Fiber Optic Measurement Techniques

To Erica and Andy, with Love

Fiber Optic Measurement Techniques

RONGQING HUI
MAURICE O'SULLIVAN

AMSTERDAM • BOSTON • HEIDELBERG • LONDON
NEW YORK • OXFORD • PARIS • SAN DIEGO
SAN FRANCISCO • SINGAPORE • SYDNEY • TOKYO
Academic Press is an imprint of Elsevier

Elsevier Academic Press
30 Corporate Drive, Suite 400, Burlington, MA 01803, USA
525 B Street, Suite 1900, San Diego, California 92101-4495, USA
84 Theobald's Road, London WC1X 8RR, UK

Library of Congress Cataloging-in-Publication Data
APPLICATION SUBMITTED

British Library Cataloguing in Publication Data
A catalogue record for this book is available from the British Library

ISBN: 978-0-12-373865-3

For all information on all Elsevier Academic Press publications visit our Web site at www.books.elsevier.com

Transferred to Digital Printing 2012

Contents

Preface xi
About the Author xv
Constants xvii

1 Fundamentals of Optical Devices 1

1.0 Introduction 2
1.1 Laser Diodes and LEDs 4
1.1.1 Pn Junction and Energy Diagram 5
1.1.2 Direct and Indirect Semiconductors 6
1.1.3 Carrier Confinement 7
1.1.4 Spontaneous Emission and Stimulated Emission 8
1.1.5 Light-Emitting Diodes (LEDs) 9
1.1.6 Laser Diodes (LDs) 13
1.1.7 Single-Frequency Semiconductor Lasers 26
1.2 Photodetectors 32
1.2.1 Pn-Junction Photodiodes 32
1.2.2 Responsivity and Bandwidth 34
1.2.3 Electrical Characteristics of a Photodiode 36
1.2.4 Photodetector Noise and SNR 37
1.2.5 Avalanche Photodiodes (APDs) 41
1.3 Optical Fibers 44
1.3.1 Reflection and Refraction 44
1.3.2 Propagation Modes in Optical Fibers 49
1.3.3 Optical Fiber Attenuation 62
1.3.4 Group Velocity and Dispersion 67
1.3.5 Nonlinear Effects in an Optical Fiber 77
1.4 Optical Amplifiers 85
1.4.1 Optical Gain, Gain Bandwidth, and Saturation 86
1.4.2 Semiconductor Optical Amplifiers 89
1.4.3 Erbium-Doped Fiber Amplifiers (EDFAs) 100
1.5 External Electro-Optic Modulator 115
1.5.1 Basic Operation Principle of Electro-Optic Modulators 116

1.5.2 Frequency Doubling and Duo-Binary Modulation 121
1.5.3 Optical Single-Side Modulation 123
1.5.4 Optical Modulators Using Electro-Absorption Effect 125

2 Basic Instrumentation for Optical Measurement 129

2.0 Introduction 130
2.1 Grating-Based Optical Spectrum Analyzers 131
2.1.1 General Specifications 131
2.1.2 Fundamentals of Diffraction Gratings 134
2.1.3 Basic OSA Configurations 138
2.2 Scanning FP Interferometer 146
2.2.1 Basic FPI Configuration and Transfer Function 146
2.2.2 Scanning FPI Spectrum Analyzer 153
2.2.3 Scanning FPI Basic Optical Configurations 157
2.2.4 Optical Spectrum Analyzer Using the Combination of Grating and FPI 159
2.3 Mach-Zehnder Interferometers 160
2.3.1 Transfer Matrix of a 2 × 2 Optical Coupler 161
2.3.2 Transfer Function of an MZI 162
2.3.3 MZI Used as an Optical Filter 164
2.4 Michelson Interferometers 168
2.4.1 Operating Principle of a Michelson Interferometer 169
2.4.2 Measurement and Characterization of Michelson Interferometers 172
2.4.3 Techniques to Increase Frequency Selectivity 174
2.5 Optical Wavelength Meter 179
2.5.1 Operating Principle of a Wavelength Meter Based on Michelson Interferometer 180
2.5.2 Wavelength Coverage and Spectral Resolution 183
2.5.3 Wavelength Calibration 185
2.5.4 Wavelength Meter Based on Fizeau Wedge Interferometer 186
2.6 Optical Polarimeter 188
2.6.1 General Description of Lightwave Polarization 188
2.6.2 The Stokes Parameters and the Poincare Sphere 190
2.6.3 Optical Polarimeters 193
2.7 Measurement Based on Coherent Optical Detection 196
2.7.1 Operating Principle 196
2.7.2 Receiver SNR Calculation of Coherent Detection 199

2.7.3 Balanced Coherent Detection and Polarization Diversity 202
2.7.4 Phase Diversity in Coherent Homodyne Detection 204
2.7.5 Coherent OSA Based on Swept Frequency Laser 207
2.8 Waveform Measurement 211
2.8.1 Oscilloscope Operating Principle 212
2.8.2 Digital Sampling Oscilloscopes 216
2.8.3 High-Speed Sampling of Optical Signal 219
2.8.4 High-Speed Electric ADC Using Optical Techniques 223
2.8.5 Short Optical Pulse Measurement Using an Autocorrelator 224
2.9 Optical Low-Coherent Interferometry 232
2.9.1 Optical Low-Coherence Reflectometry 232
2.9.2 Fourier-Domain Reflectometry 240
2.10 Optical Network Analyzer 246
2.10.1 S-Parameters and RF Network Analyzer 246
2.10.2 Optical Network Analyzers 249

3 Characterization of Optical Devices 259

3.0 Introduction 260
3.1 Characterization of RIN and Linewidth of Semiconductor Lasers 260
3.1.1 Measurement of Relative Intensity Noise (RIN) 261
3.1.2 Measurement of Laser Phase Noise and Linewidth 266
3.2 Measurement of Electro-Optic Modulation Response 276
3.2.1 Characterization of Intensity Modulation Response 277
3.2.2 Measurement of Frequency Chirp 282
3.2.3 Time-Domain Measurement of Modulation-Induced Chirp 292
3.3 Wideband Characterization of an Optical Receiver 296
3.3.1 Characterization of Photodetector Responsivity and Linearity 297
3.3.2 Frequency Domain Characterization of Photodetector Response 299
3.3.3 Photodetector Bandwidth Characterization Using Source Spontaneous-Spontaneous Beat Noise 301
3.3.4 Photodetector Characterization Using Short Optical Pulses 304

3.4 Characterization of Optical Amplifiers 306
3.4.1 Measurement of Amplifier Optical Gain 306
3.4.2 Measurement of Static and Dynamic Gain Tilt 311
3.4.3 Optical Amplifier Noise 314
3.4.4 Optical Domain Characterization of ASE Noise 316
3.4.5 Impact of ASE Noise in Electrical Domain 318
3.4.6 Noise Figure Definition and Its Measurement 323
3.4.7 Time-Domain Characteristics of EDFA 327
3.5 Characterization of Passive Optical Components 329
3.5.1 Fiber-Optic Couplers 330
3.5.2 Fiber Bragg Grating Filters 335
3.5.3 WDM Multiplexers and Demultiplexers 340
3.5.4 Characterization of Optical Filter Transfer Functions 345
3.5.5 Optical Isolators and Circulators 353

4 Optical Fiber Measurement 365

4.0 Introduction 366
4.1 Classification of Fiber Types 367
4.1.1 Standard Optical Fibers for Transmission 367
4.1.2 Specialty Optical Fibers 370
4.2 Measurement of Fiber Mode-Field Distribution 374
4.2.1 Near-Field, Far-Field, and Mode-Field Diameter 375
4.2.2 Far-Field Measurement Techniques 378
4.2.3 Near-Field Measurement Techniques 380
4.3 Fiber Attenuation Measurement and OTDR 382
4.3.1 Cutback Technique 382
4.3.2 Optical Time-Domain Reflectometers 384
4.3.3 Improvement Considerations of OTDR 391
4.4 Fiber Dispersion Measurements 394
4.4.1 Intermodal Dispersion and Its Measurement 395
4.4.2 Chromatic Dispersion and Its Measurement 400
4.5 Polarization Mode Dispersion (PMD) Measurement 409
4.5.1 Representation of Fiber Birefringence and PMD Parameter 409
4.5.2 Pulse Delay Method 413
4.5.3 The Interferometric Method 415
4.5.4 Poincare Arc Method 418
4.5.5 Fixed Analyzer Method 420
4.5.6 The Jones Matrix Method 424
4.5.7 The Mueller Matrix Method 431

4.6 Determination of Polarization-Dependent Loss 438
4.7 PMD Sources and Emulators 442
4.8 Measurement of Fiber Nonlinearity 446
4.8.1 Measurement of Stimulated Brillouin Scattering Coefficient 447
4.8.2 Measurement of the Stimulated Raman Scattering Coefficient 453
4.8.3 Measurement of Kerr effect nonlinearity 459

5 Optical System Performance Measurements 481

5.0 Introduction 482
5.1 Overview of Fiber-Optic Transmission Systems 483
5.1.1 Optical System Performance Considerations 484
5.1.2 Receiver *BER* and Q 486
5.1.3 System Q Estimation Based on Eye Diagram Parameterization 494
5.1.4 Bit Error Rate Testing 499
5.2 Receiver Sensitivity Measurement and OSNR Tolerance 508
5.2.1 Receiver Sensitivity and Power Margin 509
5.2.2 OSNR Margin and Required OSNR (R-OSNR) 514
5.2.3 BER vs. Decision Threshold Measurement 521
5.3 Waveform Distortion Measurements 524
5.4 Jitter Measurement 527
5.4.1 Basic Jitter Parameters and Definitions 527
5.4.2 Jitter Detection Techniques 532
5.5 In-situ Monitoring of Linear Propagation Impairments 537
5.5.1 *In Situ* Monitoring of Chromatic Dispersion 537
5.5.2 *In Situ* PMD Monitoring 541
5.5.3 *In Situ* PDL Monitoring 551
5.6 Measurement of Nonlinear Crosstalk in Multi-Span WDM systems 556
5.6.1 XPM-Induced Intensity Modulation in IMDD Optical Systems 556
5.6.2 XPM-induced Phase Modulation 572
5.6.3 FWM-Induced Crosstalk in IMDD Optical Systems 575
5.6.4 Characterization of Raman Crosstalk with Wide Channel Separation 581

5.7 Modulation Instability and Its Impact in WDM Optical Systems 590
5.7.1 Modulation-instability and Transfer Matrix Formulation 590
5.7.2 Impact of Modulation Instability in Amplified Multispan Fiber Systems 600
5.7.3 Characterization of Modulation Instability in Fiber-Optic Systems 601
5.8 Optical System Performance Evaluation Based On Required OSNR 606
5.8.1 Measurement of R-SNR Due to Chromatic Dispersion 607
5.8.2 Measurement of R-SNR Due to Fiber Nonlinearity 610
5.8.3 Measurement of R-OSNR Due to Optical Filter Misalignment 615
5.9 Fiber-Optic Recirculating Loop 616
5.9.1 Operation Principle of a Recirculating Loop 617
5.9.2 Measurement Procedure and Time Control 618
5.9.3 Optical Gain Adjustment in the Loop 622

Index **631**

Preface

Modern fiber-optic communications date back to the early 1960s when Charles Kao theoretically predicted that high-speed messages could be transmited long distances over a narrow glass waveguide, which is now commonly referred to as an optical fiber. In 1970, a team of researchers at Corning successfully fabricated optical fibers using fused-silica with a loss of less than 20dB/km at 633nm wavelength. The Corning breakthrough was the most significant step toward the practical application of fiber-optic communications. Over the following several years, fiber losses dropped dramatically, aided both by improved fabrication methods and the shift to longer wavelengths, where fibers have inherently lower attenuation.

Meanwhile, the prospect of long distance fiber-optic communication intensified research and development efforts in semiconductor lasers and other related optical devices. Near-infrared semiconductor lasers and LEDs operating at 810nm, 1320nm and 1550nm wavelengths were developed to fit into the low loss windows of silica optical fibers. The bandwidth in the 1550nm wavelength window alone can be as wide as 80nm, which is approximately 10THz. In order to make full and efficient use of this vast bandwidth, many innovative technologies have been developed, such as single frequency and wavelength tunable semiconductor lasers, dispersion shifted optical fibers, optical amplifiers, wavelength division multiplexing as well as various modulation formats and signal processing techniques.

In addition to optical communications, fiber-optics and photonic technologies have found a variety of other applications ranging from precision metrology, to imaging, to photonic sensors. Various optical measurement techniques have been proposed and demonstrated in research, development, maintenance and trouble-shooting of optical systems. Different optical systems demand different measurement and testing techniques based on the specific application and the key requirements of each system. Over the years, fiber-optic measurement has become a stand-alone research discipline, which is both interesting and challenging.

In general, optical measurements can be categorized into instrumentation and measurement methodology. In many cases, the measurement capability and accuracy are limited by the instruments used. Therefore, a good understanding of operation principles and performance limitations of basic optical instruments is essential in the design of experimental setups and to achieve the desired measurement speed and accuracy. From methodology point of view, a familiarity with various basic measurement system configurations and topologies is necessary, which helps in determining how to make the most efficient use of the available instrumentations, how to extract useful signals, and how to interpret and process the results.

The focus of this book is the measurement techniques related to fiber-optic systems, subsystems and devices. Since both optical systems and optical instruments are built upon various optical components, basic optical devices are discussed in chapter 1, which includes semiconductor lasers and LEDs, photodetectors, fundamental properties of optical fibers, optical amplifiers and optical modulators. Familiarity with the characteristics of these individual building blocks is essential for the understanding of optical measurement setups and optical instrumentation. Chapter 2 introduces basic optical instrumentation, such as optical spectrum analyzers, optical wavelength meters, Fabry-Perot, Mach-zehnder and Michelson interferometers, optical polarimeters, high-speed optical and RF oscilloscopes and network analyzers. Since coherent optical detection is a foundation for an entire new category of optical instrumentation, the fundamental principle of coherent detection is also discussed in this chapter, which helps in the understanding of linear optical sampling and vectorial optical network analyzer. In chapter 3, we discuss techniques of characterizing optical devices such as semiconductor lasers, optical receivers, optical amplifiers and various passive optical devices. Optical and optoelectronic transfer functions, intensity and phase noises and modulation characteristics are important parameters to investigate. Chapter 4 discusses measurement of optical fibers, including attenuation, chromatic dispersion, polarization mode dispersion and optical nonlinearity. Finally, chapter 5 is dedicated to the discussion of measurement issues related to optical communication systems.

Instead of describing performance and specification of specific instruments, the major purpose of this book is to outline the fundamental principles behind each individual measurement technique. Most of them described here can be applied to various different application fields. A good understanding of fundamental principles behind these measurement techniques is a key to making the best use of available instrumentation, to obtain the best possible results and to develop new and innovative measurement techniques and instruments.

ACKNOWLEDGEMENTS

First, I would like to thank Mr. Tim Pitts of Academic Press. The idea of developing this book began in early 2006 when Tim and I met at the Optical Fiber Communications Conference (OFC) where he asked if I would consider writing a book in fiber-optic measurements. I particularly like this topic because

I have taught a graduate course in this subject for a few years and always have a great interest in this research area.

I would also like to thank my coauthor, Dr. Maurice O'Sullivan, who contributed materials for a number of sections in Chapter 5, where his expertise in fiber-optic communication systems research and development has proven to be a great asset.

My colleague professor Chris Allen at the University of Kansas carefully read through the entire manuscript and provided invaluable corrections. My research associate Dr. Jianfeng Jiang has provided many useful suggestions in in-situ polarization measurements in fiber-optic systems.

Finally, I would like to thank my wife Jian. Without her support and tolerance it would not be possible for me to accomplish this task.

Rongqing Hui

About the Author

Rongqing Hui is a Professor of Electrical Engineering & Computer Science at the University of Kansas. He received B.S. and M.S. degrees from Beijing University of Posts & Telecommunications in China, and Ph.D degree from Politecnico di Torino in Italy, all in Electrical Engineering. Before joining the faculty of the University of Kansas in 1997, Dr. Hui was a Member of Scientific Staff at Bell-Northern Research and then Nortel, where he worked in research and development of high-speed optical transport networks. He served as a Program Director at the National Science Foundation for two years from 2006 to 2007, where he was in charge of research programs in photonics and optoelectronics. As an author or co-author, Dr. Hui has published widely in the area of fiber-optic systems and devices and holds 12 US patents. He served as a topic editor for *IEEE Transactions on Communications* from 2001 to 2007 and is currently serving as an associate editor for *IEEE Journal of Quantum Electronics* since 2006.

Maurice O'Sullivan has worked for Nortel for a score of years, at first in the optical cable business, developing factory-tailored metrology for optical fiber, but, in the main, in the optical transmission business developing, modeling and verifying physical layer designs & performance of Nortel's line and highest rate transmission product including OC-192, MOR, MOR+, LH1600G, eDCO and eDC40G. He holds a Ph.D. in physics (high resolution spectroscopy) from the University of Toronto, is a Nortel Fellow and has been granted more than 30 patents.

Constants

Physics Constants

Constant	Name	Value	Unit
κ	Boltzmann's constant	1.28×10^{-23}	$J\text{-}K^{-1}$
h	Planck's constant	6.626×10^{-34}	J-s
q	Electron charge	1.6023×10^{-19}	Coulomb
c	Speed of light in free space	2.99792×10^{8}	m/s
T	Absolute temperature	273 K = 0 °C	Kelvin
ε_0	Permittivity in free space	8.854×10^{-12}	F/m
μ_0	Permeability in free space	12.566×10^{-7}	N/A^2

Conversion table

Symbol [unit]	Parameter 1	Symbol [unit]	Parameter 2	Conversion
λ [nm]	Vacuum wavelength	f (Hz)	Frequency	$f = c/\lambda$
$\Delta\lambda$ [nm]	Wavelength difference	Δf (Hz)	Frequency difference	$\Delta f = -\frac{c}{\lambda^2}\Delta\lambda$
α [Neper/km]	Attenuation	α_{dB} [dB/km]	Attenuation	$\alpha_{dB} = 4.343\alpha$
D [ps/nm/km]	Dispersion parameter	β_2 [ps^2/nm]	Dispersion parameter	$D = -\frac{2\pi c}{\lambda^2}\beta_2$
E_g [eV]	Photon energy	λ [nm]	Wavelength	$\lambda = hc/E_g$
Λ [cm^{-1}]	Wave number	λ [nm]	Wavelength	$\Lambda = 10^7/\lambda$
T_C [°C]	Temperature	T_F [°F]	Temperature	$T_C = \frac{5}{9}(T_F - 32)$
Meter	Length	Inch	Length	1 meter = 39.37 inch

Chapter 1

Fundamentals of Optical Devices

1.0. Introduction
1.1. Laser Diodes and LEDs
    1.1.1. Pn Junction and Energy Diagram
    1.1.2. Direct and Indirect Semiconductors
    1.1.3. Carrier Confinement
    1.1.4. Spontaneous Emission and Stimulated Emission
    1.1.5. Light-Emitting Diodes (LEDs)
    1.1.6. Laser Diodes (LDs)
    1.1.7. Single-Frequency Semiconductor Lasers
1.2. Photodetectors
    1.2.1. Pn-Junction Photodiodes
    1.2.2. Responsivity and Bandwidth
    1.2.3. Electrical Characteristics of a Photodiode
    1.2.4. Photodetector Noise and SNR
    1.2.5. Avalanche Photodiodes (APDs)
1.3. Optical Fibers
    1.3.1. Reflection and Refraction
    1.3.2. Propagation Modes in Optical Fibers
    1.3.3. Optical Fiber Attenuation
    1.3.4. Group Velocity and Dispersion
    1.3.5. Nonlinear Effects in an Optical Fiber
1.4. Optical Amplifiers
    1.4.1. Optical Gain, Gain Bandwidth, and Saturation
    1.4.2. Semiconductor Optical Amplifiers
    1.4.3. Erbium-Doped Fiber Amplifiers (EDFAs)
1.5. External Electro-Optic Modulator
    1.5.1. Basic Operation Principle of Electro-Optic Modulators
    1.5.2. Frequency Doubling and Duo-Binary Modulation
    1.5.3. Optical Single-Side Modulation
    1.5.4. Optical Modulators Using Electro-Absorption Effect

1.0 INTRODUCTION

In an optical communication system, information is delivered by optical carriers. The signal can be encoded into optical intensity, frequency, and phase for transmission and be detected at the receiver. As illustrated in Figure 1.0.1, the simplest optical system has an optical source, a detector, and various optical components between them, such as an optical coupler and optical fiber. In this simple system, the electrical signal is modulated directly onto the light source such that the intensity, the wavelength, or the phase of the optical carrier is encoded by the electrical signal. This modulated optical signal is coupled into an optical fiber and delivered to the destination, where it is detected by an optical receiver. The optical receiver detects the received optical signal and recovers the electrical signal encoded on the optical carrier.

Simple optical systems like the one shown in Figure 1.0.1 are usually used for low data rate and short-distance optical transmission. First, direct modulation on a semiconductor laser source has a number of limitations such as frequency chirp and poor extinction ratio. Second, the attenuation in the transmission optical fiber limits the system reach between the transmitter and the receiver. In addition, although the low-loss window of a single mode fiber can be as wide as hundreds of Terahertz, the transmission capacity is limited by the modulation speed of the transmitter and the receiver electrical bandwidth.

To overcome these limitations, a number of new technologies have been introduced in modern optical communication systems, including single-frequency semiconductor lasers, external electro-optic modulators, wavelength division multiplexing (WDM) technique, and optical amplifiers.[1] Figure 1.0.2 shows the block diagram of a WDM optical system with N wavelength channels. In this system, each data channel is modulated onto an optical carrier with a

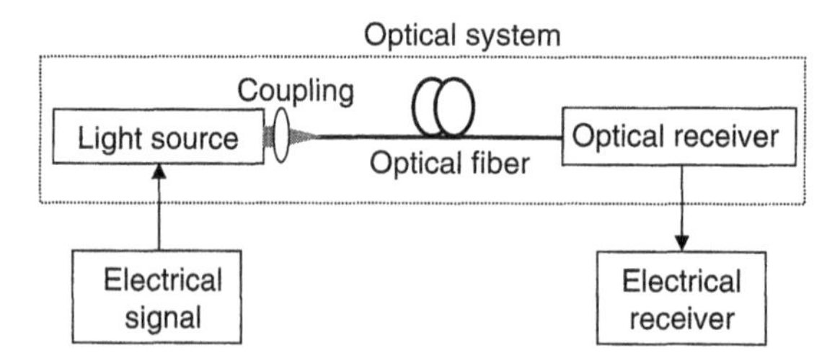

Figure 1.0.1 Configuration of the simplest optical communication system.

[1] *Optical Fiber Telecommunications V* (A and B), Ivan Kaminow, Tingye Li, and Alan Willner, eds., Academic Press, 2008, ISBN-13:978-0-12-374171.

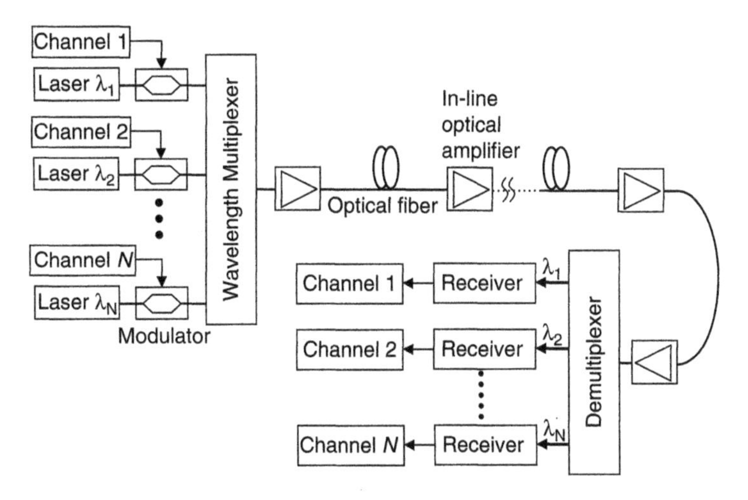

Figure 1.0.2 Block diagram of a WDM optical system using external modulation.

specific wavelength through an external electro-optic modulator. All the optical carriers are combined into a single output fiber through a wavelength division multiplexer. The power of the combined optical signal is boosted by an optical fiber amplifier and sent to the transmission optical fiber. Along the fiber transmission line, the optical signal is periodically amplified by in-line optical amplifiers to overcome the transmission loss of the optical fiber. At the destination, a wavelength division demultiplexer is used to separate optical carriers at different wavelengths, and each wavelength channel is detected separately to recover the data carried on each channel. In this system, the WDM configuration allows the full use of the wide band offered by the optical fiber, and optical amplification technology significantly extended the overall reach of the optical system.

Meanwhile, because of the dramatically increased optical bandwidth, transmission distance, and optical power levels in the fiber, other sources of performance impairments start to become significant, such as chromatic dispersion, fiber nonlinearity, polarization-mode dispersion, and amplified spontaneous emission noise generated by optical amplifiers. Understanding and characterization of these impairments have become important parts of fiber-optic system research. Another very important area in fiber-optic systems research is optical modulation format and optoelectronic signal processing. In addition to intensity modulation, optical systems based on phase modulation are gaining momentum, demonstrating excellent performance. Furthermore, due to the availability and much improved quality of single-frequency tunable semiconductor lasers, optical receivers based on coherent detection are becoming more and more popular. Many signal processing techniques previously developed for radio frequency systems can now be applied to optical systems such as orthogonal frequency-division multiplexing (OFDM) and code division multiple access (CDMA). Obviously, the overall system performance is not only determined by the system architecture but also depends on the characteristics of each individual optical device.

On the other hand, from an optical testing and characterization point of view, the knowledge of functional optical devices as building blocks and optical signal processing techniques are essential for the design and construction of measurement setup and optical instrumentation. It also helps to understand the capabilities and limitations of optical systems. This chapter introduces fundamental optical devices that are often used in fiber-optic systems and optical instrumentations.

The physics background and basic properties of each optical device will be discussed. Section 1.1 introduces optical sources such as semiconductor lasers and light-emitting diodes (LEDs). Basic properties, including optical power, spectral width, and optical modulation will be discussed. Section 1.2 presents optical detectors. Responsivity, optical and electrical bandwidth, and signal-to-noise-ratio are important characteristics. Section 1.3 reviews the basic properties of optical fibers, which include mode-guiding mechanisms, attenuation, chromatic dispersion, polarization mode dispersion, and nonlinear effects.

Section 1.4 discusses optical amplifiers, which include semiconductor optical amplifiers (SOA) and erbium-doped fiber amplifiers (EDFA). Though SOAs are often used for optical signal processing based on their high-speed dynamics, EDFAs are more convenient for application as in-line optical amplifiers in WDM optical systems due to their slow carrier dynamics and low crosstalk between high-speed channels. The last section in this chapter is devoted to the discussion of external electro-optic modulators, which are widely used in high-speed optical transmission systems. Both LiNbO3 based Mach-zehnder modulators and electro-absorption modulators will be discussed.

1.1 LASER DIODES AND LEDs

In optical systems, signals are carried by photons and therefore an optical source is an essential part of every optical system. Although there are various types of optical sources, semiconductor-based light sources are most popular in fiber-optic systems because they are small, reliable, and, most important, their optical output can be rapidly modulated by the electrical injection current, which is commonly referred to as *direct modulation*. Semiconductor lasers and LEDs are based on forward-biased pn junctions, and the output optical powers are proportional to the injection electric current.

1.1.1 Pn Junction and Energy Diagram

Figure 1.1.1 shows a homojunction between a p-type and an n-type semiconductor. For standalone n-type and p-type semiconductor materials, the Fermi level is closer to the conduction band in n-type semiconductors and it is closer to the valence band in p-type semiconductors, as illustrated in Figure 1.1.1(a). Figure 1.1.1(b) shows that once a pn junction is formed, under thermal equilibrium the Fermi level will be unified across the structure. This happens because high-energy free electrons diffuse from n-side to p-side and low-energy holes

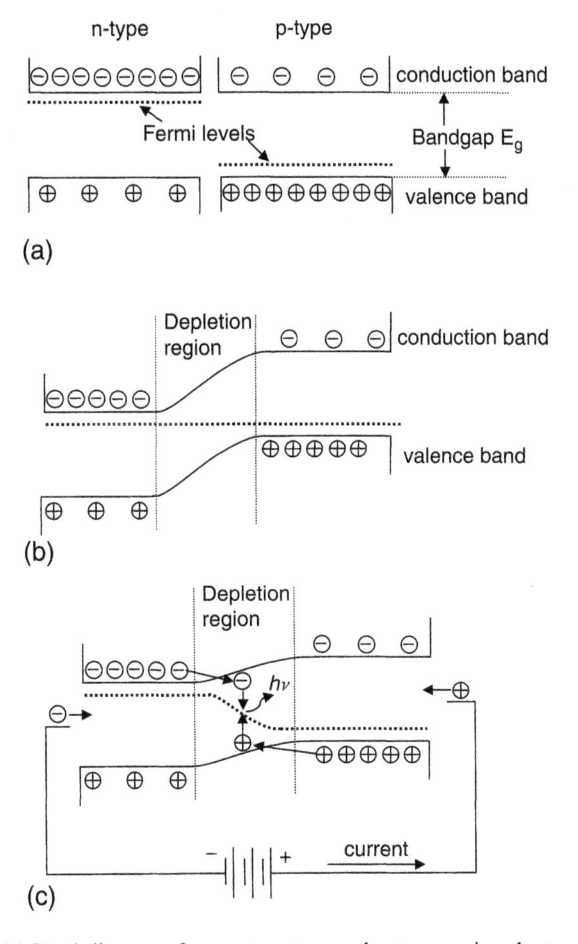

Figure 1.1.1 (a) Band diagram of separate n-type and p-type semiconductors, (b) band diagram of a pn junction under equilibrium, and (c) band diagram of a pn junction with forward bias.

diffuse in the opposite direction; as a result, the energy level of the p-type side is increased compared to the n-type side. Meanwhile, because free electrons migrate from n-type side to p-type-side, uncovered protons left over at the edge of the n-type semiconductor create a positively charged layer on the n-type side. Similarly, a negatively charged layer is created at the edge of the p-type semiconductor due to the loss of holes. Therefore built-in electrical field and thus a potential barrier is created at the pn junction, which pulls the diffused free electrons back to the n-type side and holes back to the p-type-side, a process commonly referred to as *carrier drift*. Because of this built-in electrical field, neither free electrons nor holes exist at the junction region, and therefore this region is called the *depletion region* or *space charged region*. Without an external bias, there is no net carrier flow across the pn junction due to the exact balance between carrier diffusion and carrier drift.

When the pn-junction is forward-biased as shown in Figure 1.1.1(c), excess electrons and holes are injected into the n-type side and the p-type sections, respectively. This carrier injection reduces the potential barrier and pushes excess electrons and holes to diffuse across the junction area. In this process, excess electrons and holes recombine inside the depletion region to generate photons. This is called *radiative recombination.*

1.1.2 Direct and Indirect Semiconductors

One important rule of radiative recombination process is that both energy and momentum must be conserved. Depending on the shape of their band structure, semiconductor materials can be generally classified as having direct bandgap or indirect bandgap, as illustrated in Figure 1.1.2, where E is the energy and k is the momentum.

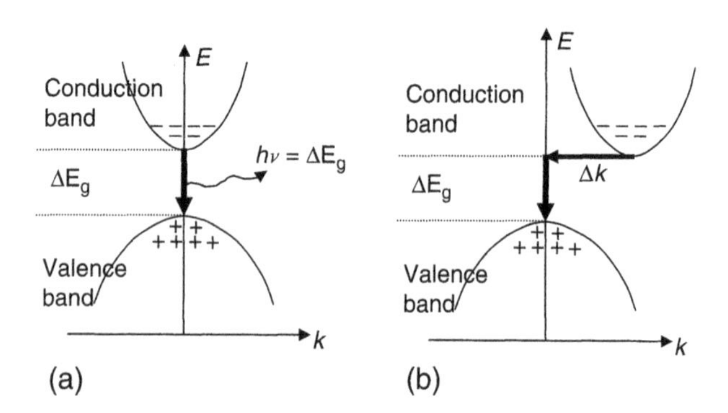

Figure 1.1.2 Illustration of direct bandgap (a) and indirect bandgap (b) of semiconductor materials.

For direct semiconductors, holes at the top of the valence band have the same momentum as the electrons at the bottom of the conduction band. In this case, electrons directly recombine with the holes to emit photons, and the photon energy is equal to the bandgap.

For indirect semiconductors, on the other hand, holes at the top of the valence band and electrons at the bottom of the conduction band have different momentum. Any recombination between electrons in the conduction band and holes in the valence band would require significant momentum change. Although a photon can have considerable energy $h\nu$, its momentum $h\nu/c$ is much smaller, which cannot compensate for the momentum mismatch between the electrons and the holes. Therefore radiative recombination is considered impossible in indirect semiconductor materials unless a third particle (for example, a phonon created by crystal lattice vibration) is involved which provides the required momentum.

1.1.3 Carrier Confinement

In addition to the requirement of using direct bandgap semiconductor material, another important requirement for semiconductor light sources is the carrier confinement. In early diode lasers, materials with the same bandgap were used at both sides of the pn junction, as shown in Figure 1.1.1. This is referred to as *homojunction*. In this case, carrier recombination happens over the entire depletion region with the width of $1 \sim 10$ μm depending on the diffusion constant of the electrons and the holes. This wide depletion region makes it difficult to achieve high carrier concentration. To overcome this problem, double heterojunction was introduced in which a thin layer of semiconductor material with a slightly smaller bandgap is sandwiched in the middle of the junction region between the p-type and the n-type sections. This concept is illustrated in Figure 1.1.3.

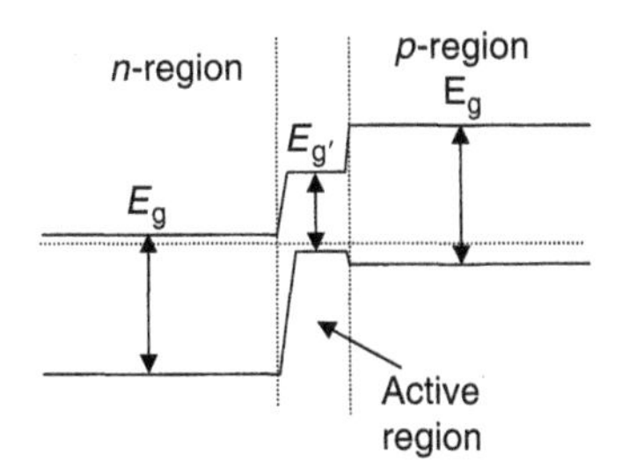

Figure 1.1.3 Illustration of semiconductor double heterostructure.

In this structure, the thin layer has slightly smaller bandgap, which attracts the concentration of carriers when the junction is forward biased; therefore this layer is referred to as the *active region* of the device. The strong carrier confinement is a result of band gap discontinuity. The sandwich layer can be controlled to be on the order of ~0.1 μm, which is several orders of magnitude thinner than the depletion region of a homojunction; therefore very high levels of carrier concentration can be realized at a certain injection current.

In addition to providing carrier confinement, another advantage of using double heterostructure is that it also provides useful photon confinement. By using a material with slightly higher refractive index for the sandwich layer, a dielectric waveguide is formed. This dielectric optical waveguide provides a mechanism to confine photons within the active layer, and therefore very high photon density can be achieved.

1.1.4 Spontaneous Emission and Stimulated Emission

As discussed, radiative recombination between electrons and holes creates photons, but this is a random process. The energy is conserved in this process, which determines the frequency of the emitted photon as $\nu = \Delta E/h$, where ΔE is the energy gap between the conduction band electron and the valence band hole that participated in the process. h is Planck's constant. However, the phase of the emitted lightwave is not predictable. Indeed, since semiconductors are solids, energy of carriers are not on discrete levels; instead they are continuously distributed within energy bands following the Fermi-Dirac distribution, as illustrated by Figure 1.1.4.

Figure 1.1.4(a) shows that different electron-hole pairs may be separated by different energy gaps and ΔE_i might not be equal to ΔE_j. Recombination

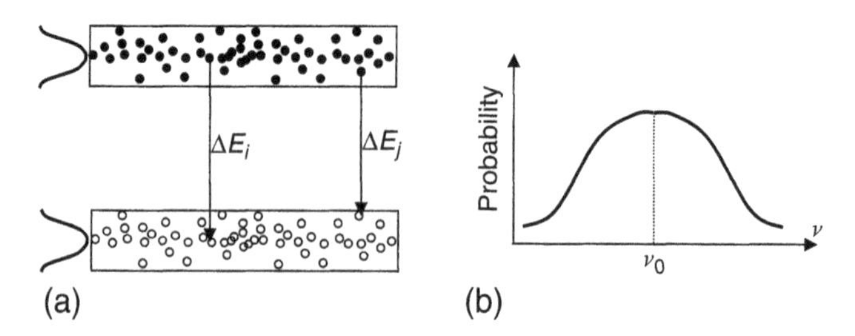

Figure 1.1.4 Illustration of an energy band in semiconductors and the impact on the spectral width of radiative recombination. (a) Energy distributions of electrons and holes. (b) Probability distribution of the frequency of emitted photons.

of different electron-hole pairs will produce emission at different wavelengths. The spectral width of the emission is determined by the statistic energy distribution of the carriers, as illustrated by Figure 1.1.4(b).

Spontaneous emission is created by the spontaneous recombination of electron-hole pairs. The photon generated from each recombination event is independent, although statistically the emission frequency falls into the spectrum shown in Figure 1.1.4(b). The frequencies, the phases, and the direction of propagation of the emitted photons are not correlated. This is illustrated in Figure 1.1.5(a).

Stimulated emission, on the other hand, is created by stimulated recombination of electron-hole pairs. In this case the recombination is induced by an incoming photon, as shown in Figure 1.1.5(b). Both the frequency and the phase of the emitted photon are identical to those of the incoming photon. Therefore photons generated by the stimulated emission process are coherent, which results in narrow spectral linewidth.

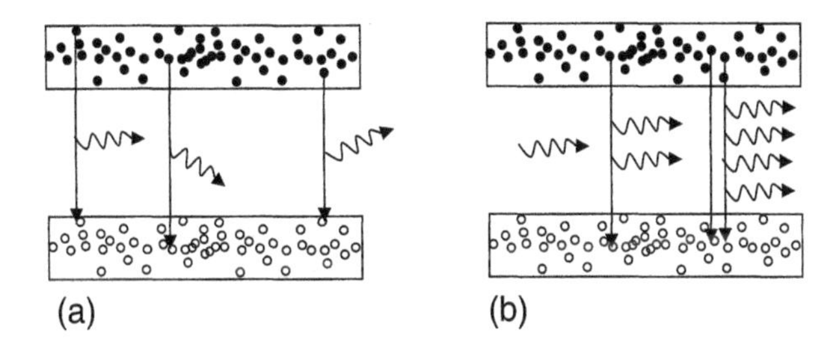

Figure 1.1.5 Illustration of spontaneous emission (a) and stimulated emission (b).

1.1.5 Light-Emitting Diodes (LEDs)

Light emission in an LED is based on the spontaneous emission of forward-biased semiconductor pn junction. The basic structures are shown in Figure 1.1.6 for surface-emitting and edge-emitting LEDs.

For surface-emitting diodes, light emits in the perpendicular direction of the active layer. The active area of the surface is equally bright and the emission angle is isotopic, which is called *Lambertian*. Optical power emission patterns can be described as $P(\theta) = P_0 \cos\theta$, where θ is the angle between the emitting direction and the surface normal and P_0 is the optical power viewed from the direction of surface normal. By properly design the shape of the bottom metal contact, the active emitting area can be made circular to maximize the coupling efficiency to optical fiber.

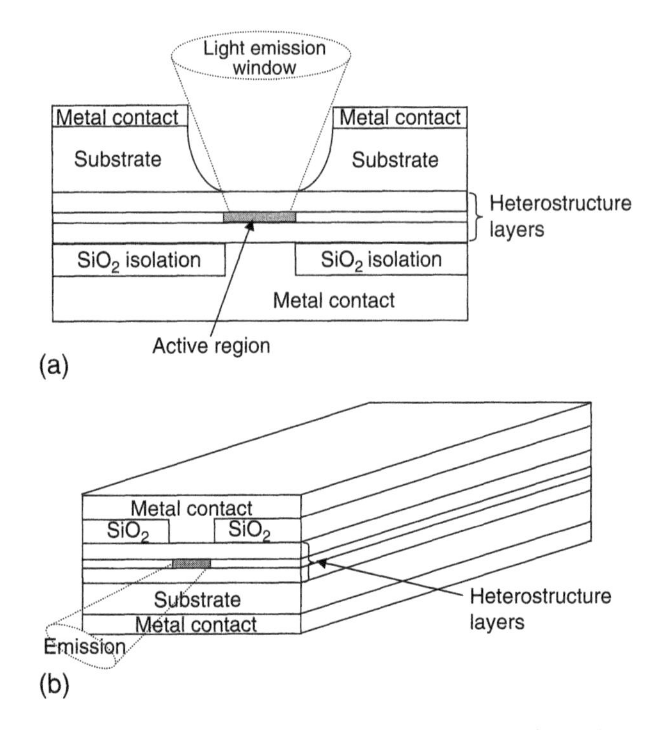

Figure 1.1.6 Illustration of surface emission (a) and edge emission (b) LEDs.

1.1.5.1 P~I Curve

For edge-emitting diodes, on the other hand, light emits in the same direction as the active layer. In this case, a waveguide is usually required in the design, where the active layer has slightly higher refractive index than the surrounding layers. Compared to surface-emitting diodes, the emitting area of edge-emitting diodes is usually much smaller, which is determined by the width and thickness of the active layer.

For an LED, the emitted optical power is linearly proportional to the injected electrical current, as shown in Figure 1.1.7. This is commonly referred to as the *P~I curve*. In the idea case, the recombination of each electron-hole pair generates a photon. If we define the power efficiency d*P*/d*I* as the ratio between the emitted optical power P_{opt} and the injected electrical current *I*, we have:

$$dP/dI = \frac{h\nu}{q} = \frac{hc}{\lambda q} \tag{1.1.1}$$

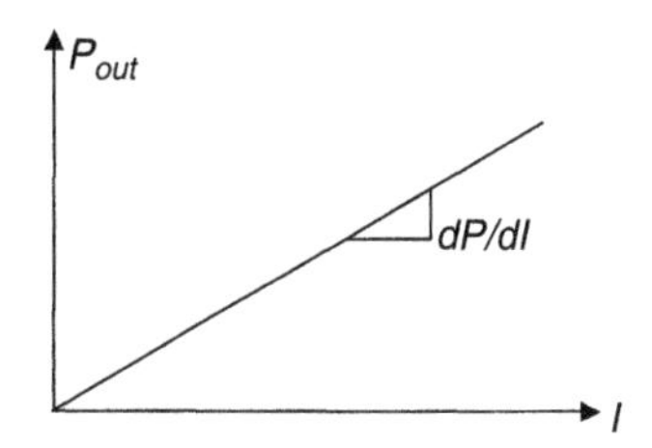

Figure 1.1.7 LED emitting power is linearly proportional to the injection current.

where q is the electron change, h is Plank's constant, c is the speed of light, and λ is the wavelength. In practical devices, in addition to radiative recombination, there is also nonradiative recombination, which does not produce photons. The internal quantum efficiency is defined as:

$$\eta_q = \frac{R_r}{R_r + R_{nr}} \tag{1.1.2}$$

where R_r and R_{nr} are the rates of radiative and nonradiative recombination, respectively. Another factor that reduces the slope of the P~I curve is that not all the photons generated through radiative recombination are able to exit the device. Various effects contribute to this efficiency reduction, such as internal material loss, interface reflection, and emitting angle. The external efficiency is defined by:

$$\eta_{ext} = \frac{R_{emit}}{R_r} \tag{1.1.3}$$

where R_{emit} is the rate of the generated photons that actually exit the LED.

Considering both internal quantum efficiency and external efficiency, the slope of the P~I curve should be:

$$dP/dI = \eta_q \eta_{ext} \frac{hc}{\lambda q} \tag{1.1.4}$$

Because of the linear relationship between output optical power and the injection current, the emitted optical power of an LED is:

$$P_{opt} = \eta_q \eta_{ext} \frac{hc}{\lambda q} I \tag{1.1.5}$$

In general, the internal quantum efficiency of an LED can be in the order of 70 percent. However, since an LED is based on spontaneous emission and the photon emission is isotropic, its external efficiency is usually less than 5 percent. As a rule of thumb, for $\eta_q = 75\%$, $\eta_{ext} = 2\%$, and the wavelength of $\lambda = 1550$ nm, the output optical power efficiency is approximately 12 µW/mA.

1.1.5.2 Modulation Dynamics

In an LED active layer, the increase of carrier population is proportional to the rate of external carrier injection minus the rate of carrier recombination. Therefore, the rate equation of carrier population N_t is:

$$\frac{dN_t(t)}{dt} = \frac{I(t)}{q} - \frac{N_t(t)}{\tau} \tag{1.1.6}$$

where τ is referred to as *carrier lifetime*. In general, τ is a function of carrier density and the rate equation does not have a closed-form solution. To simplify, if we assume τ is a constant, Equation 1.1.6 can be easily solved in the frequency domain as:

$$\tilde{N}_t(\omega) = \frac{\tilde{I}(\omega)\tau/q}{1 + j\omega\tau} \tag{1.1.7}$$

where $\tilde{N}_t(\omega)$ and $\tilde{I}(\omega)$ are the Fourier transforms of N_t(t) and I(t), respectively. Equation 1.1.7 demonstrates that carrier population can be modulated through injection current modulation and the 3-dB modulation bandwidth is proportional to $1/\tau$. Because the photons are created by radiative carrier recombination and the optical power is proportional to the carrier density, the modulation bandwidth of the optical power is also proportional to $1/\tau$.

$$P_{opt}(\omega) = \frac{P_{opt}(0)}{\sqrt{1 + (\omega\tau)^2}} \tag{1.1.8}$$

where $P_{opt}(0)$ is the optical power at DC. Typical carrier lifetime of an LED is in the order of nanoseconds and therefore the modulation bandwidth is in the 100 MHz $\sim$ 1 GHz level, depending on the structure of the LED.

It is worth noting that since the optical power is proportional to the injection current, the modulation bandwidth can be defined as either electrical or optical. The following is a practical question: To fully support an LED with an optical bandwidth of B_{opt}, what electrical bandwidth B_{ele} is required for the driving circuit? Since optical power is proportional to the electrical current, $P_{opt}(\omega) \propto I(\omega)$, and the driver electrical power is proportional to the square of the injection current, $P_{ele}(\omega) \propto I^2(\omega)$, therefore driver electrical power is proportional to the square of the LED optical power $P_{ele}(\omega) \propto P_{opt}^2(\omega)$, that is:

$$\frac{P_{ele}(\omega)}{P_{ele}(0)} = \frac{P_{opt}^2(\omega)}{P_{opt}^2(0)} \tag{1.1.9}$$

If at a given frequency the optical power is reduced by 3 dB compared to its DC value, the driver electrical power is supposed to be reduced 6 dB at that frequency. That is, 3 dB optical bandwidth is equivalent to 6 dB electrical bandwidth.

Example 1.1

Consider an LED emitting at $\lambda = 1550$ nm wavelength window. The internal quantum efficiency is 70%, the external efficiency is 2%, the carrier lifetime is $\tau = 20$ ns, and the injection current is 20 mA. Find:

1. The output optical power of the LED
2. The 3 dB optical bandwidth and the required driver electrical bandwidth

Solution:

1. Output optical power is:

$$P_{opt} = \eta_q \eta_{ext} \frac{hc}{\lambda q} I = 0.7 \times 0.02 \times \frac{6.63 \times 10^{-34} \times 3 \times 10^8 \times 20 \times 10^{-3}}{1550 \times 10^{-9} \times 1.6 \times 10^{-19}} = 0.225\ mW$$

2. To find the 3 dB optical bandwidth, we use:

$$P_{opt}(\omega) = \frac{P_{opt}(0)}{\sqrt{1 + \omega^2 \tau^2}}$$

For 3dB optical bandwidth $10\log\left\{\frac{P_{opt}(\omega_{opt})}{P_{opt}(0)}\right\} = -3\ dB$, that is:

$$10\log\left(1 + \omega_{opt}^2 \tau^2\right) = 6\ dB$$

Therefore, the angular frequency of optical bandwidth is $\omega_{opt} = 86.33$ Mrad./s, which corresponds to a circular frequency $f_{opt} = 86.33/2\pi \approx 13.7\ MHz$.

For 3 dB electrical bandwidth, $10\ \log\left\{\frac{P_{ele}(\omega_{ele})}{P_{ele}(0)}\right\} = -3\ dB$. This is equivalent to $10\log\left\{\frac{P_{opt}^2(\omega_{ele})}{P_{opt}^2(0)}\right\} = -3\ dB$. Therefore $\omega_{ele} = 1/\tau = 50\ MHz$ and $f_{ele} \approx 7.94\ MHz$.

1.1.6 Laser Diodes (LDs)

Semiconductor laser diodes are based on the stimulated emission of forward-biased semiconductor pn junction. Compared to LEDs, LDs have higher spectral purity and higher external efficiency because of the spectral and spatial coherence of stimulated emission.

One of the basic requirements of laser diodes is optical feedback. Consider an optical cavity of length L as shown in Figure 1.1.8, where the semiconductor

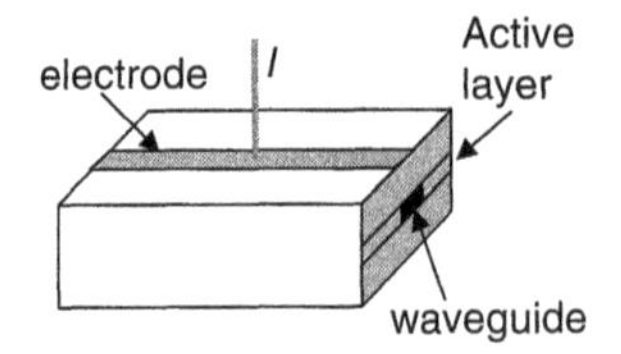

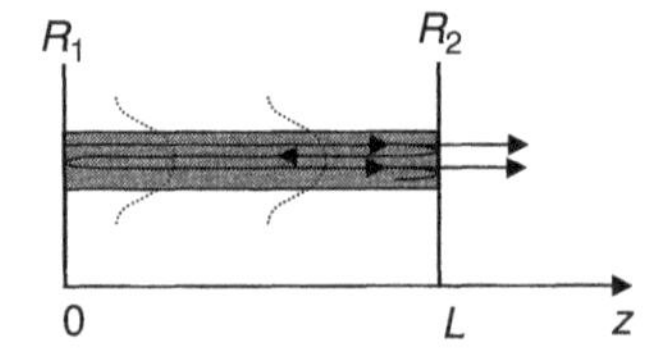

Figure 1.1.8 A laser cavity with facet reflectivity R_1 and R_2, optical gain g, optical loss α, and refractive index n.

material in the cavity provides an optical gain g and an optical loss α per unit length, the refractive index of the material in the cavity is n, and the reflectivity of the facets are R_1 and R_2.

The lightwave travels back and forth in the longitudinal ($\pm z$) direction in the cavity. After each roundtrip in the cavity, the optical field change is:

$$\frac{E_{i+1}}{E_i} = \sqrt{R_1}\sqrt{R_2}\exp(\Delta G + j\Delta\Phi) \tag{1.1.10}$$

where the total phase shift of the optical field is:

$$\Delta\Phi = \frac{2\pi}{\lambda}2nL \tag{1.1.11}$$

and the net optical gain coefficient of the optical field is:

$$\Delta G = (\Gamma g - \alpha)2L \tag{1.1.12}$$

where g is the optical gain coefficient in $[cm^{-1}]$ and α is the material absorption coefficient, also in $[cm^{-1}]$. $0<\Gamma<1$ is a confinement factor. Since not the entire optical field is confined within the active region of the waveguide, Γ is defined as the ratio between the optical field in the active region and the total optical field.

To support a self-sustained oscillation, the optical field has to repeat itself after each roundtrip. Therefore:

$$\sqrt{R_1}\sqrt{R_2}\exp(\Delta G + j\Delta\Phi) = 1 \tag{1.1.13}$$

This is a necessary condition of oscillation, which is also commonly referred to as the *threshold condition*. Equation 1.1.13 can be further decomposed into phase condition and threshold gain condition.

The phase condition is that after each roundtrip the optical phase change must be multiples of 2π, $\Delta\Phi = 2m\pi$, where m is an integer. One important implication of this phase condition is that it can be satisfied by multiple wavelengths,

$$\lambda_m = \frac{2nL}{m} \tag{1.1.14}$$

This explains the reason that a laser may emit at multiple wavelengths, which are generally referred to as *multiple longitudinal modes*.

Example 1.2

For an InGaAsP semiconductor laser operating in 1550 nm wavelength window, if the effective refractive index of the waveguide is $n \approx 3.5$ and the laser cavity length is $L = 300\ \mu m$, find the wavelength spacing between adjacent longitudinal modes.

Solution:

Based on Equation 1.1.14, the wavelength spacing between the m-th mode and the $(m+1)$-th modes can be found as $\Delta\lambda \approx \lambda_m^2/(2nL)$. Assume $\lambda_m = 1550\ nm$, this mode spacing is $\Delta\lambda = 1.144\ nm$, which corresponds to a frequency separation of approximately $\Delta f = 143\ GHz$.

The threshold gain condition is that after each roundtrip the amplitude of optical field does not change—that is, $\sqrt{R_1}\sqrt{R_2}\exp\{(\Gamma g_{th} - \alpha)2L\} = 1$, where g_{th} is the optical field gain at threshold. Therefore, in order to achieve the lasing threshold, the optical gain has to be high enough to compensate for both the material attenuation and the optical loss at the mirrors,

$$\Gamma g_{th} = \alpha - \frac{1}{4L}\ln(R_1 R_2) \tag{1.1.15}$$

In semiconductor lasers, the optical field gain coefficient is a function of the carrier density in the laser cavity, which depends on the rate of carrier injection,

$$g(N) = a(N - N_0) \tag{1.1.16}$$

In this expression, N is carrier density in $[cm^{-3}]$ and N_0 is the carrier density required to achieve material transparency. a is the differential gain coefficient in $[cm^{-2}]$; it indicates the gain per unit length along the laser cavity per unit carrier density.

In addition, due to the limited emission bandwidth, the differential gain is a function of the wavelength, which can be approximated as parabolic,

$$a(\lambda) = a_0\left\{1 - \left(\frac{\lambda - \lambda_0}{\Delta\lambda_g}\right)^2\right\} \tag{1.1.17}$$

for $|\lambda - \lambda_0| \ll \Delta\lambda_g$, where a_0 is the differential gain at central wavelength $\lambda = \lambda_0$ and $\Delta\lambda_g$ is the spectral bandwidth of the material gain.

It must be noted that material gain coefficient g is not equal to the actual gain of the optical power. When an optical field travels along the active waveguide,

$$E(z,t) = E_0 e^{(\Gamma g-\alpha)z} e^{j(\omega t-\beta z)} \tag{1.1.18}$$

where E_0 is the optical field at $z = 0$, ω is optical frequency, and $\beta = 2\pi/\lambda$ is the propagation constant. The envelope optical power along the waveguide is then

$$P(z) = |E(z)|^2 = P_0 e^{2(\Gamma g-\alpha)z} \tag{1.1.19}$$

Combine Equations 1.1.16, 1.1.17, and 1.1.19, and we have

$$P(z,\lambda) = |E(z)|^2 = P(z,\lambda_0)\exp\left\{-2\Gamma g_0\left(\frac{\lambda-\lambda_0}{\Delta\lambda_g}\right)^2 z\right\} \tag{1.1.20}$$

where $P(z,\lambda_0) = P(0,\lambda_0)e^{2(\Gamma g-\alpha)z}$ is the peak optical power at z.

As shown in Figure 1.1.9, although there are a large number of longitudinal modes that all satisfy the phase condition, the threshold gain condition can be reached only by a small number of modes near the center at the gain peak.

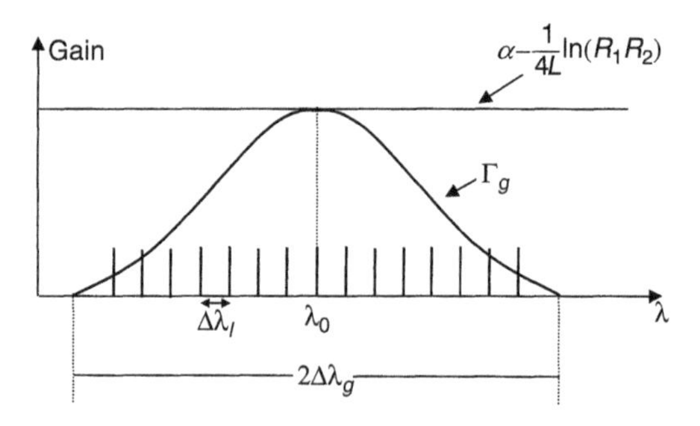

Figure 1.1.9 Gain and loss profile. Vertical bars show the wavelengths of longitudinal modes.

1.1.6.1 Rate Equations

Rate equations describe the nature of interactions between photons and electrons in the active region of a semiconductor laser. Useful characteristics such as output optical power versus injection current, modulation response, and spontaneous emission noise can be found by solving the rate equations.

$$\frac{dN(t)}{dt} = \frac{J}{qd} - \frac{N(t)}{\tau} - 2\Gamma v_g a(N - N_0)P(t) \quad (1.1.21)$$

$$\frac{dP(t)}{dt} = 2\Gamma v_g a(N - N_0)P(t) - \frac{P(t)}{\tau_{ph}} + R_{sp} \quad (1.1.22)$$

where $N(t)$ is the carrier density and $P(t)$ is the photon density within the laser cavity, they have the same unit $[cm^{-3}]$. J is the injection current density in $[A/cm^2]$, d is the thickness of the active layer, v_g is the group velocity of the lightwave in $[cm/s]$, and τ and τ_{ph} are electron and photon lifetimes, respectively. R_{sp} is the rate of spontaneous emission; it represents the density of spontaneously generated photons per second that coupled into the lasing mode, so the unit of R_{sp} is $[cm^{-3}s^{-1}]$.

On the left side of Equation 1.1.21, the first term is the number of electrons injected into each cubic centimeter within each second time window; the second term is the electron density reduction per second due to spontaneous recombination; the third term represents electron density reduction rate due to stimulated recombination, which is proportional to both material gain and the photon density.

The same term $2\Gamma v_g a(N - N_0)P(t)$ also appears in Equation 1.1.22 due to the fact that each stimulated recombination event will generate a photon, and therefore the first term at the left side of Equation 1.1.22 is the rate of photon density increase due to stimulated emission. The second term in Equation 1.1.22 is the photon density decay rate due to both material absorption and photon leakage from the two mirrors. If we distribute mirror losses into the cavity, an equivalent mirror loss coefficient in the unit of $[cm^{-1}]$ can be defined as

$$\alpha_m = \frac{\ln(R_1 R_2)}{4L} \quad (1.1.23)$$

In this way, the photon lifetime can be expressed as

$$\tau_{ph} = \frac{1}{2v_g(\alpha + \alpha_m)} \quad (1.1.24)$$

where α is the material attenuation coefficient. Using this photon lifetime expression, the photon density rate equation (1.22) can be simplified as

$$\frac{dP(t)}{dt} = 2v_g[\Gamma g - (\alpha - \alpha_m)]P(t) + R_{sp} \quad (1.1.25)$$

where g is the material gain as defined in Equation 1.1.16.

Equations 1.1.21 and 1.1.22 are coupled differential equations, and generally they can be solved numerically to predict static as well as dynamic behaviors of semiconductor lasers.

1.1.6.2 Steady State Solutions of Rate Equations

In the steady state, $d/dt = 0$, rate Equations 1.1.21 and 1.1.22 can be simplified as

$$\frac{J}{qd} - \frac{N}{\tau} - 2\Gamma v_g a(N - N_0)P = 0 \quad (1.1.26)$$

$$2\Gamma v_g a(N - N_0)P - \frac{P}{\tau_{ph}} + R_{sp} = 0 \quad (1.1.27)$$

With this simplification, the equations can be solved analytically, which will help understand some basic characteristics of semiconductor lasers.

1.1.6.3 Threshold Carrier Density

Assume that R_{sp}, τ_{ph} and α are constants. Equation 1.1.27 can be expressed as

$$P = \frac{R_{sp}}{1/\tau_{ph} - 2\Gamma v_g a(N - N_0)} \quad (1.1.28)$$

Equation 1.1.28 indicates that when the value of $2\Gamma v_g a(N - N_0)$ approaches that of $1/\tau_{ph}$, the photon density would approach infinite, and this operation point is called *threshold*. Therefore the threshold carrier density is defined as

$$N_{th} = N_0 + \frac{1}{2\Gamma v_g a \tau_{ph}} \quad (1.1.29)$$

Because the photon density should always be positive, $2\Gamma v_g a(N - N_0) < 1/\tau_{ph}$ is necessary, which requires $N < N_{\mathrm{th}}$. Practically, carrier density N can be increased to approach the threshold carrier density N_{th} by increasing the injection current density. However, the threshold carrier density level can never be reached. After the carrier density increases to a certain level, photon density will be increased dramatically and the stimulated recombination becomes significant, which, in turn, reduces the carrier density. Figure 1.10 illustrates the relationships among carrier density, photon density, and the injection current density.

1.1.6.3.1 Threshold Current Density

As shown in Figure 1.1.10, for a semiconductor laser, carrier density linearly increases with the increase of injection current density to a certain level. After that level the carrier density increase is suddenly saturated due to significant contribution of stimulated recombination. The current density corresponding to that saturation point is called *threshold current density*. Generally, about threshold the laser output is dominated by stimulated emission. However, below

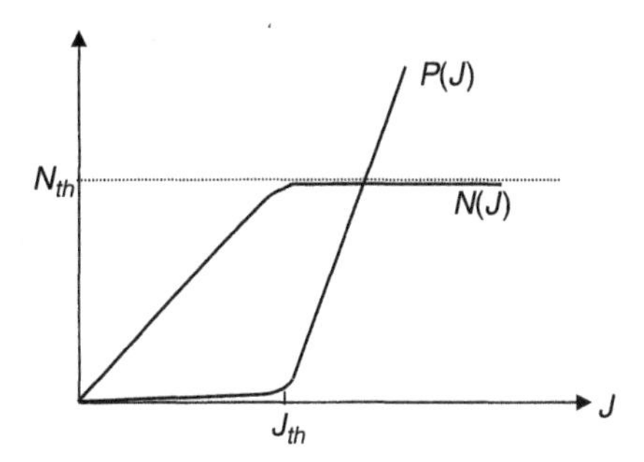

Figure 1.1.10 Photon density *P*(*J*) and Carrier density *N*(*J*) as functions of injection current density *J*. J_{th} is the threshold current density and N_{th} is the threshold carrier density.

threshold spontaneous emission is the dominant mechanism similar to that in an LED and the output optical power is usually very small. In this case, stimulated recombination is negligible, therefore Equation 1.1.26 can be simplified as $J/qd = N/\tau$ for below threshold operation. At threshold point, $N = N_{th}$, as expressed in Equation 1.1.29; therefore, the corresponding threshold current density is

$$J_{th} = \frac{qd}{\tau} = \frac{qd}{\tau}\left(N_0 + \frac{1}{2\Gamma v_g a \tau_{ph}}\right) \tag{1.1.30}$$

1.1.6.3.2 P~J Relationship About Threshold

In general, the desired operation region of a laser diode is above threshold, where high-power coherent light is generated by stimulated emission. Combining Equations 1.1.26 and 1.1.27, we have

$$\frac{J}{qd} = \frac{N}{\tau} + \frac{P}{\tau_{ph}} - R_{sp} \tag{1.1.31}$$

As shown in Figure 1.1.10, in the above threshold regime, carrier density is approximately equal to its threshold value ($N \approx N_{th}$). In addition, since $N_{th}/\tau = J_{th}/qd$, we have

$$\frac{J}{qd} = \frac{J_{th}}{qd} + \frac{P}{\tau_{ph}} - R_{sp} \tag{1.1.32}$$

Therefore, the relationship between photon density and current density above threshold is

$$P = \frac{\tau_{ph}}{qd}(J - J_{th}) + \tau_{ph} R_{sp} \tag{1.1.33}$$

Apart from a small spontaneous emission contribution $\tau_{ph}R_{sp}$, the photon density is linearly proportional to the injection current density for $J > J_{th}$ and the slop is $dP/dJ = \tau_{ph}/qd$.

One question is how to relate the photon density to the output optical power? Assume that the laser active waveguide has a length l, width w, and thickness d, as shown in Figure 1.1.11.

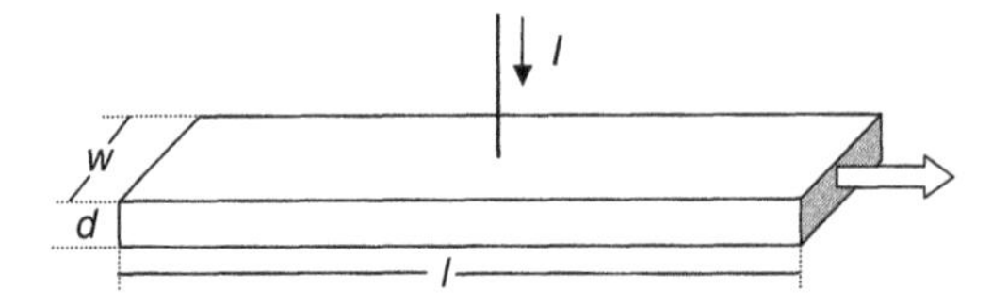

Figure 1.1.11 Illustration of the dimension of a laser cavity.

The output optical power is the flow of photons through the facet, which can be expressed as

$$P_{opt} = P \cdot (lwd)h\nu 2\alpha_m v_g \tag{1.1.34}$$

where $P \cdot (lwd)$ is the total photon number and $P \cdot (lwd)h\nu$ is the total optical power within the cavity. α_m is the mirror loss in $[cm^{-1}]$, which is the percentage of photons that escape from each mirror, and $\alpha_m v_g$ represents the rate of photon escape per second. The factor 2 means photons travel in both directions along the cavity and escape through two end mirrors.

Neglecting the contribution of spontaneous emission in Equation 1.1.33, combining Equations 1.1.34 and 1.1.24, and with Equation 1.1.33, and considering that the injection current is related to current density by $I = J \cdot wl$, we have

$$P_{opt} = \frac{(I - I_{th})h\nu}{q} \cdot \frac{\alpha_m}{\alpha_m + \alpha} \tag{1.1.35}$$

This is the total output optical power exit from both of the two laser facets. Since α is the rate of material absorption and α_m is the photon escape rate through facet mirrors, $\alpha_m/(\alpha + \alpha_m)$ represents an external efficiency.

1.1.6.3.3 Mode Suppression Ratio (MSR)

As illustrated in Figure 1.1.9, in a laser diode the phase condition can be satisfied by multiple wavelengths, which are commonly referred to as *multiple longitudinal modes*. The gain profile has a maximum in the middle, and one of the longitudinal modes is closest to the threshold gain condition. This mode usually has the highest power, which is the *main mode*. However, the power in the modes adjacent to the main mode may not be negligible for many applications that require single-mode operation. To take into account the multimodal effect in a laser diode, the rate equation for the photon density of the mth mode can be written as

$$\frac{dP_m(t)}{dt} = 2v_g\Gamma_m g_m(N)P_m(t) - \frac{P_m(t)}{\tau_{ph}} + R_{sp} \tag{1.1.36}$$

where the subscript m represents the parameter for the m-th mode. Since all the longitudinal modes share the same pool of carrier density, the rate equation for the carrier density is

$$\frac{dN(t)}{dt} = \frac{J}{qd} - \frac{N(t)}{\tau} - \sum_k 2\Gamma_k v_g g_k(N)P_k(t) \tag{1.1.37}$$

Using parabolic approximation for the material gain,

$$g(N,\lambda) = g_0(N)\left\{1 - [(\lambda - \lambda_0)/\Delta\lambda_g]^2\right\}$$

and let $\lambda_m = \lambda_0 + m\Delta\lambda_l$, where $\Delta\lambda_l$ is the mode spacing as shown in Figure 1.1.9. There should be approximately $2M + 1$ modes since there are M modes on each side of the main mode ($-M < m < M$), where $M \approx \Delta\lambda_g/\Delta\lambda_l$.

Therefore, the field gain for the m-th mode can be expressed as a function of the mode index m as

$$g_m(N) = g_0(N)\left\{1 - \left(\frac{m}{M}\right)^2\right\} \tag{1.1.38}$$

The steady state solution of the photon density rate equation of the m-th mode is

$$P_m = \frac{R_{sp}}{1/\tau_{ph} - 2\Gamma v_g g_m(N)} \tag{1.1.39}$$

The gain margin for the main mode ($m = 0$) is defined as

$$\delta = \frac{1}{\tau_{ph}} - 2\Gamma_0 v_g g_0(N) = \frac{R_{sp}}{P_0} \tag{1.1.40}$$

where, P_0 is the photon density of the main mode. Substituting main mode gain margin in Equation 1.1.40 into Equation 1.1.39, the photon density of the m-th mode is

$$P_m = \frac{R_{sp}}{\delta - 2\Gamma v_g[g_0(N) - g_m(N)]} = \frac{R_{sp}}{\delta - 2\Gamma v_g g_0(N)(m^2/M^2)} \tag{1.1.41}$$

The power ratio between the m-th mode and the main mode is then

$$MSR = \frac{P_m}{P_0} = \frac{\delta - 2\Gamma v_g g_0(N)(m^2/M^2)}{\delta} = 1 + \frac{P_0}{R_{sp}} 2\Gamma v_g g_0(N)(m^2/M^2) \tag{1.1.42}$$

Equation 1.1.42 indicates that first of all, the mode suppression ratio is proportional to m^2 because high index modes are far away from the main mode and the gain is also farther down from the threshold. In addition, the mode suppression ratio is proportional to the photon density of the main mode. The reason is that at a high photon density level, stimulated emission is predominantly higher than the spontaneous emission; thus side modes that benefited from spontaneous emission become weaker compared to the main mode.

1.1.6.3.4 Turn-on Delay

In directly modulated laser diodes, when the injection current is suddenly switched on from below to above threshold, there is a time delay between the signal electrical pulse and the output optical pulse. This is commonly referred to as *turn-on delay*.

Turn-on delay is mainly caused by the slow response of the carrier density below threshold. It needs a certain amount of time for the carrier density to build up and to reach the threshold level.

To analyze this process, we have to start from the rate equation at the low injection level J_1 below threshold, where photon density is very small and the stimulated recombination term is negligible in Equation 1.1.21:

$$\frac{dN}{dt} = \frac{J(t)}{qd} - \frac{N(t)}{\tau} \tag{1.1.43}$$

Suppose J(t) jumps from J_1 to J_2 at time $t = 0$. If J_2 is above the threshold current level, the carrier density is supposed to be switched from N_1 to a level very close to the threshold N_{th}, as shown in Figure 1.1.12. Equation 1.1.43

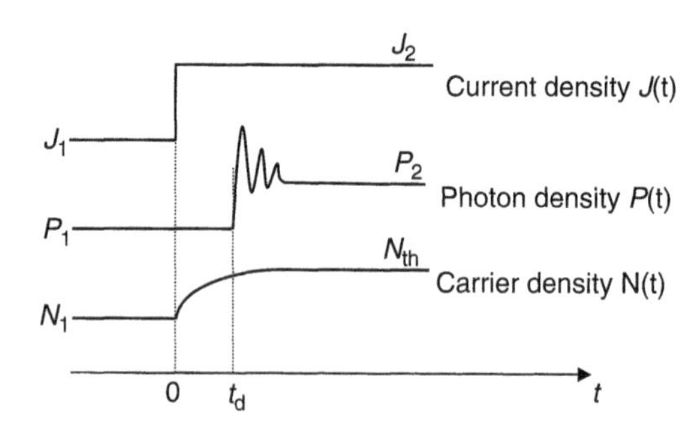

Figure 1.1.12 Illustration of the laser turn-delay. Injection current is turned on at $t = 0$ from J_1 to J_2, but both photon density and carrier density will require a certain time delay to build up toward their final values.

can be integrated to find the time required for carrier density to increase from N_1 to N_{th}:

$$t_d = \int_{N_1}^{N_{th}} \left[\frac{J(t)}{qd} - \frac{N(t)}{\tau}\right]^{-1} dN = \tau \ln\left[\frac{J - J_1}{J - J_{th}}\right] \tag{1.1.44}$$

where $J_1 = qdN_1/\tau$ and $J_{th} = qdN_{th}/\tau$. Because laser threshold is reached only for $t \geq t_d$, the actual starting time of the laser output optical pulse is at $t = t_d$. In practical applications, this time delay t_d may limit the speed of optical modulation in optical systems. Since t_d is proportional to τ, a laser diode with shorter spontaneous emission carrier lifetime may help reduce the turn-on delay. Another way to reduce turn-on delay is to bias the low-level injection current J_1 very close to the threshold. However, this may result in poor extinction ratio of the output optical pulse.

1.1.6.3.5 Small-Signal Modulation Response

In a semiconductor laser, both its optical power and its operation wavelength can be modulated by injection current. In Section 1.1.5 we show that the modulation speed of an LED is inversely proportional to the carrier lifetime. For a laser diode operating above threshold, the modulation speed is expected to be much faster than that of an LED thanks to the contribution of stimulated recombination. When a laser is modulated by a small current signal $\delta J(t)$ around a static operation point J_s: $J = J_s + \delta J(t)$, the carrier density will be $N = N_s + \delta N(t)$, where N_s and $\delta N(t)$ are the static and small signal response, respectively, of the carrier density. Rate Equation 1.1.21 can be linearized for the small-signal response as

$$\frac{d\delta N(t)}{dt} = \frac{\delta J(t)}{qd} - \frac{\delta N(t)}{\tau} - 2\Gamma v_g a P \delta N(t) \tag{1.1.45}$$

Here for simplicity we have assumed that the impact of photon density modulation is negligible. Equation 1.1.45 can be easily solved in frequency domain as

$$\delta\tilde{N}(\omega) = \frac{1}{qd} \frac{\delta\tilde{J}(\omega)}{\left(j\omega + 1/\tau + 2\Gamma v_g a P\right)} \tag{1.1.46}$$

where $\delta\tilde{J}(\omega)$ and $\delta\tilde{N}(\omega)$ are Fourier transforms of $\delta J(t)$ and $\delta N(t)$, respectively. If we define an effective carrier lifetime τ_{eff} such that

$$\frac{1}{\tau_{eff}} = \frac{1}{\tau} + 2\Gamma v_g a P \tag{1.1.47}$$

the 3-dB modulation bandwidth of the laser will be proportional to $1/\tau_{eff}$. For a laser diode operating well above threshold, stimulated recombination is much stronger than spontaneous recombination, i.e., $2\Gamma v_g aP \gg 1/\tau$, and therefore, $\tau_{eff} \ll \tau$. This is the major reason that the modulation bandwidth of a laser diode is much larger than that of an LED.

In this simplified modulation response analysis, we have assumed that photon density is a constant and therefore there is no coupling between the carrier density rate equation and the photon density rate equation. A more precise analysis has to solve coupled rate equations. A direct consequence of coupling between the carrier density and the photon density is that for a sudden increase of injection current, the carrier density will first increase, which will increase the photon density. But the photon density increase tends to reduce carrier density through stimulated recombination. Therefore there could be an oscillation of both carrier density and photon density immediately after the injection current is switched on. This is commonly referred to as *relaxation oscillation*. Detailed analysis of laser modulation can be found in [1].

A unique characteristic of a semiconductor laser is that, in addition to direct intensity modulation, its oscillation frequency can also be modulated by injection current. This frequency modulation is originated from the carrier density-dependent refractive index of the material within the laser cavity. Since refractive index is a parameter of the laser phase condition shown in Equation 1.1.14, change of refractive index will change resonance wavelength of the laser cavity. A direct modulation on a laser diode by injection current will introduce both intensity modulation and the phase modulation. This optical phase modulation is usually referred to as *chirp*. The ratio between emitting optical field phase change rate and the normalized photon density change rate is defined by a well-known linewidth enhancement factor α_{lw} as [2].

$$\alpha_{lw} = 2P\frac{d\phi/dt}{dP/dt} \tag{1.1.48}$$

And therefore, optical frequency shift is related to photon density modulation as

$$\delta f = \frac{d\phi}{dt} = \frac{\alpha_{lw}}{2P}\frac{dP}{dt} \tag{1.1.49}$$

α_{lw} is an important parameter of laser diode, which is determined both by the semiconductor material and by the laser cavity structure. For intensity modulation-based optical systems, lasers with smaller chirp are desired to minimize the spectral width of the modulated optical signal. On the other hand, for optical frequency modulation-based systems such as frequency-shift key (FSK), lasers with large chirp will be beneficial.

1.1.6.3.6 *Laser Noises*

1.1.6.3.6.1 *Relative Intensity Noise* (RIN)

In semiconductor lasers, output optical power may fluctuate due to the existence of spontaneous emission, thus producing intensity noise. Since the quality of a laser output depends on the ratio between the noise power and the total optical power, a commonly used measure of laser intensity noise is the relative intensity noise (RIN), which is defined as

$$RIN = \frac{S_P(\omega)}{P_{opt}^2} \tag{1.1.50}$$

where $S_P(\omega)$ is the intensity noise power spectral density and P_{opt} is the total optical power. Obviously, $S_P(\omega)$ increases with the increase of the total optical power. *RIN* is a convenient way to characterize a laser quality. Generally, *RIN* is a function of frequency; it peaks around relaxation oscillation frequency because of the interaction between carrier density and photon density. The unit of *RIN* is $[Hz^{-1}]$ or $[dB/Hz]$, as a relative measure.

1.1.6.3.6.2 *Phase Noise*

Phase noise is a measure of spectral purity of the laser output. Figure 1.1.13 shows that a spontaneous emission event not only generates intensity variation but also produces phase variation. The spectral width caused by phase noise is commonly referred to as *spectral linewidth*, which is proportional to the rate of spontaneous emission and inversely proportional to the photon density: $\Delta\omega \propto R_{sp}/(2P)$.

In contrast to other types of lasers, a unique feature of the semiconductor laser is the dependency of the optical phase on the photon density through the linewidth enhancement factor α_{lw} as shown by Equation 1.1.48. The photon density variation introduced by each spontaneous emission event will cause a change in the optical phase through the change of the carrier density.

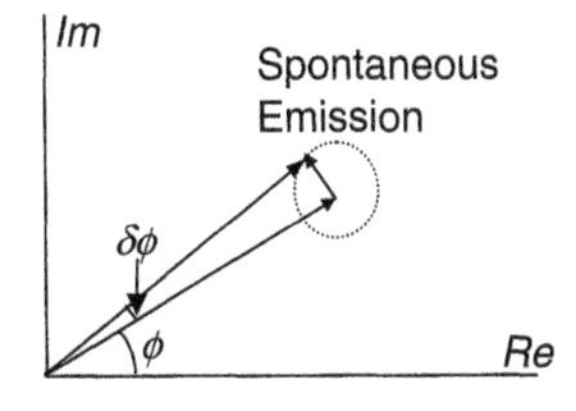

Figure 1.1.13 Optical field vector diagram. Illustration of optical phase noise generated due to spontaneous emission events.

This effect turns out to be much stronger than the direct phase noise process illustrated in Figure 1.1.13, and the overall linewidth expression of a laser diode is

$$\Delta\omega = \frac{R_{sp}}{2P}(1 + \alpha_{lw}^2) \tag{1.1.51}$$

where the second term is the contribution of the photon density-dependent refractive index. This is where the term *linewidth enhancement factor* came from. For typical semiconductor lasers, the value of α_{lw} varies between 2 and 6; therefore, it enhances the laser linewidth by 4 to 36 times [2].

1.1.6.3.6.3 Mode Partition Noise

The output from a semiconductor laser can have multiple longitudinal modes as shown in Figure 1.1.9 if the material gain profile is wide enough. All these longitudinal modes compete for carrier density from a common pool. Although several different modes may have similar gain, the winning mode will consume most of the carrier density and thus the power of other modes will be suppressed. Since the values of gain seen by different modes are not very different, spontaneous emission noise, external reflection, or temperature change may introduce a switch from one mode to another mode. This mode hopping is random and is usually associated with intensity fluctuation. In addition, if the external optical system has wavelength-dependent loss, this mode hopping will inevitably introduce additional intensity noise for the system.

1.1.7 Single-Frequency Semiconductor Lasers

So far we have only considered the laser diode where the resonator consists of two parallel mirrors. This simple structure is called a *Fabry-Perot resonator* and the lasers made with this structure are usually called *Fabry-Perot lasers*, or simply *FP lasers*. An FP laser diode usually operates with multiple longitudinal modes because a phase condition can be met by a large number of wavelengths and the reflectivity of the mirrors is not wavelength selective. In addition to mode partition noise, multiple longitudinal modes occupy wide optical bandwidth, which results in poor bandwidth efficiency and low tolerance to chromatic dispersion of the optical system.

The definition of a single-frequency laser can be confusing. An absolute single-frequency laser does not exist because of phase noise and frequency noise. A single-frequency laser diode may simply be a laser diode with a single longitudinal mode. A more precise definition of single-frequency laser is a laser that not only has a single mode but that mode also has very narrow spectral linewidth. To achieve single-mode operation, the laser cavity has to have a special wavelength selection mechanism. One way to introduce wavelength selectivity is to add a grating along the active layer, which is called *distributed feedback* (DFB). The other way is to add an additional mirror outside the laser cavity, which is referred to as the *external cavity*.

1.1.7.1 DFB Laser Diode

A DFB laser diode is a very popular device that is widely used in optical communication systems. Figure 1.1.14 shows the structure of a DFB laser, where a corrugating grating is written just outside the active layer, providing a periodic refractive index perturbation [3]. Similar to what happens in an FP laser, the lightwave resonating within the cavity is composed of two counter-propagating waves, as shown in Figure 1.1.14(b). However, in the DFB structure, the index grating creates a mutual coupling between the two waves propagating in opposite directions, and therefore mirrors on the laser surface are no longer needed to provide the optical feedback.

Because the grating is periodic, constructive interference between the two waves happens only at certain wavelengths, which provides a mechanism of wavelength selection for the laser cavity. To resonate, the wavelength has to match the grating period, and the resonant condition of a DFB laser is thus,

$$\lambda_g = 2n\Lambda \tag{1.1.52}$$

This is called the *Bragg wavelength*, where Λ is the grating pitch and n is the effective refractive index of the optical waveguide. For wavelengths away from the Bragg wavelength, the two counter-propagated waves do not enhance each other along the way; therefore, self-oscillation cannot be sustained for these wavelengths.

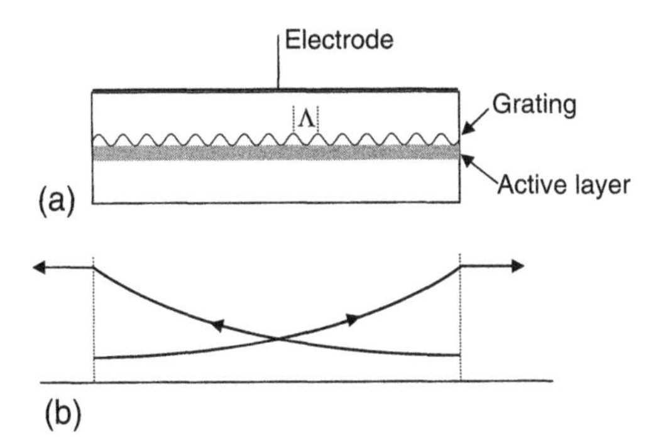

Figure 1.1.14 (a) Structure of a DFB laser with a corrugating grating just outside the active layer, and (b) an illustration of two counter-propagated waves in the cavity.

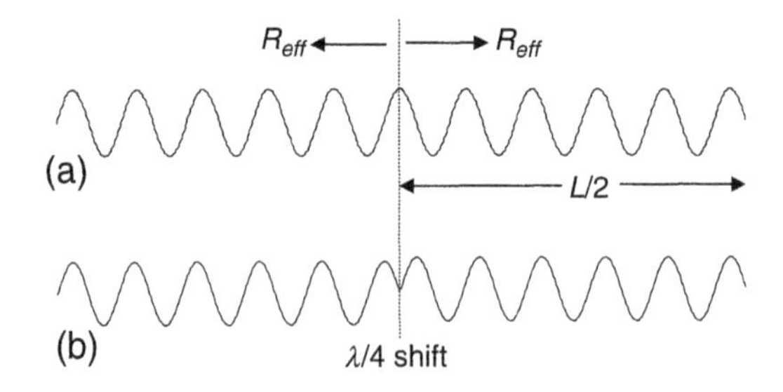

Figure 1.1.15 (a) A uniform DFB grating and (b) a DFB grating with a quarter-wave shift in the middle.

Another way to understand this distributed feedback is to treat the grating as an effective mirror. As shown in Figure 1.1.15, from the reference point in the middle of the cavity looking left and right, the effect of the grating on each side can be viewed as an equivalent mirror, as in a FP laser cavity, with the effective reflectivity R_{eff}. This effective reflectivity is frequency-dependent, as

$$R_{eff} \propto \left| 1 - \left(\frac{\sin x}{x} \right)^2 \right| \tag{1.1.53}$$

where $x = \pi Lc(\lambda - \lambda_g)/(2v_g\lambda_g^2)$, v_g is the group velocity, c is the speed of light, and L is the cavity length.

If the grating is uniform, this effective reflectivity has two major resonance peaks separated by a deep stop band. As a result, a conventional DFB laser generally has two degenerate longitudinal modes and the wavelength separation between these two modes is $\Delta\lambda = 4v_g\lambda_g^2/Lc$. Although residual reflectivity from laser facets may help to suppress one of the two degenerate modes by breaking up the symmetry of transfer function, it is usually not reliable enough for mass production. A most popular technique to create single-mode operation is to add a quarter-wave shift in the middle of the Bragg grating, as shown in Figure 1.1.15(b). This $\lambda/4$ phase shift introduces a phase discontinuity in the grating and results in a strong reflection peak at the middle of the stop-band, as shown in Figure 1.1.16(b). This ensures single longitudinal mode operation in the laser diode at the Bragg wavelength.

1.1.7.2 External Cavity Laser Diode

The operation of a semiconductor laser is sensitive to external feedback [4]. Even a −40 dB optical feedback is enough to bring a laser from single-frequency operation into chaos. Therefore an optical isolator has to be used at the output of a laser diode to prevent optical feedback from

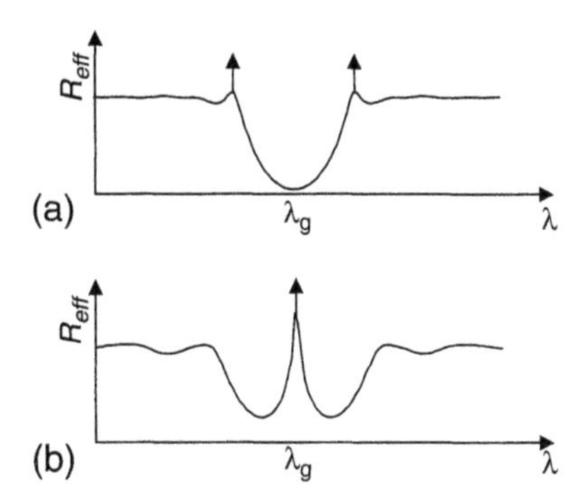

Figure 1.1.16 (a) Structure of a DFB laser with a corrugating grating just outside the active layer, and (b) an illustration of two counter-propagated waves in the cavity.

external optical interfaces. On the other hand, precisely controlled external optical feedback can be used to create wavelength-tunable lasers with very narrow spectral linewidth.

The configuration of a grating-based external cavity laser is shown in Figure 1.1.17, where laser facet reflectivities are R_1 and R_2 and the external grating has a wavelength-dependent reflectivity of $R_3(\omega)$. In this complex cavity configuration, the reflectivity R_2 of the facet facing the external cavity has to be replaced by an effective reflectivity R_{eff} as shown in Figure 1.1.17, as

$$R_{eff}(\omega) = \left\{ \sqrt{R_2} + (1 - R_2)\sqrt{R_3(\omega)} \sum_{m=1}^{\infty} (R_2 R_3(\omega))^{\frac{m-1}{2}} e^{j\omega\tau_e} \right\}^2 \tag{1.1.54}$$

If external feedback is small enough ($R_3 \ll 1$), only one roundtrip needs to be considered in the external cavity. Then Equation 1.1.54 can be simplified as

$$R_{eff}(\omega) \approx R_2 \left\{ 1 + \frac{(1 - R_2)\sqrt{R_3(\omega)}}{\sqrt{R_2}} \right\}^2 \tag{1.1.55}$$

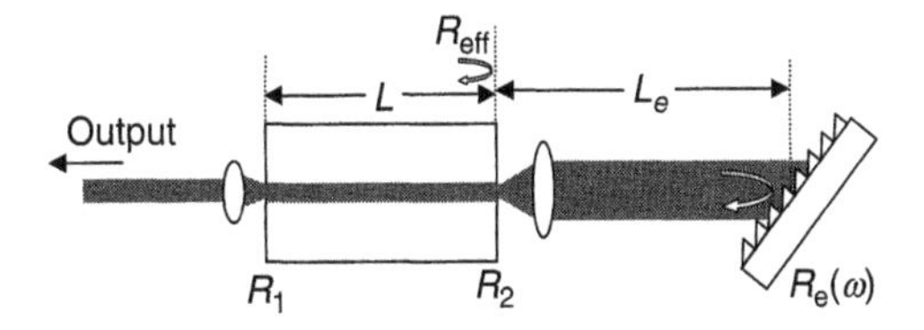

Figure 1.1.17 Configuration of an external cavity semiconductor laser, where the external feedback is provided by a reflective grating.

Then the mirror loss α_m shown in Equation 1.1.23 can be modified by replacing R_2 with R_{eff}. Figure 1.1.18 illustrates the contributions of various loss terms: α_1 is the reflection loss of the grating, which is wavelength selective, and α_2 and α_3 are resonance losses between R_1 and R_2 and between R_2 and R_3, respectively. Combining these three contributions, the total wavelength-dependent loss α_m has only one strong low loss wavelength, which determines the lasing wavelength. In practical external cavity laser applications, an antireflection coating is used on the laser facet facing the external cavity to reduce R_2, and the wavelength dependency of both α_1 and α_2 can be made very small compared to that of the grating; therefore, a large wavelength tuning range can be achieved by rotating the angle of the grating while maintaining single longitudinal mode operation.

External optical feedback not only helps to obtain wavelength tuning, it also changes the spectral linewidth of the emission. The linewidth of an external cavity laser can be expressed as,

$$\Delta v = \frac{\Delta v_0}{1 + k\cos(\omega_0 \tau_e + \tan^{-1}\alpha_{lw})} \tag{1.1.56}$$

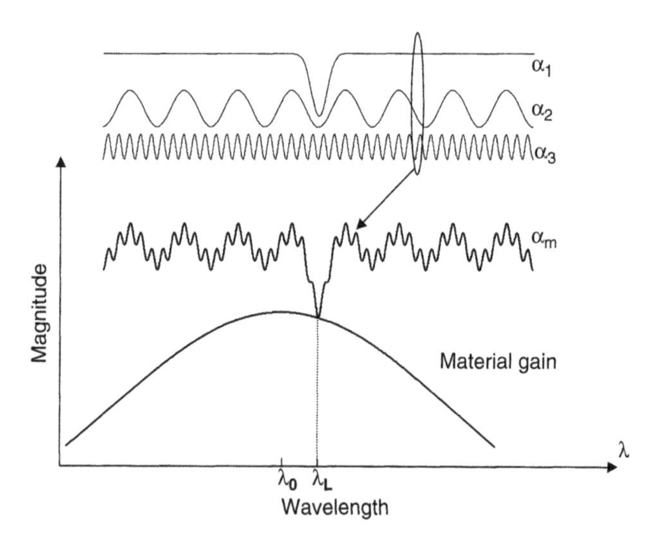

Figure 1.1.18 Illustration of resonance losses between R_1 and R_2 (α_2) and between R_2 and R_3 (α_3). α_1 is the reflection loss of the grating and α_m is the combined mirror loss. Lasing threshold is reached only by one mode at λ_L.

where $\Delta\nu_0$ is the linewidth of the laser diode without external optical feedback, ω_0 is the oscillation angular frequency, α_{lw} is the linewidth enhancement factor, and k represents the strength of the optical feedback. When the feedback is not very strong, this feedback strength can be expressed as

$$k = \frac{\tau_e}{\tau} \frac{(1 - R_2)\sqrt{R_3}}{\sqrt{R_2}} \sqrt{1 + \alpha_{lw}^2} \tag{1.1.57}$$

$\tau = 2nL/c$, $\tau_e = 2n_e L_e/c$ are roundtrip delays of the laser cavity and the external cavity, respectively, with L and L_e the lengths of the laser cavity and the external cavity. n and n_e are refractive indices of the two cavities.

Equation 1.1.56 shows that the linewidth of the external cavity laser depends on the phase of the external feedback. To obtain narrow linewidth, precise control of the external cavity length is critical; a mere $\lambda/2$ variation in the length of external cavity can change the linewidth from the minimum to the maximum. This is why an external cavity has to have a very stringent mechanical stability requirement.

An important observation from Equation 1.1.57 is that the maximum linewidth reduction is proportional to the ratio of the cavity length ratio L_e/L. This is because there is no optical propagation loss in the external cavity and the photon lifetime is increased by increasing the external cavity length. In addition, when photons travel in the external cavity, there is no power-dependent refractive index; this is the reason for including the factor $\sqrt{1 + \alpha_{lw}^2}$ in Equation 1.1.57.

In fact, if the antireflection coating is perfect such that $R_2 = 0$, this ideal external cavity laser can be defined as an *extended-cavity* laser because it becomes a two-section one, with one of the sections passive. With $R_2 = 0$, α_2 and α_3 in Figure 1.1.18 will be wavelength independent. In this case, the laser operation will become very stable and the linewidth is no longer a function of the phase of the external optical feedback. The extended-cavity laser diode linewidth can simply be expressed as [5]

$$\Delta\nu = \frac{\Delta\nu_0}{1 + \frac{\tau_e}{\tau}\sqrt{1 + \alpha_{lw}^2}} \tag{1.1.58}$$

Grating-based external cavity lasers are commercially available and they are able to provide >60 nm continuous wavelength tuning range in a 1550 nm wavelength window <100 kHz spectral linewidth. These lasers are important tools for optical measurement using coherent detection technique, which is discussed in later sections.

1.2 PHOTODETECTORS

The photodetector is the key device in an optical receiver that converts the incoming optical signal into an electrical signal [6, 7]. Semiconductor photodetectors, commonly referred to as *photodiodes*, are the predominant detectors used in optical communication systems and for optical measurements because they are small and have fast detection speed and high detection efficiency. Similar to laser diodes, photodiodes are also based on pn junctions; however, unlike laser diodes where the pn junctions are forward biased, in photodiodes the pn junctions are reverse biased. Although the basic structure of a photodiode is a pn junction, which is usually called a *PN* detector, practical photodiodes use *PIN* structure to enhance quantum efficiency, where an intrinsic layer is sandwiched between the *p*-type and *n*-type layers. An avalanche photodiode (*APD*) is another type of often used detector that introduces photon multiplication through avalanche gain when the bias voltage is high enough. In this section, we discuss the basic structures and useful parameters of these photodetectors.

1.2.1 Pn-Junction Photodiodes

Consider a homojunction between a *p*-type and an *n*-type semiconductor, as shown in Figure 1.2.1. Under reverse bias, the potential barrier is increased and the width of the depletion region is expanded. In the depletion region, there are no free electrons and holes because the electrical field in this region is strong. At this time, when photons are launched into this region and if the photon energy is higher than the material bandgap ($h\nu > E_g$), they may break up initially neutral electron-hole pairs into free electrons and holes. Then under the influence of the electric field, the holes (electrons) will move to the right (left) and create an electrical current flow, called a *photocurrent*.

Since the *p*-region and the *n*-regions are both highly conductive and the electrical field is built up only within the depletion region where the photons are absorbed to generate photocurrent, it is desirable to have a thick depletion layer so that the photon absorption efficiency can be improved. This is usually accomplished by adding an undoped (or very lightly doped) intrinsic layer between the *p*-type and the *n*-type layers to form the so-called PIN structure. In this way, the depletion region can be thick, which is determined in the fabrication. Typically, the thickness of the intrinsic layer is on the order of 100-μm. Another desired property of the photodiode is a sufficient optical window to accept the incoming photons. Practical photodiode geometry is shown in Figure 1.2.2, where the optical signal comes from the *p*-type side of the wafer and the optical window size can be large, independent of the thickness of the intrinsic layer.

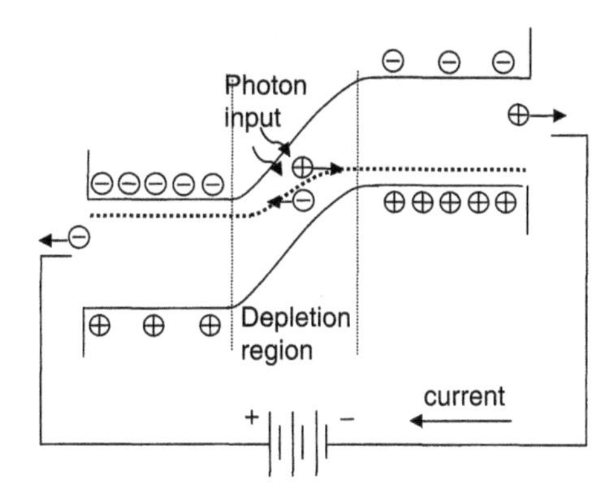

Figure 1.2.1 Band diagram of a pn junction with reverse bias.

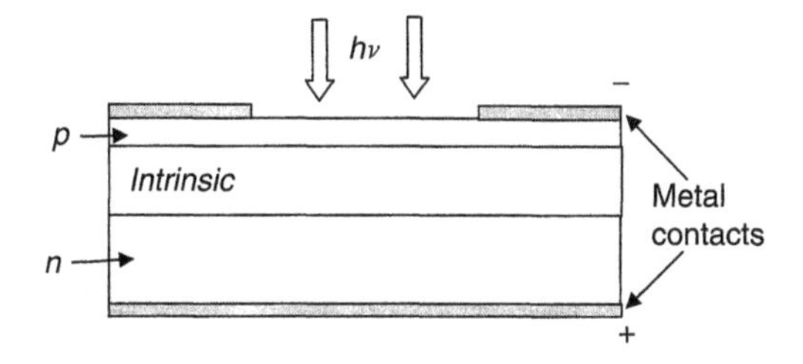

Figure 1.2.2 Geometry of a typical PIN photodetector, where the optical signal is injected from the *p*-side.

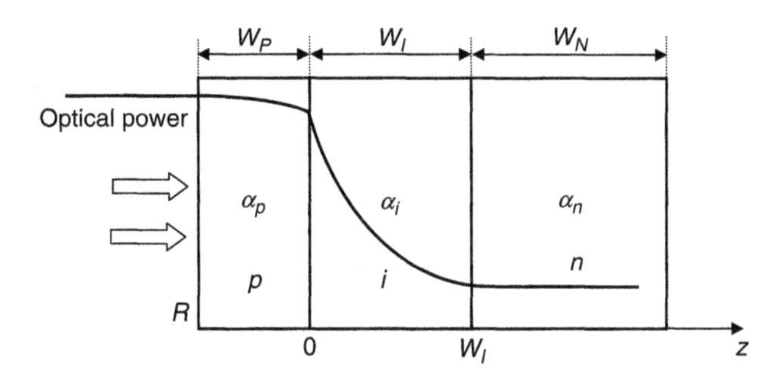

Figure 1.2.3 Illustration of photon absorption in a PIN structure.

The photon absorption process in a PIN structure is illustrated in Figure 1.2.3, where W_P, W_N, and W_I are the thicknesses of *p*-type, *n*-type, and the intrinsic layers, respectively, and the absorption coefficients are α_p, α_n, and α_i for these three regions.

Assume that each photon absorbed within the intrinsic layer produces an electrical carrier, whereas the photons absorbed outside the intrinsic layer are lost; the quantum efficiency η can then be defined by the ratio between the number of electrical carriers generated and the number of photons injected:

$$\eta = (1 - R)\exp(-\alpha_p W_P)[1 - \exp(-\alpha_i W_I)] \tag{1.2.1}$$

where R is the surface reflectivity of the device. Obviously, a necessary condition for Equation 1.2.1 is that each incoming photon has a higher energy than the bandgap ($hv > E_g$), otherwise, η is equal to zero because no carriers can be generated.

1.2.2 Responsivity and Bandwidth

As we have discussed previously that quantum efficiency is defined as the number of photons generated for every incoming photon, responsivity $\Re$, on the other hand defines how many mA photocurrent that can be generated for every mW of signal optical power. If each photon generates a carrier, the responsivity will be $\Re = q/hv = q\lambda/hc$, where q is the electron charge hv is the photon energy, c is the speed of light and λ is the signal wavelength. Considering the non-ideal quantum efficiency η, the photodiode responsivity will be,

$$\Re = \frac{I(mA)}{P(mW)} = \eta\frac{q}{hv} = \eta\frac{q\lambda}{hc} \tag{1.2.2}$$

It is interesting to note that the responsivity is linearly proportional to the wavelength of the optical signal. With the increase of wavelength, the energy per photon becomes smaller, and each photon is still able to generate a carrier but with a lower energy. Therefore the responsivity becomes higher at longer wavelengths. However, when the wavelength is too long and the photon energy is too low, the responsivity will suddenly drop to zero because the necessary condition $hv > E_g$ is not satisfied, as shown in Figure 1.2.4. The longest wavelength to which a photodiode can still have nonzero responsivity is called *cutoff wavelength*, which is, $\lambda_c = hc/E_g$, where E_g is the bandgap of the semiconductor material used to make the photodiode. Typically, a silicon-based photodiode has a cutoff wavelength at about 900 nm and an InGaAs-based photodiode can extend the wavelength to approximately 1700 nm. For that reason, in optical communication systems at 1550 nm wavelengths, Si photodiodes cannot be used.

For example, for a photodiode operating at 1550 nm wavelength, if the quantum efficiency is $\eta = 0.65$, the responsivity can be easily found, using Equation 1.2.2, as $\Re \approx 0.81[mA/mW]$.

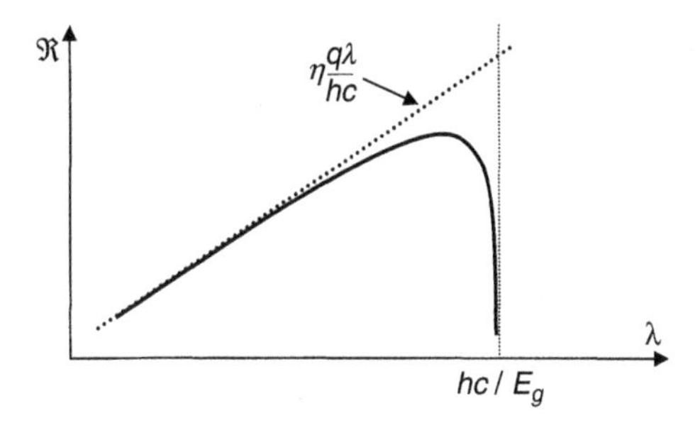

Figure 1.2.4 Photodiode responsivity versus wavelength.

The responsivity can be improved using an antireflection coating at the surface to reduce the reflectivity R. Another way to improve the responsivity is to increase the thickness of the intrinsic layer so that fewer photons can escape into the n-type layer, as shown in Figure 1.2.3. However, this can result in a slower response speed of the photodiode. Now let's see why.

Suppose the input optical signal is amplitude modulated as

$$P(t) = P_0\left(1 + k_m e^{j\omega t}\right) \tag{1.2.3}$$

where P_0 is the average optical power, ω is the modulation angular frequency, and k_m is the modulation index. The generated electrical carriers are then distributed along the z-direction within the intrinsic region, as shown in Figure 1.2.3. Assume that carrier drift velocity is v_n under the electrical field; the photocurrent density distribution will be

$$J(z,t) = J_0[1 + k_m \exp j\omega(t - z/v_n)] \tag{1.2.4}$$

where J_0 is the average current density. The total electrical current will be the collection of contributions across the intrinsic layer thickness:

$$I(t) = \int_0^{W_I} J(z,t)dz = J_0 W_I\left[1 + \frac{k_m}{j\omega\tau_t}\left(e^{j\omega\tau_t} - 1\right)e^{j\omega_m t}\right] \tag{1.2.5}$$

where $\tau_t = W_I/v_n$ is the carrier drift time across the intrinsic region. Neglecting the DC parts in Equations 1.2.3 and 1.2.5, the photodiode response can be obtained as

$$H(\omega) = \frac{I_{ac}}{P_{ac}} = \frac{\Re}{j\omega\tau_t}\left(e^{j\omega\tau_t} - 1\right) = \Re\exp\left(\frac{j\omega\tau_t}{2}\right)\cdot\mathrm{sinc}\left(\frac{\omega\tau_t}{2\pi}\right) \tag{1.2.6}$$

where $\Re = J_0/P_0$ is the average responsivity. The 3-dB bandwidth of $|H(\omega)|^2$ can be easily found as

$$\omega_c = \frac{2.8}{\tau_t} = 2.8\frac{v_n}{W_I} \tag{1.2.7}$$

Clearly, increasing the thickness of the intrinsic layer will slow the response speed of the photodiode. Carrier drift speed is another critical parameter of detector speed. v_n increases with the increase of external bias voltage, but it saturates at approximately $8 \times 10^6\ cm/s$ in silicon, when the field strength is about $2 \times 10^4\ V/cm$. As an example, for a silicon-based photodiode with 10 μm intrinsic layer thickness, the cutoff bandwidth will be about $f_c = 3.6\ GHz$. The carrier mobility in InGaAs is usually much higher than that in silicon; therefore, the ultrahigh-speed photodiodes can be made.

Another important parameter affecting the speed of a photodiode is the junction capacitance, which can be expressed as

$$C_j = \frac{\varepsilon_i A}{W_I} \tag{1.2.8}$$

where ε_i is the permittivity of the semiconductor and A is the junction area. Usually, large area photodiodes have lower speed due to their large junction capacitance.

1.2.3 Electrical Characteristics of a Photodiode

The terminal electrical characteristic of a photodiode is similar to that of a conventional diode; its current-voltage relationship is shown in Figure 1.2.5. The diode equation is

$$I_j = I_D\left[\exp\left(\frac{V}{xV_T}\right) - 1\right] \tag{1.2.9}$$

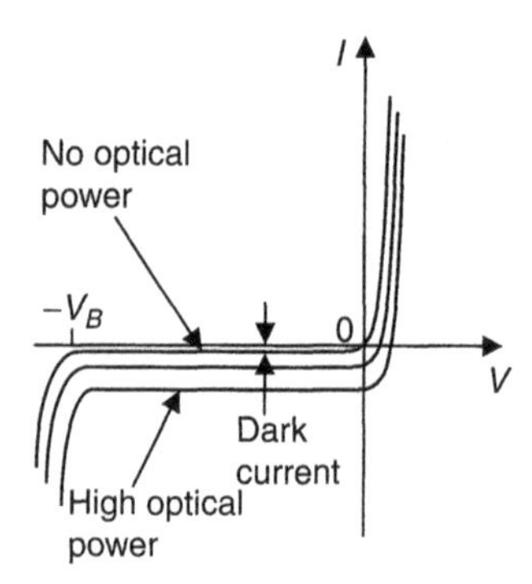

Figure 1.2.5 Photodiode current-voltage relationship.

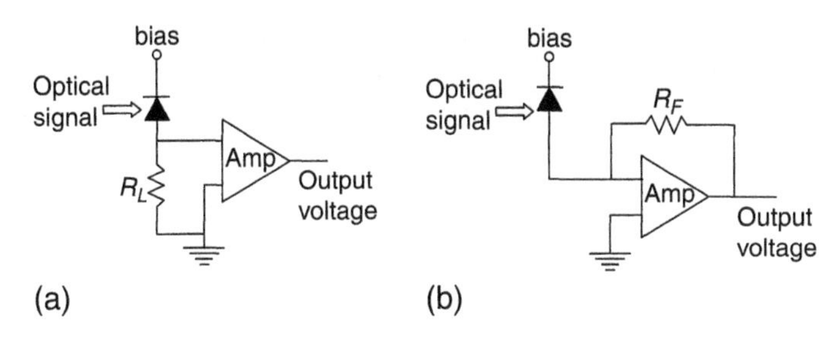

Figure 1.2.6 Two often used pre-amplifier circuits: (a) a voltage amplifier and (b) a transimpedance amplifier.

where $V_T = kT/q$ is the thermal voltage ($V_T \approx 25\ mV$ at room temperature), $2 > x > 1$ is a device structure-related parameter, and I_D is the reverse saturation current, which may range from pico-ampere to nano-ampere, depending on the structure of the device. When a photodiode is forward biased (please do not try this; it could easily damage the photodiode), current flows in the forward direction, which is exponentially proportional to the bias voltage. On the other hand, when the photodiode is reversely biased, the reverse current is approximately equal to the reverse saturation current I_D when there is no optical signal received by the photodiode. With the increase of the signal optical power, the reverse current linearly increases as described by Equation 1.2.2. Reverse bias also helps increase the detection speed as described in the last section. This is the normal operation region of a photodiode. When the reverse bias is too strong ($V \leq -V_B$), the diode may break down, where V_B is the breakdown voltage.

To construct an optical receiver, a photodiode has to be reverse biased and the photocurrent has to be amplified. Figure 1.2.6 shows two typical electrical circuit examples of optical receivers. Figure 1.2.6(a) is a voltage amplifier, where the load resistance seen by the photodiode can be high. High load resistance makes the amplifier highly sensitive; we will see later that the thermal noise is relatively small in this case. However, the frequency bandwidth of the amplifier, which is inversely proportional to the parasitic capacitance and the load resistance, may become narrow. On the other hand, in the transimpedance amplifier shown in Figure 1.2.6(b), the equivalent load resistance seen by the photodiode can be low. This configuration is usually used in optical receivers, which require high speed and wide bandwidth.

1.2.4 Photodetector Noise and SNR

In optical communication systems and electro-optical measurement setups, signal quality depends not only on the signal level itself but also on the signal-to-noise ratio (SNR). To achieve a high SNR, the photodiode must have high quantum efficiency and low noise. We have already discussed the quantum efficiency and the responsivity of a photodiode previously; in this section, we discuss noise sources associated with photodetection.

It is straightforward to find the expression of a signal photocurrent as

$$I_s(t) = \Re P_s(t) \tag{1.2.10}$$

where I_s(t) is the signal photocurrent, P_s(t) is the signal optical power, and $\Re$ is the photodiode responsivity as defined in Equation 1.2.2.

Major noise sources in a photodiode can be categorized as thermal noise, shot noise, and dark current noise. Because of the random nature of the noises, the best way to specify them is to use their statistical values, such as spectral density, power, and bandwidth.

Thermal noise is generated by the load resistor, which is a white noise. Within an electrical bandwidth B, the mean-square thermal noise current can be expressed as

$$\langle i_{th}^2 \rangle = \frac{4kTB}{R_L} \tag{1.2.11}$$

where R_L is the load resistance, k is the Boltzmann's constant, and T is the absolute temperature. Large load resistance helps reducing thermal noise; however, as we discussed in the last section, receiver bandwidth will be reduced by increasing the RC constant. In most high-speed optical receivers, 50Ω is usually used as a standard load resistance.

Shot noise arises from the statistic nature of photodetection. For example, if 1 μW optical power at 1550 nm wavelength is received by a photodiode, it means that statistically about 7.8 trillion photons hit the photodiode every second. However, these photons are not synchronized and they come randomly. The generated photocurrent will fluctuate as the result of this random nature of photon arrival. Shot noise is also a wideband noise and the mean-square shot noise current is proportional to the signal photocurrent detected by the photodiode. Within a receiver bandwidth B, the mean-square shot noise current is

$$\langle i_{sh}^2 \rangle = 2qI_sB = 2q\Re P_sB \tag{1.2.12}$$

Dark current noise is a constant current that exists when no light is incident on the photodiode. This is the reason it is called *dark* current. As shown in Figure 1.2.5, this dark current is the same thing as the reverse saturation current. Because of the statistical nature of the carrier generation process, this dark current also has a variance and its mean-square value is

$$\langle i_{dk}^2 \rangle = 2qI_DB \tag{1.2.13}$$

where I_D is the average dark current of the photodiode.

After discussing signal and noise photocurrents, now we can put them together to discuss SNR. SNR is usually defined as the ratio of signal electrical power and noise electrical power. This is equivalent to the ratio of their mean-square currents,

$$SNR = \frac{\langle I_s^2(t)\rangle}{\langle i_{th}^2\rangle + \langle i_{sh}^2\rangle + \langle i_{dk}^2\rangle + \langle i_{amp}^2\rangle} \tag{1.2.14}$$

where $\langle i_{amp}^2\rangle$ is the equivalent mean-square noise current of the electrical pre-amplifier.

In many optical systems, the optical signal is very weak when it reaches the receiver. In these cases, thermal noise is often the dominant noise. If we only consider thermal noise and neglect other noises, the SNR will be simplified as

$$SNR_{thermal} = \frac{\Re^2 R_L}{4kTB} P_s^2 \propto P_s^2 \tag{1.2.15}$$

Here SNR is proportional to the square of the incident optical power.

On the other hand, if the incident optical signal is very strong, shot noise will become the dominant noise source. If we consider only shot noise and neglect other noises, the SNR will be

$$SNR_{shot} = \frac{\Re^2 P_s^2}{2q\Re P_s B} = \frac{\Re P_s}{2qB} \propto P_s \tag{1.2.16}$$

In this case, SNR is linearly proportional to the incident optical power.

Example 1.3

For a photodiode operating in a 1550 nm wavelength window with the following parameters: $\eta = 0.85$, $R_L = 50\Omega$, $I_D = 5\text{nA}$, $T = 300K$ and $B = 1$ GHz, find the output SNR versus incident optical power when separately considering various noise sources.

Solution:

This problem can be solved using Equations 1.2.10—1.2.13. As a result, Figure 1.2.7 shows the calculated SNR versus the input optical power in dBm. In this example, the thermal noise remains the limiting factor to SNR when the signal optical power is low. When the signal power is higher than approximately 0 dBm, shot noise becomes the limiting factor.

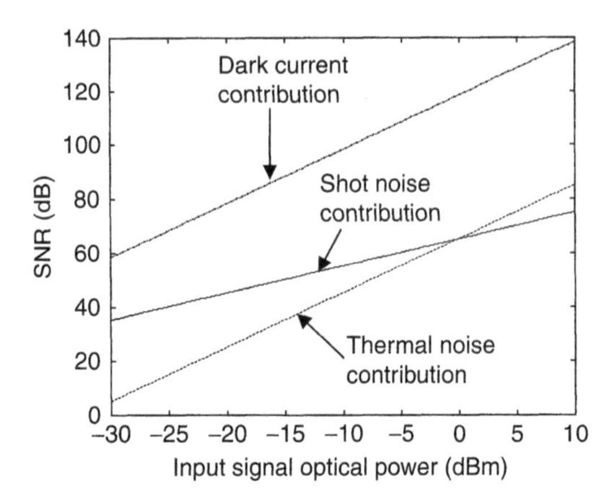

Figure 1.2.7 Decomposing SNR by its three contributing factors.

Also, thermal noise-determined SNR has a slope of 2 dB/1dB, whereas shot noise-determined SNR has a slope of 1 dB/1dB, as discussed in Equations 1.2.15 and 1.2.16. In this example, the impact of dark current noise is always negligible. In general, dark current noise is only significant in low-speed optical receivers where load resistance R_L is very high and the thermal noise level is therefore very low.

1.2.4.1 Noise-Equivalent Power (NEP)

NEP is another useful parameter to specify a photodetector. *NEP* is defined as the minimum optical power required to obtain a unity SNR in a 1Hz bandwidth. Usually only thermal noise is considered in the definition of NEP. From Equation 1.2.15, if we let $\Re^2 R_L P_s^2/(4kTB) = 1$, it will require

$$NEP = \frac{P_s}{\sqrt{B}} = \sqrt{\frac{4kT}{\Re^2 R_L}} \tag{1.2.17}$$

According to this definition, the unit of NEP is in $[W/\sqrt{Hz}]$. Some manufacturers specify NEP for their photodiode products. Obviously, small NEP is desired for high- quality photodiodes. As an example at room temperature, for a photodetector operating in a 1550 nm wavelength window with quantum efficiency $\eta = 0.85$ and load resistance $R_L = 50\Omega$, the NEP value is approximately $NEP = 1.72 \times 10^{-11} \left[W/\sqrt{Hz}\right]$. In an optical system requiring the measurement of low levels of optical signals, NEP of the optical receiver has to be low enough to guarantee the required SNR and thus the accuracy of the measurement.

1.2.5 Avalanche Photodiodes (APDs)

The typical responsivity of a PIN photodiode is limited to the level of approximately 1mA/mW because the quantum efficiency η cannot be higher than 100 percent. To further increase the detection responsivity, avalanche photodiodes were introduced in which photocurrents are internally amplified before going to the load. A very high electrical field is required to initiate the carrier multiplication; therefore, the bias voltage an APD needed can be as high as 100V.

Under a very strong electrical field, an electron is able to gain sufficient kinetic energy and it can knock several electrons loose from neutral electron-hole pairs; this is called *ionization*. These newly generated free electrons will also be able to gain sufficient kinetic energy under the same electrical field and to create more free carriers; this is commonly referred to as the *avalanche effect*. Figure 1.2.8 illustrates this carrier multiplication process, where one input electron generates five output electrons and four holes.

Figure 1.2.9 shows a commonly used APD structure, where a highly resistive intrinsic semiconductor material is deposited as an epitaxial layer on a heavily doped *p*-type (p^+) substrate. Then a *p*-type implantation or diffusion is made on top of the intrinsic layer to create a thin *p*-type layer. Finally, another heavily doped *n*-type (n^+) epitaxial layer is deposited on the top. Because the charge density at the n^+p junction suddenly changes the sign, the built-in electrical field is very strong. This electrical field density is further intensified when a high reverse biasing across the device is applied.

When a photon is absorbed in the intrinsic region, it generates a free carrier. The electron moves toward the avalanche region under the applied electrical field. Once it enters the avalanche region, it is quickly accelerated by the very strong electrical field within the region to initiate the ionization and avalanche

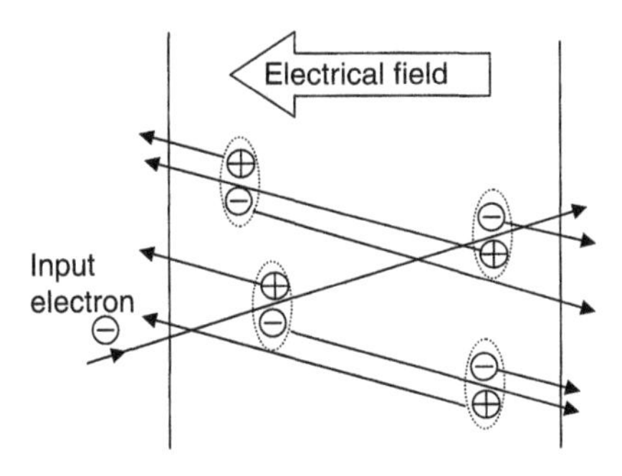

Figure 1.2.8 Illustration of the carrier multiplication process.

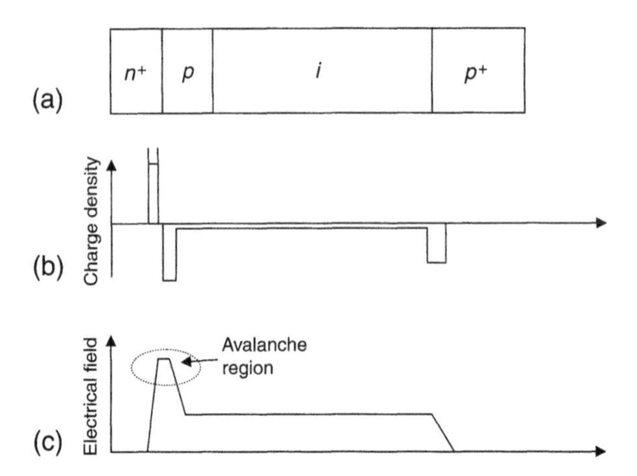

Figure 1.2.9 (a) APD layer structure, (b) charge density distribution, and (c) electrical field density profile.

process. The result of the avalanche process is that each input photon may generate multiple free electrons and holes. In this case, the responsivity expression of the photodiode needs to be modified as

$$\Re_{APD} = M_{APD}\Re_{PIN} = M_{APD}\eta\frac{q\lambda}{hc} \tag{1.2.18}$$

where M_{APD} is defined as the APD gain. Since the avalanche process depends on the applied external voltage, the APD gain strongly depends on the reverse bias. An often used simple expression of APD gain is

$$M_{APD} = \frac{1}{1-(V/V_B)^{n_B}} \tag{1.2.19}$$

where n_B is a parameter that depends on the device structure and the material. V is the applied reverse bias voltage and V_B is defined as the breakdown voltage of the APD, and Equation 1.2.19 is valid only for $V < V_B$. Obviously, when the reverse bias voltage approaches the breakdown voltage V_B, the APD gain approaches infinite.

In addition to APD gain, another important parameter in an APD is its frequency response. In general the avalanche process slows the response time of the APD, and this bandwidth reduction is proportional to the APD gain. A simplified equation describing the frequency response of APD gain is

$$M_{APD}(\omega) = \frac{M_{APD,0}}{\sqrt{1+(\omega\tau_e M_{APD,0})^2}} \tag{1.2.20}$$

where $M_{APD,0} = M_{APD}(0)$ is the APD gain at DC as shown in Equation 1.2.19 and τ_e is an effective transient time, which depends on the thickness of the avalanche region and the speed of the carrier drift. Therefore, the 3-dB bandwidth of APD gain is

$$f_c = \frac{1}{2\pi\tau_e M_{APD,0}} \tag{1.2.21}$$

In practical applications, the frequency bandwidth requirement has to be taken into account when choosing APD gain.

Due to carrier multiplication, signal photocurrent will be amplified by a factor M_{APD} as

$$I_{s,APD}(t) = \Re M_{APD} P_s(t) \tag{1.2.22}$$

As far as the noises are concerned, since the thermal noise is generated in the load resistor R_L, it is not affected by the APD gain. However, both shot noise and the dark current noise are generated within the photodiode, and they will be enhanced by the APD gain. Within a receiver bandwidth B, the mean-square shot noise current in an APD is

$$\langle i_{sh,APD}^2 \rangle = 2q\Re P_s B M_{APD}^2 F(M_{APD}) \tag{1.2.23}$$

The dark current noise in an APD is

$$\langle i_{dk,APD}^2 \rangle = 2qI_D B M_{APD}^2 F(M_{APD}) \tag{1.2.24}$$

In both Equations 1.2.23 and 1.2.14, $F(M_{APD})$ is a noise figure associated with the random nature of carrier multiplication process in the APD. This noise figure is proportional to the APD gain M_{APD}. The following simple expression is found to fit well with measured data for most practical APDs:

$$F(M) = (M_{APD})^x \tag{1.2.25}$$

where $0 \leq x \leq 1$, depending on the material. For often used semiconductor materials, $x = 0.3$ for Si, $x = 0.7$ for InGaAs, and $x = 1$ for Ge avalanche photodiodes.

From a practical application point of view, APD has advantages compared to conventional PIN when the received optical signal is very weak and the receiver SNR is limited by thermal noise. In quantum noise limited optical receivers, such as coherent detection receivers, APD should, in general, not be used, because it would only increase noise level and introduce extra limitations in the electrical bandwidth.

1.3 OPTICAL FIBERS

Optical fiber is the most important component in fiber-optic communication systems as well as in many fiber-based optical measurement setups. The basic structure of an optical fiber is shown in Figure 1.3.1, which has a central core, a cladding, and an external coating to protect and strengthen the fiber.

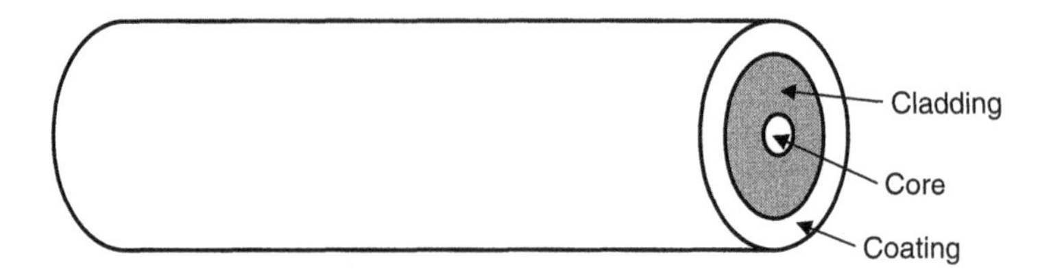

Figure 1.3.1 Basic structure of an optical fiber.

Light guiding and propagation in an optical fiber is based on total internal reflection between the core and the cladding. Because of the extremely low attenuation in silica, optical signals can transmit for very long distance along an optical fiber without significant loss in optical power. To understand the mechanism of optical propagation in fiber, we start by discussing fundamental concept of light wave reflection and refraction at an optical interface.

1.3.1 Reflection and Refraction

An optical interface is generally defined as a plane across which optical property discontinues. For example, water surface is an optical interface because the refractive indices suddenly change from $n = 1$ in the air to $n = 1.3$ in the water. To simplify our discussion, the following assumptions have been made:

1. Plane wave propagation
2. Linear medium
3. Isotropic medium
4. Smooth planar optical interface

As illustrated in Figure 1.3.2, an optical interface is formed between two optical materials with refractive indices of n_1 and n_2, respectively. A plane optical wave is projected onto the optical interface at an incident angle θ_1 (with respect to the surface normal). The optical wave is linearly polarized, and its field amplitude vector can be decomposed into two orthogonal components, $E^i_{//}$ and $E^i_{\perp}$ parallel and perpendicular to the incidence plane. At the optical interface, part of the energy is reflected back to the same side of the interface and the other part is refracted across the interface.

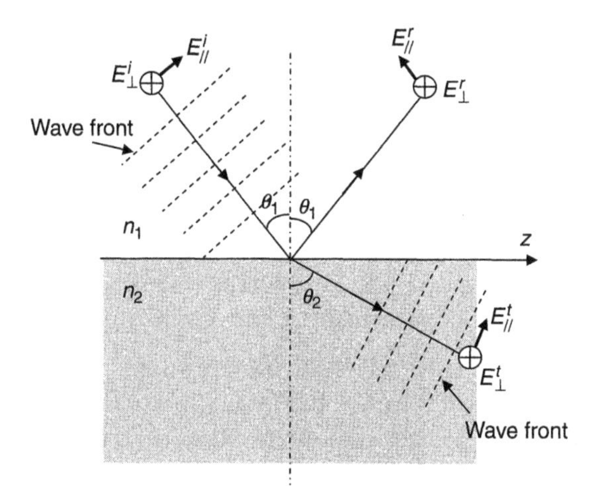

Figure 1.3.2 Plane wave reflection and refraction at an optical interface.

Snell's Law:

$$n_2 \sin\theta_2 = n_1 \sin\theta_1 \tag{1.3.1}$$

tells the propagation direction of the refracted light wave with respect to that of the incident wave. The wave propagation direction change is proportional to the refractive index difference across the interface.

Snell's Law is derived based on the fact that phase velocity along z-direction should be continuous across the interface. Since the phase velocities in the z-direction are $v_{p1} = \frac{c}{n_1}\frac{1}{\sin\theta_1}$ and $v_{p2} = \frac{c}{n_2}\frac{1}{\sin\theta_2}$ at the two sides of the interface, Equation 1.3.1 can be obtained with $v_{p1} = v_{p2}$.

Because Snell's Law was obtained without any assumption of light wave polarization state and wavelength, it is independent of these parameters. An important implication of Snell's Law is that $\theta_2 > \theta_1$ with $n_2 < n_1$.

1.3.1.1 Fresnel Reflection Coefficients

To find out the strength and the phase of the optical field that is reflected back to the same side of the interface, we have to treat optical field components $E^i_{//}$ and $E^i_\perp$ separately. An important fact is that optical field components parallel to the interface must be continuous at both sides of the interface.

Let's first consider the field components $E^i_{//}$, $E^t_{//}$ and $E^r_{//}$ (they are all parallel to the incident plane but not to the interface). They can be decomposed into components parallel with and perpendicular to the interface; the parallel components are, $E^i_{//}\cos\theta_1$, $E^t_{//}\cos\theta_2$, and $-E^r_{//}\cos\theta_1$, respectively, which can be derived from Figure 1.3.2. Because of the field continuity across the interface, we have

$$\left(E^i_{//} - E^r_{//}\right)\cos\theta_1 = E^t_{//}\cos\theta_2 \tag{1.3.2}$$

At the same time, the magnetic field components associated with $E^i_{//}$, $E^t_{//}$ and $E^r_{//}$ have to be perpendicular to the incident plane, and they are $H^i_\perp = \sqrt{\varepsilon_1/\mu_1}E^i_{//}$, $H^t_\perp = \sqrt{\varepsilon_2/\mu_2}E^t_{//}$ and $H^r_\perp = \sqrt{\varepsilon_1/\mu_1}E^r_{//}$, respectively, where ε_1 and ε_2 are electrical permittivities and μ_1 and μ_2 are magnetic permittivities of the optical materials at two sides of the interface. Since $H^i_\perp$, $H^t_\perp$ and $H^r_\perp$ are all parallel to the interface (although perpendicular to the incident plane), magnetic field continuity requires $H^i_\perp + H^r_\perp = H^t_\perp$. Assume that $\mu_1 = \mu_2$, $\sqrt{\varepsilon_1} = n_1$ and $\sqrt{\varepsilon_2} = n_2$, and we have

$$n_1 E^i_{//} + n_1 E^r_{//} = n_2 E^t_{//} \tag{1.3.3}$$

Combine Equations 1.3.2 and 1.3.3 we can find the field reflectivity:

$$\rho_{//} = \frac{E^r_{//}}{E^i_{//}} = \frac{n_1\cos\theta_2 - n_2\cos\theta_1}{n_1\cos\theta_2 + n_2\cos\theta_1} \tag{1.3.4}$$

Using Snell's Law, Equation 1.3.4 can also be written as

$$\rho_{//} = \frac{-n_2^2\cos\theta_1 + n_1\sqrt{\left(n_2^2 - n_1^2\sin^2\theta_1\right)}}{n_2^2\cos\theta_1 + n_1\sqrt{\left(n_2^2 - n_1^2\sin^2\theta_1\right)}} \tag{1.3.5}$$

where variable θ_2 is eliminated.

Similar analysis can also find the reflectivity for optical field components perpendicular to the incident plane as

$$\rho_\perp = \frac{E^r_\perp}{E^i_\perp} = \frac{n_1\cos\theta_1 - n_2\cos\theta_2}{n_1\cos\theta_1 + n_2\cos\theta_2} \tag{1.3.6}$$

or, equivalently,

$$\rho_\perp = \frac{n_1\cos\theta_1 - \sqrt{\left(n_2^2 - n_1^2\sin^2\theta_1\right)}}{n_1\cos\theta_1 + \sqrt{\left(n_2^2 - n_1^2\sin^2\theta_1\right)}} \tag{1.3.7}$$

Power reflectivities for parallel and perpendicular field components are therefore

$$R_{//} = \left|\rho_{//}\right|^2 = \left|E^r_{//}/E^i_{//}\right|^2 \tag{1.3.8}$$

and

$$R_{\perp} = \left|\rho_{\perp}\right|^2 = \left|E^r_{\perp}/E^i_{\perp}\right|^2 \tag{1.3.9}$$

Then, according to energy conservation, the power transmission coefficients can be found as

$$T_{//} = \left|E^t_{//}/E^i_{//}\right|^2 = 1 - \left|\rho_{//}\right|^2 \tag{1.3.10}$$

and

$$T_{\perp} = \left|E^t_{\perp}/E^i_{\perp}\right|^2 = 1 - \left|\rho_{\perp}\right|^2 \tag{1.3.11}$$

In practice, for an arbitrary incidence polarization state, the input field can always be decomposed into $E_{//}$ and $E_{\perp}$ components. Each can be treated independently.

1.3.1.2 Special Cases

1.3.1.2.1 Normal Incidence

This is when the light is launched perpendicular to the material interface. In this case, $\theta_1 = \theta_2 = 0$ and $\cos\theta_1 = \cos\theta_2 = 1$, the field reflectivity can be simplified as

$$\rho_{//} = \rho_{\perp} = \frac{n_1 - n_2}{n_1 + n_2} \tag{1.3.12}$$

Note that there is no phase shift between incident and reflected field if $n_1 > n_2$ (the phase of both $\rho_{//}$ and $\rho_{\perp}$ is zero). On the other hand, if $n_1 < n_2$, there is a π phase shift for both $\rho_{//}$ and $\rho_{\perp}$ because they both become negative.

With normal incidence, the power reflectivity is

$$R_{//} = R_{\perp} = \left|\frac{n_1 - n_2}{n_1 + n_2}\right|^2 \tag{1.3.13}$$

This is a very often used equation to evaluate optical reflection. For example, reflection at an open fiber end is approximately 4 percent. This is because the refractive index in the fiber core is $n_1 \approx 1.5$(silica fiber) and refractive index of air is $n_1 = 1$. Therefore:

$$R = \left|\frac{n_1 - n_2}{n_1 + n_2}\right|^2 = \left|\frac{1.5 - 1}{1.5 + 1}\right|^2 = 0.2^2 = 0.04 \approx -14\ dB$$

In practical optical measurement setups using optical fibers, if optical connectors are not properly terminated, the reflections from fiber-end surface can potentially cause significant measurement errors.

1.3.1.2.2 Critical Angle

Critical angle is defined as an incident angle θ_1 at which total reflection happens at the interface. According to Fresnel Equations 1.3.5 and 1.3.7, the only possibility that $|\rho_{//}|^2 = |\rho_{\perp}|^2 = 1$ is to have $n_2^2 - n_1^2 \sin^2\theta_1 = 0$ or

$$\theta_1 = \theta_c = \sin^{-1}(n_2/n_1) \tag{1.3.14}$$

where θ_c is defined as the critical angle. Obviously the necessary condition to have a critical angle depends on the interface condition.

First, if $n_1 < n_2$, there is no real solution for $\sin^2\theta_1 = (n_2/n_1)^2$. This means that when a light beam goes from a low index material to a high index material, total reflection is not possible.

Second, if $n_1 > n_2$, a real solution can be found for $\theta_c = \sin^{-1}(n_2/n_1)$ and therefore total reflection can only happen when a light beam launches from a high index material to a low index material.

It is important to note that at a larger incidence angle $\theta_1 > \theta_c$, $n_2^2 - n_1^2 \sin^2\theta_1 < 0$ and $\sqrt{n_2^2 - n_1^2 \sin^2\theta_1}$ becomes imaginary. Equations 1.3.5 and 1.3.7 clearly show that if $\sqrt{n_2^2 - n_1^2 \sin^2\theta_1}$ is imaginary, both $|\rho_{//}|^2$ and $|\rho_{\perp}|^2$ are equal to 1.

The important conclusion is that for all incidence angles satisfying $\theta_1 > \theta_c$ total internal reflection will happen with $R = 1$.

1.3.1.3 Optical Field Phase Shift Between the Incident and the Reflected Beams

(a) When $\theta_1 < \theta_c$ (partial reflection and partial transmission), both $\rho_{//}$ and $\rho_{\perp}$ are real and therefore there is no phase shift for the reflected wave at the interface.

(b) When total internal reflection happens, $\theta_1 > \theta_c$, $\sqrt{n_2^2 - n_1^2 \sin^2\theta_i}$ is imaginary. Fresnel Equations 1.3.5 and 1.3.7 can be written as

$$\rho_{//} = \frac{-n_2^2 \cos\theta_1 + jn_1\sqrt{(n_1^2 \sin^2\theta_1 - n_2^2)}}{n_2^2 \cos\theta_1 + jn_1\sqrt{(n_1^2 \sin^2\theta_1 - n_2^2)}} \tag{1.3.15}$$

$$\rho_{\perp} = \frac{n_1 \cos\theta_1 - j\sqrt{(n_1^2 \sin^2\theta_1 - n_2^2)}}{n_1 \cos\theta_1 + j\sqrt{(n_1^2 \sin^2\theta_1 - n_2^2)}} \tag{1.3.16}$$

Therefore phase shift for the parallel and the perpendicular electrical field components are, respectively,

$$\Delta\Phi_{//} = \arg\left(\frac{E^r_{//}}{E^i_{//}}\right) = -2\tan^{-1}\left(\frac{\sqrt{n_1^2 \sin^2\theta_1 - n_2^2}}{n_1 \cos\theta_1}\right) \tag{1.3.17}$$

$$\Delta\Phi_{\perp} = \arg\left(\frac{E^r_{\perp}}{E^i_{\perp}}\right) = -2\tan^{-1}\left(\frac{n_1\sqrt{(n_1^2/n_2^2)\sin^2\theta_1 - 1}}{n_2 \cos\theta_1}\right) \tag{1.3.18}$$

This optical phase shift happens at the optical interface, which has to be considered in optical waveguide design, as will be discussed later.

1.3.1.4 Brewster Angle (Total Transmission $\rho = 0$)

Consider a light beam launching onto an optical interface. If the input electrical field is parallel to the incidence plane, there exists a specific incidence angle θ_B at which the reflection is equal to zero. Therefore the energy is totally transmitted across the interface. This angle θ_B is defined as the *Brewster angle*.

Consider the Fresnel Equation 1.3.5 for parallel field components. If we solve this equation for $\rho_{//} = 0$ and use θ_1 as a free parameter, the only solution is $\tan\theta_1 = n_2/n_1$ and therefore the Brewster angle is defined as

$$\theta_B = \tan^{-1}(n_2/n_1) \tag{1.3.19}$$

Two important points we need to note: (1) The Brewster angle is only valid for the polarization component which has the electrical field vector parallel to the incidence plane. For the perpendicular polarized component, no matter how you choose θ_1, total transmission will never happen. (2) $\rho_{//} = 0$ happens only at one angle $\theta_1 = \theta_B$. This is very different from the critical angle where total reflection happens for all incidence angles within the range of $\theta_c < \theta_1 < \pi/2$.

The Brewster angle is often used to minimize the optical reflection and it can also be used to select the polarization. Figure 1.3.3 shows an example of optical field reflectivities $\rho_{//}$ and $\rho_{\perp}$, and their corresponding phase shifts $\Delta\Phi_{//}$ and $\Delta\Phi_{\perp}$ at an optical interface of two materials with $n_1 = 1.5$ and $n_2 = 1.4$. In this example, the critical angle is $\theta_c \approx 69°$ and the Brewster angle is $\theta_B \approx 43°$.

1.3.2 Propagation Modes in Optical Fibers

As mentioned earlier in this chapter, an optical fiber is a cylindrical glass bar with a core, a cladding, and an external coating, as shown in Figure 1.3.1. To confine and guide the lightwave signal within the fiber core, a total internal

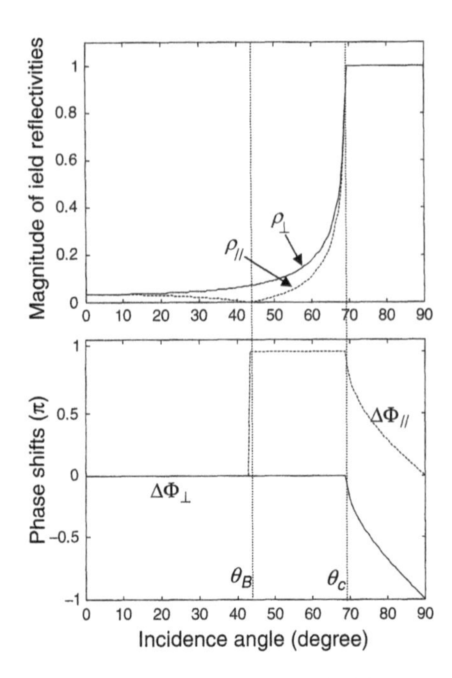

Figure 1.3.3 Field reflectivities and phase shifts versus incidence angle. Optical interface is formed with $n_1 = 1.5$ and $n_2 = 1.4$.

reflection is required at the core-cladding interface. According to what we have discussed in Section 1.3.1, this requires the refractive index of the core to be higher than that of the cladding.

Practical optical fibers can be divided into two categories: step-index fiber and graded-index fiber. The index profiles of these two types of fibers are shown in Figure 1.3.4. In a step-index fiber the refractive index is n_1 in the core and n_2 in the cladding; there is an index discontinuity at the core-cladding interface. A lightwave signal is bounced back and forth at this interface, which forms guided modes propagating in the longitudinal direction. On the other hand, in a graded-index fiber, the refractive index in the core gradually reduces its value along the radius. A generalized Fresnel equation indicates that in a medium with a continual index profile, a light trace would always bend toward high refractive areas. In fact this graded-index profile creates a self-focus effect within the fiber core to form an optical waveguide [8]. Although graded-index fibers form a unique category, they are usually made for multimode applications. The popular single-mode fibers are made with step-index profiles. Because of their popularity and simplicity, we will focus our analysis on step-index fibers.

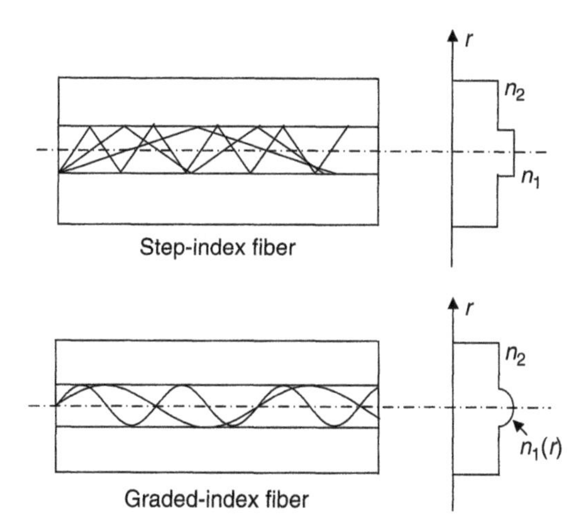

Figure 1.3.4 Illustration and index profiles of step-index and graded-index fibers. Used with permission.

Rigorous description of wave propagation in optical fibers requires solving Maxwell's equations and applying appropriate boundary conditions. In this section, we first use geometric ray trace approximation to provide a simplified analysis, which helps us understand the basic physics of wave propagation. Then we present electromagnetic field analysis, which provides precise mode cutoff conditions.

1.3.2.1 Geometric Optics Analysis

In this geometric optics analysis, different propagation modes in an optical fiber can be seen as rays traveling at different angles. There are two types of light rays that can propagate along the fiber: skew rays and meridional rays. Figure 1.3.5(a) shows an example of skew rays, which are not confined to any particular plane along the fiber. Although skew rays represent a general case of fiber modes, they are difficult to analyze. A simplified special case is the meridional rays shown in Figure 1.3.5(b), which are confined to the meridian plane, which contains the symmetry axis of the fiber. The analysis of meridional rays is relatively easy and provides a general picture of ray propagation along an optical fiber.

Consider meridional rays as shown in Figure 1.3.5(b). This is a two-dimensional problem where the optical field propagates in the longitudinal

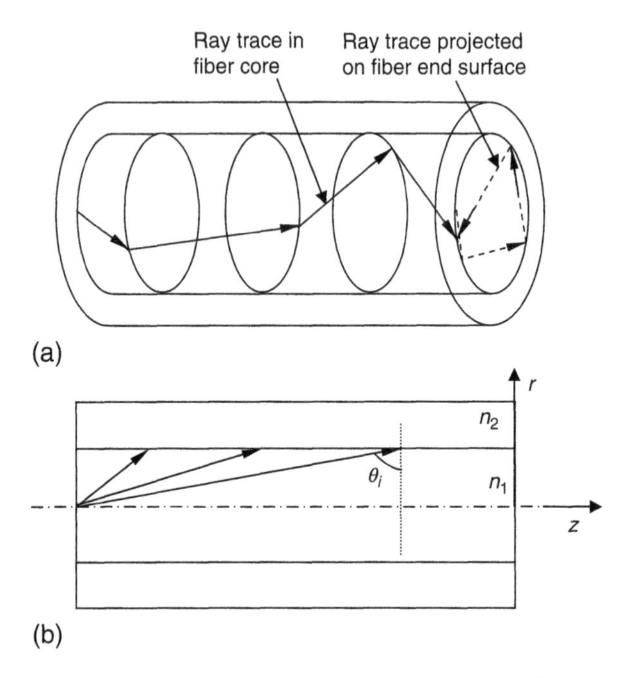

Figure 1.3.5 Illustration of fiber propagation modes in geometric optics: (a) skew ray trace and (b) meridional ray trace.

direction z and its amplitude varies over the transversal direction r. We define $\beta_1 = n_1\omega/c = 2\pi n_1/\lambda$ (in *rad./m*) as the propagation constant in a homogeneous core medium with a refraction index of n_1. Each fiber mode can be explained as a light ray that travels at a certain angle, as shown in Figure 1.3.6. Therefore, for i^{th} mode propagating in $+z$ direction, the propagation constant can be decomposed into a longitudinal component β_{zi} and a transversal component k_{i1} such that

$$\beta_1^2 = \beta_{zi}^2 + k_{i1}^2 \tag{1.3.21}$$

Then the optical field vector of the ith mode can be expressed as

$$\vec{E}_i(r,z) = \vec{E}_{i0}(r,z)\exp\{j(\omega t - \beta_{zi}z)\} \tag{1.3.22}$$

where $\vec{E}_{i0}(r,z)$ is the field amplitude of the mode.

Since the mode is propagating in the fiber core, both k_{i1} and β_{zi} must be real. First, for k_i to be real in the fiber core, we must have

$$k_{i1}^2 = \beta_1^2 - \beta_{zi}^2 \geq 0 \tag{1.3.23}$$

The physical meaning of this real propagation constant in the transversal direction is that the light wave propagates in the transverse direction but is bounced

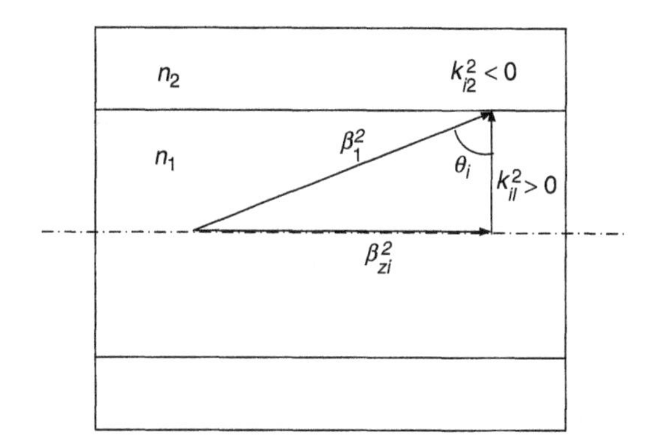

Figure 1.3.6 Decompose propagating vector β_1 into longitudinal and transversal components.

back and forth between the core-cladding interfaces. This creates a standing wave pattern in the transverse direction, like a resonant cavity. In addition, k_{i1} can only have discrete values because the standing wave pattern in the transversal direction requires phase matching. This is the reason that propagating optical modes in a fiber have to be discrete.

Now let's look at what happens in the cladding. Because the optical mode is guided in the fiber core, there should be no energy propagating in the transversal direction in the cladding (otherwise optical energy would be leaked). Therefore, k_i has to be imaginary in the cladding, that is,

$$k_{i2}^2 = \beta_2^2 - \beta_{zi}^2 < 0 \tag{1.3.24}$$

where subscript 2 represents parameters in the cladding and $\beta_2 = n_2\omega/c = 2\pi n_2/\lambda$ is the propagation constant in the homogeneous cladding medium.

Note that since the optical field has to propagate with the same phase velocity in the z-direction both in the core and in cladding, β_{zi} has the same value in both Equations 1.3.23 and 1.3.24.

Equations 1.3.23 and 1.3.24 can be simplified as

$$\beta_{zi}/\beta_1 \leq 1 \tag{1.3.25}$$

and

$$\beta_{zi}/\beta_2 > 1 \tag{1.3.26}$$

Bringing equations 1.3.25 and 1.3.26 together with $\beta_2 = \beta_1 n_2/n_1$, we can find the necessary condition for a propagation mode,

$$1 \geq \frac{\beta_{zi}}{\beta_1} > \frac{n_2}{n_1} \tag{1.3.27}$$

It is interesting to note that in Figure 1.3.6, θ_i is, in fact, the incidence angle of the ith mode at the core-cladding interface. The triangle in Figure 1.3.6 clearly shows that $\beta_{zi}/\beta_1 = \sin\theta_i$. This turns Equation 1.3.27 into $1 \geq \sin\theta_i > n_2/n_1$, which is the same as the definition of the critical angle as given by Equation 1.3.14.

The concept of discrete propagation modes comes from the fact that the transversal propagation constant k_{i1} in the fiber core can only take discrete values to satisfy standing wave conditions in the transverse direction. Since β_1 is a constant, the propagation constant in the z-direction $\beta_{zi}^2 = \beta_1^2 - k_i^2$ can only take discrete values as well. Or equivalently the ray angle θ_i can only take discrete values within the range defined by $1 \geq \sin\theta_i > n_2/n_1$.

The geometric optics description given here is simple and it qualitatively explains the general concept of fiber modes. However, it is not adequate to obtain quantitative mode field profiles and cutoff conditions. Therefore electromagnetic field theory has to be applied by solving Maxwell's equations and using appropriate boundary conditions, which we discuss next.

1.3.2.2 Mode Analysis Using Electromagnetic Field Theory

Mode analysis in optical fibers can be accomplished more rigorously by solving Maxwell's equations and applying appropriate boundary conditions defined by fiber geometries and parameters. We start with classical Maxwell's Equations,

$$\nabla \times \vec{E} = -\mu \frac{\partial}{\partial t} \tag{1.3.28}$$

$$\nabla \times \vec{H} = \varepsilon \frac{\partial}{\partial t} \tag{1.3.29}$$

The complex electrical and the magnetic fields are represented by their amplitudes and phases,

$$\vec{E}(t,\vec{r}) = \vec{E}_0 \exp\{-j(\omega t - \vec{k}\cdot\vec{r})\} \tag{1.3.30}$$

$$\vec{H}(t,\vec{r}) = \vec{H}_0 \exp\{-j(\omega t - \vec{k}\cdot\vec{r})\} \tag{1.3.31}$$

Since fiber material is passive and there is no generation source within the fiber,

$$\left(\nabla\cdot\vec{E}\right) = 0 \tag{1.3.32}$$

$$\nabla\times\nabla\times\vec{E} \equiv \nabla\left(\nabla\cdot\vec{E}\right) - \nabla^2\vec{E} = -\nabla^2\vec{E} \tag{1.3.33}$$

Combining Equations 1.3.28–1.3.33 yields,

$$\nabla \times \nabla \times \vec{E} = j\omega\mu\left(\nabla \times \vec{H}\right) = j\omega\mu(-j\omega\varepsilon) \tag{1.3.34}$$

And the Helmholtz equation,

$$\nabla^2\vec{E} + \omega^2\mu\varepsilon\vec{E} = 0 \tag{1.3.35}$$

Similarly, a Helmholtz equation can also be obtained for the magnetic field $\vec{H}$:

$$\nabla^2\vec{H} + \omega^2\mu\varepsilon\vec{H} = 0 \tag{1.3.36}$$

The next task is to solve Helmholtz equations for the electrical and the magnetic fields. Because the geometric shape of an optical fiber is cylindrical, we can take advantage of this axial symmetry to simplify the analysis by using cylindrical coordinates. In cylindrical coordinates, the electrical field can be decomposed into radial, azimuthal and longitudinal components: $\vec{E} = \vec{a}_r E_r + \vec{a}_\phi E_\phi + \vec{a}_z E_z$ and $\vec{H} = \vec{a}_r H_r + \vec{a}_\phi H_\phi + \vec{a}_z H_z$ where $\vec{a}_r$, $\vec{a}_\phi$, and $\vec{a}_z$ are unit vectors. With this separation, the Helmholtz Equations 1.3.35 and 1.3.36 can be decomposed into separate equations for E_r, E_ϕ, E_z, H_r, H_ϕ, and H_z, respectively. However, these three components are not completely independent. In fact, classic electromagnetic theory indicates that in cylindrical coordinate the transversal field components E_r, E_ϕ, H_r, and H_ϕ can be expressed as a combination of longitudinal field components E_z and H_z. This means that E_z and H_z need to be determined first and then we can find all other field components.

In cylindrical coordinates, the Helmholtz equation for E_z is:

$$\frac{\partial^2 E_z}{\partial r^2} + \frac{1}{r}\frac{\partial E_z}{\partial r} + \frac{1}{r^2}\frac{\partial^2 E_z}{\partial \phi^2} + \frac{\partial^2 E_z}{\partial z^2} + \omega^2\mu\varepsilon E_z = 0 \tag{1.3.37}$$

Since $E_z = E_z(r, \phi, z)$ is a function of both r, ϕ, and z, Equation 1.3.37 cannot be solved analytically. We assume a standing wave in the azimuthal direction and a propagating wave in the longitudinal direction, then the variables can be separated as

$$E_z(r, \phi, z) = E_z(r)e^{jl\phi}e^{j\beta_z z} \tag{1.3.38}$$

where $l = 0, \pm 1, \pm 2, \ldots.$ is an integer. Substituting 1.3.38 into 1.3.37, we can obtain a one-dimensional wave equation:

$$\frac{\partial^2 E_z(r)}{\partial r^2} + \frac{1}{r}\frac{\partial E_z(r)}{\partial r} + \left(\frac{n^2\omega^2}{c^2} - \beta_z^2 - \frac{l^2}{r^2}\right)E_z(r) = 0 \tag{1.3.39}$$

This is commonly referred to as a *Bessel equation* because its solutions can be expressed as Bessel functions.

For a step-index fiber with a core radius a, its index profile can be expressed as,

$$n = \begin{cases} n_1 & (r \leq a) \\ n_2 & (r > a) \end{cases} \tag{1.3.40}$$

We have assumed that the diameter of the cladding is infinite in this expression. The Bessel Equation 1.3.39 has solutions only for discrete β_z values, which correspond to discrete modes. The general solutions of Bessel Equation 1.3.39 can be expressed in Bessel functions as,

$$E_z(r) = \begin{cases} AJ_l(U_{lm}r) + A'Y_l(U_{lm}r) & (r \leq a) \\ CK_l(W_{lm}r) + C'I_l(W_{lm}r) & (r > a) \end{cases} \tag{1.3.41}$$

where $U_{lm}^2 = \beta_1^2 - \beta_{z,lm}^2$ and $W_{lm}^2 = \beta_{z,lm}^2 - \beta_2^2$ represent equivalent transversal propagation constants in the core and cladding, respectively, with $\beta_1 = n_1\omega/c$ and $\beta_2 = n_2\omega/c$ as defined before. $\beta_{z,lm}$ is the propagation constant in the z-direction. This is similar to the vectorial relation of propagation constants shown in Equation 1.3.21 in geometric optics analysis. However, we have two mode indices here, l and m. The physical meanings of these two mode indices are the amplitude maximums of the standing wave patterns in the azimuthal and the radial directions, respectively.

In Equation 1.3.41, J_l and Y_l are the first and the second kind of Bessel functions of the lth order, and K_l and I_l are the first and the second kind of *modified* Bessel functions of the lth order. Their values are shown in Figure 1.3.7. A, A', C, and C' in Equation 1.3.41 are constants that need to be defined using appropriate boundary conditions.

The first boundary condition is that the field amplitude of a guided mode should be finite at the center of the core ($r = 0$). Since the special function $Y_l(0) = -\infty$, one must set $A' = 0$ to ensure that $E_z(0)$ has a finite value.

The second boundary condition is that the field amplitude of a guided mode should be zero far away from the core ($r = \infty$). Since $I_l(\infty) \neq 0$, one must set $C' = 0$ to ensure that $E_z(\infty) = 0$. Consider $A' = C' = 0$ Equation 1.3.41 can be simplified, and for the mode index of (l, m), Equation 1.3.38 becomes:

$$E_{z,lm}(r,\phi,z) = \begin{cases} AJ_l(U_{lm}r)e^{jl\phi} \cdot e^{j\beta_{z,lm}z} & (r \leq a) \\ CK_l(W_{lm}r)e^{jl\phi} \cdot e^{j\beta_{z,lm}z} & (r > a) \end{cases} \tag{1.3.42}$$

Mathematically, the modified Bessel function fits well to an exponential characteristic $K_l(W_{lm}r) \propto \exp(-W_{lm}r)$, so that $K_l(W_{lm}r)$ represents an exponential decay of optical field over r in the fiber cladding. For a propagation mode, $W_{lm} > 0$ is required to ensure that energy does not leak through the cladding. In the fiber core, the Bessel function $J_l(U_{lm}r)$ oscillates as shown in Figure 1.3.7, which represents a standing-wave pattern in the core over the radius direction. For a propagating mode, $U_{lm} \geq 0$ is required to ensure this standing-wave pattern in the fiber core.

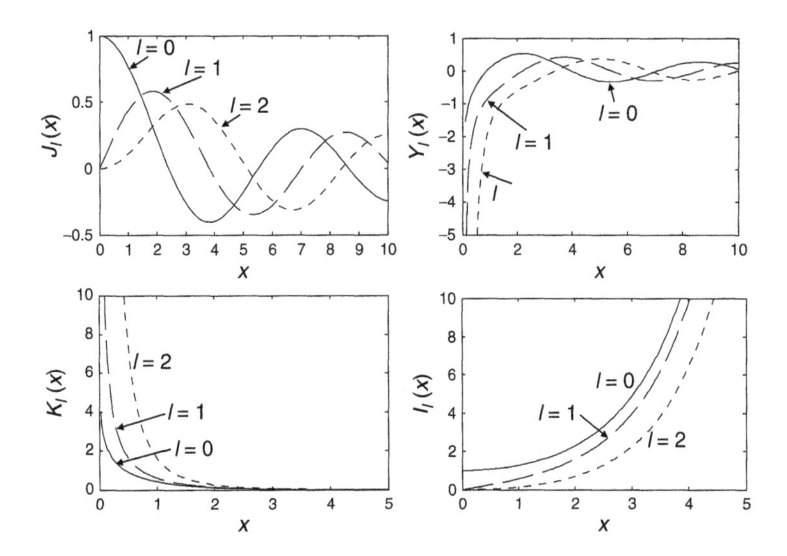

Figure 1.3.7 Bessel function (top) and modified Bessel functions (bottom).

It is interesting to note that based on the definitions of $U_{lm}^2 = \beta_1^2 - \beta_{z,lm}^2$ and $W_{lm}^2 = \beta_{z,lm}^2 - \beta_2^2$, the requirement of $W_{lm} > 0$ and $U_{lm} \geq 0$ is equivalent to $\beta_2^2 < \beta_{z,lm}^2 \leq \beta_1^2$ or $(n_2/n_1) < \beta_{z,lm}^2/\beta_1 \leq 1$. This is indeed equivalent to the mode condition (1.3.27) derived by the ray optics.

There are a few often used definitions to categorize the propagation modes in the fiber:

Transverse electric-field mode (TE mode): $E_z = 0$
Transverse magnetic-field mode (TM mode): $H_z = 0$
Hybrid mode (HE mode) $E_z \neq 0$ and $H_z \neq 0$

V-number is an important parameter of a fiber, which is defined as

$$V = a\sqrt{U_{lm}^2 + W_{lm}^2} \tag{1.3.43}$$

since

$$U_{lm}^2 = \beta_1^2 - \beta_{z,lm}^2 = \left(\frac{2\pi n_1}{\lambda}\right)^2 - \beta_{z,lm}^2$$

and

$$W_{lm}^2 = \beta_{z,lm}^2 - \beta_2^2 = \beta_{z,lm}^2 - \left(\frac{2\pi n_2}{\lambda}\right)^2$$

V-number can be expressed as

$$V = a\sqrt{U_{lm}^2 + W_{lm}^2} = \frac{2\pi a}{\lambda}\sqrt{n_1^2 - n_2^2} \tag{1.3.44}$$

In an optical fiber with large core size and large core-cladding index difference, it will support a large number of propagating modes. Approximately, the total number of guided modes in a fiber is related to the V-number as [7]

$$M \approx V^2/2 \tag{1.3.45}$$

In a multimode fiber, the number of guided modes can be on the order of several hundred. Imagine that a short optical pulse is injected into a fiber and the optical energy is carried by many different modes. Because different modes have different propagation constants $\beta_{z,lm}$ in the longitudinal direction and they will arrive at the output of the fiber in different times, the short optical pulse at the input will become a broad pulse at the output. In optical communications systems, this introduces signal waveform distortions and bandwidth limitations. This is the reason single-mode fiber is required in high-speed long distance optical systems.

In a single-mode fiber, only the lowest-order mode is allowed to propagate; all higher-order modes are cut off. In a fiber, the lowest-order propagation mode is HE_{11}, whereas the next lowest modes are TE_{01} and TM_{01} ($l = 0$ and $m = 1$). In fact, TE_{01} and TM_{01} have the same cutoff conditions: (1) $W_{01} = 0$ so that these two modes radiate in the cladding, and (2) $J_0(U_{01}a) = 0$ so that the field amplitude at core/cladding interface ($r = a$) is zero. Under the first condition ($W_{01} = 0$), we can find the cutoff V-number $V = a\sqrt{U_{01}^2 + W_{01}^2} = aU_{01}$, whereas under the second condition, we can find $J_0(aU_{01}) = J_0(V) = 0$, which implies that $V = 2.405$ as the first root of $J_0(V) = 0$.

Therefore, the single-mode condition is

$$V = \frac{2\pi a}{\lambda}\sqrt{n_1^2 - n_2^2} < 2.405 \tag{1.3.46}$$

1.3.2.3 Numerical Aperture

Numerical aperture is a parameter that is often used to specify the acceptance angle of a fiber. Figure 1.3.8 shows an azimuthal cross-section of a step-index fiber and a light ray that is coupled into the fiber from the left side end surface.

For the light to be coupled into the guided mode in the fiber, total internal reflection has to occur inside the core and $\theta_i > \theta_c$ is required, as shown in Figure 1.3.8, where $\theta_c = \sin^{-1}(n_2/n_1)$ is the critical angle of the core-cladding interface. With this requirement on θ_i, there is a corresponding requirement on

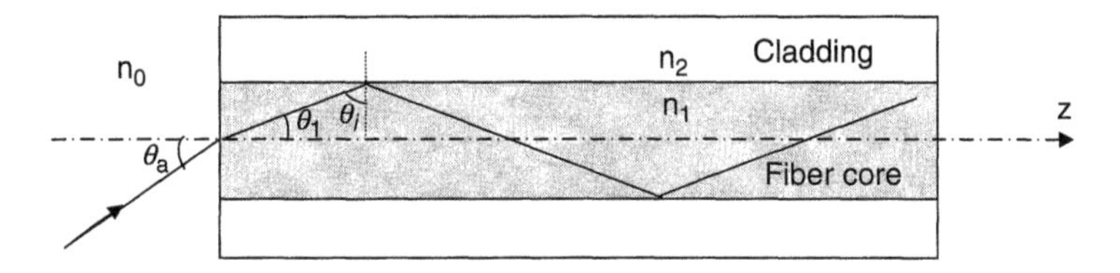

Figure 1.3.8 Illustration of light coupling into a step-index fiber.

incident angle θ_a at the fiber end surface. It is easy to see from the drawing that $\sin\theta_1 = \sqrt{1 - \sin^2\theta_i}$, and by Snell's Law,

$$n_0 \sin\theta_a = n_1\sqrt{1 - \sin^2\theta_i}$$

If total reflection happens at the core-cladding interface, which requires $\theta_i \geq \theta_c$, then $\sin\theta_i \geq \sin\theta_c = n_2/n_1$. This requires the incidence angle θ_a to satisfy the following condition:

$$n_0 \sin\theta_a \leq \sqrt{n_1^2 - n_2^2} \tag{1.3.47}$$

The definition of *numerical aperture* is

$$NA = \sqrt{n_1^2 - n_2^2} \tag{1.3.48}$$

For weak optical waveguide like a single-mode fiber, the difference between n_1 and n_2 is very small (not more than 1 percent). Use $\Delta = (n_1 - n_2)/n_1$ to define a normalized index difference between core and cladding, then Δ must also be very small ($\Delta \ll 1$). In this case, the expression of numerical aperture can be simplified as

$$NA \approx n_1\sqrt{2\Delta} \tag{1.3.49}$$

In most cases fibers are placed in air and $n_0 = 1$. $\sin\theta_a \approx \theta_a$ is valid when $\sin\theta_a \ll 1$ (weak waveguide); therefore, Equation 1.3.47 reduces to

$$\theta_a \leq n_1\sqrt{2\Delta} = NA \tag{1.3.50}$$

From this discussion, the physical meaning of numerical aperture is very clear. Light entering a fiber within a cone of acceptance angle, as shown in Figure 1.3.9, will be converted into guided modes and will be able to propagate along the fiber. Outside this cone, light coupled into fiber will radiate into the cladding. Similarly, light exits a fiber will have a divergence angle also defined by the numerical aperture. This is often used to design focusing optics if a collimated beam is needed at the fiber output.

Typically parameters of a single-mode fiber are $NA \approx 0.1 \sim 0.2$ and $\Delta \approx 0.2\% \sim 1\%$. Therefore, $\theta_a \approx \sin^{-1}(NA) \approx 5.7^\circ \sim 11.5^\circ$. This is a very small

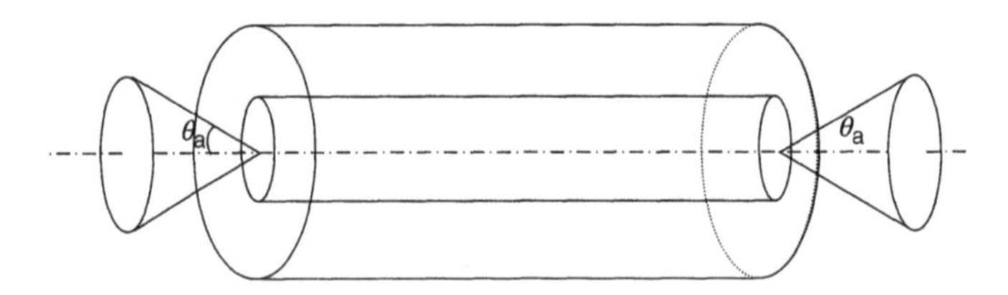

Figure 1.3.9 Light can be coupled to an optical fiber only when the incidence angle is smaller than the numerical aperture.

angle and it makes difficult to couple light into a single-mode fiber. Not only that, the source spot size has to be small (~80 μm²) too while the angle has to be within ±10 degrees.

With the definition of the numerical aperture in Equation 1.3.48, the V-number of a fiber can be expressed as a function of NA:

$$V = \frac{2\pi a}{\lambda} NA$$

Another important fiber parameter is the cutoff wavelength λ_c. It is defined such that the second lowest mode ceases to exist when the signal wavelength is longer than λ_c, and therefore when $\lambda < \lambda_c$ a single-mode fiber will become multimode. According to Equation 1.3.46, cutoff wavelength is

$$\lambda_c = \frac{\pi d}{2.405} NA$$

where d is the core diameter of the step-index fiber. As an example, for a typical standard single-mode fiber with, $n_1 = 1.47$, $n_2 = 1.467$, and $d = 9$ μm, the numerical aperture is

$$NA = \sqrt{n_1^2 - n_2^2} = 0.0939$$

The maximum incident angle at the fiber input is

$$\theta_a = \sin^{-1}(0.0939) = 5.38°$$

and the cutoff wavelength is

$$\lambda_c = \pi d \cdot NA/2.405 = 1.1\ \mu m$$

Example 1.4

To reduce the Fresnel reflection, the end surface of a fiber contactor can be made nonperpendicular to the fiber axis. This is usually referred to as APC (angle-polished connector) contactor. If the fiber has the core index $n_1 = 1.47$ and cladding index $n_2 = 1.467$, what is the minimum angle ϕ such that the Fresnel reflection by the fiber end facet will not become the guided fiber mode?

Solution:

To solve this problem, we use ray trace method and consider three extreme light beam angles in the fiber. The angle has to be designed such that after reflection at the fiber end surface, all these three light beams will not be coupled into fiber-guided mode in the backward propagation direction.

As illustrated in Figure 1.3.10(a), first, for the light beam propagating in the fiber axial direction (z-direction), the direction of the reflected beam from the end surface has an angle θ with respect to the surface normal of the fiber sidewall: $\theta = \pi/2 - 2\phi < \theta_c$. In order for this reflected light beam not to become the guided mode of the fiber, $\theta < \theta_c$ is required, where θ_c is the critical angle defined by Equation 1.3.14. Therefore, the first requirement for ϕ is

$$\phi > \pi/4 - \theta_c/2$$

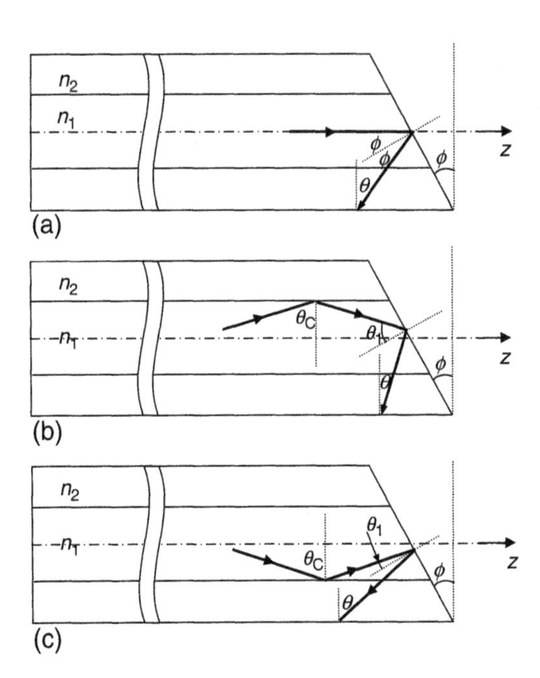

Figure 1.3.10 Illustration of an angle-polished fiber surface.

Second, for the light beam propagating at the critical angle of the fiber as shown in Figure 1.3.10(b), the beam has an angle θ_1 with respect to the surface normal of the fiber end surface, which is related to ϕ by $\theta_1 = (\pi/2 - \theta_c) + \phi$. Then the θ angle of the reflected beam from the end surface with respect to the fiber sidewall can be found as $\theta = \pi - \theta_c - 2\theta_1 = \theta_c - 2\phi$. This angle also has to be smaller than the critical angle, that is, $\theta_c - 2\phi < \theta_c$, or

$$\phi > 0$$

In the third case as shown in Figure 1.3.10(c), the light beam propagates at the critical angle of the fiber but at the opposite side as compared to the ray trace shown in Figure 1.3.10(b). This produces the smallest θ_1 angle, which is, $\theta_1 = \phi - (\pi/2 - \theta_c)$. This corresponds to the biggest θ angle, $\theta = \pi - \theta_c - 2\theta_1 = 2\pi - 3\theta_c - 2\phi$. Again, this θ angle has to be smaller than the critical angle, $\theta < \theta_c$, that is,

$$\phi > \pi - 2\theta_c$$

Since in this example

$$\theta_c = \sin^{-1}\left(\frac{n_2}{n_1}\right) = \sin^{-1}\left(\frac{1.467}{1.47}\right) = 86.34^\circ$$

The three constraints given above become $\phi > 1.83^\circ$, $\phi > 0$, and $\phi > 7.3^\circ$, respectively. Obviously, in order to satisfy all these three specific conditions, the required surface angle is $\phi > 7.3^\circ$. In fact, as an industry standard, commercial APC connectors usually have the surface tilt angle of approximately $\phi = 8^\circ$.

1.3.3 Optical Fiber Attenuation

Optical fiber is an ideal medium that can be used to carry optical signals over long distances. Attenuation is one of the most important parameters of an optical fiber; it, to a large extent, determines how far an optical signal can be delivered at a detectable power level. There are several sources that contribute to fiber attenuation, such as absorption, scattering, and radiation.

Material absorption is mainly caused by photo-induced molecular vibration, which absorbs signal optical power and turns it into heat. Pure silica molecules absorb optical signals in ultraviolet (UV) and infrared (IR) wavelength bands. At the same time, there are often impurities inside silica material such as OH^-

ions, which may be introduced in the fiber perform fabrication process. These impurity molecules create additional absorption in the fiber. Typically, OH^- ions have high absorptions around 700 nm, 900 nm, and 1400 nm, which are commonly referred to as *water absorption peaks.*

Scattering loss arises from microscopic defects and structural inhomogeneities. In general, the optical characteristics of scattering depend on the size of the scatter in comparison to the signal wavelength. However, in optical fibers, the scatters are most likely much smaller than the wavelength of the optical signal, and in this case the scattering is often characterized as *Rayleigh scattering.* A very important spectral property of Rayleigh scattering is that the scattering loss is inversely proportional to the fourth power of the wavelength. Therefore, Rayleigh scattering loss is high in a short wavelength region.

In the last few decades, the loss of optical fiber has been decreased significantly by reducing the OH^- impurity in the material and eliminating defects in the structure. However, absorption by silica molecules in the UV and IR wavelength regions and Rayleigh scattering still constitute fundamental limits to the loss of silica-based optical fiber. Figure 1.3.11 shows the typical absorption spectra of silica fiber. The dotted line shows the attenuation of old fibers that were made before the 1980s. In addition to strong water absorption peaks, the attenuation is generally higher than new fibers due to material impurity and waveguide scattering. Three wavelength windows have been used since the 1970s for optical communications in 850 nm, 1310 nm, and 1550 nm where optical attenuation has local minimums. In the early days of optical communication, the first wavelength window in 850 nm was used partly because of the availability of GaAs-based laser sources, which emit in that wavelength

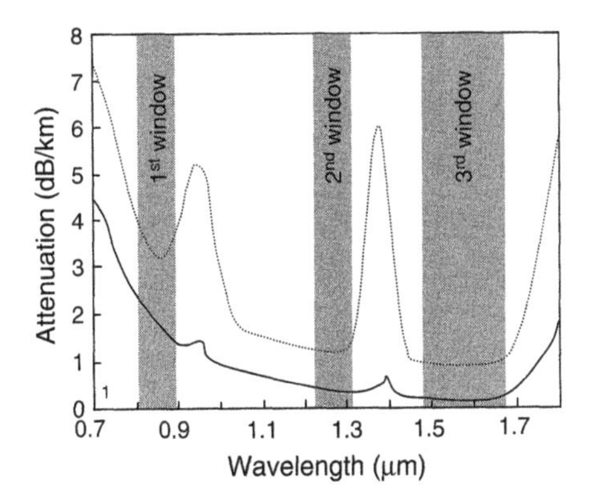

Figure 1.3.11 Attenuation of old (dotted line) and new (solid line) silica fibers. The shaded regions indicate the three telecommunication wavelength windows.

window. The advances in longer wavelength semiconductor lasers based on InGaAs and InGaAsP pushed optical communications toward the second and the third wavelength windows in 1310 nm and 1550 nm where optical losses are significantly reduced and optical systems can reach longer distances without regeneration.

Another category of optical loss that may occur in optical fiber cables is radiation loss. It is mainly caused by fiber bending. Micro bending, usually caused by irregularities in the drawing process, may introduce coupling between the fundamental optical mode and high-order radiation modes and thus creating losses. On the other hand, macro bending, often introduced by cabling process and fiber handling, causes the spreading of optical energy from fiber core into the cladding. For example, for a standard single-mode fiber, bending loss starts to be significant when the bending diameter is smaller than approximately 30 cm.

Mathematically, the complex representation of an optical wave propagating in z-direction is

$$E(z,t) = E_0 \exp\left[-j(\omega t - kz)\right] \tag{1.3.51}$$

where E_0 is the complex amplitude, ω is the optical frequency and k is the propagation constant. Considering attenuation in the medium, the propagation constant should be complex:

$$k = \beta + j\frac{\alpha}{2} \tag{1.3.52}$$

where $\beta = 2\pi n/\lambda$ is the real propagation constant and α is the power attenuation coefficient. By separating the real and the imaginary parts of the propagation constant, Equation 1.3.51 can be written as

$$E(z,t) = E_0 \exp[-j(\omega t - \beta z)] \cdot \exp\left(-\frac{\alpha}{2}\right)z \tag{1.3.53}$$

The average optical power can be simply expressed as

$$P(z) = P_0 e^{-\alpha z} \tag{1.3.54}$$

where P_0 is the input optical power. Note here the unit of α is in Neper per meter.

This attenuation coefficient α of an optical fiber can be obtained by measuring the input and the output optical power:

$$\alpha = \frac{1}{L}\ln\left[\frac{P_0}{P(L)}\right] \tag{1.3.55}$$

where L is the fiber length and $P(L)$ is the optical powers measured at the output of the fiber.

However, engineers like use decibel (*dB*) to describe fiber attenuation and use *dB/km* as the unit of attenuation coefficient. If we define α_{dB} as the attenuation coefficient which has the unit of *dB/m*, then the optical power level along the fiber length can be expressed as

$$P(z) = P_0 \times 10^{-\frac{\alpha_{dB}}{10} z} \tag{1.3.56}$$

Similar to Equation 1.3.55, for a fiber of length L, α_{dB} can be estimated using

$$\alpha_{dB} = \frac{1}{L} 10\log\left[\frac{P_0}{P(L)}\right] \tag{1.3.57}$$

Comparing Equations 1.3.55 and 1.3.57, the relationship between α and α_{dB} can be found as

$$\frac{\alpha_{dB}}{\alpha} = \frac{10\log[P_0/P(L)]}{\ln[P_0/P(L)]} = 10\log(e) = 4.343 \tag{1.3.58}$$

or simply, $\alpha_{dB} = 4.343\alpha$.

α_{dB} is a simpler parameter to use for evaluation of fiber loss. For example, for an 80 km-long fiber with $\alpha_{dB} = 0.25$ dB/km attenuation coefficient, the total fiber loss can be easily found as $80 \times 0.25 = 20$ *dB*. On the other hand, if complex optical field expression is required to solve wave propagation equations, α needs to be used instead. In practice, people always use the symbol α to represent optical-loss coefficient, no matter in [Neper/m] or in [dB/km]. But one should be very clear which unit to use when it comes to finding numerical values.

The following is an interesting example that may help to understand the impact of fiber numerical aperture and attenuation.

Example 1.5

A very long step-index, single-mode fiber has a numerical aperture $NA = 0.1$ and a core refractive index $n_1 = 1.45$. Assume that the fiber end surface is ideally antireflection coated and fiber loss is only caused by Rayleigh scattering. Find the reflectivity of the fiber $R_{ref} = P_b/P_i$ where P_i is the optical power injected into the fiber and P_b is the optical power that is reflected back to the input fiber terminal, as illustrated in Figure 1.3.12(a).

Solution:

In this problem, Rayleigh scattering is the only source of attenuation in the fiber. Each scattering source scatters the light into all directions uniformly, which fills the entire 4π solid angle. However, only a small part of the scattered light with the angle within the numerical aperture can be converted into guided mode within the fiber core and travels back to

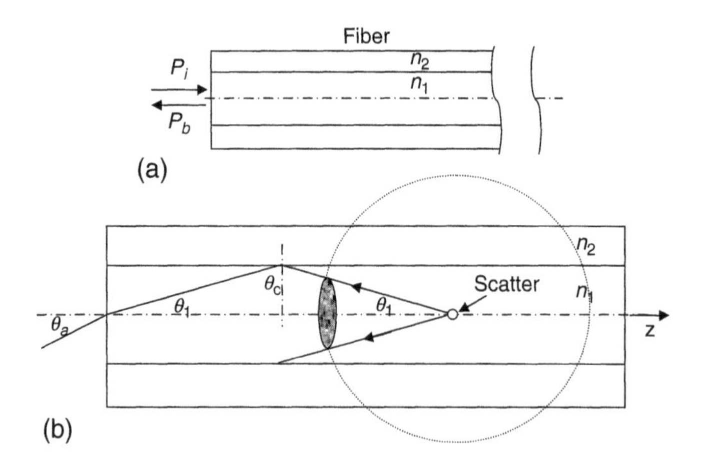

Figure 1.3.12 Illustrations of fiber reflection (a) and scattering in fiber core (b).

the input. The rest will be radiated into cladding and lost. For simplicity, we assume that scatters are only located along the center of the fiber core, as shown in Figure 1.3.12(b). Therefore the conversion efficiency from the scattered light to that captured by the fiber is

$$\eta = \frac{2\pi(1 - \cos\theta_1)}{4\pi} \tag{1.3.59}$$

where the numeritor is the area of the spherical cap shown as the shaded area in Figure 1.3.12(b), whereas the denominator 4π is the total area of the unit sphere. θ_1 is the maximum trace angle of the guided mode with respect to the fiber axis. Since the numerical aperture is defined as the maximum acceptance angle θ_a at the fiber entrance as given in Equation 1.3.50, θ_1 can be expressed as a function of NA as

$$\theta_1 = \frac{n_0}{n_1} NA = \frac{1}{1.45} \times 0.1 = 0.069 \tag{1.3.60}$$

where $n_0 = 1$ is used for the index of air. Substitute the value of θ_1 into Equation 1.3.59, the ratio between the captured and the total scattered optical power can be found as $\eta = 1.19 \times 10^{-3} = -29\ dB$.

Now let's consider a short fiber section Δz located at position z along the fiber. The optical power at this location is $P(z) = P_i e^{-\alpha z}$, where α is the fiber attenuation coefficient. The optical power loss within this section is

$$\Delta P(z) = \frac{dP(z)}{dz} \Delta z = -\alpha P_i e^{-\alpha z} \Delta z$$

Since we assumed that the fiber loss is only caused by Rayleigh scattering, this power loss $\Delta P(z)$ should be equal to the total scattered power within the short section. $\eta \Delta P(z)$ is the amount of scattered power that is turned into the guided mode that travels back to the fiber input.

However, during the traveling from location z back to the fiber input, attenuation also applies, and the total amount of power loss is, again, $e^{-\alpha z}$. Considering that the fiber is composed of many short sections and adding up the contributions from all sections, the total reflected optical power is

$$P_b = \sum \eta |\Delta P(z)| e^{-\alpha z} = \int_0^\infty \eta \alpha P_i e^{-2\alpha z} dz = \frac{\eta P_i}{2} \tag{1.3.61}$$

Therefore the reflectivity is

$$R_{ref} = \frac{P_b}{P_i} = \frac{\eta}{2} = 5.95 \times 10^{-4} = -32\ dB \tag{1.3.62}$$

This result looks surprisingly simple. The reflectivity, sometimes referred to as return loss, only depends on the fiber numerical aperture and is independent of the fiber loss. The physical explanation is that since Rayleigh scattering is assumed to be the only source of loss, increasing the fiber loss will increase both scattered signal generation and its attenuation. In practical single-mode fibers, this approximation is quite accurate, and the experimentally measured return loss in a standard single-mode fiber is between –31 dB and –34 dB.

1.3.4 Group Velocity and Dispersion

When an optical signal propagates along an optical fiber, not only is the signal optical power attenuated but also different frequency components within the optical signal propagate in slightly different speeds. This frequency-dependency of propagation speed is commonly known as the *chromatic dispersion.*

1.3.4.1 Phase Velocity and Group Velocity

Neglecting attenuation, the electric field of a single-frequency, plane optical wave propagating in z-direction is often expressed as

$$E(z,t) = E_0 \exp[-j\Phi(t,z)] \tag{1.3.63}$$

where $\Phi(t,z) = (\omega_0 t - \beta_0 z)$ is the optical phase, ω_0 is the optical frequency, and $\beta_0 = 2\pi n/\lambda = n\omega_0/c$ is the propagation constant.

The phase front of this light wave is the plane where the optical phase is constant:

$$(\omega_0 t - \beta_0 z) = constant \tag{1.3.64}$$

The propagation speed of the phase front is called *phase velocity*, which can be obtained by differentiating both sides of Equation 1.3.64:

$$v_p = \frac{dz}{dt} = \frac{\omega_0}{\beta_0} \tag{1.3.65}$$

Now consider that this single-frequency optical wave is modulated by a sinusoid signal of frequency $\Delta\omega$. Then the electrical field is

$$E(z,t) = E_0 \exp[-j(\omega_0 t - \beta z)] \cos(\Delta\omega t) \tag{1.3.66}$$

This modulation splits the signal frequency light wave into two frequency components. At the input ($z = 0$) of the optical fiber, the optical field is

$$E(0,t) = E_0 e^{-j\omega_0 t} \cos(\Delta\omega t) = \frac{1}{2} E_0 \left(e^{-j(\omega_0+\Delta\omega)t} + e^{-j(\omega_0-\Delta\omega)t} \right) \tag{1.3.67}$$

Since wave propagation constant $\beta = n\omega/c$ is linearly proportional to the frequency of the optical signal, the two frequency components at $\omega_0 \pm \Delta\omega$ will have two different propagation constants $\beta_0 \pm \Delta\beta$. Therefore, the general expression of the optical field is

$$\begin{aligned} E(z,t) &= \frac{1}{2} E_0 \left\{ e^{-j[(\omega_0+\Delta\omega)t-(\beta_0-\Delta\beta)z]} + e^{-j[(\omega_0-\Delta\omega)t-((\beta_0+\Delta\beta)z]} \right\} \\ &= E_0 e^{-j(\omega_0 t-\beta_0 z)} \cos(\Delta\omega t - \Delta\beta z) \end{aligned} \tag{1.3.68}$$

where $E_0 e^{-j(\omega_0 t-\beta_0 z)}$ represents an optical carrier, which is identical to that given in Equation 1.3.63, whereas $\cos(\Delta\omega t - \Delta\beta z)$ is an envelope that is carried by the optical carrier. In fact, this envelope represents the information that is modulated onto the optical carrier. The propagation speed of this information-carrying envelope is called *group velocity*. Similar to the derivation of phase velocity, one can find group velocity by differentiating both sides of $(\Delta\omega t - \Delta\beta z) = constant$, which yields

$$v_g = dz/dt = \Delta\omega/\Delta\beta$$

With infinitesimally low modulation frequency, $\Delta\omega \rightarrow d\omega$ and $\Delta\beta \rightarrow d\beta$, so the general expression of group velocity is

$$v_g = \frac{d\omega}{d\beta} \tag{1.3.69}$$

In a nondispersive medium, the refractive index n is a constant that is independent of the frequency of the optical signal. In this case, the group velocity is equal to the phase velocity: $v_g = v_p = n/c$. However, in many practical optical materials, the refractive index $n(\omega)$ is a function of the optical frequency and therefore $v_g \neq v_p$ in these materials.

Over a unit length, the propagation phase delay is equal to the inverse of the phase velocity:

$$\tau_p = \frac{1}{v_p} = \frac{\beta_0}{\omega_0} \tag{1.3.70}$$

And similarly, the propagation group delay over a unit length is defined as the inverse of the group velocity:

$$\tau_g = \frac{1}{v_g} = \frac{d\beta}{d\omega} \tag{1.3.71}$$

1.3.4.2 Group Velocity Dispersion

To understand group velocity dispersion, we consider that two sinusoids with the frequencies $\Delta\omega \pm \delta\omega/2$ are modulated onto an optical carrier of frequency ω_0. When propagating along a fiber, each modulating frequency will have its own group velocity; then over a unit fiber length, the group delay difference between these two frequency components can be found as

$$\delta\tau_g = \frac{d\tau_g}{d\omega}\delta\omega = \frac{d}{d\omega}\left(\frac{d\beta}{d\omega}\right)\delta\omega = \frac{d^2\beta}{d\omega^2}\delta\omega \tag{1.3.72}$$

Obviously, this group delay difference is introduced by the frequency dependency of the propagation constant. In general, the frequency-dependent propagation constant $\beta(\omega)$ can be expended in a Taylor series around a central frequency ω_0:

$$\begin{aligned}\beta(\omega) &= \beta(\omega_0) + \left.\frac{d\beta}{d\omega}\right|_{\omega=\omega_0}(\omega-\omega_0) + \left.\frac{1}{2}\frac{d^2\beta}{d\omega^2}\right|_{\omega=\omega_0}(\omega-\omega_0)^2 + \ldots\ldots \\ &= \beta(\omega_0) + \beta_1(\omega-\omega_0) + \frac{1}{2}\beta_2(\omega-\omega_0)^2 + \ldots\ldots\end{aligned} \tag{1.3.73}$$

where

$$\beta_1 = \frac{d\beta}{d\omega} \tag{1.3.74}$$

represents the group delay and

$$\beta_2 - \frac{d^2\beta}{d\omega^2} \tag{1.3.75}$$

is the group delay dispersion parameter.

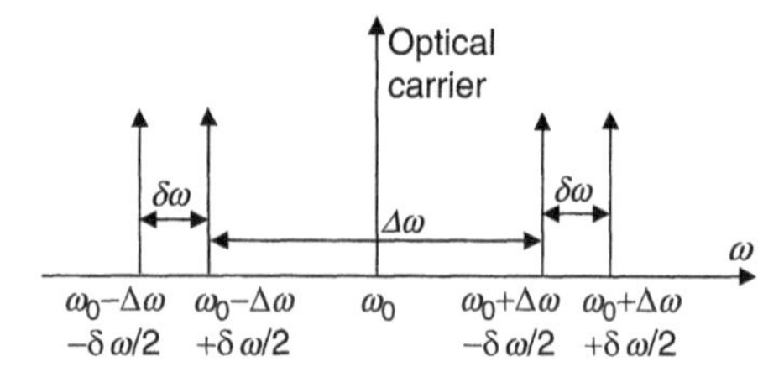

Figure 1.3.13 Spectrum of two-tone modulation on an optical carrier, where ω_0 is the carrier frequency and $\Delta\omega \pm \delta\omega/2$ are the modulation frequencies.

If the fiber is uniform with length L, use Equation 1.3.72, we can find the relative time delay between two frequency components separated by $\delta\omega$ as

$$\Delta\tau_g = \beta_2 \cdot L \cdot \delta\omega \tag{1.3.76}$$

Sometimes it might be convenient to use wavelength separation $\delta\lambda$ instead of frequency separation $\delta\omega$ between the two frequency (or wavelength) components. In this case, the relative delay over a unit fiber length can be expressed as

$$\delta\tau_g = \frac{d\tau_g}{d\lambda}\delta\lambda \equiv D\delta\lambda \tag{1.3.77}$$

where $D = d\tau_g/d\lambda$ is another group delay dispersion parameter. The relationship between the two dispersion parameters D and β_2 can be found as

$$D = \frac{d\tau_g}{d\lambda} = \frac{d\omega}{d\lambda} \cdot \frac{d\tau_g}{d\omega} = -\frac{2\pi c}{\lambda^2}\beta_2 \tag{1.3.78}$$

For a fiber of length L, we can easily find the relative time delay between two wavelength components separated by $\delta\lambda$ as

$$\Delta\tau_g = D \cdot L \cdot \delta\lambda \tag{1.3.79}$$

In practical fiber-optic systems, the relative delay between different wavelength components is usually measured in picoseconds; wavelength separation is usually expressed in nanometers and fiber length is usually measured in kilometers. Therefore, the most commonly used units for β_1, β_2, and D are [*s/m*], [*ps*^{2}*/nm*], and [*ps/nm-km*], respectively.

1.3.4.3 Sources of Chromatic Dispersion

The physical reason of chromatic dispersion is the wavelength-dependent propagation constant $\beta(\lambda)$. Both material property and waveguide structure may contribute to the wavelength dependency of $\beta(\lambda)$, which are referred to as material dispersion and waveguide dispersion, respectively.

Material dispersion is originated by the wavelength dependent material refractive index $n = n(\lambda)$; thus the wavelength dependent propagation constant is $\beta(\lambda) = 2\pi n(\lambda)/\lambda$.

For a unit fiber length, the wavelength-dependent group delay is

$$\tau_g = \frac{d\beta(\lambda)}{d\omega} = -\left(\frac{\lambda^2}{2\pi}\right)\frac{d\beta(\lambda)}{d\lambda} = \frac{1}{c}\left[n(\lambda) - \lambda\frac{dn(\lambda)}{d\lambda}\right] \quad (1.3.80)$$

The group delay dispersion between two wavelength components separated by $\delta\lambda$ is then

$$\delta\tau_g = \frac{d\tau_g}{d\lambda}\delta\lambda = \frac{-1}{c}\left[\lambda\frac{d^2n(\lambda)}{d\lambda^2}\right]\delta\lambda \quad (1.3.81)$$

Therefore material induced dispersion is proportional to the second derivative of the refractive index.

Waveguide dispersion can be explained as the wavelength-dependent angle of the light ray propagating inside the fiber core, as illustrated by Figure 1.3.6. In the guided mode, the projection of the propagation constant in z-direction β_z has to satisfy $\beta_2^2 < \beta_z^2 \leq \beta_1^2$, where $\beta_1 = kn_1$, $\beta_2 = kn_2$ and $k = 2\pi/\lambda$. This is equivalent to

$$0 < \frac{(\beta_z/k)^2 - n_2^2}{n_1^2 - n_2^2} < 1 \quad (1.3.82)$$

If we define

$$b = \frac{(\beta_z/k)^2 - n_2^2}{n_1^2 - n_2^2} \approx \frac{(\beta_z/k) - n_2}{n_1 - n_2} \quad (1.3.83)$$

as a normalized propagation constant, the actual propagation constant in the z-direction, β_z can be expressed as a function of b as

$$\beta_z(\lambda) = kn_2(b\Delta + 1) \quad (1.3.83)$$

where $\Delta = (n_1 - n_2)/n_2$ is the normalized index difference between the core and the cladding. Then the group delay can be found as

$$\tau_g = \frac{d\beta_z(\lambda)}{d\omega} = \frac{n_2}{c}\left(1 - b\Delta - k\frac{db}{dk}\Delta\right)$$

Figure 1.3.14 shows the variation of b as a function of V for different modes. Since V is inversely proportional to λ, as defined by Equation 1.3.44, τ_g varies with λ as well.

In general, material dispersion is difficult to modify, because doping other materials into silica might introduce excess attenuation. On the other hand, waveguide dispersion can be modified by index profile design.

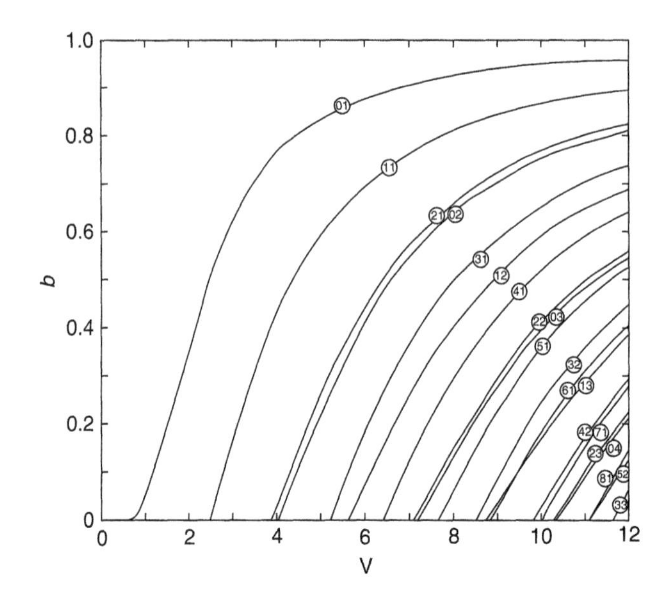

Figure 1.3.14 Normalized propagation constant b for various fiber modes. [9]

The overall chromatic dispersion in an optical fiber is the combination of material dispersion and waveguide dispersion. In general, different types of fiber have different dispersion characteristics. However, for a standard single-mode fiber, the dispersion parameter D can usually be described by a Sellmeier equation [8]:

$$D(\lambda) = \frac{S_0}{4}\left(\lambda - \frac{\lambda_0^4}{\lambda^3}\right) \tag{1.3.84}$$

where S_0 is the dispersion slope, which ranges from 0.08 to 0.09 ps/nm^2-km and λ_0 is the zero-dispersion wavelength, which is around 1315 nm.

Figure 1.3.15(a) shows the dispersion parameter D versus wavelength for standard single-mode fiber, which has dispersion slope $S_0 = 0.09$ ps/nm^2-km and zero-dispersion wavelength $\lambda_0 = 1315$ nm. $D(\lambda)$ is generally nonlinear; however, if we are only interested in a relatively narrow wavelength window, it is often convenient to linearize this parameter. For example, if the central frequency of an optical signal is at $\lambda = \lambda_a$, then $D(\lambda)$ can be linearized in the vicinity of λ_a as

$$D(\lambda) \approx D(\lambda_a) + S(\lambda_a) \cdot (\lambda - \lambda_a) \tag{1.3.85}$$

where

$$D(\lambda_a) = \frac{S_0}{4}\left(\lambda - \frac{\lambda_0^4}{\lambda_a^3}\right) \tag{1.3.86}$$

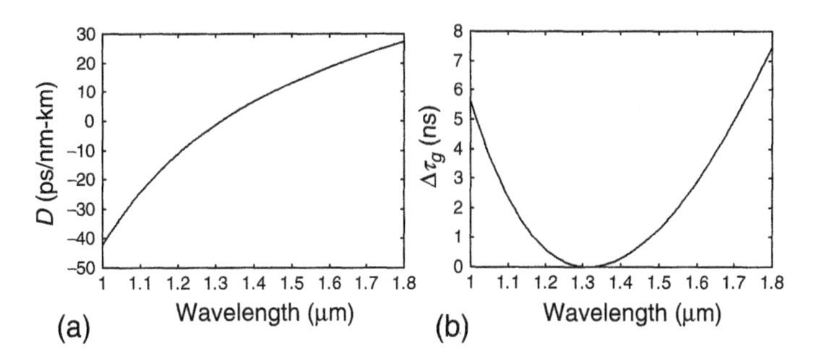

Figure 1.3.15 (a) Chromatic dispersion D versus wavelength and (b) relative group delay versus wavelength. S_0 = 0.09 ps/nm^2-km, λ_0 = 1315 nm.

and the local dispersion slope at λ_a is

$$S(\lambda_a) = \frac{S_0}{4}\left(1 + \frac{3\lambda_0^4}{\lambda_a^4}\right) \tag{1.3.87}$$

In general, $S(\lambda_a) \neq S_0$, except when the optical signal is near the zero-dispersion wavelength $\lambda_a = \lambda_0$.

As a consequence of wavelength-dependent dispersion parameter, the group delay is also wavelength-dependent. Considering the definition of dispersion $D = d\tau_g/d\lambda$ as given by Equation 1.3.78, the wavelength-dependent group delay $\tau_g(\lambda)$ can be found by integrating $D(\lambda)$ over wavelength. Based on Equation 1.3.84, the group delay can be derived as

$$\tau_g(\lambda) = \int D(\lambda)d\lambda = \tau_0 + \frac{S_0}{8}\left(\lambda - \frac{\lambda_0^2}{\lambda}\right)^2 \tag{1.3.88}$$

Figure 1.3.15(a) shows the relative group delay $\Delta\tau_g(\lambda) = \tau_g(\lambda) - \tau_0$ versus wavelength. The group delay is not sensitive to wavelength change around $\lambda = \lambda_0$ because where the dispersion is zero.

1.3.4.4 Modal Dispersion

Chromatic dispersion discussed earlier specifies wavelength-dependent group velocity within one optical mode. If a fiber has more than one mode, different modes will also have different propagation speeds; this is called *model dispersion*. In a multimode fiber, the effect of model dispersion is typically much stronger than the chromatic dispersion within each mode; therefore chromatic dispersion is usually neglected.

Modal dispersion depends on the number of propagation modes that exist in the fiber, which, in turn, is determined by the fiber core size and the index difference between the core and the cladding. By using geometric optics analysis as described in Section 1.3.2, we can find the delay difference between the fastest and the slowest propagation modes. Obviously, the fastest mode is the one that travels along the fiber longitudinal axis, whereas the ray trace of the slowest mode has the largest angle with respect to the fiber longitudinal axis or the smallest θ_i shown in Figure 1.3.5(b). This smallest angle θ_i is limited by the condition of total reflection at the core-cladding interface, $\theta_i > \sin^{-1}(n_2/n_1)$. Since the group velocity of the fastest ray trace is c/n_1 (here we assume n_1 is a constant), the group velocity of the slowest ray trace should be $(c/n_1)\sin\theta_i = (cn_2/n_1^2)$. Therefore, for a fiber of length L, the maximum group delay difference is approximately

$$\delta T_{\max} = \frac{n_1 L}{c}\left(\frac{n_1 - n_2}{n_2}\right) \tag{1.3.89}$$

1.3.4.5 Polarization Mode Dispersion (PMD)

Polarization mode dispersion is a special type of modal dispersion that exists in single mode fibers. It is worth noting that there are actually two fundamental modes that coexist in a single mode fiber. As shown in Figure 1.3.16, these two modes are orthogonally polarized. In an optical fiber with perfect cylindrical symmetry, these two modes have the same cutoff condition and they are referred to as *degenerate modes.*

However, practical optical fibers might not have perfect cylindrical symmetry due to birefringence; therefore these two fundamental modes may propagate in different speeds. Birefringence in an optical fiber is usually caused by small perturbations of the structure geometry as well as the anisotropy of the refractive index. The sources of the perturbations can be categorized as intrinsic and extrinsic. *Intrinsic perturbation* refers to permanent structural perturbations of fiber geometry, which are often caused by errors in the manufacturing process.

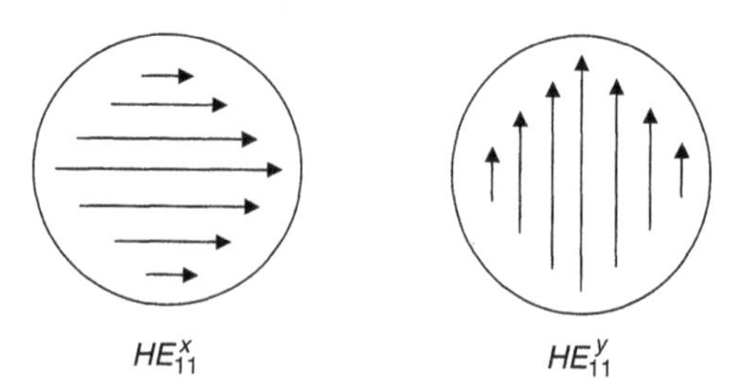

Figure 1.3.16 Illustration of optical field vector across the core cross-section of a single-mode fiber. Two degenerate modes exist in a single-mode fiber.

The effect of intrinsic perturbation includes (1) noncircular fiber core, which is called *geometric birefringence* and (2) nonsymmetric stress originated from the non-ideal perform, which is usually called *stress birefringence.* On the other hand, *extrinsic perturbation* usually refers to perturbations due to random external forces in the cabling and installation process. Extrinsic perturbation also causes both geometric and stress birefringence.

The effect of birefringence is that the two orthogonal polarization modes HE_{11}^{x} and HE_{11}^{y} experience slightly different propagation constants when they travel along the fiber; therefore their group delays become different. Assuming that the effective indices in the core of a birefringence fiber are n_x and n_y for the two polarization modes, their corresponding propagation constants will be $\beta_x = \omega n_x/c$ and $\beta_y = \omega n_y/c$, respectively. Due to birefringence, β_x and β_y are not equal and their difference is

$$\Delta\beta = (\beta_x - \beta_y) = \frac{\omega}{c}\Delta n_{eff} \tag{1.3.90}$$

where $\Delta n_{eff} = n_{//} - n_{\perp}$ is the effective differential refractive index of the two modes.

For a fiber of length L, the relative group delay between the two orthogonal polarization modes is

$$\Delta\tau_g = \frac{(n_{//} - n_{\perp})}{c}L = \frac{L\Delta n_{eff}}{c} \tag{1.3.91}$$

This is commonly referred to as *differential group delay* (DGD).

As a result of fiber birefringence, the state of polarization of the optical signal will rotate while propagating along the fiber because of the accumulated relative phase change $\Delta\Phi$ between the two polarization modes:

$$\Delta\Phi = \frac{\omega\Delta n_{eff}}{c}L \tag{1.3.92}$$

According to Equation 1.3.92, when an optical signal is launched into a birefringence fiber, its polarization evolution can be caused by the changes of either the fiber length L, the differential refractive index Δn_{eff}, or the lightwave signal frequency ω.

At a certain fiber length $L = L_p$, if the state of polarization of the optical signal completes a full $\Delta\Phi = 2\pi$ rotation, L_p is therefore defined as the birefringence beat length. On the other hand, at a fixed fiber length L, the polarization state of the optical signal can also be varied by changing the frequency. For a complete polarization rotation, the change of optical frequency should be

$$\Delta\omega_{cycle} = \frac{2\pi c}{L\Delta n_{eff}} = \frac{2\pi}{\Delta\tau_g} \tag{1.3.93}$$

In modern high-speed optical communications using single-mode fiber, polarization mode dispersion has become one of the most notorious sources of transmission performance degradation. Due to the random nature of the perturbations that cause birefringence, polarization mode dispersion in an optical fiber is a stochastic process. Its characterization and measurement are discussed in great detail in Chapter 4.

Example 1.6

A 1550 nm optical signal from a multilongitudinal mode laser diode has two discrete wavelength components separated by 0.8 nm. There are two pieces of optical fiber; one of them is a standard single-mode fiber with chromatic dispersion parameter $D = 17$ ps/nm/km at 1550 nm wavelength, and the other is a step-index multimode fiber with core index $n_1 = 1.48$, cladding index $n_2 = 1.46$, and core diameter $d = 50$ μm. Both of these two fibers have the same length of 20 km. Find the allowed maximum signal data rate that can be carried by each of these two fibers.

Solution:

For the single-mode fiber, chromatic dispersion is the major source of pulse broadening of the optical signal. In this example, the chromatic dispersion-induced pulse broadening is

$$\Delta t_{SMF} = D \cdot L \cdot \Delta\lambda = 17 \times 20 \times 0.8 = 272ps$$

For the multimode fiber, the major source of pulse broadening is modal dispersion. Using Equation 1.3.89, this pulse broadening is

$$\Delta t_{MMF} \approx \frac{n_1 L}{c}\left(\frac{n_1 - n_2}{n_2}\right) = 1.35\mu s$$

Obviously, multimode fiber introduces pulse broadening more than three orders of magnitude higher than the single-mode fiber. The data rate of the optical signal can be on the order of 1 Gb/s if the single-mode fiber is used, whereas it is limited to less than 1 Mb/s with the multimode fiber.

1.3.5 Nonlinear Effects in an Optical Fiber

Fiber parameters we have discussed so far, such as attenuation, chromatic dispersion, and modal dispersion, are all linear effects. The values of these parameters do not change with the change in the signal optical power. On the other hand, the effects of fiber nonlinearity depend on the optical power density inside the fiber core. The typical optical power carried by an optical fiber may not seem very high, but since the fiber core cross-section area is very small, the power density can be high enough to cause significant nonlinear effects. For example, for a standard single-mode fiber, the cross-section area of the core is about 80 $\mu m.^2$ If the fiber carries 10 mW of average optical power, the power density will be as high as 12.5 $kW/cm.^2$ Stimulated Brillouin scattering (SBS), stimulated Raman scattering (SRS), and the Kerr effect are the three most important nonlinear effects in silica optical fibers.

1.3.5.1 Stimulated Brillouin Scattering

Stimulated Brillouin scattering (SBS) is originated by the interaction between the signal photons and the traveling sound waves, also called *acoustic phonons* [10]. It is similar to blowing air into an open-ended tube; a sound wave is often generated. Because of the SBS, the signal light wave is modulated by the traveling sound wave. Stokes photons are generated in this process, and the frequency of the Stokes photons is downshifted from that of the original optical frequency. The amount of this frequency shift can be estimated approximately by

$$\Delta f = 2f_0 \frac{V}{(c/n_1)} \tag{1.3.94}$$

where n_1 is the refractive index of the fiber core, c/n_1 is the group velocity of the light wave in the fiber, V is the velocity of the sound wave, and f_0 is the original frequency of the optical signal. In silica-based optical fibers, this frequency shift is about 11 GHz.

SBS is highly directional and narrowband. The generated Stokes photons only propagate in the opposite direction of the original photons, and therefore the scattered energy is always counter-propagating with respect to the signal. In addition, since SBS relies on the resonance of a sound wave, which has very narrow spectral linewidth, the effective SBS bandwidth is as narrow as 20 MHz. Therefore, SBS efficiency is high only when the linewidth of the optical signal is narrow.

When the optical power is high enough, SBS turns signal optical photons into frequency-shifted Stokes photons that travel in the opposite direction. If another optical signal travels in the same fiber, in the same direction and at the same frequency of this Stokes wave, it can be amplified by the SBS process. Based on this, the SBS effect has been used to make optical amplifiers; however, the narrow amplification bandwidth nature of SBS limits their applications. On the other hand, in optical fiber communications, the effect of SBS introduces an extra loss for the optical signal and sets an upper limit for the amount of optical power that can be used in the fiber. In commercial fiber-optic systems, an effective way to suppress the effect of SBS is to frequency-modulate the optical carrier and increase the spectral linewidth of the optical source to a level much wider than 20 MHz.

1.3.5.2 Stimulated Raman Scattering

Stimulated Raman scattering (SRS) is an inelastic process where a photon of the incident optical signal (pump) stimulates molecular vibration of the material and loses part of its energy. Because of the energy loss, the photon reemits in a lower frequency [11]. The introduced vibrational energy of the molecules is referred to as an *optical phonon*. Instead of relying on the acoustic vibration as in the case of SBS, SRS in a fiber is caused by the molecular-level vibration of the silica material. Consequently, through the SRS process, pump photons are progressively absorbed by the fiber, and new photons, called *Stokes photons*, are created at a downshifted frequency.

Unlike SBS, where the Stokes wave only propagates in the backward direction, the Stokes waves produced by the SRS process propagate in both forward and backward directions. Therefore SRS can be used to amplify both co- and counterpropagated optical signals if their frequencies are within the SRS bandwidth. Also, the spectral bandwidth of SRS is much wider than that of SBS. As shown in Figure 1.3.17, in silica-based optical fibers, the maximum Raman efficiency happens at a frequency shift of about 13.2 THz, and the bandwidth

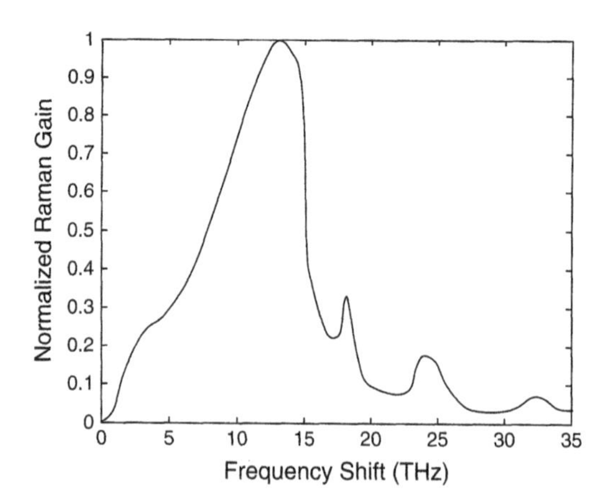

Figure 1.3.17 A normalized Raman gain spectrum of a silica fiber.

can be as wide as 10 THz. Optical amplifiers based on the SRS effect has become popular in recent years because of their unique characteristics compared to other types of optical amplifiers.

On the other hand, SRS may create interchannel crosstalk in wavelength-division multiplexed (WDM) optical systems. In a fiber carrying multiple wavelength channels, SRS effect may create energy transfer from short wavelength (higher-frequency) channels to long wavelength (lower-frequency) channels.

1.3.5.3 Kerr Effect Nonlinearity and Nonlinear Schrödinger Equation

Kerr effect nonlinearity is introduced by the fact that the refractive indices of an optical material is often a weak function of the optical power density:

$$n = n_0 + n_2 \frac{P}{A_{eff}} \tag{1.3.95}$$

where n_0 is the linear refractive index of the material, n_2 is the nonlinear index coefficient, P is the optical power, and A_{eff} is the effective cross-section area of the optical waveguide. P/A_{eff} represents optical power density.

Considering both linear and nonlinear effects, a nonlinear differential equation is often used to describe the envelope of optical field propagating along an optical fiber [1]:

$$\frac{\partial A(t,z)}{\partial z} + \frac{i\beta_2}{2}\frac{\partial^2 A(t,z)}{\partial t^2} + \frac{\alpha}{2}A(t,z) - i\gamma|A(t,z)|^2 A(t,z) = 0 \tag{1.3.96}$$

This equation is known as the *nonlinear Schrödinger (NLS) equation*. $A(t,z)$ is the complex amplitude of the optical field. The fiber parameters β_2 and α are group delay dispersion and attenuation, respectively. γ is defined as the *nonlinear parameter*:

$$\gamma = \frac{n_2 \omega_0}{c A_{eff}} \tag{1.3.97}$$

On the left side of Equation 1.3.96 the second term represents the effect of chromatic dispersion; the third term is optical attenuation; and the last term represents a nonlinear phase modulation caused by the Kerr effect. To understand the physical meaning of each term in the nonlinear Schrödinger equation, we can consider dispersion and nonlinearity separately.

First, we only consider the dispersion effect and assume

$$\frac{\partial A}{\partial z} + \frac{i\beta_2}{2}\frac{\partial^2 A}{\partial t^2} = 0 \tag{1.3.98}$$

This equation can easily be solved in Fourier domain as

$$\tilde{A}(\omega, L) = \tilde{A}(\omega, 0)\exp\left(-j\frac{\omega^2}{2}\beta_2 L\right) \tag{1.3.99}$$

where L is the fiber length, $\tilde{A}(\omega, L)$ is the Fourier transform of $A(t, L)$, and $\tilde{A}(\omega, 0)$ is the optical field at the fiber input. Consider two frequency components separated by $\delta\omega$; their phase difference at the end of the fiber is

$$\delta\Phi = \frac{\delta\omega^2}{2}\beta_2 L \tag{1.3.100}$$

If we let $\delta\Phi = \omega_c\delta t$, where δt is the arrival time difference at the fiber output between these two frequency components and ω_c is the average frequency, δt can be found as $\delta t \approx \delta\omega\beta_2 L$. Then if we convert β_2 into D using Equation 1.3.78, we find

$$\delta t \approx D \cdot L \cdot \delta\lambda \tag{1.3.101}$$

where $\delta\lambda = \delta\omega\lambda^2/(2\pi c)$ is the wavelength separation between these two components. In fact, Equation 1.3.101 is identical to Equation 1.3.79.

Now let's neglect dispersion and only consider fiber attenuation and Kerr effect nonlinearity. Then the nonlinear Schrödinger equation is simplified to

$$\frac{\partial A(t, z)}{\partial z} + \frac{\alpha}{2}A(t, z) = i\gamma|A(t, z)|^2 A(t, z) \tag{1.3.102}$$

If we start by considering the optical power, $P(z, t) = |A(z, t)|^2$, Equation 1.3.102 gives $P(z, t) = P(0, t)e^{-\alpha z}$. Then we can use a normalized variable $E(z, t)$ such that

$$A(z, t) = \sqrt{P(0, t)}\exp\left(\frac{-\alpha z}{2}\right)E(z, t) \tag{1.3.103}$$

Equation 1.3.102 becomes

$$\frac{\partial E(z, t)}{\partial z} = i\gamma P(0, t)e^{-\alpha z}E(z, t) \tag{1.3.104}$$

And the solution is

$$E(z, t) = E(0, t)\exp[j\Phi_{NL}(t)] \tag{1.3.105}$$

where

$$\Phi_{NL}(t) = \gamma P(0, t)\int_0^L e^{-\alpha z}dz = \gamma P(0, t)L_{eff} \tag{1.3.106}$$

with

$$L_{eff} = \frac{1 - e^{-\alpha L}}{\alpha} \approx \frac{1}{\alpha} \tag{1.3.107}$$

known as the nonlinear length of the fiber, which only depends on the fiber attenuation (where $e^{-\alpha L} \ll 1$ is assumed). For a standard single-mode fiber operating in a 1550 nm wavelength window, the attenuation is about 0.25 dB/km (or 5.8×10^{-5} Np/m) and the nonlinear length is approximately 17.4 km.

According to Equation 1.3.106, the nonlinear phase shift $\Phi_{NL}(t)$ follows the time-dependent change of the optical power. The corresponding optical frequency change can be found by

$$\delta f(t) = \frac{1}{2\pi}\gamma L_{eff}\frac{\partial}{\partial t}[P(0,t)] \tag{1.3.108}$$

Or the corresponding signal wavelength modulation:

$$\delta\lambda(t) = -\frac{\lambda^2}{2\pi c}\gamma L_{eff}\frac{\partial}{\partial t}[P(0,t)] \tag{1.3.109}$$

Figure 1.3.18 illustrates the waveform of an optical pulse and the corresponding nonlinear phase shift. This phase shift is proportional to the signal waveform, an effect known as *self-phase modulation* (SPM) [12].

If the fiber has no chromatic dispersion, this phase modulation alone would not introduce optical signal waveform distortion if optical intensity is detected at the fiber output. However, if the fiber chromatic dispersion is considered, wavelength deviation created by SPM at the leading edge and the falling edge of the optical pulse, as shown in Figure 1.3.18, will introduce group delay mismatch between these two edges of the pulse, therefore creating waveform distortion. For example, if the fiber has anomalous dispersion ($D > 0$), short wavelength components will travel faster than long wavelength components. In this case, the blue-shifted pulse falling edge travels faster than the red-shifted leading edge; therefore the pulse will be squeezed by the SMP process. On the

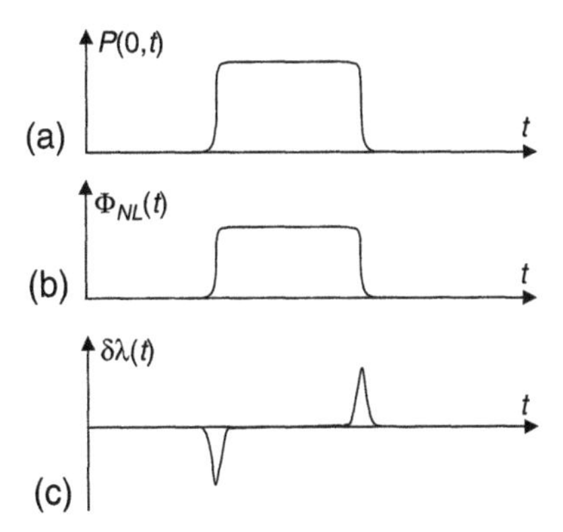

Figure 1.3.18 (a) Optical pulse waveform, (b) nonlinear phase shift, and (c) wavelength shift introduced by self-phase modulation.

other hand, if the fiber dispersion is normal (D < 0), the blue-shifted pulse falling edge travels slower than the red-shifted leading edge and this will result in pulse spreading.

In the discussion of self-phase modulation, we have only considered one wavelength channel in the fiber, and its optical phase is affected by its own intensity. If there is more than one wavelength channel traveling in the same fiber, the situation becomes more complicated and crosstalk between channels will be created by Kerr effect nonlinearity.

Now let's consider a system with only two wavelength channels; the combined optical field is

$$A(z,t) = A_1(z,t)e^{-j\theta_1} + A_2(z,t)e^{-j\theta_2} \tag{1.3.110}$$

where A_1 and A_2 are the optical field amplitude of the two wavelength channels and $\theta_1 = n\omega_1/c$ and $\theta_2 = n\omega_2/c$ are optical phases of these two optical carriers. Substituting Equation 1.3.110 into the nonlinear Schrödinger Equation 1.3.96 and collecting terms having $e^{-j\theta_1}$ and $e^{-j\theta_2}$, respectively, will result in two separate equations:

$$\frac{\partial A_1}{\partial z} + \frac{i\beta_2}{2}\frac{\partial^2 A_1}{\partial t^2} + \frac{\alpha}{2}A_1 = j\gamma|A_1|^2 A_1 + j2\gamma|A_2|^2 A_1 + j\gamma A_1^2 A_2^* e^{j(\theta_1-\theta_2)} \tag{1.3.111}$$

$$\frac{\partial A_2}{\partial z} + \frac{i\beta_2}{2}\frac{\partial^2 A_2}{\partial t^2} + \frac{\alpha}{2}A_2 = j\gamma|A_2|^2 A_2 + j2\gamma|A_1|^2 A_2 + j\gamma A_2^2 A_1^* e^{j(\theta_2-\theta_1)} \tag{1.3.112}$$

Each of these two equations describes the propagation characteristics of an individual wavelength channel. On the right side of each equation, the first term represents the effect of self-phase modulation as we have described; the second term represents *cross-phase modulation* (XPM); and the third term is responsible for another nonlinear phenomenon called *four-wave mixing* (FWM).

XPM is originated from the nonlinear phase modulation of one wavelength channel by the optical power change of the other channel [13]. Similar to SPM, it requires chromatic dispersion of the fiber to convert this nonlinear phase modulation into waveform intensity distortion. Since signal waveforms carried by these wavelength channels are usually not synchronized with each other, the precise time-dependent characteristic of crosstalk is less predictable. Statistical analysis is normally used to estimate the effect of XPM-induced crosstalk.

FWM can be better understood as two optical carriers copropagating along an optical fiber; the beating between the two carriers modulates the refractive index of the fiber at the frequency difference between them. Meanwhile, a third optical carrier propagating along the same fiber is phase modulated by this index modulation and then creates two extra modulation sidebands [14, 15].

If the frequencies of three optical carriers are ω_j, ω_k, and ω_l, the new frequency component created by this FWM process is

$$\omega_{jkl} = \omega_j + \omega_k - \omega_l = \omega_j - \Delta\omega_{kl} \tag{1.3.113}$$

where $l \neq j$, $l \neq k$ and $\Delta\omega_{kl}$ is the frequency spacing between channel k and channel l. If there are only two original optical carriers involved as in our example shown in Figure 1.3.19, the third carrier is simply one of the two original carriers ($j \neq k$), and this situation is known as *degenerate FWM*. Figure 1.3.19 shows the wavelength relations of degenerate FWM where two original carriers at ω_j *and* ω_k beat in the fiber, creating an index modulation at the frequency of $\Delta\omega_{jk} = \omega_j - \omega_k$. Then the original optical carrier at ω_j is phase-modulated at the frequency $\Delta\omega_{jk}$, creating two modulation sidebands at $\omega_i = \omega_j - \Delta\omega_{jk}$ and $\omega_k = \omega_j + \Delta\omega_{jk}$. Similarly, the optical carrier at ω_k is also phase-modulated at the frequency $\Delta\omega_{jk}$ and creates two sidebands at $\omega_j = \omega_k - \Delta\omega_{jk}$ and $\omega_l = \omega_k + \Delta\omega_{jk}$. In this process, the original two carriers become four, and this is probably where the name "four-wave mixing" came from.

FWM is an important nonlinear phenomenon in optical fiber; it introduces interchannel crosstalk when multiple wavelength channels are used. The buildup of new frequency components generated by FWM depends on the phase match between the original carriers and the new frequency component when they travel along the fiber. Therefore, the efficiency of FWM is very sensitive to the dispersion-induced relative walk-off between the participating wavelength components. In general, the optical field amplitude of the FWM component created in a fiber of length L is

$$A_{jkl}(L) = j\gamma\sqrt{P_j(0)P_k(0)P_l(0)}\int_0^L e^{-\alpha z}\exp\left(j\Delta\beta_{jkl}z\right)dz \tag{1.3.114}$$

where $P_j(0)$, $P_k(0)$, and $P_l(0)$ are the input powers of the three optical carriers and

$$\Delta\beta_{jkl} = \beta_j + \beta_k - \beta_l - \beta_{jkl} \tag{1.3.115}$$

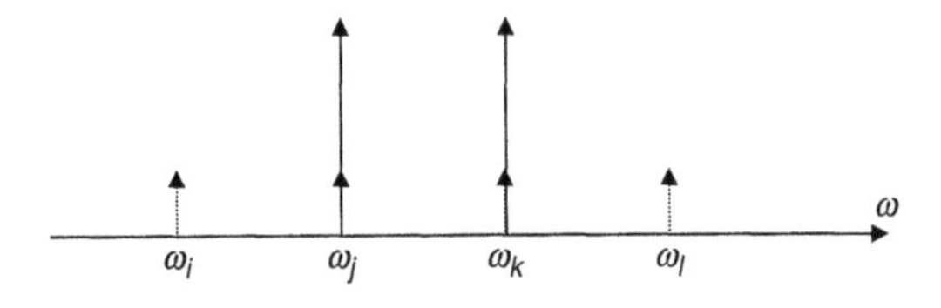

Figure 1.3.19 Degenerate four-wave mixing, where two original carriers at ω_j and ω_k create four new frequency components at ω_i, ω_j, ω_k, and ω_l.

is a propagation constant mismatch, $\beta_j = \beta(\omega_j)$, $\beta_k = \beta(\omega_k)$, $\beta_l = \beta(\omega_l)$, and $\beta_{jkl} = \beta(\omega_{jkl})$. Expanding β as $\beta(\omega) = \beta_0 + \beta_1(\omega - \omega_0) + (\beta_2/2)(\omega - \omega_0)^2$ and using the frequency relation given in Equation (1.3.113), we can find

$$\Delta\beta_{jkl} = -\beta_2(\omega_j - \omega_l)(\omega_k - \omega_l)$$

Here for simplicity we neglected the dispersion slope and considered that the dispersion value is constant over the entire frequency region. This propagation constant mismatch can also be expressed as the functions of the corresponding wavelengths:

$$\Delta\beta_{jkl} = \frac{2\pi c D}{\lambda^2}(\lambda_j - \lambda_l)(\lambda_k - \lambda_l) \tag{1.3.116}$$

where dispersion parameter β_2 is converted to D using Equation 1.3.78. The integration of Equation 1.3.114 yields

$$A_{jkl}(L) = j\gamma\sqrt{P_j(0)P_k(0)P_l(0)}\frac{e^{(j\Delta\beta_{jkl}-\alpha)L} - 1}{j\Delta\beta_{jkl} - \alpha} \tag{1.3.117}$$

The power of the FWM component is then

$$P_{jkl}(L) = \eta_{FWM}\gamma^2 L_{eff} P_j(0)P_k(0)P_l(0) \tag{1.3.118}$$

where $L_{eff} = (1 - e^{-\alpha L})/\alpha$ is the fiber nonlinear length and

$$\eta_{FWM} = \frac{\alpha^2}{\Delta\beta_{jkl}^2 + \alpha^2}\left[1 + \frac{4e^{-\alpha L}\sin^2(\Delta\beta_{jkl}L/2)}{(1 - e^{-\alpha L})^2}\right] \tag{1.3.119}$$

is the FWM efficiency. In most of the practical cases, when the fiber is long enough, $e^{-\alpha L} \ll 1$ is true. The FWM efficiency can be simplified as

$$\eta_{FWM} \approx \frac{\alpha^2}{\Delta\beta_{jkl}^2 + \alpha^2} \tag{1.3.120}$$

In this simplified expression, FWM efficiency is no longer dependent on the fiber length. The reason is that as long as $e^{-\alpha L} \ll 1$, for the fiber lengths far beyond the nonlinear length, the optical power is significantly reduced and thus the nonlinear contribution. Consider the propagation constant mismatch given in Equation 1.3.116: The FWM efficiency can be reduced either by the increase of fiber dispersion or by the increase of channel separation. Figure 1.3.20 shows the FWM efficiency for several different fiber dispersion parameters calculated with Equations 1.3.116 and 1.3.120, where the fiber loss is $\alpha = 0.25\ dB/km$ and the operating wavelength is 1550 nm. Note in these calculations, the unit of attenuation α has to be in Np/m when using eq.s (1.3.117) – (1.3.120). As an example, if two wavelength channels with 1 nm channel spacing, the

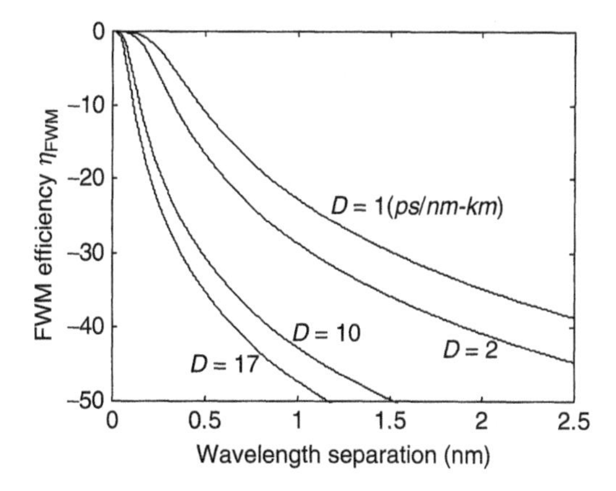

Figure 1.3.20 Four-wave mixing efficiency η_{FWM}, calculated for α =0.25 dB/km, λ = 1550 nm.

FWM efficiency increases for approximately 25 dB when the chromatic dispersion is decreased from 17ps/nm/km to 1ps/nm/km. As a consequence, in WDM optical systems, if dispersion-shifted fiber is used, interchannel crosstalk introduced by FWM may become a legitimate concern, especially when the optical power is high.

1.4 OPTICAL AMPLIFIERS

The introduction of the optical amplifier is one of the most important advances in optical fiber communications. Linear optical amplifiers are often used to compensate losses in optical communication systems and networks due to fiber attenuation, optical power splitting, and other factors. Optical amplifiers can also be used to perform nonlinear optical signal processing and waveform shaping when they are used in a nonlinear regime.

Figure 1.4.1 shows a typical multispan point-to-point optical transmission system in which optical amplifiers are used to perform various functions. The post-amplifier (post-amp) is used at the transmitter to boost the optical power. Sometimes the post-amp is integrated with the transmitter to form a powerful "extended" optical transmitter. The post-amp is especially useful if the power of a transmitter has to be split into a number of broadcasting outputs in an optical network to compensate the splitting loss. The basic requirement for a post-amp is to be able to supply high enough output optical power.

In-line optical amplifiers (line-amps) are used along the transmission system to compensate for the attenuation caused by optical fibers. In high-speed optical

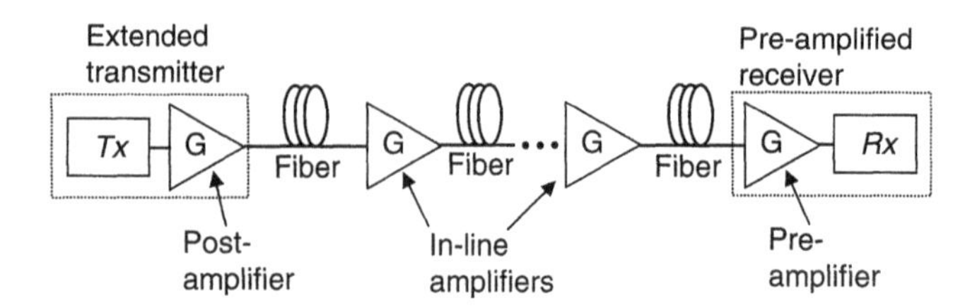

Figure 1.4.1 Illustration of a point-to-point optical transmission system and the functions of optical amplifiers. *Tx*: transmitter, *Rx*: receiver.

transmission systems, line-amps are often spaced periodically along the fiber link, one for approximately every 80 km of transmission distance. The basic requirement for a line-amp is to provide high enough optical gain. In addition, to support wavelength division multiplexed (WDM) optical systems, a line-amp needs to have wide optical bandwidth and flat optical gain within the bandwidth. Gain equalization techniques are often used in line-amps. Also, the gain of a line-amp has to be linear to prevent nonlinear crosstalk between different wavelength channels.

A pre-amplifier is often used immediately before the photodiode in an optical receiver to form the so called pre-amplified optical receiver. In addition to the requirement of high optical gain, the most important qualification of a pre-amplifier is that the noise should be low. The sensitivity of a pre-amplified optical receiver is largely dependent on the noise characteristic of the pre-amplifier. Because of their different applications, various types of optical amplifiers will generally have to be designed and optimized differently for best performance.

The most popular optical amplifiers used for optical communications and other electro-optic systems are semiconductor optical amplifiers (SOAs) and erbium-doped fiber amplifiers (EDFAs). In this section, we first discuss the common characteristics of optical amplifiers, such as optical gain, gain bandwidth, and saturation. Then we will discuss the specific characteristics and applications of SOAs and EDFs separately.

1.4.1 Optical Gain, Gain Bandwidth, and Saturation

As in a laser, the most important component of an optical amplifier is the gain medium. An optical amplifier differs from a laser in that it does not require optical feedback and the optical signal passes through the gain medium only once. The optical signal is amplified through the stimulated emission process in the gain medium where carrier density is inverted, as discussed in Section 1.1.4. At the same time, spontaneous emission also exists in the gain medium, creating optical noise added to the amplified signal. In continuous wave operation, the propagation equation of the optical field E in the gain medium along the longitudinal direction z can be expressed as

$$\frac{dE}{dz} = \left[\frac{g}{2} + jn\right]E + \rho_{sp} \tag{1.4.1}$$

where g is the optical power gain coefficient at position z, n is the refractive index of the optical medium, and ρ_{sp} is the spontaneous emission factor.

In general, the local gain coefficient g is a function of optical frequency due to the limited optical bandwidth. It is also a function of optical power because of the saturation effect. For homogeneously broadened optical systems, gain saturation does not affect the frequency dependency of the gain; therefore we can use the following expression

$$g(f, P) = \frac{g_0}{1 + 4\left(\frac{f - f_0}{\Delta f}\right)^2 + \frac{P}{P_{sat}}} \tag{1.4.2}$$

where g_0 is the linear gain, Δf is the FWHM bandwidth, f_0 is the central frequency of the material gain, and P_{sat} is the saturation optical power.

Neglecting the spontaneous emission term, optical power amplification can be easily evaluated using Equation 1.4.1:

$$\frac{dP(z)}{dz} = g(f, P)P(z) \tag{1.4.3}$$

When signal optical power is much smaller than the saturation power ($P \ll P_{sat}$), the system is linear and the solution of equation (1.4.3) is

$$P(L) = P(0)G_0 \tag{1.4.4}$$

where L is the amplifier length, $P(0)$ and $P(L)$ are the input the output optical power, respectively, and G_0 is the small-signal optical gain.

$$G_0 = \frac{P(L)}{P(0)} = \exp\left[\frac{g_0 L}{1 + 4(f - f_0)^2/\Delta f^2}\right] \tag{1.4.5}$$

It is worth noting that the FWHM bandwidth of optical gain G_0 can be found as

$$B_0 = \Delta f \sqrt{\frac{\ln 2}{g_0 L - \ln 2}} \tag{1.4.6}$$

which is different from the material gain bandwidth Δf. Although Δf is a constant, optical gain bandwidth is inversely proportional to the peak optical gain $g_0 L$.

If the signal optical power is strong enough, nonlinear saturation has to be considered. To simplify, if we only consider the peak gain at $f = f_0$, optical power evolution along the amplifier can be found by

$$\frac{dP(z)}{dz} = \frac{g_0 P(z)}{1 + P(z)/P_{sat}} \tag{1.4.7}$$

This equation can be converted into

$$\int_{P(0)}^{P(L)} \left(\frac{1}{P} + \frac{1}{P_{sat}}\right) dP = g_0 L \tag{1.4.8}$$

And its solution is

$$\ln\left[\frac{P(L)}{P(0)}\right] + \frac{P(L) - P(0)}{P_{sat}} = g_0 L \tag{1.4.9}$$

We know that $G_0 = \exp(g_0 L)$ is the small-signal gain, if we define $G = P(L)/P(0)$ as the large-signal gain, Equation 1.4.9 can be written as

$$\frac{\ln(G/G_0)}{1 - G} = \frac{P(0)}{P_{sat}} \tag{1.4.10}$$

With the increase of the signal optical power, the large-signal optical gain G will decrease. Figure 1.4.2 shows the large-signal gain versus the input optical signal power when the small-signal optical gain is 30 dB and the saturation optical power is 3 dBm.

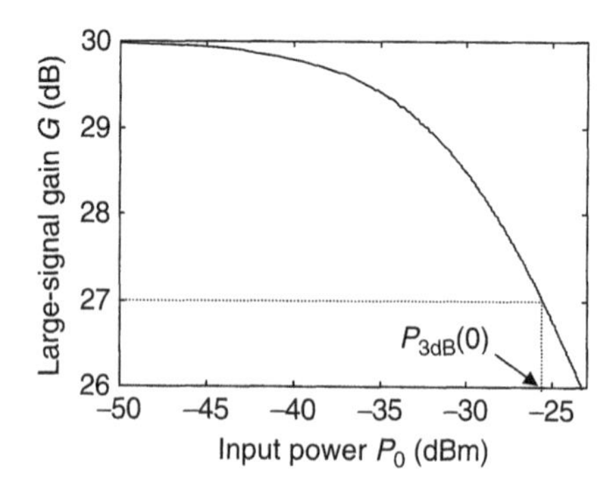

Figure 1.4.2 Optical signal gain versus input power calculated with Equation 1.4.10. $G_0 = 30$ dB, $P_{sat} = 3$ dBm.

The 3-dB saturation input power $P_{3dB}(0)$ is defined as the input power at which the large-signal optical gain is reduced by 3 dB. From Equation 1.4.10, $P_{3dB}(0)$ can be found as

$$P_{3dB}(0) = P_{sat} \frac{2\ln(2)}{G_0 - 2} \tag{1.4.11}$$

In the example shown in Figure 1.4.2, $P_{3dB}(0)$ can be found as –25.6 dBm.

Obviously, the 3-dB saturation input power depends on the small-signal optical gain G_0. When G_0 is high, $P_{3dB}(0)$ will become smaller. In a similar way, we can also define a 3-dB saturation output power $P_{3dB}(L)$, which will be less dependent on the small-signal optical gain. Since $P_{3dB}(L) = GP_{3dB}(0)$,

$$P_{3dB}(L) \approx P_{sat}\ln(2) \tag{1.4.12}$$

where we have assumed that $G_0 \gg 2$.

1.4.2 Semiconductor Optical Amplifiers

A semiconductor optical amplifier (SOA) is similar to a semiconductor laser operating below threshold. It requires an optical gain medium and an optical waveguide, but it does not require an optical cavity [16, 17, 18]. An SOA can be made by a laser diode with antireflection coating on each facet; an optical signal passes through the gain medium only once and therefore it is also known as a *traveling-wave optical amplifier*. Because of the close similarity between an SOA and a laser diode, the analysis of SOA can be based on the rate Equation 1.1.21 we have already discussed in Section 1.1.

$$\frac{dN(z,t)}{dt} = \frac{J}{qd} - \frac{N(z,t)}{\tau} - 2\Gamma v_g a(N - N_0)P(z,t) \tag{1.4.13}$$

where $N(z, t)$ is the carrier density and $P(z, t)$ is the photon density within the SOA optical waveguide, J is the injection current density, d is the thickness of the active layer, v_g is the group velocity of the lightwave, τ is the spontaneous emission carrier lifetime, a is the differential optical field gain, N_0 is the transparency carrier density, and Γ is the optical confinement factor of the waveguide.

1.4.2.1 Steady-State Analysis

Let's start with a discussion of steady-state characteristics of an SOA. In the steady state, $d/dt = 0$ and therefore

$$\frac{J}{qd} - \frac{N(z,t)}{\tau} - 2\Gamma v_g a(N - N_0)P(z,t) = 0 \tag{1.4.14}$$

If the optical signal is very small ($P(z,t) \approx 0$), the small-signal carrier density can be found as $N = J\tau/qd$ and the small-signal material gain can be found as

$$g_0 = 2a(N - N_0) = 2a\left(\frac{J\tau}{qd} - N_0\right) \tag{1.4.15}$$

We can then use this small-signal gain g_0 as a parameter to investigate the general case where the signal optical power may be large. Thus Equation 1.4.14 can be written as

$$\frac{g_0}{a} - (N - N_0) - 2\Gamma v_g \tau a(N - N_0)P(z,t) = 0 \tag{1.4.16}$$

In large-signal case, if we define $g = 2a(N - N_0)$, then Equation 1.4.16 yields,

$$g = \frac{g_0}{1 + P(z,t)/P_{sat}} \tag{1.4.17}$$

where

$$P_{sat} = \frac{1}{2\Gamma v_g \tau a} \tag{1.4.18}$$

is the saturation photon density, and it is an OSA parameter that is independent of the actual signal optical power. The saturation photon density can also be found to be related to the saturation optical power by

$$P_{sat,opt} = P_{sat} w d v_g hf \tag{1.4.19}$$

where w is the waveguide width (wd is the cross section area) and hf is the photon energy. Therefore, the corresponding saturation optical power is

$$P_{sat,opt} = \frac{wdhf}{2\Gamma \tau a} \tag{1.4.20}$$

As an example, for an SOA with waveguide width $w = 2\ \mu m$ and thickness $d = 0.25\ \mu m$, optical confinement factor $G = 0.3$, differential field gain $a = 5 \times 10^{-20} m^2$, and carrier lifetime $\tau = 1\ ns$, the saturation optical power is approximately 22 mW.

We need to point out that in the analysis of semiconductor lasers, we often assume that the optical power along the laser cavity is uniform; this is known as *mean-field approximation*. For an SOA, on the other hand, the signal optical power at the output side is much higher than that at the input side, and they differ by the amount of the SOA optical gain. In this case, mean-field application is no longer accurate. Optical evolution along the SOA has to be modeled by a z-dependent equation,

$$\frac{dP_{opt}(z)}{dz} = [\Gamma g(P_{opt}) - \alpha_{int}]P_{opt}(z) \tag{1.4.21}$$

where α_{int} is the internal loss of the waveguide. Because the optical power varies significantly along the waveguide, the material gain is also a function of location parameter z due to power-induced saturation. Similar to Equation 1.4.2, the optical gain can be expressed as a function of the optical power,

$$g(f,P,z) = \frac{g_0}{1 + 4\left(\frac{f - f_0}{\Delta f}\right)^2 + \frac{P_{opt}(z)}{P_{sat,opt}}} \tag{1.4.22}$$

Therefore, at $f = f_0$ and neglecting the waveguide loss α_{int}, the large-signal optical gain can be obtained using the equation similar to Equation 1.4.10:

$$\frac{\ln(G/G_0)}{1 - G} = \frac{P_{opt}(0)}{P_{sat,opt}} \tag{1.4.23}$$

where $P_{opt}(0)$ is the input signal optical power and $G_0 = \exp(\Gamma g_0 L)$ is the small-signal optical gain.

1.4.2.2 Gain Dynamics of OSA

Due to its short carrier lifetime, optical gain in an SOA can change quickly. Fast-gain dynamics is one of the unique properties of semiconductor optical amplifiers as opposed to fiber amplifiers, which are discussed later in this chapter. The consequence of fast gain dynamics is twofold: (1) It introduces cross-talk between different wavelength channels through cross-gain saturation and (2) it can be used to accomplish all-optical switch based on the same cross-gain saturation mechanism.

Using the same carrier density rate equation as Equation 1.4.13 but converting photon density into optical power,

$$\frac{dN(z,t)}{dt} = \frac{J}{qd} - \frac{N(z,t)}{\tau} - \frac{2\Gamma a}{hfA}(N - N_0)P_{opt}(z,t) \tag{1.4.24}$$

where A is the waveguide cross-section area. To focus on the time domain dynamics, it would be helpful to eliminate the spatial variable z. This can be achieved by investigating the dynamics of the total carrier population N_{tot},

$$N_{tot}(t) = \int_0^L N(z,t)dz$$

Then Equation 1.4.24 becomes

$$\frac{dN_{tot}(t)}{dt} = \frac{I}{qd} - \frac{N_{tot}(t)}{\tau} - \frac{2\Gamma a}{hfA}\int_0^L [N(z,t) - N_0]P_{opt}(z,t)dz \tag{1.4.25}$$

where $I = JL$ is the total injection current. Considering that relationship between the optical power and the optical gain as shown in Equation 1.4.21, the last term of Equation 1.4.25 can be expressed by

$$2\Gamma a\int_0^L [N(z,t) - N_0]P_{opt}(z,t)dz = P_{opt}(L,t) - P_{opt}(0,t) \tag{1.4.26}$$

Here $P_{opt}(0,t)$ and $P_{opt}(L,t)$ are the input and the output optical waveforms, respectively. We have neglected the waveguide loss α_{int}. Also, we have assumed that the power distribution along the amplifier does not change, which implies that the rate of optical power change is much slower than the transit time of the amplifier. For example, if the length of an SOA is $L = 300\ \mu m$ and the refractive index of the waveguide is $n = 3.5$, the transit time of the SOA is approximately 3.5ps. Under these approximations, Equation 1.4.25 can be simplified as

$$\frac{dN_{tot}(t)}{dt} = \frac{I}{qd} - \frac{N_{tot}(t)}{\tau} - \frac{P_{opt}(0,t)}{hfA}[G(t) - 1] \tag{1.4.27}$$

where $G(t) = P_{opt}(L,t)/P_{opt}(0,t)$ is the optical gain. On the other hand, since the optical gain of the amplifier can be expressed as

$$G(t) = \exp\left\{2\Gamma a\int_0^L (N(z,t) - N_0)dz\right\} = \exp\{2\Gamma a(N_{tot}(t) - N_0)L\} \tag{1.4.28}$$

That is,

$$N_{tot}(t) = \frac{\ln G(t)}{2\Gamma a} + N_0 L$$

Or

$$\frac{dN_{tot}(t)}{dt} = \frac{1}{2\Gamma aG(t)}\frac{dG(t)}{dt} \tag{1.4.29}$$

Equations 1.4.27 and 1.4.29 can be used to investigate how an SOA responds to a short optical pulse. When a short and intense optical pulse is injected into an SOA with the pulse width $T \ll \tau$, within this short time interval of optical pulse duration, the contributions of current injection I/qd and spontaneous recombination N_{tot}/τ are very small, which can be neglected. With this simplification, Equation 1.4.27 becomes

$$\frac{dN_{tot}(z)}{dt} \approx -\frac{P_{opt}(0,t)}{hfA}[G(t) - 1] \tag{1.4.30}$$

Equations 1.4.29 and 1.4.30 can be combined to eliminate dN_{tot}/dt and then the equation can be integrated over the pulse duration from $t = 0$ to $t = T$,

$$\ln\left[\frac{1 - 1/G(T)}{1 - 1/G(0)}\right] + \frac{1}{\tau P_{sat,opt}}\int_0^T P_{opt}(0,t)dt = 0 \qquad (1.4.31)$$

where $P_{sat,opt}$ is the saturation power as defined in Equation 1.4.20. $G(0)$ and $G(T)$ are SOA optical gains immediately before and immediately after the optical pulse, respectively.

Equation 1.4.31 can be written as

$$G(T) = \frac{1}{1 - \left(1 - \frac{1}{G_0}\right)\exp\left[-\frac{W_{in}(T)}{\tau P_{sat,opt}}\right]} \qquad (1.4.32)$$

where $W_{in}(T) = \int_0^T P_{opt}(t,0)dt$ is the input signal optical pulse energy.

An SOA is often used in nonlinear applications utilizing its fast cross-gain saturation effect, where an intense optical pulse at one wavelength can saturate the optical gain of the amplifier and therefore all-optically modulate optical signals of other wavelengths passing through the same SOA. Figure 1.4.3 illustrates the short pulse response of an SOA. Assume that the input optical

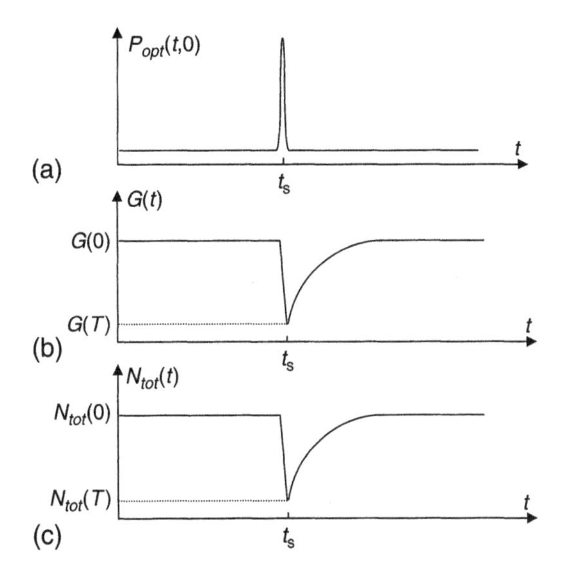

Figure 1.4.3 Illustration of short pulse response of optical gain and carrier population of an SOA: (a) short optical pulse, (b) optical gain of the SOA, and (c) carrier population.

pulse is very short so we do not need to consider its width. Then, immediately after the signal pulse is over, the optical gain of the amplifier can be calculated by Equation 1.4.32. As an example, for an SOA with a saturation power $P_{sat,opt} = 110\,mW$ and a carrier lifetime $\tau = 1\ ns$, the optical gain suppression ratio versus input pulse energy is shown in Figure 1.4.4 for three different small-signal gains of the SOA. With high small-signal optical gain, it is relatively easy to achieve a high gain suppression ratio. For example, if the input optical pulse energy is 10 pJ, to achieve 10 dB optical gain suppression, the small signal gain has to be at least 20 dB, as shown in Figure 1.4.4.

Although the optical gain suppression is fast in an SOA, which to some extent is only limited by the width of the input optical pulse, the gain recovery after the optical pulse can be quite long; this gain recovery process depends on the carrier lifetime of the SOA.

After the optical pulse passes through an SOA, the gain of the SOA starts to recover toward its small-signal value due to constant carrier injection into the SOA. Since the photon density inside the SOA is low without the input optical signal, stimulated emission can be neglected in the gain recovery process, and the carrier population rate equation is

$$\frac{dN_{tot}(t)}{dt} = \frac{I}{qd} - \frac{N_{tot}(t)}{\tau} \tag{1.4.33}$$

The solution of Equation 1.4.33 is

$$N_{tot}(t) = [N_{tot}(t_s) - N_{tot}(0)]\exp\left(-\frac{t-t_s}{\tau}\right) + N_{tot}(0) \tag{1.4.34}$$

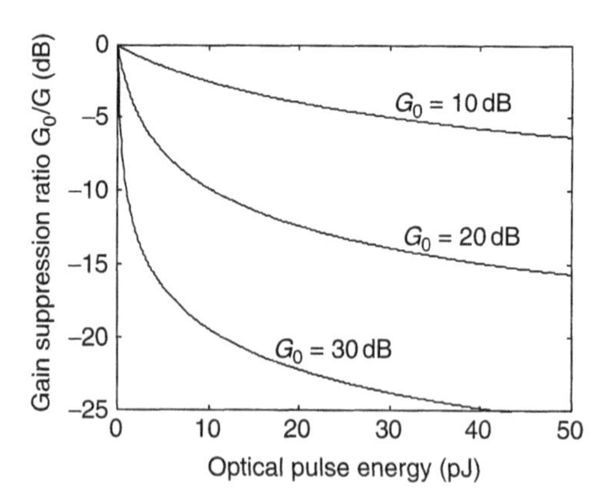

Figure 1.4.4 Gain suppression ratio versus input optical pulse energy in an SOA with 110 mW saturation optical power and 1 ns carrier lifetime.

where $N_{tot}(0) = N_{tot}(\infty)$ is the small-signal carrier population of the SOA. Since optical gain is exponentially proportional to the carrier population, as shown in Equation 1.4.28:

$$\ln\left(\frac{G(t)}{G_0}\right) = \ln\left(\frac{G(t_s)}{G_0}\right)\exp\left(-\frac{t-t_s}{\tau}\right) \quad (1.4.35)$$

Equation 1.4.35 clearly demonstrates that the dynamics of optical gain recovery primarily depend on the carrier lifetime τ of the SOA.

1.4.2.2.1 Optical Wavelength Conversion Using Cross-Gain Saturation

One of the important applications of gain saturation effect in SOA is all-optical wavelength conversion as schematically shown in Figure 1.4.5. In this application, a high-power intensity-modulated optical signal at wavelength λ_2 is combined with a relatively weak CW signal at λ_1 before they are both injected to an SOA. The optical gain of the SOA is then modulated by the signal at λ_2 through SOA gain saturation effect as illustrated in Figure 1.4.5(b). At the same time, the weak CW optical signal at wavelength λ_1 is amplified by the time-varying gain of the SOA, and therefore its output waveform is complementary to that of the optical signal at λ_2. At the output of the SOA, an optical

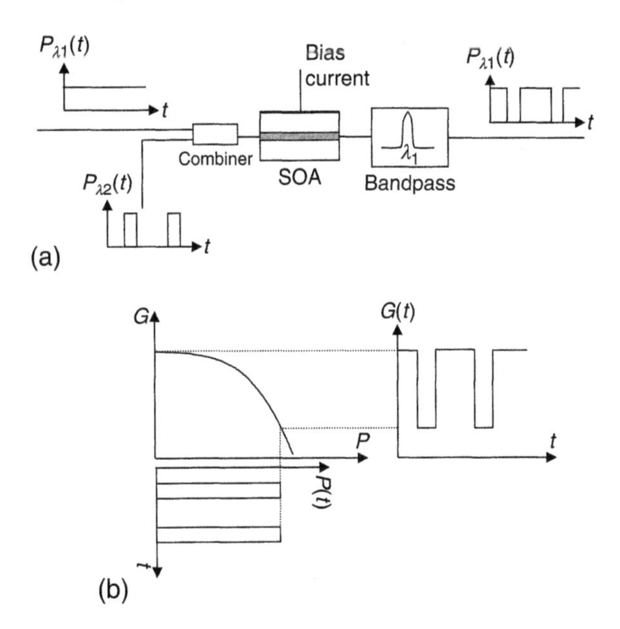

Figure 1.4.5 All-optical wavelength conversion using cross gain saturation effect of SOA: (a) system configuration and (b) illustration of principle.

bandpass filter is used to select the wavelength component at λ_1 and reject that at λ_2. In this way, the wavelength of the optical carrier is converted from λ_2 to λ_1. This process is all-optical and without the need for electrical modulation.

1.4.2.2.2 Wavelength Conversion Using FWM in SOA

In the last section, we discussed four-wave mixing (FWM) in an optical fiber, where two optical carriers at different frequencies, ω_1 and ω_2, copropagate in the fiber, causing refractive index modulation at the frequency $\Omega = \omega_2 - \omega_1$. New frequency components of FWM are created due to this phase modulation in the optical fiber. The concept of FWM in an SOA is similar to that in an optical fiber except the nonlinear mechanism is different. In an SOA, two optical carriers generate a modulation of carrier population N_{tot} and therefore the optical gain is modulated at the difference frequency between the two carriers. Rigorous analysis of FWM in an SOA can be performed by numerically solving the rate equations. However, to understand the physical mechanism of FWM, we can simplify the analysis using mean field approximation. With this approximation, the carrier population rate Equation 1.4.25 can be written as

$$\frac{dN_{tot}(t)}{dt} = \frac{I}{qd} - \frac{N_{tot}(t)}{\tau} - \frac{2\Gamma a}{hfA}[N_{tot}(t) - N_{0,tot}]P_{ave}(t)dz \tag{1.4.36}$$

where P_{ave} is the spatially averaged optical power and $N_{0,tot}$ is the transparency carrier population. Since there are two optical carriers that co-exist in the SOA, the total optical power inside the SOA is the combination of these two carriers,

$$P_{ave}(t) = \left|E_1 e^{j\omega_1 t} + E_2 e^{j\omega_2 t}\right|^2 \tag{1.4.37}$$

where ω_1 and ω_2 are the optical frequencies and E_1 and E_2 are the field amplitudes of the two carriers, respectively.

Equation 1.4.37 can be expanded as

$$P_{ave}(t) = P_{DC} + 2E_1E_2 \cos \Delta\omega t \tag{1.4.38}$$

where $P_{DC} = E_1^2 + E_2^2$ is the constant part of the power and $\Delta\omega = \omega_2 - \omega_1$ is the frequency difference between the two carriers. As a consequence of this optical power modulation at frequency $\Delta\omega$ as shown in equation (1.4.38), the carrier population will also be modulated at the same frequency due to gain saturation:

$$N_{tot}(t) = N_{DC} + \Delta N \cos \Delta\omega t \tag{1.4.39}$$

where N_{DC} is the constant part of carrier population and ΔN is the magnitude of carrier population modulation. Substituting Equations 1.4.38 and 1.4.39 into the carrier population rate Equation 1.4.36 and separating the DC and time-varying terms, we can determine the values of ΔN and N_{DC} as

$$\Delta N = \frac{2(N_{DC} - N_{0,tot})E_1E_2}{P_{sat,opt}(1 + P_{DC}/P_{sat,opt} + j\Delta\omega\tau)} \quad (1.4.40)$$

And

$$N_{DC} = \frac{I\tau/q + N_{0,tot}P_{DC}/P_{sat,opt}}{1 + P_{DC}/P_{sat,opt}} \quad (1.4.41)$$

Since the optical gain of the amplifier is exponentially proportional to the carrier population as $G \propto \exp(N_{tot}L)$, the magnitude of carrier population modulation reflects the efficiency of FWM. Equation 1.4.40 indicates that FWM efficiency is low when the carrier frequency separation $\Delta\omega$ is high and the effective FWM bandwidth is mainly determined by the carrier lifetime τ of the SOA. In a typical SOA, the carrier lifetime is on the order of nanosecond and therefore this limits the effective FWM bandwidth to the order of Gigahertz. However, efficient FWM in SOA has been experimentally observed even when the frequency separation between the two optical carriers was as wide as in the Terahertz range, as illustrated in Figure 1.4.6.

The wide bandwidth of FWM measured in SOAs indicated that the gain saturation is not only determined by the lifetime of interband carrier recombination τ; other effects, such as intraband carrier relaxation and carrier heating, have to be considered to explain the ultrafast carrier recombination mechanism. In Figure 1.4.6, the normalized FWM efficiency, which is defined as the ratio between the power of the FWM component and that of the initial optical carrier, can be estimated as a multiple time constant system [19]:

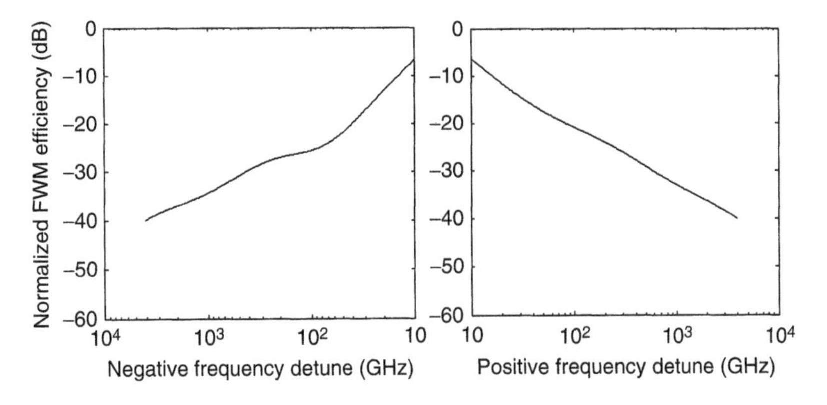

Figure 1.4.6 Normalized FWM signal power versus frequency detune.

$$\eta_{FWM} \propto \left| \sum_n \frac{k_i}{1 + i\Delta\omega\tau_n} \right|^2 \tag{1.4.42}$$

where τ_n and k_n are the time constant and the importance of each carrier recombination mechanism. To fit the measured curve, three time constants had to be used, with $k_1 = 5.65e^{-1.3i}$, $k_2 = 0.0642e^{1.3i}$, $k_3 = 0.0113e^{1.53i}$, $\tau_1 = 0.2ns$, $\tau_2 = 650fs$, and $\tau_3 = 50fs$ [19].

The most important application of FWM in SOA is frequency conversion, as illustrated in Figure 1.4.7. A high-power CW optical carrier at frequency ω_1, which is usually referred to as *pump*, is injected into an SOA together with an optical signal at frequency ω_2, which is known as the *probe*. Due to FWM in SOA, a new frequency component is created at $\omega_{FWM} = \omega_2 - 2\omega_1$. A bandpass optical filter is then used to select this FWM frequency component. In addition to performing frequency conversion from ω_2 to ω_{FWM}, an important advantage of this frequency conversion is frequency conjugation, as shown in Figure 1.4.7(b). The upper (lower) modulation sideband of the probe signal is translated into the lower (upper) sideband of the wavelength converted FWM component. This frequency conjugation effect has been used to combat chromatic dispersion in fiber-optical systems.

Suppose the optical fiber has an anomalous dispersion. High-frequency components travel faster than low-frequency components. If an SOA is added in the middle of the fiber system, performing wavelength conversion and frequency conjugation, the fast-traveling upper sideband over the first half of the system will be translated into the lower sideband, which will travel in a slower speed over the second half of the fiber system. Therefore, signal waveform distortion due to fiber chromatic dispersion can be canceled by this midspan frequency conjugation.

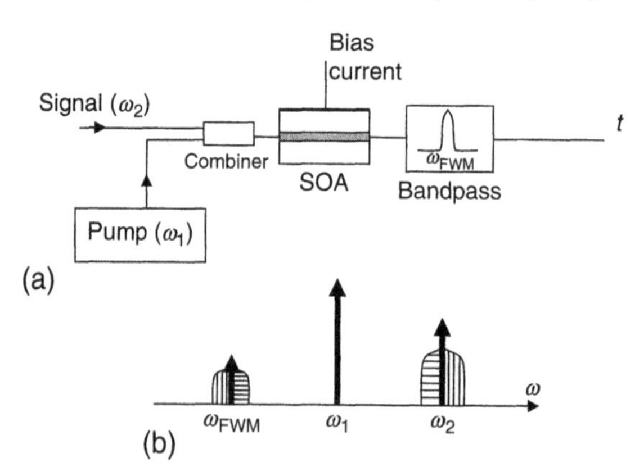

Figure 1.4.7 Wavelength conversion using FWM in an SOA: (a) system configuration and (b) illustration of frequency relationship of FWM, where $\omega_{FWM} = \omega_2 - 2\omega_1$

1.4.2.2.3 Optical Phase Modulation in an SOA

As we discussed in Section 1 of this chapter, in a semiconductor laser, a carrier density change not only changes the optical gain, it also changes the refractive index of the material and thus introduces an optical phase change. This is also true in semiconductor optical amplifiers. We know that photon density P of an optical amplifier can be related to the material gain g as, $P = P_{in}\exp(gL)$, where P_{in} is the input photon density and L is the SOA length. If there is a small change in the carrier density that induces a change of the material gain by Δg, the corresponding change in the photon density is $\Delta P = P \cdot \Delta gL$. At the same time, the optical phase will be changed by $\Delta\phi = \omega\Delta nL/c$, where Δn is the change of the refractive index and ω is the signal optical frequency. According to the definition of linewidth enhancement factor α_{lw} in Equation 1.1.48, the gain change and the phase change are related by:

$$\alpha_{lw} = 2P\frac{\Delta\phi}{\Delta P} = \frac{2\Delta\phi}{\Delta gL} \tag{1.4.43}$$

If the phase is changed from ϕ_1 to ϕ_2, the corresponding material gain is changed from g_1 to g_2, and the corresponding optical gain is changed from $G_1 = \exp(g_1L)$ to $G_2 = \exp(g_2L)$, Equation 1.4.43 can be used to find their relationship as

$$\phi_2 - \phi_1 = -\frac{\alpha_{lw}}{2}\ln\left(\frac{G_2}{G_1}\right) \tag{1.4.44}$$

Optical phase modulation has been used to achieve an all-optical switch. A typical system configuration is shown in Figure 1.4.8, where two identical SOAs are used in a Mach-zehnder interferometer setting [20]. A control optical beam at wavelength λ_2 is injected into one of the two SOAs to create an imbalance of the interferometer through optical power-induced phase change. A $\pi/2$ phase change in an interferometer arm will change the interferometer operation from constructive to destructive interference, thus switching off the optical signal at wavelength λ_1.

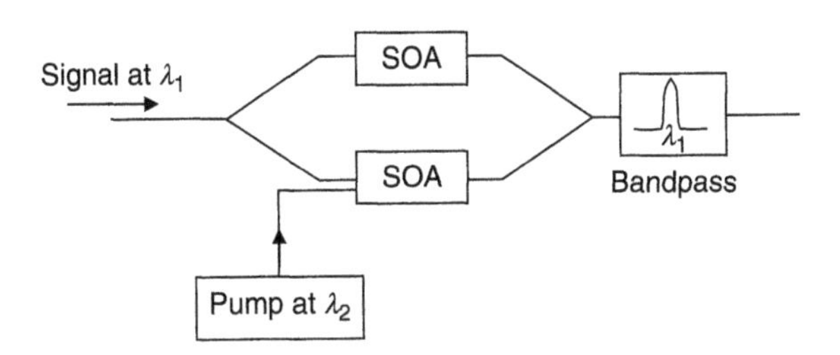

Figure 1.4.8 All-optical switch using SOAs in a Mach-zehnder configuration.

The major advantage of an all-optical switch using phase modulation in SOA is that the extinction ration can be made very high by carefully balancing the amplitude and the phase of the two interferometer arms. While in gain saturation-based optical switch, a very strong pump power is required to achieve a reasonable level of gain saturation, as illustrated in Figure 1.4.5(b). In addition, phase modulation is usually less wavelength-sensitive compared to gain saturation, which allows a wider wavelength separation between the signal and the pump.

To summarize, a semiconductor optical amplifier has a structure similar to that of a semiconductor laser except it lacks optical feedback. The device is small and can be integrated with other optical devices such as laser diodes and detectors in a common platform. Because of its fast gain dynamics, an SOA can often be used for all-optical switches and other optical signal processing purposes. However, for the same reason, an SOA is usually not suitable to be used as a line amplifier in a multiwavelength optical system. Cross-gain modulation in the SOA will create significant crosstalk between wavelength channels. For that purpose, the optical fiber amplifier is a much better choice.

1.4.3 Erbium-Doped Fiber Amplifiers (EDFAs)

An erbium-doped fiber amplifier is one of the most popular optical devices in modern optical communication systems as well as in fiber-optic instrumentation [21]. EDFAs provide many advantages over SOAs in terms of high gain, high optical power, low crosstalk between wavelength channels, and easy optical coupling from and to optical fibers. The basic structure of an EDFA shown in Figure 1.4.9, is composed of an erbium-doped fiber, a pump laser, an optical isolator, and a wavelength-division multiplexer.

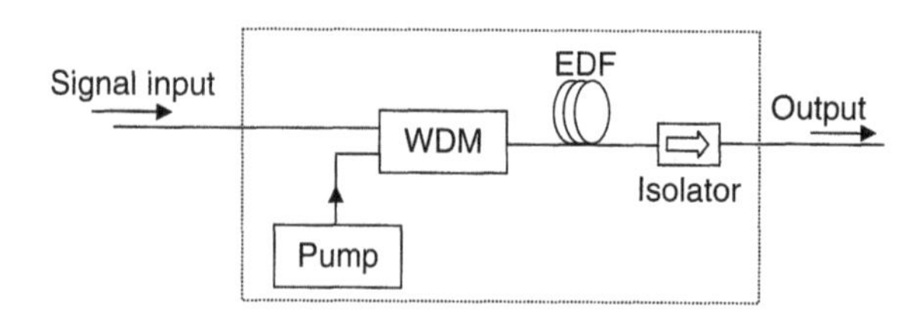

Figure 1.4.9 Configuration of an EDFA. *EDF:* erbium-doped fiber, *WDM:* wavelength-division multiplexer.

In contrast to SOAs, an EDFA is optically pumped and therefore it requires a pump source, which is usually a high-power semiconductor laser. The wavelength-division multiplexer (WDM) is used to combine the short wavelength pump with the longer wavelength signal. The reason to use a WDM combiner instead of a simple optical coupler is to avoid the combination loss. The optical isolator is used to minimize the impact of optical reflections from interfaces of optical components. Since an EDFA may provide a significant amount of optical gain, even a small amount of optical reflection may be able to cause oscillation, and therefore degrade EDFA performance. Some high-gain EDFAs may also employ another optical isolator at the input side.

The optical pumping process in an erbium-doped fiber is usually described by a simplified three-level energy system as illustrated in Figure 1.4.10(a). The bandgap between the ground state and the excited state is approximately 1.268eV; therefore pump photons at 980 nm wavelength are able to excite ground state carriers to the excited state and create population inversion. The carriers stay in the excited state for only about 1μs, and after that they decay into a metastable state through a nonradiative transition. In this process, the energy loss is turned into mechanical vibrations in the fiber. The energy band of the metastable state extends roughly from 0.8eV to 0.84eV, which correspond to a wavelength band from 1480 nm to 1550 nm. Within the metastable energy band, carriers tend to move down from the top of the band to somewhere near the bottom through a process known as *intraband relaxation.*

Finally, radiative recombination happens when carriers step down from the bottom of the metastable state to the ground state and emit photons in the 1550 nm wavelength region. The carrier lifetime in the metastable state is on

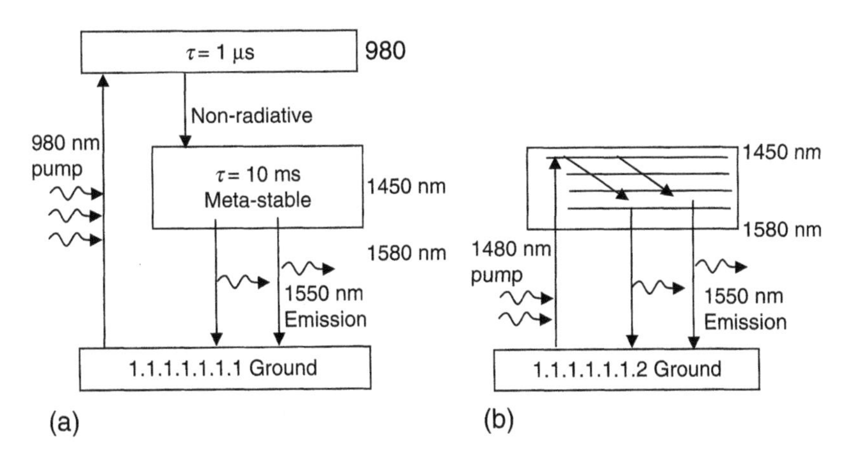

Figure 1.4.10 Simplified energy band diagrams of erbium ions in silica: (a) three-level system and (b) two-level system.

the order of 10 ms, which is four orders of magnitude longer than the carrier lifetime in the excited state. Therefore, with constant optical pumping at 980 nm, almost all the carriers will be accumulated in the metastable state. Thus, the three-level system can be simplified into two levels for most of the practical applications.

An erbium-doped fiber can also be pumped at 1480 nm wavelength, which corresponds to the bandgap between the top of the metastable state and the ground state. In this case, 1480 nm pump photons excite carriers from ground state to the top of metastable state directly. Then these carriers relax down to the bottom part of the metastable band. Typically, 1480 nm pumping is more efficient than 980 nm pumping because it does not involve the nonradiative transition from the 980 nm to the 1480 nm band. Therefore, 1480 nm pumping is often used for high-power optical amplifiers. However, amplifiers with 1480 nm pumps usually have higher noise figures than 980 nm pumps, which will be discussed later.

1.4.3.1 Absorption and Emission Cross-Sections

Absorption and emission cross-sections are two very important properties of erbium-doped fibers. Although the name *cross-section* may seem to represent a geometric size of the fiber, it does not. The physical meanings of absorption and emission cross-sections in an erbium-doped fiber are absorption and emission efficiencies as the functions of the wavelength.

Now let's start with the pump absorption efficiency W_p, which is defined as the probability of each ground state carrier that is pumped to the metastable state within each second. If the pump optical power is P_p, within each second the number of incoming pump photons is P_p/hf_p, where f_p is the optical frequency of the pump. Then, pump absorption efficiency is defined as

$$W_p = \frac{\sigma_a P_p}{hf_p A} \tag{1.4.45}$$

where A is the effective fiber core cross-section area and σ_a is the absorption cross-section. In another words, the absorption cross-section is defined as the ratio of pump absorption efficiency and the density of pump photon flow rate:

$$\sigma_a = W_p / \left(\frac{P_p}{hf_p A} \right) \tag{1.4.46}$$

Since the unit of W_p is $[s^{-1}]$ and the unit of the density of the pump photon flow rate $P_p/(hf_p A)$ is $[s^{-1}\, m^2]$, the unit of absorption cross-section is $[m^2]$. This is probably where the term absorption *cross-section* comes from. As we mentioned, it has nothing to do with the geometric cross-section of the fiber. We also need to note that the *absorption* cross-section does not mean attenuation; it indicates the efficiency of energy conversion from photons to the excited carriers.

Similarly, we can define an emission cross-section as

$$\sigma_e = W_s / \left(\frac{P_s}{hf_s A} \right) \quad (1.4.47)$$

where P_s and f_s are the emitted signal optical power and frequency, respectively. W_s is the stimulated emission efficiency, which is defined as the probability of each carrier in the metastable state that recombines to produce a signal photon within each second. It has to be emphasized that both absorption and emission cross-sections are properties of the erbium-doped fiber itself. They are independent of operation conditions of the fiber, such as pump and signal optical power levels.

At each wavelength both absorption and emission exist because photons can be absorbed to generate carriers through a stimulated absorption process; at the same time new photons can be generated through the stimulated emission process. Figure 1.4.11 shows an example of absorption and emission cross-sections; obviously both of them are functions of wavelength. If the carrier densities in the ground state and the metastable state are N_1 and N_2, respectively, the net stimulated emission rate per cubic meter is $R_e = W_s(\lambda_s)N_2 - W_p(\lambda_s)N_1$, where $W_s(\lambda_s)$ and $W_p(\lambda_s)$ are the emission and absorption efficiencies of the erbium-doped fiber at the signal wavelength λ_s. Considering the definitions of absorption and emission cross-sections, this net emission rate can be expressed as

$$R_e = \frac{\Gamma_s \sigma_e(\lambda_s) P_s}{hf_s A} \left(N_2 - \frac{\sigma_a(\lambda_s)}{\sigma_e(\lambda_s)} N_1 \right) \quad (1.4.48)$$

where a field confinement factor Γ_s is introduced to take into account the overlap fact between the signal optical field and the erbium-doped area in the fiber core. The physical meaning of this net emission rate is the number of photons that are generated per second per cubic meter.

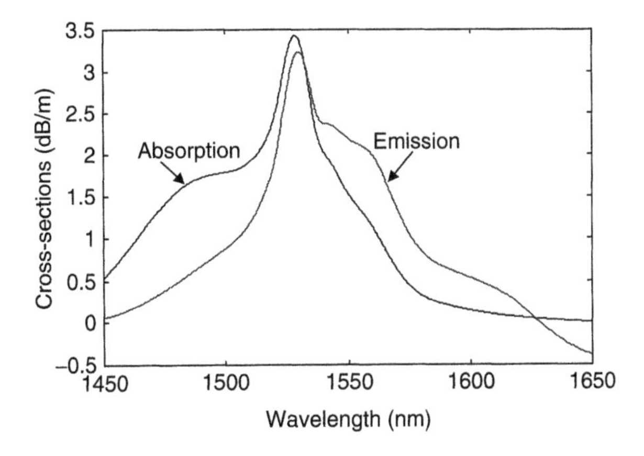

Figure 1.4.11 Absorption and emission cross-sections of HE980 erbium-doped fiber.

1.4.3.2 Rate Equations

The rate equation for the carrier density in the metastable state N_2 is

$$\frac{dN_2}{dt} = \frac{\Gamma_p \sigma_a(\lambda_p) P_p}{hf_p A} N_1 - \frac{\Gamma_s \sigma_e(\lambda_s) P_s}{hf_s A} \left(N_2 - \frac{\sigma_a(\lambda_s)}{\sigma_e(\lambda_s)} N_1 \right) - \frac{N_2}{\tau} \quad (1.4.49)$$

On the right-hand side of this equation, the first term is the contribution of carrier generation by stimulated absorption of the pump (Γ_p is the overlap factor at pump wavelength). The second term represents net stimulated recombination which consumes the upper-level carrier density. The last term is spontaneous recombination, in which upper-level carriers spontaneously recombine to generate spontaneous emission. τ is the spontaneous emission carrier lifetime. For simplicity we neglected the emission term at pump wavelength because usually $\sigma_e(\lambda_p) \ll \sigma_a(\lambda_s)$.

In Equation 1.4.49, both lower (N_1) and upper (N_2) state population densities are involved. However, these two carrier densities are not independent and they are related by

$$N_1 + N_2 = N_T \quad (1.4.50)$$

where N_T is the total erbium ion density that is doped into the fiber. The energy state of each erbium ion stays either in the ground level or in the metastable level.

In the steady state ($d/dt = 0$), if only one signal wavelength and one pump wavelength are involved, Equation 1.4.49 can be easily solved with the help of Equation 1.4.50:

$$N_2 = \frac{\Gamma_p \sigma_a(\lambda_p) f_s P_p + \Gamma_s \sigma_a(\lambda_s) f_p P_s}{\Gamma_p \sigma_a(\lambda_p) f_s P_p + \Gamma_s f_p P_s \left(\sigma_e(\lambda_s) + \sigma_a(\lambda_s) \right) + hf_s f_p A/\tau} N_T \quad (1.4.51)$$

Note that since the optical powers of both the pump and the signal change significantly along the fiber, the upper-level carrier density is also a function of the location parameter z along the fiber $N_2 = N_2(z)$. To find the z-dependent optical powers of the pump and the signal, propagation equations need to be established for them.

The propagation equations for the optical signal at wavelength λ_s and for the pump at wavelength λ_p are

$$\frac{dP_s(\lambda_s)}{dz} = g(z, \lambda_s) P_s(z) \quad (1.4.52)$$

$$\frac{dP_p(\lambda_p)}{dz} = \alpha(z, \lambda_p) P_p(z) \quad (1.4.53)$$

where

$$g(z, \lambda_s) = \Gamma_s [\sigma_e(\lambda_s) N_2(z) - \sigma_a(\lambda_s) N_1(z)] \quad (1.4.54)$$

is the effective gain coefficient at the signal wavelength and

$$\alpha(z, \lambda_p) = \Gamma_p[\sigma_e(\lambda_p)N_2(z) - \sigma_a(\lambda_p)N_1(z)] \tag{1.4.55}$$

is the effective absorption coefficient at the pump wavelength.

These propagation equations can be easily understood by looking at, for example, the term $\sigma_e(\lambda_s)N_2(z)P_s(z) = W_s hf_s AN_2(z)$. It is the optical power generated per unit length at the position z. Similarly, $\sigma_a(\lambda_s)N_1(z)P_s(z) = W_a hf_s AN_1(z)$ is the optical power absorbed per unit length at position z. Note that the unit of both g and α is in $[m^{-1}]$.

The two propagation Equations 1.4.52 and 1.4.53 are mutually coupled through carrier density $N_2(z)$ and $N_1(z) = N_T - N_2(z)$. Once the position-dependent carrier density $N_2(z)$ is found, the performance of the EDFA will be known and the overall signal optical gain of the amplifier can be found as

$$G(\lambda_s) = \exp\left\{\Gamma\left[(\sigma_e(\lambda_s) + \sigma_a(\lambda_s))\int_0^L N_2(z)dz - \sigma_a(\lambda_s)N_T L\right]\right\} \tag{1.4.56}$$

It depends on the accumulated carrier density along the fiber length.

In practice, since both the pump and the signal are externally injected into the EDF and they differ only by the wavelength. There could be multiple pump channels and multiple signal channels. Therefore, Equations 1.4.52 and 1.4.53 can be generalized as

$$\frac{dP_k(z)}{dz} = g_k(z, \lambda_k)P_k(z) \tag{1.4.57}$$

where the subscript k indicates the kth channel with the wavelength λ_k, and the gain coefficient for the kth channel is

$$g_k(z) = \Gamma_k[\sigma_e(\lambda_k)N_2(z) - \sigma_a(\lambda_k)N_1(z)] \tag{1.4.58}$$

In terms of the impact on the carrier density, the only difference between the pump and the signal is that at signal wavelength $\sigma_e > \sigma_a$, whereas at the pump wavelength we usually have $\sigma_a \gg \sigma_e$. Strictly speaking, both signal and pump participate in the emission and the absorption processes. In the general approach, the upper-level carrier density depends on the optical power of all the wavelength channels, and the rate equation of N_2 is

$$\frac{dN_2}{dt} = -\sum_j \left\{\frac{P_j(z)}{Ahf_j}\Gamma_j\left[\sigma_e(\lambda_j)N_2 - \sigma_a(\lambda_j)N_1\right]\right\} - \frac{N_2}{\tau} \tag{1.4.59}$$

Considering expressions in Equations 1.4.57 and 1.4.58, Equation 1.4.59 can be simplified as

$$\frac{dN_2(t,z)}{dt} = -\frac{N_2}{\tau} - \sum_j \left\{ \frac{1}{Ahf_j} \frac{dP_j(z)}{dz} \right\} \tag{1.4.60}$$

In the steady state, $d/dt = 0$, Equation 1.4.60 can be written as

$$N_2(z) = -\tau \sum_j \left\{ \frac{1}{Ahf_j} \frac{dP_j(z)}{dz} \right\} \tag{1.4.61}$$

In addition, Equation 1.4.57 can be integrated on both sides:

$$\int_{P_{k,in}}^{P_{k,out}} \frac{1}{P_k(z)} dP_k(z) = \int_0^L g_k(z)dz = \ln(P_{k,out}) - \ln(P_{k,in}) = \ln\left(\frac{P_{k,out}}{P_{k,in}}\right) \tag{1.4.62}$$

Since

$$\begin{aligned} g_k(z) &= \Gamma_k\{\sigma_e(\lambda_k)N_2(z) - \sigma_a(\lambda_k)[N_T - N_2(z)]\} \\ &= \Gamma_k\{[\sigma_e(\lambda_k) + \sigma_a(\lambda_k)]N_2(z) - \sigma_a(\lambda_k)N_T\} \end{aligned}$$

Using Equation 1.4.61, we have

$$g_k(z) = -\Gamma_k \left\{ [\sigma_e(\lambda_k) + \sigma_a(\lambda_k)]\tau \left[\sum_j \frac{1}{Ahf_j} \frac{dP_j(z)}{dz} \right] + \sigma_a(\lambda_k)N_T \right\} \tag{1.4.63}$$

$g_k(z)$ can be integrated over the erbium-doped fiber length L:

$$\begin{aligned} \int_0^L g_k(z)dz &= -\Gamma_k \left\{ [\sigma_e(\lambda_k) + \sigma_a(\lambda_k)]\tau \left[\sum_j \frac{1}{Ahf_j} \int_0^L \frac{dP_j(z)}{dz} dz \right] + \sigma_a(\lambda_k)N_T L \right\} \\ &= -\Gamma_k \left\{ [\sigma_e(\lambda_k) + \sigma_a(\lambda_k)]\tau \left[\sum_j \frac{(P_{j,out} - P_{j,in})}{Ahf_j} \right] + \sigma_a(\lambda_k)N_T L \right\} \end{aligned}$$

Define

$$a_k = \Gamma_k \sigma_a(\lambda_k) N_T \tag{1.4.64}$$

$$S_k = \frac{\Gamma_k \tau[\sigma_e(\lambda_k) + \sigma_a(\lambda_k)]}{A} \tag{1.4.65}$$

Then we have

$$\int_0^L g_k(z)dz = -\left\{ S_k \left[\sum_j \frac{(P_{j,out} - P_{j,in})}{hf_j} \right] + a_k L \right\} \tag{1.4.66}$$

The right side of Equation 1.4.66 can be rewritten as

$$S_k\left[\sum_j \frac{(P_{j,out} - P_{j,in})}{hf_j}\right] + a_k L = S_k\left[\sum_j \frac{P_{j,out}}{hf_j} - \sum_j \frac{P_{j,in}}{hf_j}\right] + a_k L$$

$$= S_k[Q_{OUT} - Q_{IN}] + a_k L$$

where

$$Q_{OUT} = \sum_j \frac{P_{j,out}}{hf_j}$$

and

$$Q_{IN} = \sum_j \frac{P_{j,in}}{hf_j}$$

are total photon flow rate at the output and the input, respectively (including both signals and pump).

Combining Equations 1.4.62 and 1.4.66, we have [22]

$$P_{k,out} = P_{k,in} e^{-a_k L} \cdot e^{-S_k(Q_{OUT} - Q_{IN})} \tag{1.4.67}$$

In this equation, both a_k and S_k are properties of the EDF. Q_{IN} is the total photon flow rate at the input, which is known. However, Q_{OUT} is the total photon flow rate at the output, which is *not* known. To find Q_{OUT}, one can solve the following equation iteratively:

$$Q_{OUT} = \sum_{k=1}^{N} \left\{ \frac{1}{hf_k} P_{k,in} \exp(S_K Q_{IN} - a_k L) \exp(-S_k Q_{OUT}) \right\} \tag{1.4.68}$$

In this equation, everything is known except Q_{OUT}. A simple numerical algorithm should be enough to solve this iterative equation and find Q_{OUT}. Once Q_{OUT} is obtained, optical gain for each channel can be found.

Although Equation 1.4.68 provides a semi-analytical formulation that can be used to investigate the performance of an EDFA, it has several limitations: (1) In this calculation, we have neglected the carrier saturation effect caused by amplified spontaneous emission (ASE) noise. In fact, when the optical gain is high enough, ASE noise may be significant and it contributes to saturate the EDFA gain. Therefore, the quasi analytical formulation is only accurate when the EDFA has relatively low gain. (2) It only predicts the relationship between the input and the output optical power levels; power evolution P(z) along the EDF is not given. (3) It only calculates the accumulated carrier population $\int_0^L N_2(z)dz$ in the EDF; the actual carrier density distribution is not provided. Since ASE noise is generated and amplified in a distributive manor along the EDF, we are not able to accurately calculate the ASE noise at the EDFA output using this semi-analytical method. Therefore, although this semi-analytical method serves as a quick evaluation tool, it does not give accurate enough results when the optical gain is high. Precise modeling of EDFA performance has to involve numerical simulations. The model has to consider optical pumps and optical signals of various wavelengths. Because the spontaneous emission noise generated along the erbium fiber propagates both in the forward direction and the backward direction, they have to be considered as well.

Figure 1.4.12 illustrates the evolution of optical signals and ASE noises. $S^f_{in}(\lambda)$ is the combination of the input optical signal and the input pump. After propagating through the fiber, the optical signal is amplified and the optical pump is depleted. $S^f_{out}(\lambda)$ is the combination of output optical signal and the remnant pump. Because of carrier inversion, spontaneous emission is generated along the fiber, and it is amplified in both directions.

$SP^f(\lambda)$ is the ASE in the forward direction, which is zero at the fiber input and becomes nonzero at the fiber output. Similarly, $SP^b(\lambda)$ is the ASE in the backward direction, which is zero at the fiber output and nonzero at the fiber input. In the steady state, the propagation equations for the optical signal (including the pump), the forward ASE and the backward ASE are, respectively,

$$\frac{dP^f_s(\lambda, z)}{dz} = P^f_s(\lambda, z)g(\lambda, z) \tag{1.4.69}$$

$$\frac{dP^f_{SP}(\lambda, z)}{dz} = P^f_{SP}(\lambda, z)g(\lambda, z) + 2n_{sp}(z)g(\lambda, z) \tag{1.4.70}$$

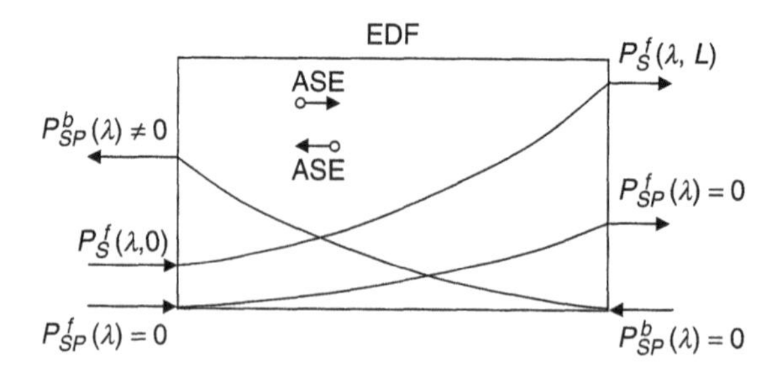

Figure 1.4.12 Illustration of signal and ASE noise in an EDFA: $P^f_s(\lambda, 0)$ and $P^f_s(\lambda, L)$ represent input and output optical signal (include pump) in the forward direction, $P^f_{SP}(\lambda)$ and $P^b_{SP}(\lambda)$ represent forward and backward propagated ASE.

$$\frac{dP_{SP}^{b}(\lambda, z)}{dz} = -P_{SP}^{b}(\lambda, z)g(\lambda, z) - 2n_{sp}(z)g(\lambda, z) \quad (1.4.71)$$

where n_{sp} is the spontaneous emission factor, which depends on the carrier inversion level N_2 as

$$n_{sp}(z) = \frac{N_2(z)}{N_2(z) - N_1(z)} = \frac{N_2(z)}{2N_2(z) - N_T} \quad (1.4.72)$$

The carrier density N_2 is described by a steady-state carrier density rate equation:

$$\frac{N_2}{\tau} + \int_{\lambda} \left[\frac{P_s^f(\lambda, z)}{Ahc/\lambda} + \frac{P_{SP}^f(\lambda, z)}{Ahc/\lambda} + \frac{P_{SP}^b(\lambda, z)}{Ahc/\lambda} \right] g(\lambda, z)d\lambda = 0 \quad (1.4.73)$$

where

$$g(\lambda, z) = \Gamma(\lambda)\{[\sigma_e(\lambda) + \sigma_a(\lambda)]N_2(z) - \sigma_a(\lambda)N_T\} \quad (1.4.74)$$

is the gain coefficient and $\Gamma(\lambda)$ is the wavelength-dependent confinement factor.

Equations 1.4.69 through 1.4.73 are coupled through $N_2(z)$ and can be solved numerically. A complication in this numerical analysis is that the ASE noise has a continuous spectrum that varies over the wavelength. In the numerical analysis, the ASE optical spectrum has to be divided into narrow slices, and within each wavelength slice, the noise power can be assumed to be in a single wavelength. Therefore, if the ASE spectrum is divided into m wavelength slices, Equations 1.4.70 and 1.4.71 each must be split into m equations. Since the carrier density $N_2(z)$ is a function of z, the calculation usually has to divide the erbium-doped fiber into many short sections, and within each short section N_2 can be assumed to be a constant.

As shown in Figure 1.4.12, if the calculation starts from the input side of the fiber, we know that the forward ASE noise at the fiber input ($z = 0$) is zero, and we also know the power levels of the input optical signal and the pump. However, we do not know the level of the backward ASE noise at $z = 0$. To solve the rate equations, we have to assume a value of $P_{SP}^{b}(\lambda)$ at $z = 0$. Then rate equations can be solved section by section along the fiber. At the end of the fiber, an important boundary condition has to be met—that is, the backward ASE noise has to be zero ($P_{SP}^{b}(\lambda) = 0$). If this condition is not satisfied, we have to perform the calculation again using a modified initial guess of $P_{SP}^{b}(\lambda)$ at z = 0. This process usually has to be repeated several times until the backward ASE noise at the fiber output is lower than a certain level (say, –40 dBm); then the results are accurate.

Figure 1.4.13 shows an example of EDFA optical gain versus wavelength at different input optical signal power levels. This example was obtained using

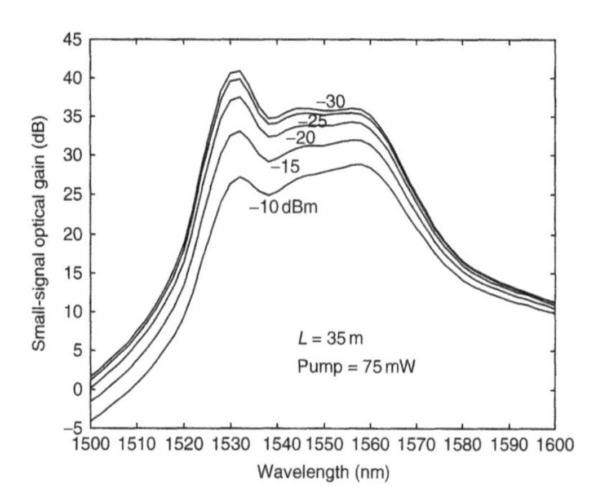

Figure 1.4.13 Example of EDFA optical gain versus wavelength at different input optical signal power levels.

a 35 m Lucent HE20P EDF and a 75 mW forward pump power at 980 nm wavelength. First of all, the gain spectrum of an EDFA is not uniform across the bandwidth and its shape cannot be represented by a simple mathematic formula. There is typically a narrow gain peak around 1530 nm and a relatively flat region between 1540 nm and 1560 nm. The gain and spectral shape of an EDFA depend on the emission and absorption cross-sections of the EDF, the pump power level, and the signal power. Figure 1.4.13 shows that with the input signal optical power increased from –30 dBm to –10 dBm, the optical gain at 1550 nm decreases from 36 dB to 27 dB, which is caused by gain saturation. In addition, the spectral shape of the optical gain also changes with the signal optical power. With the increase of signal optical power, the gain tends to tilt toward the longer wavelength side and the short wavelength peak around 1530 nm is suppressed. This phenomenon is commonly referred to as *dynamic gain tilt*.

1.4.3.3 EDFA Design Considerations

1.4.3.3.1 Forward Pumping and Backward Pumping

EDFA performance, to a large extent, depends on the pumping configuration. In general, an EDFA can have either a forward pump or a backward pump; some EDFAs have both forward and backward pumps. As illustrated in Figure 1.4.14, *forward pumping* refers to pumps propagating in the same direction as the signals, whereas in backward pumping, the pump travels in the opposite direction of the optical signal.

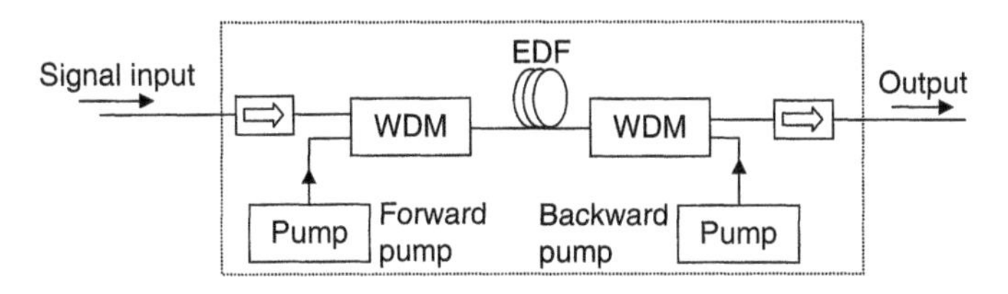

Figure 1.4.14 Configuration of an EDFA with both a forward pump and a backward pump.

An EDFA with a forward pump alone usually has relatively low signal output optical power, whereas at the same time the forward ASE noise level is also relatively low. However, the backward ASE noise level at the input side of the erbium-doped fiber may become very high. The reason is that although the pump power can be strong at the fiber input, its level is significantly reduced at the fiber output because the pump energy is converted to amplify the signal along the fiber. Because of the weak pump at the EDF output, the carrier inversion level is low as well; therefore the output optical signal level is limited. On the other hand, since the pump power and thus the carrier inversion level is very strong near the fiber input side, the backward propagated spontaneous emission noise meets high amplification in that region; therefore the backward ASE noise power level is strong at the fiber input side.

With backward pumping, the EDFA can provide a higher signal optical power at the output, but the forward ASE noise level can also be high. The backward ASE noise level in a backward pumping configuration is relatively low at the input side of the erbium-doped fiber. In this case, the pump power is strong at the fiber output side and its level is significantly reduced when it reaches the input side. Because of the strong pump power at the EDF output and the high carrier inversion, the output optical signal power can be much higher than the forward pumping configuration. Figure 1.4.15 illustrates the power levels of pump, signal, forward ASE, and backward ASE along the erbium-doped fiber for forward-pumping and backward-pumping configurations.

1.4.3.3.2 EDFAs with AGC and APC

Automatic gain control (AGC) and automatic power control (APC) are important features in practical EDFAs that are used in optical communication systems and networks. Since the optical gain of an EDFA depends on the signal optical power, system performance will be affected by signal optical power fluctuation and add/drop of optical channels. Therefore, AGC and APC are usually used in in-line optical amplifiers to regulate the optical gain and the output signal optical power of an EDFA. Because both the optical gain and the output

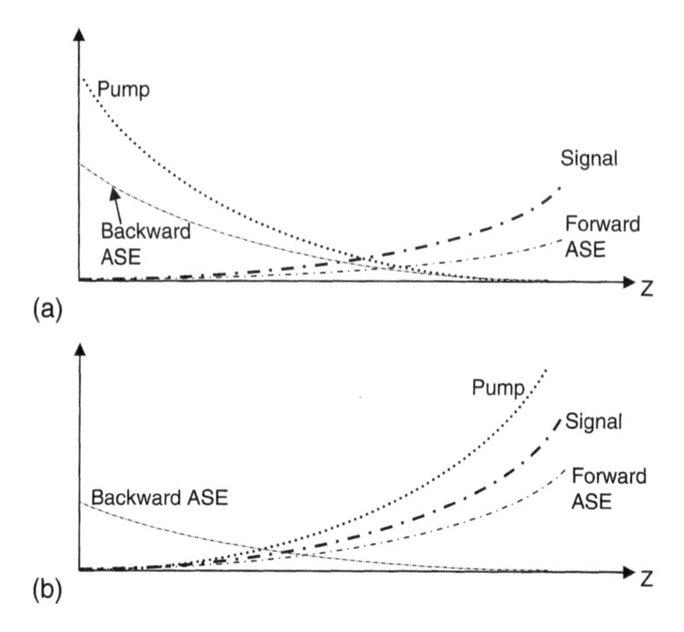

Figure 1.4.15 Illustration of power levels of pump, signal, forward ASE, and backward ASE along the erbium-doped fiber for forward pumping (a) and backward pumping (b) configurations.

signal optical power are dependent on the power of the pump, the automatic control can be accomplished by adjusting the pump power.

Figure 1.4.16 shows an EDFA design with AGC in which two photodetectors, PDI and PDO, are used to detect the input and the output signal optical powers. An electronic circuit compares these two power levels, calculates the optical gain of the EDFA, and generates an error signal to control the injection current of the pump laser if the EDFA gain is different from the target value. If the EDFA carries multiple WDM channels and we assume that the responsivity of each photodetector is wavelength independent within the EDFA bandwidth, this AGC configuration controls the overall gain of the total optical power such that

$$G = \frac{\sum_{j=1}^{N} P_{j,out}}{\sum_{j=1}^{N} P_{j,in}} = \text{constant} \tag{1.4.75}$$

where N is the number of wavelength channels. In this case, although the gain of the total optical power is fixed by the feedback control, the optical gain of each

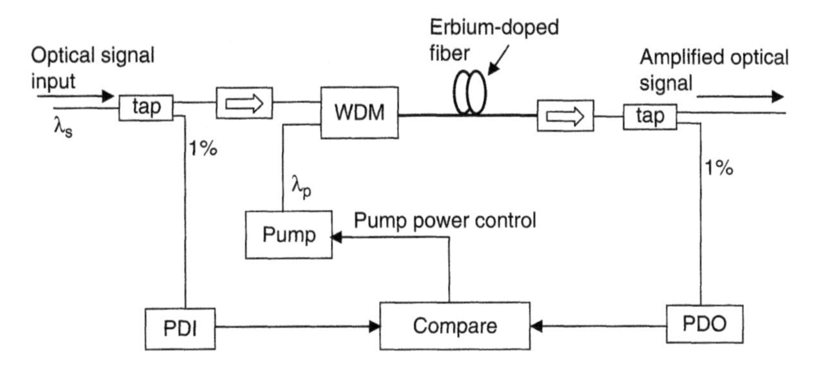

Figure 1.4.16 Configuration of an EDFA with both automatic gain control. *PDI:* input photodiode, *PDO:* output photodiode.

individual wavelength channel may vary, depending on the gain flatness over the EDFA bandwidth.

In long distance optical transmission, many in-line optical amplifiers may be concatenated along the system. If the optical gain of each EDFA is automatically controlled to a fixed level, any loss fluctuation along the fiber system will make the output optical power fluctuate and therefore the optical system may become unstable.

APC, on the other hand, regulates the total output optical power of an EDFA, as shown in Figure 1.4.17. In this configuration, only one photodetector is used to detect the total output signal optical power. The injection current of the pump laser is controlled to ensure that this measured power is equal to the desired level such that

$$\sum_{j=1}^{N} P_{j,out} = \text{constant} \tag{1.4.76}$$

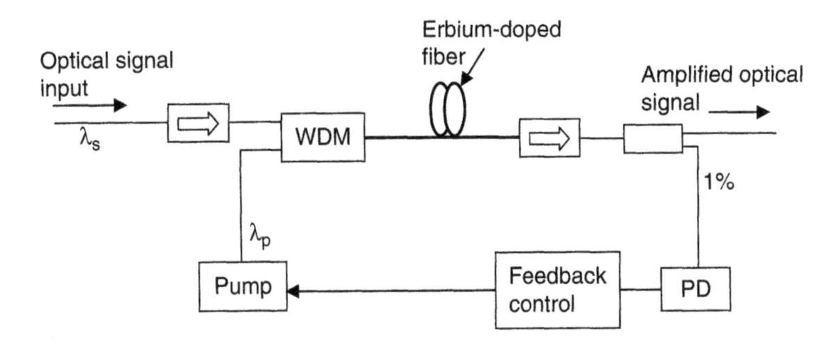

Figure 1.4.17 Configuration of an EDFA with automatic power control.

The advantage of APC is that it isolates optical power fluctuations along the system, because the variation of fiber loss in one span does not affect other spans. However, since the total power is regulated, channel add/drop in WDM systems will affect the optical power of each individual channel. Therefore, in advanced optical communication systems, EDFAs are controlled by intelligent systems taking into account the number of wavelength channels, optical signal-to-noise ratio, and other important system parameters.

1.4.3.4 EDFA Gain Flattening

As shown in Figure 1.4.13, the optical gain of an EDFA changes with the wavelength within the gain bandwidth. In a WDM optical system, the wavelength-dependent gain characteristic of EDFAs makes transmission performance different from channel to channel, and thus greatly complicates system design and provisioning. In addition, dynamic gain tilt in EDFA may also affect system design and transmission performance, considering signal power change and channel add/drop.

In high-end optical amplifiers, the gain variation versus wavelength can be compensated by passive optical filters, as shown in Figure 1.4.18(a). To flatten the gain spectrum, an optical filter with the transfer function complementing to the original EDFA gain spectrum is used with the EDFA so that the overall optical gain does not vary with the wavelength and this is illustrated in Figure 1.4.19. However, the passive gain flattening is effective only with one particular optical gain of the amplifier. Any change of EDFA operation condition and the signal optical power level would change the gain spectrum and therefore require a different filter.

To address the dynamic gain tilt problem, a spatial light modulator (SLM) is used as shown in Figure 1.4.18(b). In this configuration, the output of the

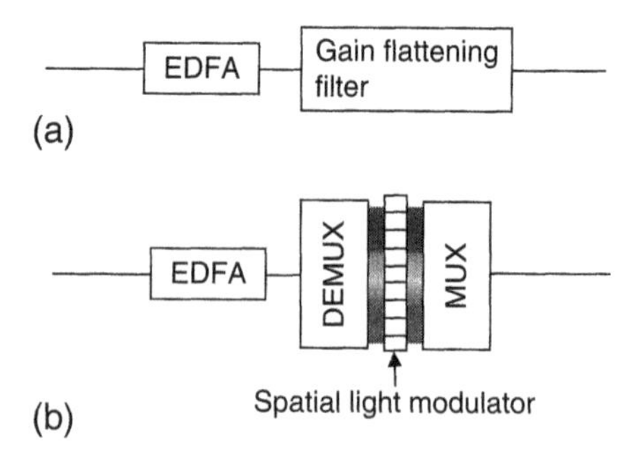

Figure 1.4.18 EDFA gain flattening using (a) a passive filter and (b) a spatial light modulator.

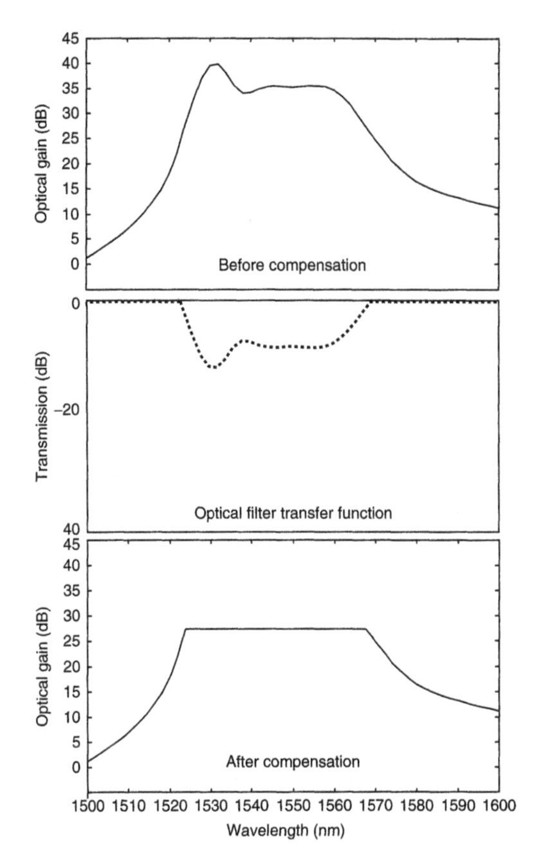

Figure 1.4.19 Illustration of EDFA gain flattening using an optical filter.

EDFA is demultiplexed, and different wavelength components are dispersed into the spatial domain. Then a multipixel, one-dimensional SLM is used to attenuate each wavelength component individually to equalize the optical gain across the EDFA bandwidth. With the rapid advance of liquid crystal technology, SLM is commercially available. Since each wavelength component is addressed independently and dynamically in this configuration, the effect of dynamic gain tilt can be reduced.

1.5 EXTERNAL ELECTRO-OPTIC MODULATOR

In an optical transmitter, encoding electrical signals into optical domains can be accomplished either by direct injection current modulation of laser diode or by electro-optic modulation using an external modulator. The speed of direct modulation is usually limited by carrier dynamics. In addition, the direct modulation-induced frequency chirp may significantly degrade the transmission performance of the optical signal in high-speed systems. On the other hand, external modulation can potentially achieve >100 GHz modulation bandwidth with well-controlled frequency chirp. Therefore external modulation is widely adopted in long distance and high-speed optical communication systems. From an application point of view, an electro-optic modulator is typically used after a laser source which provides a constant optical power. The external modulator simply manipulates the optical signal according to the applied voltage, thus converting the electrical signal into the optical domain. Various types of external modulators have been developed to perform intensity modulation as well as phase modulation.

1.5.1 Basic Operation Principle of Electro-Optic Modulators

In some crystal materials, the refractive indices are functions of the electrical field applied on them and the index change is linearly proportional to the applied field magnitude as $\delta n = \alpha_{EO} E$, where E is the electrical field and α_{EO} is the linear electro-optic coefficient. The most popular electro-optical material used so far for electro-optical modulator is $LiNbO_3$, which has a relatively high electro-optic coefficient. Electro-optic phase modulator is the simplest external modulator, which is made by a $LiNbO_3$ optical waveguide and a pair of electrodes, as shown in Figure 1.5.1. If the length of the electrode is L, the separation between the two electrodes is d, and the applied voltage is V, the optical phase change introduced by the linear electro-optic effect is

$$\phi(V) = \left(\frac{2\pi\alpha_{EO}L}{\lambda d}\right) V \tag{1.5.1}$$

The modulation efficiency, defined as $d\phi/dV$, is directly proportional to the length of the electrode L and inversely proportional to the electrodes separation d. Increasing the length and reduce the separation of the electrodes would certainly increase the modulation efficiency but will inevitably increase the parasitic

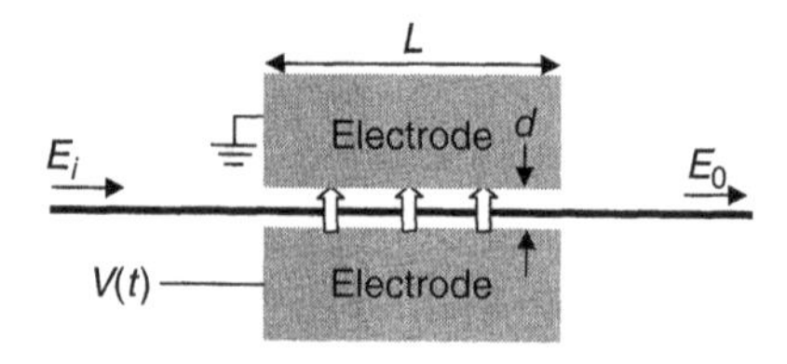

Figure 1.5.1 Electro-optic phase modulator.

capacitance and thus reduce the modulation speed. Another speed limitation is due to the device transit time $t_d = L/v_p$, where $v_p = c/n$ is the propagation speed in the waveguide of refractive index n. For a 10 mm-long $LiNbO_3$ waveguide with $n = 2.1$, the transit time is approximately 35 ps, which sets a speed limit to the modulation. In very high-speed electro-optic modulators, traveling-wave electrodes are usually used, where the RF modulating signal propagates in the same direction as the optical signal. This results in a phase matching between the RF and the optical signal and thus eliminates the transit time induced speed limitation.

Based on the same electro-optic effect, an optical intensity modulator can also be made. A very popular external intensity modulator is made by planar waveguide circuits designed in a Mach-zehnder interferometer (MZI) configuration as shown in Figure 1.5.2.

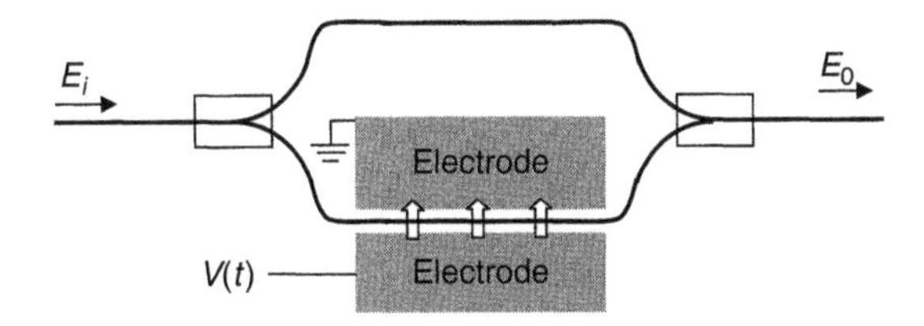

Figure 1.5.2 Electro-optic modulator based on MZI configuration.

In this configuration, the input optical signal is equally split into two interferometer arms and then recombined. Similar to the phase modulator, electrical field is applied across one of the two MZI arms to introduce an additional phase delay and thus control the differential phase between the two arms.

If the phase delays of the two MZI arms are ϕ_1 and ϕ_2, respectively, the combined output optical field is

$$E_0 = \frac{1}{2}(e^{j\phi_1} + e^{j\phi_2})E_i \tag{1.5.2}$$

where E_i is the complex field of the input optical signal. Then Equation 1.5.2 can be rewritten into

$$E_0 = \cos\left(\frac{\Delta\phi}{2}\right)e^{j\phi_c/2}E_i \tag{1.5.3}$$

where $\phi_c = \phi_1 + \phi_2$ is the average (common-mode) phase delay and $\Delta\phi = \phi_1 - \phi_2$ is the differential phase delay of the two arms.

The input-output power relationship of the modulator is then

$$P_0 = \cos^2\left(\frac{\Delta\phi}{2}\right)P_i \tag{1.5.4}$$

where $P_i = |E_i|^2$ and $P_0 = |E_0|^2$ are the input and the output powers, respectively. Obviously, in this intensity modulator transfer function, the differential phase delay between the two MZI arms plays a major role. Again, if we use L and d for the length and the separation of the electrodes, and α_{OE} for the linear electro-optic coefficient, the differential phase shift will be

$$\Delta\phi(V) = \phi_0 + \left(\frac{2\pi}{\lambda}\right)\frac{\alpha_{EO}V}{d}L \tag{1.5.5}$$

where ϕ_0 is the initial differential phase without the applied electrical signal. Its value may vary from device to device and may change with temperature mainly due to practical fabrication tolerance. In addition, if the driving electrical voltage has a DC bias and an AC signal, the DC bias can be used to control this initial phase ϕ_0.

A convenient parameter to specify the efficiency of an electro-optic intensity modulator is V_π, which is defined as the voltage required to change the optical power transfer function from the minimum to the maximum. From Equations 1.5.4 and 1.5.5, V_π can be found as

$$V_\pi = \frac{\lambda d}{2\alpha_{EO}L} \tag{1.5.6}$$

V_π is obviously a device parameter depending on the device structure as well as the material electro-optic coefficient. With the use of V_π, the power transfer function of the modulator can be simplified as

$$T(V) = \frac{P_0}{P_i} = \cos^2\left[\phi_0 + \frac{\pi V}{2V_\pi}\right] \tag{1.5.7}$$

Figure 1.5.3 illustrates the relationship between the input electrical voltage waveform and the corresponding output optical signal waveform. V_{b0} is the *DC* bias voltage, which determines the initial phase ϕ_0 in Equation 1.5.7. The DC bias is an important operational parameter because it determines the electrical-to-optical (E/O) conversion efficiency. If the input voltage signal is bipolar, the modulator is usually biased at the quadrature point, as shown in Figure 2.3.6. This corresponds to the initial phase $\phi_0 = \pm\pi/4$ depending on the selection of the positive or the negative slope of the transfer function. With this DC bias, the E/O transfer function of the modulator has the best linearity and allows the largest swing of the signal voltage, which is $\pm V_\pi/2$. In this case, the output optical power is

$$P_0(t) = \frac{P_i}{2}\left\{1 \pm \sin\left(\frac{\pi V(t)}{V_\pi}\right)\right\} \tag{1.5.8}$$

Although this transfer function is nonlinear, it is not a significant concern for binary digital modulation, where the electrical voltage switches between

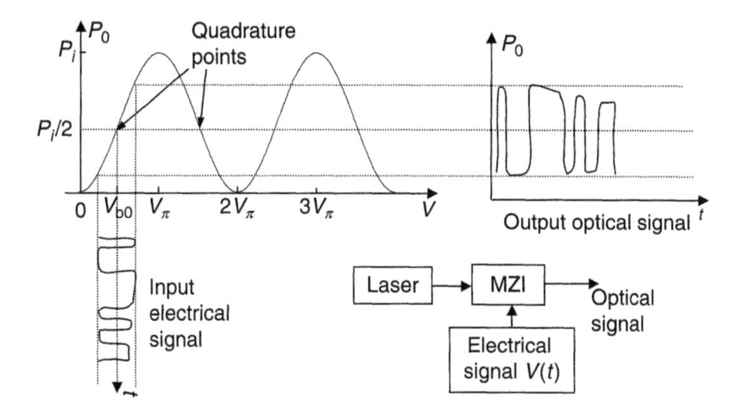

Figure 1.5.3 Electro-optic modulator transfer function and input (electrical)/output (optical) waveforms.

$-V_\pi/2$ and $+V_\pi/2$ and the output optical power switches between zero and P_i. However, for analog modulation, the nonlinear characteristic of the modulator transfer function may introduce signal waveform distortion. For example, if the modulating electrical signal is a sinusoid, $V(t) = V_m \cos(\Omega t)$, and the modulator is biased at the quadrature point with $\phi_0 = m\pi - \pi/4$, where m is an integer, the output optical power can be expanded in a Bessel series:

$$P_0(t) = \frac{P_i}{2} + P_i J_1(x)\cos(\Omega t) - P_i J_3(x)\cos(3\Omega t) + P_i J_5(x)\cos(5\Omega t) + \ldots. \tag{1.5.9}$$

where $J_n(x)$ is the n-th order Bessel function with $x = \pi V_m / V_\pi$. In this equation, the first term is the average output power, the second term is the result of linear modulation, and the third and the fourth terms are high-order harmonics caused by the nonlinear transfer function of the modulator. To minimize nonlinear distortion, the amplitude of the modulating voltage signal has to be very small, such that $V_m \ll V_\pi/2$, and therefore the high-order terms in the Bessel series can be neglected.

It is noticed that so far, we have only discussed the intensity transfer function of the external modulator as given by Equation 1.5.7, whereas the optical phase information has not been discussed. In fact, an external optical modulator may also have a modulating chirp similar to the direct modulation of semiconductor lasers but originated from a different mechanism and with a much smaller chirp parameter. To investigate the frequency chirp in an external modulator, the input/output optical field relation given by Equation 1.5.3 has to be used.

If the optical phase ϕ_2 is modulated by an external electrical signal from its static value ϕ_{20} to $\phi_{20} + \delta\phi(t)$, we can write $\phi_c = \phi_1 + \phi_{20} + \delta\phi(t)$ and $\Delta\phi = \phi_1 - \phi_{20} - \delta\phi(t)$. Equation 1.5.3 can be modified as

$$E_0 = \cos\left(\frac{\phi_0 - \delta\phi(t)}{2}\right) e^{j(\phi_{c0}+\delta\phi(t))/2} E_i \tag{1.5.10}$$

where $\phi_{c0} = \phi_1 + \phi_{20}$ and $\phi_0 = \phi_1 - \phi_{20}$ are the static values of the common mode and the differential phases. Obviously, the phase of the optical signal is modulated, which is represented by the factor $e^{j\delta\phi(t)/2}$.

As defined in Equation 1.1.48, the chirp parameter is the ratio between the phase modulation and the intensity modulation. It can be found from Equation 1.5.10 that

$$\frac{dP_0}{dt} = \frac{-P_i}{2} \sin(\phi_0 - \delta\phi(t)) \frac{d\delta\phi(t)}{dt} \tag{1.5.11}$$

Therefore the equivalent chirp parameter is

$$\alpha_{lw} = 2P_0 \left.\frac{d\delta\phi(t)/dt}{dP_0/dt}\right|_{\delta\phi(t)=0} = \frac{1+\cos(\phi_0)}{\sin(\phi_0)} \tag{1.5.12}$$

Not surprisingly, the chirp parameter is only the function of the DC bias. The reason is that although the phase modulation efficiency $d\delta\phi(t)/dt$ is independent of the bias, the efficiency of the normalized intensity modulation is a function of the bias. At $\phi_0 = (2m+1)\pi$, the chirp parameter $\alpha_{lw} = 0$ because the output optical power is zero, whereas at $\phi_0 = 2m\pi \pm \pi/2$ the chirp parameter $\alpha_{lw} = \infty$ because the intensity modulation efficiency is zero. It is also interesting to note that the sign of the chirp parameter can be positive or negative depending on the DC bias on the positive or negative slopes of the power transfer function. This adjustable chirp of external modulator in system designs, for example, may be used to compensate for the chromatic dispersion in the optical fiber.

However, for most of the applications, chirp is not usually desired; therefore external modulators with zero chirp have been developed. These Zero-chirp modulators can be built based on a balanced MZI configuration and antisymmetric driving of the two MZI arms, as shown in Figure 1.5.4. In this case, the two MZI arms have the same physical length, but the electrical fields are applied in the opposite directions across the two arms, creating an antisymmetric phase modulation.

Recall that in Equation 1.5.3, the common-mode phase delay, which determines the chirp, is $\phi_c = \phi_1 + \phi_2$. If both ϕ_1 and ϕ_2 are modulated by the same amount $\delta\phi$(t) but with opposite signs $\phi_1 = \phi_{10} + \delta\phi$ and $\phi_2 = \phi_{20} - \delta\phi$, then $\phi_c = \phi_{10} + \phi_{20}$ will not be time-dependent and therefore no optical phase modulation is introduced. For the differential phase delay, $\Delta\phi = \phi_{10} - \phi_{20} + 2\delta\phi(t)$, which doubles the intensity modulation efficiency.

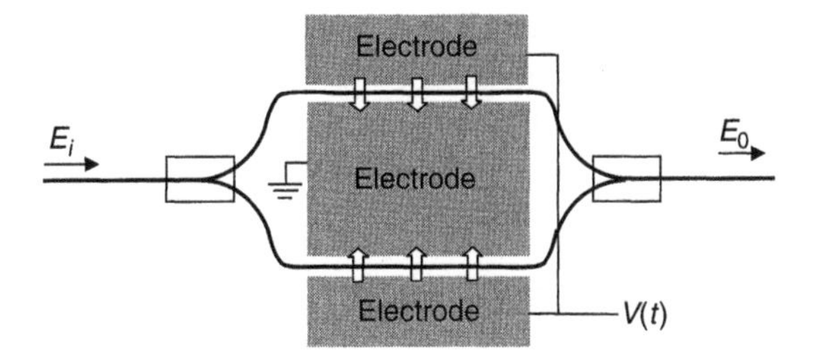

Figure 1.5.4 Electro-optic modulator based on an MZI configuration.

1.5.2 Frequency Doubling and Duo-Binary Modulation

Electro-optical modulators based on an MZI configuration have been used widely in high-speed digital optical communications due to their high modulation bandwidth and low frequency chirp. In addition, due to their unique bias-dependent modulation characteristics, electro-optic modulators can also be used to perform advanced electro-optic signal processing, such as frequency doubling and single-sideband modulation. These signal processing capabilities are useful not only for optical communications but also for optical measurements and instrumentation.

Frequency doubling is relatively easy to explain, as illustrated in Figure 1.5.5. In order for the output intensity-modulated optical signal to double the frequency compared to the driving RF signal, the modulator should be biased at

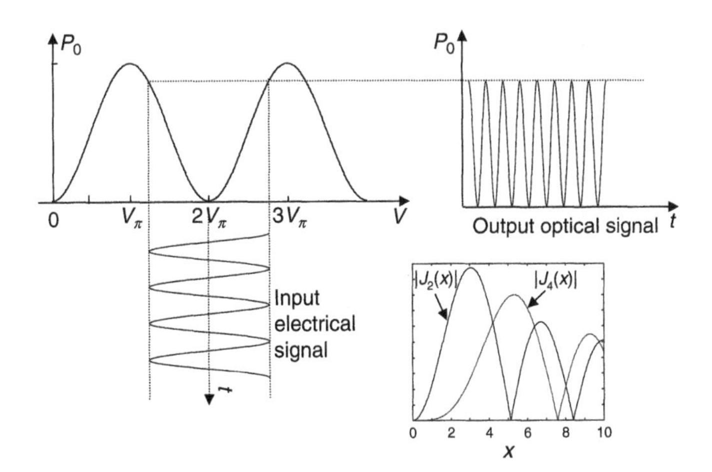

Figure 1.5.5 Electro-optic modulator transfer function and input (electrical)/output (optical) waveforms.

either the minimum or the maximum transmission point. Referring to the power transfer function shown in Equation 1.5.7, the bias should make the initial phase $\phi_0 = m\pi \pm \pi/2$. Therefore:

$$P_0 = P_i \frac{1 \pm \cos(\pi V(t)/V_\pi)}{2} \tag{1.5.13}$$

If the input RF signal is a sinusoid, $V(t) = V_m \cos(\Omega t)$, then the output is

$$P_0 = \frac{P_i}{2}\left[1 \pm \cos\left(\frac{\pi V_m}{V_\pi}\cos(\Omega t)\right)\right] \tag{1.5.14}$$

This can be expanded into a Bessel series as:

$$P_0 = \frac{P_i}{2}\left[1 \pm J_0\left(\frac{\pi V_m}{V_\pi}\right) \mp 2J_2\left(\frac{\pi V_m}{V_\pi}\right)\cos(2\Omega t) \pm 2J_4\left(\frac{\pi V_m}{V_\pi}\right)\cos(4\Omega t) \mp \ldots\right] \tag{1.5.15}$$

In this optical output, the fundamental frequency component is 2Ω, which doubles the input RF frequency. The absolute values of $J_2(x)$ and $J_4(x)$ shown in the inset of Figure 1.5.5 for convenience indicate the relative amplitudes of the second- and fourth-order harmonics. When the modulating amplitude is small enough, $\pi V_m / V_\pi \ll 1$, Bessel terms higher than the second order can be neglected, resulting in only a DC and a frequency-doubled component. This modulation technique can also be used to generate quadruple frequency by increasing the amplitude of the modulating RF signal such that $\pi V_m / V_\pi \approx 5$. But in practice, a large-amplitude RF signal at high frequency is usually difficult to generate.

It is important to notice the difference between the power transfer function and the optical field transfer function of an external modulator, as shown in Figure 1.5.6. Because the power transfer function is equal to the square of the field transfer function, the periodicity is doubled in this squaring operation. As a consequence, if the modulator is biased at the minimum power transmission point, the field transfer function has the best linearity. Although the optical intensity waveform is frequency-doubled compared to the driving RF signal, the waveform of the output optical field is quasi-linearly related to the input RF waveform. In this case, in the signal optical spectrum, the separation between each modulation sideband and the optical carrier is equal to the RF modulating frequency Ω rather than 2Ω. However, if this optical signal is detected by a photodiode, only optical intensity is received while the phase information is removed. Therefore, the received signal RF spectrum in the electrical domain will have a discrete frequency at 2Ω, which is twice the RF modulating frequency.

This unique property of MZI-based electro-optic modulator has been used to create duo-binary optical modulation format in high-speed optical transmission

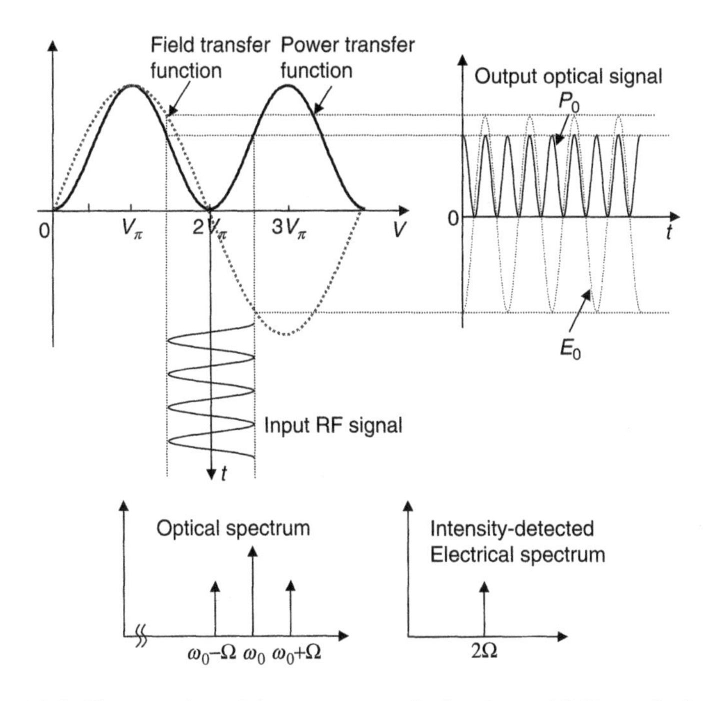

Figure 1.5.6 Electro-optic modulator power transfer function and field transfer function.

systems, which requires only half the optical bandwidth compared with direct amplitude modulation at the same data rate [Penninckx 1997].

1.5.3 Optical Single-Side Modulation

In general, under a sinusoid modulation, an electro-optic modulation generates two sidebands—one on each side of the optical carrier, which is usually referred to as *double-sideband modulation.* Since these two sidebands carry redundant information, removing one of them will not affect signal transmission. Single sideband optical signals occupy narrower spectral bandwidth and thus result in better bandwidth efficiency in multiwavelength WDM systems. Also, optical signals with narrower spectral bandwidth suffer less from chromatic dispersion. These are the major reasons that optical single sideband (OSSB) modulation is attractive [24, 25].

An easy way to generate OSSB is to use a notch optical filter, which directly removes one of the two optical sidebands. But this technique requires stringent wavelength synchronization between the notch filter and the transmitter. OSSB can also be generated using an electro-optic modulator with a balanced MZI structure and two electrodes, as shown in Figure 1.5.7.

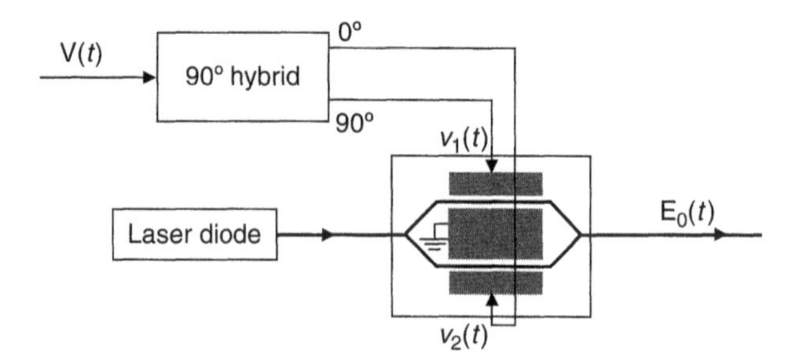

Figure 1.5.7 Optical single-sideband modulation using a dual-electrode MZI electro-optic modulator.

If we assume that the frequency of the optical carrier is ω_0 and the RF modulating signal is sinusoid at frequency Ω, the real function representation of the output optical field is

$$E_0(t) = \frac{E_i}{2}[\cos(\omega_0 t + \phi_1(t)) + \cos(\omega_0 t + \phi_2(t))] \tag{1.5.16}$$

where $\phi_1(t)$ and $\phi_2(t)$ are the phase delays of the two MZI arms. Now we suppose that these two phase delays can be modulated independently by two RF signals, $v_1(t)$ and $v_2(t)$. These two RF signals have the same amplitude and the same frequency, but there is a relative RF phase difference θ between them: $v_1(t) = V_b + V_m \cos \Omega t$ and $v_2(t) = V_m(\cos \Omega t + \theta)$, where V_b is a DC bias and V_m is the amplitude. Therefore the two phase terms are

$$\phi_1(t) = \phi_{01} + \beta \cos \Omega t \text{ and } \phi_2(t) = \beta \cos(\Omega t + \theta)$$

where $\beta = \pi V_m/(2V_\pi)$ is the normalized phase modulation efficiency. Then the output optical field can be expressed as

$$E_0(t) = \frac{E_i}{2}\left\{\cos[\omega_0 t + \phi_{01} + \beta \cos \Omega t] + \cos[\omega_0 t + \beta \cos(\Omega t + \theta)]\right\} \tag{1.5.17}$$

If the modulator is biased at the quadrature point (for example, $\phi_{01} = 3\pi/2$), Equation 1.5.17 becomes

$$E_0(t) = \frac{E_i}{2}\left\{\sin[\omega_0 t + \beta \cos \Omega t] + \cos[\omega_0 t + \beta \cos(\Omega t + \theta)]\right\} \tag{1.5.18}$$

We will see that the relative RF phase shift between the signals coming to the two electrodes have dramatic effect in the spectral properties of the modulated optical signal.

In first case, if the relative RF phase shift is 180°, that is, $\theta = (2m+1)\pi$, Equation 1.5.18 becomes

$$E_0(t) = \frac{E_i}{2}(\sin\omega_0 t + \cos\omega_0 t)[\sin(\beta\cos\Omega t) + \cos(\beta\cos\Omega t)] \tag{1.5.19}$$

Using Bessel expansion, Equation 1.5.19 can be written as

$$E_0(t) = \frac{E_i}{2} J_0(\beta)(\sin\omega_0 t + \cos\omega_0 t) + J_1(\beta)[\cos(\omega_0 - \Omega)t + \cos(\omega_0 + \Omega)t] \\ + J_1(\beta)[\sin(\omega_0 - \Omega)t + \sin(\omega_0 - \Omega)t] + \ldots.. \tag{1.5.20}$$

Obviously, this is a typical double-sideband spectrum because the three major frequency components are ω_0, $\omega_0 - \Omega$ and $\omega_0 + \Omega$.

In the second case, if the relative RF phase shift is 90°, that is, $\theta = 2m\pi + \pi/2$, Equation 1.5.18 becomes

$$E_0(t) = \frac{E_i}{2}\Big\{\sin\omega_0 t[\cos(\beta\cos\Omega t) + \sin(\beta\sin\Omega t)] + \cos\omega_0 t[\cos(\beta\sin\Omega t) \\ + \sin(\beta\cos\Omega t)]\Big\} \tag{1.5.21}$$

Again, using Bessel expansion, Equation 1.5.21 becomes

$$E_0(t) = \frac{E_i}{2} J_0(\beta)(\sin\omega_0 t + \cos\omega_0 t) + 2J_1(\beta)\cos(\omega_0 - \Omega)t + \ldots\ldots \tag{1.5.22}$$

This optical spectrum only has the optical carrier component at ω_0 and a lower modulation sideband at $\omega_0 - \Omega$. Similarly, if the RF phase shift is –90°, that is, $\theta = 2m\pi - \pi/2$. The optical spectrum will only have the carrier component at ω_0 and a higher modulation sideband at $\omega_0 + \Omega$.

1.5.4 Optical Modulators Using Electro-Absorption Effect

As discussed in the last section, electro-optic modulators made of $LiNbO_3$ in an MZI configuration have good performances in terms of high modulation speed and low frequency chirp. However, these modulators cannot be integrated with semiconductor lasers due to the difference in the material systems. As a consequence, $LiNbO_3$-based external modulators are always standalone devices with input and output pigtail fibers, and the insertion loss is typically on the order of 3 to 5 dB. Difficulties in high-speed packaging and input/output optical coupling make $LiNbO_3$ expensive; therefore they are usually used in high-speed, long-distance optical systems. In addition, $LiNbO_3$-based electro-optic modulators are generally sensitive to the state of polarization of the input optical signal; therefore, a polarization-maintaining fiber has to be used to connect the modulator to the source laser. This prevents this type of modulators to perform remodulation in locations far away from the laser source.

Electro-absorption (EA) modulation can be made from the same type of semiconductors, such as group III-V materials, as are used for semiconductor lasers. By appropriate doping, the bandgap of the material can be engineered such that it does not absorb the signal photons and therefore the material is transparent to the input optical signal. However, when an external reverse electrical voltage is applied across the pn junction, its bandgap will be changed to the same level of the signal photon energy, mainly through the well known Stark effect.

Then the material starts to absorb the signal photons and converts them into photocurrent, similar to what happens in a photodiode. Therefore, the optical transmission coefficient through this material is a function of the applied voltage. Because electro-absorption modulators are made by semiconductor materials, they can usually be monolithically integrated with semiconductor lasers, as illustrated in Figure 1.5.8. In this case, the DFB laser section is driven by a constant injection current I_C, thus producing a constant optical power. The EA section is separately controlled by a reverse-bias voltage that determines the strength of absorption.

The optical field transfer function of an EA modulator can be expressed as

$$E_0(t) = E_i \exp\left\{ -\frac{\Delta\alpha[V(t)]}{2} L - j\Delta\beta[V(t)]L \right\} \tag{1.5.23}$$

where $\Delta\alpha[V(t)]$ is the bias voltage-dependent power attenuation coefficient, which is originated from the electro-absorption effect in reverse-biased semiconductor pn junctions. $\Delta\beta[V(t)]$ is the voltage-dependent phase coefficient, which is introduced by the electro-optic effect. L is the length of the EA modulator. In general, both $\Delta\alpha$ and $\Delta\beta$ are strongly nonlinear functions of the applied voltage V, as shown in Figure 1.5.9, which are determined by the material bandgap structure as well as the specific device configuration.

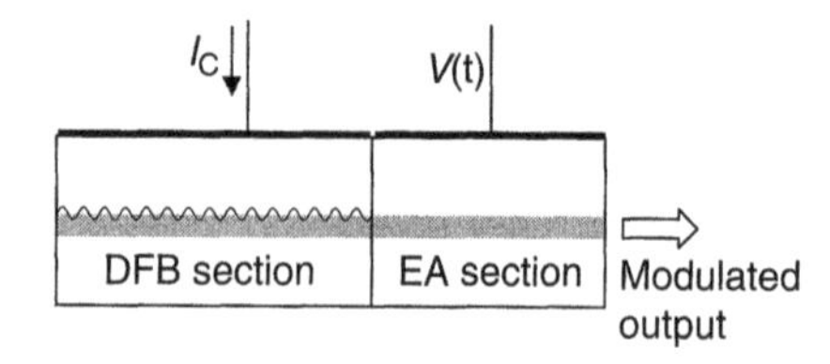

Figure 1.5.8 EA modulator integrated with a DFB laser source.

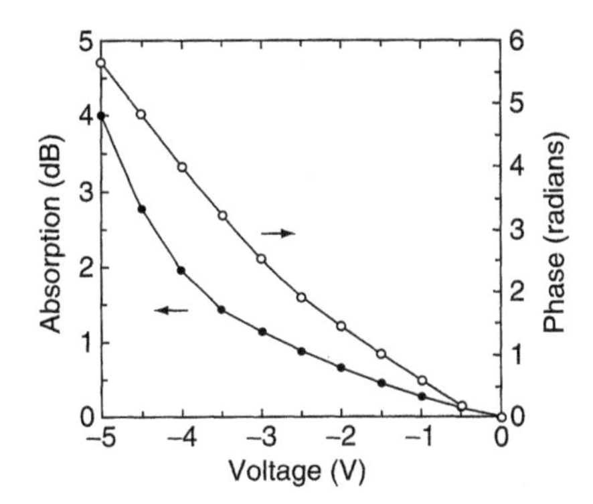

Figure 1.5.9 Phase and absorption coefficients of an EA modulator [26]. Used with permission.

From an application point of view, EA modulators can be monolithically integrated with semiconductor lasers on the same chip; therefore, they are relatively low cost and more compact. On the other hand, because both the absorption and the chirp parameters of an EA modulator are nonlinear functions of the bias voltage, the overall performance of optical modulation is generally not as controllable as using a $LiNbO_3$-based external optical modulator. In addition, since there is generally no optical isolator between the semiconductor laser source and the integrated EA modulator, optical reflection from an EA modulator often affects the wavelength and the phase of the laser itself [27]. From a measurement point of view, the characterization of a DFB laser-EA modulator unit is relatively complicated due to the nonlinear absorption and phase transfer functions of the modulator, as well as the mutual interaction between the laser and the modulator.

REFERENCES

[1] G. P. Agrawal, *Nonlinear Fiber Optics*, Academic Press, 3rd ed., 2001.

[2] C. H. Henry, "Theory of the linewidth of semiconductor lasers," *IEEE J. Quantum Electronics,* Vol. 18, p. 259, 1982.

[3] H. Kogelink and C. V. Shank, "Coupled-wave theory of distributed feedback lasers," *J. Appl. Phys.* 43 (5), 2327 (1972).

[4] R. Lang and K. Kobayashi, "External optical feedback effects on semiconductor injection laser properties," *IEEE J. Quantum Electronics*, Vol. 16, No. 3, pp. 347–355, 1980.

[5] R. Hui and S. Tao, "Improved rate-equation for external cavity semiconductor lasers," *IEEE J. Quantum Electronics*, Vol. 25, No. 6, p. 1580, 1989.

[6] K. Schneider and H. Zimmermann, *Highly Sensitive Optical Receivers*, Springer, 2006.

[7] Gerd Keiser, *Optical Fiber Communications*, 3rd ed., McGraw-Hill, 2000.

[8] E. G. Neumann, *Single-Mode Fibers Fundamentals*, Springer Series in Optical Science, Vol. 57, Springer-Verlag, 1988, ISBN 3-54018745-6.
[9] D. Gloge, "Weakly Guiding Fibers," *Applied Optics*, Vol. 10, No. 10, pp. 2252–2258, 1971.
[10] Robert Boyd, *Nonlinear Optics*. Academic Press, 1992.
[11] R. G. Smith, "Optical power handling capacity of low loss optical fibers as determined by stimulated Raman and Brillouin scattering," *Appl. Opt.* 11 (11), 2489 (1972).
[12] R. H. Stolen and C. Lin, "Self-phase modulation in silica optical fibers," *Phys. Rev. A* 17 (4), 1448 (1978).
[13] M. N. Islam et al., "Cross-phase modulation in optical fibers," *Opt. Lett.* 12 (8), 625 (1987).
[14] K. O. Hill, D. C. Johnson, B. S. Kawasaki, and R. I. MacDonald, "cw three-wave mixing in single-mode optical fibers," *J. Appl. Phys.* 49, 5098 (1978).
[15] K. Inoue, "Polarization effect on four-wave mixing efficiency in a single-mode fiber," *IEEE J. Quantum Electron.* 28, 883–894 (1992).
[16] M. J. Connelly, *Semiconductor Optical Amplifiers*, Kluwer Academic Publishers, (2002).
[17] S. Shimada and H. Ishio, eds., *Optical amplifiers and their applications*, John Wiley (1992).
[18] N. A. Olsson, "Semiconductor optical amplifiers," *IEEE Proceedings*, Vol. 80, pp. 375–382, 1992.
[19] Jianhui Zhou, et al., "Terahertz four-wave mixing spectroscopy for study of ultrafast dynamics in a semiconductor optical amplifier," *Applied Physics Letters*, Vol. 63, No. 9, pp. 1179–1181, 1993.
[20] Jade P. Wang, et al., "Efficient performance optimization of SOA-MZI devices," *Optics Express*, Vol. 16, No. 5, pp. 3288–3292, 2008.
[21] E. Desurvire, *Erbium-doped fiber amplifiers: Principles and applications*, John Wiley (1994).
[22] A. A. M. Saleh, R. M. Jopson, J. D. Evankow, and J. Aspell, "Accurate modeling of gain in erbium-doped fiber amplifiers," *IEEE Photon. Technol. Lett.*, Vol. 2, pp. 714–717, Oct. 1990.
[23] D. Penninckx, M. Chbat, L. Pierre, and J. P. Thiery, "The Phase-Shaped Binary Transmission (PSBT): a new technique to transmit far beyond the chromatic dispersion limit," *IEEE Photon. Technol. Lett.* Vol. 9, pp. 259–261, 1997.
[24] G. H. Smith, D. Novak, and Z. Ahmed, "Overcoming chromatic dispersion effects in fiber-wireless systems incorporating external modulators," *IEEE Trans. Microwave Technol.*, Vol. 45, pp. 1410–1415, 1997.
[25] R. Hui, B. Zhu, R. Huang, C. Allen, K. Demarest, and D. Richards, "Subcarrier multiplexing for high-speed optical transmission," *IEEE Journal of Lightwave Technology*, Vol. 20, pp. 417–427, 2002.
[26] J. C. Cartledge, "Comparison of Effective Parameters for Semiconductor Mach Zehnder Optical Modulators," *J. Lightwave Technology*, Vol. 16, pp. 372–379, 1998.
[27] J.-I. Hashimoto, Y. Nakano, and K. Tada, "Influence of facet reflection on the performance of a DFB laser integrated with an optical amplifier/modulator," *IEEE J. Quantum Electronics*, Vol. 28, No. 3, pp. 594–603, 1992.

Chapter 2

Basic Instrumentation for Optical Measurement

2.0. Introduction
2.1. Grating-Based Optical Spectrum Analyzers
    2.1.1. General Specifications
    2.1.2. Fundamentals of Diffraction Gratings
    2.1.3. Basic OSA Configurations
2.2. Scanning FP Interferometer
    2.2.1. Basic FPI Configuration and Transfer Function
    2.2.2. Scanning FPI Spectrum Analyzer
    2.2.3. Scanning FPI Basic Optical Configurations
    2.2.4. Optical Spectrum Analyzer Using the Combination of Grating and FPI
2.3. Mach-Zehnder Interferometers
    2.3.1. Transfer Matrix of a 2 × 2 Optical Coupler
    2.3.2. Transfer Function of an MZI
    2.3.3. MZI Used as an Optical Filter
2.4. Michelson Interferometers
    2.4.1. Operating Principle of a Michelson Interferometer
    2.4.2. Measurement and Characterization of Michelson Interferometers
    2.4.3. Techniques to Increase Frequency Selectivity
2.5. Optical Wavelength Meter
    2.5.1. Operating Principle of a Wavelength Meter Based on Michelson Interferometer
    2.5.2. Wavelength Coverage and Spectral Resolution
    2.5.3. Wavelength Calibration
    2.5.4. Wavelength Meter Based on Fizeau Wedge Interferometer
2.6. Optical Polarimeter
    2.6.1. General Description of Lightwave Polarization
    2.6.2. The Stokes Parameters and the Poincare Sphere
    2.6.3. Optical Polarimeters
2.7. Measurement Based on Coherent Optical Detection
    2.7.1. Operating Principle
    2.7.2. Receiver SNR Calculation of Coherent Detection
    2.7.3. Balanced Coherent Detection and Polarization Diversity
    2.7.4. Phase Diversity in Coherent Homodyne Detection
    2.7.5. Coherent OSA Based on Swept Frequency Laser

2.8. Waveform Measurement
2.8.1. Oscilloscope Operating Principle
2.8.2. Digital Sampling Oscilloscopes
2.8.3. High-Speed Sampling of Optical Signal
2.8.4. High-Speed Electric ADC Using Optical Techniques
2.8.5. Short Optical Pulse Measurement Using an Autocorrelator
2.9. Optical Low-Coherent Interferometry
2.9.1. Optical Low-Coherence Reflectometry
2.9.2. Fourier-Domain Reflectometry
2.10. Optical Network Analyzer
2.10.1. S-Parameters and RF Network Analyzer
2.10.2. Optical Network Analyzers

2.0 INTRODUCTION

Characterization and testing of optical devices and systems require essential tools and instrumentations. Good understanding of basic operation principles of various test instruments, their general specifications, advantages, and limitations will help us design experimental setups, optimize test procedures, and minimize measurement errors. Optical measurement and testing is a fast-moving field due to the innovations and rapid advances in optical devices and systems as well as emerging applications. The complexity and variety of optical systems require high levels of flexibility in the testing and characterization. Realistically, it is not feasible for a single laboratory to possess all the specialized instrumentations to satisfy all the measurement requirements. It is important to be flexible in configuring measurement setups based on the available tools and instruments. In doing so, it is also important to understand the implications of the measurement accuracy, efficiency and their tradeoffs.

This chapter introduces a number of basic optical measurement instruments and functional devices. The focus is on the discussion of the basic operating principles, general specifications, unique advantages, and limitations of each measurement tool. Section 2.1 discusses grating-based optical spectrum analyzers (OSAs), the most popular instruments in almost every optics laboratory. Section 2.2 introduces scanning Fabry-Perot interferometers, which can usually provide better spectral resolution than grating-based OSA but with narrower range of wavelength coverage. Sections 2.3 and 2.4 describe Mach-zehnder and Michelson interferometers. Their optical configurations are similar but the analysis and specific applications may differ. Section 2.5 introduces optical wavelength meters based on a Michelson interferometer. Although the wavelength of an optical signal can be measured by an OSA, a wavelength meter is application specific and usually provides much better wavelength measurement accuracy and efficiency.

Section 2.6 presents various designs of optical polarimeters, which are used to characterize polarization states of optical signals passing through optical devices or systems. Section 2.7 introduces optical measurement techniques based on coherent detection. In recent years, high-quality single-frequency semiconductor tunable lasers become reliable and affordable. This makes coherent detection a viable candidate for many high-end measurement applications that require high spectral resolution and detection sensitivity.

Section 2.8 discusses various optical waveform measurement techniques, including analog and sampling oscilloscopes, linear and nonlinear optical sampling, and optical auto-correlation. Section 2.9 is devoted to the discussion of optical low-coherent interferometers, which are able to accurately allocate optical interfaces and reflections. The same technique has also been widely used for biomedical imaging, where it is known as optical coherence tomography (OCT). The last section of this chapter, Section 2.10, discusses optical network analyzers. Based on the similar idea of an RF network analyzer which operates in electronics domain, an optical network analyzer characterizes complex transfer functions of a device or system, but in optical domain.

The purpose of this chapter is to set up a foundation and basic understanding of optical measurement methodologies and their theoretical background. Similar operation principles and methodologies can be extended to create various specialized instrumentations, as discussed in later chapters.

2.1 GRATING-BASED OPTICAL SPECTRUM ANALYZERS

An optical spectrum analyzer is an instrument that is used to measure the spectral density of a lightwave signal at different wavelengths. It is one of the most useful pieces of instruments in fiber-optic system and device measurement, especially when wavelength division multiplexing is introduced into the systems where different data channels are carried by different wavelengths. In addition to being able to identify the wavelength of an optical signal, an optical spectrum analyzer is often used to find optical signal power level at each wavelength channel, evaluate optical signal-to-noise-ratio and optical crosstalk, and check the optical bandwidth when an optical carrier is modulated.

2.1.1 General Specifications

The most important parameter an OSA provides is the optical spectral density versus wavelength. The unit of optical spectral density is usually expressed in watts per Hertz [W/Hz], which is defined as the optical power within a bandwidth of a Hertz measured at a certain wavelength. The most important qualities of an OSA can be specified by the following parameters [1, 2]:

1. *Wavelength range*. It is the maximum wavelength range the OSA can cover while guaranteeing the specified performance. Although a large wavelength range is desired, practical limitation comes from the applicable wavelength window of optical filters, photodetectors, and other optical devices. The typical wavelength range of a commercially available grating-based OSA covers is from 400 nm to 1700 nm.
2. *Wavelength accuracy*. Specifies how accurately the OSA measures the wavelength. Most commercial OSAs separately specify absolute wavelength accuracy and relative wavelength accuracy. Absolute wavelength accuracy specifies how accurate is the measured absolute wavelength value, which is often affected by the wavelength calibration. Relative wavelength accuracy tells how accurate is the measured wavelength separation between two optical signals, which is mainly determined by the nonlinearity of the optical filters. In typical OSAs, wavelength accuracy of less than 0.1 nm can be achieved.
3. *Resolution bandwidth*. It defines how fine an OSA slices the signal optical spectrum during the measurement. As an OSA measures signal optical spectral density, which is the total optical power within a specified bandwidth, a smaller-resolution bandwidth means a more detailed characterization of the optical signal. However, the minimum resolution bandwidth of an OSA is usually limited by the narrowest bandwidth of the optical system the OSA can provide and limited by the lowest detectable optical power of the receiver. The finest optical resolution bandwidth of a grating-based commercial OSA ranges from 0.1 nm to 0.01 nm.
4. *Sensitivity*. Specifies the minimum measurable signal optical power before it reaches the background noise floor. Therefore the detection sensitivity is basically determined by the noise characteristic of the photodiode used inside the OSA. For a short-wavelength OSA, the detection sensitivity is generally better due to the use of silicon photodiode, which covers the wavelength from 400 nm to 1000 nm. For the wavelength from 1000 nm to 1700 nm, an InGaAs photodiode has to be used, and the noise level is generally high and the detection sensitivity is therefore poor compared to the short-wavelength OSAs. Commercially available OSAs can provide a detection sensitivity of –120 dBm in the 400–1700 nm wavelength range.
5. *Maximum power*. The maximum allowable signal optical power before the OSA detection system is saturated. An OSA can generally tolerate signal optical power levels in the order of 20 dBm or higher.

6. *Calibration accuracy.* Specifies how accurate the absolute optical power reading is in the measurement. Typically a calibration accuracy of less than 0.5 dB can be achieved in a commercial OSA.
7. *Amplitude stability.* Specifies the maximum allowable fluctuation of the power reading over time, when the actual input signal optical power is constant. Typical amplitude stability of a commercial OSA is less than 0.01 dB per minute.
8. *Dynamic range.* The maximum distinguishable amplitude difference between two optical signals with their wavelengths a certain number of nanometers apart. This becomes a concern because in practical OSAs, if a weak optical signal is located in close vicinity to a strong optical signal, the weak signal may become un-measurable because the receiver is overwhelmed by the strong signal; obviously, this effect depends on the wavelength separation between these two optical signals. Typical dynamic range of a commercial OSA is about 60 dB for a 0.8 nm wavelength interval and 52 dB for a 0.2 nm wavelength interval.
9. *Frequency sweep rate.* Specifies the speed of an OSA sweeping over the wavelength during the measurement. It depends on the measurement wavelength span and the resolution bandwidth used. Generally, the number of measurement samples an OSA takes in each sweep is equal to the span width divided by the resolution bandwidth. In practical applications, choosing sweep speed also depends on the power levels of the optical signal. At low power levels, the average may have to be made in the detection, and that may slow down the sweep speed.
10. *Polarization dependence.* Specifies the maximum allowable fluctuations of the power reading while changing the state of polarization of the optical signal. Polarization dependence of an OSA is usually caused by the polarization-dependent transmission of the optical system, such as gratings and optical filters used in the OSA. Commercial OSAs usually have less than 0.1 dB polarization dependence.

An OSA measures signal optical power within the resolution bandwidth at various wavelengths; obviously this can be accomplished by a tunable narrowband optical filter. Although there are various types of optical filters, considering the stringent specifications, such as wide wavelength range, fine optical resolution, high dynamic range, and so on, the option is narrowed down and the most commonly used optical filter for OSA applications is diffraction grating. High-quality gratings can have wide wavelength range and reasonably good resolution, and they can be mechanically tuned to achieve swept wavelength selection. In this section, we discuss the operation principle and various configurations of the grating-based optical spectrum analyzer.

2.1.2 Fundamentals of Diffraction Gratings

Most of the commercial optical spectrum analyzers are based on diffraction gratings because of their superior characteristics in terms of resolution, wavelength range, and simplicity [1]. A diffraction grating is a mirror onto which very closely spaced thin groove lines are etched, as illustrated in Figure 2.1.1. Important parameters defining a diffraction grating include the grating period d, which is the distance between adjacent grooves, grating normal $\vec{n}_G$, which is perpendicular to the grating surface, and groove normal $\vec{n}_V$, which is perpendicular to the groove surface.

As illustrated in Figure 2.1.2, when a collimated optical signal is projected onto a grating at an incident angle α (with respect to the grating normal $\vec{n}_G$) and a receiver collects the diffracted lightwave at an angle of β, the path length difference between adjacent ray traces can be expressed as

$$\Delta = d(\sin\beta - \sin\alpha) \tag{2.1.1}$$

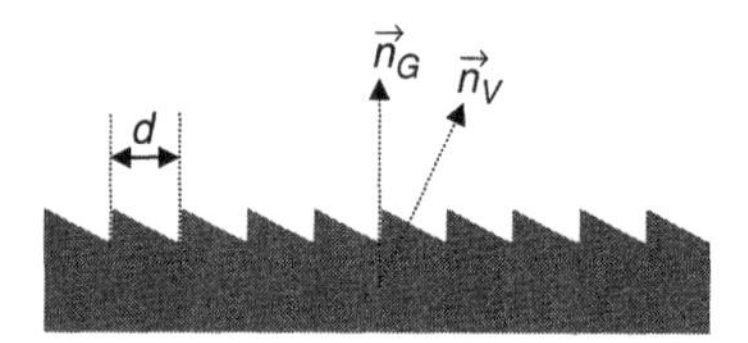

Figure 2.1.1 Illustration of a diffraction grating with period d, grating normal $\vec{n}_G$, and groove normal $\vec{n}_V$.

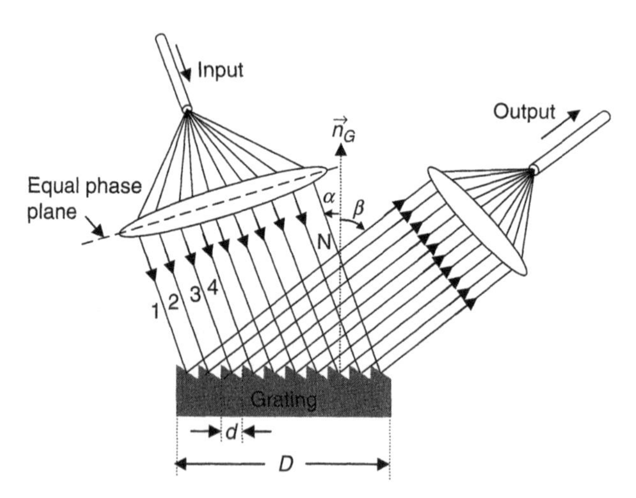

Figure 2.1.2 Lightwave incident and diffracted from a diffraction grating.

Assume that the input field amplitude of each ray is A_0 (the total field amplitude within a spatial width of a grating slot) and there are N lightwave rays project onto the grating. Then at the receiver, the field of all the ray traces will add up and the total electrical field is

$$A(\beta) = A_0 U_0(\beta)\left\{1 + \exp\left[j\frac{2\pi}{\lambda}\Delta\right] + \exp\left[j\frac{2\pi}{\lambda}2\Delta\right] + \exp\left[j\frac{2\pi}{\lambda}3\Delta\right] + \ldots\ldots + \exp\left[j\frac{2\pi}{\lambda}N\Delta\right]\right\}$$
$$= A_0 U_0(\beta)\frac{1 - \exp\left[j\frac{2\pi}{\lambda}N\Delta\right]}{1 - \exp\left[j\frac{2\pi}{\lambda}\Delta\right]} \tag{2.1.2}$$

Here $U_0(\beta)$ is the angle-dependent diffractive efficiency of the field within each grating slot:

$$U_0(\beta) = \frac{1}{d}\int_0^d \exp\left[j\frac{2\pi}{\lambda}x\sin\beta\right]dx = A_0\exp\left(j\frac{\pi d}{\lambda}\sin\beta\right)\frac{\sin\left(\frac{\pi d}{\lambda}\sin\beta\right)}{\left(\frac{\pi d}{\lambda}\sin\beta\right)} \tag{2.1.3}$$

Since

$$\frac{1 - \exp(jNx)}{1 - \exp(jx)} = \frac{\sin(Nx/2)}{\sin(x/2)}\exp\left(j\frac{N-1}{2}x\right) \tag{2.1.4}$$

the overall optical field transfer function from the input to the output is

$$H(\lambda, \beta) = \frac{A(\lambda, \beta)}{NA_0} = \frac{\sin\left(\frac{\pi}{\lambda}d\sin\beta\right)}{\left(\frac{\pi}{\lambda}d\sin\beta\right)}\frac{\sin\left(\frac{\pi N}{\lambda}\Delta\right)}{N\sin\left(\frac{\pi}{\lambda}\Delta\right)}\exp j\left(\frac{N-1}{\lambda}\pi\Delta + \frac{\pi d}{\lambda}\sin\beta\right) \tag{2.1.5}$$

and the corresponding power transfer function is

$$T(\lambda, \beta) = |\, H(\lambda, \beta)\,|^2 = \text{sinc}^2\left(\frac{\pi}{\lambda}d\sin\beta\right)\frac{\sin^2\left[\frac{\pi N}{\lambda}\Delta\right]}{N^2\sin^2\left[\frac{\pi}{\lambda}\Delta\right]} \tag{2.1.6}$$

Obviously, the power transfer function reaches to its peak values when $\Delta = m\lambda$, where the grating order m is an integer number. So a grating equation is defined as

$$m\lambda = d(\sin\beta - \sin\alpha) \tag{2.1.7}$$

Although this grating equation defines the wavelengths and the diffraction angles whereby the power transmission finds its maxima, Equation (2.1.6) is a general equation describing the overall grating transfer function versus wavelength. To better understand this grating transfer function, let's look at a few special cases.

2.1.2.1 Measure the Diffraction Angle Spreading When the Input Only has a Single Frequency

In this case, we can find the spatial resolution of the grating by measuring the output optical intensity versus the diffraction angle β.

For an m-th order grating, one of the power transmission maxima happens at a diffraction angle β_m, which satisfies the grating equation $d \sin \beta_m = m\lambda + d \sin\alpha$, or equivalently,

$$Nd\sin\beta_m = mN\lambda + Nd\sin\alpha \tag{2.1.8}$$

Suppose that the adjacent transmission minimum happens at an angle $\beta_m + \Delta\beta$, we can write

$$Nd\sin(\beta_m + \Delta\beta) = (mN+1)\lambda + Nd\sin\alpha \tag{2.1.9}$$

so that

$$\sin(\beta_m + \Delta\beta) - \sin\beta_m = \frac{\lambda}{Nd} \tag{2.1.10}$$

For a high-resolution grating, suppose this angle $\Delta\beta$ is small enough, Equation 2.1.10 can be linearized as

$$\sin(\beta + \Delta\beta) - \sin\beta \approx \Delta\beta\cos\beta \tag{2.1.11}$$

Then we can find the value of $\Delta\beta$, which is the half-angular width of the transmission peak:

$$\Delta\beta = \frac{\lambda}{Nd\cos\beta} \tag{2.1.12}$$

$\Delta\beta$ specifies the angular spreading of the diffracted light when the input has a single wavelength at λ. For high-resolution OSAs, the value of $\Delta\beta$ should be small. High angular resolution can be achieved using a large effective area grating because the total width of the grating illuminated by the optical signal is $D = (N-1)d \approx Nd$.

2.1.2.2 Sweep the Signal Wavelength While Measuring the Output at a Fixed Diffraction Angle

In this case we can find the frequency resolution of the grating by measuring the output optical intensity as the function of the input signal wavelength. Here β is fixed and λ is varying.

For an m-th order grating and at the output diffraction angle β, the transmission maximum happens when the input signal is tuned to a wavelength λ_m. The relationship between β and λ_m is given by the grating equation

$$Nd(\sin\beta - \sin\alpha) = Nm\lambda_m \tag{2.1.13}$$

To evaluate the width of this transmission peak, it is useful to find the position of the transmission minimum around the peak. Assume that a transmission minimum happens at $(\lambda_m - \Delta\lambda)$, according to the grating equation

$$Nd(\sin\beta - \sin\alpha) = (mN + 1)(\lambda_m - \Delta\lambda) \tag{2.1.14}$$

Combining Equations 2.1.13 and 2.1.14, we find

$$\Delta\lambda = \frac{\lambda}{mN + 1} \approx \frac{\lambda}{mN} \tag{2.1.15}$$

$\Delta\lambda$ specifies the wavelength resolution of the grating-based optical filter when the output is measured at a fixed diffraction angle. Equation 2.1.15 reveals that a high-resolution OSA requires a large number of grating grooves N to be illuminated by the input signal. For a certain input beam size D, this requires a high groove density or a small groove period d. Commercially available gratings usually have groove densities ranging from 400 to 1200 lines per millimeter. Gratings with higher groove densities suffer from larger polarization sensitivities and narrow wavelength range. A high diffraction order m also helps to improve the wavelength resolution. However, at high diffraction angles, the grating efficiency is usually decreased, although there are special designs to enhance the grating efficiency at a specific diffraction order.

Example 2.1

To achieve a spectral resolution of 0.08 nm at a signal wavelength of 1550 nm, use a second-order grating ($m = 2$ in eq.(2.1.7)) with a groove density of 800 lines/mm. What is the minimum diameter of the collimated input beam?

Solution

According to Equation 2.1.15, to achieve a resolution bandwidth of 0.08 nm, the minimum number of grooves illuminated by the input optical signal is

$$N = \frac{\lambda}{m\Delta\lambda} = \frac{1550}{2 \times 0.08} = 9688$$

Since the grating has a groove density of 800 lines/mm, the required beam diameter is

$$D = \frac{9688}{800} = 12.1\ mm$$

Another often used specification for a grating is the *grating dispersion*, which is defined as the diffraction angle change induced by a unit wavelength change of the optical signal.

Using the grating equation given by Equation 2.1.7 and assuming that the input angle is fixed, differentiating both sides of the equation, we have $m\Delta\lambda = d\cos\beta\Delta\beta$. Therefore, the grating dispersion can be expressed as

$$Disp = \frac{\Delta\beta}{\Delta\lambda} = \frac{m}{d\cos\beta} \tag{2.1.16}$$

2.1.3 Basic OSA Configurations

The heart of a grating-based optical spectrum analyzer is a monochromator as schematically illustrated in Figure 2.1.3. In this basic optical configuration, the input optical signal is converted into a collimated beam through a lens

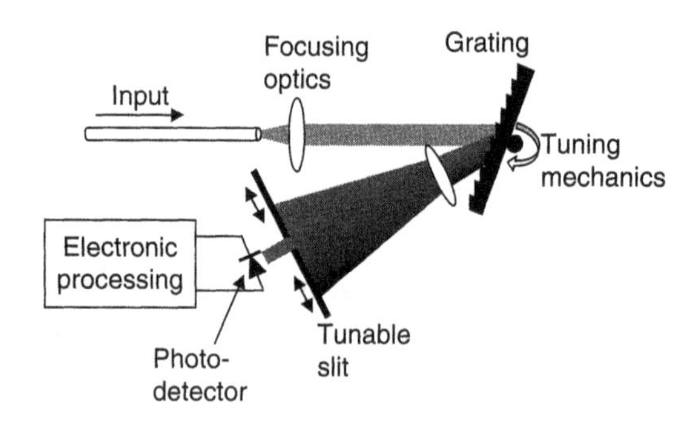

Figure 2.1.3 Basic configuration of an optical spectrum analyzer.

system and launched onto a grating. The grating diffracts and disperses wavelength components of the optical signal into different diffraction angles. A tunable mechanical aperture (slit) is used to select various wavelength components within the signal spectrum and a photodiode converts the selected wavelength component into electrical signal. The electric signal is then amplified, digitized and recorded, or displayed on the OSA screen.

The photodiode measures the total optical power within the spectral bandwidth determined by the width of the slit. Taking into account the grating transfer function and the width of the optical aperture, the measured value can be converted into an optical power spectral density in the unit of mW/nm, or dBm/nm.

In a practical OSA, there is usually a button to select the desired resolution bandwidth for each measurement, which is accomplished by selecting the width of the optical aperture. Reducing the width of the aperture allows to cut fine slices of the optical spectrum, however, the minimum wavelength resolution is determined by the grating quality and the beam size as described in Equation 2.1.15. In addition, focusing optics also needs to be designed properly to minimize aberration while maintaining a large beam size on the grating.

2.1.3.1 OSA Based on a Double Monochromator

Spectral resolution and measurement dynamic range are the two most important parameters of an OSA, and both of them depend on the grating transfer function. These parameters can be improved by (1) increasing groove-line density of the grating and (2) increasing the size of the beam that launches onto the grating. Practically, gratings with groove densities higher than 1200/mm are difficult to make and they usually suffer from high polarization sensitivity and limited wavelength coverage. On the other hand, increased beam size demands very high-quality optic systems and high uniformity of the grating. It also proportionally increases the physical size of the OSA.

One often used method to improve OSA performance is to let the optical signal diffract on the grating twice, as shown in Figure 2.1.4. In this double-pass configuration, if the two gratings are identical, the power transfer function will be

$$T_2(\lambda, \beta) = | T_1(\lambda, \beta) |^2 \tag{2.1.17}$$

where $T_1(\lambda, \beta)$ is the power transfer function of a single grating as given by Equation 2.1.6.

Figure 2.1.5 shows an example of grating power transfer function versus the relative diffraction angle β. The grating groove density is 1000 lines/mm and the beam size that launched on the grating is 10 mm. Therefore, there

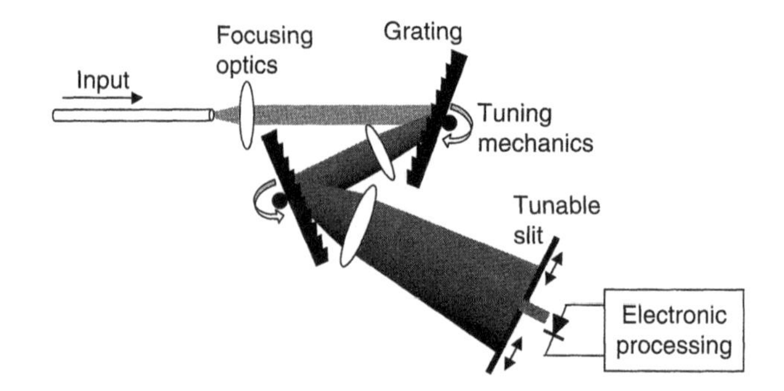

Figure 2.1.4 Optical spectrum analyzer based on a double-pass monochromator.

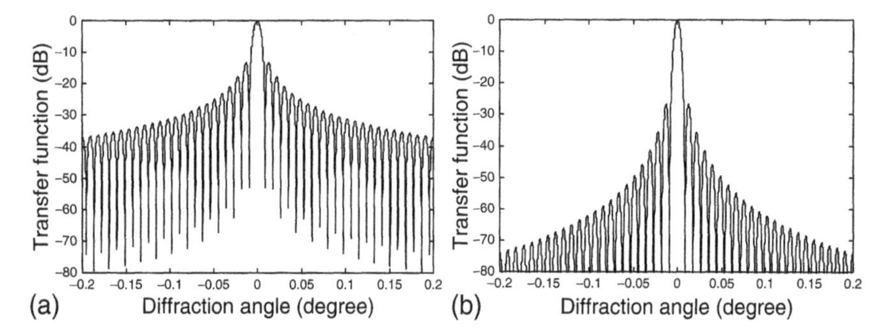

Figure 2.1.5 Transfer functions of (a) single-pass and (b) and double-pass grating with $d = 1$ µm, $N = 10000$, and $\lambda = 1550$ nm.

are 10,000 groove lines involved in the interference. The signal wavelength is $\lambda = 1550$ nm. Using double-pass configuration, the spectral resolution defined by Equations 2.1.12 and 2.1.15 does not change; however, it does improve the sideband suppression and thus increase the dynamic range of the OSA.

2.1.3.2 OSA with Polarization Sensitivity Compensation

For practical OSA applications, especially when the input is from an optical fiber, the signal polarization state is not typically controlled and it can vary randomly. This requires an OSA to operate independent of polarization. However, diffraction gratings are typically polarization sensitive. The diffraction efficiency of a grating is usually different for input optical field polarizations parallel (s-plane) or perpendicular (p-plane) to the direction of the groove lines. This polarization sensitivity is especially pronounced when the groove density is high.

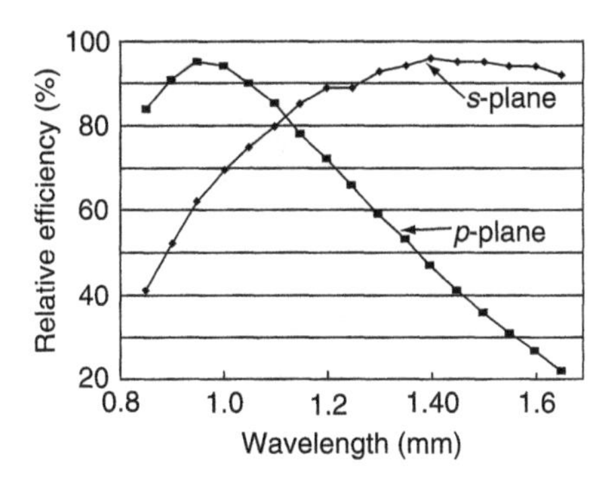

Figure 2.1.6 Example of diffraction efficiency of a plane-ruled reflectance grating versus wavelength for optical signal polarized on *s*-plane and *p*-plane [3].

Figure 2.1.6 shows an example of a plane-ruled reflectance grating with 1000 line/nm groove density. The diffraction efficiencies for signal polarized on the *s*-plane and the *p*-plane are generally different, and this difference is a function of wavelength. Obviously this polarization sensitivity needs to be compensated. There are a number of techniques that can be used to design an OSA. The most popular ones are polarization diversity and polarization average.

In polarization diversity configuration, two photodiodes are used to detect polarization components in the *p*-plane and the *s*-plane separately, as shown in Figure 2.1.7. Since the grating diffraction efficiencies for these two polarizations can be measured precisely and they are deterministic, each detector only needs to compensate for the grating efficiency in its corresponding polarization. Adding the calibrated electrical signals from the two detectors, polarization sensitivity can be easily minimized through electronic signal processing.

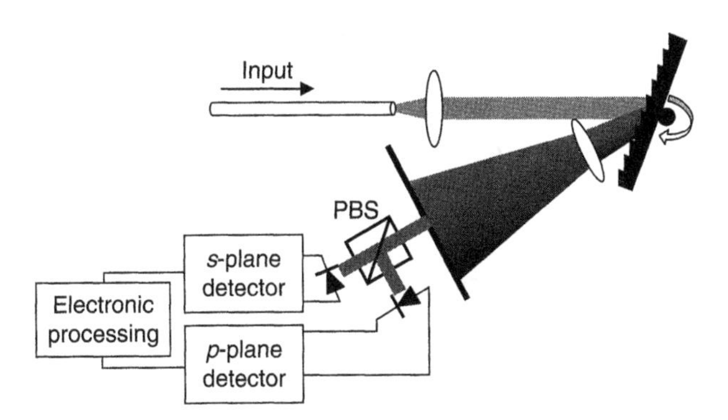

Figure 2.1.7 Optical spectrum analyzer with polarization diversity detection. *PBS*: polarization beam splitter.

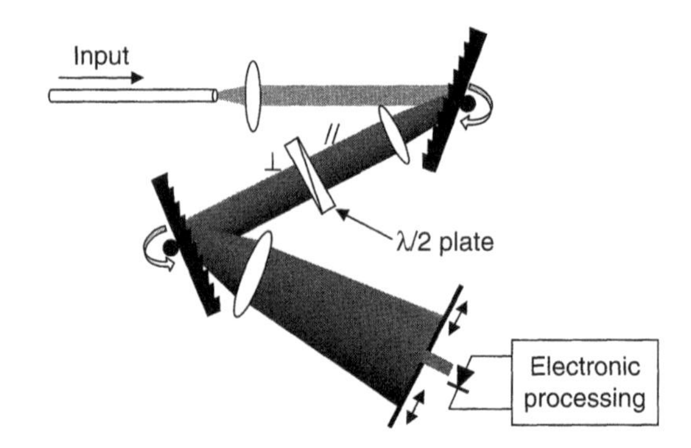

Figure 2.1.8 Optical spectrum analyzer with polarization averaging.

Polarization average is a technique that can be used in double-pass monochromators, as shown in Figure 2.1.8, where a halfwave plate is used to rotate the polarization by 90°. In this way, a signal entering the first grating on the *p*-plane will be rotated to the *s*-plane for the second grating. If the grating diffraction efficiencies for the *p*-plane and the *s*-plane are $A_p(\lambda)$ and $A_s(\lambda)$, respectively, the overall efficiency after a double-pass will be $A(\lambda) = A_p(\lambda) \cdot A_s(\lambda)$, which is an efficiency independent of the polarization state of the optical signal.

There are various commercial OSAs, each with its unique design. However, their operations are based on similar principles, as described previously. Figure 2.1.9 shows an example of practical optical system design of a double-pass OSA. In this design, the optical signal is folded back and forth in the optical system and diffracted by the same grating twice. A $\lambda/2$ waveplate is used to rotate the polarization state of the signal by 90° between the two diffractions. This OSA can also be used as a tunable optical filter by selecting the optical output instead of the electrical signal from photodiode 1. To better explain the operation mechanism, the lightpass sequence is marked by 1, 2, 3, and 4 on Figure 2.1.9, each with a direction arrow.

2.1.3.3 Consideration of Focusing Optics

In an OSA, although the performance is largely determined by the grating characteristics, focusing optics also plays an important role in ensuring that the optimum performance is achieved. We must consider several key parameters in the design of an OSA focusing optics system:

1. *Size of the optical input aperture and the collimating lens.* The collimated beam quality depends largely on the optical input aperture size and the divergence angle. In principle, if the input signal is from a point source with infinitesimal spot size, it can always be converted into an ideally collimated beam

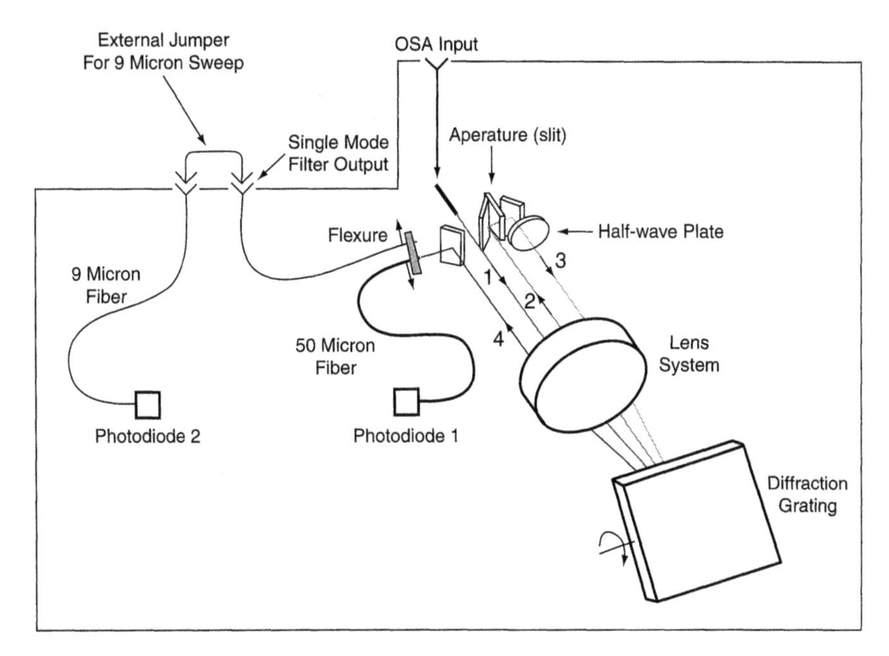

Figure 2.1.9 A practical optical system configuration of a double-pass OSA with polarization averaging using a $\lambda/2$ plate. This OSA can also be used as a tunable optical filter. [2]

through an optical lens system. However, if the optical input is from a standard single-mode fiber that has a core diameter of 9 μm and a maximum divergence angle of $\pm 10°$, the lens system will not be able to convert this into an ideally collimated beam. This idea is illustrated in Figure 2.1.10. Since the quality of an OSA depends on the precise angular dispersion of the grating, any beam divergence will create degradations in the spectral resolution.

2. *Output focusing lens.* After the optical signal is diffracted off the grating, different wavelength components within the optical signal will travel at different angles, and the output focusing lens will then focus them to different locations

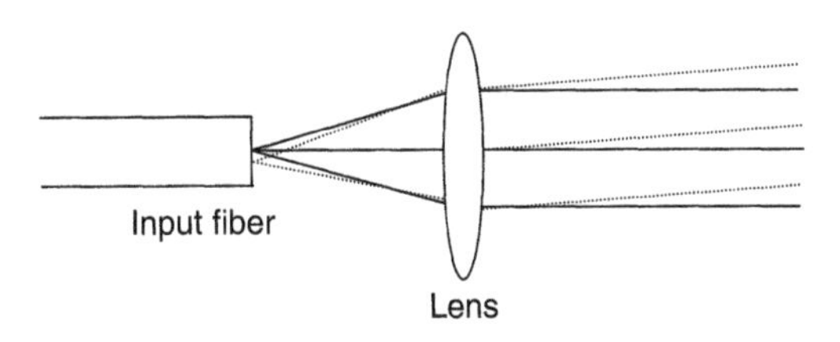

Figure 2.1.10 Illustration of beam divergence when the input aperture is big.

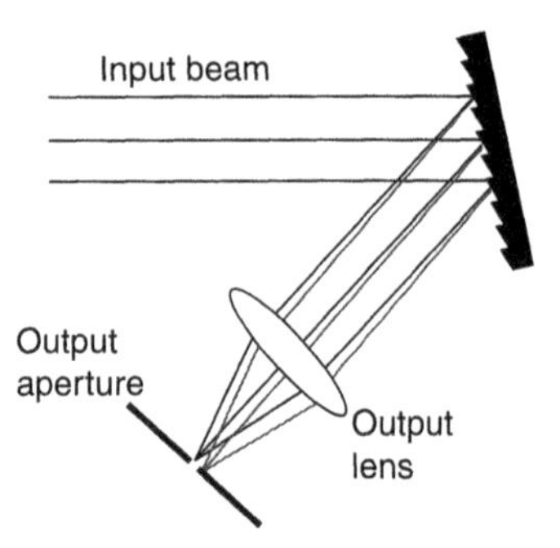

Figure 2.1.11 Illustration of beam divergence when the input aperture is large.

at the output aperture, as shown in Figure 2.1.11. Ideally the focusing spot size has to be infinitesimal to ensure the optimum spectral resolution. However, the minimum focusing spot size is limited by the diffraction of the lens system, which is usually called an *Airy disk*. This diffraction-limited spot size depends on the focus length f and the diameter of the lens D as

$$w = \frac{2\lambda f}{\pi D} \tag{2.1.18}$$

Large collimated beam size and a large lens are preferred to increase the spectral resolution and reduce the diffraction-limited spot size. However, this will increase the size of the OSA.

3. *Aberration induced by lens.* Aberration increases the focusing spot size and therefore degrades resolution. Aberration is directly proportional to the beam diameter. Therefore the beam diameter cannot be too large.
4. *Output slit or pin-hole.* To select diffracted light of different wavelength, a smaller exit pinhole gives better spectral resolution but less signal energy (poor SNR). Therefore, usually longer averaging time has to be used when a small resolution bandwidth is chosen.
5. *Photodetection.* Requires photodiodes with wide spectral coverage, or sometimes multiple photodiodes (each covers a certain wavelength range). Response calibration must be made and the noise level has to be low.

2.1.3.4 Optical Spectral Meter Using Photodiode Array

Traditional grating-based OSA uses a small optical aperture for spectral selection. To measure an optical spectrum, one has to mechanically scan the angle of the grating, or the position of the optical aperture. Mechanical moving parts potentially make the OSA less robust and reliable. Like an automobile, after running for years, the engine will be worn out. For some applications requiring continuous monitoring of optical spectrum, this reliability issue would be an important concern.

In recent years, miniaturized optical spectrometers have been used in WDM optical systems to monitor signal quality and optical signal-to-noise ratio. In this type of optical communication system applications, the spectrometers have to operate continuously for many years, and therefore mechanical moving parts have to be eliminated.

A simple spectrometer design without mechanical moving parts is to use a one-dimensional photodiode array as the detector. The configuration is shown in Figure 2.1.12, where each pixel in the photodiode array measures a slice of the optical spectrum. Parallel electronic processing can be used to perform data acquisition and spectrum analysis. A most often used detector array is the one-dimensional charge coupled device (CCD). Silicon-based CCD have been used widely for imaging sensors, they are reliable with low noise, high sensitivity, and most important, low cost. However, silicon-based CCDs only cover wavelength regions from visible to approximately 1 μm. For long wavelength applications such as in the 1550 nm optical communication window, InGaAs-based linear arrays are commercially available with the pixel numbers ranging from 512 to 2048 and the pixel width ranging from 10 μm to 50 μm. In addition to higher noise levels compared to silicon-based CCDs, the cost of an InGaAs detector array is much higher than its silicon counterpart. This high cost, to a large extent, limits the applications of OSA-based optical performance monitors in commercial communication networks.

For a commercially available compact OSA operating in a 1550 nm wavelength window using a InGaAs linear array, the spectral resolution is typically on the order of 0.2 nm and the wavelength coverage can be in both C-band (1530 nm – 1562 nm) and L-band (1565 nm – 1600 nm). This is made primarily for all-optical performance monitoring of optical networks. The limitation in the spectral resolution is mainly doe to the limited grating dispersion and the small beam size, as well as the limited pixel numbers of the diode array.

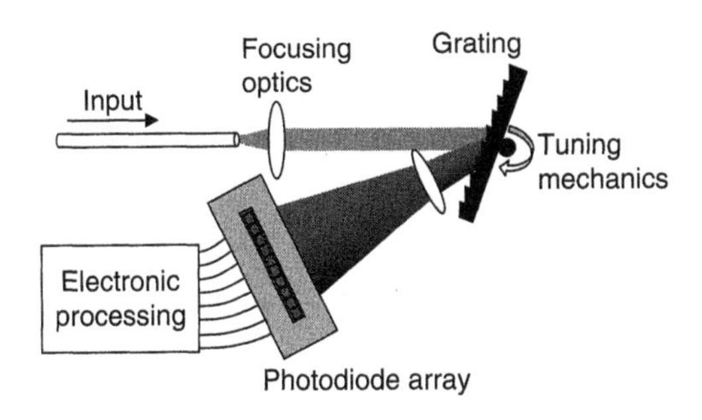

Figure 2.1.12 Photodiode array-based optical spectrometer.

2.2 SCANNING FP INTERFEROMETER

In the last section, we discussed optical spectrum analyzers based on diffraction gratings. The major advantage of grating-based OSAs is the wide wavelength coverage. A commercial OSA usually covers a wavelength range from 400 nm to 1700 nm. However, it is not easy to achieve a very fine spectral resolution using grating-based OSA, which is limited by the groove-line densities and the maximum optical beam diameter. For example, for a first-order grating with line density of 1200/mm operating in a 1550 nm wavelength window, to achieve a frequency resolution of 100 MHz, which is equivalent to 0.0008 nm wavelength resolution, the required optical beam diameter has to be approximately 1.6 m; this is simply too big to be realistic. On the other hand, an ultra-high-resolution optical spectrum analyzer can be made using a scanning Fabry-Perot interferometer (FPI). In this section, we discuss the operating principle and realization of FPI-based optical spectrum analyzers.

2.2.1 Basic FPI Configuration and Transfer Function

The basic configuration of a FPI is shown in Figure 2.2.1, where two parallel mirrors, both having power reflectivity R, are separated by a distance d [4, 5]. If a light beam is launched onto the mirrors at an incident angle α, a part of the light will penetrate through the left mirror and propagates to the right mirror at point A. At this point, part of the light will pass through the mirror and the other part will be reflected back to the left mirror at point B. This process will be repeated many times until the amplitude is significantly reduced due to the multiple reflection losses.

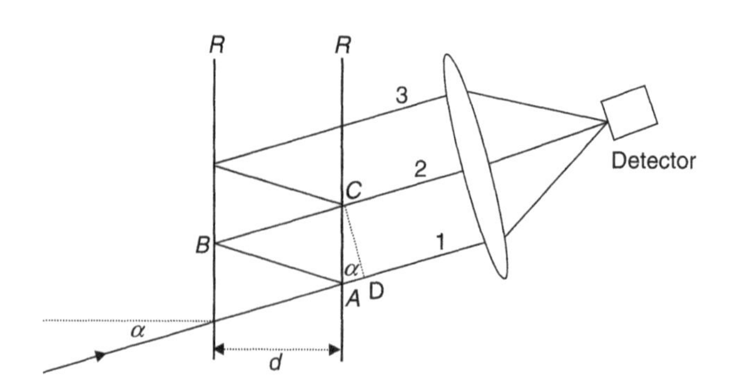

Figure 2.2.1 Illustration of a Febry-Perot Interferometer with two partially reflecting mirrors separated by a distance d.

The propagation phase delay can be found after each roundtrip between the two mirrors. The path length difference between the adjacent output traces, as illustrated in Figure 2.2.1, is

$$\Delta L = AB + BC - AD = 2d\cos\alpha \tag{2.2.1}$$

After passing through the second mirror, the optical field amplitudes of light rays 1, 2, and 3 can be expressed as, respectively,

$$E_0(f) = E_i(f)\sqrt{1-R}\exp(j\beta n x_0)\sqrt{1-R}$$

$$E_1(f) = E_i(f)\sqrt{1-R}\exp(j\beta n x_0)\exp(j\beta n\Delta L)R\sqrt{1-R}$$

and

$$E_2(f) = E_i(f)\sqrt{1-R}\exp(j\beta n x_0)\exp(j2\beta n\Delta L)R^3\sqrt{1-R}$$

where $E_i(f)$ is the input optical field, f is the optical frequency, x_0 is the propagation delay of the beam 1 shown in Figure 2.2.1, $\beta = \frac{2\pi}{\lambda} = \frac{2\pi f}{c}$ is the propagation constant, and n is the refractive index of the material between the two mirrors.

The general expression of the optical field, which experienced N roundtrips between the two mirrors can be expressed as

$$E_N(f) = E_i(f)\sqrt{1-R}\exp(j\beta n x_0)\exp(jN\beta n\Delta L)R^{2N+1}\sqrt{1-R} \tag{2.2.2}$$

Adding all the components together at the FPI output, the total output electrical field is

$$\begin{aligned} E_{out}(f) &= E_i(f)(1-R)\exp(j\beta n x_0)\sum_{m=0}^{\infty}\exp(jm\beta n\Delta L)R^m \\ &= E_i(f)\frac{(1-R)\exp(j\beta n x_0)}{1-\exp(j\beta n\Delta L)R} \end{aligned} \tag{2.2.3}$$

Because of the coherent interference between various output ray traces, the transfer function of an FPI becomes frequency-dependent. The field transfer function of the FPI is then

$$H(f) = \frac{E_{out}(f)}{E_i(f)} = \frac{(1-R)\exp\left(j\frac{2\pi f}{c}n x_0\right)}{1-\exp\left(j\frac{2\pi f}{c}n\Delta L\right)R} \tag{2.2.4}$$

The power transfer function is the square of the field transfer function:

$$T_{FP}(f) = |H(f)|^2 = \frac{(1-R)^2}{(1-R)^2 + 4R\sin^2\left(\frac{2\pi f}{2c}n\Delta L\right)} \tag{2.2.5}$$

Equivalently, this power transfer function can also be expressed using a signal wavelength as the variable:

$$T_{FP}(\lambda) = \frac{(1-R)^2}{(1-R)^2 + 4R\sin^2\left(\frac{2\pi dn\cos\alpha}{\lambda}\right)} \tag{2.2.6}$$

For a fixed signal wavelength, Equation 2.2.6 is a periodic transfer function of the incidence angle α. For example, if a point light source is illuminated on an FPI, as shown in Figure 2.2.2, a group of bright rings will appear on the screen behind the FPI. The diameters of the rings depend on the thickness of the FPI as well as on the signal wavelength. To provide a quantitative demonstration, Figure 2.2.3 shows the FPI power transfer function versus the incidence angle at the input. To obtain Figure 2.2.3, the power reflectivity of the mirror $R = 0.5$, mirror separation $d = 300\ \mu m$, and media refractive index $n = 1$ were used. The solid line shows the transfer function with the signal wavelength at $\lambda = 1540.4$ nm. At this wavelength the center of the screen is dark. When the signal wavelength is changed to 1538.5 nm, as shown by the dashed line, the center of the screen becomes bright. Note too that with the increase of the mirror separation d, the angular separation between transmission peaks will become smaller and the rings on the screen will become more crowded.

In fiber-optic FPI applications, because of the small numerical aperture of the single-mode fiber, collimated light is usually used, and one can assume that

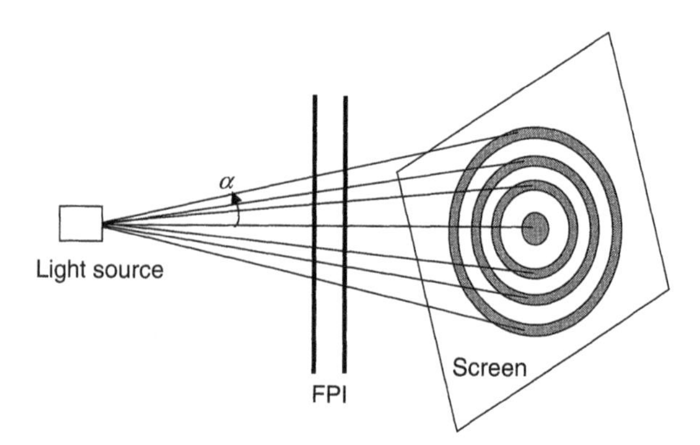

Figure 2.2.2 Illustration of a circular fringes pattern when a noncollimated light source is launched onto a screen through an FPI.

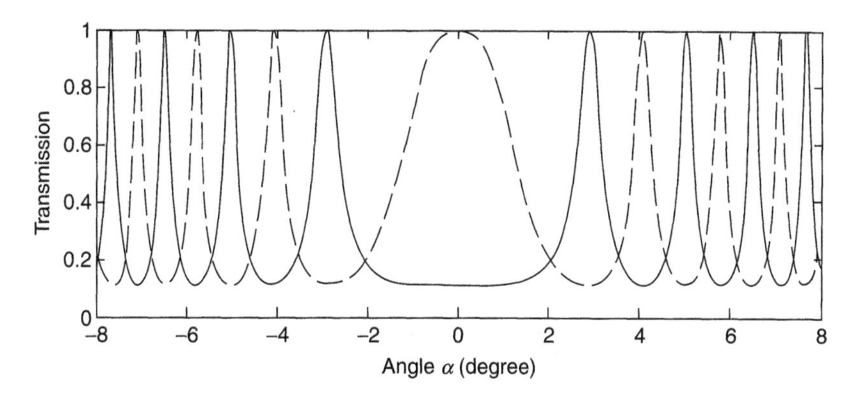

Figure 2.2.3 Transmission versus the beam incident angle to the FPI with mirror reflectivity $R = 0.5$, mirror separation $d = 300$ μm and media refractive index $n = 1$. Calculations were made at two wavelengths: $\lambda = 1538.5$ nm (dashed line) and $\lambda = 1540.4$ nm (solid line).

the incidence angle is approximately $\alpha = 0$. This is known as a *collinear configuration*. With $\alpha = 0$, Equation 2.2.6 can be simplified to

$$T_{FP}(\lambda) = \frac{(1-R)^2}{(1-R)^2 + 4R\sin^2(2\pi nd/\lambda)} \tag{2.2.7}$$

In this simple case, the power transmission is a periodic function of the signal optical frequency. Figure 2.2.4 shows two examples of power transfer functions in a collinear FPI configuration where the mirror separation is $d = 5$ mm, the media refractive index is $n = 1$, and the mirror reflectivity is $R = 0.95$ for Figure 2.2.4(a) and $R = 0.8$ for Figure 2.2.4(b). With a higher mirror reflectivity, the transmission peaks become narrower and the transmission minima become lower. Therefore the FPI has better frequency selectivity with high mirror reflectivity.

In the derivation of FPI transfer function, we have assumed that there is no loss within the cavity, but in practice, optical loss always exists. After each roundtrip in the cavity, the reduction in the optical field amplitude should be ηR instead of just R. The extra loss factor η may be introduced by cavity material absorption and beam misalignment. The major cause of beam misalignment is that the two mirrors that form the FPI are not exactly parallel. As a consequence, after each roundtrip the beam exit from the FPI only partially overlaps with the previous beam; therefore the strength of interference between them is reduced. For simplicity, we still use R to represent the FPI mirror reflectivity; however, we have to bear in mind that this is an effective reflectivity which includes the effect of cavity loss. The following are a few parameters that are often used to describe the properties of an FPI.

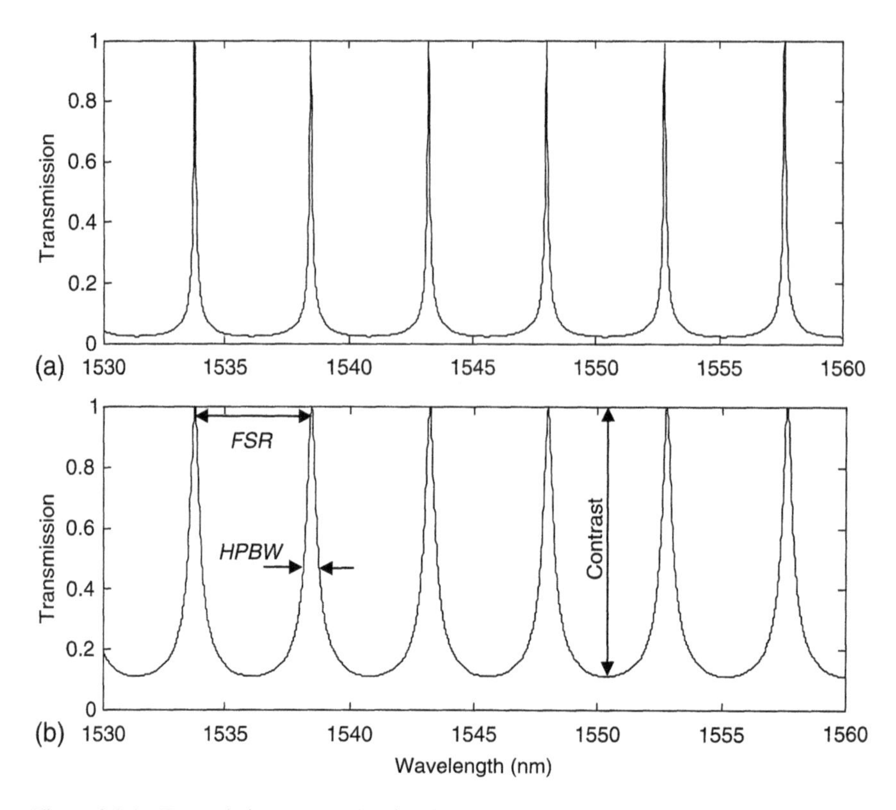

Figure 2.2.4 Transmission versus the signal wavelength with mirror separation $d = 5$ mm, media refractive index $n = 1$, and mirror reflectivity $R = 0.95$ (a) and $R = 0.8$ (b).

2.2.1.1 Free Spectral Range (FSR)

FSR is the frequency separation Δf between adjacent transmission peaks of an FPI. For a certain incidence angle α, the FSR can be found from transfer function 2.2.6 as

$$FSR = \Delta f = \frac{c}{2nd\cos\alpha} \tag{2.2.8}$$

FSR is inversely proportional to the cavity optical length *nd*. For a simple Febry-Perot type laser diode, if the cavity length is $d = 300$ μm and the refractive index is $n = 3.5$, the FSR of this laser diode is approximately 143 GHz. If the laser operates in a 1550 nm wavelength window, this FSR is equivalent to 1.14 nm. This is the mode spacing of the laser, as we discussed in Section 1.1.

2.2.1.2 Half-Power Bandwidth (HPBW)

HPBW is the width of each transmission peak of the FPI power transfer function, which indicates the frequency selectivity of the FPI. From Equation 1.2.6, if we assume at $f = f_{1/2}$ the transfer function is reduced to $^1/_2$ its peak value, then

$$4R\sin^2\left(\frac{2\pi f_{1/2} nd\cos\alpha}{c}\right) = (1-R)^2$$

Assuming that the transmission peak is narrow enough, which is the case for most of the FPIs, $\sin(2\pi f_{1/2} nd\cos\alpha/c) \approx (2\pi f_{1/2} nd\cos\alpha/c)$ and $f_{1/2}$ can be found as

$$f_{1/2} = \frac{(1-R)c}{4\pi nd\sqrt{R}\cos\alpha}$$

Therefore the full width of the transmission peak is

$$HPBW = 2f_{1/2} = \frac{(1-R)c}{2\pi nd\sqrt{R}\cos\alpha} \qquad (2.2.9)$$

In most applications, large FSR and small HPBW are desired for good frequency selectivity. However, these two parameters are related by a FPI quality parameter known as *finesse*.

2.2.1.3 Finesse

Finesse of an FPI is related to the percentage of the transmission window within the free spectral range. It is defined by the ratio between the FSR and HPBW:

$$F = \frac{FSR}{HPBW} = \frac{\pi\sqrt{R}}{1-R} \qquad (2.2.10)$$

Finesse is a quality measure of FPI that depends only on the effective mirror reflectivity R. Technically, very high R is hard to obtain, because the effective reflectivity not only depends on the quality of mirror itself, it also depends on the mechanical alignment between the two mirrors. Current state-of-the-art technology can provide finesse of up to a few thousand.

2.2.1.4 Contrast

Contrast is the ratio between the transmission maximum and the transmission minimum of the FPI power transfer function. It specifies the ability of wavelength discrimination if the FPI is used as an optical filter. Again, from the transfer function 2.2.6, the highest transmission is $T_{max} = 1$ and the minimum transmission is $T_{min} = (1 - R)^2/[(1 - R)^2 + 4R]$. Therefore the contrast of the FPI is

$$C = \frac{T_{max}}{T_{min}} = 1 + \frac{4R}{(1-R)^2} = 1 + \left(\frac{2F}{\pi}\right)^2 \tag{2.2.11}$$

The physical meanings of FSR, HPBW, and contrast are illustrated in Figure 2.2.4(b).

Example 2.2

The reflection characteristics of a Fabry-Perot filter can be used as a narrowband notch filter, as illustrated in Figure 2.2.5. To achieve a power rejection ratio of 20dB, what is the maximally allowed loss of the mirrors?

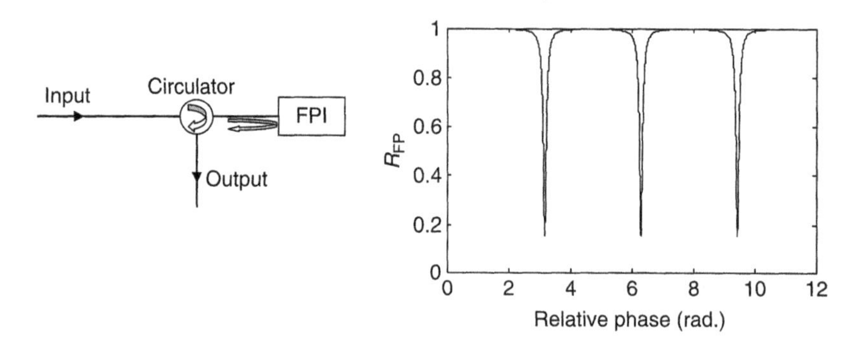

Figure 2.2.5 Using FPI as a notch filter.

Solution:

If the FP interferometer is ideal and the mirror has no excess loss, the wavelength-dependent power transmission of an FP filter is given by Equation 2.2.7 as

$$T(\lambda) = \frac{(1-R)^2}{(1-R)^2 + 4R\sin^2(2\pi nd/\lambda)}$$

and the reflection of the FPI is

$$R_{FP}(\lambda) = 1 - T(\lambda) = \frac{4R\sin^2(2\pi nd/\lambda)}{(1-R)^2 + 4R\sin^2(2\pi nd/\lambda)}$$

If this is true, the notch filter would be ideal because $R_{FP}(\lambda) = 0$ whenever $(2nd/\lambda)$ is an integer. However, in real devices, absorption, scattering, and no-ideal beam collimation would contribute to reflection losses; therefore the wavelength-dependent power transmission of an FP filter becomes

$$T(\lambda) = \frac{(1-R)^2}{(1-R\eta)^2 + 4R\eta\sin^2(2\pi nd/\lambda)}$$

where $\eta < 1$ is the excess power loss of each reflection on the mirror. Then the FPI power reflectivity is

$$R_{FP}(\lambda) = \frac{(1-R\eta)^2 - (1-R)^2 + 4R\eta\sin^2(2\pi nd/\lambda)}{(1-R\eta)^2 + 4R\eta\sin^2(2\pi nd/\lambda)}$$

Obviously, the minimum reflection of the FPI happens at wavelengths where $\sin^2(2\pi nd/\lambda) = 0$; therefore the value of minimum reflection is

$$R_{FP}(\lambda) = 1 - \frac{(1-R)^2}{(1-R\eta)^2}$$

In order for the minimum reflectivity to be –20 dB (0.01), the reflection loss has to be

$$\eta \geq \frac{1}{R}\left[1 - \frac{(1-R)}{\sqrt{0.99}}\right]$$

Suppose that $R = 0.9$. The requirement for the excess reflection loss of the mirror is $\eta > 0.9994$. That means it allows for only 0.12 percent power loss in each roundtrip in the FP cavity, which is not usually easy to achieve.

2.2.2 Scanning FPI Spectrum Analyzer

Optical spectrum analyzer based on scanning FPI is a popular optical instrument for its superior spectral resolution. Typically, the resolution of a grating-based OSA is on the order of 0.08 nm, which is about 10 GHz in the 1550 nm wavelength window. A commercial scanning FPI can easily provide spectral resolution better than 10 MHz, which is approximately 0.08 pm in a 1550 nm wavelength window. However, the major disadvantage of an FPI-based OSA is its relatively narrow wavelength coverage. Since in fiber-optic systems, collinear configuration is often used ($\alpha = 0$), we only consider this case in the following description; the simplified FPI transfer function is given by Equation 2.2.7.

The power transfer function of an FPI is periodic in the frequency domain. From Equation 2.2.7, the wavelength λ_m corresponding to the mth transmission peak can be found as

$$\lambda_m = \frac{2nd}{m} \tag{2.2.12}$$

By changing the length of the cavity, this peak transmission wavelength will movc. When the cavity length is scanned by an amount of

$$\delta d = \frac{\lambda_m}{2n} \tag{2.2.13}$$

the mth transmission peak frequency will be scanned over one entire free spectral range. This is the basic mechanism of making an OSA using FPI. To measure the signal spectral density over a wavelength band corresponding to an FSR, we only need to sweep the cavity length for approximately half the wavelength. This mechanical scanning can usually be accomplished using a voltage-controlled piezo-electric transducer (PZT); therefore the mirror displacement (or equivalently, the change in cavity length *d*) is linearly proportional to the applied voltage.

Figure 2.2.6 shows a block diagram of optical spectrum measurement using a scanning FPI. The FPI is driven by a sawtooth voltage waveform, and therefore the FPI mirror displacement is linearly scanned. As a result, the peak transmission frequency of the FPI transfer function is also linearly scanned. A photodiode is used at the FPI output to convert the optical signal into electrical waveform, which is then displayed on an oscilloscope. To synchronize with the mirror scanning, the oscilloscope is triggered by the sawtooth waveform. As shown in Figure 2.2.7, if the swing of the driving voltage is between V_0 and V_1, the frequency scan of the transmission peak will be from f_0 and f_1 with

$$f_1 - f_0 = \eta_{PZT}(V_1 - V_0)\frac{2n \cdot FSR}{\lambda} \tag{2.2.14}$$

where η_{PZT} is the PZT efficiency, which is defined as the ratio between mirror displacement and the applied voltage.

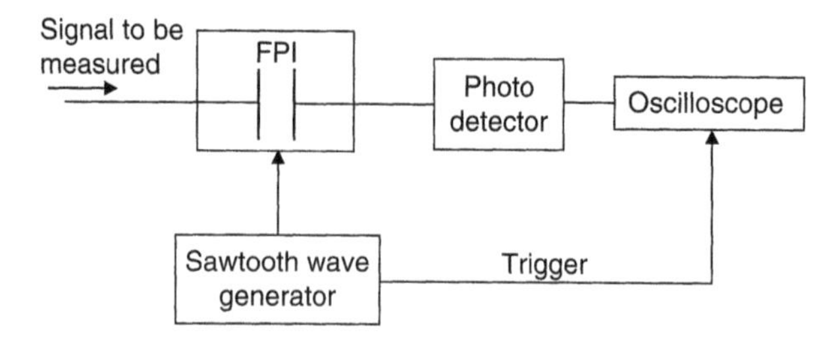

Figure 2.2.6 Block diagram of optical spectrum measurement using a scanning FPI.

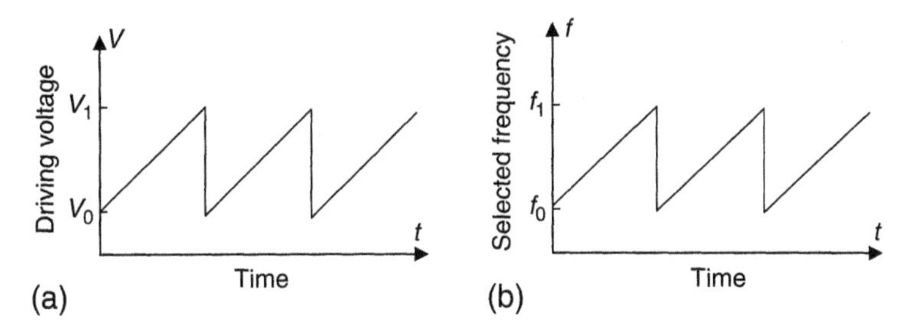

Figure 2.2.7 (a) FPI driving voltage waveform and (b) transmission peak frequency versus time.

In this spectrum measurement system, the spectral resolution is determined by the width of the FPI transmission peak and the frequency coverage is limited by the FSR of the FPI. Because of the periodic nature of the FPI transfer function, the spectral content of the optical signal to be measured has to be limited within one FSR of the FPI. All the spectral components outside an FSR will be folded together, thus introducing measurement errors. This concept is illustrated in Figure 2.2.8, where frequency scanning of three FPI transmission peaks is displayed. The two signal spectral components are separated wider than an FSR, and they are selected by two different transmission peaks of the FPI. As a result, they are folded together in the measured oscilloscope trace. Because of the limited frequency coverage, a scanning FPI is often used to measure narrow optical spectra where high spectral resolution is required.

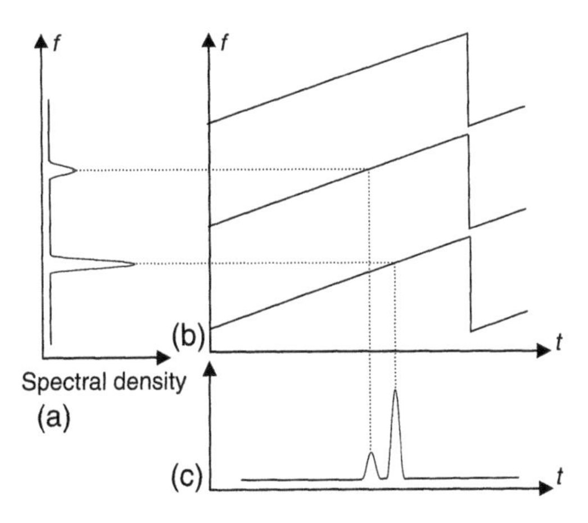

Figure 2.2.8 Illustration of spectrum folding when signal spectrum exceeds an FSR. (a) Original signal optical spectrum, (b) FPI response, and (c) measured oscilloscope trace.

To convert the measured oscilloscope waveform $p(t)$ into an optical spectrum $p(f)$, a precise calibration is required to determine the relationship between time t and frequency f. Theoretically, this can be done using Equation 2.2.14 and the knowledge of the sawtooth waveform V(t). In practice, this calibration can easily be done experimentally. As shown in Figure 2.2.9, we can apply a sawtooth waveform with a high swing voltage so that the frequency of each FPI transmission peak swings wider than an FSR. A single-wavelength source needs to be used in the measurement. Since the FPI is scanning for more than an FSR, two signal peaks can be observed on the oscilloscope trace corresponding to the spectrum measured by two adjacent FPI transmission peaks.

Obviously the frequency separation between these two peaks on the oscilloscope trace should correspond to one free spectral range of the FPI. If the measured time separation between these two peaks is Δt on the oscilloscope, the conversion between the time scale of the oscilloscope and the corresponding frequency should be

$$f = \frac{FSR}{\Delta t} t \tag{2.2.15}$$

We need to point out that what we have just described is a *relative* calibration of the frequency scale. This allows us to determine the frequency scale for each time division measured on the oscilloscope. However, this does not provide absolute frequency calibration. In practice, since the frequency of an FPI transmission peak can swing across an entire FSR with the cavity length change of merely half a wavelength, absolute wavelength calibration is not a simple task. Cavity length change on the order of a micrometer can easily be introduced by a temperature change or a change in the mechanical stress.

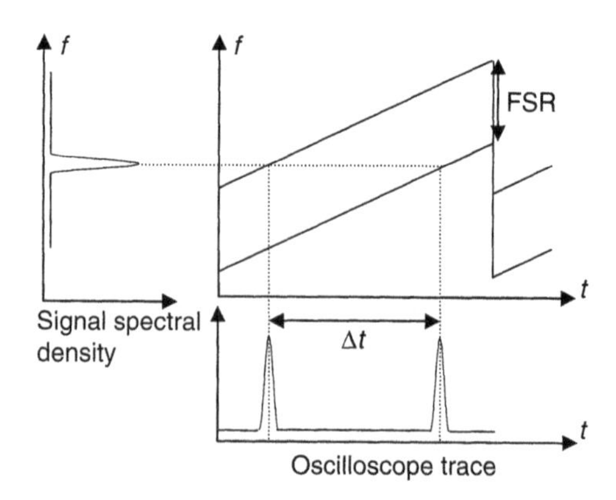

Figure 2.2.9 Calibration procedure converting time to frequency.

However, the relative frequency measurement can still be very accurate because FSR will not be affected by micrometer-level cavity length changes.

Although absolute wavelength calibration can be done by adding a fixed wavelength reference, FPI-based optical spectrometers are traditionally not built to make accurate measurement in the absolute wavelength of the optical signal. Rather, they are often used to measure source linewidth, modulation sidebands, and other situations in which high spectral resolution is required.

2.2.3 Scanning FPI Basic Optical Configurations

There are various types of commercially available scanning FPIs, such as plano-mirror design, confocal-mirror design, and all-fiber design.

Plano-mirror design uses a pair of flat mirrors that are precisely parallel to each other. As shown in Figure 2.2.10, one of the two mirrors is mounted on a PZT so that its position can be controlled by an electrical voltage. The other mirror is attached to a metal ring that is mounted on a mechanical translation stage to adjust the angle of the mirror. The major advantage of this plano-mirror design is its flexibility. The cavity length can be adjusted by sliding one of the mirrors along the horizontal bars of the frame.

However, in this configuration, the optical alignment between the two mirrors is relatively difficult. The requirement for the beam quality of the optical signal is high, especially when the cavity length is long. The finesse of this type of FPI is usually less than 100.

The confocal-mirror design uses a pair of concave mirrors whose radii of curvature are equal to half their separation, resulting in a common focus point in the middle of the cavity, as shown in Figure 2.2.11. An FPI using this confocal configuration usually has much higher finesse compared to plano-mirror configuration. The reason is that the focusing of the incident beam reduces possible

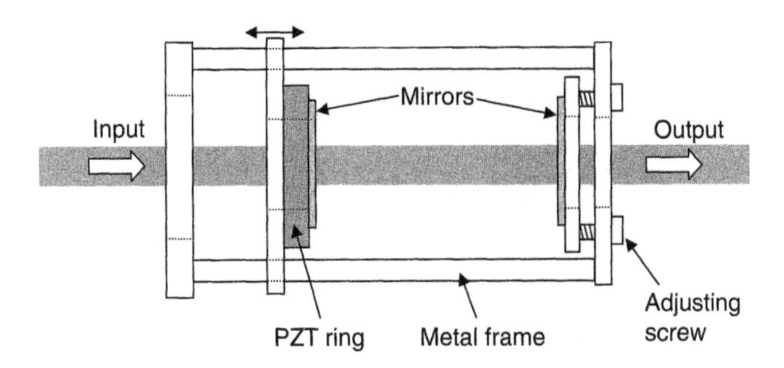

Figure 2.2.10 Configuration of a scanning FPI using plano mirrors.

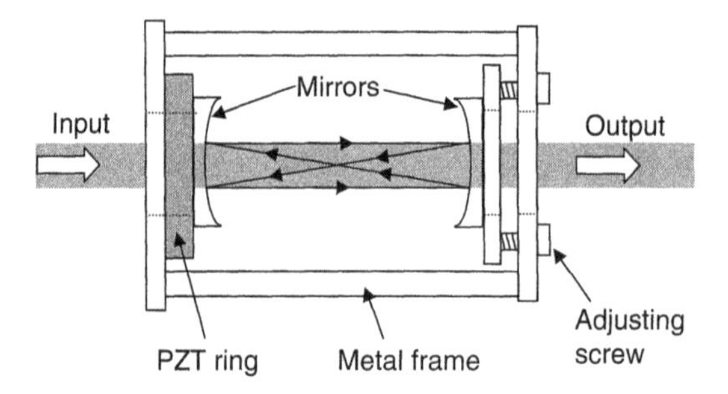

Figure 2.2.11 Configuration of a scanning FPI using concave mirrors.

finesse degradation due to mirror surface imperfections. This configuration also has better tolerance to the quality of the incident beam. However, the cavity length of the FPI is determined by the curvature of the mirror, which cannot be adjusted. Therefore, FPI using confocal mirror design has a fixed free spectral range and is not as flexible as that using plano mirrors.

For an FPI using free-space optics, beam and mirror alignments are usually not easy tasks. Even a slight misalignment may result in significant reduction of the cavity finesse. For applications in fiber-optic systems, an FPI based on all-fiber configuration is attractive for its simplicity and compactness. An example of an all-fiber FPI is shown in Figure 2.2.12 [6]. In this configuration, micro mirrors are made in the fiber and a small air gap between fibers makes it possible for the scanning of cavity length. Antireflection (AR) coating is applied on each side of the air gap to eliminate the unwanted reflection. The entire package of an all-fiber FPI can be made small enough to mount onto a circuit board.

Because the optical signal within the cavity is mostly confined inside single-mode optical fiber and it does not need additional optical alignment, an all-fiber FPI can provide much higher finesse compared to free-space PFIs. A state-of-the-art all-fiber FPI can provide a finesse of as high as several thousand.

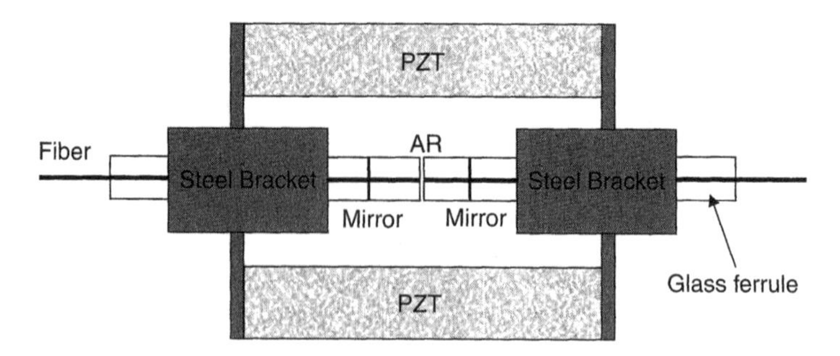

Figure 2.2.12 All-fiber FPI.

2.2.4 Optical Spectrum Analyzer Using the Combination of Grating and FPI

In the last two sections, we have discussed OSAs based on diffractive gratings and FPIs. A grating-based OSA has wide wavelength coverage, but the spectral resolution is limited by the groove-line density of the grating and the size of the collimated optical beam on the grating. A desktop OSA can usually provide 0.05 nm spectral resolution, and a miniature OSA using photodiode array provides about 0.25 nm resolution in a 1550 nm wavelength window. On the other hand, a scanning FPI-based OSA can have much higher spectral resolution by using long cavity length. However, since the transfer function of an FPI is periodic, the spectral coverage is limited to a free spectral range. The ratio between the wavelength coverage and spectral resolution is equal to the finesse.

To achieve both high spectral resolution and wide wavelength coverage, we can design an OSA that combines an FPI and a grating.

The operation principle of the high-resolution OSA using the combination of an FPI and a grating is illustrated in Figure 2.2.13. The optical signal is transformed into discrete narrowband slices by the FPI with the wavelength separation equal to FSR between each other. After the transmission grating, each wavelength slice is dispersed into a light beam at a certain spatial angle with the angular width of $\Delta\phi$. Theoretically, $\Delta\phi$ is determined by the convolution between spectral bandwidth of the FPI and the angle resolution of the grating. Because of this angular width, the beam has a width ΔL when it reaches the surface of the photo-diode array.

It is assumed that each photodiode has a width of d, and then there will be $n = \Delta L/d$ photodiodes simultaneously illuminated by this light beam. Adding up the photocurrents generated by all the n associated photodiodes,

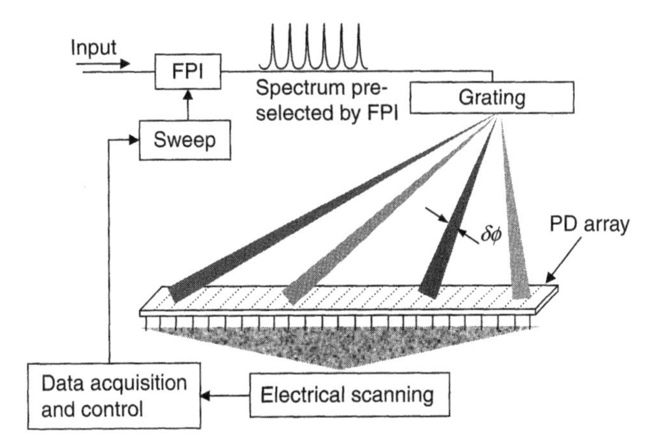

Figure 2.2.13 High-resolution OSA design combining an FPI and a grating.

an electrical signal can be obtained that is linearly proportional to the optical power within the optical bandwidth selected by each transmission peak of the FPI. As shown in the example of Figure 2.2.13, there are m light beams (the figure shows $m = 4$), each corresponding to a specific transmission peak of the FPI transfer function. Therefore, for each FPI setting, the signal optical spectral density can be measured simultaneously at m separate wavelengths. If each transmission peak of the FPI is linearly swept across an FSR, the measurement will be able to cover a continuous wavelength range of m times FSR. In this OSA, the frequency sweep of FPI is converted into angular sweep of the light beams at the output of the grating and thus the spatial position on the surface of the photodiode array. With proper signal processing, this OSA is able to provide a high spectral resolution, as well as a wide spectral coverage [7, 8].

2.3 MACH-ZEHNDER INTERFEROMETERS

The Mach-zehnder interferometer (MZI) is one of the oldest optical instruments, in use for more than a century. The basic configuration of an MZI is shown in Figure 2.3.1, which consists of two optical beam splitters (combiners) and two mirrors to alter beam directions.

The beam splitter splits the incoming optical signal into two equal parts. After traveling through two separate arms, these two beams recombine at the beam combiner. The nature of beam interference at the combiner depends, to a large extent, on the coherence length of the optical signal which will be discussed later in sec.2.5 (e.g. eq.(2.5.7)). If the path length difference between these two arms is shorter than the coherent length of the optical signal, the two beams interfere with each other at the combiner. If the two beams are in phase at the combiner, the output optical power is equal to the input power; otherwise, if they are antiphase, the output optical power is equal to zero. In fiber-optical systems, the beam splitter and the combiner can be replaced by fiber couplers; therefore all-fiber MZIs can be made. Because of the wave-guiding mechanism, fiber-based MZI can be made much more compact than an MZI using free space optics. In each case, the basic components to construct an MZI are optical couplers and optical delay lines.

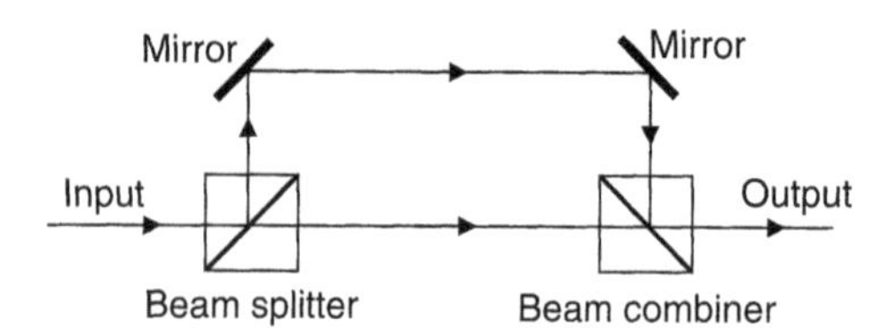

Figure 2.3.1 Basic configuration of a Mach-zehnder interferometer.

2.3.1 Transfer Matrix of a 2 × 2 Optical Coupler

In Figure 2.3.1, the beam splitter and the beam combiner are essentially the same device and they can be generalized as 2 × 2 optical couplers. As shown in Figure 2.3.2, a 2 × 2 optical coupler has two inputs and two outputs.

The input/output relationship of both free-space and fiber-based 2 × 2 couplers shown in Figure 2.3.2 can be represented as a transfer matrix [9, 10]:

$$\begin{bmatrix} b_1 \\ b_2 \end{bmatrix} = \begin{bmatrix} s_{11} & s_{12} \\ s_{21} & s_{22} \end{bmatrix} \begin{bmatrix} a_1 \\ a_2 \end{bmatrix} \tag{2.3.1}$$

where a_1, a_2 and b_1, b_2 are electrical fields at the two input and two output ports, respectively. This device is reciprocal, which means that the transfer function will be identical if the input and the output ports are exchanged; therefore

$$s_{12} = s_{21} \tag{2.3.2}$$

We also assume that the coupler has no loss so that energy conservation applies; therefore the total output power is equal to the total input power:

$$|b_1|^2 + |b_2|^2 = |a_1|^2 + |a_2|^2 \tag{2.3.3}$$

Equation 2.3.1 can be expanded into

$$|b_1|^2 = |s_{11}|^2|a_1|^2 + |s_{12}|^2|a_2|^2 + s_{11}s_{12}^*a_1a_2^* + s_{11}^*s_{12}a_1^*a_2 \tag{2.3.4}$$

and

$$|b_2|^2 = |s_{21}|^2|a_1|^2 + |s_{22}|^2|a_2|^2 + s_{21}s_{22}^*a_1a_2^* + s_{21}^*s_{22}a_1^*a_2 \tag{2.3.5}$$

Where * indicates complex conjugate. Using the energy conservation condition, Equations 2.3.4 and 2.3.5 yield

$$|s_{11}|^2 + |s_{12}|^2 = 1 \tag{2.3.6}$$

$$|s_{21}|^2 + |s_{22}|^2 = 1 \tag{2.3.7}$$

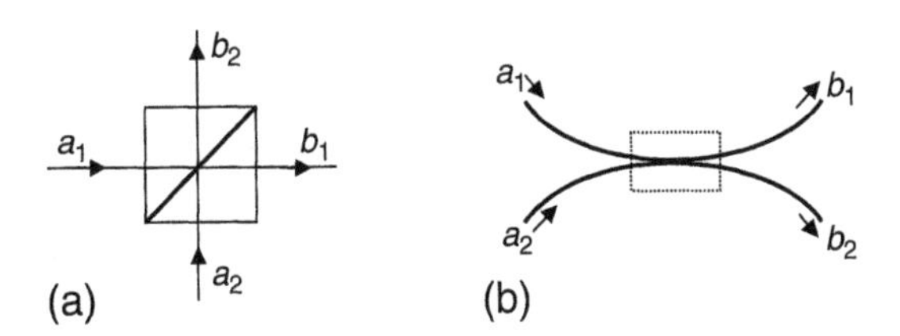

Figure 2.3.2 2 x 2 optical couplers with (a) free-space optics and (b) fiber optics.

and

$$s_{11}s_{12}^* + s_{21}s_{22}^* = 0 \tag{2.3.8}$$

Since $s_{12} = s_{21}$, Equations 2.3.6 and (2.3.7) give

$$|s_{11}|^2 = |s_{12}|^2 \tag{2.3.9}$$

For a coupler, if ε is the fraction of the optical power coupled from input port 1 to output port 2, the same coupling ratio will be from input port 2 to output port 1. For the optical field, this coupling ratio is $\sqrt{\varepsilon}$. Then the optical field coupling from input port 1 to output port 1 or from input port 2 to output port 2 will be $\sqrt{1-\varepsilon}$. If we assume $s_{11} = \sqrt{1-\varepsilon}$ is real, $s_{22} = \sqrt{1-\varepsilon}$ should also be real because of the symmetry assumption. Then we can assume there is a phase shift for the cross-coupling term, $s_{12} = s_{21} = \sqrt{\varepsilon}e^{j\Phi}$. The transfer matrix is then

$$\begin{bmatrix} b_1 \\ b_2 \end{bmatrix} = \begin{bmatrix} \sqrt{1-\varepsilon} & e^{j\Phi}\sqrt{\varepsilon} \\ e^{j\Phi}\sqrt{\varepsilon} & \sqrt{1-\varepsilon} \end{bmatrix} \begin{bmatrix} a_1 \\ a_2 \end{bmatrix} \tag{2.3.10}$$

Using Equation 2.3.8, we have

$$e^{-j\Phi} + e^{j\Phi} = 0 \tag{2.3.11}$$

The solution of Equation 2.3.11 is $\Phi = \pi/2$ and $e^{j\Phi} = j$. Therefore the transfer matrix of a 2×2 optical coupler is

$$\begin{bmatrix} b_1 \\ b_2 \end{bmatrix} = \begin{bmatrix} \sqrt{1-\varepsilon} & j\sqrt{\varepsilon} \\ j\sqrt{\varepsilon} & \sqrt{1-\varepsilon} \end{bmatrix} \begin{bmatrix} a_1 \\ a_2 \end{bmatrix} \tag{2.3.12}$$

An important conclusion of Equation 2.3.12 is that there is a 90-degree relative phase shift between the direct pass (s_{11}, s_{22}) and cross-coupling (s_{12}, s_{21}). This property is important for several applications including phase-balanced detection in coherent receivers which we discuss in Section 2.7.

2.3.2 Transfer Function of an MZI

An MZI is composed of two optical couplers and optical delay lines. It can be made by free space optics as well as guided-wave optics. Using 2×2 couplers, a general MZI may have two inputs and two outputs, as shown in Figure 2.3.3.

The transfer function of an MZI can be obtained by cascading the transfer functions of two optical couplers and that of the optical delay line. Suppose the optical lengths of the delay lines in *arm1* and *arm2* are n_1L_1 and n_2L_2, respectively, the transfer matrix of the two delay lines is simply

$$\begin{bmatrix} c_1 \\ c_2 \end{bmatrix} = \begin{bmatrix} \exp(-j\phi_1) & 0 \\ 0 & \exp(-j\phi_2) \end{bmatrix} \begin{bmatrix} b_1 \\ b_2 \end{bmatrix} \tag{2.3.13}$$

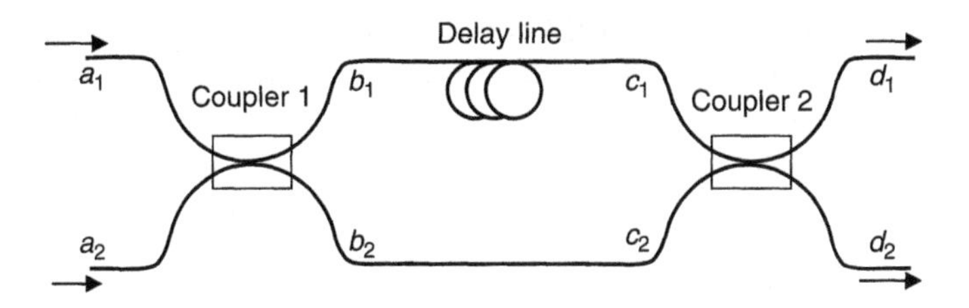

Figure 2.3.3 Illustration of an MZI with two inputs and two outputs.

where $\phi_1 = (2\pi/\lambda)n_1L_1$ and $\phi_2 = (2\pi/\lambda)n_2L_2$ are the phase delays of the two delay lines. If we further assume that the two optical couplers are identical with a power splitting ratio of ε,

$$\begin{bmatrix} d_1 \\ d_2 \end{bmatrix} = \begin{bmatrix} \sqrt{1-\varepsilon} & j\sqrt{\varepsilon} \\ j\sqrt{\varepsilon} & \sqrt{1-\varepsilon} \end{bmatrix} \begin{bmatrix} e^{-j\phi_1} & 0 \\ 0 & e^{-j\phi_2} \end{bmatrix} \begin{bmatrix} \sqrt{1-\varepsilon} & j\sqrt{\varepsilon} \\ j\sqrt{\varepsilon} & \sqrt{1-\varepsilon} \end{bmatrix} \begin{bmatrix} a_1 \\ a_2 \end{bmatrix} \quad (2.3.14)$$

To achieve the highest extinction ratio, most MZIs use a 50 percent power splitting ratio for the optical couplers. In that case, $\varepsilon = 0.5$ and the equation can be simplified as,

$$\begin{bmatrix} d_1 \\ d_2 \end{bmatrix} = \frac{1}{2} \begin{bmatrix} (e^{-j\phi_1} - e^{-j\phi_2}) & j(e^{-j\phi_1} + e^{-j\phi_2}) \\ j(e^{-j\phi_1} + e^{-j\phi_2}) & -(e^{-j\phi_1} - e^{-j\phi_2}) \end{bmatrix} \begin{bmatrix} a_1 \\ a_2 \end{bmatrix} \quad (2.3.15)$$

If the input optical signal is only at the input port 1 and the input port 2 is disconnected ($a_2 = 0$), the optical field at the two output ports will be

$$d_1 = \frac{1}{2}(e^{-j\phi_1} - e^{-j\phi_2})a_1 = -e^{-j\phi_0} \sin\left(\frac{\Delta\phi}{2}\right) a_1 \quad (2.3.16)$$

and

$$d_2 = \frac{j(e^{-j\phi_1} + e^{-j\phi_2})}{2} a_1 = je^{-j\phi_0} \cos\left(\frac{\Delta\phi}{2}\right) a_1 \quad (2.3.17)$$

where $\phi_0 = (\phi_1 + \phi_2)/2$ is the average phase delay and $\Delta\phi = (\phi_1 - \phi_2)$ is the differential phase shift of the two MZI arms. Then the optical power transfer function from input port 1 to output port 1 is

$$T_{11} = \left.\frac{d_1}{a_1}\right|_{a_2=0} = \sin^2\left[\frac{\pi f}{c}(n_2L_2 - n_1L_1)\right] \quad (2.3.18)$$

and the optical power transfer function from input port 1 to output port 2 is

$$T_{12} = \left.\frac{d_2}{a_1}\right|_{a_2=0} = \cos^2\left[\frac{\pi f}{c}(n_2L_2 - n_1L_1)\right] \quad (2.3.19)$$

where $f = c/\lambda$ is the signal optical frequency and c is the speed of light. Obviously, the optical powers coming out of the two output ports are complementary such that $T_{11} + T_{12} = 1$, which results from energy conservation, since we assumed that there is no loss.

The transmission efficiencies T_{11} and T_{12} of an MZI are the functions of both the wavelength λ and the differential optical length $\Delta l = n_2 L_2 - n_1 L_1$ between the two arms. Therefore, an MZI can usually be used in two different categories: optical filter and electro-optic modulator.

2.3.3 MZI Used as an Optical Filter

The application of MZI as electro-optic modulator was discussed in Chapter 1, where the differential arm length can be modulated to modulate the optical transfer function. However, a more classic application of MZI is as an optical filter, where wavelength-dependent characteristics of T_{11} and T_{12} are used, as illustrated in Figure 2.3.4(a). For example, if we assume that the two arms have the same refractive index $n_1 = n_2 = 2.387$ and the physical length difference between the two arms is $L_1 - L_2 = 0.5$ mm, the wavelength spacing between adjacent transmission peaks of each output arm is

$$\Delta\lambda \approx \frac{\lambda^2}{n_2 L_2 - n_1 L_1} = 2\ nm$$

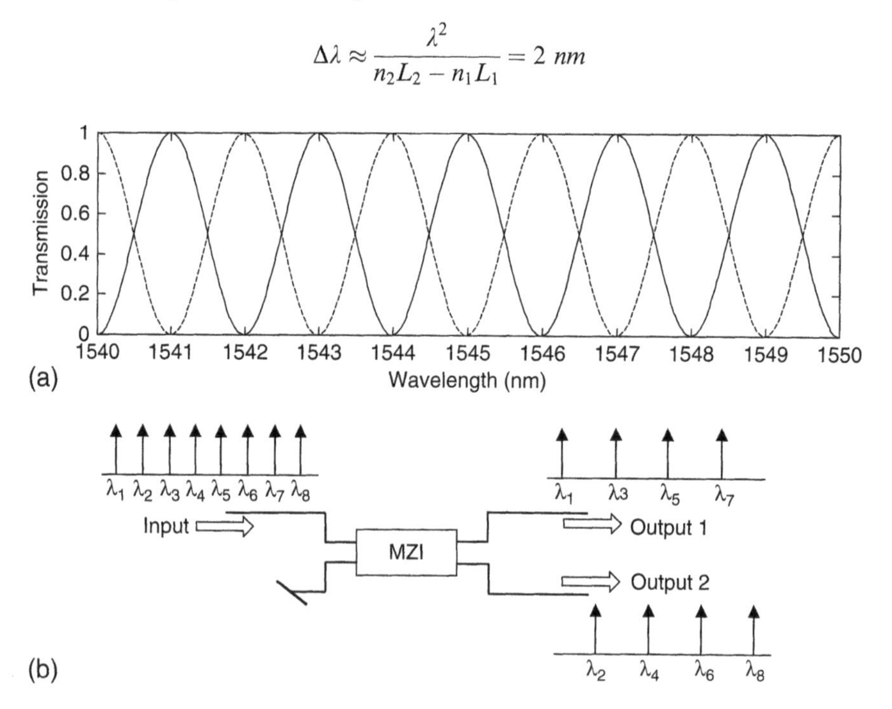

Figure 2.3.4 (a) MZI power transfer functions T_{11} (solid line) and T_{12} (dashed line) with $n_1 = n_2 = 2.387$, $L_1 - L_2 = 0.5$ mm. (b) An MZI is used as a wavelength interleaver.

whereas the wavelength spacing between adjacent transmission peaks of the two output ports is equal to $\Delta\lambda/2 = 1\ nm$. In addition to using an MZI as an optical filter, it is also often used as a wavelength interleaver, as shown in Figure 2.3.4(b) [11, 12]. In this application, a wavelength division multiplexed optical signal with narrow channel spacing can be interleaved into two groups of optical signals each with a doubled channel spacing compared to the input.

Example 2.3

The fiber ring resonator shown in Figure 2.3.5 is made with a 2 × 2 fiber directional coupler. Suppose that the power-splitting ratio of the fiber coupler is ε, the length of the fiber ring is L, and the refractive index of the fiber is n:

1. Derive the power transfer function $T(f) = P_{out}(f)/P_{in}(f)$ for the resonator.
2. In order to have high finesse, what coupling ratio α of the 2 × 2 fiber coupler you would like to choose?
3. Discuss the similarity and difference between this transfer function and that of a Fabry-Perot filter.

Solution:

Based on Equation 2.3.12, the transfer function of a 2 × 2 directional coupler is

$$b_1 = a_1\sqrt{1-\varepsilon} + ja_2\sqrt{\varepsilon}$$
$$b_2 = ja_1\sqrt{\varepsilon} + a_2\sqrt{1-\varepsilon}$$

The fiber ring then connects the output port b_2 to the input port a_2 and therefore

$$a_2 = \eta b_2 e^{j\varphi}$$

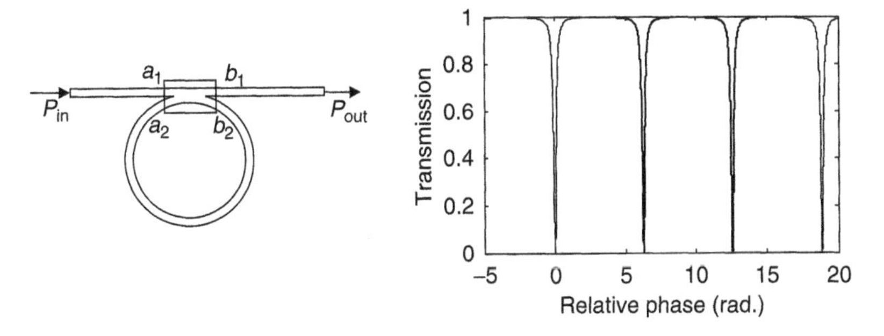

Figure 2.3.5 Fiber-based ring resonator (left) and transfer function (right).

where, η is the transmission coefficient of the fiber, $\varphi = 2\pi f\tau$ is the phase delay introduced by the fiber ring, and $\tau = nL/c$ is the time delay of the ring. Combine these three equations:

$$T(\varphi) = \left|\frac{b_1}{a_1}\right|^2 = \left|\sqrt{1-\varepsilon} - \frac{\varepsilon\eta e^{j\varphi}}{1 - e^{j\varphi}\eta\sqrt{1-\varepsilon}}\right|^2$$

It is easy to find that if the fiber loss in the ring is neglected, $\eta = 1$, the power transfer function is independent of the phase delay of the ring, $T(\varphi) \equiv 1$. In practice, any transmission media would have some loss, and $\eta < 1$, even though it is usually very close to unity.

Figure 2.3.5 shows the transfer function with fiber coupler power splitting ratio $\varepsilon = 10$ percent and the transmission coefficient $\eta = 99.9$ percent. Obviously, this is a periodic notch filter. The transmission stays near to 100 percent for most of the phases except at the resonance frequencies of the ring, where the power transmission is minimized. These transmission minima correspond to the ring phase delay of $\varphi = 2m\pi$, where m is an integer, and therefore the FSR of this ring resonator filter is $\Delta f = nL/c$. The value of the minimum transmission can be found easily as

$$T_{\min} = \frac{1 - \varepsilon - 2\eta\sqrt{1-\varepsilon} + \eta^2}{1 - 2\eta\sqrt{1-\varepsilon} + \eta^2(1-\varepsilon)}$$

From a notch filter application point of view, a high extinction ratio requires the minimum transmission $T_{\min}$ to be as small as possible. Figure 2.3.6 shows the minimum power transmission of a ring resonator-based filter as the function of the power-splitting ratio of the fiber coupler.

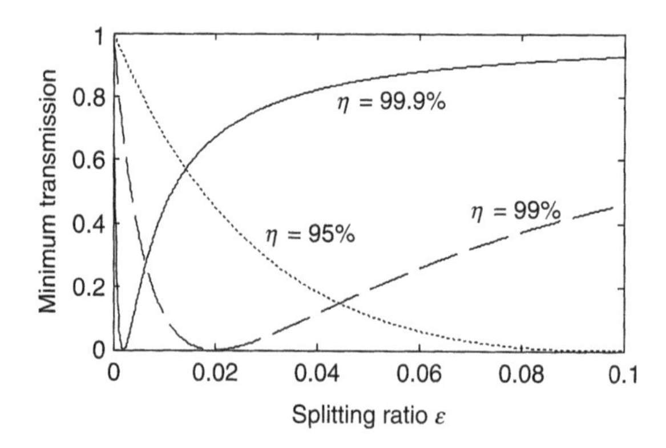

Figure 2.3.6 Minimum power transmission of a ring resonator-based filter.

Three different loss values of the fiber ring are used. Obviously the optimum power-splitting ratio depends on the loss of the fiber ring. In fact, in order for the minimum transmission to be zero, $\eta = \sqrt{1-\varepsilon}$ is required.

Similar to the characterization of an FPI, finesse is also an important parameter to specify a ring resonator. The finesse, also known as the Q-factor in a ring resonator, is defined by the FSR divided by the FWHM width of the transmission notch. If we use the optimized value $\eta = \sqrt{1-\varepsilon}$, the ring resonator transfer function will be simplified to

$$T(\varphi) = \left| \frac{\eta(1 - e^{j\varphi})}{1 - \eta^2 e^{j\varphi}} \right|^2$$

Assuming that at a certain phase, φ_Δ, this power transfer function is $T(\varphi_\Delta) = 0.5$, we can find that

$$\varphi_\Delta = cos^{-1}\left(\frac{1 - 4\eta^2 + \eta^4}{2\eta^2}\right)$$

Since the FSR corresponds to 2π in phase, the finesse of the resonator can easily be found as

$$\text{Finesse} = \frac{2\pi}{2\varphi_\Delta} = \frac{\pi}{\cos^{-1}\left(\frac{1 - 4\eta^2 + \eta^4}{2\eta^2}\right)}$$

Figure 2.3.7 shows the calculated finesse as the function of the round-trip loss of the fiber ring. Apparently high finesse requires low loss of the optical ring. In practice this loss can be introduced either by the fiber loss or by the excess loss of the directional coupler.

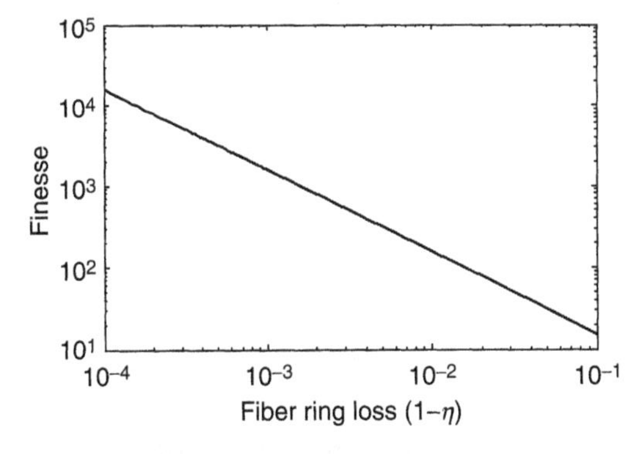

Figure 2.3.7 Calculated finesse as a function of the fiber ring loss.

From an application point of view, the biggest problem with the fiber-based ring resonator is probably the birefringence in the fiber. If there is a polarization rotation after a roundtrip propagation in the ring, the overall transfer function will be affected. Therefore, planar lightwave circuit technology might be more appropriate to make ring-based resonators compared to all-fiber devices.

2.4 MICHELSON INTERFEROMETERS

The optical configuration of a Michelson interferometer is similar to that of an MZI except that only one beam splitter is used and the optical signal passes bidirectionally in the interferometer arms. The basic optical configuration of a Michelson interferometer is shown in Figure 2.4.1. It can be made by free-space optics as well as fiber-optic components.

In a free-space Michelson interferometer, a partial reflection mirror with 50 percent reflectivity is used to split the input optical beam into two parts. After traveling a distance, each beam is reflected back to recombine at the partial reflection mirror. The interference between these two beams is detected by the photodetector. One of the two end mirrors is mounted on a translation stage so that its position can be scanned to vary the interference pattern. The configuration of a fiber-optics-based Michelson interferometer compares to the free-space version except that the partial reflection mirror is replaced by a 3-dB fiber coupler and the light beams are guided within optical fibers. The scan of the arm length can be accomplished by stretching the fiber in one of the two arms using a piezo-electrical transducer. Because the fiber

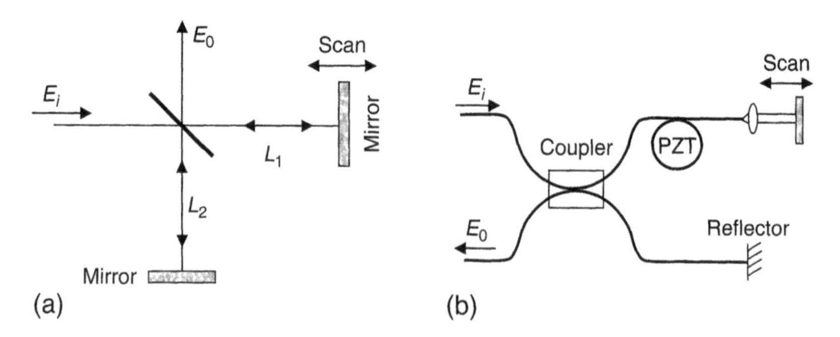

Figure 2.4.1 Basic configurations of (a) free-space optics and (b) a fiber-optic Michelson interferometer. *PZT:* Piezo-electric transducer.

can be easily bent and coiled, the size of a fiber-optic interferometer can be much smaller.

In free-space Michelson interferometers there is no birefringence in the air and polarization rotation is generally not a problem. However, for fiber-based Michelson interferometers, random birefringence in the fiber causes polarization rotation in each arm of the interferometer. The optical signals reflected from the mirrors in the two arms might not have the same polarization state when they combine at the fiber coupler. This polarization mismatch may cause significant reduction of the extinction ratio and degrade the performance of the interferometer. This problem caused by random birefringence of the fiber can be solved using Faraday mirrors, as shown in Figure 2.4.2 [13]. The Faraday mirror is composed of a 45° Faraday rotator and a total reflection mirror. The polarization of the optical signal rotates 45° in each direction; therefore the total polarization rotation is 90° after reflected from the Faraday mirror. Because of their applications in optical measurement and optical sensors, standard Faraday mirrors with fiber pigtails are commercially available.

2.4.1 Operating Principle of a Michelson Interferometer

For the fiber directional coupler using the notations in Figure 2.4.2, suppose the power splitting ratio is α; the optical field transfer matrix is

$$\begin{bmatrix} \vec{E}_C \\ \vec{E}_D \end{bmatrix} = \begin{bmatrix} \sqrt{1-\alpha} & j\sqrt{\alpha} \\ j\sqrt{\alpha} & \sqrt{1-\alpha} \end{bmatrix} \begin{bmatrix} \vec{E}_A \\ \vec{E}_B \end{bmatrix} \tag{2.4.1}$$

Optic fields are generally vectorial and their polarization states may change along the fiber because of the birefringence. Due to the effects of fiber propagation delays in the two arms and their polarization rotation, the relationship between reflected fields ($E_{C'}$, $E_{D'}$) and input fields (E_C, E_D) can be expressed in a transfer matrix form. First,

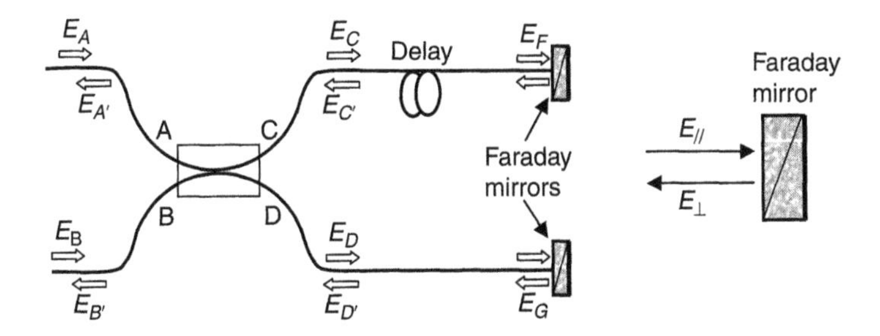

Figure 2.4.2 Illustration of a fiber Michelson interferometer using Faraday mirrors.

$$\begin{bmatrix} \vec{E}_F \\ \vec{E}_G \end{bmatrix} = \begin{bmatrix} [U^{//}]\exp(-j\omega\tau_1 - j\varphi_1) & 0 \\ 0 & [V^{//}]\exp(-j\omega\tau_2 - j\varphi_2) \end{bmatrix} \begin{bmatrix} \vec{E}_C \\ \vec{E}_D \end{bmatrix} \quad (2.4.2)$$

where τ_1 and τ_2 are propagation delays of the two fiber arms, ω is the optical frequency, φ_1 and φ_2 are initial optical phases, and $U^{//}$ and $V^{//}$ are tensors representing the birefringence effect in the two fiber arms seen by the optical signal in the // polarization. At the end of each fiber arm, a Faraday mirror with the reflectivity of R_1 and R_2, respectively, reflects the optical signal and also rotates its polarization state by 90°. The backward propagating optical field will see the birefringence of the fiber in the perpendicular polarization compared to the forward propagating signal. The transfer matrix for the backward propagated optical fields is

$$\begin{bmatrix} \vec{E}_{C'} \\ \vec{E}_{D'} \end{bmatrix} = \begin{bmatrix} [U^{\perp}]R_1\exp(-j\omega\tau_1 - j\varphi_1) & 0 \\ 0 & [V^{\perp}]R_2\exp(-j\omega\tau_2 - j\varphi_2) \end{bmatrix} \begin{bmatrix} \vec{E}_F \\ \vec{E}_G \end{bmatrix} \quad (2.4.3)$$

where $U^{\perp}$ and $V^{\perp}$ are tensors representing the birefringence effect in the two fiber arms seen by the optical signal in the $\perp$ polarization. Combining Equations 2.4.2 and 2.4.3, the roundtrip transmission matrix is

$$\begin{bmatrix} \vec{E}_{C'} \\ \vec{E}_{D'} \end{bmatrix} = \begin{bmatrix} [U^{//}U^{\perp}]R_1\exp(-j2\omega\tau_1 - j\varphi_1) & 0 \\ 0 & [V^{//}V^{\perp}]R_2\exp(-j2\omega\tau_2 - j\varphi_2) \end{bmatrix} \begin{bmatrix} \vec{E}_C \\ \vec{E}_D \end{bmatrix} \quad (2.4.4)$$

Because of the reciprocity principle, after a roundtrip, both $[U^{//}U^{\perp}]$ and $[V^{//}V^{\perp}]$ will be independent of the actual birefringence of the fiber. Therefore, in the following analysis, the polarization effect will be neglected and $[U^{//}U^{\perp}] = [V^{//}V^{\perp}] = 1$ will be assumed. Finally, the backward-propagated optical fields passing through the same fiber directional coupler are

$$\begin{bmatrix} \vec{E}_{A'} \\ \vec{E}_{B'} \end{bmatrix} = \begin{bmatrix} \sqrt{1-\alpha} & j\alpha \\ j\sqrt{\alpha} & \sqrt{1-\alpha} \end{bmatrix} \begin{bmatrix} \vec{E}_C \\ \vec{E}_D \end{bmatrix} \quad (2.4.5)$$

Therefore, the overall transfer matrix from the input (E_A, E_B) to the output ($E_{A'}$, $E_{B'}$) is

$$\begin{bmatrix} \vec{E}_{A'} \\ \vec{E}_{B'} \end{bmatrix} = \begin{bmatrix} \sqrt{1-\alpha} & j\alpha \\ j\sqrt{\alpha} & \sqrt{1-\alpha} \end{bmatrix} \begin{bmatrix} R_1\exp(-j2\omega\tau_1 - j2\varphi_1) & 0 \\ 0 & R_2\exp(-j2\omega\tau_2 - j2\varphi_2) \end{bmatrix} \begin{bmatrix} \sqrt{1-\alpha} & j\alpha \\ j\sqrt{\alpha} & \sqrt{1-\alpha} \end{bmatrix} \begin{bmatrix} \vec{E}_A \\ \vec{E}_B \end{bmatrix} \quad (2.4.6)$$

For unidirectional application, this fiber Michelson interferometer can be used as an optical filter. Letting $E_B = 0$, the filter transfer function is

$$T = \left|\frac{\vec{E}_{B'}}{\vec{E}_A}\right|^2 = \alpha(1-\alpha)|\ R_1\exp(-j2\omega\tau_1 - j2\varphi_1) + R_2\exp(-j2\omega\tau_2 - j2\varphi_2)\ |^2 =$$
$$= \alpha(1-\alpha)\{R_1^2 + R_2^2 + 2R_1R_2\cos - (2\omega\Delta\tau + 2\varphi)\} \tag{2.4.7}$$

where $\Delta\tau = \tau_1 - \tau_2$ is the differential roundtrip delay and $\varphi = (\varphi_1 - \varphi_2)/2$ is the initial phase difference between the two arms. If we further assume that the two Faraday mirrors have the same reflectivity and, neglecting the propagation loss of the fiber, $R_1 = R_2 = R$, then

$$T = 4\alpha(1-\alpha)R^2\cos^2[\omega(\tau_1 - \tau_2) + \varphi] = 4\alpha(1-\alpha)R^2\cos^2(2\pi f\Delta\tau + \varphi) \tag{2.4.8}$$

Except for the transmission loss term $4\alpha(1 - \alpha)$, Equation 2.4.8 is equivalent to an ideal MZI transfer function, and the infinite extinction ratio is due to the equal roundtrip loss in the two arms $R_1 = R_2$. Figure 2.4.3 shows the transmission loss of the Michelson interferometer as the function of the power-splitting ratio of the coupler. Even though the minimum loss is obtained with a 3dB (50 percent) coupling ratio of the fiber coupler, the extinction ratio of the Michelson interferometer transfer function is independent of the power-splitting ratio of

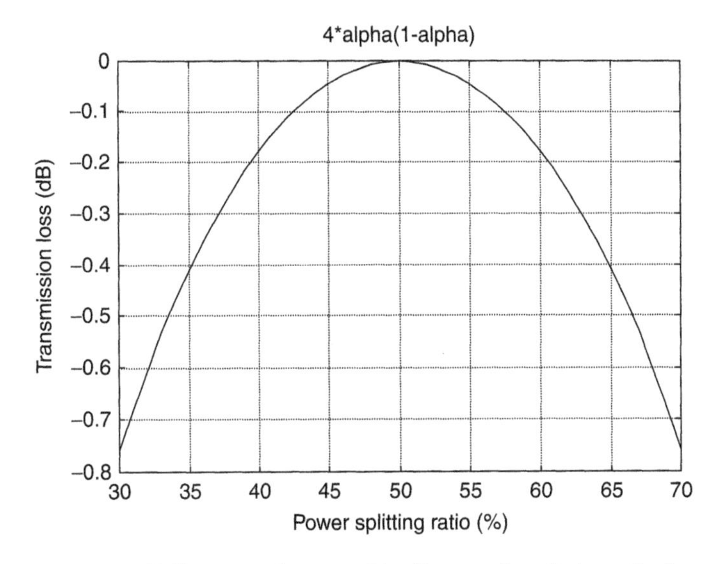

Figure 2.4.3 MZ filter power loss caused by fiber coupler splitting ratio change.

the fiber coupler. This relaxes the coupler selection criterion, and relatively inexpensive couplers can be used. Figure 2.4.3 indicates that even with ± 10 percent variation of the power-splitting ratio in the fiber coupler, the additional insertion loss for the filter is less than 0.2 dB.

Figure 2.4.4 shows a typical power transfer function of a Michelson interferometer-based optical filter with a 1 cm optical path length difference between the two arms, which corresponds to a free spectrum range of 10 GHz. The four curves in Figure 2.4.4 were obtained with $R_1^2 = 98$ percent while $R_2^2 = 98$ percent, 88 percent, 78 percent, and 68 percent, respectively. Obviously, when the reflectivity difference between the two arms increases, the filter extinction ratio decreases.

Figure 2.4.5 systemically shows the extinction ratio of the transfer function versus the power reflectivity ratio $(R_1/R_2)^2$ between the two arms. To guarantee the extinction ratio of the transfer function is higher than 30 dB, for example, the difference of power reflectivities between the two arms must be less than 10 percent.

2.4.2 Measurement and Characterization of Michelson Interferometers

The characterization of the Michelson interferometer power transfer function is straightforward using an optical spectrum analyzer, as shown in Figure 2.4.6. In this measurement setup, a wideband light source is used to provide the

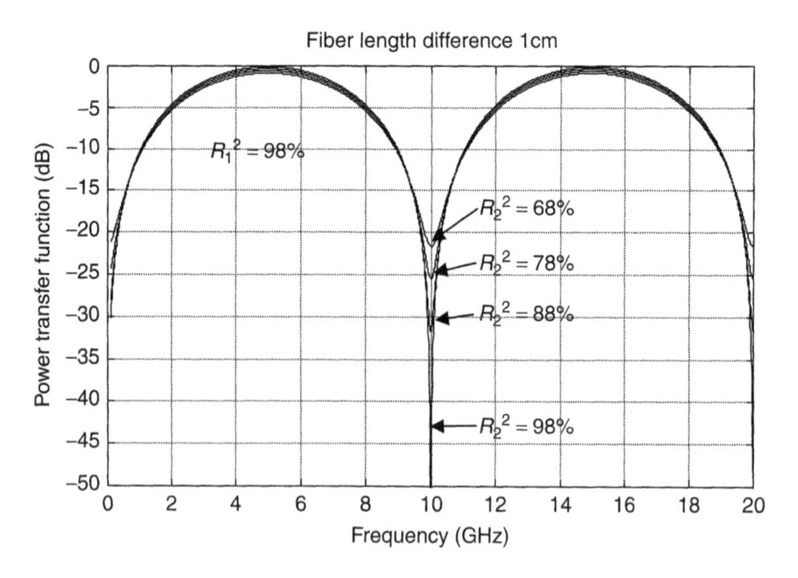

Figure 2.4.4 Power transfer function for different reflectivities of Faraday mirrors.

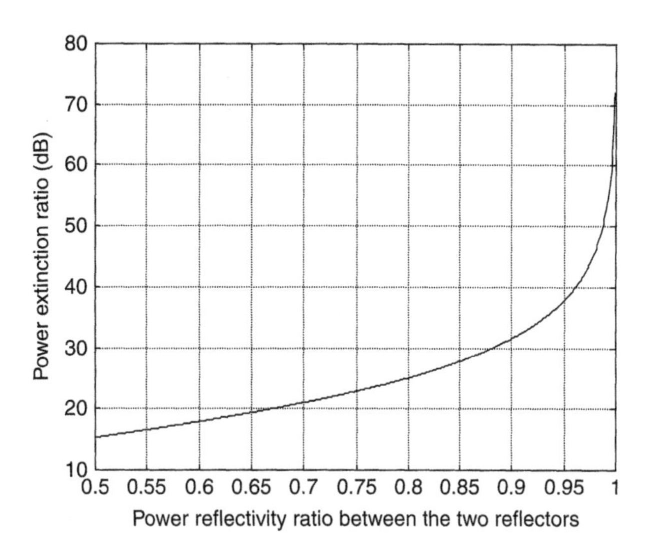

Figure 2.4.5 Extinction ratio of the power transfer function as the function of power reflectivity ratio R_1^2/R_2^2 between the two arms.

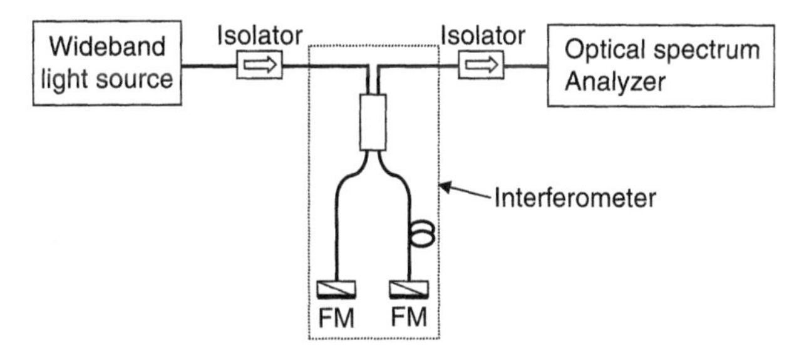

Figure 2.4.6 Experimental setup to measure the transfer function of a Michelson interferometer. *FM:* Faraday mirror.

probing signal, which can simply be an EDFA without the input. Since the optical spectrum from the wideband source is flat, the measurement of optical spectrum at the interferometer output is equivalent to the filter power transfer function. For accurate measurement, optical isolators have to be used at both the input and the output side of the interferometer to prevent unwanted reflections, which may potentially introduce wavelength-dependent errors.

Figure 2.4.7 shows an example of the optical spectrum at the output of a Michelson interferometer measured by an optical spectrum analyzer with 0.05 nm spectral resolution. The free spectral range of this interferometer is

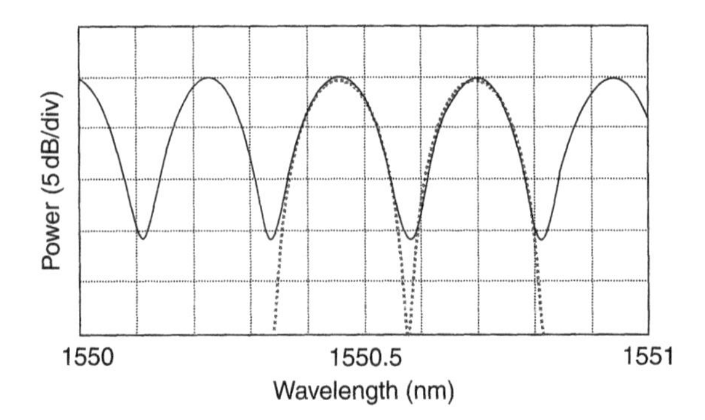

Figure 2.4.7 Optical spectrum measured by an optical spectrum analyzer with 0.05 nm spectral resolution. Dashed curve is the theoretical fit.

approximately 0.25 nm, and the extinction ratio shown in Figure 2.4.7 is only approximately 16 dB. Part of the reason for the poor extinction ratio measured by a grating-based OSA is due to the insufficient spectral resolution. The theoretically sharp rejection band shown by the dashed cure in Figure 2.4.7 is smoothed out by the spectrum analyzer. Therefore, a scanning Fabry-Perot interferometer with high spectral resolution may be more appropriate for this measurement.

2.4.3 Techniques to Increase Frequency Selectivity

As indicated in Equation 2.4.7, the theoretical transfer function of a Michelson interferometer is a simple sinusoid. High rejection is only provided in very narrow frequency windows, as shown in Figure 2.4.4. Therefore an optical filter based on a Michelson interferometer is only suitable to be used as a notch filter to reject narrowband optical signals. Used as a transmission filter, the frequency selectivity of a Michelson interferometer is relatively poor.

On the other hand, a FP interferometer has sharp transmission peaks as shown in Figure 2.2.4. FP optical filters are useful to select optical signal channels in DWDM optical systems. However, due to the slow rolling-off of the transmission peak in a FP interferometer transfer function, the rejection ratio of the adjacent channel is usually small. The combination of a FP and a Michelson interferometer, as shown in Figure 2.4.8, may enhance the frequency selectivity of the power transfer function. An optical isolator has to be inserted between the FPI and the Michelson interferometer to prevent multiple reflections and the interference between these two devices.

Figure 2.4.9 shows a typical power transfer function of an optical filter made by the concatenation of an FPI and a Michelson interferometer. The FPI filter

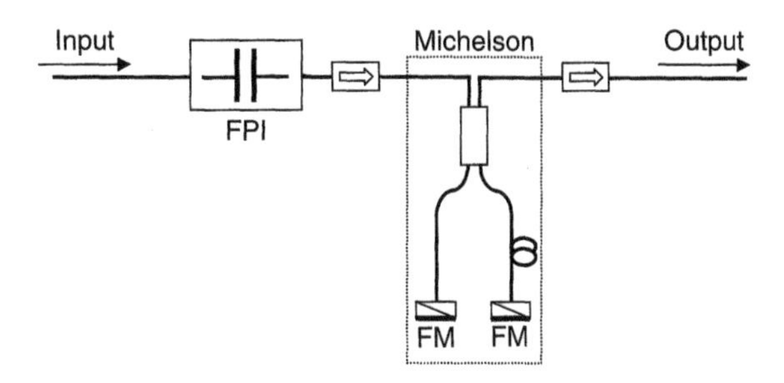

Figure 2.4.8 The combination of an FPI and a Michelson interferometer.

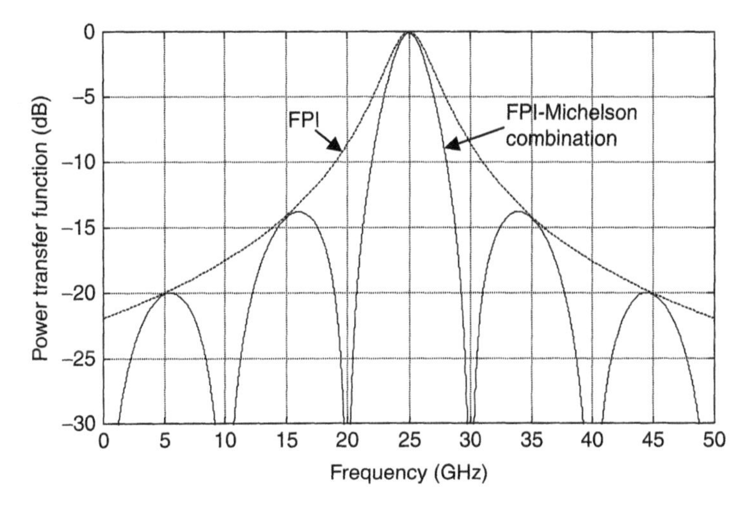

Figure 2.4.9 Power transfer function of an FPI alone (dashed curve) and an FPI-Michelson combination (solid line).

has a finesse of 100 and a FWHM bandwidth of 4 GHz. The Michelson interferometer has a free-spectral range of 20 GHz. Obviously, the Michelson interferometer helps to increase the rejection at the edge of the passband significantly, which increases the channel selectivity in DWDM optical systems, especially when a high-power crosstalk channel exists nearby. In this application, since both the FPI filter and the Michelson interferometer are polarization independent, the overall transfer function can be polarization insensitive. Furthermore, since both of them can be made tunable using thermal or PZT, their combination is also tunable.

Another possibility for increasing frequency selectivity of Michelson interferometers is to concatenate multiple units together, as shown schematically in

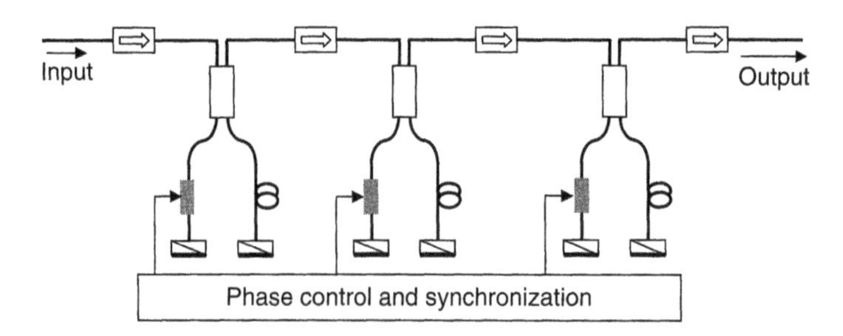

Figure 2.4.10 Concatenate several Michelson interferometers to increase frequency selectivity.

Figure 2.4.10. By carefully adjusting the transfer function of each interferometer, the overall transfer function of the concatenated filters can be optimized for particular applications. For example, Figure 2.4.11 shows the transfer function of three concatenated Michelson interferometers. One of them has the free-spectral range (FSR) of 10 GHz and the other two have the same FSR of 20 GHz. In this particular case, the phases of these three filters are tuned away from each other by approximately $\pi/6$ to achieve the narrowest transmission peaks. Obviously, the concatenation effectively increases the overall finesse of the filter and thus the frequency selectivity.

Phase control and synchronization are important issues to ensure the proper transfer function as desired. Phase control in a Michelson interferometer can be

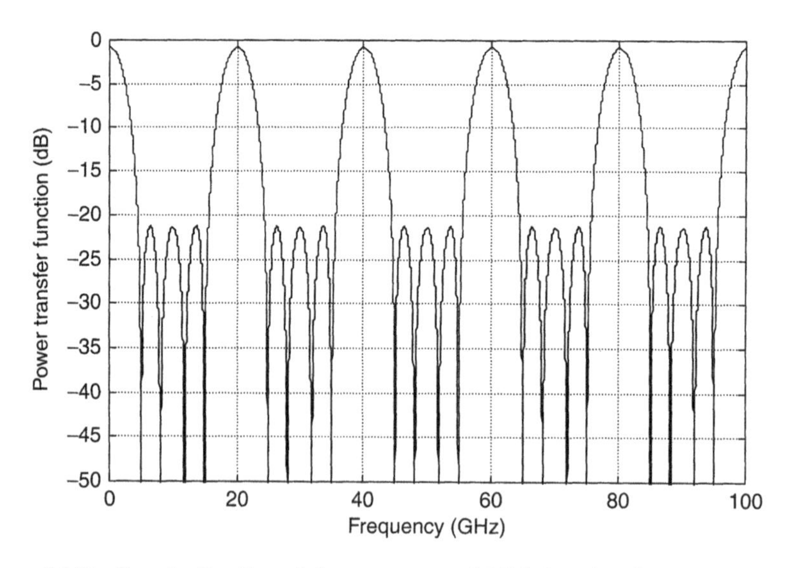

Figure 2.4.11 Transfer function of three concatenated Michelson interferometers. One has 10 GHz FSR and the other two have 20 GHz FSR.

accomplished through thermal effect by adding an electrical heater at one of the two interferometer arms. In practice, this can be done by spattering resistive metal coating on the fiber or planar optical waveguides. The heat generated by electrical current flowing through the resistive metal introduces a local temperature change, so that the refractive index can be adjusted through thermal-induced index change of silica. A feedback control is usually necessary to dynamically optimize the phase synchronization.

Michelson interferometers can also be connected in a tree configuration, as shown in Figure 2.4.12. In this configuration, optical isolators between fiber couplers are not required. Using the notations marked in the figure, the general power transfer function of this composite filter is

$$\begin{aligned} T &= \left| \frac{\vec{E}_{out}}{\vec{E}_{in}} \right|^2 \\ &= R^2\alpha(1-\alpha) \left| \begin{array}{l} \exp(-j\omega\Delta\tau_0)[(1-\alpha_1)\exp(-j\omega\Delta\tau_1) - \alpha_1\exp(j\omega\Delta\tau_1)] + \\ +\exp(j\omega\Delta\tau_0)[(1-\alpha_2)\exp(-j\omega\Delta\tau_2) - \alpha_2\exp(j\omega\Delta\tau_2)] \end{array} \right|^2 \end{aligned} \quad (2.4.9)$$

where

$$\Delta\tau_0 = \left(\frac{\tau_{11}+\tau_{12}}{2} - \frac{\tau_{21}+\tau_{22}}{2} + \tau_1 - \tau_2\right)$$

$$\Delta\tau_1 = (\tau_{11} - \tau_{12})$$

and

$$\Delta\tau_2 = (\tau_{21} - \tau_{22})$$

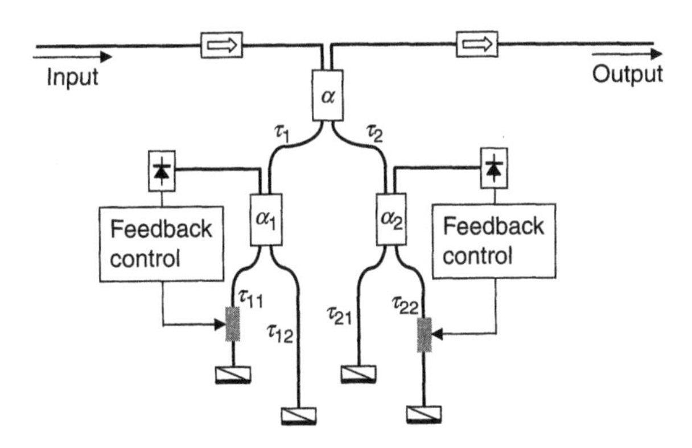

Figure 2.4.12 Tree configuration of three Michelson interferometers to increase frequency selectivity.

τ_1, τ_2, τ_{11}, τ_{12}, τ_{21}, and τ_{22}, are roundtrip time delays of the respective optical arms indicated in Figure 2.4.12. All the reflectivities of the Faraday mirrors are assumed to be equal to R. In the simplest case, suppose all the fiber couplers have a 50 percent power-splitting ratio, that is, $\alpha = \alpha_1 = \alpha_2 = 0.5$, then Equation 2.4.9 can be simplified as

$$T = \left|\frac{\vec{E}_{out}}{\vec{E}_{in}}\right|^2 = \frac{R^2}{4}\{\sin^2(\omega\Delta\tau_1) + \sin^2(\omega\Delta\tau_2) + 2\sin(\omega\Delta\tau_1)\sin(\omega\Delta\tau_2)\cos(2\omega\Delta\tau_0)\} \quad (2.4.10)$$

In this configuration, the bottom two interferometers operate as frequency-selective reflectors for the top interferometer. To obtain the best operation characteristics, these two frequency-selective reflectors need to have identical characteristics. This can be understood as in a simple Michelson interferometer: equal mirror reflectivity $R_1 = R_2 = R$ gives the best extinction ratio. Figure 2.4.13 shows an example of the power transfer functions of three Michelson interferometers in a tree configuration (solid curve), where all the three interferometers have identical relative delay: $\Delta\tau_1 = \Delta\tau_2 = \Delta\tau_0$. This transfer function is equivalent to the synchronous tuning of two identical single-stage filters. Therefore the width of the passband is decreased. The transfer function of

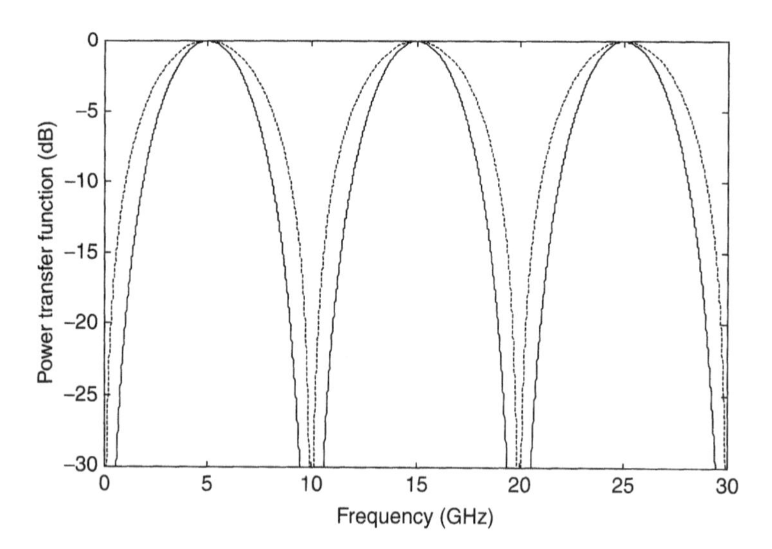

Figure 2.4.13 Transfer functions of tree configuration of Michelson interferometers (solid line). Three filters are identical and synchronized in phase with $\Delta\tau_1 = \Delta\tau_2 = \Delta\tau_0$. The dashed line is the transfer function of a single Michelson interferometer.

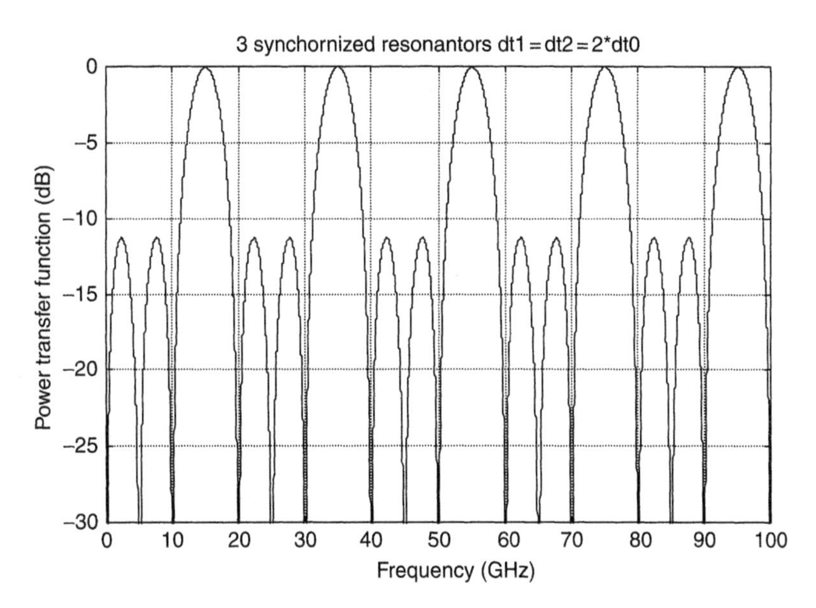

Figure 2.4.14 Transfer functions of tree configuration of Michelson interferometers with asynchronized tuning. Three filters are identical but with $\Delta\tau_1 = \Delta\tau_2 = \Delta\tau_0$.

each individual Michelson interferometer filter is shown as the dashed curve in the same figure for comparison.

Figure 2.4.14 shows the result of asynchronized tuning of three Michelson interferometers with $\Delta\tau_1 = \Delta\tau_2 = 2\Delta\tau_0$. In fact, this transfer function is equivalent to the asynchronized tuning of two concatenated Michelson interferometers shown in Figure 2.4.10 but without the need for an optical isolator between the interferometers.

2.5 OPTICAL WAVELENGTH METER

In optical communication and optical sensor systems, the wavelength of the laser source is an important parameter, especially when multiple wavelengths are used in the system. The wavelength of an optical signal can be measured by an OSA; the measurement accuracy is often limited by the spectral resolution of the OSA, which is usually in the order of 0.1 nm for a compact version. The most popular optical wavelength meter is based on a Michelson interferometer, which is structurally simpler than an OSA and can easily be made into a small package using fiber-optic components [14, 15]. A Michelson interferometer-based wavelength meter can typically provide a measurement accuracy of better than 0.001 nm.

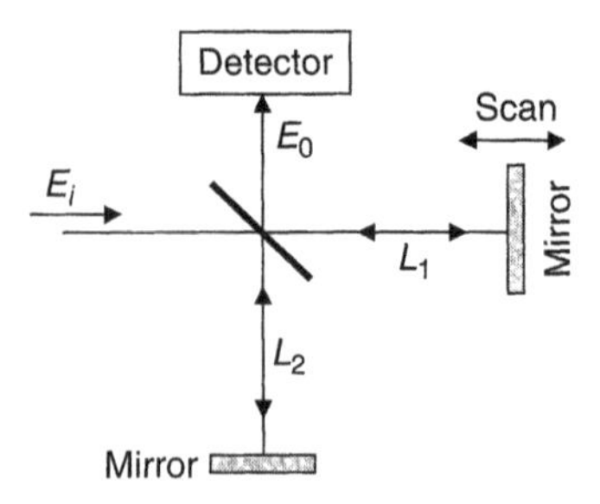

Figure 2.5.1 Basic configurations of a wavelength meter based on a Michelson interferometer.

An optical wavelength meter can be built from a simple Michelson interferometer by scanning the optical length of one of the two interferometer arms. Based on the Michelson interferometer configuration shown in Figure 2.5.1, assuming that the input optical field is E_i, the power-splitting ratio of the coupler is α, and the optical lengths are L_1 and L_2 for the two arms, then the optical field reaches the receiving photodetector is

$$\begin{aligned} E_0 &= \sqrt{1-\alpha}\sqrt{\alpha}E_i[\exp j(2\beta L_1 + \varphi_1) + \exp j(2\beta L_2 + \varphi_2)]e^{j\omega t} \\ &= 2\sqrt{\alpha(1-\alpha)}E_i \exp j\left(\beta(L_1 + L_2) + \frac{\varphi_1 + \varphi_2}{2}\right)\cos\left(\beta\Delta L + \frac{\varphi_1 - \varphi_2}{2}\right)e^{j\omega t} \end{aligned} \quad (2.5.1)$$

where $\beta = 2\pi n/\lambda$ is the propagation constant, n is the refractive index, $\Delta L = L_1 - L_2$ is the length difference between the two arms, and ω is the optical frequency. φ_1 and φ_2 are the initial phase shift of the two arms, which can be introduced by the beam splitter as well as the reflecting mirrors.

Since the photodiode is a square-law detection device, the photo-current generated by the photodiode is directly proportional to the optical power:

$$I = \Re|E_0|^2 = 2\alpha(1-\alpha)P_i\Re\{1 + \cos(2\beta\Delta L + \Delta\varphi)\} \quad (2.5.2)$$

where $\Re$ is the detector responsivity and $\Delta\varphi = (\varphi_1 - \varphi_2)$ is the initial phase difference between the two arms when $\Delta L = 0$. Obviously, if the path length of one of the two arms is variable, the photo-current changes with ΔL.

2.5.1 Operating Principle of a Wavelength Meter Based on Michelson Interferometer

In the simplest case, the optical signal has a single wavelength λ. Equation 2.5.2 indicates that when the length of ΔL scans, the photocurrent I passes through maximums and minimums alternatively. By analyzing the photocurrent I as a function of ΔL, the signal wavelength λ can be determined.

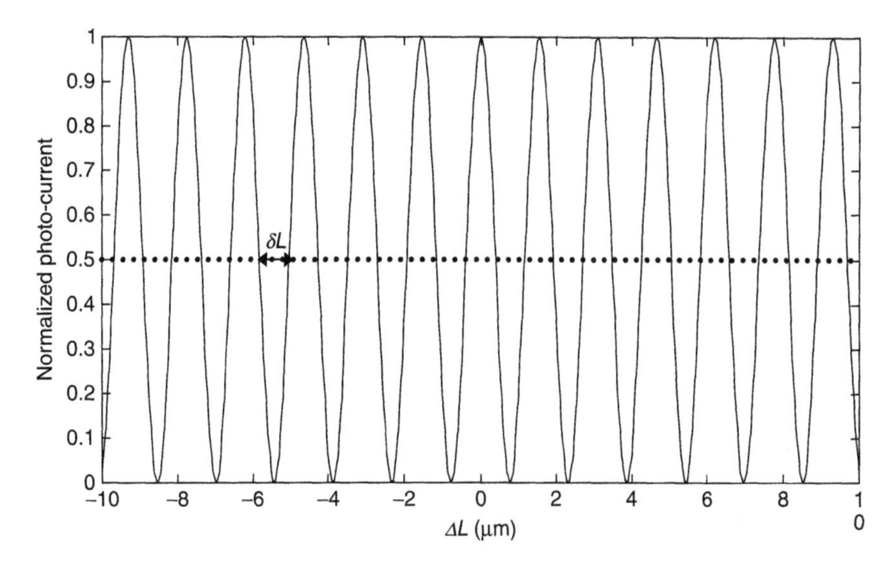

Figure 2.5.2 Normalized photocurrent versus differential arm length. Signal wavelength $\lambda = 1550$ nm.

For example, every time the photocurrent crosses the half-maximum point, Equation 2.5.2 becomes, $\cos(4\pi\Delta L/\lambda + \Delta\varphi) = 0$. The interval between adjacent crossings δL, as shown in Figure 2.5.2, can be recorded, satisfying the following equation:

$$4\pi\frac{\delta L}{\lambda} = \pi$$

and therefore the signal wavelength can be found as $\lambda = 4\delta L$.

In principle, to find the signal wavelength, the arm length of the interferometer only needs to scan for a half wavelength. However, in practice, because of the measurement uncertainty and noise associated with the optical signal, the arm length usually has to scan for much longer than a wavelength. This long scan length provides multiple measures of δL, and accurate wavelength estimation can be obtained by averaging the measured δL values.

The wavelength measurement technique we described is also known as *fringe counting*. The key to this measurement technique is the knowledge of the exact differential length variation of the interferometer arm. The initial phase difference $\Delta\varphi$ in Equation 2.5.2 does not make a real impact as long as it does not change with time.

So far, we have discussed the simplest case where the signal has only a single wavelength. Now let's look at the case when the optical signal has two wavelength components. Assuming that these two wavelength components have an equal amplitude $E_i/2$ and their frequencies are ω_1 and ω_2, respectively, the combined field at the photodiode is

$$E_0 = 0.5E_i \left\{ \begin{array}{l} \exp[j\beta_1(L_1 + L_2)]\exp(j\omega_1 t)\cos(\beta_1 \Delta L) + \\ \exp[j\beta_2(L_1 + L_2)]\exp(j\omega_2 t)\cos(\beta_2 \Delta L) \end{array} \right\} \tag{2.5.3}$$

where the initial phase difference $\Delta\varphi$ is neglected for simplicity and a 50 percent splitting ratio of the optical coupler is assumed. The photo-current can then be found as

$$I = \Re P_i \left\{ 1 + \frac{\cos(2\beta_1 \Delta L) + \cos(2\beta_2 \Delta L)}{2} \right\} \tag{2.5.4}$$

In obtaining Equation 2.5.4 we have removed the cross-term which contains a time-dependent factor $\exp[j(\omega_1 - \omega_2)t]$. This fast-changing random phase can be averaged out because the much slower moving speed of the interferometer arms.

Figure 2.5.3 shows the normalized photocurrent versus differential arm length ΔL where the two wavelength components in the optical signal are $\lambda_1 = 1350$ nm and $\lambda_2 = 1550$ nm. Obviously, in this case, a simple fringe counting would not be sufficient to determine the wavelengths in the optical signal.

However, Figure 2.5.3 shows distinct features of the mixing between the two wavelength components. Applying a Fast Fourier Transformation (FFT) on the $I(\Delta L)$ characteristic given by Equation 2.5.4, we should be able to determine the two distinct wavelengths within the optical signal.

In general, to measure an optical signal with multiple wavelengths using a Michelson interferometer, FFT is always used to determine the wavelengths

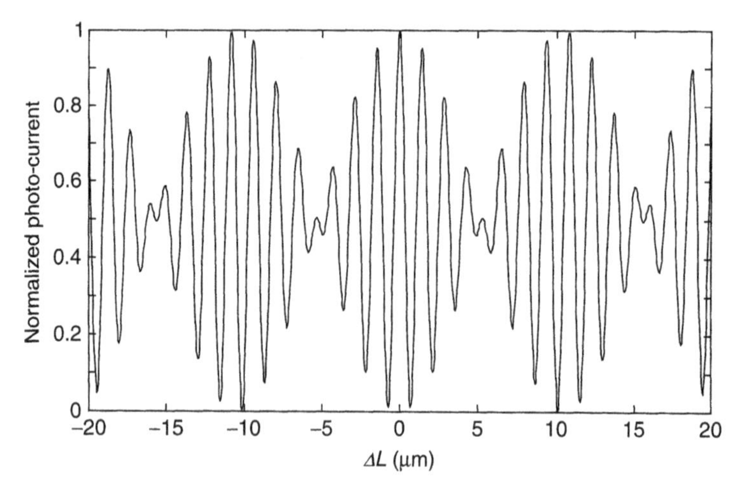

Figure 2.5.3 Normalized photocurrent versus differential arm length. The optical signal has two equal-amplitude wavelength components at $\lambda_1 = 1350$ nm and $\lambda_2 = 1550$ nm.

that contained in the optical signal. This technique is indeed equivalent to analyzing the optical spectrum and therefore, in principle, it can be used to construct an optical spectrum analyzer. However, an OSA made by a Michelson interferometer would only be suitable to measure optical signals with discrete wavelength components; it cannot be used to measure power spectral densities of optical noises. This will be discussed in the next section.

2.5.2 Wavelength Coverage and Spectral Resolution

In a wavelength meter based on the Michelson interferometer, the accuracy of determining the signal wavelength, to a large extent, depends on how accurately the differential length ΔL is measured during the scanning. Since FFT is used to convert the measured data from the length (ΔL) domain into the frequency (f) domain, the wavelength coverage and the spectral resolution of the wavelength meter will be determined by the step size and the total range of the scanning interferometer arm, respectively.

2.5.2.1 Wavelength Coverage

A good wavelength meter has to be able to cover a large wavelength range. For example, in an WDM system C-band extends from 1530 nm to 1560 nm. To measure all wavelength components within this 30 nm bandwidth, the wavelength coverage of the wavelength meter has to be wide enough. Because the wavelengths are determined by the Fourier transformation of the measured photocurrent versus differential arm length, a smaller step size of ΔL will result in a wider wavelength coverage of the measurement.

It is well known that to perform FFT on a time domain waveform $\cos(\omega t)$, if the time domain sampling interval is δt as illustrated in Figure 2.5.4, the corresponding frequency domain coverage will be $f_{range} = 1/(2\delta t)$ after FFT. It can be understood because a smaller time domain-sampling interval picks up fast variation features of the signal and it is equivalent to a wider frequency bandwidth.

Similarly for the wavelength meter, the photocurrent is proportional to cos $(2\omega\Delta L/c)$ for each frequency component, as shown in Equation 2.5.2. If the sampling interval (step size) of the differential arm length ΔL is δl, after FFT the frequency-domain coverage will be

$$f_{range} = \frac{c}{4dl} \tag{2.5.5}$$

For example, in order for the wavelength meter to cover the entire C-band from 1530 nm to 1560 nm, f_{range} has to be as wide as 3750 GHz. In this case, the step size of the differential arm length has to be $\delta l \leq 20\ \mu m$.

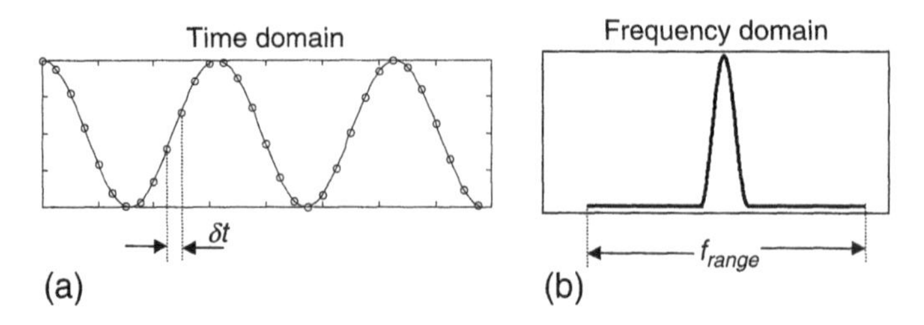

Figure 2.5.4 Relationship between (a) time-domain sampling interval δt and (b) frequency-domain coverage f_{range}.

2.5.2.2 Spectral Resolution

Another important parameter of a wavelength meter is its spectral resolution, which specifies its ability to identify two closely spaced wavelength components. Again, because the wavelengths are determined by the Fourier transformation of the measured photocurrent versus differential arm length, a larger traveling distance of the mirror will result in a finer wavelength resolution of the measurement.

In general, for a time-domain waveform $\cos(\omega t)$, if the total sampling window in the time domain is T, as illustrated in Figure 2.5.5, the corresponding sampling interval in the frequency domain is $f_{resolution} = 1/T$ after FFT. Obviously a longer time domain window helps to pick up slow variation features of the signal and it is equivalent to a finer-frequency resolution.

For a wavelength meter, the photocurrent is proportional to $\cos(2\omega\Delta L/c)$ for each frequency component. If the differential arm length of the interferometer scans for a total amount of L_{span}, after FFT the sampling interval in the frequency domain will be

$$f_{resolution} = \frac{c}{2L_{span}} \tag{2.5.6}$$

For example, in order for the wavelength meter to have a spectral resolution of 5 GHz, which is approximately 0.04 nm in a 1550 nm wavelength window, the interferometer arm has to scan for a total distance of $L_{span} \geq 3\ cm$.

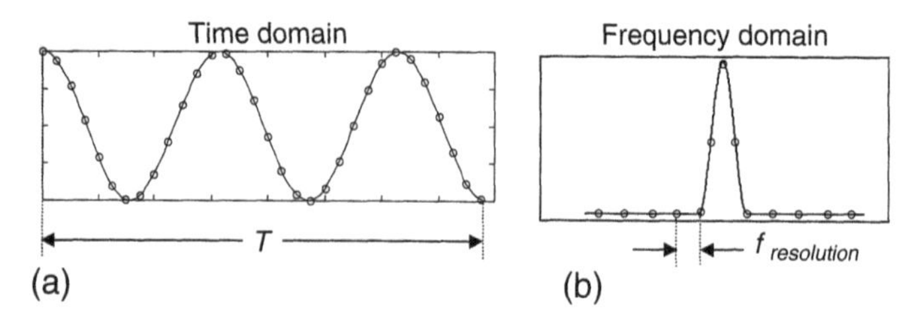

Figure 2.5.5 Relationship between (a) time-domain sampling window T and (b) frequency-domain resolution $f_{resolution}$.

2.5.2.3 Effect of Signal Coherence Length

A wavelength meter based on Michelson interferometer utilizes beam interference between its two arms. One fundamental requirement for interference is that the two participating beams have to be mutually coherent. This requires that the path length difference between the two interferometer arms be shorter than the coherence length of the optical signal to be measured. For an optical signal with linewidth of Δf_{lw}, its coherence length is defined as

$$L_{coherent} = \frac{c}{n\Delta f_{lw}} \tag{2.5.7}$$

where, n is the refractive index of the material. In the derivation of Michelson interferometer Equation 2.5.2, we have assumed that the phase difference between the optical signal traveled through the two arms is fixed and not varying with time: $\Delta\phi = \text{constant}$. This is true only when the arm's-length difference is much less than the signal coherence length. If the arm's-length difference is longer than the signal coherence length, the coherent interference will not happen and the photocurrent will become a random function of time. This restricts the distance that the mirror can scan such that

$$L_{span} \ll \frac{c}{n\Delta f_{lw}} \tag{2.5.8}$$

Then, since the traveling distance of the mirror is limited by the coherence length of the optical signal, according to Equation 2.5.6, the achievable spectral resolution will be limited to

$$f_{resolution} \gg \frac{n\Delta f_{lw}}{2} \tag{2.5.9}$$

Since the available spectral resolution has to be much wider than the spectral linewidth of the optical signal, Michelson interferometer-based wavelength meter is better suited to measure optical signals with discrete wavelength components. In other words, a wavelength meter may not be suitable to measure power spectral density of random optical noise, such as amplified spontaneous emission generated by optical amplifiers. This is one of the major differences between a wavelength meter and an OSA.

2.5.3 Wavelength Calibration

Throughout our discussion so far, it is important to note that the accuracy of wavelength measurement relies on the accurate estimation of the differential arm length ΔL. In practice, absolute calibration of mirror displacement is not easy to accomplish mechanically. Therefore, wavelength calibration using a reference light source is usually used in commercial wavelength meters.

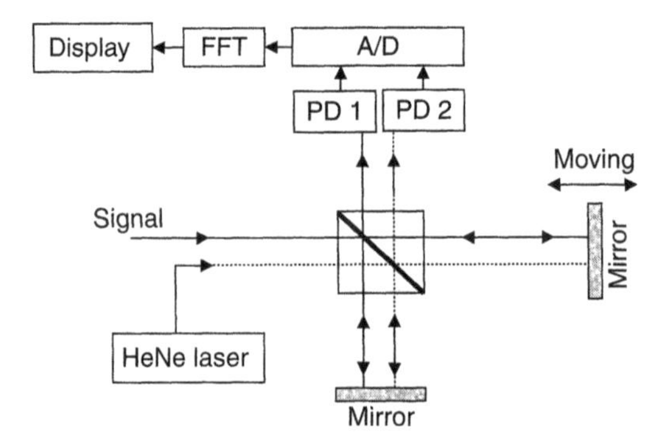

Figure 2.5.6 Calibration of wavelength meter using a wavelength reference laser.

Figure 2.5.6 shows the block diagram of a typical wavelength meter with an *in-situ* calibration using a HeNe laser as the wavelength reference. The reference light is launched into the same system simultaneously with the optical signal to be measured. Two independent photodiodes are used to detect the signal and the reference separately. The outputs from both photodetectors are digitized through an A/D converter before performing the FFT operation.

Since the wavelength of the reference is precisely known (for example $\lambda_{ref} = 632.8$ nm for a HaNe laser), this reference wavelength can be used to calibrate the differential arm length ΔL of the interferometer. Then this ΔL is used as a determining parameter in the FFT of the signal channel to precisely predict the wavelength of the unknown signal.

2.5.4 Wavelength Meter Based on Fizeau Wedge Interferometer

In a Michelson interferometer-based optical wavelength meter as described in the previous sections, the scan of interference fringe is accomplished by mechanically scanning the position of one of the two mirrors. A Fizeau wedge interferometer-based wavelength meter equivalently spreads the interference fringe in the spatial domain and uses a photodiode array to simultaneously acquire fringe pattern. Therefore no moving part is needed.

As illustrated in Figure 2.5.7, in a Fizeau wedge interferometer-based wavelength meter, the incoming optical signal is collimated and launched onto a Fizeau wedge whose front surface is partially reflective and the back facet is a total reflector. The two reflective surfaces are not parallel and there is a small angle ϕ between them. Therefore the light beams reflected from the front and the back facets converge at the surface of the photodiode array, and the angle

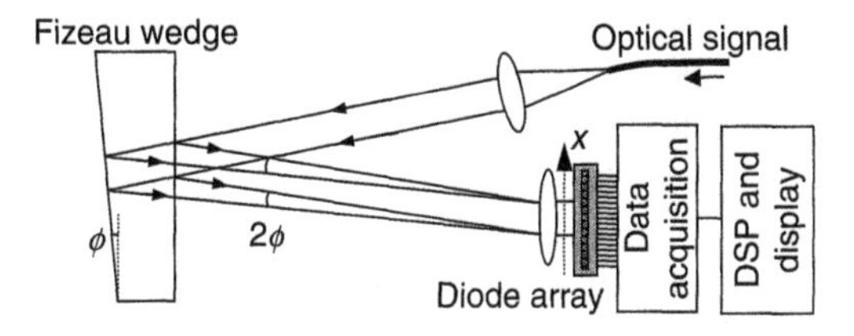

Figure 2.5.7 Optical configuration of a Fizeau wedge interferometer-based wavelength meter.

between these two reflected beams is 2ϕ. From simple geometric optics analysis, the normalized interference pattern formed on the one-dimensional photodiode array can be obtained as the function of the position x:

$$P(x) = 1 + \cos\left(\frac{2\pi}{\Lambda}x + \varphi_0\right) \tag{2.5.10}$$

where $\Lambda = \lambda/(2\phi)$ is the fringe spacing on the diode array and φ_0 is a constant phase that represents the phase at $x = 0$. Since the angle ϕ of the Fizeau wedge is fixed, a Fourier transform on the measured interference pattern $P(x)$ determines the wavelength of the optical signal.

Similar to the Michelson interferometer-based wavelength meter, the spectral resolution of the Fizeau wedge-based wavelength meter is determined by the total length of the diode array L and the angle of the Fizeau wedge ϕ:

$$f_{resolution} = \frac{c}{2\phi L} \tag{2.5.11}$$

The overall range of wavelength coverage is determined by the width of each pixel l of the diode array

$$f_{range} = \frac{c}{4\phi l} \tag{2.5.12}$$

For example, if a diode array has 1024 pixels and the width of each pixel is 25 μm, the angle of the Fizeau wedge is $\phi = 10°$, and the frequency resolution is about 33 GHz, which is approximately 0.26 nm in the 1550 nm wavelength window. The wavelength coverage range is about 138 nm.

It is important to point out the difference between the spectral resolution and the wavelength measurement accuracy. The spectral resolution specifies the resolving power. For example, if the optical signal has two frequency components and their frequency difference is δf, to distinguish these two frequency components the spectral resolution of the wavelength meter has to satisfy $f_{resolution} < \delta f$. On the other hand, the wavelength measurement accuracy depends on both the frequency resolution and the noise level of the measurement system. Although the signal-to-noise ratio can be enhanced by averaging, it will inevitably slow the measurement.

Since a Fizeau wedge-based wavemeter is based on the measurement of the spatial interference pattern introduced by the optical wedge, the optical flatness has to be guaranteed. In addition, the wavefront curvature of the incident beam may also introduce distortions on the fringe pattern, which has to be compensated for. Furthermore, if the optical wedge is made by a solid silica glass, chromatic dispersion may also be a potential source of error, which causes different wavelength component to have different optical path length inside the wedge. A number of designs have been proposed and demonstrated to solve these problems [16, 17]. In comparison to Michelson interferometer-based wavelength meter, the biggest advantage of a Fizeau wedge-based wavemeter is its potentially compact design, which does not need any moving part. The disadvantage is the spectral resolution, which is mainly limited by the size of the linear diode array and the total differential optical delay through the wedge.

2.6 OPTICAL POLARIMETER

It is well known that an electromagnetic wave has four basic parameters, including amplitude, frequency, phase, and the state of polarization. For an optical signal, its amplitude is represented by the brightness and its frequency is represented by the color. While the brightness can be easily measured by an optical power meter and the wavelength can be measured by a wave meter or an optical spectrum analyzer, the state of polarization of an optical signal is less intuitive and the measurement techniques are not as well-known. In optical communications and optical sensors, the effect of optical polarization can be very significant and must be considered.

As an additional degree of freedom, the state of polarization in a lightwave signal can be used to carry information in an optical system. At the same time the polarization effect may also degrade the performance of an optical fiber transmission system through polarization mode dispersion. The polarimeter is an instrument that can be used to measure the state of polarization of a lightwave signal; it is useful in many applications, such as optical system monitoring and characterization.

2.6.1 General Description of Lightwave Polarization

A propagating lightwave can be represented by electromagnetic field vectors in the plane that is perpendicular to the wave propagation direction. The complex electrical field envelope can be expressed as [18, 19]

$$\vec{E} = \vec{a}_x E_{x0} \cos(\omega t - kz) + \vec{a}_y E_{y0} \cos(\omega t - kz - \phi) \tag{2.6.1}$$

where $\vec{a}_x$ and $\vec{a}_y$ are unit vectors and φ is the relative phase delay between the two orthogonal field components. The polarization of the optical signal is determined by the pattern traced out on the transverse plane of the electrical field vector over time.

Figure 2.6.1 illustrates two-dimensional descriptions of various polarization states of polarized lights, such as linear polarization, circular polarization, and elliptical polarization. If we rewrite the optical field vector as shown in Equation 2.6.1 into two scalar expressions,

$$E_x = E_{x0} \cos(\omega t - kz) \tag{2.6.2a}$$

$$E_y = E_{y0} \cos(\omega t - kz - \phi) \tag{2.6.2b}$$

The pattern of the optical field traced on a fixed xy-plane can be described by the polarization ellipse,

$$\frac{E_x^2}{E_{x0}^2} + \frac{E_y^2}{E_{y0}^2} - 2\frac{E_x E_y}{E_{x0} E_{y0}} \cos\phi = \sin^2\phi \tag{2.6.3}$$

Generally, an optical signal may not be fully polarized. That means that the x-component and the y-component of the light are not completely correlated with each other; in other words, the phase difference ϕ between them may

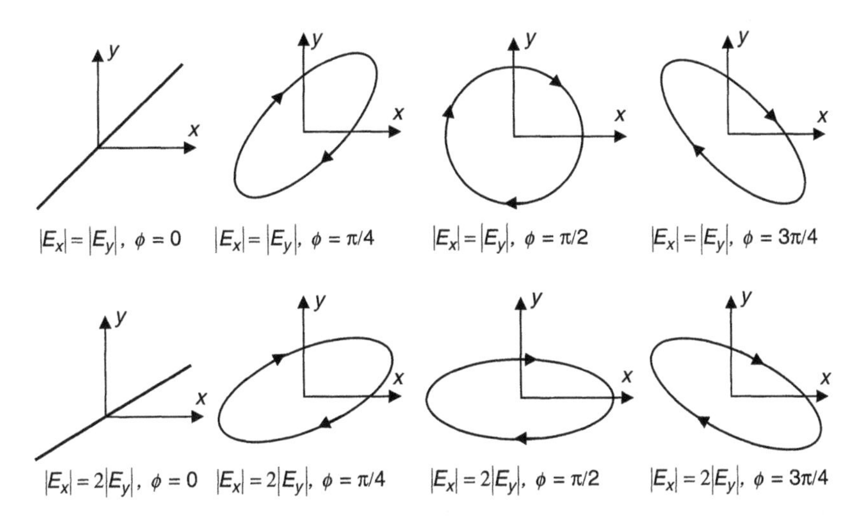

Figure 2.6.1 Two-dimensional view of polarization states of a polarized light.

be random. Many light sources in nature are unpolarized, such as sunlight, whereas optical signals from lasers are mostly polarized. An optical signal can usually be divided into a fully polarized part and a completely unpolarized part. The *degree of polarization* (DOP) is often used to describe the polarization characteristics of a partially polarized light, which is defined as

$$DOP = \frac{P_{polarized}}{P_{polarized} + P_{unpolarized}} \tag{2.6.4}$$

where $P_{polarized}$ and $P_{unpolarized}$ are the powers of the polarized part and unpolarized part, respectively. DOP is therefore the ratio between the power of the polarized part and the total power. DOP is equal to zero for an unpolarized light and is equal to unity for a fully polarized light.

2.6.2 The Stokes Parameters and the Poincare Sphere

The state of polarization of a lightwave signal is represented by the maximum amplitudes E_{x0}, E_{y0}, in the x and y directions, and a relative phase ϕ between them, as shown in Equation 2.6.1. In addition, if the signal is not completely polarized, the DOP also has to be considered. Because of the multiple parameters, the representation and interpretation of polarization states are often confusing. A Stokes vector is one of the most popular tools to describe the state of polarization of an optical signal. A Stokes vector is determined by four independent Stokes parameters, which can be represented by optical powers in various specific reference polarization states.

Considering an optical signal with the electrical field given by Equation 2.6.1, the four Stokes parameters are defined as

$$S_0 = P \tag{2.6.5}$$

$$S_1 = |E_{x0}|^2 - |E_{y0}|^2 \tag{2.6.6}$$

$$S_2 = 2|E_{x0}||E_{y0}|\cos\phi \tag{2.6.7}$$

$$S_3 = 2|E_{x0}||E_{y0}|\sin\phi \tag{2.6.8}$$

where P is the total power of the optical signal. For an ideally polarized lightwave signal, since there are only three free variables, E_{x0}, E_{y0}, and ϕ, in Equation 2.6.1, the four Stokes parameters should not be independent and the total optical power is related to S_1, S_1 and S_1 by

$$S_1^2 + S_2^2 + S_3^2 \equiv P \tag{2.6.9}$$

S_0 represents the total power of the optical signal, $S_0 = |E_{x0}|^2 + |E_{y0}|_2 = P$. However, if the optical signal is only partially polarized, $S_1^2 + S_2^2 + S_3^2$ only represents the power of the polarized part $P_{polarized}$, which is smaller than the total optical power because $S_0 = P = P_{polarized} + P_{unpolarized}$. In this case, the Stokes parameters can be normalized by the total power S_0, and the normalized Stokes parameters have three elements:

$$s_1 = \frac{S_1}{S_0} \quad s_2 = \frac{S_2}{S_0} \quad s_3 = \frac{S_3}{S_0} \tag{2.6.10}$$

According to the definition of DOP given by Equation 2.6.4,

$$\sqrt{s_1^2 + s_2^2 + s_3^2} = DOP \tag{2.6.11}$$

These normalized Stokes parameters can be displayed in a real three-dimensional space. Any state of polarization can be represented as a vector $s = [s_1\ s_2\ s_3]$, as shown in Figure 2.6.2.

The value of each normalized Stokes parameter ranges between –1 and +1. If the optical signal is fully polarized, the normalized Stokes vector endpoint is always on a sphere with unit radius, which is commonly referred to as *Poincare sphere*. On the other hand, if the optical signal is partially polarized, the endpoint of the normalized Stokes vector should be inside the unit sphere and the length of the vector is equal to the DOP of the optical signal, as defined by Equation 2.6.11. To convert variables from a three-dimensional Cartesian coordinate to a spherical polar coordinate, an azimuthal angle $0 < \psi < 2\pi$ and a polar angle $-\pi/2 < \zeta < \pi/2$ must be introduced with

$$s_1 = |s| \cos\zeta \cos\psi \quad s_2 = |s| \cos\zeta \sin\psi \quad s_3 = |s| \sin\zeta \tag{2.6.12}$$

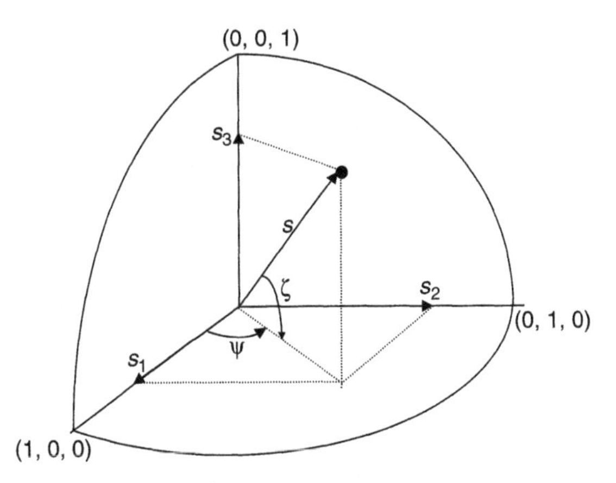

Figure 2.6.2 Three-dimensional representation of a normalized Stokes vector.

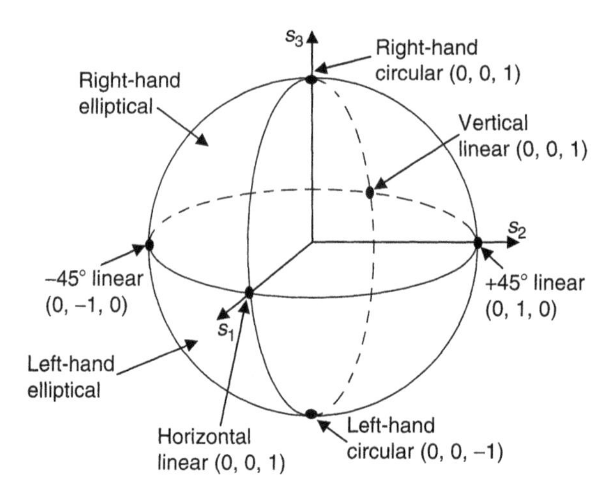

Figure 2.6.3 Polarization states represented on a Poincare sphere.

Since each polarization state of an optical signal can be represented by a specific location on the Poincare sphere, it is useful to identify a few special locations on the Poincare sphere, as shown in Figure 2.6.3, and their relationships with the polarization states described in Figure 2.6.1.

For *circular* polarization with $|E_{x0}| = |E_{y0}|$ and $\phi = \pm\pi/2$, Equations 2.6.6–2.6.10 indicate that $s_1 = 0$, $s_2 = 0$ and $s_3 = \pm|s|$. Therefore, the North Pole and the South Pole on the Poincare sphere represent circular polarizations of the optical signal. Specifically, the North and South Poles represent right and left circular polarization, respectively.

For a *linear* polarization, $\phi = 0$ so that $s_3 = 0$. Therefore, the Stokes vector endpoint of an optical signal with linear polarization has to be located on the equator of the Poincare sphere. As examples, for a horizontal linear polarization $|E_{y0}| = 0$, $s_1 = |s|$ and $s_2 = s_3 = 0$, whereas for a vertical linear polarization $|E_{x0}| = 0$, $s_1 = -|s|$ and $s_2 = s_3 = 0$. These two polarization states are represented at (1, 0, 0) and (–1, 0, 0), respectively, on the equator. For optical signals with $\pm 45°$ linear polarizations, $|E_x| = |E|$ and $\phi = 0$ or π so that $s_1 = s_3 = 0$ and $s_2 = \pm|s|$. These two polarization states are therefore located at $(0, \pm 1, 0)$ on the equator.

$(\pm S_0, 0, 0)$ $[S_1 = S_0, S_2 = 0$ and $S_3 = 0]$ represents either horizontal or vertical polarized light (because either $|E_x| = |E_y|$ or $|E_y| = |E_x|$ to make $S_2 = 0$ and $S_3 = 0$).

In general, right and left elliptical polarization states occupy the northern and the southern hemisphere, respectively, on the Poincare sphere. The azimuthal angle ψ indicates the spatial tilt of the ellipse, and the polar angle ζ is related to the differential optical phase ϕ between E_x and E_x by $\sin\zeta = \sin 2\alpha \sin\phi$ where $\alpha = \tan^{-1}(|E_{x0}|/|E_{y0}|)$.

2.6.3 Optical Polarimeters

An optical polarimeter is an instrument used to measure the Stokes parameters of optical signals [20, 21, 22, 23]. According to the definition of Stokes parameters given by Equations 2.6.5–2.6.8, they can be obtained by measuring the optical powers after the optical signal passes through polarization-sensitive optical devices such as linear polarizer and retarders. If we define the following seven measurements:

P = total optical power

$P_{0°}$ = optical power measured after a linear horizontal polarizer

$P_{90°}$ = optical power measured after a linear vertical polarizer

$P_{+45°}$ = optical power measured after a linear +45° polarizer

$P_{-45°}$ = optical power measured after a linear –45° polarizer

$P_{+45°,\lambda/4}$ = optical power measured after a $\lambda/4$ retarder and a linear +45° polarizer

$P_{-45°,\lambda/4}$ = optical power measured after a $\lambda/4$ retarder and a linear –45° polarizer

then the Stokes parameters can be obtained as

$$S_0 = P_{0°} + P_{90°} \tag{2.6.13}$$

$$S_1 = P_{0°} - P_{90°} \tag{2.6.14}$$

$$S_2 = P_{+45°} - P_{-45°} \tag{2.6.15}$$

$$S_3 = P_{+45°,\lambda/4} - P_{-45°,\lambda/4} \tag{2.6.16}$$

However, these seven measurements are not independent. In fact, only four measurements are necessary to determine the Stokes parameters. For example, use $P_{0°}$, $P_{90°}$, $P_{+45°}$ and $P_{+45°,\lambda/4}$ we can find S_2 and S_3 by

$$S_2 = P_{+45°} - P_{-45°} = 2P_{+45°} - S_0 \tag{2.6.17}$$

$$S_3 = P_{+45°,\lambda/4} - P_{-45°,\lambda/4} = 2P_{+45°,\lambda/4} - S_0 \tag{2.6.18}$$

Based on these relations, most polarimeters rely on four independent power measurements, either in parallel or sequential. As shown in Figure 2.6.4, a parallel measurement technique uses a 1×4 beam splitter that splits the input optical signal into four equal parts. A polarizer is used in each of the first three branches, with 0°, 90°, and 45° orientation angles. The fourth branch has a $\lambda/4$ waveplate and a polarizer oriented at 45°. The principle axis of the waveplate is aligned with 0° so that it introduces an additional $\pi/2$ phase shifts

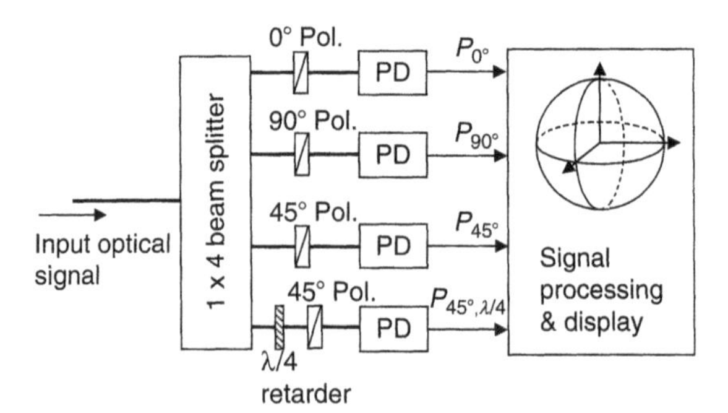

Figure 2.6.4 Block diagram of a polarimeter with parallel detection. *Pol.:* polarizer, *PD:* photodetector.

between the x and y components of the optical field. A photodetector detects the optical power of each branch, providing the required parameters of $P_{0°}$, $P_{90°}$, $P_{+45°}$ and $P_{+45°,\lambda/4}$, respectively. These optical power values are then analyzed by a signal processing system, and Stokes parameters can be calculated using Equations 2.6.13, 2.6.14, 2.6.17, and 2.6.18. Most polarimeters have graphic interfaces that allow the display of the measured Stokes parameters on a Poincare sphere.

Another type of polarimeter uses sequential measurement, which requires only one set of polarizers and waveplates, as shown in Figure 2.6.5. The simplest measurement setup is shown in Figure 2.6.5(a), where a $\lambda/4$ waveplate and a polarizer are used and their principle axis can be rotated manually. The power of optical signal passing through these devices is measured and recorded with a photodetector followed by an electronic circuit. The four measurements of $P_{0°}$, $P_{90°}$, $P_{+45°}$ and $P_{+45°,\lambda/4}$ as described previously can be performed one at a time by properly adjusting the orientations (or removing, if required) of the $\lambda/4$ waveplate and the polarizer. This measurement setup is simple; however, one has to make sure that the polarization state of the optical signal does not change during each set of measurements. Since manual adjustment is slow, the determination of Stokes parameters could take as long as several minutes. It is not suitable for applications in fiber-optical systems where polarization variation may be fast.

To improve the speed of sequential measurement, electrically switchable retarders can be used as shown in Figure 2.6.5(b). In this configuration, the orientation of the optical axis can be switched between 90° and 135° for the first retarder and between 135° (equivalent to –45°) and 0° for the second retarder. The optical axis of the polarizer is fixed at 0°.

Table 2.6.1 shows the state of polarizations (SOAs) at locations A, B, and C (referring to Figure 2.6.5(b)) with different combinations of retarders optic-axis

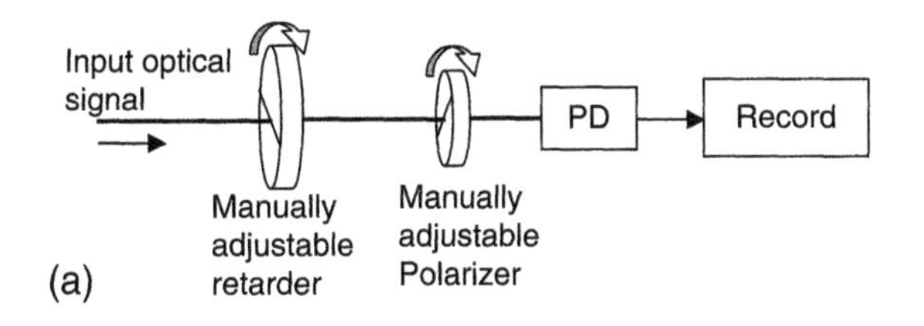

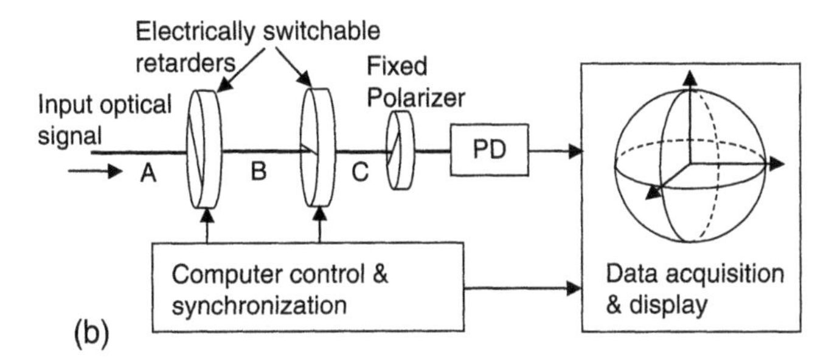

Figure 2.6.5 Block diagram of polarimeters with sequential detection. (a) Manual adjustment (b) automated measurement.

Table 2.6.1

State of Polarizations (SOPs) at Locations A, B, and C in Figure 2.6.5(b) with Different Sets of Retarder Optic-Axis Orientations; *RHC:* Right-Hand Circular

SOP at Point A	First Retarder Optic Axis	SOP at Point B	Second Retarder Optic Axis	SOP at Point C	Equivalent PD Power
0°	90°	0°	0°	0°	$P_{0°}$
90°	135°	RHC	135°	0°	$P_{90°}$
45°	90°	RHC	135°	0°	$P_{+45°}$
RHC	135°	0°	0°	0°	$P_{+45°,\ \lambda/4}$

orientations. Four specific combinations of retarder optic-axis orientations can convert input SOPs of 0° linear, 90° linear, 45° linear, and right-hand circular into 0° linear at the output, which is then selected by the fixed polarizer. Therefore, the optical powers measured after the polarizer should be equivalent to $P_{0°}$, $P_{90°}$, $P_{+45°}$ and $P_{+45°,\lambda/4}$ as defined previously. Thanks to the rapid development of liquid crystal technology, fast switchable liquid crystal retarders are commercially available with submillisecond-level switching speed, which allows the Stokes parameters to be measured using the sequential method. As shown in Figure 2.6.5, a computerized control system is usually used to synchronize data acquisition with instantaneous retarder orientation status. Therefore, the measurement can be fast and fully automatic.

2.7 MEASUREMENT BASED ON COHERENT OPTICAL DETECTION

Coherent detection is a novel detection technique that is useful both in fiber-optic communication systems [24] and in optical measurement and sensors. In a direct detection optical receiver, thermal noise is usually the dominant noise, especially when the power level of an optical signal is low when it reaches the receiver.

In coherent detection, a strong local oscillator is used, mixing with the optical signal at the receiver and effectively amplifying the weak optical signal. Therefore, compared to direct detection, coherent detection has much improved detection sensitivity. In addition, coherent detection also provides superior frequency selectivity because of the wavelength tunability of local oscillator [Betti 1995].

The coherent detection technique was investigated extensively in the 1980s to increase receiver sensitivity in optical communication systems. However, the introduction of EDFA in the early 1990s made coherent detection less attractive, mainly for two reasons: (1) EDFA provides sufficient optical amplification without the requirement of an expensive local oscillator and (2) EDFA has wide gain bandwidth to support multichannel WDM optical systems. Nevertheless, coherent detection still has its own unique advantages because optical phase information is preserved, which is very important in many practical applications such as optical measurement, instrumentation, and optical signal processing.

2.7.1 Operating Principle

Coherent detection originates from radio communications, where a local carrier mixes with the received RF signal to generate a product term. As a result, the received RF signal can be demodulated or frequency translated.

A block diagram of coherent detection is shown in Figure 2.7.1. In this circuit, the received signal $m(t)\cos(\omega_{sc}t)$ has an information-carrying amplitude $m(t)$ and an RF carrier at frequency ω_{sc}, whereas the local oscillator has a single

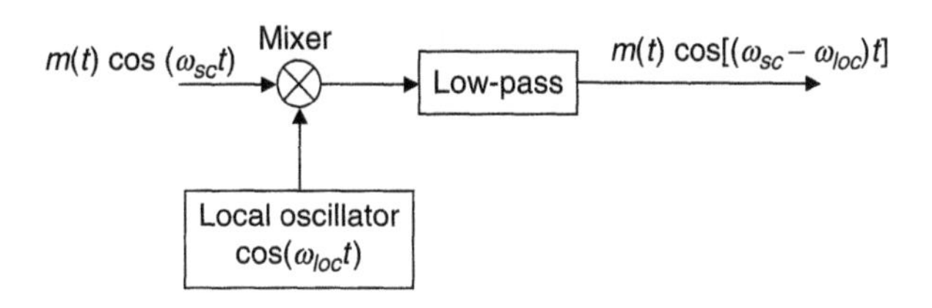

Figure 2.7.1 Block diagram of coherent detection in radio communications.

frequency at ω_{loc}. The RF signal multiplies with the local oscillator in the RF mixer, generating the sum and the difference frequencies between the signal and the local oscillator. The process can be described by the following equation:

$$m(t)\cos(\omega_{sc}t) \times \cos(\omega_{loc}t) = \frac{m(t)}{2}\{\cos[(\omega_{sc} + \omega_{loc})t] + \cos[(\omega_{sc} - \omega_{loc})t]\} \tag{2.7.1}$$

A lowpass filter is usually used to eliminate the sum frequency component and thus the baseband signal can be recovered if the frequency of the local oscillator is equal to that of the signal ($\omega_{loc} = \omega_{sc}$). When the RF signal has multiple and very closely spaced frequency channels, excellent frequency selection in coherent detection can be accomplished by tuning the frequency of the local oscillator. This technique has been used in ratio communications for many years, and high-quality RF components such as RF mixers and amplifiers are standardized.

For coherent detection in lightwave systems, although the fundamental principle is similar, its operating frequency is many orders of magnitude higher than the radio frequencies; thus the required components and circuit configurations are different. In a lightwave coherent receiver, mixing between the input optical signal and the local oscillator is done in a photodiode, which is a square-law detection device. A typical schematic diagram of coherent detection in a lightwave system is shown in Figure 2.7.2, where the incoming optical signal and the optical local oscillator are combined in an optical coupler. The optical coupler can be made by a partial reflection mirror in free space or by a fiber directional coupler in guided wave optics. To match the polarization state between the input optical signal and the local oscillator (LO), a polarization controller is used; it can be put either at the LO or at the input optical signal. A photodiode converts the composite optical signal into an electrical domain and performs mixing between the signal and LO.

Consider the field vector of the incoming optical signal as

$$\vec{E}_s(t) = \vec{A}_s(t)\exp(j\omega_s t + j\phi_s(t)) \tag{2.7.2}$$

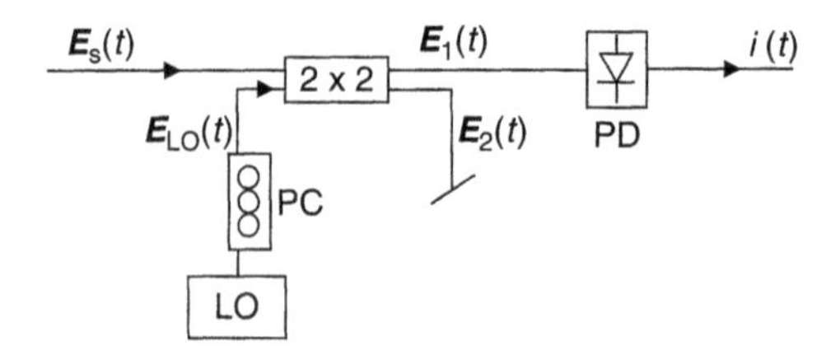

Figure 2.7.2 Block diagram of coherent detection in a lightwave receiver. *PC:* polarization controller, *PD:* photodiode, *LO:* local oscillator.

and the field vector of the local oscillator

$$\vec{E}_{LO}(t) = \vec{A}_{LO}\exp(j\omega_{LO}t + j\phi_{LO}) \tag{2.7.3}$$

where $\vec{A}_s(t)$ and $\vec{A}_{LO}(t)$ are the real amplitudes of the incoming signal and the LO, respectively, ω_s and ω_{LO} are their optical frequencies, and $\phi_s(t)$ and ϕ_{LO} are their optical phases. According to the optical transfer function of the 2×2 optical coupler,

$$\begin{bmatrix}\vec{E}_1\\ \vec{E}_2\end{bmatrix} = \begin{bmatrix}\sqrt{1-\varepsilon} & j\sqrt{\varepsilon}\\ j\sqrt{\varepsilon} & \sqrt{1-\varepsilon}\end{bmatrix}\begin{bmatrix}\vec{E}_S\\ \vec{E}_{LO}\end{bmatrix} \tag{2.7.4}$$

where ε is the power-coupling coefficient of the optical coupler. In this simple coherent receiver configuration, only one of the two outputs of the 2×2 coupler is used and the other output is terminated. The composite optical signal at the coupler output is

$$\vec{E}_1(t) = \sqrt{1-\varepsilon}\,\vec{E}_s(t) + j\sqrt{\varepsilon}\,\vec{E}_{Lo}(t) \tag{2.7.5}$$

As discussed in Section 1.2, a photodiode has a square-law detection characteristic and the photocurrent is proportional to the square of the input optical signal, that is,

$$\begin{aligned} i(t) &= \Re\,|\,\vec{E}_1(t)|^2\\ &= \Re\left\{(1-\varepsilon)|\,A_s(t)\,|^2 + \varepsilon|\,A_{LO}\,|^2 + 2\sqrt{\varepsilon(1-\varepsilon)}\,\vec{A}_s(t)\cdot\vec{A}_{LO}\cos(\omega_{IF}t + \Delta\phi(t))\right\} \end{aligned} \tag{2.7.6}$$

where $\Re$ is the responsivity of the photodiode, $\omega_{IF} = \omega_s - \omega_{LO}$ is the frequency difference between the signal and the LO, which is referred to as the *intermediate frequency* (IF), and $\Delta\phi(t) = \phi_s(t) - \phi_{LO}$ is their relative phase difference. We have neglected the sum-frequency term in Equation 2.7.6 because $\omega_s + \omega_{LO}$ will still be in the optical domain and will be eliminated by the RF circuit that follows the photodiode.

The first and the second terms on the right side of Equation 2.7.6 are the direct detection components of the optical signal and the LO. The last term is the coherent detection term, which is the result of mixing between the optical signal and the LO.

Typically, the LO is a laser operating in continuous waves. Suppose the LO has no intensity noise; $A_{LO} = \sqrt{P_{LO}}$ is a constant, where P_{LO} is the optical power of the LO. In general, optical power of the LO is much stronger than that of the optical signal such that $|A_s(t)|^2 \ll |\vec{A}_s(t) \cdot \vec{A}_{LO}|$. Therefore the only significant component in Equation 2.7.6 is the last term, and thus

$$i(t) \approx 2\Re\sqrt{\varepsilon(1-\varepsilon)}\cos\theta\sqrt{P_s(t) \cdot P_{LO}}\cos(\omega_{IF}t + \Delta\phi(t)) \tag{2.7.7}$$

where $\cos\theta$ results from the dot product of $\vec{A}_s(t) \cdot \vec{A}_{LO}$, which is the angle of polarization state mismatch between the optical signal and the LO. $P_s(t) = |A_s(t)|^2$ is the signal optical power. Equation 2.7.7 shows that the detected electrical signal level depends also on the coefficient of the optical coupler ε. Since $2\sqrt{\varepsilon(1-\varepsilon)}$ reaches the maximum value of 1 when $\varepsilon = ½$, 3-dB couplers are usually used in the coherent detection receiver. In addition, in the ideal case, if the polarization state of the LO matches the incoming optical signal, $\cos\theta = 1$, therefore the photocurrent of coherent detection is simplified as

$$i(t) \approx \Re\sqrt{P_s(t) \cdot P_{LO}}\cos(\omega_{IF}t + \Delta\phi) \tag{2.7.8}$$

Equation 2.7.8 shows that coherent detection shifts the spectrum of the optical signal from the optical carrier frequency ω_s to an intermediate frequency ω_{IF}, which can be handled by an RF circuit. In general, the coherent detection can be categorized as homodyne detection if $\omega_{IF} = 0$ or heterodyne detection if $\omega_{IF} \neq 0$.

2.7.2 Receiver SNR Calculation of Coherent Detection

2.7.2.1 Heterodyne and Homodyne Detection

In coherent detection, the photocurrent is proportional to the field of the optical signal and thus the phase information of the optical signal is preserved. In the simplest coherent detection case, assuming the matched polarization states between the signal and the LO, the photocurrent is

$$i(t) = \Re A_s(t)A_{LO}\cos(\omega_{IF}t + \Delta\phi(t)) \tag{2.7.9}$$

In a heterodyne configuration, the intermediate frequency ω_{IF} is usually much higher than the bandwidth of the signal carried by P_s(t). To convert the IF signal into baseband, two basic techniques can be used. One of these techniques, shown in Figure 2.7.3(a), uses an RF bandpass filter and an RF detector.

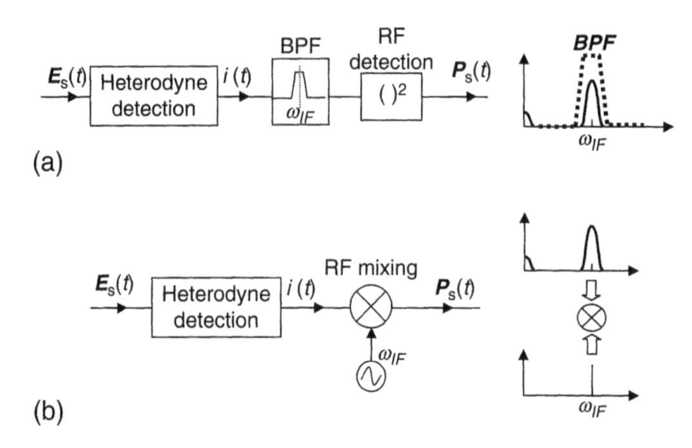

Figure 2.7.3 (a) Heterodyne optical detection and direct RF detection and (b) heterodyne optical detection and RF coherent detection. *PC:* polarization controller, *PD:* photodiode, *LO:* local oscillator.

This is usually referred to as *RF direct detection.* Another technique, shown in Figure 2.7.3(b), uses an RF local oscillator at the intermediate frequency ω_{IF}, which mixes with the heterodyne signal to recover the baseband signal. This second technique is called *RF coherent detection.*

Although both of these two techniques recover baseband signals from the intermediate frequency, each has its unique characteristics. Direct RF detection is relatively insensitive to the IF frequency variation because the width of the bandpass filter can be chosen wider than the signal bandwidth, but on the other hand, a wider filter bandwidth also increases the noise level. RF coherent detection has higher detection efficiency compared to direct RF detection because a strong RF local oscillator effectively amplifies the signal level. In addition, the signal-to-noise ratio of RF coherent detection is 3-dB higher than the direct RF detection because only the in-phase noise component is involved and the quadrature noise component can be eliminated. Obviously, RF coherent detection is sensitive to the variation of the intermediate frequency, and typically a frequency locked loop needs to be used.

In coherent homodyne detection, the intermediate frequency is set to $\omega_{IF} = 0$, that is, the frequency of the LO is equal to that of the signal optical carrier. Therefore baseband information carried in the input optical signal is directly obtained in the homodyne detection as

$$i(t) = \Re A_s(t) A_{LO} \cos\Delta\phi(t) \tag{2.7.10}$$

Although homodyne detection seems to be simpler than homodyne detection, it requires both frequency locking and phase locking between the LO and the optical signal because a random phase $\Delta\phi(t)$ in Equation 2.7.10 may also introduce signal fading.

2.7.2.2 Signal-to-Noise-Ratio in Coherent Detection Receivers

In an optical receiver with coherent detection, the power of LO is usually much stronger than the received optical signal; therefore the SNR is mainly determined by the shot noise caused by the LO. If we consider only thermal noise and shot noise, the receiver SNR is

$$SNR = \frac{\langle i^2(t)\rangle}{\langle i_{th}^2\rangle + \langle i_{sh}^2\rangle} \tag{2.7.11}$$

where

$$\langle i^2(t)\rangle = \frac{\Re^2 P_s(t)\cdot P_{LO}}{2} \tag{2.7.12}$$

is the RF power of the signal;

$$\langle i_{th}^2\rangle = \frac{4kTB}{R_L} \tag{2.7.13}$$

is the thermal noise power, and

$$\langle i_{sh}^2\rangle = q\,\Re[P_s(t) + P_{LO}]B \tag{2.7.14}$$

is the short noise power, where B is the receiver electrical bandwidth, R_L is the load resistance, and q is the electron charge. Then the SNR can be expressed as

$$SNR = \frac{1}{2B}\cdot\frac{\Re^2 P_s(t)P_{LO}}{4kT/R_L + q\,\Re[P_s(t) + P_{LO}]} \tag{2.7.15}$$

Figure 2.7.4(a) shows the SNR comparison between direct detection and coherent detection, where we assume that the receiver bandwidth is $B = 10\,\text{GHz}$, the load resistance is $R_L = 50\Omega$, the photodiode responsivity is $\Re = 0.75A/W$, the temperature is $T = 300K$, and the power of the LO is fixed at $P_{LO} = 20$ dBm. Obviously, coherent detection significantly improves the SNR compared to direct detection, especially when the signal optical power level is low. In fact, if the LO power is strong enough, the shot noise caused by the LO can be significantly higher than the thermal noise; therefore Equation 2.7.15 can be simplified into

$$SNR_{coh} \approx \frac{\Re}{2Bq}P_s(t) \tag{2.7.16}$$

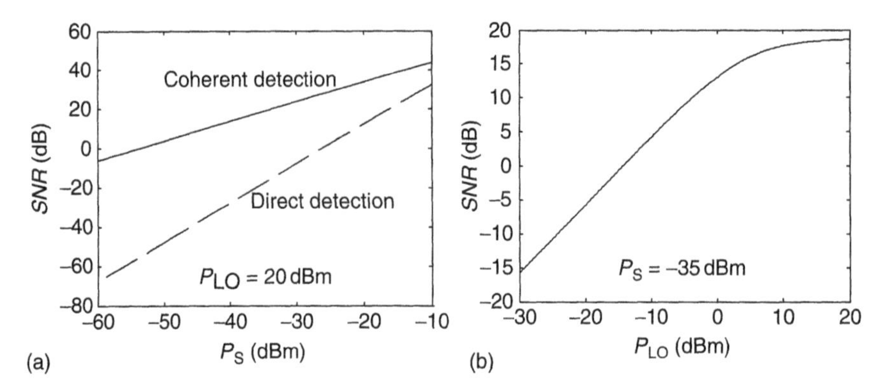

Figure 2.7.4 (a) Comparison between direct detection and coherent detection when P_{LO} is 20 dBm and (b) effect of LO power when P_s is fixed at -35 dBm. Other parameters: $B = 10$ GHz, $R_L = 50$W, R = 75% and $T = 300$K.

In this approximation, we also assumed that $P_{LO} \gg P_s$, which is usually true. Therefore *SNR* in a coherent detection receiver is linearly proportional to the power of the input optical signal P_s. In contrast, in a direct detection receiver, $SNR \propto P_s^2$.

Figure 2.7.4(b) shows that SNR in coherent detection also depends on the power of the LO. If the LO power is not strong enough, the full benefit of coherent detection is not achieved and the SNR is a function of the LO power. When LO power is strong enough, the SNR no longer depends on LO because strong LO approximation is valid and the SNR can be accurately represented by Equation 2.7.16.

2.7.3 Balanced Coherent Detection and Polarization Diversity

Coherent detection using the simple configuration shown in Figure 2.7.2 has both the mixed-frequency term and direct-detection terms shown in Equation 2.7.6. In the analysis of the last section, we assumed that the local oscillator has no intensity noise and that the direct-detection term of the signal is negligible. In practice, however, the direct-detection term may overlap with the IF component and introducing crosstalk. In addition, since the optical power of the LO is significantly higher than the optical signal, any intensity noise of the LO would introduce the excessive noise in coherent detection.

Another important concern is that a strong LO may saturate the electrical pre-amplifier after the photodiode. Therefore, the first two direct detection terms in Equation 2.7.6 may degrade the performance of coherent detection.

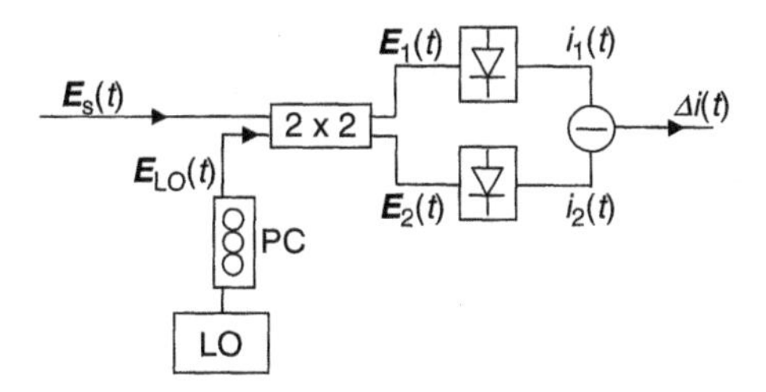

Figure 2.7.5 Block diagram of balanced coherent detection.

To solve this problem and eliminate the effect of direct detection, a balanced coherent detection configuration is often used. The schematic diagram of a balanced coherent detection is shown in Figure 2.7.5. Instead of using one photodetector as shown in Figure 2.7.2, two head-to-toe photodetectors are used in this balanced coherent detection configuration.

Based on the transfer function of the 2 × 2 optical coupler shown in Equation 2.7.4, the optical fields at the two output ports are

$$\vec{E}_1(t) = \left[\vec{E}_s(t) + j\vec{E}_{Lo}(t)\right]/\sqrt{2} \tag{2.7.17}$$

$$\vec{E}_2(t) = j\left[\vec{E}_s(t) - j\vec{E}_{Lo}(t)\right]/\sqrt{2} \tag{2.7.18}$$

and the photocurrents generated by the two photodiodes are

$$i_1(t) = \frac{1}{2}\Re\left[|A_s(t)|^2 + |A_{LO}|^2 + 2A_s(t)\cdot A_{LO}\cos(\omega_{IF}t + \Delta\phi)\cos\theta\right] \tag{2.7.19}$$

$$i_2(t) = \frac{1}{2}\Re\left[|A_s(t)|^2 + |A_{LO}|^2 - 2A_s(t)\cdot A_{LO}\cos(\omega_{IF}t + \Delta\phi)\cos\theta\right] \tag{2.7.20}$$

Therefore the direct-detection components can be eliminated by subtracting these two photocurrents and we obtain

$$\Delta i(t) = 2\Re A_s(t)A_{LO}\cos(\omega_{IF}t + \Delta\phi)\cos\theta \tag{2.7.21}$$

In the derivation of Equations 2.7.18–2.7.21, we assumed that the power coupling coefficient of the optical coupler is $\varepsilon = 0.5$ (3-dB coupler) and the two photodetectors have identical responsivities. In practical applications, paired photodiodes with identical characteristics are commercially available for this purpose.

In Equation 2.7.21, although direct detection components are eliminated, the effect of polarization mismatch, represented by cos θ, still affects coherent detection efficiency. Since the polarization state of the optical signal is usually not predictable after the transmission over long distance, the photocurrent term may fluctuate over time and thus causing receiver fading. Polarization diversity is a technique that overcomes polarization mismatch-induced receiver fading in

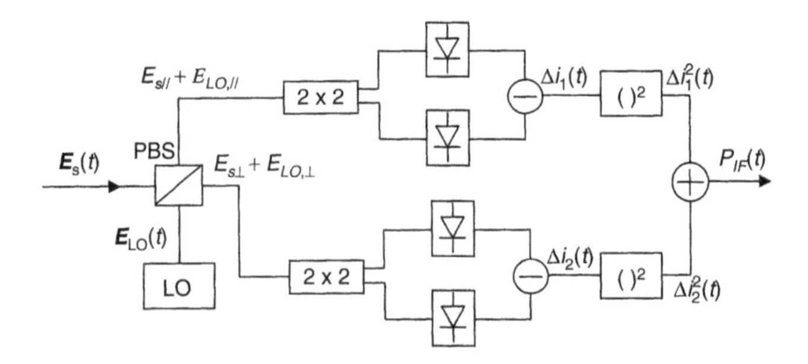

Figure 2.7.6 Block diagram of balanced coherent detection.

coherent systems. The block diagram of polarization diversity in a coherent receiver is shown in Figure 2.7.6, where a polarization beam splitter (PBS) is used to separate the input optical signal into horizontal ($E_{s//}$) and vertical ($E_{s\perp}$) polarization components. The polarization state of the local oscillator is aligned midway between the two principle axis of the PBS such that optical power of the LO is equally split between the two outputs of the PBS, that is, $E_{LO,//} = E_{LO,\perp} = E_{LO}/\sqrt{2}$. As a result, the photocurrents at the output of the two balanced detection branches are

$$\Delta i_1(t) = \sqrt{2}\,\Re A_s(t) A_{LO} \cos(\omega_{IF} t + \Delta\phi) \cos\theta \tag{2.7.22}$$

$$\Delta i_2(t) = \sqrt{2}\,\Re A_s(t) A_{LO} \cos(\omega_{IF} t + \Delta\phi) \sin\theta \tag{2.7.23}$$

where θ is the angle between the polarization state of the input optical signal and the principle axis of the PBS, $E_{s//} = E_s \cos\theta$ and $E_{s\perp} = E_s \sin\theta$.

Both photocurrents are squared by RF power detectors before they combine to produce the RF power of the coherently detected signal,

$$P_{IF}(t) = \Delta i_1^2 + \Delta i_2^2 = 2\Re^2 P_s(t) P_{LO} \cos^2(\omega_{IF} t + \Delta\phi) \tag{2.7.24}$$

This RF power is independent of the polarization state of the input optical signal. A lowpass filter can then be used to select the baseband information carried by the input optical signal.

2.7.4 Phase Diversity in Coherent Homodyne Detection

In coherent homodyne detection, the photocurrent is proportional to the field of the optical signal and thus the phase information of the optical signal is preserved. At the same time, phase noise in the LO and the received optical signal also plays an important role because $i(t) = \Re A_s(t) \cdot A_{LO} \cos\Delta\phi(t)$, as

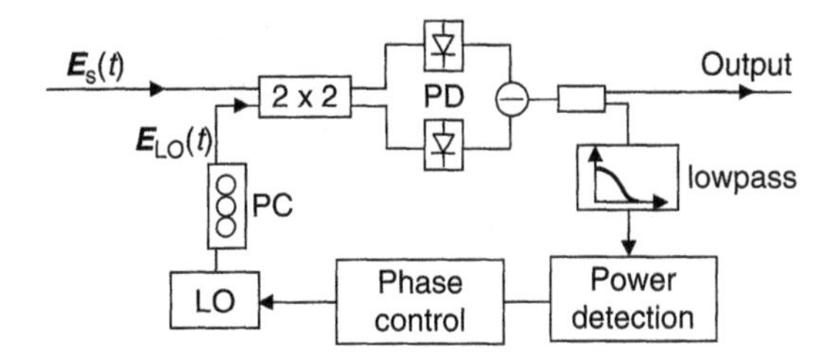

Figure 2.7.7 Coherent homodyne detection with a phase locked feedback loop.

given by Equation 2.7.10. If $\Delta\phi(t)$ varies randomly with time, the photocurrent will also fluctuate randomly and cause signal fading.

Traditionally, a phase locked loop can be used to overcome this signal fading problem. Assuming that the variation of $\Delta\phi(t)$ is much slower than the data rate on the optical carrier, the $\cos\Delta\phi(t)$ term can be seen as a slow varying envelope. Figure 2.7.7 schematically shows a phase locked loop, which includes a lowpass filter that selects the slow varying envelope, a power detect, and a phase control unit. The optical phase of the LO is adjusted by the feedback from the phase control unit to maximize the power level such that $\cos\Delta\phi(t) = 1$.

In practice, a coherent homodyne system requires narrow linewidth lasers for both the transmitter and the LO for the low phase noise. The implementation of adaptive phase locked loop is also expensive, which limits the application of coherent homodyne receiver.

Another technique to minimize phase noise induced signal fading in coherent homodyne system is *phase diversity*. As shown in Figure 2.7.8, phase diversity coherent receiver is based on a 90° optical hybrid. It is well known that the transfer matrix of a conventional 3-dB, 2 × 2 optical fiber coupler is given by

$$\begin{bmatrix} \vec{E}_1 \\ \vec{E}_2 \end{bmatrix} = \frac{1}{\sqrt{2}} \begin{bmatrix} 1 & j \\ j & 1 \end{bmatrix} \begin{bmatrix} \vec{E}_S \\ \vec{E}_{LO} \end{bmatrix} \tag{2.7.25}$$

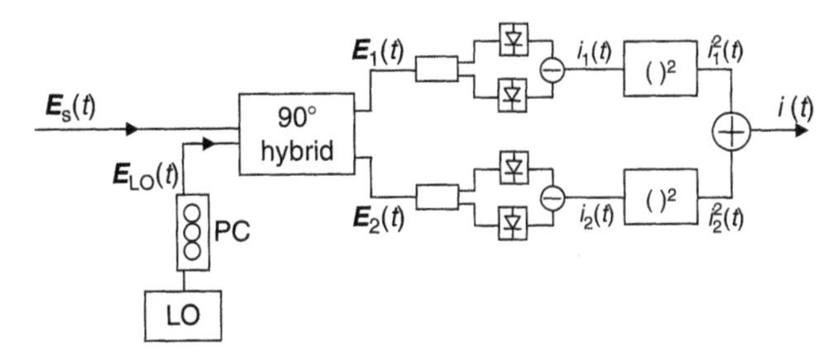

Figure 2.7.8 Block diagram of a phase diversity coherent homodyne receiver.

In the balanced detection coherent receiver shown in Figure 2.7.5 using a 3-dB, 2×2 fiber coupler, the cross-products are 180° apart in the two output arms after photodetectors, as shown in Equation 2.7.19 and 2.7.20. Therefore the conventional 2×2 fiber coupler can be classified as a 180° hybrid.

An ideal 90° optical hybrid should have a transfer matrix as

$$\begin{bmatrix} \vec{E}_1 \\ \vec{E}_2 \end{bmatrix} = \frac{1}{\sqrt{2}} \begin{bmatrix} 1 & \exp(j\pi/4) \\ \exp(j\pi/4) & 1 \end{bmatrix} \begin{bmatrix} \vec{E}_S \\ \vec{E}_{LO} \end{bmatrix} \tag{2.7.26}$$

Applying this 90° optical hybrid in the receiver shown in Figure 2.7.8 and assuming that the matched polarization states between the input optical signal and the LO,

$$i_1(t) = \Re A_s(t) \cdot A_{LO} \cos\left(\Delta\phi(t) - \frac{\pi}{4}\right) \tag{2.7.27}$$

$$i_2(t) = \Re A_s(t) \cdot A_{LO} \sin\left(\Delta\phi(t) - \frac{\pi}{4}\right) \tag{2.7.28}$$

where for simplicity we have neglected the direct detection components and assumed that the input optical signal and the LO have the same polarization state. Also, $\omega_{IF} = 0$ is assumed because of the homodyne detection. Obviously, in this circuit, there is a quadrature relationship between $i_1(t)$ and $i_2(t)$, and

$$i_1^2(t) + i_2^2(t) = \Re^2 P(t) \cdot P_{LO} \tag{2.7.29}$$

which is independent of the differential phase $\Delta\phi(t)$.

It is clear that in the phase diversity receiver configuration shown in Figure 2.7.8, the key device is the 90° optical hybrid. Unfortunately, the transfer matrix shown in Equation 2.7.26 cannot be provided by a simple 2×2 fiber coupler. In fact, a quick test shows that the transfer function of Equation 2.7.26 does not even satisfy the energy conservation principle, because $|E_1|^2 + |E_2|^2 \neq |E_s|^2 + |E_{LO}|^2$.

A number of optical structures have been proposed to realize the 90° optical hybrid, but the most straightforward way is to use a specially designed 3×3 fiber coupler with the following transfer matrix [26]:

$$\begin{bmatrix} E_1 \\ E_2 \\ E_3 \end{bmatrix} = \begin{bmatrix} \sqrt{0.2} & \sqrt{0.4}\exp\left(j\frac{3\pi}{4}\right) & \sqrt{0.4}\exp\left(j\frac{3\pi}{4}\right) \\ \sqrt{0.4}\exp\left(j\frac{3\pi}{4}\right) & \sqrt{0.2} & \sqrt{0.4}\exp\left(j\frac{3\pi}{4}\right) \\ \sqrt{0.4}\exp\left(j\frac{3\pi}{4}\right) & \sqrt{0.4}\exp\left(j\frac{3\pi}{4}\right) & \sqrt{0.2} \end{bmatrix} \begin{bmatrix} E_s \\ E_{LO} \\ 0 \end{bmatrix} \tag{2.7.30}$$

where only two of the three ports are used at each side of the coupler and therefore it is used as a 2 × 2 coupler. On the input side, the selected two ports connect to the input optical signal and the LO, whereas the first two output ports provide

$$E_{01} = \left(\sqrt{0.2}E_s(t) + \sqrt{0.4}\exp\left(j\frac{3\pi}{4}\right)E_{LO}\right) \quad (2.7.31)$$

$$E_{02} = \left(\sqrt{0.4}\exp\left(j\frac{3\pi}{4}\right)E_s(t) + \sqrt{0.2}E_{LO}\right) \quad (2.6.32)$$

After photodetection and neglecting the direct detection components, the AC parts of these two photocurrents are

$$i_1(t) = 2\Re\sqrt{0.08}A_s(t)A_{LO}\cos\left(\Delta\phi(t) - \frac{3\pi}{4}\right) \quad (2.7.33)$$

$$i_2(t) = 2\Re\sqrt{0.08}A_s(t)A_{LO}\sin\left(\Delta\phi(t) - \frac{3\pi}{4}\right) \quad (2.7.34)$$

Therefore, after squaring and combining, the receiver output is

$$i_1^2(t) + i_2^2(t) = 0.32\Re^2 P(t) \cdot P_{LO} \quad (2.7.35)$$

Compare this result with Equation 2.7.29, where an ideal 90° hybrid was used, there is a signal RF power reduction of approximately 5 dB using the 3 × 3 coupler. Because one of the three output ports is not used, obviously a portion of the input optical power is dissipated through this port.

2.7.5 Coherent OSA Based on Swept Frequency Laser

As has been discussed previously, coherent detection not only has high detection sensitivity, it also has excellent frequency selectivity because the coherent detection process converts the optical spectrum into RF domain. Optical spectrum analyzing using tunable laser and coherent detection has become practical because in recent years tunable semiconductor lasers are widely available with wide continuous tuning range and fast wavelength sweeping speed. Because of the narrow linewidth of tunable lasers in the submegahertz level, unprecedented spectral resolution can be obtained with coherent OSAs [25].

Figure 2.7.9 shows a block diagram of a coherent OSA; basically it is a balanced coherent detection receiver discussed in the previous section, except that the LO is a wavelength-swept laser that sweeps across the wavelength window of interest. The purpose of using balanced coherent detection is to minimize the direct detection component and improve the measurement signal-to-noise

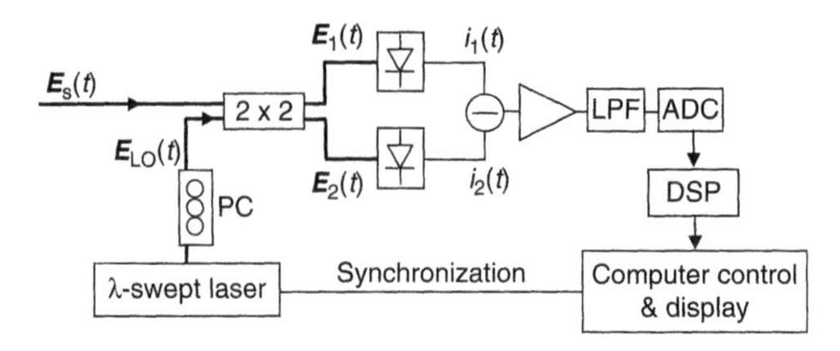

Figure 2.7.9 Block diagram of a coherent OSA. *LPF*: lowpass filter, *ADC*: analog-to-digital converter, *DSP*: digital signal processing.

ratio. Because the differential photocurrent is linearly proportional to the optical field of the input optical signal $E_s(t)$, this setup allows the measurement of the incident optical field strength.

First, consider that the input is a monochromatic optical signal:

$$E_s(t) = A_s(t)\exp[j2\pi\upsilon_s t + j\phi_s(t)] \tag{2.7.36}$$

where $A_s(t)$ is the amplitude of the incident light, υ_s is the mean optical frequency, and $\phi_s(t)$ is phase. The LO is a swept frequency laser and its optical field is

$$E_{LO}(t) = A_{LO}(t)\exp\left[j2\pi\int_0^t \upsilon_{LO}(\tau)d\tau + j\phi_{LO}(t)\right] \tag{2.7.37}$$

where $A_{LO}(t)$ and $\phi_{LO}(t)$ are the amplitude and phase of the local oscillator and $\int_0^t \upsilon_{LO}(\tau)d\tau = \int_0^t(\upsilon_0 + \gamma\tau)d\tau = \upsilon_0 t + \frac{1}{2}\gamma t^2$ is the instantaneous scanning phase of the local oscillator, υ_0 is the initial scanning frequency, and γ is the rate of frequency change with time. At the balanced coherent receiver, the differential photocurrent detected by the photodiodes is

$$\Delta i(t) = 2\Re|A_s(t)||A_{LO}(t)|\cos(\pi\gamma\, t^2 + \psi) \tag{2.7.38}$$

where $\Re$ is the responsivity of the two photodiodes and ψ is the combined phase of LO and the input optical signal. Assuming that the temporal response of the lowpass electrical filter is Gaussian, then after the filter the differential photocurrent becomes

$$\Delta i(t) = 2\Re|A_s(t)||A_{LO}(t)|\exp\left(\frac{t^2}{\tau}\right)\cos(\pi\gamma\, t^2 + \psi) \tag{2.7.39}$$

where τ is equal to the filter bandwidth divided by γ.

Figure 2.7.10 shows an example of the measured photocurrent as a function of time, which represents the signal optical spectrum. In this setup, a 1550 nm tunable external-cavity semiconductor laser was used as the local oscillator

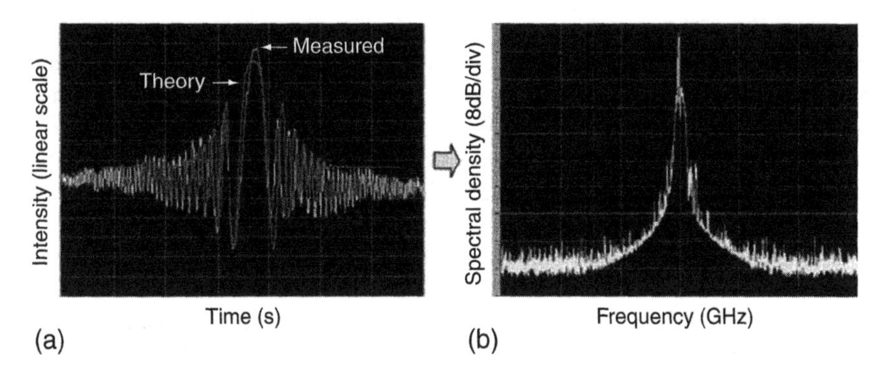

Figure 2.7.10 (a) A detected coherent detection photocurrent of an external-cavity laser (yellow solid line) and theoretical calculation result (red solid line). (b) Optical spectrum of the external-cavity laser.

with 100 KHz spectral linewidth. A balanced photodiode pair was used, each connected to an output of the 3-dB coupler, which ensures the subtraction of the direct detection intensity noise in this heterodyne coherent receiver. The wavelength of the local oscillator was linearly swept under a computer control. In this experiment, the wavelength scanning rate was 0.5 nm/s, which is equivalent to approximately 62.5 GHz/s. After the balanced coherent detection, a lowpass electrical filter was used with a bandwidth of approximately 500 kHz. the electrical signal was magnified by a logarithmic amplifier with 90 dB dynamic range. A 5 MHz, 12-bit A/D converter was used for data acquisition, the speed of which was 2 Mega-samples per second. Therefore, the sampling interval was equivalent to 31.25 kHz, which defined the frequency resolution due to data acquisition. It is important to note that the actual spectral resolution of this OSA was determined by the spectral linewidth of the local oscillator, which was 100 kHz in this case.

As the frequency of the LO sweeps, the incident optical field E_s is effectively sampled in the frequency domain by coherent detection. Therefore, the spectral density of the signal optical field is represented by the average fringe envelope described by Equation 2.7.39, and the power spectral density can be calculated from the square of the average envelope amplitude. In Figure 2.7.10, the input optical signal is a single-frequency laser with narrow linewidth; therefore the beating with the swept-frequency LO simply produces a photocurrent that is proportional to $\cos(\pi\gamma t^2 + \psi)$, as shown in Figure 2.7.10(a), where the measured photocurrent agrees well with the calculated result based on Equation 2.7.39.

Ideally, if the local oscillator has a linear phase, the phase of the input optical signal can be extracted from this measurement. However, a practical LO using an external cavity semiconductor laser may have phase discontinuities during frequency sweeping, especially when the frequency tuning range is large.

Therefore this phase measurement may not be reliable. In the future, if the LO phase continuity can be guaranteed during a frequency sweep, this coherent SOA technology may be used to measure phase distortion of the optical signal in an optical system introduced by fiber chromatic dispersion.

In a coherent SOA using frequency-swept LO, an important aspect is to translate the measured time-dependent photocurrent waveform into an optical power spectrum; this translation depends on the speed of the LO frequency sweep. Note that the resolution of the coherent OSA is determined by the linewidth of the local oscillator and the bandwidth of the receiver filter. Selecting the bandwidth of the lowpass filter should consider the frequency scanning rate of the LO. Assuming that the LO frequency scanning rate is γ GHz/s and the desired coherent OSA frequency resolution is Δf, then the scanning time interval across a resolution bandwidth Δf is

$$\Delta t = \frac{\Delta f}{\gamma} \tag{2.7.40}$$

Equivalently, to obtain the signal spectral density within a time interval Δt, the electrical filter bandwidth B should be wide enough such that

$$B \geq \frac{1}{\Delta t} = \frac{\gamma}{\Delta f} \tag{2.7.41}$$

In this measurement example, $\gamma = 62.5 GHz/s$ and B = 500 kHz, and the spectral resolution is approximately 125 kHz, according to Equation 2.7.41. Since the LO linewidth is only 100 kHz, it has a negligible effect on the spectral resolution of this OSA. Although a wider bandwidth of the filter gives higher spectral resolution of the OSA, it will certainly introduce more wideband noise and thus degrade receiver signal-to-noise ratio.

To measure a wideband optical signal, the coherent detection is equivalent to a narrowband optical filter. Continuously scanning the frequency of the LO across the frequency range of interest, the signal power spectral density can be obtained. Because of the excellent spectral resolution and detection sensitivity, a coherent OSA can be used as an optical system performance monitor, which is able to tell modulation data rates as well as modulation formats by measuring the optical spectral characteristics using the coherent OSA. As an example, Figure 2.7.11(a) and (b) show the measured optical spectrum of a 10 Gb/s optical system with RZ and NRZ modulation formats. Similarly, Figure 2.7.12(a) and (b) demonstrate the measured spectrum of a 2.5 Gb/s optical system with RZ and NRZ modulation formats. Clearly the coherent OSA revealed the detailed spectral features of the optical signals. This enables precise identification of datarate and modulation format of the signal in the optical domain, which cannot be achieved by conventional grating-based OSAs.

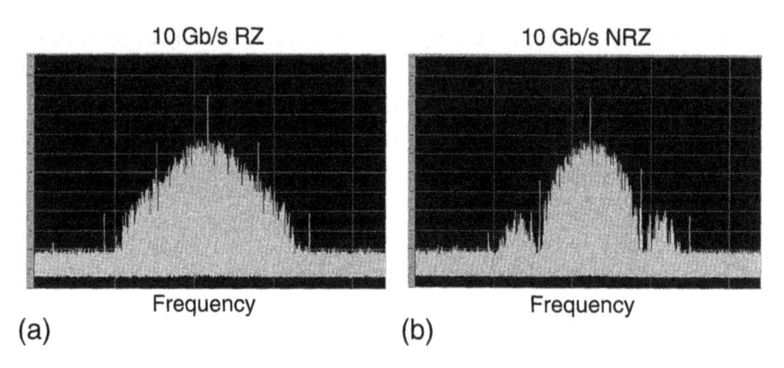

Figure 2.7.11 Optical spectrum of 10 Gb/s RZ (left) and 10 Gb/s NRZ (right) signals measured by a COSA. Horizontal: 15.6 GHz/div. Vertical: 8 dB/div.

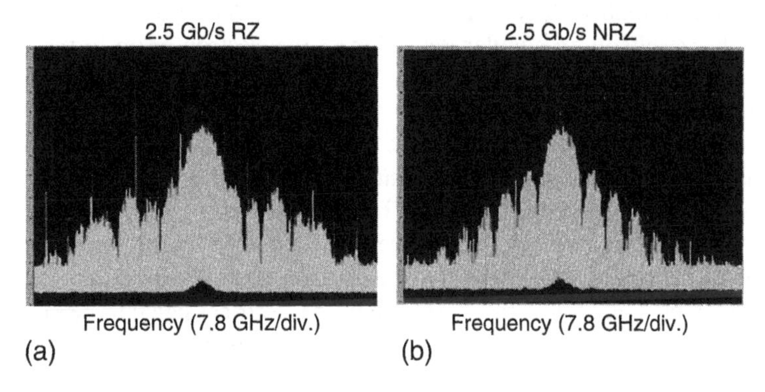

Figure 2.7.12 Optical spectrum of (a) 2.5 Gb/s RZ and (b) 2.5 Gb/s NRZ signals measured by a COSA. Horizontal: 7.8 GHz/div. Vertical: 8 dB/div.

2.8 WAVEFORM MEASUREMENT

Waveform measurement is a very important aspect of both electronic systems and optical systems. An oscilloscope can usually be used to characterize signal electric waveform, transient effect, and distortion of system response as well as eye-diagrams in digital communication systems. To display high-speed time-domain signal waveforms, an oscilloscope should have wide enough bandwidth, low noise, and excellent response linearity. Although traditionally an oscilloscope is based on analog detection and linear display, thanks to the rapid development of high-speed silicon technology and digital electronic circuits, digital oscilloscopes are becoming more and more popular. In addition, the combination with computerized digital signal processing and display has greatly improved the capability and flexibility of oscilloscopes. In this section, we will discuss the operating principles of various types of oscilloscopes, including analog oscilloscopes, digital oscilloscopes, and sampling oscilloscopes. The application of oscilloscopes in characterizing digital optical communication systems is also discussed.

2.8.1 Oscilloscope Operating Principle

The traditional oscilloscope is based on the analog format, in which the input time-domain electrical signal is linearly amplified and displayed on the screen.

Figure 2.8.1 shows the operation principle of an analog oscilloscope based on the cathode ray tube (CRT). The input signals to be measured are linearly amplified and applied directly to a pair of vertical deflection plates inside the CRT. This results in an angular deflection of the electron beam in the vertical direction, which is linearly proportional to the amplitude of the electrical signal. The time scale is introduced by a saw-tooth voltage signal inside the oscilloscope, which sweeps at a uniform rate and is applied to a pair of horizontal deflection plates of the CRT. This generates a repetitive and uniform left-to-right horizontal motion on the screen. The direction of the high-speed electrons generated by the electron gun inside the CRT is guided by the vertical and the horizontal deflectors and, therefore, the electron beam projected on the phosphor screen precisely traces the waveform of the electrical signal to be measured.

Figure 2.8.2 shows the block diagram of an analog oscilloscope. Each input signal channel passes through a pre-amplifier, a delay line, and a power amplifier before it is connected to the CRT to drive the vertical deflection plates. A trigger can be selected from either a signal channel or a separate external trigger. The purpose of triggering is to provide a timing to start each horizontal sweep of the CRT.

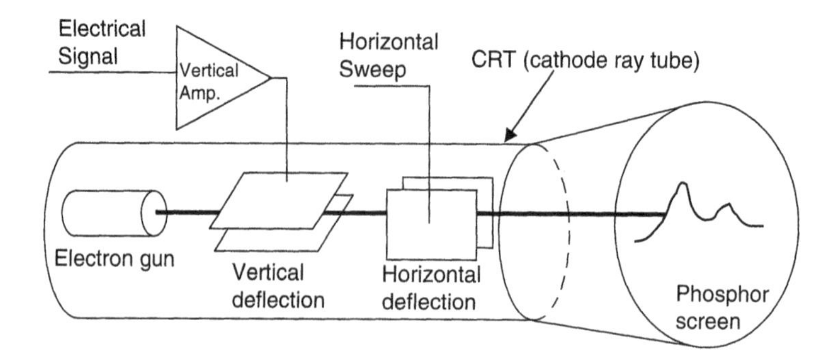

Figure 2.8.1 Illustration of an analog oscilloscope using a cathode ray tube.

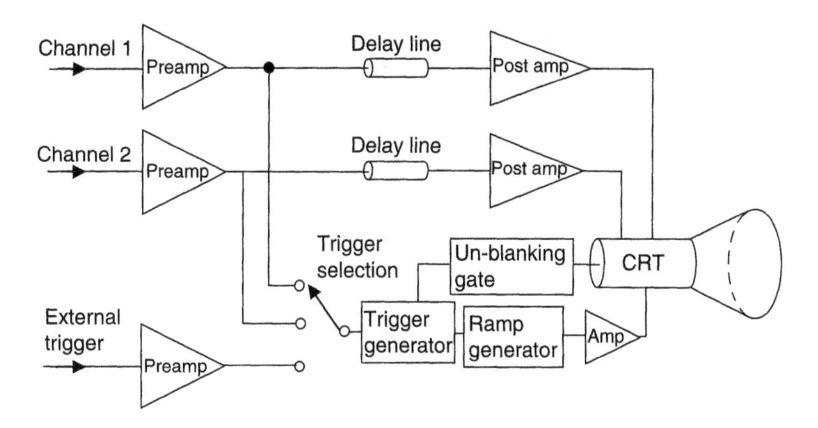

Figure 2.8.2 Block diagram of an analog oscilloscope.

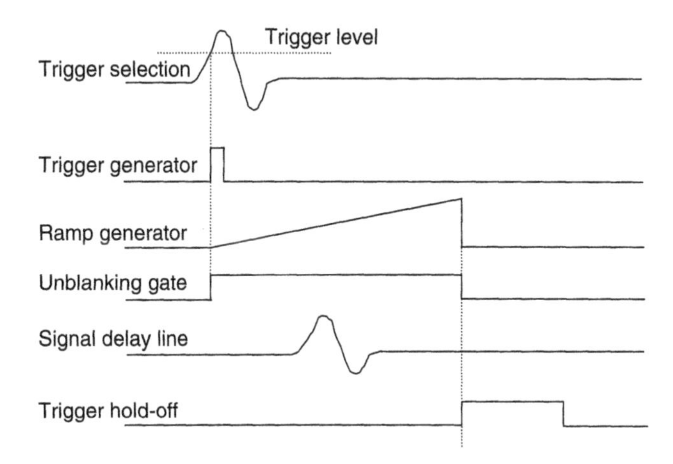

Figure 2.8.3 Waveforms and their synchronization at different parts of the oscilloscope.

Therefore, for periodic electrical signals, the period of the trigger should be either the same as or the exact multiple of the period length of the signal.

A trigger level can be set by the trigger generator and compare with the selected trigger signal. At the moment when the trigger signal amplitude reaches the trigger level, a pulse is created by the trigger generator, which sets the starting time for each horizontal sweep. A linear ramp signal is used to drive the horizontal sweep and an amplifier is used to adjust the horizontal driving amplitude.

The electron beam inside the CRT is normally blocked unless an unblanking signal is applied. As shown in Figure 2.8.3, the unblanking gate is activated only during the ramp period, which determines the display time window. An electrical delay line is used in each signal channel to allow the horizontal adjustment of the waveform on the oscilloscope screen. Finally, a trigger hold-off pulse appears to mark the end of the display window.

In recent years, rapid advances in digital electronic technologies, such as microprocessors, digital signal processing, and memory storage, have made digital oscilloscopes more and more popular. The combination between an oscilloscope and a computer enables many new capabilities and functionalities that could not be achieved previously. Figure 2.8.4 shows the block diagram of a basic digital oscilloscope.

In a digital oscilloscope, an A/D converter samples the signal of each channel and converts the analog voltage into a digital value. This sampled digital signal is stored in memory. The sampling frequency and timing are determined by the time base of a crystal oscillator. At a later time, the series of sampled digital values can be retrieved from memory and the graph of signal voltage versus time can be reconstructed and displayed on the screen. The graphic display of the digital oscilloscope is quite flexible using digital display technologies such as liquid crystal panel or even on a computer screen. Similar to analog oscilloscopes, triggering is also necessary in digital oscilloscopes to determine the exact instant in time to begin signal capturing. Two different modes of triggering are usually used in a digital oscilloscope.

The most often used trigger mode is *edge triggering*. As shown in Figure 2.8.5, in this mode a trigger pulse is generated at the moment the input signal passes through a threshold voltage, which is called a *trigger level*. The trigger pulse provides the exact timing to start each data acquisition process. To ensure the time accuracy of the trigger pulse, the trigger level can be adjusted

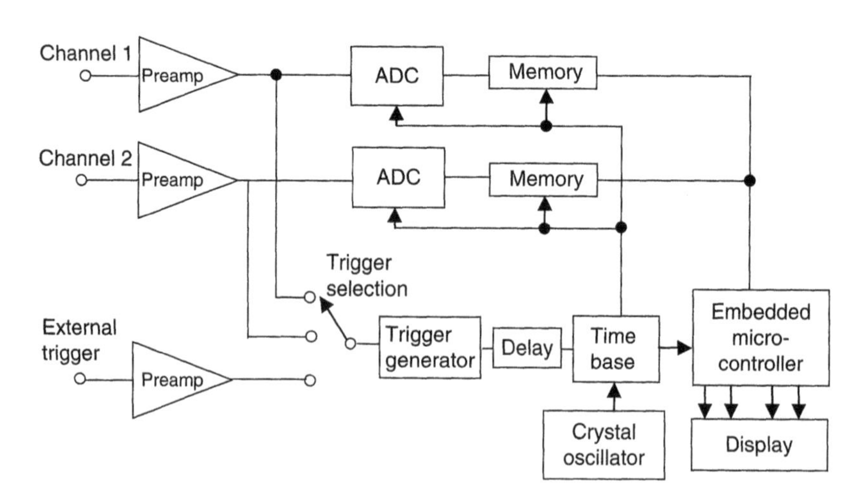

Figure 2.8.4 Block diagram of a digital oscilloscope.

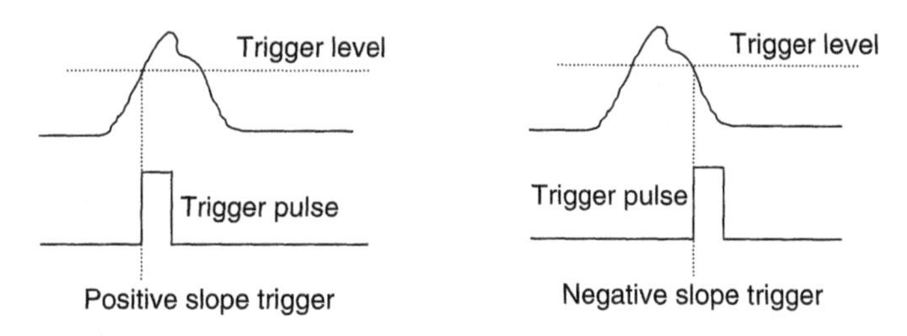

Figure 2.8.5 Illustration of edge triggering in a digital oscilloscope.

such that the slope of the signal at the trigger level is steep enough. Usually an oscilloscope also has an option that allows the selection of either a positive slope trigger or a negative slope trigger, as illustrated in Figure 2.8.5. The general rules of trigger-level selection are (a) it has to be within the signal amplitude range, (b) near the trigger level the signal should have the smallest time jitter, and (3) at the trigger level the signal should have the steepest slope.

Another often used trigger mode in a multichannel digital oscilloscope is *pattern triggering*. Instead of selecting a trigger level, the trigger condition for pattern triggering is defined as a set of states of the multichannel input. A trigger pulse is produced at the time when the defined pattern appears. Figure 2.8.6 shows an example of pattern triggering in a four-channel digital oscilloscope; the selected trigger pattern is high-low-high-low (HLHL). In this case, as soon as the preselected pattern (which is high, low, high, and low for channels 1, 2, 3, and 4, respectively) appears, a trigger pulse is generated. Pattern triggering is useful in digital systems and can be used in isolating simple logical patterns in digital systems.

In a digital oscilloscope, the speed of horizontal sweeps is determined by the selection of the time base. The sweep speed control labeled *time/division* is usually set in 1, 2, 5 sequence, such as 1 ms/division, 2 ms/division, 5 ms/division, 10 ms/division, and so on. To be able to accurately evaluate the transient time of

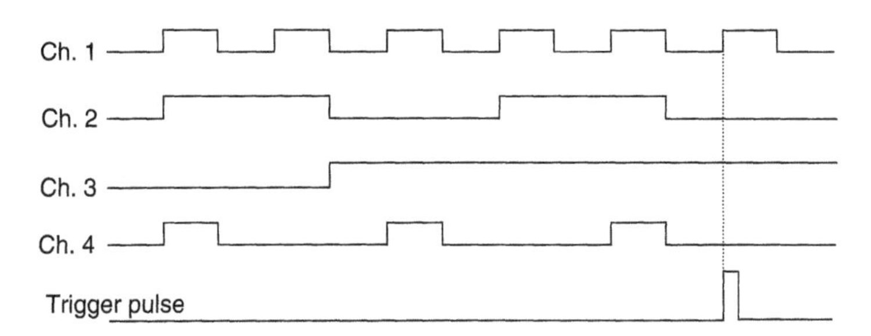

Figure 2.8.6 Pattern triggering in a four-channel digital oscilloscope using HLHL as the trigger pattern.

the measured waveform, horizontal sweeping speed of the oscilloscope needs to be precise. In a digital oscilloscope, the precision of the time base is mainly determined by the accuracy of the crystal oscillator.

The sweep control in an oscilloscope allows the selection of single sweep or normal sweep. If the *single sweep* button is pushed, when the trigger event is recognized the signal is acquired once and displayed on the screen. After that, the signal acquisition stops until the single sweep button is pressed again. If we choose *normal sweep*, the oscilloscope will operate in the continuous sweep mode. In this mode, as long as trigger pulses are generated regularly, the signal data is continually acquired and the graph on the display screen is continually updated. However, if the trigger condition is not met, the sweep stops automatically.

To avoid the stop of horizontal sweeping, *auto-trigger* mode may be used. This is similar to normal sweep mode except that if the trigger condition is not met for a certain time (called *trigger timeout*), a forced sweep is initiated by an artificial trigger. Therefore, in auto-trigger mode, sweeping never really stops, even if the trigger condition is not met. However, the disadvantage of auto-trigger is that the forced sweep is not synchronized to the signal. Unless the periodic signal has the same time base as the oscilloscope, the phase of the measured waveform will not be stable. In addition to the sweeping speed selection, the horizontal phase shift of the measured waveform can be adjusted by changing the relative delay between the signal and the trigger pulse. As shown in Figure 2.8.4, this is accomplished by a tunable trigger delay.

2.8.2 Digital Sampling Oscilloscopes

As discussed in the last section, data acquisition in a digital oscilloscope is accomplished by digital sampling and data storage. To capture enough details of the waveform, the sampling has to be continuous at a fast enough rate compared to the time scale of the waveform. This is referred to as *real-time sampling*. According to the Nyquist criterion, $f_s \geq 2f_{\mathrm{sig,max}}$ is required to characterize the waveform, where f_s is the sampling frequency and $f_{\mathrm{sig,max}}$ is the maximum signal frequency. For a sinusoid signal waveform, the maximum frequency is equal to its fundamental frequency, and sampling at Nyquist frequency, that is two samples per period, is sufficient to reproduce the waveform. However, this is assuming that we know that the waveform was sinusoid before the measurement.

In general, for an unknown waveform, the maximum frequency $f_{\mathrm{sig,max}}$ can be much higher than the fundamental frequency f_{sig}. Therefore, to capture waveform details, the sampling frequency has to be much higher than f_{sig}. Typically $f_s \geq 5\, f_{sig}$ is necessary for an accurate measurement. That means that in order to characterize a signal with 10 GHz fundamental frequency, the sampling frequency has to be higher than 50 GHz, which is a very challenging task for amplifiers and electronic circuits in the oscilloscope. The purpose of a sampling oscilloscope is to be able to measure high-speed signals using relatively low sampling rates, thus relaxing the requirements for the electronic circuits.

The operating principle of a sampling oscilloscope is based on the undersampling of repetitive waveforms. Although only a small number of samples might be acquired within each period of the waveform, with the sampling of over many periods and combining the data together, the waveform can be reconstructed with a large number of sampling points. This method is also called *equivalent-time sampling* and there are two basic requirements: (1) the waveform must be repetitive and (2) a stable trigger and precisely controlled relative delay must be available.

Figure 2.8.7(a) illustrates the principle of waveform sampling. Since the signal waveform must be repetitive, a periodic trigger pulse train is used and the synchronization between the first trigger pulse and the signal waveform is determined by the trigger-level selection. A discrete data point is sampled at the moment of each trigger pulse. By setting the trigger period to be slightly longer than the signal period by ΔT, data sampling happens at different positions within each period of the signal waveform. Since only one data point is acquired within each trigger period, by definition this is *sequential sampling*.

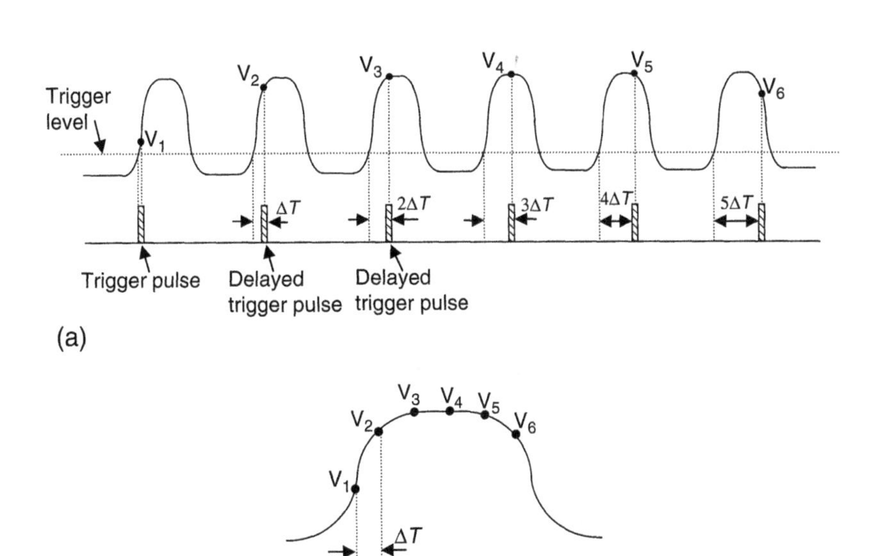

Figure 2.8.7 Illustration of (a) sequential sampling method and (b) the waveform reconstructed after sequential sampling.

If the total number of sampling points required to reconstruct the waveform on the screen is N and the time window of the waveform to display is ΔT_{dsp}, the sequential delay of each sampling event within a signal period should be

$$\Delta T = \frac{\Delta T_{dsp}}{N-1} \tag{2.8.1}$$

Figure 2.8.7(b) shows the reconstructed waveform and obviously ΔT represents the time domain resolution of the sampling. Here ΔT can be chosen very short to have high resolution while still using a relatively low sampling frequency. A compromise has to be made, though, when choosing to use high resolution; data acquisition becomes longer because a large number of signal periods will have to be involved. A major disadvantage is that since only one sampling point is allowed for each signal period, when the signal period is long, the measurement time can be very long.

Random sampling is an alternative sampling algorithm, as illustrated in Figure 2.8.8. In random sampling, the signal is continuously sampled at a frequency that is independent of the trigger frequency. Therefore, within a trigger period, more than one or less than one data point(s) may be sampled, depending on the frequency difference between the sampling and the triggering. In the measurement process, the data is stored in memory with the sampled voltage

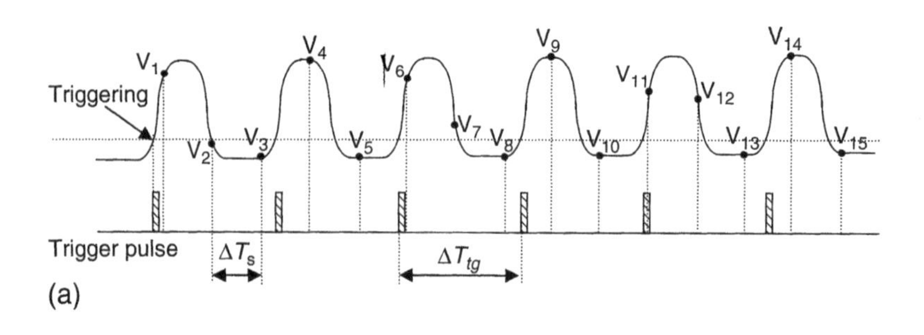

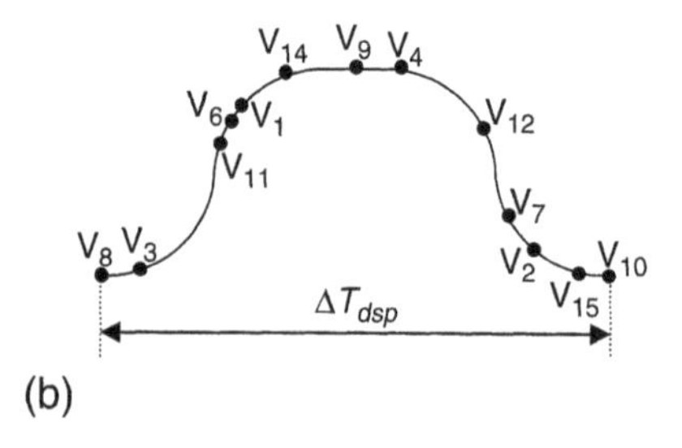

Figure 2.8.8 Illustration of (a) random sampling method and (b) the waveform reconstructed after random sampling.

value of the signal and the time at which the voltage was taken. The time difference between each sampling moment and the nearest trigger is collected and arranged to reconstruct the signal waveform on the display screen.

Assume that the sampling period is ΔT_s, the trigger period is ΔT_{tg} and the time window of signal to display on screen is ΔT_{dsp}. If $\Delta T_{dsp} = \Delta T_{tg}$, then, on average there are $N_{ave} = \Delta T_{tg}/\Delta T_s$ data points sampled within every trigger period. Therefore, in order to accumulate N sampling points for graphic display, the total time required for data acquisition is $T_N = N\Delta T_s = N_{ave}\Delta T_{tg}$. Using random sampling, the sampling rate is not restricted to one sample per trigger period as in sequential sampling, and thus the sampling efficiency can be higher, limited only by the sampling speed the oscilloscope can offer. The major disadvantage is that since the sampling has a different frequency compared to the trigger, the reconstructed sampling points within the display window might not be equally spaced, as shown in Figure 2.8.8(b). Therefore more samples are usually required compared to sequential sampling to reconstruct the original signal waveform.

It is interesting to note that for all the oscilloscopes and sampling methods discussed so far, a trigger is always required to start the sweeping process and synchronize data sampling with the signal. In recent years the capability of high-speed digital signal processing has been greatly expanded, bringing benefits to many technical areas, including instrumentation. *Microwave transition analysis* is another sampling technique that does not require a separate trigger. It repetitively samples the signal and uses DSP algorithms such as FFT to determine the fundamental frequency and high order harmonics (or subharmonics) of the signal. Once the fundamental period ΔT_s of the signal is determined, measurement can be performed similar to sequential sampling.

2.8.3 High-Speed Sampling of Optical Signal

The measurement of optical waveforms in the time domain is important in the characterization of optical devices, systems, and networks. The simplest and most often used way to measure an optical waveform is to use an oscilloscope equipped with an optical detector, as shown in Figure 2.8.9.

To measure high-speed and typically weak optical signals, the photodiode has to have wide bandwidth and low noise. Suppose that the photodiode has 1A/W responsivity and 50Ω impedance, a –20 dBm (10 μW) optical power produces only 0.5 mV electric voltage at the photodetector output. Therefore, a low-noise pre-amplifier is usually required before sending the signal into the electric oscilloscope. In modern optical communication and optical signal processing systems, short optical pulses in the picosecond and even femtosecond levels are often used, which would require extremely wide bandwidth for photodiodes and amplifiers in order to characterize these waveforms.

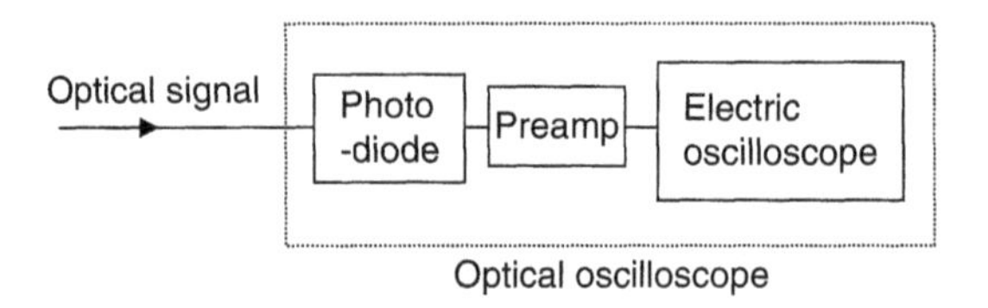

Figure 2.8.9 Block diagram of an optical oscilloscope.

2.8.3.1 Nonlinear Optical Sampling

Direct optical sampling as shown in Figure 2.8.10 has been used to measure ultrafast optical waveforms without the need for an ultrawideband photodetector. In this configuration, the optical signal to be tested is mixed with a short pulse train in a nonlinear optical mixer. The mixer has output only at the times when the reference pulses exist; therefore the optical signal is sampled by these short pulses. The speed of the photodiode only needs to be faster than the repetition rate of the short pulses, whereas its bandwidth can be much narrower than that of the optical signal to be tested. An optical sampling oscilloscope can be built based on this sampling technique, with the same operating principle and time alignments as for electrical sampling oscilloscopes.

The key device in the optical sampling is the nonlinear optical mixer, which can be realized using nonlinear optical components such as semiconductor optical amplifiers, nonlinear optical fibers, and periodically polled $LiNbO_3$ (PPLN) [27, 28]. A simple example of this nonlinear mixer, shown in Figure 2.8.11, is based on four-wave mixing (FWM) in a nonlinear medium, which can be either an SOA or a piece of highly nonlinear fiber. Assume that the wavelengths of the short pulse mode-locked laser and the optical signal to be measured are λ_p and λ_s, respectively. Degenerate FWM between these two wavelength components creates a frequency conjugated wavelength component at $\lambda_c = \lambda_p\lambda_s/(2\lambda_s - \lambda_p)$. This conjugate wave can then be selected by an optical bandpass filter (OBPF) before it is detected by a photodiode. As discussed in Section

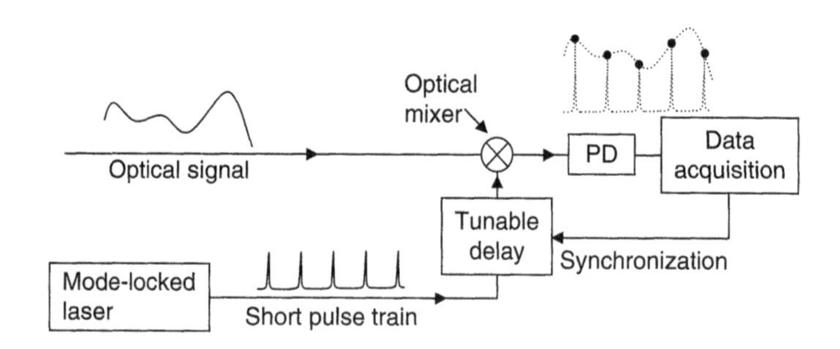

Figure 2.8.10 Direct sampling of high-speed optical waveform.

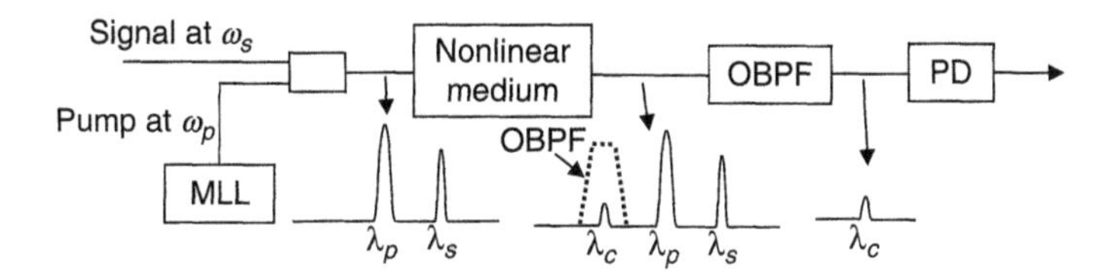

Figure 2.8.11 Illustration of optical sampling in a nonlinear medium using FWM. *OBPF:* Optical bandpass filter, *MLL:* mode-locked laser, *PD:* photodiode.

1.3.5, the power of the FWM component is proportional to the power product of the pump and the signal in the time domain, $P_{FWM}(t) \propto P_p^2(t)P_s(t)$. This process samples the value of the signal optical power precisely at the instant of each short optical pulse.

One of the challenges in optical sampling is the low nonlinear coefficient in photonic devices and optical fibers; therefore both the optical signal and the sampling pulses have to be optically amplified prior to mixing. Another challenge is that the dispersion in optical devices or fibers creates a walk-off between the signal and the sampling pulse, causing a relatively narrow optical bandwidth to be available for the optical signal. Nevertheless, optical sampling has been a topic of research for many years, and various novel techniques have been proposed to increase the sampling efficiency and increase the optical bandwidth [29, 30, 31, 32, 33].

Although nonlinear optical sampling is a powerful technique that allows the characterization of ultrafast optical waveforms, it is only capable of measuring the optical intensity while the phase information carried by the optical signal is usually lost. In many cases optical phase information is important, especially in phase-modulated optical systems. Therefore alternative techniques are required to characterize complex fields of the optical signal.

2.8.3.2 Linear Optical Sampling

It is well known that coherent detection preserves phase information of the optical signal; the principle of coherent detection is presented in Section 2.7. In fact, linear optical sampling can be understood as modified coherent homodyne detection which uses a short pulse source as the local oscillator [34, 35, 36].

The circuit configuration of linear optical sampling is shown in Figure 2.8.12. A 90° optical hybrid is used to combine the optical signal with a mode-locked short pulse laser. Although the 90° optical hybrid may be constructed in various ways, the following discussion provides the understanding of the general operating principle.

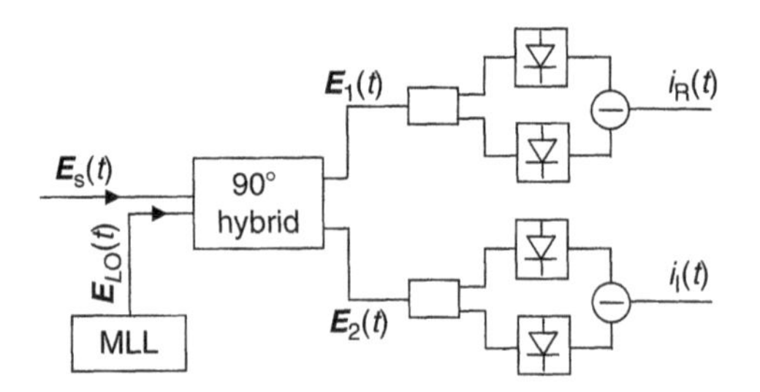

Figure 2.8.12 Block diagram of linear optical sampling.

Consider the field of the optical signal to be measured:

$$E_s(t) = A_s(t)\exp(j\omega_s t + j\phi_s(t)) \tag{2.8.2}$$

and the field of the mode-locked short pulse train:

$$E_{LO}(t) = A_{LO}(t)\exp(j\omega_{LO} t + j\phi_{LO}) \tag{2.8.3}$$

where $A_s(t)$ and $A_{LO}(t)$ are the real amplitude of the incoming signal and the LO, respectively, ω_s and ω_{LO} are their optical frequencies, and ϕ_s(t) and ϕ_{LO} are their optical phases. Since the LO is a short pulse train, $A_{LO}(t) \neq 0$ only happens at discrete time windows of $t_N = t_0 + NT$, where t_0 is an arbitrary starting time, T is the pulse period, and N is an integer.

Using the circuit shown in Figure 2.8.12 and following the derivation of coherent homodyne detection with phase diversity discussed in Section 2.8.4, we can derive expressions similar to Equations 2.7.27 and 2.7.28:

$$i_R(t) = \Re A_s(t)\cdot A_{LO}(t)\cos\left(\Delta\phi(t) - \frac{\pi}{4}\right) \tag{2.8.4}$$

$$i_I(t) = \Re A_s(t)\cdot A_{LO}(t)\sin\left(\Delta\phi(t) - \frac{\pi}{4}\right) \tag{2.8.5}$$

where $\Delta\phi(t) = \phi_s(t) - \phi_{LO}$ is the phase difference between the optical signal to be measured and that of the LO. In this linear sampling system the detection speed of the photodiode should be higher than the pulse repetition rate but can be much lower than both the input signal modulation speed and the speed determined by the width of the LO short pulses; therefore the photodiodes perform integration within each pulse window. As a consequence, there is only one effective data point acquired for each pulse period. Therefore, Equations 2.8.4 and 2.8.5 can be written as

$$i_R(t_N) = \mathrm{Re}\left[\tilde{A}(t_N)\right] \tag{2.8.6}$$

$$i_I(t_N) = \mathrm{Im}\left[\tilde{A}(t_N)\right] \tag{2.8.7}$$

where

$$\tilde{A}(t_N) = \Re \cdot A_s(t_N)e^{j\phi_{s,N}} \cdot A_{LO}(t_N)e^{j(\phi_{LO}-\pi/4)} \tag{2.8.8}$$

If we assume that the amplitude of the LO pulses is constant over time and the phase of the LO does not change in the measurement time window, $\tilde{A}(t_N)$ is, in fact, the complex field of the input optical signal sampled at $t = t_N$. In practice, if the variation of the LO optical phase is small enough, the measurement can be corrected via digital signal processing.

The measurement results of linear sampling can be represented on a constellation diagram, as illustrated in Figure 2.8.13(a). The complex optical field sampled by each LO short pulse is represented as a dot on this complex plan, with the radius representing the field amplitude and the polar angle representing the optical phase.

Figure 2.8.13(b) shows an example of a constellation diagram in a 10 Gb/s BPSK signal generated by a Mach-zehnder modulator [36]. In this example, the signal optical phase is switched between 0 and π while the amplitude has negligible change.

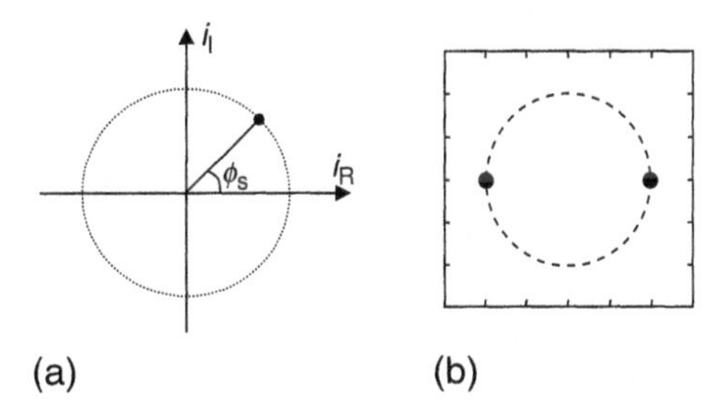

Figure 2.8.13 (a) Illustration of constellation diagram to represent complex optical field. (b) An example of a 10 Gb/s BPSK signal generated by a MZ-modulator [36].

2.8.4 High-Speed Electric ADC Using Optical Techniques

As discussed in the last section, a sampling oscilloscope can only be used to measure repetitive signals because it relies on the sparse sampling and waveform reconstruction. The measurement of ultrafast and nonrepetitive electric signals has been a challenging task. In recent years, digital signal processing (DSP) has become very popular, where an analog-to-digital converter (ADC) is a key component that translates the analog signal into a digital domain for processing. To sample a high-speed, nonrepetitive analog signal, the sampling rate

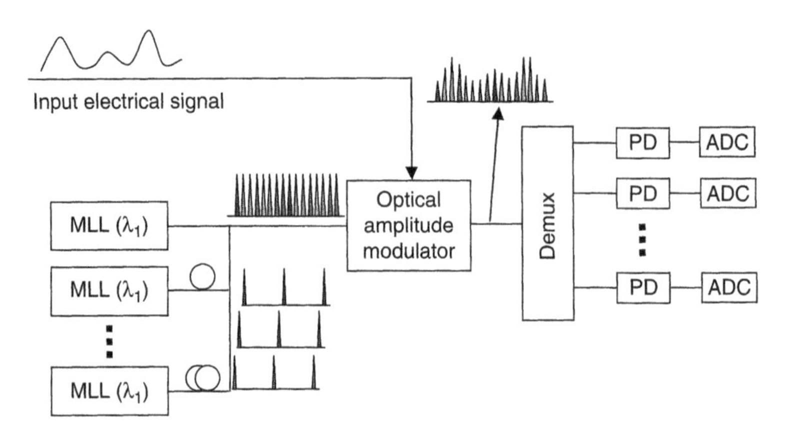

Figure 2.8.14 ADC using optical sampling.

has to be high enough to satisfy the well-known Nyquist Criterion, and the speed of electrical sampling is often a limiting factor.

Optical techniques can help increase the sampling speed of electrical signals and realize ultrahigh-speed ADC. Figure 2.8.14 shows an example of sampling ultrafast electrical signals with the help of optical techniques [37, 38]. A group of short-pulse mode-locked lasers is used; each operates in a different central wavelength. The low-repetition rate pulse trains from different lasers are interleaved to form a high-repetition rate pulse train. This pulse train passes through an electro-optic modulator, which is driven by the electrical signal to be sampled. Therefore the amplitude of the high-repetition rate optical pulse train is linearly proportional to the modulating electrical signal. Then this modulated optical signal is split by a wavelength division demultiplexer into various wavelength components, each of which is a low-repetition rate pulse train. At this point, each subrate pulse train can be easily detected by a relatively low-speed optical receiver and converted into a digital signal by a low-speed ADC. These channels can then be reconstructed in the digital domain to recover the original waveform.

2.8.5 Short Optical Pulse Measurement Using an Autocorrelator

Time-domain characterization of ultrashort optical pulses is usually challenging because it requires very high speed of the photodetector, the amplifying electronics, and the oscilloscope. In recent years, ultrafast optics are becoming very important in optical communication, optical signal processing, and optical measurement and testing. Optical autocorrelators are popular instruments for characterizing ultrashort optical pulses.

The fundamental operation principle of an autocorrelator is based on the correlation of the optical pulse with itself. Experimentally, autocorrelation can be accomplished in a Michelson interferometer configuration as shown in Figure 2.8.15. In this setup, the short optical pulse is split into two paths and recombines at a focal point. A key device in this setup is a second order harmonic generation (SGH) nonlinear crystal, which is an optical "mixer." In general, this mixing process generates a second order harmonic wave E_{SGH}, which is proportional to the product of the two input optical signals E_1 and E_2. The frequency of the second order harmonic is equal to the sum frequency of the two participating optical signals:

$$E_{SGH}(\omega_1 + \omega_2) \propto E_1(\omega_1) \cdot E_2(\omega_2) \tag{2.8.9}$$

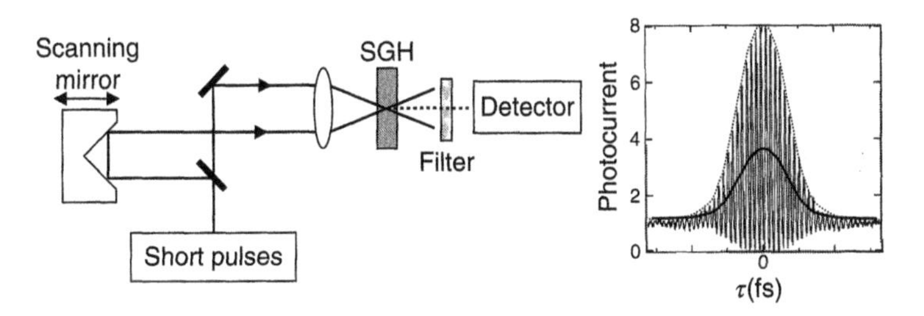

Figure 2.8.15 Optical autocorrelation using second harmonic generation.

In the specific case of autocorrelation, as shown in Figure 2.8.15, the mixing happens between the original signal and its time-delayed replica, and the SGH process simply produces the frequency-doubled output. In the time domain, if the relative delay between the two pulses is τ, the SGH output will be:

$$E_{SGH}(t, \tau) \propto E(t) \cdot E(t - \tau) \tag{2.8.10}$$

where $E(t)$ and $E(t - \tau)$ represent the original optical signal and its time-delayed replica, respectively.

Note that the fundamental frequency of $E_{SGH}(t, \tau)$ is frequency-doubled compared to the input optical signal, and the purpose of the short-pass optical filter after the SGH in Figure 2.8.15 is to block the original optical frequency ω and allow the frequency-doubled output 2ω from the SGH to be detected. Based on square-law detection, the photocurrent generated at the photodetector is

$$I_{SGH}(t, \tau) \propto | E(t) \cdot E(t - \tau) |^2 \tag{2.8.11}$$

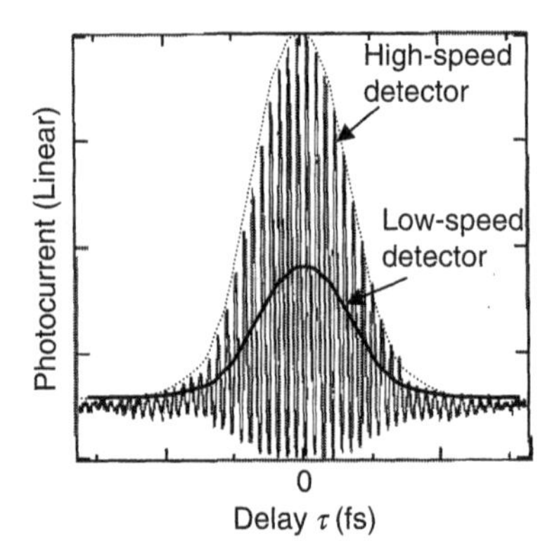

Figure 2.8.16 Illustration of the measured waveforms of the autocorrelator by a high-speed detector or by a low-speed detector.

If the speed of the photodetector is high enough, the optical phase walk-off between the two pulses can be observed and the envelope of the measured waveform, shown as the dotted line in Figure 2.8.16, represents the convolution between the two pulses. On the other hand, if the speed of the photodetector is low, the impact of the optical phase cannot be observed and the measured waveform is simply the convolution of the intensity of the two pulses:

$$I_{SGH}(\tau) \propto \int_{-\infty}^{\infty} P(t) \cdot P(t-\tau) dt \tag{2.8.12}$$

where, $P(t) = |E(t)|^2$ is the optical intensity of the pulse.

Figure 2.8.17 shows the optical circuit of the autocorrelator. The collimated input optical beam is split by a beam splitter (partial reflecting mirror). One part of the beam (path 1) goes through a fixed delay formed by mirror 1 and a retroreflector. It is reflected back into the nonlinear crystal through the beam splitter and the concave mirror. The other part (path 2) of the beam is directed into a scanning delay line that consists of a parallel pair of mirrors (mirror 2 and mirror 3) mounted on a spinning wheel and a fixed mirror (mirror 4). The optical signal reflected back from this scanning delay line is also sent to the nonlinear crystal through the beam splitter and the concave mirror. The optical beams from path 1 and path 2 overlap in the nonlinear crystal, creating a nonlinear wave mixing between them. In this case, the SHG crystal operates noncollinearly; therefore it is relatively easy to spatially separate the second order harmonic from the fundamental waves. The mixing product is directed onto a photodetector. Because of the frequency doubling of this nonlinear mixing process, the photodetector should be sensitive to the wavelength which is

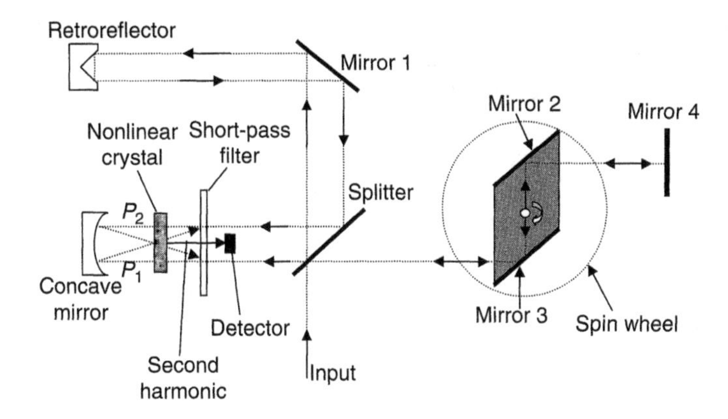

Figure 2.8.17 Optical configuration of an autocorrelator.

approximately half of the input optical signal wavelength. A short-pass optical filter is usually placed in front of the photodetector to further block the optical signal of the original wavelength.

Due to the autocorrelation process we discussed previously, if the input optical signal is a periodic pulse train, the SHG component can measure only optical pulses from the two paths overlapping in time at the nonlinear crystal. Therefore, the waveform of the optical pulses measured by an autocorrelator is the self-convolution of the original optical pulses. The measured waveform can be displayed as a function of the differential time delay τ between the two paths in the autocorrelator, as shown in Figure 2.8.18. As the result of autocorrelation, one disadvantage of the autocorrelator is that the actual optical pulse width cannot be precisely predicted unless the pulse shape is known. For example, for the pulses produced by a Ti:sapphire laser, the pulse shape is assumed to be sech^2, and the pulse width measured by an autocorrelator (with low-

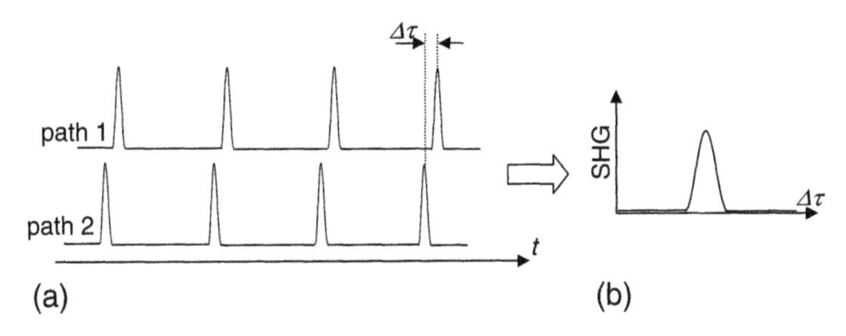

Figure 2.8.18 SGH is generated by the cross-correlation between the optical signals arriving from the two paths: (a) optical pulse train and (b) second harmonic signal.

speed photodetector) is approximately 1.543 times the original pulse width. This number is 1.414 for Gaussian pulses.

There are several types of nonlinear crystals for SHG. For example, BBO (Beta-Barium Borate) crystals are often used for autocorrelation; they have relatively high second-order coefficients. Recently it was reported that an unbiased light-emitting diode (LED) could also be used as the nonlinear medium for autocorrelation through the two-photon absorption effect [39].

The linear scan of the differential delay τ between the two paths is accomplished by the rotating wheel, as illustrated in Figure 2.8.19(a). If the separation between the parallel mirrors on the optical path is D, the path length difference should be a function of the rotating angle ϕ as

$$\Delta L = 2D\sin(\phi) \approx 2D\phi \tag{2.8.13}$$

where we have assumed that only within a small rotation angle φ of the wheel the light beam can be effectively reflected back. Usually, beyond this angular range the light beam will not reach to mirror 4 and therefore it is referred to as the *dead zone*.

Suppose the wheel rotation frequency is f; the differential delay can be expressed as a function of time:

$$\tau(t) = \frac{\Delta L}{c} \approx \frac{2D\phi}{c} = \frac{4\pi f D}{c} t \tag{2.8.14}$$

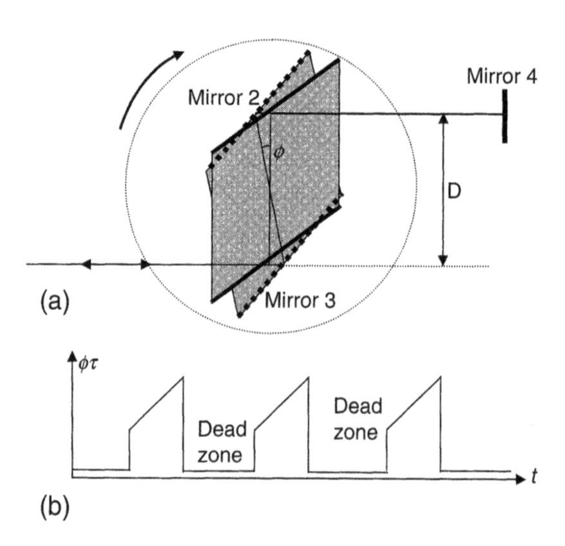

Figure 2.8.19 (a) Illustration of optical path length change introduced by rotating the wheel and (b) waveform of differential delay versus time.

It is a periodic ramp signal and there is a dead time slot within each period when the optical beam is outside the mirror coverage, as shown in Figure 2.8.19(b).

At the time when τ is zero or the exact multiples of the period of the optical pulse train, pulses from the two paths overlap, producing strong second order harmonic signals. Because of the nature of autocorrelation, the measured pulse waveform in the SHG is the self-convolution of the original optical signal pulses. The actual pulse width of the original optical signal under test can be calculated from the measured SHG with a correcting factor depending on the pulse shape.

Figure 2.8.20 shows the typical experimental setup using an autocorrelator to characterize short optical pulses. The second order harmonic signal generated by the detector in the autocorrelator is amplified and displayed by an oscilloscope. A periodic electrical signal provided by the autocorrelator, which is related to the spin rate of the rotation wheel, is used to trigger the oscillator and to provide the time base. With the measured waveform on the oscilloscope

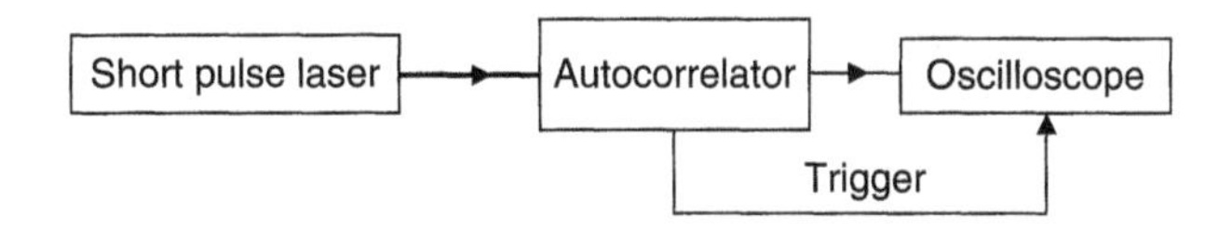

Figure 2.8.20 Measurement setup using an autocorrelator.

$V(t)$, the corresponding autocorrelation function $I_{SGH}(\tau)$ can be found by converting time t to the differential delay τ using Equation 2.8.11. However, a more precise calibration can be accomplished by adjusting the position of the retroreflection mirror (see Figure 2.8.17). Obviously, shifting the retroreflection mirror by a distance Δx, the corresponding differential delay change should be $\Delta\tau = 2\Delta x/c$. If the pulse peak of a measured oscilloscope trace is shifted by a time ΔT, we can find a calibration factor of $\tau = \frac{2\Delta x}{c\Delta T}t$

Another very important issue to note is that due to the nature of autocorrelation, the relationship between the measured pulse width and the actual pulse width of the optical signal depends on the actual time-domain shape of the pulse. For example, if the pulse is Gaussian,

$$P(t) = P_0 \exp\left[-2.77\left(\frac{t}{\sigma_{actual}}\right)^2\right] \qquad (2.8.15)$$

where σ_{actual} is the FWHM of the pulse; the FWHM of the measured pulse using an autocorrelator is $\sigma_{measured} = 1.414\sigma_{actual}$.

If the optical pulse is hyperbolic secant,

$$P(t) = P_0 \operatorname{Sech}^2\left(1.736\frac{t}{\sigma_{actual}}\right) \tag{2.8.16}$$

the FWHM of the measured pulse using an autocorrelator is $\sigma_{measured} = 1.543\sigma_{actual}$.

For the autocorrelator discussed so far, the measurement has been made only for the intensity of the optical pulse, whereas the phase information is lost. In addition, since the measurement is based on autocorrelation, detailed time-domain features of the optical pulses are smoothed out. Frequency-resolved optical gating, known as FROG, has been proposed to solve these problems [Trebino, 1997]; it provides the capability of characterizing both the intensity and the phase of an optical pulse.

Figure 2.8.21 shows the basic optical configuration of FROG based on SGH, where the key difference from a conventional autocorrelator is the added spectral analysis of the second harmonic signal and the signal analysis to recover the intensity and the phase information of the optical pulse.

As indicated in Equation 2.8.10, in the SHG process, if we neglect the constant multiplication factor, the field of the second-order harmonic signal is $E_{SGH}(t, \tau) = E(t) \cdot E(t - \tau)$ and its power spectral density can be obtained as the square of a one-dimensional Fourier transform:

$$S_{FROG}(\omega, \tau) = \left| \int_{-\infty}^{\infty} E(t)E(t-\tau)\exp(-j\omega t)dt \right|^2 \tag{2.8.17}$$

Through the measurement using setup shown in Figure 2.8.21, the FROG trace $S_{FROG}(\omega, \tau)$ can be obtained experimentally, which is a two-dimensional data array representing the spectrum of the SHG signal at different differential delay τ. Then the next step is to determine the actual optical field of the signal pulse $E(t)$ based on the knowledge of $S_{FROG}(\omega, \tau)$ [40]. This is a numerical process based on the phase-retrieval algorithm known as iterative Fourier transform [41].

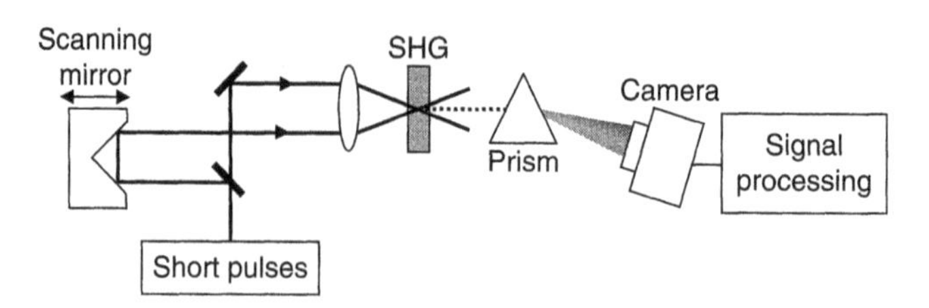

Figure 2.8.21 Optical configuration of FROG based on SHG.

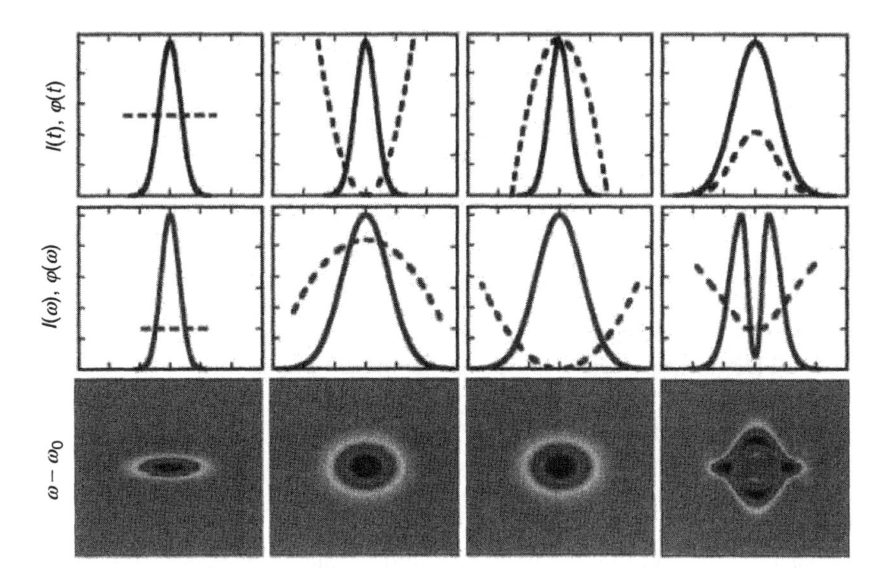

Figure 2.8.22 Example of the FROG measurements. Top row: pulse intensity (solid line) and phase (dashed line). Bottom row: measured 2-D FROG traces [42]. Used with permission.

Figure 2.8.22 shows an example of FROG traces of optical pulses with various phase characteristics, where $I(t) = |E(t)|^2$ is the pulse intensity and $\varphi(t)$ is the relative phase shift of the pulse. The first and second rows show the time-domain and frequency-domain descriptions of the pulse, respectively. The bottom row in the figure shows the measured FROG traces in which the horizontal axis indicates the differential time delay τ and the vertical axis is the frequency domain information measured with the spectrometer. Among the different phase shapes shown in this example, the first column in Figure 2.8.22 is a transform-limited pulse with constant phase, the second and the third columns are optical pulses with negative and positive chirps, respectively, and the third column is an optical pulse with significant self-phase modulation so that the pulse spectrum is split into two peaks. It is evident that the FROG spectrograms predict most of the pulse features, both intensity and phase; however, the SHG-based FROG is not able to directly predict the sign of the frequency chirp as shown in the second and third columns. There are a number of FROG techniques based on other nonlinear mechanisms which are able to distinguish the sign of the chirp. More detailed description of these techniques can be found in an excellent review paper [42].

2.9 OPTICAL LOW-COHERENT INTERFEROMETRY

Optical reflection as a function of reflector distance is an important characteristic of optical systems as well as in composite optical components. The ability to precisely locate optical reflections is often required in system characterization and troubleshooting. An optical time-domain interferometer (OTDR) is often used for this purpose, in which a short optical pulse is launched to the optical system. By measuring the timing and the magnitude of the echo reflected from the optical system, the location and the strength of optical reflection can be evaluated. In principle, the spatial resolution $R_{resolution}$ of an OTDR measurement depends on the optical pulses width τ by $R_{resolution} = 0.5 v_g \tau$, where v_g is the group velocity of the optical pulse. For example, if the pulse width in an OTDR is $\tau = 100$ ns, the spatial resolution is approximately 10 m in the fiber. For some applications which require better spatial resolutions and accuracy, interferometer-based optical instrumentation has to be used.

Generally there are two categories of optical interferometers that are often used to measure optical reflections and find the precise locations of optical interfaces from which the reflections come. One of them is commonly known as optical low-coherent reflectometry (OLCR), which uses a wideband low-coherent light source and a tunable optical delay line to characterize optical reflections. The other one is known as Fourier-domain reflectometry, which uses a wavelength swept laser source; the magnitudes and the locations of the optical reflections can be found through Fourier analysis of the received interference signal.

Because of its superior spatial resolution, OLCR has also been used in biomedical applications to provide three-dimensional images with high resolution, where this technique is known as optical coherent tomography (OCT) [43].

2.9.1 Optical Low-Coherence Reflectometry

OLCR was developed to measure optical reflectivity as a function of distance [44]. OLCR has demonstrated both higher spatial resolution and detection sensitivity compared to OTDR. The operation of an OLCR system is based on a Michelson interferometer configuration, as shown in Figure 2.9.1.

In this simplified OLCR configuration, the wideband (and thus low-coherence) optical source is divided evenly between the reference and test arms using a 3-dB fiber coupler. The optical delay in the reference arm can then be varied by moving the scanning mirror. The reflected signals from each arm travel back through the coupler, where they are recombined and received at the photodiode. Due to the nature of the directional coupler, half the reflected power will be

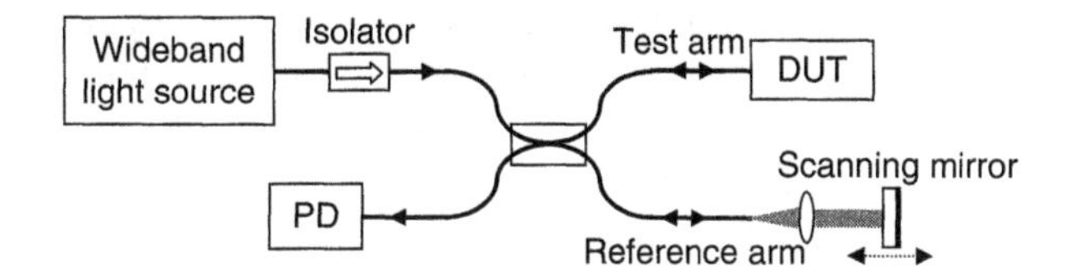

Figure 2.9.1 simplified OLCR optical configuration.

directed back to the source, where it is attenuated by the isolator and the other half goes to the photodiode. From the arrangement shown in Figure 2.9.1, coherent interference will appear at the photodiode if the difference in optical length between the reference and test arms is less than the source coherence length. The coherence length of the light source is determined by its spectral width $\Delta\lambda$ according to the equation $L_c = \lambda^2/(n\Delta\lambda)$, where n is the refractive index of the material and λ is the average source wavelength. The incident optical power on the photodiode leads to a photocurrent which is described by

$$I = \Re\left[P_{REF} + P_{DUT} + 2\sqrt{P_{REF}P_{DUT}}\cos(\phi_{REF}(t) - \phi_{DUT}(t))\right] \qquad (2.9.1)$$

where $\Re$ is the responsivity of the photodiode, P_{REF} is the reflected reference signal with optical phase of $\phi_{REF}(t)$, and P_{DUT} is the reflected signal from the test arm with phase $\phi_{DUT}(t)$.

One assumption made in using the photocurrent expression is the matched polarization states between the reference and test-arm signals incident upon the photodetector. This matched polarization state maximizes the interference on the photodiode. Conversely, signals received at the photodiode having orthogonal states of polarization will create no interference signal, even when the optical length of the two arms is within the coherence length, L_C, of the source.

An illustration of the interference signal received at the detector is shown in Figure 2.9.2. As shown, the signal is a function of the differential length z, with the peak occurring when the optical lengths of the two arms are equal ($z = 0$).

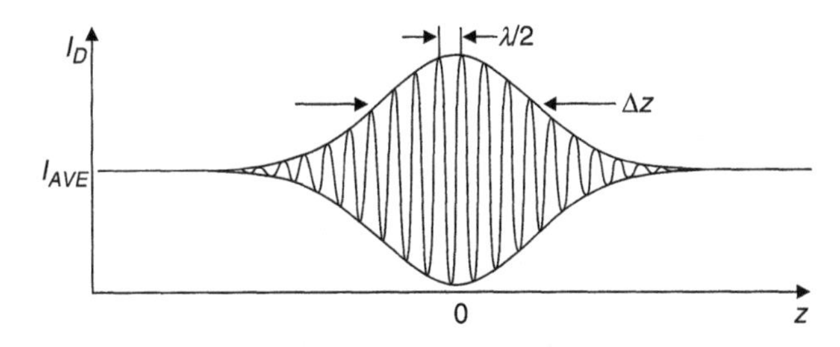

Figure 2.9.2 Example of detected interference signal current.

The DC value, I_{AVE}, is a result of the constant powers, P_{REF} and P_{DUT}, reflected from each arm. The sinusoidal wave represents the interference between the two reflected signals received at the photodiode when the differential delay is small. As the reference arm length changes by one-half the average source wavelength, $\lambda/2$, the sinusoidal interference signal completes one period. This is a result of the signal in the reference arm traveling twice across the variable delay, in effect doubling the distance.

According to Equation 2.9.1, when the differential delay between the two arms is greater than the coherence length, the phases of the two signals are uncorrelated, varying randomly with respect to one another. Since the bandwidth of the photodetector is much slower than the optical frequency at which the phase difference varies, a constant current will be observed at the photodiode output. However, once the differential delay decreases to less than L_C, the phase difference, $\phi_{REF}(t) - \phi_{DUT}(t)$, does not average to zero. Therefore, even though the photodetector bandwidth is unable to observe the details of the resulting sinusoidal interference signal, it will recognize the signal envelope shown in Figure 2.9.2. The 3-dB width of this envelope, Δz, is the spatial resolution of the OLCR setup. The resolution is ultimately determined by the coherence length of the source,

$$\Delta z \approx \frac{L_C}{2} = A\frac{\lambda^2}{2n\Delta\lambda} \tag{2.9.2}$$

where the factor A is governed by the low-coherence source spectral shape. For example, $A = 0.44$ for Lorentzian, $A = 0.88$ for Gaussian, and $A = 1.2$ for rectangular line shape [45].

Figure 2.9.3 shows a typical OLCR spatial resolution as a function of the source spectral bandwidth for $A = 1$ and $n = 1.5$. As shown, a spatial resolution of less than 10 μm is achieved by implementing a source with approximately 100 nm spectral bandwidth. The measurement resolution is also of interest in determining the ability of an OLCR system to detect multiple reflections in the test arm. To be observed, the spacing of adjacent reflections must be greater than the spatial resolution, Δz. In this manner, any number of reflections can be detected as long as they lie within the system dynamic range.

In practical applications, the optical reflection from the device under test (DUT) is usually much weaker than that from the reference arm. Therefore the Michelson interferometer is strongly unbalanced. As a consequence, although the average photocurrent is strong, the signal contrast ratio is weak. To minimize the DC component in the photocurrent, a balanced photodetection is often used, as shown in Figure 2.9.4. In addition, an optical phase modulation in the reference arm can help improve measurement signal-to-noise ratio. This can be accomplished using a PZT, which stretches the fiber. This phase modulation at frequency f_m converts the self-homodyne detection into a heterodyne detection

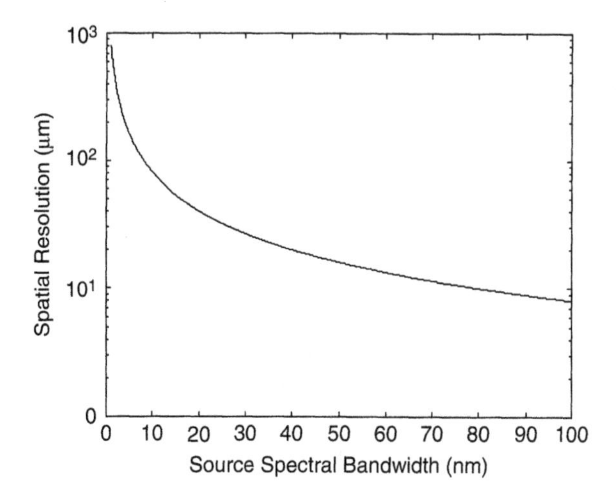

Figure 2.9.3 OLCR spatial resolution versus source spectral bandwidth, assuming $A = 1$ and $n = 1.5$.

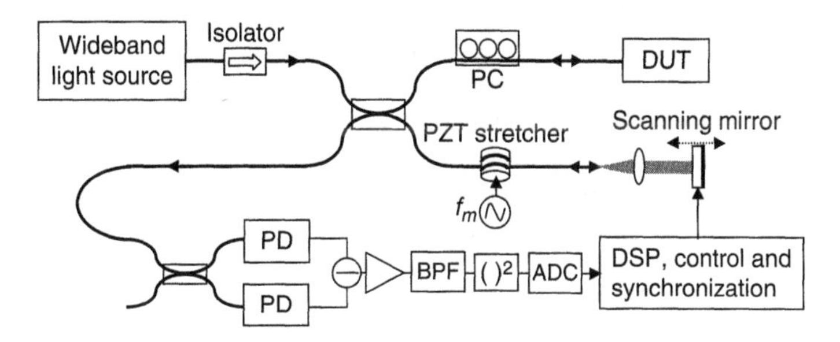

Figure 2.9.4 Practical OLCR using balanced detection and phase modulation.

with the IF frequency at f_m. It shifts the signal electrical spectrum away from DC, thus reducing the low-frequency noise introduced in the receiver electronics. A polarization controller can be inserted in either the reference or the test arm to match the polarization states and maximize the detection efficiency.

Another limitation of an OLCR is that the maximum measurement range is usually limited to a few centimeters, mainly determined by the length coverage of the scanning optical delay line. Range extension has been proposed using a pair of retroreflectors in the optical delay line [46]. By letting the light bouncing back and forth between two retroreflectors for N times, the length coverage range of the delay line is increased N times. In addition to extra insertion loss in the delay line, another disadvantage of this method is that both motor step size and mechanical errors in the translation stage (which controls the position of the retroreflectors in the scanning delay line) will be amplified N times, degrading the length resolution of the measurement.

Another difficulty for an OLCR is the polarization state mismatch between the signals reflected from the two interferometer arms. Due to the random nature of polarization-mode coupling in an optical fiber, polarization state mismatch may occur, causing temporary fading of the coherent beating signal at the optical receiver. To avoid this fading, a polarization controller is usually required in one of the two interferometer arms. However, because the birefringence of an optical fiber is very sensitive to temperature and mechanical variations, the polarization controller has to be adjusted endlessly, complicating the measurement procedure. This approach is not feasible for many sensor applications in which long-term measurements are required.

A polarization-independent OLCR has been demonstrated [47] for which the low coherent light source was first polarized and then launched into an interferometer consists of polarization-maintaining (PM) fibers, thus eliminating the polarization sensitivity.

Figure 2.9.5 shows an example of extending the measurement length coverage using an FBG array and eliminating polarization mismatch-induced signal fading using polarization spreading in the optical receiver. This makes it possible for an OLCR to be used in long-term monitoring of fiber length variations, which has a wide range of applications such as system troubleshooting and fiber-optic sensors. In this configuration, a number of fiber-Bragg gratings (FBG) are used and connected in tandem, each reflecting at a different wavelength. A scanning optical delay line with the maximum tuning range l_t is used

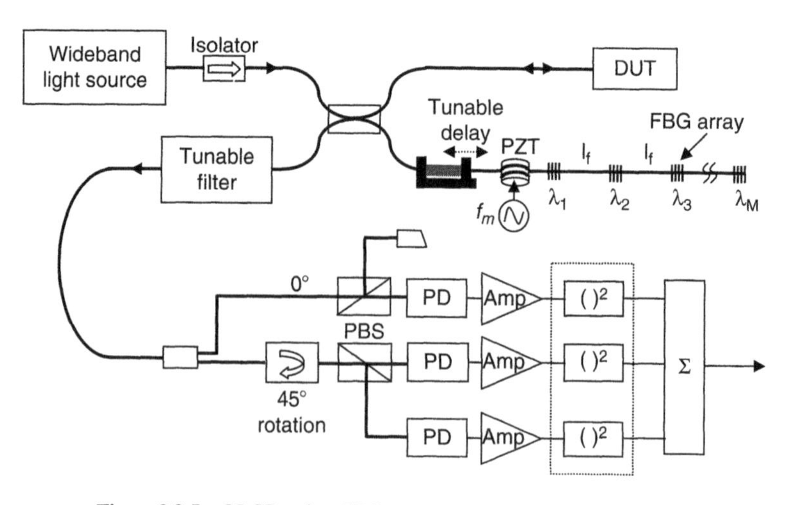

Figure 2.9.5 OLCR using FBG array and polarization spreading.

to vary the length of the reference arm. A tunable optical filter is used at the receiver side for wavelength selection. At a certain wavelength, the length of the reference arm is determined by the location of the FBG reflecting at that particular wavelength. Although the maximum tuning range, l_t, of the optical delay line may be limited by physical constrains, the overall measurable length coverage is approximately $L = Ml_f$, where M is the total number of FBGs. As a practical concern, to ensure continuous length coverage, the distance between adjacent FBGs, l_f, should be slightly shorter than l_t. The spatial resolution of this OLCR is determined by the bandwidth of the optical filter and fiber gratings. Basically, this technique effectively trades off the source spectral bandwidth with the length coverage of the measurement. To create an ultrawideband source, it is possible to combine a number of wideband light sources, such as SLEDs that each has a different central wavelength, by a WDM MUX.

To implement a polarization-spreading technique, the combined optical signal reflected from the two interferometer arms is first split by a 3-dB polarization-maintaining beam splitter. One of the beam splitter outputs is connected to a linear horizontal polarizer and then detected by a photodiode. The signal from the other output port of the beam splitter is connected to a polarization beam splitter (PBS) via a 45° polarization rotation patch cord, and then the two outputs from the PBS are each detected by a photodiode. After amplification and envelope detection, low-frequency electrical signals from the three photodiode branches are added together to form the output.

Assume that the two optical signal vectors (reflected from the two arms of the interferometer) at the input of the receiver are $\vec{E}_1 = \vec{a}_x E_{x1} e^{j\varphi_{xy1}} + \vec{a}_y E_{y1}$ and $\vec{E}_2 = \vec{a}_x E_{x2} e^{j\varphi_{xy2}} + \vec{a}_y E_{y2}$, where φ_{xy1} and φ_{xy2} are the relative phases between horizontal and vertical optical field components. Without losing generality, we can assume that the two optical fields have the same amplitude E_0: $E_{x1} = E_0 \cos(\Phi_1)$ and $E_{y1} = E_0 \sin(\Phi_1)$, $E_{x2} = E_0 \cos(\Phi_2)$, and $E_{y2} = E_0 \sin(\Phi_2)$, where $0 < \Phi_1, \Phi_2 < \pi$ denotes the orientation of signal polarization states. If these two optical fields are directly combined and detected by a photodiode, the photocurrent of the beating component is

$$\begin{aligned} I_s &= \Re \left| E_{x1} E_{x2} e^{j(\varphi_{xy1}+\varphi_{xy2})} + E_{y1} E_{y2} \right| \\ &= \Re |E_0|^2 \left| \cos(\Phi_1)\cos(\Phi_2) e^{j(\varphi_{xy1}+\varphi_{xy2})} + \sin(\Phi_2)\sin(\Phi_1) \right| \end{aligned} \tag{2.9.3}$$

where $\Re$ is the photodiode responsivity. Assuming both φ_{xy1}, φ_{xy2}, Φ_1 and Φ_2 are random and each has a uniform distribution between 0 and π, then the photocurrent I_s will be random. Using a Monte Carlo simulation, a statistical distribution of I_s can be found as shown in Figure 2.9.6(a), which displays 5000 simulated samples. Figure 2.9.6(a) demonstrates a high probability of severe signal fading with direct detection using a single photodiode.

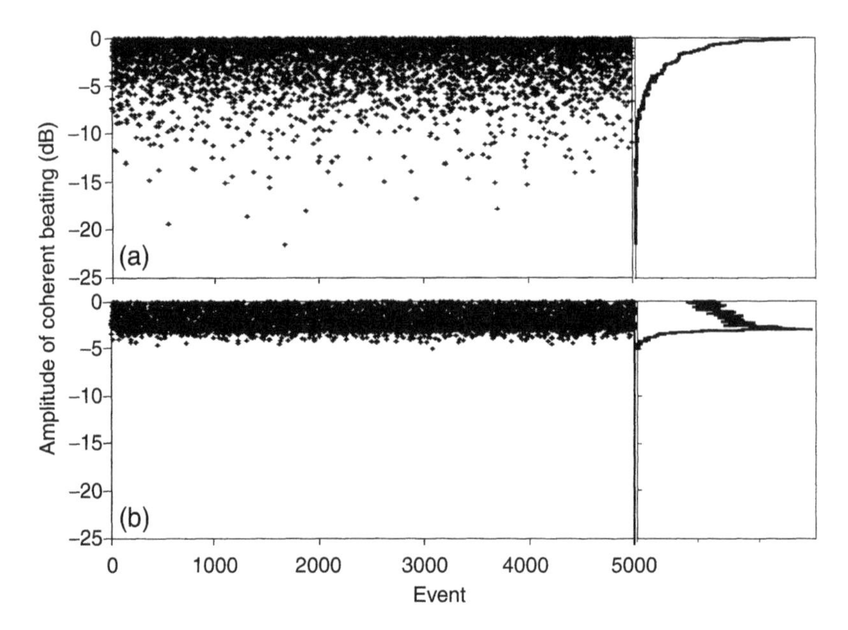

Figure 2.9.6 Monte Carlo simulation of signal fading induced by polarization state mismatch. (a) Using a conventional receiver with a single photodiode, and (b) using the proposed polarization spreading in the optical receiver.

Using the polarization spreading optical receiver as shown in Figure 2.9.5, the photocurrents produced at three photodiodes can be obtained by a Jones Matrix analysis:

$$I_1 = 0.5\,\Re|E_0|^2|\cos(\Phi_2)\cos(\Phi_1)| \tag{2.9.4a}$$

$$I_2 = 0.25\,\Re|E_0|^2|(\cos\Phi_1 e^{j\varphi_{xy1}} + \sin\Phi_1)(\cos\Phi_2 e^{j\varphi_{xy2}} + \sin\Phi_2)| \tag{2.9.4b}$$

$$I_3 = 0.25\,\Re|E_0|^2|(\cos\Phi_1 e^{j\varphi_{xy1}} - \sin\Phi_1)(\cos\Phi_2 e^{j\varphi_{xy2}} - \sin\Phi_2)| \tag{2.9.4c}$$

Although the sum of these three photocurrents is still a function of φ_{xy1}, φ_{xy2}, Φ_1, and Φ_2, it is not possible for these three photocurrent components to fade simultaneously. Figure 2.9.6(b) shows 5000 simulated samples and the statistical distribution of the total photocurrent $I = I_1 + I_2 + I_3$. In this case, the maximum variation of the total photocurrent I is approximately 5 dB. For many applications in the measurement of long-term fiber length variations, a polarization spreading optical receiver can simplify the measurement procedure by eliminating the necessity for endless polarization adjustments.

For example, Figure 2.9.7 shows the results of continuous fiber length monitoring for about three days [48]. In this experiment four FBGs were used with

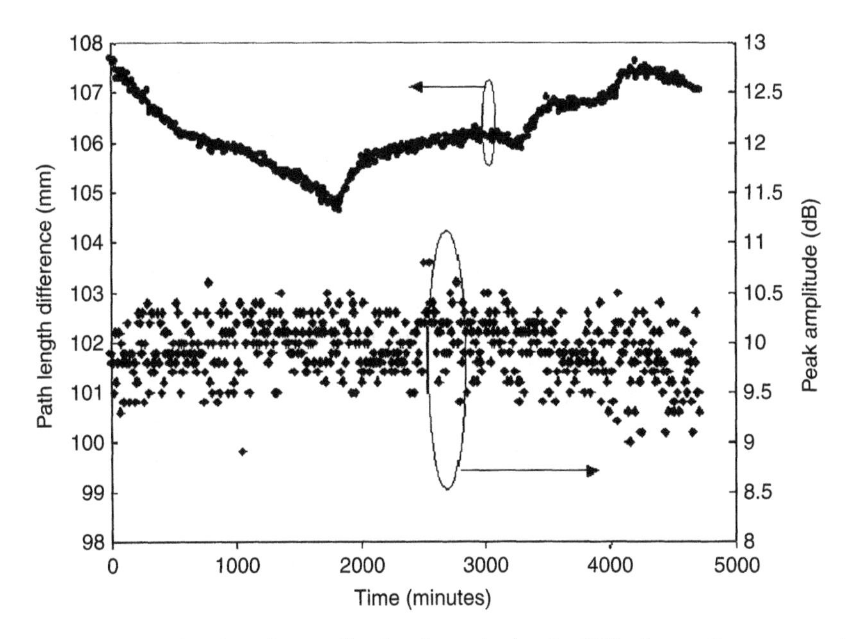

Figure 2.9.7 Results of continuous fiber length monitoring for 5000 minutes. Dots: measured fiber length versus time; diamonds: coherent interference peak amplitude versus time.

peak reflection wavelengths at 1550 nm, 1553 nm, 1556 nm, and 1559 nm, respectively. The optical bandwidth of each FBG is approximately 1.2 nm and the step motor-controlled tunable optical filter at the receiver has an optical bandwidth of 1 nm. The four FBGs were spliced together in tandem, and the separations between them are 94 mm, 75 mm, and 72 mm, respectively. The step-motor-driven scanning optical delay line used in the experiment has a maximum tuning range of 160 mm in the air. Considering the refractive index difference between air and fiber ($n = 1.5$), this tuning range is equivalent to approximately 106 mm in the fiber. The FBG array extended the total measurement length coverage to 347 mm. The effective optical bandwidth of the cascaded FBG and tunable optical filter is approximately 1 nm, which is much smaller than the total bandwidth of the optical source. The spatial resolution of the measurement can then be estimated as 0.8 mm. If a higher spatial resolution is required, the optical bandwidths of FBGs and the optical filter will have to be increased. This may require a wider bandwidth of the optical source, which can be achieved by wavelength multiplexing.

In this measurement, each of the two interferometer arms is made by a 1-km LEAF fiber cable with a dispersion of 2.5 ps/nm-km in the 1550 nm wavelength window. Figure 2.9.7 indicates that the relative length change between the two fiber arms is approximately 3 mm during the 5000 minutes of measurement, which is mainly caused by the relative temperature difference between the two fiber spools. Figure 2.9.7 also shows the measured coherent interference peak amplitude versus time. Although no polarization controller is used in the setup, significant signal fading never occurred during the three days, although the interference peak amplitude varied by about 3 dB.

Although there have been numerous demonstrations of OLCR for many applications, the major limitation of OLCT technique is the requirement of a swept optical delay line. Since the sweep of optical delay is usually accomplished by mechanical means, the sweeping speed is generally slow and the long-term reliability may not be guaranteed. To overcome this problem, Fourier-domain reflectometry was proposed based on the fact that scanning the length of the optical delay line can be replaced by sweeping the wavelength of the optical source. Since wavelength sweeping in lasers can be made fast and without the need of mechanical tuning, Fourier-domain reflectometry has the potential to be more practical.

2.9.2 Fourier-Domain Reflectometry

As shown in Figure 2.9.8, Fourier-domain reflectometry (FDR) is also based on a Michelson interferometer, which is similar to the OLCR, but a wavelength-swept laser is used instead of a low-coherent light source. In this configuration a fixed length of the reference arm is used. Although the wavelength-swept laser source needs a wide wavelength tuning range, at each point of time its spectral linewidth is narrow. The interference between optical signals from the test and the reference arms is detected by a photodiode, and the reflection in the DUT as a function of distance can be obtained by a Fourier-domain analysis of the interference signal.

The operation principle of FDR is illustrated in Figure 2.9.9, in which the optical frequency of the laser source linearly sweeps from f_1 to f_2 during a time interval τ, and the source optical bandwidth is therefore $B = |f_2 - f_1|$. After being reflected from the DUT, the signal from the test arm is mixed with that reflected from the reference mirror at the optical receiver. Similar to the case of FM radar (Kachelmyer, 1988), in this process the signal is "dechirped." In

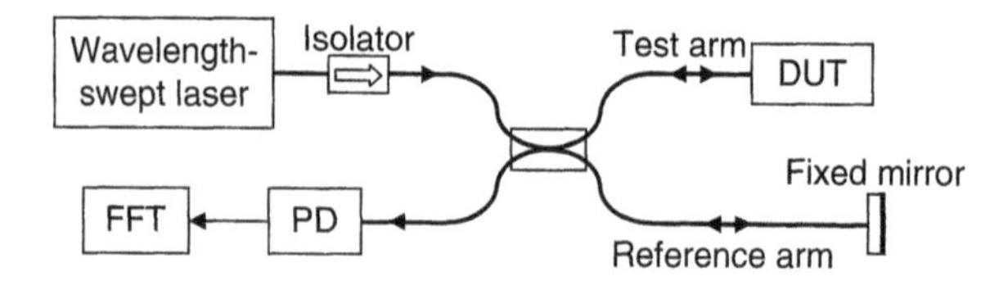

Figure 2.9.8 Simplified Fourier-domain reflectometry optical configuration.

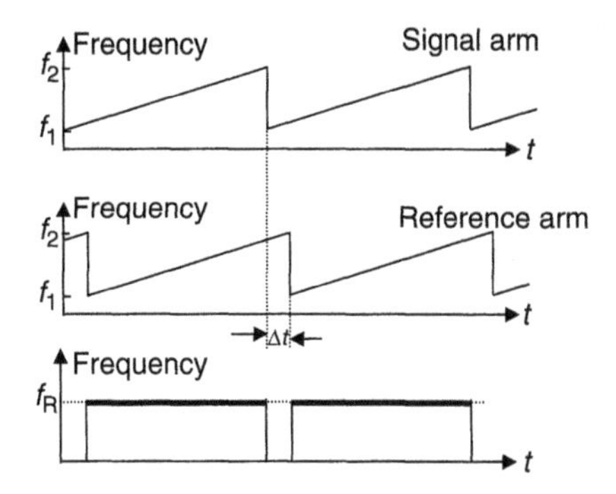

Figure 2.9.9 Illustration of the Fourier-domain reflectometry operation principle.

the simplest case, if the DUT has only one discrete reflection interface, the mixed signal will produce only one constant frequency at the receiver, which indicates the location of the reflection within the DUT. In general, if the DUT has continuous reflections with distributed intensity, a Fourier analysis can be used to translate the measured frequency components to the longitudinal distribution of reflections at various locations within the DUT.

The spatial resolution in the longitudinal dimension in a FDR is inversely proportional to the source optical bandwidth B as $\delta x = c/(2nB)$, or equivalently, $\delta x = \lambda_0^2/(2n\Delta\lambda)$, where $\Delta\lambda$ is the width of the region in which the source wavelength sweeps, n is the refractive index of the material, and λ_0 is the central wavelength of the swept-wavelength source. For example, if the source wavelength is in the 1550 nm window, a wavelength sweep over 50nm bandwidth can provide a spatial resolution of approximately 17 μm in an optical fiber, with $n = 1.45$.

Based on the principle of coherent detection, if the 3-dB coupler is ideal, the photocurrent obtained at the output of the photodetector is

$$I_{sig} = 2\Re\{\sqrt{P_{ref}}\exp[j2\pi f(t)\cdot t]\}\left\{\sqrt{P_{test}}\exp[-j2\pi f(t-\Delta t)\cdot t]\right\} \tag{2.9.5}$$

where Δt is the roundtrip propagation delay difference between the reference and the test arms. The direct detection component is neglected for simplicity. P_{ref} and P_{test} are the optical powers that come from the reference and the test arms, respectively. Due to wavelength sweep of the light source, the signal optical frequency linearly changes from f_1 to f_2 in a time interval τ, and thus the optical frequency can be expressed as

$$f(t) = f_1 - (f_2 - f_1)t/\tau \tag{2.9.6}$$

Dechirping takes place in the photodiode due to the mixing between the $\{\sqrt{P_{ref}}\exp[j2\pi f(t)\cdot t]\}$ and $\{\sqrt{P_{test}}\exp[-j2\pi f(t-\Delta t)\cdot t]\}$ in the coherent detection process. The beating frequency can be determined by a Fourier analysis of the photocurrent signal as

$$f_R = f(t) - f(t - \Delta t) = \left(\frac{f_2 - f_1}{\tau}\right)\Delta t \tag{2.9.7}$$

Therefore the location of the reflector inside the DUT can be easily calculated by

$$x = \frac{c\Delta t}{2n} = \frac{cf_R\tau}{2n(f_2 - f_1)} \tag{2.9.8}$$

In contrast to OLCR, where low coherent light source was used, the wavelength-swept light source used in FDR is highly coherent. In fact, in FDR the range coverage of the measurement Δx depends on the coherence length of the laser source, which is inversely proportional to the spectral linewidth of the laser $\delta\lambda$ by $\Delta x = \lambda_0^2/(2n\delta\lambda)$. This means that the optical signals reflected from the test and the reference arms should always be mutually coherent so that their path length difference can be found by fast Fourier transform. For example, suppose the spectral linewidth of a swept laser source is 2 GHz, which is equivalent to 0.016 nm in a 1550 nm wavelength window, it will provide an approximately 51 mm spatial coverage. This length coverage is not enough to diagnose failures in fiber-optical systems, but it is enough to characterize the structures of optical components.

It is important to note that the same issues affecting OLCR will also affect the performance of an FDR, such as polarization state mismatch and power unbalance between signals from the two interferometer arms. Similar approaches can be used to overcome these problems, including active polarization control, polarization diversity, and balanced-detection receivers, as illustrated in the last section.

One unique challenge in FDR, though, is the lack of high-quality wavelength-swept laser sources, especially for high-speed wavelength sweep. Although small-footprint distributed Bragg reflector (DBR) semiconductor lasers can provide >20 nm wavelength tuning range with about 100 MHz linewidth, wavelength sweep of a DBR laser is usually not continuous across the entire 20 nm wavelength band; therefore it cannot be used for FDR. In practical FDR systems, fiber-based ring lasers are often used with an intercavity wavelength selection filter as shown in Figure 2.9.10.

In this fiber laser configuration, a semiconductor optical amplifier (SOA) is used as the gain medium. Two optical isolators are used to ensure the unidirectionality of the optical signal and prevent unwanted optical reflections. A swept narrowband optical filter is used to select the oscillation wavelength in this fiber ring cavity. In general, the spectral-width of the laser, $\Delta\lambda$, is determined by the SOA used in the ring. Because of the saturation effect of the laser, this

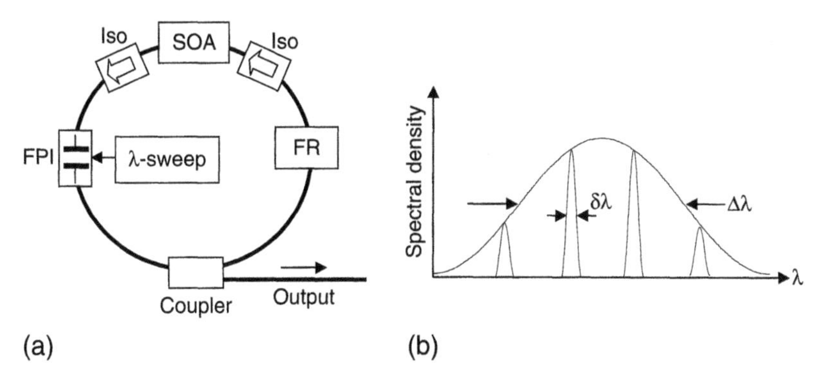

Figure 2.9.10 (a) Configuration of a wavelength-swept fiber ring laser. *FR:* Faraday rotator, *FPI:* Fabry-Perot interferometer, *ISO:* isolator, *SOA:* semiconductor optical amplifier. (b) Illustration of the optical bandwidth $\Delta\lambda$ and the linewidth $\delta\lambda$ of the laser.

spectral bandwidth is in general wider than the small-signal 3-dB bandwidth of the SOA itself. Commercial fiber-pigtailed SOAs in the 1320 nm region usually have >50 nm bandwidth, which can potentially provide an approximately >70 nm ring laser bandwidth. The spectral linewidth of the laser output $\delta\lambda$, on the other hand, is determined by the narrowband optical filter in the loop. A commercially available fiber Febry-Perot tunable filter can have a finesse of 2000. Therefore, if one chooses the free spectral range (FSR) of 100 nm, the width of the passband can be on the order of 0.05 nm. Generally, because of the mode competition mechanism in the homogenous-broadening gain medium, the linewidth of the ring laser may be slightly narrower than the 3-dB passband of the tunable filter. A Faraday rotator in the ring cavity is used to minimize polarization sensitivity of the lasing mode.

For practical FDR applications, the wavelength sweeping rate of the laser source is an important parameter that, to a large extent, determines the speed of measurement. In a fiber ring laser, after the bandpass optical filter selects a certain wavelength, it needs a certain time for the laser to build up the lasing mode at that wavelength. Theoretically, when the lasing threshold is achieved, the optical power should reach its saturation value P_{sat}. If the net optical gain of each roundtrip is g (including amplifier gain and cavity loss) and the ASE noise selected by the narrowband filter is δP_{ASE}, it requires N roundtrips to build up the ASE to the saturation power, and N can be found from,

$$P_{sat} = \delta P_{ASE} \cdot g^N \tag{2.9.9}$$

where $\delta P_{ASE} = P_{ASE}\,(\delta\lambda_f/\Delta\lambda)$, P_{ASE} is the total ASE noise power within the bandwidth $\Delta\lambda$ of the SOA, and $\delta\lambda_f$ is the bandwidth of the optical filter. Since the time required for each roundtrip in the ring cavity is $\tau_{cavity} = Ln/c$, with L the cavity length, the time required to build the lasing mode is

$$\delta t = \frac{Ln}{c \cdot \log(g)} \log\left(\frac{P_{sat}\Delta\lambda}{P_{ASE}\delta\lambda_f}\right) \tag{2.9.10}$$

For example, if $P_{ASE} = 1\ mW$, $P_{sat} = 10\ mW$, $\Delta\lambda = 120\ nm$, $\delta\lambda_f = 0.135\ nm$, $L = 2.4\ m$, $g = 15\ dB$, and n $= 1.46$, we can find $\delta t \approx 30.73\ ns$. Then the maximum wavelength sweeping speed v_{tune} can be determined such that with the time interval δt the wavelength tuning should be less than the bandwidth of the filter, and in this example, $v_{\max} \leq \delta\lambda_f / \delta t = 4.39\ nm/\mu s$. Consider the total SOA bandwidth of $\Delta\lambda = 120\ nm$; the limitation on the repetition rate of wavelength sweeping is approximately

$$f_{\max} \leq \frac{1}{\pi} \frac{v_{\max}}{\Delta\lambda} \approx 11.65\ kHz \tag{2.9.11}$$

where the factor $1/\pi$ accounts for the speed of a sinusoidal sweep compared to a linear, unidirectional sweep, assuming the filter sweep occurs over the same spectral range [49]. If the tuning frequency is higher than f_{max}, the lasing mode would not have enough time to establish itself, and the output power would be weak and not stable.

One technique to increase wavelength-sweeping speed is to synchronize the repetition time of wavelength sweeping with the roundtrip time of the ring cavity, that is, the narrowband optical filter is tuned periodically at the roundtrip time or a harmonic of the roundtrip time of the ring cavity. This allows the delay line in the ring cavity to store the wavelength-swept optical signal. This signal reenters the narrowband filter and the SOA at the exact time when the filter is tuned to the particular wavelength of the optical signal. In this case, lasing mode does not have to build up from the spontaneous emission at each wavelength; therefore a much higher sweeping speed of up to a few hundred kilohertz is permitted [50, 51].

The Fourier-domain reflectrometer eliminates the requirement of a mechanical tunable delay line at the reference arm of the interferometer; however, it shifts the tenability requirement to the wavelength of the laser source. Without a tunable laser source, an alternative technique is to use a fixed wideband light source in the Fourier-domain reflectrometer but to employ an optical spectrometer at the detection side, as shown in Figure 2.9.11. In this case, if the target has only one reflection interface, the optical power spectral density measured at the end of the interferometer will be similar to that shown by Equation 2.4.8:

$$I(f) \propto R_{ref} + R_{DUT} + 2R_{ref}R_{DUT}\cos(4\pi fcx) \tag{2.9.12}$$

where R_{ref} and R_{DUT} are power reflectivities of the reference arm and the DUT, respectively, and x is the differential length between the reference and the test

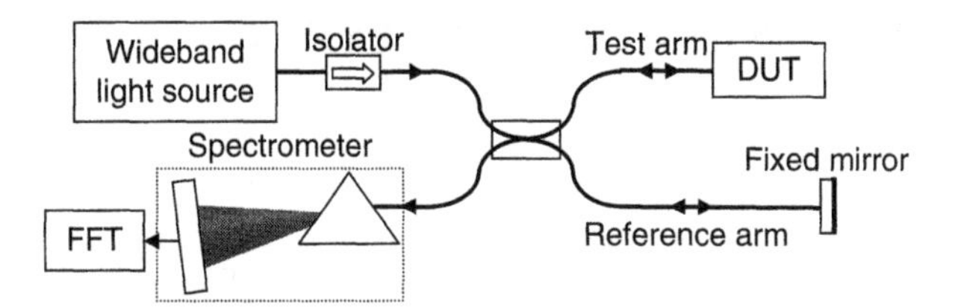

Figure 2.9.11 Fourier-domain reflectometry using a fixed wideband source and an optical spectrometer for optical length determination.

arms. Obviously the value of x can be predicted by a Fourier transform of the measured optical spectrum.

Using the setup shown in Figure 2.9.11, the spatial resolution of the measurement is determined by the spectral bandwidth $\Delta\lambda$ of the light source $\delta x = \lambda_0^2/(2n\Delta\lambda)$, whereas the length coverage is determined by the spectral resolution $\delta\lambda$ of the optical spectrometer $\Delta x = \lambda_0^2/(2n\delta\lambda)$, where λ_0 is the central wavelength of the source and n is the refractive index of the DUT material. Although this measurement technique does not need a tunable laser, it has stringent requirements on the optical spectrometer. For the measurement requiring fast speed, grating and a CCD array-based optical spectrometer are often used, allowing the parallel processing of signal wavelength components simultaneously; also there is no moving part required in the spectrometer, as discussed at the end of Section 2.1. Practically speaking, commercially available, compact CCD array-based spectrometers usually have low-spectral resolution that is typically not better than 0.5 nm and therefore the length coverage is less than 1.6 mm if operating in the 1550 nm wavelength window. Another problem is that a CCD array is available in silicon material which is sensitive only up to 1 μm wavelength. For long wavelength operations such as 1320 nm or 1550 nm, we must use InGaAs-based diode arrays, which are very expensive, especially with a large number of pixels in the array.

Although tunable filter-based optical spectrometers, such scanning FPI, can be used in this setup, the detection efficiency is rather poor. The reason is that the optical source emits energy that simultaneously covers a wide wavelength band, whereas the tunable filter in the spectrometer only picks up a narrow slice of the spectrum at any given time. Therefore most of the optical energy provided by the source is wasted in the process, and thus the detection efficiency is low.

2.10 OPTICAL NETWORK ANALYZER

Network analyzer is one of the basic instruments in the research and development of RF, microwave, and recently lightwave technologies. An optical network analyzer is based on the same concept as an RF network analyzer except for their difference in the operation frequencies. Therefore our discussion should start from RF network analyzers.

2.10.1 S-Parameters and RF Network Analyzer

An RF network analyzer performs accurate measurement of the ratios of the transmitted signal to the incident signal, and of the reflected signal to the incident signal. Typically an RF network analyzer has a frequency swept source and a detector that allows the measurement of both the amplitude and the phase responses at each frequency that produce complex transfer functions of RF devices and systems.

In general, the linear characteristics of a two-port RF network can be defined by its *S*-parameters, as illustrated in Figure 2.10.1. By definition, the *S*-parameter relates the input, the output, and the reflected fields by

$$\begin{bmatrix} b_1 \\ b_2 \end{bmatrix} = \begin{bmatrix} s_{11} & s_{12} \\ s_{21} & s_{22} \end{bmatrix} \begin{bmatrix} a_1 \\ a_2 \end{bmatrix} \tag{2.10.1}$$

where s_{11}, s_{12}, s_{21} and s_{22} are matrix elements, each providing a specific relation between the input and the output,

$$s_{11} = \frac{\text{reflected from port 1}}{\text{incident into port 1}} = \left.\frac{b_1}{a_1}\right|_{a_2=0} \tag{2.10.2}$$

$$s_{12} = \frac{\text{transmitted from port 2 to port 1}}{\text{incident into port 2}} = \left.\frac{b_1}{a_2}\right|_{a_1=0} \tag{2.10.3}$$

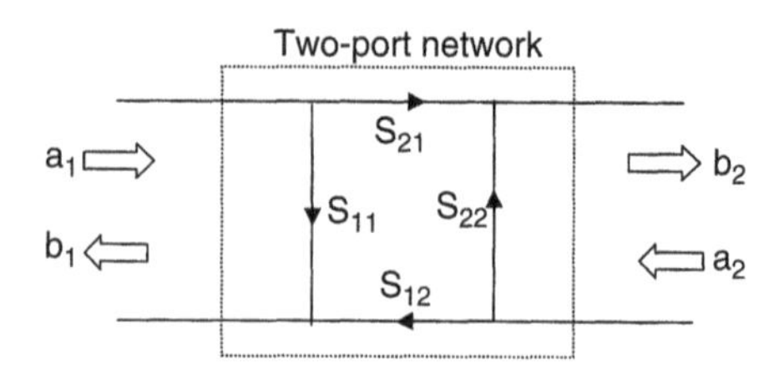

Figure 2.10.1 *S*-parameter of a two-port network.

$$s_{21} = \frac{\text{transmitted from port 1 to port 2}}{\text{incident into port 1}} = \left.\frac{b_2}{a_1}\right|_{a_2=0} \tag{2.10.4}$$

$$s_{22} = \frac{\text{reflected from port 2}}{\text{incident into port 2}} = \left.\frac{b_2}{a_2}\right|_{a_1=0} \tag{2.10.5}$$

In general, each element of the S-parameter is both wavelength-dependent and complex, representing both the amplitude and the phase of the network properties.

Figure 2.10.2 shows the block diagram of a transmission/reflection (T/R) test set, which can be used to measure the S-parameters of RF components. In this test set, a swept synthesizer is used as the signal source, which supplies the stimulus for the stimulus-response test system. We can either sweep the frequency of the source or sweep its power level, depending on the measurement requirement. Part of the signal from the oscillator is sent to the device under test (DUT) and part of it is split out through a directional coupler as the reference. To measure s_{11}, the signal reflected from DUT is directed to an RF coherent receiver, where it is mixed with the reference signal that serves as the local oscillator. An ADC is used to convert the mixed intermediate frequency (IF) signal into digital format and send for signal processing and display. Since RF coherent detection is used, both the amplitude and the phase information of the reflected signal from port 1 of DUT can be determined, which yields s_{11}. Similarly, the signal transmitted from port 1 to port 2 can also be coherently detected by setting the switch to the other RF mixer on the right; this measures the parameter s_{21}.

Note that in this T/R test set, the stimulus RF power always comes into the test port 1 and the test port 2 is always connected to a receiver in the analyzer. To measure s_{12} and s_{22}, the stimulus RF source has to be connected to the output of the DUT and therefore one must disconnect the DUT, turn it around,

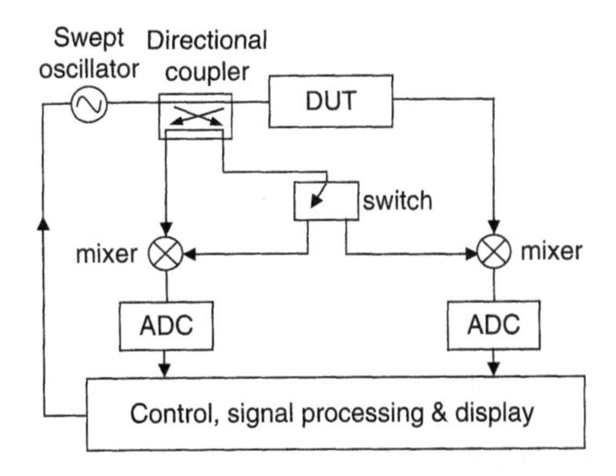

Figure 2.10.2 Block diagram of a transmission/reflection test set.

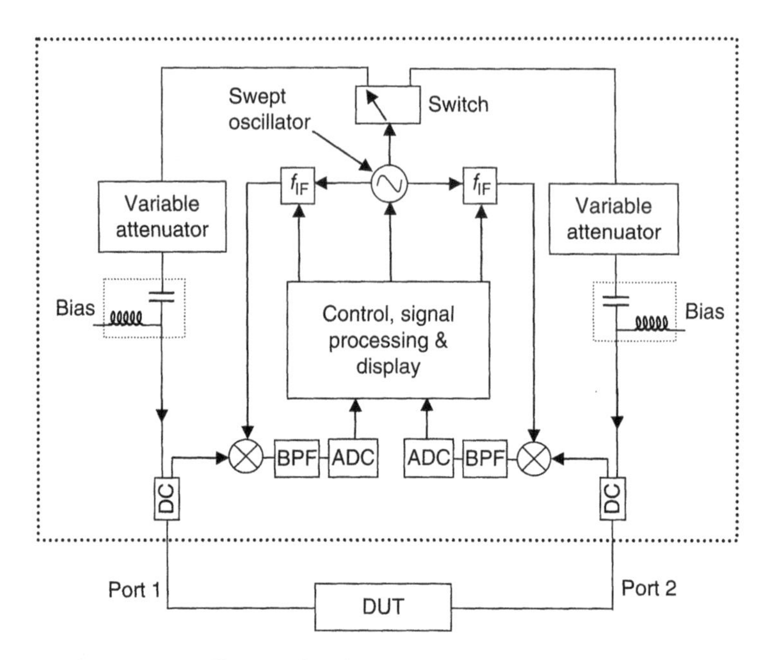

Figure 2.10.3 Block diagram of an RF network analyzer. *DC:* directional coupler.

and reconnect it to the test set. In doing this, the test set may need to be recalibrated, which often degrades the measurement accuracy.

Figure 2.10.3 shows the block diagram of an RF network analyzer. It allows both forward and reverse measurements on the DUT, which are needed to characterize all the four *S*-parameters. The variable attenuators are used to set the stimulus signal power levels, and the bias-Tees are used to add a CW bias level to the signal, if necessary. The directional couplers (DCs) separate the stimulus and the reflected signals from the DUT. In the coherent receiver, the reflected (transmitted) signal from (through) the DUT is mixed with the frequency-shifted local oscillator and selected by narrow bandpass filters (BPFs) before been sent to analog-to-digital converters (ADCs). The constant frequency shift f_{IF} in the local oscillator ensures the heterodyne RF detection with the intermediate frequency at f_{IF}. The central frequency of the BPF is determined by the intermediate frequency f_{IF}. To achieve high detection sensitivity, the bandwidth of BPF is usually very narrow. A central electronic control circuit synchronizes the swept-frequency oscillator and the signal processing unit to find the *S*-parameters of the DUT.

The network analyzer can be used to characterize both RF devices and complicated RF systems. For example, an optical transmission system can be regarded as a two-port RF system and the transmission performance can be characterized by a network analyzer, as illustrated in Figure 2.10.4. In an optical transmission system, the transmitter (Tx) converts the RF input into an

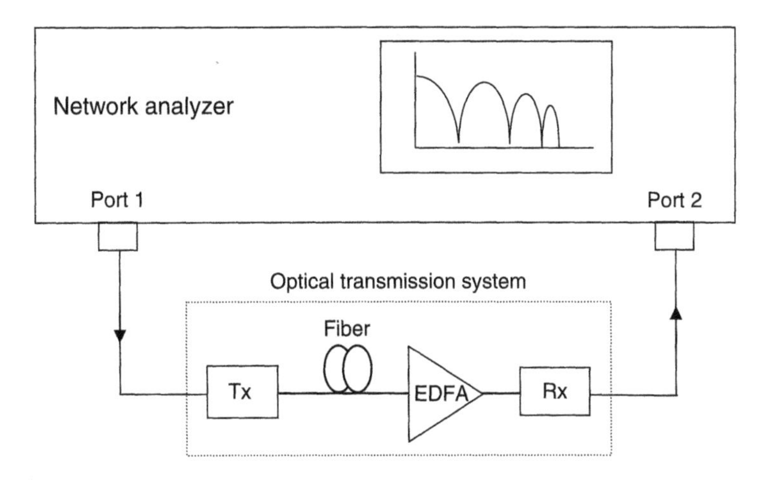

Figure 2.10.4 Measure transmission characteristics using an RF network analyzer.

optical signal, optical fibers and optical amplifiers deliver the optical signal to the destination, and the receiver (Rx) converts the received optical signal back to the RF domain. Therefore the frequency response of the overall system can be measured by the complex S_{21} parameter.

In this measurement, an important detail we need to pay attention to is the time delay of the transmission system. For example, for a 100 km long optical fiber, the propagation delay is approximately 500 μs. Since the local oscillator in the network analyzer is linearly swept, during this long time delay the mixed frequency between the transmitted signal over the system and the local oscillator will likely be outside the bandwidth of the BPF; as a consequence, no response will be measured by the network analyzer. In most commercial RF network analyzers, the "step sweep" function is available, in which the local oscillator stays in each frequency for a certain amount of time before jumping to the next frequency. By choosing the dwell time of step sweep longer than the system delay, the BPF in the coherent receiver of the network analyzer will be able to capture the heterodyne signal.

2.10.2 Optical Network Analyzers

An optical network analyzer is an instrument to characterize the transfer function of optical components and systems. Although the transfer function of an optoelectronic system can be measured by an RF network analyzer, as illustrated in Figure 2.10.4, the system must have an electrical-to-optical (E/O) converter and an optical-to-electrical (O/E) converter as terminals of the optical system.

2.10.2.1 Scalar Optical Network Analyzer

To be able to characterize an optical system itself without optoelectronic conversions, an optical network analyzer can be extended from an RF network analyzer by providing calibrated O/E and E/O converters. Figure 2.10.5 shows the block diagram of a scalar optical network analyzer. The O/E converter consists of a tunable laser and a wideband external electro-optic intensity modulator, which converts an RF modulation into optical modulation. A wideband photodetector and a low noise preamplifier are used in the optical receiver, which converts the optical signal back into RF domain. An optical directional coupler is used at the transmitter port, providing the capability of measuring the reflection from the DUT. An optical switch at the receiver port selects between the measurements of transmission (S_{21}) or reflection (S_{11}). A conventional RF network analyzer is used to provide RF stimulation, detection, signal processing, and display. Since the transmitter and receiver pair is unidirectional, this optical network analyzer only measures S_{21} and S_{11}, and obviously S_{12} and S_{22} can be measured by reversing the direction of the DUT.

In fact, this optical network analyzer simply provides a calibrated electro-optic transmitter/ receiver pair, which is an interface between the RF network analyzer and the optical device or system to be tested. Integrating the wideband transmitter/receiver pair into the same equipment makes it easy to calibrate the overall transfer function including the RF network analyzer and the O/E and E/O converters. This enables an optical network analyzer to accurately characterize wideband response of optical devices and systems up to 50 GHz.

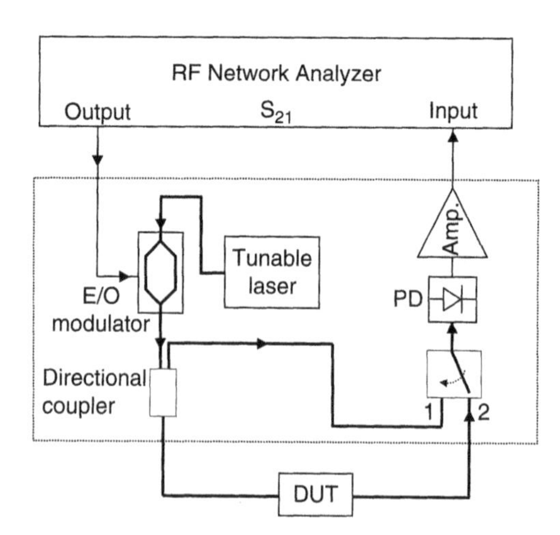

Figure 2.10.5 Block diagram of a "scalar" lightwave network analyzer.

It is important to notice that in the optical network analyzer block diagram shown in Figure 2.10.5, intensity modulation and direct detection is used for the O/E and E/O conversion. The instrument is effective in measuring the response of the DUT to the intensity modulated lightwave signal at various stimulating frequencies determined by the RF network analyzer, and the wavelengths determined by the tunable laser. However, optical phase information of the DUT cannot be determined in the measurements due to the intensity modulation/direct detection nature of the O/E and E/O conversion. For this reason, this optical network analyzer is usually referred to as a *scalar* network analyzer.

2.10.2.2 Vector Optical Network Analyzer

Generally, to be able to measure optical phase information, coherent detection will have to be used in the O/E-E/O interface box. However, another measurement technique using optical interferometric method appears to be more effective and simpler to obtain the vector transfer function of the DUT. Figure 2.10.6 shows the block diagram of a vector optical network analyzer using the interferometric technique [52, 53, 54]. A wavelength swept tunable laser is used to provide an optical stimulus. Two optical interferometers are used to provide relative optical delays. The first interferometer is made by the polarization-maintaining fiber and the polarization-maintaining coupler (PMC) equally splitting the optical signal into the two branches. At the output of this first interferometer, the polarization axis of one of the PM fiber branches is rotated by 90°, and a polarization beam splitter (PBS) is used to combine the optical signals. Since the optical signals carried by the two branches are orthogonal with each other at the PBS, there is no interference between them. Thus,

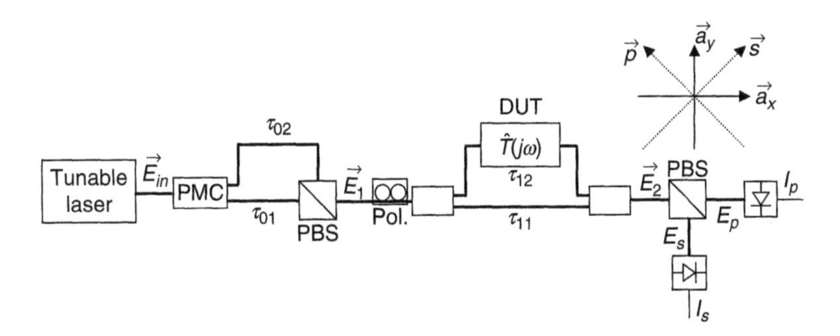

Figure 2.10.6 Block diagram of a "vector" optical network analyzer.

from a classic definition, this is not a real interferometer. Instead, it merely produces a differential delay between the two orthogonally polarized components.

Assume the field of the input optical signal is

$$\vec{E}_{in} = \vec{a}_x A e^{-j\omega t} \tag{2.10.6}$$

where $\vec{a}_x$ is a unit vector indicating the polarization orientation, A is a complex envelope, and ω is the frequency of the optical signal. Then, after the first interferometer, the optical field is

$$\vec{E}_1 = \frac{A}{\sqrt{2}} \left[\vec{a}_x \exp(-j\omega(t - \tau_{01})) + \vec{a}_y \exp(-j\omega(t - \tau_{02}))\right] \tag{2.10.7}$$

where $\vec{a}_x$ and $\vec{a}_y$ are two mutually orthogonal unit vectors and τ_{01} and τ_{02} are time delays of the two interferometer branches.

The second interferometer in Figure 2.10.6 is composed of two conventional 3-dB optical couplers. The lower branch is a simple optical delay line and the upper branch connects to the DUT. Another PBS is used at the end to split the optical signal into two orthogonal polarization components s and p before they are detected by two photodiodes. The optical field at the output of the second interferometer is composed of two components:

$$\vec{E}_2 = \vec{E}_{2L} + \vec{E}_{2U} \tag{2.10.8}$$

where $\vec{E}_{2L}$ and $\vec{E}_{2U}$ are the optical signals that come from the lower and upper branches of the second interferometer, respectively:

$$\vec{E}_{2L} = \frac{A}{2\sqrt{2}} \left[\vec{a}_x e^{-j\omega(t-\tau_{01})} + \vec{a}_y e^{-j\omega(t-\tau_{02})}\right] e^{j\omega\tau_{11}} \tag{2.10.9}$$

$$\vec{E}_{2U} = \frac{A}{2\sqrt{2}} \hat{T}(j\omega) \left[\vec{a}_x e^{-j\omega(t-\tau_{01})} + \vec{a}_y e^{-j\omega(t-\tau_{02})}\right] e^{j\omega\tau_{12}} \tag{2.8.10}$$

where τ_{11} and τ_{12} are time delays of the lower and upper branches and $\hat{T}(j\omega)$ is the vector transfer function of the DUT, which can be expressed as a 2×2 Jones matrix:

$$\hat{T}(j\omega) = \begin{bmatrix} T_{xx}(j\omega) & T_{xy}(j\omega) \\ T_{yx}(j\omega) & T_{yy}(j\omega) \end{bmatrix} \tag{2.10.11}$$

The polarization controller (Pol.) before the second interferometer is adjusted such that after passing the lower branch of the second interferometer, each field component in $\vec{E}_1$ is equally split into the s and p branches. This can be simply explained in a 45° rotation between the two orthogonal bases (a_x, a_y) and (s, p), as illustrated in the inset of Figure 2.10.6. Based on this setting, the $\vec{E}_{2L}$ component is equally split into s and p branches and their optical fields are, respectively:

$$E_{2L,s} = \frac{A}{4}\left[e^{-j\omega(t-\tau_{01}-\tau_{11})} + e^{-j\omega(t-\tau_{02}+\tau_{11})}\right] \quad (2.10.12)$$

$$E_{2L,p} = \frac{A}{4}\left[e^{-j\omega(t-\tau_{01}-\tau_{11})} - e^{-j\omega(t-\tau_{02}+\tau_{11})}\right] \quad (2.10.13)$$

Similarly, for the signal passing through the upper branch of the second interferometer, the corresponding optical fields reaching the s and p detectors are, respectively,

$$E_{2U,s} = \frac{A}{4}\left[T_{xx}e^{-j\omega(t-\tau_{12}-\tau_{01})} + T_{xy}e^{-j\omega(t-\tau_{12}-\tau_{02})} + T_{yx}e^{-j\omega(t-\tau_{12}-\tau_{01})} + T_{yy}e^{-j\omega(t-\tau_{12}-\tau_{02})}\right] \quad (2.10.14)$$

$$E_{2U,p} = \frac{A}{4}\left[T_{xx}e^{-j\omega(t-\tau_{12}-\tau_{01})} + T_{xy}e^{-j\omega(t-\tau_{12}-\tau_{02})} - T_{yx}e^{-j\omega(t-\tau_{12}-\tau_{01})} - T_{yy}e^{-j\omega(t-\tau_{12}-\tau_{02})}\right] \quad (2.10.15)$$

Then the composite optical field at the s and p photodetectors can be obtained by combining these two contributing components as

$$E_s = E_{2U,s} + E_{2L,s}$$
$$= \frac{|E_{in}|}{4}\left[(T_{xx} + T_{yx})e^{j\omega(\tau_{12}+\tau_{01})} + (T_{xy} + T_{yy})e^{j\omega(\tau_{12}+\tau_{02})} + e^{j\omega(\tau_{11}+\tau_{01})} + e^{j\omega(\tau_{11}+\tau_{02})}\right]e^{-j\omega t} \quad (2.10.16)$$

and

$$E_p = E_{2U,p} + E_{2L,p}$$
$$= \frac{|E_{in}|}{4}\left[(T_{xx} - T_{yx})e^{j\omega(\tau_{12}+\tau_{01})} + (T_{xy} - T_{yy})e^{j\omega(\tau_{12}+\tau_{02})} + e^{j\omega(\tau_{11}+\tau_{01})} - e^{j\omega(\tau_{11}+\tau_{02})}\right]e^{-j\omega t} \quad (2.10.17)$$

Since photodiodes are square-law detection devices, photocurrents in the s and p branches can be obtained by $I_s = \Re|E_s|^2$ and $I_p = \Re\,|E_p|^2$, where $\Re$ is the responsivity of the photodiode. It can be easily found that there are four discrete propagation delay components in the expression of the photocurrent in each branch: $e^{j\omega\delta\tau_0}$, $e^{j\omega\delta\tau_1}$, $e^{j\omega(\delta\tau_1-\delta\tau_0)}$, and $e^{j\omega(\delta\tau_1+\delta\tau_0)}$, where $\tau_{02} - \tau_{01} = \delta\tau_0$ is the relative delay in the first interferometer and $\tau_{12} - \tau_{11} = \delta\tau_1$ is the relative delay in the second interferometer. From a signal processing point of view, it is not generally easy to distinguish terms corresponding to different propagation delays in a photocurrent signal. However, as illustrated in Figure 2.10.7, if the tunable laser source is linearly swept, these discrete propagation delay terms will

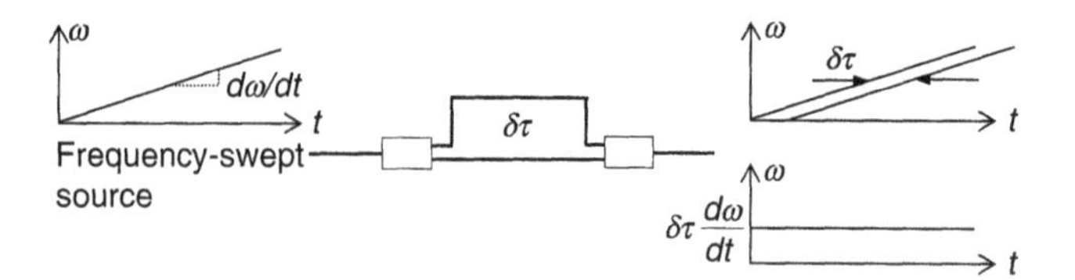

Figure 2.10.7 Illustration of the differential frequency term generated by an interferometer and a frequency-swept source. *dω/dt* is the frequency slope and $\delta\tau$ is the differential time delay between the two interferometer arms.

each have a different frequency and can be easily analyzed by a Fourier analysis.

If the tunable laser is swept at a rate of $d\omega/dt$, the four discrete delay terms $e^{j\omega\delta\tau_0}$, $e^{j\omega\delta\tau_1}$, $e^{j\omega(\delta\tau_1-\delta\tau_0)}$, and $e^{j\omega(\delta\tau_1+\delta\tau_0)}$ will become four discrete frequency components in the photocurrent $\delta\tau_0\left(\frac{d\omega}{dt}\right)$, $\delta\tau_1\left(\frac{d\omega}{dt}\right)$, $(\delta\tau_1-\delta\tau_0)\left(\frac{d\omega}{dt}\right)$, and $(\delta\tau_1+\delta\tau_0)\left(\frac{d\omega}{dt}\right)$, respectively. Among these frequency components, only $(\delta\tau_1-\delta\tau_0)\left(\frac{d\omega}{dt}\right)$ and $(\delta\tau_1+\delta\tau_0)\left(\frac{d\omega}{dt}\right)$ are useful because their coefficients are proportional to the DUT transfer matrix elements.

Simple derivation from Equations 2.10.16 and 2.10.17 yields

$$I_s = \Re|E_s|^2 \propto \left[T_{11}\exp\left(j(\delta\tau_1-\delta\tau_0)\frac{d\omega}{dt}t\right) + T_{12}\exp\left(j(\delta\tau_1+\delta\tau_0)\frac{d\omega}{dt}t\right)\right] \tag{2.10.18}$$

and

$$I_p = \Re|E_p|^2 \propto \left[T_{21}\exp\left(j(\delta\tau_1-\delta\tau_0)\frac{d\omega}{dt}t\right) + T_{22}\exp\left(j(\delta\tau_1+\delta\tau_0)\frac{d\omega}{dt}t\right)\right] \tag{2.10.19}$$

where

$$\begin{bmatrix} T_{11} & T_{12} \\ T_{21} & T_{22} \end{bmatrix} = \begin{bmatrix} (T_{xx}+T_{yx}) & (T_{xy}+T_{yy}) \\ (T_{xx}-T_{yx}) & (T_{xy}-T_{yy}) \end{bmatrix} \tag{2.10.20}$$

is the base rotated transfer function of the DUT. In practical implementations, the photocurrent signals I_s and I_p are digitized and analyzed through digital signal processing. The amplitudes at frequencies $(\delta\tau_1-\delta\tau_0)(d\omega/dt)$ and $(\delta\tau_1+\delta\tau_0)$ $(d\omega/dt)$ can be calculated by fast Fourier transformation (FFT) and the matrix elements T_{11}, T_{11}, T_{21}, T_{22} can be obtained. Finally, the vector transfer function of the DUT can be expressed as

$$\hat{T}(j\omega) = \begin{bmatrix} T_{xx} & T_{xy} \\ T_{yx} & T_{yy} \end{bmatrix} = \frac{1}{2}\begin{bmatrix} (T_{11} + T_{21}) & (T_{12} + T_{22}) \\ (T_{11} - T_{21}) & (T_{12} - T_{22}) \end{bmatrix} \tag{2.10.21}$$

This linear transfer function is a 2 × 2 Jones Matrix, and each element is generally a function of optical frequency. This transfer function contains all the information required to characterize the performance of the DUT, such as optical loss, chromatic dispersion, and polarization effects.

Note that the vector optical network analyzer shown in Figure 2.10.6 is unidirectional, which only measures the transmission characteristics of the DUT. This configuration can be modified to measure both transmission and reflection characteristics, as shown in Figure 2.10.8 [55, 56]. In this configuration, the reference signal does not go through the differential polarization delay and therefore, compared to the configuration in Figure 2.10.6, Equations 2.10.12 and 2.10.13 should be replaced by

$$E_{2L,s} = E_{2L,p} = A_2 e^{-j\omega(t-\tau_0)} \tag{2.10.22}$$

where A_2 is the reference signal amplitude at the receiver and τ_0 is the delay of the reference signal. Here we have assumed that the polarization controller (Pol.2) in the reference path is set such that the reference signal is equally split into s and p polarizations in the receivers.

Therefore, there will be only three discrete frequency components in this configuration: $(\tau_{01} - \tau_0)\left(\frac{d\omega}{dt}\right)$, $(\tau_{02} + \tau_0)\left(\frac{d\omega}{dt}\right)$, and $(\tau_{02} - \tau_{01})\left(\frac{d\omega}{dt}\right)$, where, τ_{01} and τ_{02} in this case represent overall propagation delays (from source to

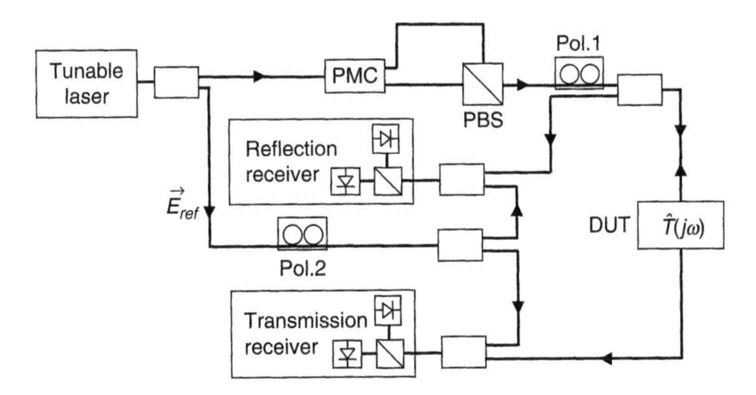

Figure 2.10.8 Optical configuration of a vector optical network analyzer that is capable of both transmission and reflection measurements [55].

detector) of optical signals passing through the lower and upper branches of the only one interferometer. Among these three frequency components, the coefficients of the first two are related to the DUT transfer functions and then similar procedures can be used to determine the vector transfer function of the DUT as described previously.

REFERENCES

[1] Erwin G. Loewen and Evgeny Popov, *Diffraction Gratings and Applications*, CRC, 1st ed., 1997, ISBN-10: 0824799232.
[2] For example, Agilent Optical Spectrum Analyzer User Guide: http://cp.literature.agilent.com/litweb/pdf/86140-90068.pdf
[3] Newport Grating product description: http://gratings.newport.com/products/table1.asp
[4] G. Hernandez, *Fabry-Perot Interferometers*, Cambridge University Press, 1986.
[5] J. M. Vaughan, *The Fabry-Perot Interferometer: History, Practice and Applications*, Bristol England, Adam Hilger, 1989.
[6] J. B. Clayton, M. A. El, L. J. Freeman, J. Lucius, and C. M. Miller, Tunable optical filter, US patent #5,073,004, 1991.
[7] R. Hui, Optical domain signal analyzer, US Patent #6,697,159, 2004.
[8] R. Hui and M. O'Sullivan, Method and apparatus for high-resolution monitoring of optical signals, US Patent #6,687,009, 2004.
[9] J. Pietzsch, "Scattering matrix analysis of 3 × 3 fiber couplers," *Journal of Lightwave Technology*, Vol. 7, No. 2, pp. 303–307, 1989.
[10] P. E. Green, *Fiber Optic Networks*, Prentice-Hall, 1991.
[11] S. Cao, J. Chen, J. N. Damask, C. R. Doerr, L. Guiziou, G. Harvey, Y. Hibino, H. Li, S. Suzuki, K.-Y. Wu, and P. Xie, "Interleaver technology: Comparisons and applications requirements," *J. Lightwave Technology*, Vol. 22, No. 1, pp. 281–289, Jan. 2004.
[12] M. Oguma, T. Kitoh, Y. Inoue, T. Mizuno, T. Shibata, M. Kohtoku, and Y. Kibino, "Compact and low-loss interleave filter employing latticeform structure and silica-based waveguide," *J. Lightwave Technology*, Vol. 22, No. 3, pp. 895–902, March 2004.
[13] A. D. Kersey, M. J. Marrone, and M. A. Davis, "Polarisation-insensitive fibre optic Michelson interferometer," *Electronics Letters*, Vol. 27, No. 6, pp. 518–520, 1991.
[14] J. J. Snyder, "Laser wavelength meters," *Laser Focus* 18, 55–61 (1982).
[15] J. P. Monchalin, M. J. Kelly, J. E. Thomas, N. A. Kurnit, A. Szoeke, F. Zernike, P. H. Lee, and A. Javan, "Accurate laser wavelength measurement with a precision two-beam scanning Michelson interferometer," *Appl. Opt.* 20, 736-757 (1981).
[16] D. F. Gray, K. A. Smith and F. B. Dunning, "Simple compact Fizeau wavemeter," *Applied Optics*, Vol. 25, No. 8, pp. 1339–1343, 1986.
[17] C. Reiser and R. B. Lopert, "Laser wavemeter with solid Fizeau wedge interferometer," *Applied Optics*, Vol. 27, No. 17, pp. 3656–3660, 1988.
[18] M. Born and E. Wolf, *Principles of Optics*, 7th ed., Cambridge University Press.
[19] Edward Collett, *Polarized Light: Fundamentals and Applications*, Dekker, 1992.
[20] R. A. Chipman, "Polarimetry," in *Handbook of Optics*, 2nd ed., Vol. 2, M. Bass, ed., New York: McGraw-Hill, 1994.
[21] C. Madsen et al., "Integrated optical spectral polarimeter for signal monitoring and feedback to a PMD compensator," *J. Opt. Netw.*, Vol. 3, No. 7, pp. 490–500, July 2004.

[22] Shawn X. Wang and Andrew M. Weiner, "A Complete Spectral Polarimeter Design for Lightwave Communication Systems," *Journal of Lightwave Technologies*, Vol. 24, No. 11, pp. 3982–3991, 2006.

[23] B. L. Heffner, "Automated measurement of polarization mode dispersion and using Jones matrix eigenanalysis," *IEEE Photon. Technol. Lett.* Vol. 4, pp. 1066–1069, 1992.

[24] S. Betti, G. De Marchis, and E. Iannone, *Coherent Optical Communications Systems,* Wiley Series in Microwave and Optical Engineering, Wiley-Interscience (1995).

[25] D. M. Baney, B. Szafraniec, and A. Motamedi, "Coherent optical spectrum analyzer," *IEEE Photonics Technology Letters*, Vol. 14, Issue 3, pp. 355–357, March 2002.

[26] R. Epworth, 3 Fibre I and Q coupler, US patent #6,859,586, 2005.

[27] D. H. Jundt, G. A. Magela, M. M. Fejer, and R. L. Byer, "Periodically poled LiNbO3 for high-efficiency second-harmonic generation," *Appt. Phys. Lett.* Vol. 59, No. 21, pp. 2657–2659, 1991.

[28] M. M. Fejer, G. A. Magel, D. H. Jundt, and R. L. Byer, "Quasi-phase-matched second harmonic generation: tuning and tolerances," *IEEE Quantum Electronics*, Vol. 28, No. 11, pp. 2631–2654, 1992.

[29] S. Diez, R. Ludwig, C. Schmidt, U. Feiste, and H. G. Weber, "160-Gb/s optical sampling by gain-transparent four-wave mixing in a semiconductor optical amplifier," *IEEE Photonics Technology Letters*, Vol. 11, No. 11, pp. 1402–1404, 1999.

[30] M. Westlund, H. Sunnerud, B.-E. Olsson, P. A. Andrekson, "Simple scheme for polarization-independent all-optical sampling," *IEEE Photonics Technology Letters*, Vol. 16, No. 9, pp. 2108–2110, 2004.

[31] J. Li, M. Westlund, H. Sunnerud, B.-E. Olsson, M. Karlsson, P. A. Andrekson, "0.5-Tb/s eye-diagram measurement by optical sampling using XPM-induced wavelength shifting in highly nonlinear fiber," *IEEE Photonics Technology Letters*, Vol. 16, No. 2, pp. 566–568, 2004.

[32] S. Nogiwa, H. Ohta, Y. Kawaguchi, Y. Endo, "Improvement of sensitivity in optical sampling system," *IEE Electronics Letters*, Vol. 35, No. 11, pp. 917–918, 1999.

[33] K. Kikuchi, F. Futami, K. Katoh, "Highly sensitive and compact cross-correlator for measurement of picosecond pulse transmission characteristics at 1550 nm using two-photon absorption in Si avalanche photodiode," *IEE Electronics Letters*, Vol. 34, No. 22, pp. 2161–2162, 1998.

[34] C. Dorrer, D. C. Kilper, H. R. Stuart, G. Raybon, and M. G. Raymer, "Linear Optical Sampling," *IEEE Photonics Technology Letters*, Vol. 15, No. 12, pp. 1746–1748, 2003.

[35] Christophe Dorrer, High-Speed Measurements for Optical Telecommunication Systems, *IEEE Journal of Selected Topics in Quantum Electronics*, Vol. 12, No. 4, pp. 843–858, 2006.

[36] Christophe Dorrer, "Monitoring of Optical Signals from Constellation Diagrams Measured with Linear Optical Sampling," *Journal of Lightwave Technology*, Vol. 24, No. 1, pp. 313–321, 2006.

[37] A. Yariv and R. G. M. P. Koumans, "Time interleaved optical sampling for ultrahigh speed AID conversion," *Electronics Letters*, Vol. 34, No. 21, pp. 2012–2013, 1998.

[38] Yan Han and Bahram Jalali, "Differential Photonic Time-Stretch Analog-to-Digital Converter," *CLEO* 2003.

[39] D. T. Reid, M. Padgett, C. McGowan, W. Sleat, and W. Sibbett, "Light-emitting diodes as measurement devices for femtosecond laser pulses," *Opt. Lett.*, Vol. 22, pp. 233–235, 1997.

[40] Kenneth W. DeLong, David N. Fittinghoff, and Rick Trebino, "Practical Issues in Ultra-short-Laser-Pulse Measurement Using Frequency-Resolved Optical Gating," *IEEE Journal of Quantum Electronics*, Vol. 32, No. 7, pp. 1253–1264, 1996.

[41] J. R. Fienup, "Phase retrieval algorithms: A comparison," *Appi. Opt.* Vol. 21, No. 15, pp. 2758–2769, 1982.

[42] Rick Trebino, Kenneth W. DeLong, David N. Fittinghoff, John N. Sweetser, Marco A. Krumbügel, Bruce A. Richman, and Daniel J. Kane, "Measuring ultrashort laser pulses in the time-frequency domain using frequency-resolved optical gating," *Rev. Sci. Instrum.*, Vol. 68, No. 9, pp. 3277–3295, 1997.

[43] D. Huang, E. A. Swanson, C. P. Lin, J. S. Schuman, W. G. Stinson, W. Chang, M. R. Hee, T. Flotte, K. Gregory, C. A. Puliafito, and J. G. Fujimoto, "Optical coherence tomography," *Science* Vol. 254, pp. 1178–1181 (1991).

[44] C. R. Giles, "Lightwave Applications of Fiber Bragg Gratings," *Journal of Lightwave Technology*, Vol. 15, No. 8, 1997.

[45] D. Derickson, *Fiber Optic Test and Measurement,* Prentice Hall PTR (1998).

[46] K. Takada, H. Yamada, Y. Hibino, S. Mitachi, "Range extension in optical low coherence reflectometry achieved by using a pair of retroreflectors," *IEEE Electronics Letters*, Vol. 31, No. 18, pp. 1565–1567, 1995.

[47] M. Kobayashi, H. Hanafusa, K. Takada and J. Noda, "Polarization-independent interferometric optical-time-domain reflectometer," *IEEE/OSA J. Lightwave Technology*, Vol. 9, No. 5, pp. 623–628, 1991.

[48] R. Hui, J. Thomas, C. Allen, B. Fu, and S. Gao, "Low-coherent WDM reflectometry for accurate fiber length monitoring," *IEEE Photonics Technol. Lett.*, Vol. 15, No. 1, pp. 96–98, 2003.

[49] R. Huber, M. Wojtkowski, K. Taira, J. G. Fujimoto, and K. Hsu, "Amplified, frequency swept lasers for frequency domain reflectometry and OCT imaging: design and scaling principles," *Optics Express*, Vol. 13, No. 9, pp. 3513–3528, 2005.

[50] R. Huber, M. Wojtkowski, and J. G. Fujimoto, "Fourier Domain Mode Locking (FDML): A new laser operating regime and applications for optical coherence tomography," *Optics Express*, Vol. 14, No. 8, pp. 3225–3237, 2006.

[51] Maciej Wojtkowski, Vivek J. Srinivasan, Tony H. Ko, James G. Fujimoto, Andrzej Kowalczyk, Jay S. Duker, "Ultrahigh-resolution, high-speed, Fourier domain optical coherence tomography and methods for dispersion compensation," *Optics Express*, Vol. 12, No. 11, pp. 2404–2422, 2004.

[52] U. Glombitza and E. Brinkmeyer, "Coherent frequency domain reflectometry for characterization of single-mode integrated optical waveguides," *J. Lightwave Technol.* Vol. 11, pp. 1377–1384, 1993.

[53] Luna Technologies, Fiber Optic Component Characterization, www.lunatechnologies.com.

[54] M. Froggatt, T. Erdogan, J. Moore, and S. Shenk, "Optical frequency domain characterization (OFDC) of dispersion in optical fiber Bragg gratings," in *Bragg Gratings, Photosensitivity, and Poling in Glass Waveguides*, OSA Technical Digest Series (Optical Society of America, Washington, DC, 1999), paper FF2.

[55] G. D. VanWiggeren, A. R. Motamedi, and D. M. Baney, "Single-scan interferometric component analyzer," *IEEE Photonics Technology Letters*, Vol. 15, pp. 263–265, 2003.

[56] A. P. Freundorfer, "A Coherent Optical Network Analyzer," *IEEE Photonics Technology Letters*, Vol. 3, No. 12, pp. 1139–1142, 1991.

Chapter 3

Characterization of Optical Devices

3.0. Introduction
3.1. Characterization of RIN and Linewidth of Semiconductor Lasers
3.1.1. Measurement of Relative Intensity Noise (RIN)
3.1.2. Measurement of Laser Phase Noise and Linewidth
3.2. Measurement of Electro-Optic Modulation Response
3.2.1. Characterization of Intensity Modulation Response
3.2.2. Measurement of Frequency Chirp
3.2.3. Time-Domain Measurement of Modulation-Induced Chirp
3.3. Wideband Characterization of an Optical Receiver
3.3.1. Characterization of Photodetector Responsivity and Linearity
3.3.2. Frequency Domain Characterization of Photodetector Response
3.3.3. Photodetector Bandwidth Characterization Using Source Spontaneous-Spontaneous Beat Noise
3.3.4. Photodetector Characterization Using Short Optical Pulses
3.4. Characterization of Optical Amplifiers
3.4.1. Measurement of Amplifier Optical Gain
3.4.2. Measurement of Static and Dynamic Gain Tilt
3.4.3. Optical Amplifier Noise
3.4.4. Optical Domain Characterization of ASE Noise
3.4.5. Impact of ASE Noise in Electrical Domain
3.4.6. Noise Figure Definition and Its Measurement
3.4.7. Time Domain Characteristics of EDFA
3.5. Characterization of Passive Optical Components
3.5.1. Fiber-Optic Couplers
3.5.2. Fiber Bragg Grating Filters
3.5.3. WDM Multiplexers and Demultiplexers
3.5.4. Characterization of Optical Filter Transfer Functions
3.5.5. Optical Isolators and Circulators

3.0 INTRODUCTION

Optical devices are building blocks of optical systems and networks. The performance and reliability of a complex optical network depend heavily on the quality of optical components. As the capability requirement of optical systems and networks has increased, various new optical devices have been created with unprecedented high quality and rich functionalities. As the complexity of optical systems increases, the requirement for qualification of optical components also becomes more and more stringent. Therefore, a good understanding of various techniques to characterize optical devices is critical for optical systems and network research, development, implementation, and trouble shooting.

This chapter is devoted to various measurement and characterization techniques of fundamental optical devices. Sections 3.1 and 3.2 discuss the measurement of semiconductor lasers, including intensity noise, phase noise, intensity, and phase modulation responses, as well as frequency chirp characterization. Section 3.3 presents techniques of testing wideband optical receivers, in both frequency domain and time domain. The characterization of optical amplifiers is discussed in Section 3.4, which includes wavelength-dependent gain, noise, gain saturation, noise figure, and dynamic properties. At the end of this chapter, Section 3.5 discusses the working principles and qualification test techniques of a number of passive optical devices, including optical fiber couplers, Bragg grating filters, WDM multiplexers and demultiplexers, optical isolators, and circulators. These basic fiber-optical components are indispensible in building optical systems as well as optical testing apparatus.

3.1 CHARACTERIZATION OF RIN AND LINEWIDTH OF SEMICONDUCTOR LASERS

The semiconductor laser is a key device in an optical transmitter; it translates electrical signals into optical domain. It is one of the most important components in lightwave communication systems and photonic sensors, as well as in optical instrumentation. When operating above threshold, the output optical power of a laser diode is linearly proportional to the injection current, the electrical current signal can therefore be converted into an optical signal through direct modulation of the injection current. However, direct modulation on a laser diode not only introduces optical intensity modulation, it also creates a frequency modulation on the optical carrier, which is known as the *frequency chirp*. Frequency chirp broadens the signal optical spectrum and introduces transmission performance degradation in high-speed optical systems due to the dispersive nature of the transmission media. To overcome this problem, external electro-optic modulators are often used in high-speed and long-distance optical systems. In this case, the laser diode operates in a continuous wave mode while the external modulator encodes the electrical signal into the optical domain.

Optical system performance, to a large extent, depends on the characteristics of the laser source. Important issues such as optical power, wavelength, spectral line width, relative intensity noise, modulation response, and modulation chirp are all practical concerns in an optical transmitter. Some of these properties have relatively simple definitions and can be measured straightforwardly; for example, the optical power of a semiconductor laser can be directly measured by an optical power meter and its wavelength can be measured by either a wavelength meter or an optical spectrum analyzer. Some other properties of semiconductor lasers are defined rather nonintuitively, and the measurements of these properties often require good understanding of their physical mechanisms and the limitations of various measurement techniques. This section is devoted to the measurements and explanations of these properties in semiconductor lasers.

3.1.1 Measurement of Relative Intensity Noise (RIN)

In semiconductor lasers operating above threshold, although stimulated emission dominates the emission process, there are still a small percentage of photons that are generated by spontaneous emission. The origin of optical intensity noise in a semiconductor laser is caused by these spontaneous emission photons. As a result, when a semiconductor laser operates in CW mode, the effect of spontaneous emission makes both carrier density and photon density fluctuate around their equilibrium values [1].

In general, the intensity noise in a laser diode caused by spontaneous emission is a wideband noise; however, this noise can be further amplified by the resonance of the relaxation oscillation in the laser cavity, which modifies the noise spectral density. As illustrated in Figure 3.1.1, each single event of spontaneous increase in the carrier density will increase the optical gain of the laser medium and thus increase the photon density in the laser cavity. However, this increased

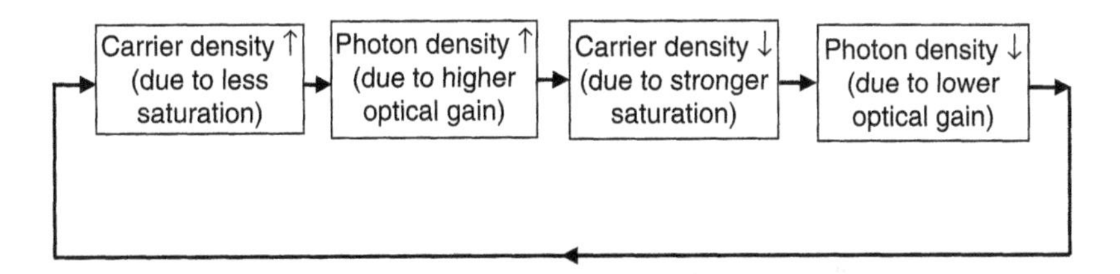

Figure 3.1.1 Explanation of a relaxation oscillation process in a semiconductor laser.

photon density will consume more carriers in the cavity, creating a gain saturation in the medium; as a result, the photon density will tend to be decreased due to the reduced optical gain. This, in turn, will increase the carrier density due to the reduced saturation effect. This resonance process is strong in a specific frequency Ω_R which is determined by the optical gain G, the differential gain dG/dN, and the photon density P in the laser cavity as

$$\Omega_R^2 = G \cdot P \cdot \frac{dG}{dN} \tag{3.1.1}$$

Due to relaxation oscillation, the intensity noise of a semiconductor laser is a function of frequency. The normalized intensity noise spectral density can be fit by the following expression:

$$H(\Omega) \propto \left| \frac{\Omega_R^2 + B\Omega^2}{j\Omega(j\Omega + \gamma) + \Omega_R^2} \right|^2 \tag{3.1.2}$$

where B and γ are damping parameters depending on specific laser structure and bias condition. Figure 3.1.2 shows examples of the normalized intensity noise spectral density with three different damping parameters. The relaxation oscillation frequency used in this figure is $\Omega_R = 2\pi \times 10\ GHz$. At frequencies much higher than the relaxation oscillation frequency, the dynamic coupling between the photon density and the carrier density is weak and therefore the intensity noise becomes less dependent on the frequency.

In addition to relaxation oscillation, external optical feedback to semiconductor lasers may also change the spectral distribution of intensity noise [2, 3]. This can be

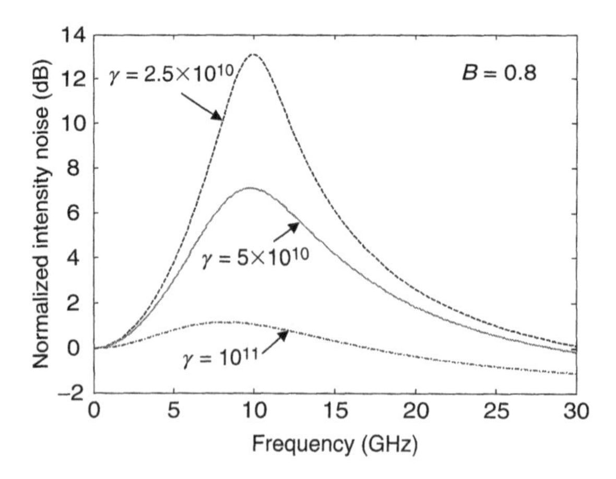

Figure 3.1.2 Normalized intensity noise spectral density with 10 GHz relaxation oscillation frequency.

caused by external optical reflectors such as connectors, or other optical devices, which form external cavities with the help of the output mirror of the semiconductor laser. The resonance nature in the external cavity may be strong enough to cause excessive carrier density fluctuation in the laser and result in strong optical power fluctuation. The distinct feature of optical feedback-induced intensity noise is that it has well-defined periodical resonance peaks in the frequency domain and the period is $\Delta f = c/(2nL)$, where L is the optical length of the external cavity, c is the speed of light, and n is refractive index of the external cavity.

Another source of optical intensity noise is a result of frequency noise to intensity noise conversion. In an optical system, frequency instabilities in the laser source can be converted into intensity noise through a frequency-dependent transfer function as well as chromatic dispersion. For example, Fabry-Perot cavities can be formed by multiple reflectors (e.g., connectors) in an optical system, which introduce frequency-dependent transmission characteristics. As illustrated in Figure 3.1.3, when laser frequency fluctuates, the transmission efficiency varies, thus causing intensity fluctuation.

In addition, if the transmission system is dispersive, different frequency components will arrive at the detector at different times; interference between them also causes frequency noise to intensity noise conversion [4].

In general, the optical intensity noise in an optical system is linearly proportional to signal optical power; therefore it is more convenient to normalize the intensity noise by the total optical power. Relative intensity noise (RIN) is defined as the ratio between the noise power spectral density and the total power. Figure 3.1.4 shows the block diagram of a RIN measurement system, which consists of a wideband photodetector, an electrical preamplifier, and an RF spectrum analyzer.

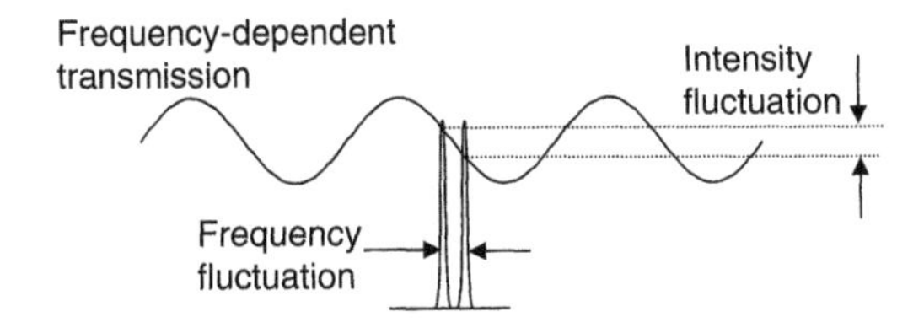

Figure 3.1.3 Illustration of frequency noise to intensity noise conversion through frequency-dependent transmission.

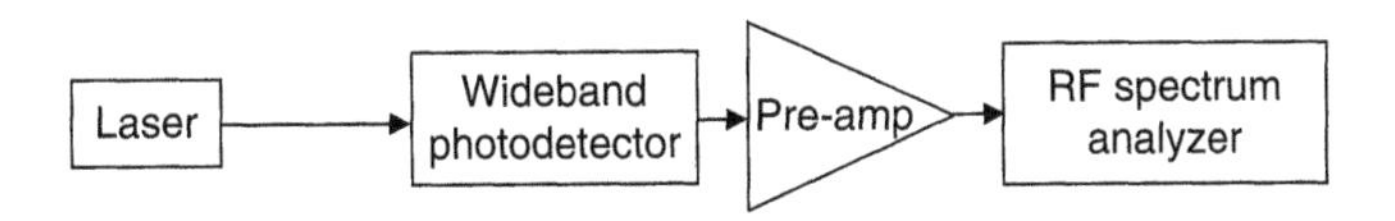

Figure 3.1.4 System block diagram to measure laser RIN.

To characterize the RIN characteristic, the laser diode to be measured is operated in a continuous wave. Obviously, without intensity noise in the laser, one would only be able to see a DC component on the RF spectrum analyzer. Due to intensity noise, the optical power fluctuation introduces photocurrent fluctuation, and the spectral density of this photocurrent fluctuation is measured by the RF spectrum analyzer. Because of the square-law detection principle, the electrical power P_{ele} generated at the output of a photodiode is proportional to the square of the received optical power P_{opt}, that is, $P_{ele} \propto P_{opt}^2$. Therefore, the RIN can be defined as the ratio between noise power spectral density and the signal power in the *electrical domain*,

$$RIN(\omega) = F\left\{\frac{\left\langle (P_{opt} - P_{opt,ave})^2 \right\rangle}{P_{opt,ave}^2}\right\} \qquad (3.1.3)$$

where $P_{opt,ave}$, is the average optical power and F denotes Fourier transformation. $\Re^2\langle(P_{opt} - P_{opt,ave})^2\rangle$ is the mean square intensity fluctuation, and its Fourier transform is the noise power spectral density in the electrical domain, where $\Re$ is the detector responsivity. $\Re P_{opt}$ is the signal photocurrent, and $\Re^2 P_{opt}^2$ is the signal electrical power. Therefore, the definition of *RIN* can also be simply expressed as

$$RIN = \frac{S_P(\omega)}{\Re^2 P_{opt,\,ave}^2} \qquad (3.1.4)$$

where $S_P(\omega)$ is the measured electrical noise power spectral density on the RF spectrum analyzer. In fact, if the detector noise is neglected, the value of photodiode responsivity does not affect the result of RIN measurement, because both the nominator and dinomintor of Equation 3.1.4 are proportional to this responsivity.

Obviously, the noise power spectral density $S_P(\omega)$ increases with the increase of the square of the average optical power $P_{opt,\,ave}^2$. RIN is a convenient way to characterize optical signal quality, and the result is not affected by the uncertainties in optical attenuation in optical systems; therefore absolute calibration of the photodetector responsivity is not required. The unit of RIN is $[Hz^{-1}]$, or $[dB/Hz]$ as a relative measure.

Practically, RIN is a convenient parameter to use in optical system performance calculation. For example, in an analog system, if RIN is the only source of noise, the signal-to-noise ratio (in electrical domain) can be easily determined by

$$SNR = \frac{m^2}{2}\frac{1}{\int_0^B RIN(f)df} \qquad (3.1.5)$$

where m is the modulation index and B is the receiver bandwidth.

Good-quality semiconductor lasers usually have the RIN levels lower than –155 dB/Hz, and it is usually challenging to accurately measure the RIN at such a low level. This requires the measurement system to have a much lower noise level. Considering thermal noise and quantum noise in the photodiode and thermal noise in the electrical preamplifier, the measured RIN can be expressed as

$$RIN_{measure} = \frac{S_p + \sigma_{shot}^2 + \sigma_{th}^2 + \sigma_{amp}^2}{\Re^2 P_{opt}^2} = RIN_{laser} + RIN_{error} \tag{3.1.6}$$

where

$$RIN_{error} = \frac{2q}{\Re P_{opt}} + \frac{kT(F_A G_A + F_{SA} - 1)/G_A}{\Re^2 P_{opt}^2} + \frac{4kT/R_L}{\Re^2 P_{opt}^2} \tag{3.1.7}$$

is the measurement error due to instrument noise. Where $\sigma_{shot}^2 = 2q\Re P_{opt}$ is the shot noise power spectral density, $\sigma_{th}^2 = 4kT/R_L$ is the thermal noise power spectral density at the photodiode and $\sigma_{amp}^2 = kT(F_A G_A + F_{SA} - 1)/G_A$ is the equivalent noise power spectral density introduced by the electrical preamplifier and the RF spectrum analyzer. F_A is the noise figure of the electrical preamplifier, F_{SA} is the noise figure of the spectrum analyzer front end, and G_A is the gain of the preamplifier.

Figure 3.1.5 shows the RIN measurement error estimated with $\Re = 0.75A/W$, $R_L = 50\Omega$, $T = 300K$, $F_A = F_{SA} = 3$ dB, and $G_A = 30$ dB. The measurement error decreases with the increase of the signal optical power, and eventually the quantum noise is the dominant noise in the high power region. In this particular case, in order to achieve 180 dB/Hz error level (which is considered much lower than the actual laser RIN of –155 dBm/Hz), the signal optical power has to be higher than 0 dBm.

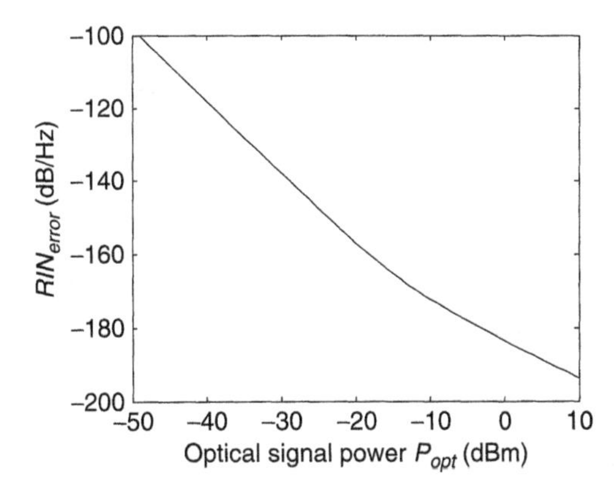

Figure 3.1.5 Error in RIN versus input optical signal power.

Example 3.1

Consider the RIN measurement system shown in Figure 3.1.4. The resolution bandwidth of the RF spectrum analyzer is set at 3 MHz. When a laser source is DC biased at an operation current of $I_B = 50$ mA, the measured noise level on the spectrum analyzer is –80 dBm at 1 GHz frequency. Then the laser source is modulated by a sinusoid at $f_m = 1$ MHz, in which the injection current swings from the threshold to 50 mA. A very narrow peak at 1 MHz is measured on the spectrum analyzer with the amplitude of –5 dBm. Find the value of the RIN of this laser at 1 GHz frequency.

Solution:

Typically the power-level reading on the RF spectrum analyzer is the RF power within a resolution bandwidth. Therefore, the reading of –80 dBm on the spectrum analyzer in this case means a power spectral density of $S_p = -80$ dBm/3 MHz = –144.78 dBm/Hz.

When the laser is analog modulated with the index m, the mean-root-square electrical power measured on the spectrum analyzer is –5 dBm. Since this is a sinusoid modulation, the spectral bandwidth at the modulation frequency should be much narrower than the resolution bandwidth of 3 MHz. Therefore the actual total electrical power converted from the optical signal should be approximately –2 dBm (at the modulation index of $m = 1$, the effective signal power is ½). Therefore the RIN of this laser at 50 mA bias should be RIN = –144.78 – (–2) = 142.78 dBm/Hz.

3.1.2 Measurement of Laser Phase Noise and Linewidth

Phase noise is an important issue in semiconductor lasers, especially when they are used in coherent optical systems where optical phase information is utilized. Phase noise in semiconductor lasers is originated from spontaneous emission. A spontaneous emission event not only generates variation in the photon density, it also produces phase variation [1]. In addition, the photon density-dependent refractive index in the laser cavity significantly enhances the phase noise, which makes the linewidth of semiconductor lasers much wider than other types of solid-state lasers [5].

Assume that the optical field in a laser cavity is

$$E(t) = \sqrt{P(t)}\exp\{j[\omega_0 t + \phi(t)]\} \tag{3.1.8}$$

where $P(t)$ is the photon density inside the laser cavity, ω_0 is the central optical frequency, and $\phi(t)$ is the time-varying part of optical phase. A differential rate equation that describes the phase variation in the time domain is [6]

$$\frac{d\phi(t)}{dt} = F_\phi(t) - \frac{\alpha_{lw}}{2P} F_P(t) \tag{3.1.9}$$

where α_{lw} is the linewidth enhancement factor of a semiconductor laser, which accounts for the coupling between intensity and phase variations. $F_\phi(t)$ and $F_P(t)$ are Langevin noise terms for phase and intensity. They are random and their statistic measures are

$$\left\langle F_P(t)^2 \right\rangle = 2R_{sp}P \tag{3.1.10}$$

and

$$\left\langle F_\phi(t)^2 \right\rangle = \frac{R_{sp}}{2P} \tag{3.1.11}$$

R_{sp} is the spontaneous emission factor of the laser.

Although we directly use Equation 3.1.9 without derivation, the physical meaning of the two terms on the right side of the equation are very clear. The first term is created directly due to spontaneous emission contribution to the phase variation. Each spontaneous emission event randomly emits a photon which changes the optical phase as illustrated in Figure 1.1.13 in Chapter 1. The second term in Equation 3.1.9 shows that each spontaneous emission event randomly emits a photon that changes the carrier density, and this carrier density variation, in turn, changes the refractive index of the material. Then this index change will alter the resonance condition of the laser cavity and thus will introduce a phase change of the emitting optical field. Equation 3.1.9 can be solved by integration:

$$\phi(t) = \int_0^t \frac{d\phi(t)}{dt} dt = \int_0^t F_\phi(t)dt - \frac{\alpha_{lw}}{2P} \int_0^t F_P(t)dt \tag{3.1.12}$$

If we take an ensemble average, the power spectral density of phase noise can be expressed as

$$\left\langle \phi(t)^2 \right\rangle = \left\langle \left[\int_0^t F_\phi(t)dt - \frac{\alpha_l w}{2P} \int_0^t F_P(t)dt \right]^2 \right\rangle = \int_0^t \left\langle F_\phi(t)^2 \right\rangle dt - \frac{\alpha_{lw}}{2P} \int_0^t \left\langle F_P(t)^2 \right\rangle dt =$$

$$\left[\frac{R_{sp}}{2P} + \left(\frac{\alpha_{lw}}{2P} \right)^2 2R_{sp}P \right] |t| = \frac{R_{sp}}{2P} [1 + \alpha_{lw}^2] |t| \tag{3.1.13}$$

Since $\phi(t)$ is a random Gaussian process

$$\left\langle e^{j\phi(t)} \right\rangle = e^{-\frac{1}{2}\langle \phi(t)^2 \rangle} \tag{3.1.14}$$

Then the optical power spectral density is

$$S_{op}(\omega) = \int_{-\infty}^{\infty} \langle E(t)E^*(0) \rangle e^{-j\omega t} dt = P \int_{-\infty}^{\infty} e^{-j(\omega - \omega_0)t} e^{-\frac{1}{2}\langle \phi(t)^2 \rangle} dt \tag{3.1.15}$$

Then the normalized optical power spectral density can be found as

$$S_{op}(\omega) = \frac{\left[\frac{R_{sp}}{4P} \left(1 + \alpha_{lw}^2\right) \right]^2}{\left[\frac{R_{sp}}{4P} \left(1 + \alpha_{lw}^2\right) \right]^2 + (\omega - \omega_0)^2} \tag{3.1.16}$$

The FWHM linewidth of this spectrum is [6]

$$\Delta v = \frac{\Delta \omega}{2\pi} = \frac{R_{sp}}{4\pi P} \left(1 + \alpha_{lw}^2\right) \tag{3.1.17}$$

This formula is commonly referred to as the modified Scholow-Towns formula because of the introduction of linewidth enhancement factor α_{lw}.

Spectral linewidth of an unmodulated laser is determined by the phase noise, which is a measure of coherence of the lightwave signal. Other measures such as coherence time and coherence length can all be related to the linewidth. Coherence time is defined as

$$t_{coh} = \frac{1}{\Delta v} \tag{3.1.18}$$

It is the time over which a lightwave may still be considered coherent. Or, in other words, it is the time interval within which the phase of a lightwave is still predictable. Similarly, coherence length is defined by

$$L_{coh} = t_{coh} v_g = \frac{v_g}{\Delta v} \tag{3.1.19}$$

It is the propagation distance over which a lightwave signal maintains its coherence, where v_g is the group velocity of the optical signal.

As a simple example, for a lightwave signal with 1 MHz linewidth, the coherence time is approximately 320 ns and the coherence length is about 95 m in free space.

In principle, the linewidth of an optical signal can be directly measured by an optical spectrum analyzer. However, a finest resolution of a conventional OSA is on the order of 0.1 nm, which is approximately 12 GHz in a 1550 nm wavelength window, which is not suitable to measure a laser linewidth. In fact, the linewidth of a commercial single longitudinal mode laser diode (such as a DFB or DBR laser) is on the order of 1–100 MHz. In this section we discuss two techniques that are commonly used to measure linewidths of semiconductor lasers: coherent detection and self-homodyne selection.

3.1.2.1 Coherent Detection

As we discussed in Section 2.6, in coherent detection (Figure 3.1.6), the incoming lightwave signal mixes with an optical local oscillator in a photodiode and downshifts the signal from an optical frequency to an intermediate frequency (IF). Since an RF spectrum analyzer can have an excellent spectral resolution, the detailed spectral shape of the frequency-downshifted IF signal can be precisely measured [7].

Similar to Equation 2.7.6 in Chapter 2, if the splitting ratio of the fiber directional coupler is 50 percent, after the input optical signal mixes with the local oscillator in the photodiode the photocurrent is

$$I(t) = \Re|E_1(t)|^2 = \frac{\Re}{2}\left\{|E_s(t)|^2 + |E_{LO}|^2 + E_s(t)E_{LO}^* + E_s^*(t)E_{LO}\right\} \quad (3.1.20)$$

where E_s and E_{LO} are the complex optical fields of the input lightwave signal and the local oscillator, respectively. Neglecting the direct detection contributions, the useful part of the photocurrent, which produces the IF frequency component, is the cross term.

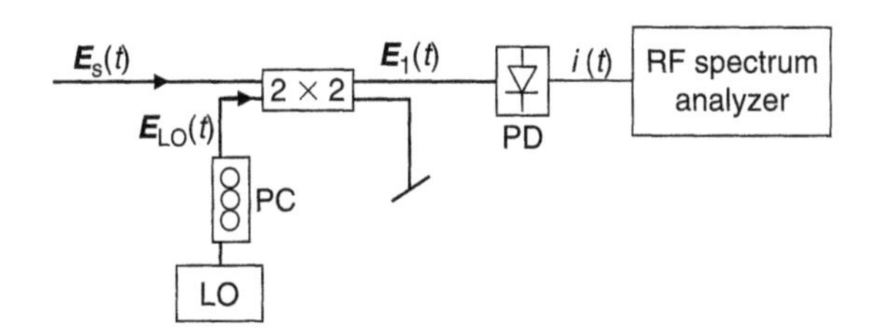

Figure 3.1.6 Linewidth measurement using coherent detection.

$$i(t) = \frac{\Re}{2}\left[E_s(t)E_{LO}^*(t) + E_s^*(t)E_{LO}(t)\right] \tag{3.1.21}$$

This time-domain multiplication between the signal and the local oscillator is equivalent to a convolution between the two optical spectra in the frequency domain. Therefore the *IF* power spectral density measured by the RF spectrum analyzer is

$$S_{IF}(\omega) \propto S_{p,s}(\omega) \otimes S_{p,LO}(-\omega) \tag{3.1.22}$$

where $S_{p,s}$ and $S_{p,LO}$ are the power spectral densities of the input lightwave signal and the local oscillator, respectively. Suppose these two power spectral densities are both Lorenzian,

$$S_{p,s}(f) = \frac{1}{1 + \left(\frac{f - f_{s0}}{\Delta\nu_s/2}\right)^2} \tag{3.1.23}$$

and

$$S_{p,LO}(f) = \frac{1}{1 + \left(\frac{f - f_{LO0}}{\Delta\nu_{LO}/2}\right)^2} \tag{3.1.24}$$

where $\Delta\nu_s$ and $\Delta\nu_{LO}$ are the FWHM spectral linewidths and f_{s0} and f_{LO0} are the central frequencies of the signal and the local oscillator, respectively. The normalized *IF* power spectral density measured by the RF spectrum analyzer should be

$$S_{IF}(f) = \frac{1}{1 + \left(\frac{f - f_{IF}}{(\Delta\nu_s + \Delta\nu_{LO})/2}\right)^2} \tag{3.1.25}$$

where $f_{IF} = |f_{s0} - f_{LO0}|$ is the central frequency of the heterodyne IF signal; obviously, the measured IF signal linewidth in the electrical domain is the sum of the linewidths of the incoming lightwave signal and the local oscillator,

$$\Delta\nu_{IF} = \Delta\nu_s + \Delta\nu_{Lo} \tag{3.1.26}$$

Figure 3.1.7 illustrates an example of frequency translation from optical domain to RF domain by coherent heterodyne detection, as well as the linewidth relationship among the signal, the local oscillator, and the IF beating note. In this case, the central frequencies of the optical signal and the local oscillator are 193500 GHz (1550.388 nm) and 193501 GHz (1550.396 nm), and their linewidths are 10 MHz and 100 kHz, respectively. As the result of coherent detection, the central RF frequency is 1 GHz and the RF linewidth is 10.1 MHz.

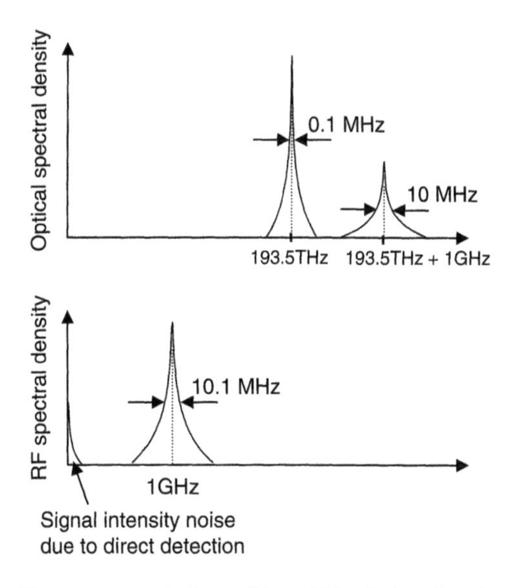

Figure 3.1.7 Frequency translation and linewidth relations in coherent detection.

Although the operation principle of linewidth measurement using coherent heterodyne detection is straightforward, a few practical issues need to be considered in the measurement. First, the linewidth of the local oscillator has to be narrow enough. In fact, $\Delta\nu_{Lo} \ll \Delta\nu_s$ is usually required to accurately measure the linewidth of an optical signal. Second, the wavelength of the local oscillator has to be continuously tunable in order to translate the coherent beating note to the measurable frequency range of an RF spectrum analyzer. Generally, grating-based external-cavity lasers can be used for this purpose, which provides both narrow spectral linewidth and wide continuous tuning range. Third, polarization states have to be aligned between the signal and the local oscillator to maximize the coherent detection efficiency. This is usually accomplished by a polarization controller.

Finally, the effect of slow frequency drifts of both signal and local oscillator can potentially make the intermediate frequency unstable. This f_{IF} fluctuation will certainly affect the accuracy of linewidth measurement (often causing an overestimation of the linewidth). Since the spectral shape of the laser line is known as Lorenzian as shown in Equation 3.1.25, one can easily measure the linewidth at –20 dB point and then calculate the linewidth at –3 dB point using Lorentzian fitting. This may significantly increase the measurement tolerance to frequency fluctuations and thus improve the accuracy.

Example 3.2

If the –20 dB linewidth of a laser diode measured by a coherent heterodyne detection is 1 GHz and assuming that the signal and the local oscillator have the same linewidth, what is the 3-dB linewidth of this laser?

Solution:

Assuming Lorenzian line shape, the RF spectrum after coherent detection is

$$S_p(f) = \frac{1}{1 + \left(\frac{f - f_0}{\Delta/2}\right)^2}$$

where Δ is the 3-dB linewidth. At the –20 dB point, the half-linewidth can be calculated with

$$1 + \left(\frac{f_{-20dB} - f_0}{\Delta/2}\right)^2 = 100$$

Since the FWHM at the –20 dB point is 1 GHz, thus $\Delta\sqrt{99}/2 = 500\ MHz$. Thus

$$\Delta = 1\ GHz/\sqrt{99} \approx 100\ MHz$$

is the FWHM linewidth of this beating note at –3 dB point. Excluding the impact of local oscillator linewidth, the 3-dB linewidth of the optical signal is about 50 MHz.

3.1.2.2 Self-homodyne detection

In coherent detection, the frequency of the input optical signal is down-converted to the RF domain through mixing with a local oscillator. Self-homodyne detection eliminates the requirement of a local oscillator, and the optical signal mixes with a delayed version of itself. Figure 3.1.8 shows the optical

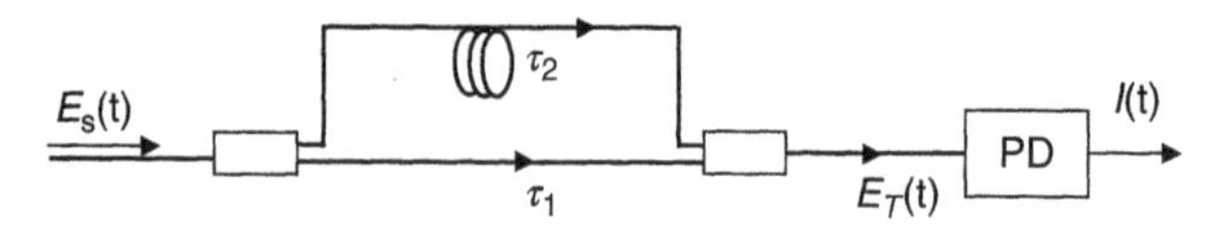

Figure 3.1.8 Self-homodyne method to measure the linewidth of a optical signal.

circuit of self-homodyne detection, where two delay lines are used between two directional couplers to form a Mach-zehnder configuration.

We can use the same expression as Equation 3.1.8 for the input signal optical field,

$$E(t) = \sqrt{P}\exp[j(\omega t + \phi(t))]$$

After the Mach-zehnder interferometer, the composite optical field is

$$E_T(t) = A_1 \exp\Big[j\Big(\omega(t-\tau_1) + \phi(t-\tau_1)\Big)\Big] + A_2 \exp\Big[j\Big(\omega(t-\tau_2) + \phi(t-\tau_2)\Big)\Big] \tag{3.1.27}$$

where τ_1 and τ_2 are the propagation delays of the two interferometer arms and A_1 and A_2 are the amplitudes of the fields emerging from these two arms. Then the photocurrent is

$$I(t) = \Re|E_T(t)|^2 = \Re\Big\{P_1 + P_2 + 2\sqrt{P_1 P_2}\cos[\omega\Delta\tau + \Delta\phi(t)]\Big\} \tag{3.1.28}$$

where P_1(t) and P_2(t) are the powers of the optical signals passing through the two arms, $\Delta\tau = \tau_2 - \tau_1$ is their differential time delay, and $\Delta\phi(t) = \phi(t-\tau_2) - \phi(t-\tau_1)$ is their differential phase. Since the phase noise has a Gaussian statistics, in a stationary process we have

$$\Delta\phi(t) = \phi(t) - \phi(t-\Delta\tau)$$

In a self-homodyne detection system, differential delay $\Delta\tau$ is an important parameter. If the differential delay is much longer than the coherence time of the optical signal ($\Delta\tau \gg t_{coh}$), the Mach-zehnder interferometer is said to operate in the *incoherent* regime. Otherwise, if $\Delta\tau \ll t_{coh}$, the interferometer will be in the *coherent* regime. Specifically, the self-homodyne linewidth measurement technique operates in the incoherent regime.

In the incoherent interference regime, optical signals pass through the two interferometer branches and then combine incoherently at the second directional coupler. In this case, the two terms in Equation 3.1.27 are not correlated with each other because the differential phase term is not deterministic, which resembles the mixing between lights from two independent laser sources with identical spectral linewidth. Therefore, the normalized power spectral density of the photocurrent in this self-homodyne setup is the autoconvolution of the signal power spectral density $S_{p,s}(f)$,

$$S_{IF}(f) = S_{p,s}(f) \otimes S_{p,s}(-f) = \frac{1}{1 + \left(\dfrac{f}{\Delta\nu_s}\right)^2} \tag{3.1.29}$$

Note that here the IF frequency is $f_{IF} = 0$ and the measured RF signal spectral linewidth is twice as wide as the actual optical signal spectral linewidth. Since the center of the RF peak is at zero frequency and the negative part of the spectrum will be flipped onto the positive side, only half the spectrum can be seen on the RF spectrum analyzer. Therefore, the width of this single-sided spectrum is equal to the linewidth of the optical signal.

Self-homodyne linewidth measurement is simpler than the coherent detection because it does not require a tunable local oscillator; however, it has a few disadvantages. First, in order for the interferometer to operate in the incoherent regime, the length difference between the two arms has to be large enough so that Equation 3.1.29 is valid. This requirement may be difficult to satisfy when the linewidth of the light source is very narrow. For example, for a light source with the linewidth of $\Delta\nu = 10\ kHz$, the coherence length is approximately $L_{coh} = c/(\mathrm{n}\Delta\nu) \approx 30\ km$ in the air. To ensure an accurate measurement of this linewidth, the length of the delay line in one of the two interferometer arms has to be much longer than 10 km. Another problem of self-homodyne measurement is that the central RF frequency is at zero, while most RF spectrum analyzers have high noise levels in this very low-frequency region. Laser intensity noise (usually strong at low frequency, such as 1/f noise) may also significantly affect the measurement accuracy.

To improve the performance of self-homodyne measurement, it would be desirable to move the intermediate frequency away from DC and let $f_{IF} > \Delta\nu$ to avoid the accuracy concern at the low–frequency region. This leads to the use of self-heterodyne measurement. Figure 3.1.9 shows the measurement setup of the self-heterodyne linewidth measurement technique, which is similar to the self-homodyne setup except an optical frequency shifter is used in one of the two interferometer arms.

As a result, the frequency of the optical signal is shifted by f_{IF} in one arm whereas in the other arm the optical frequency is not changed. The mixing of the optical signal with its frequency-shifted version in the photodiode creates an intermediate frequency at f_{IF}. Therefore the normalized RF spectral density in the electrical domain will be

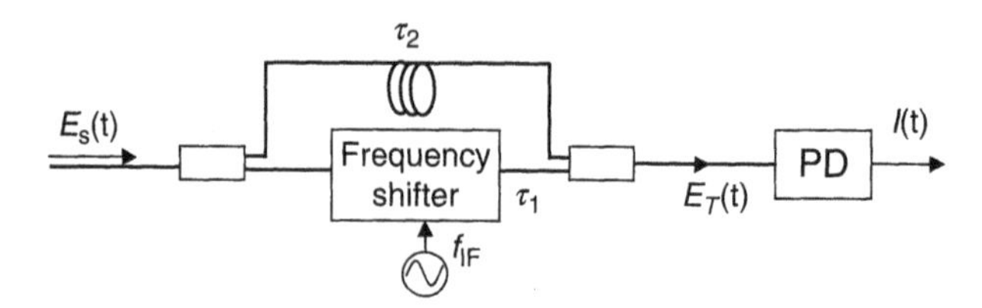

Figure 3.1.9 Self-homodyne method to measure the linewidth of an optical signal.

$$S_{IF}(f) = S_{p,s}(f - f_{IF}) \otimes S_{p,s}(-f) = \frac{1}{1 + \left(\frac{f - f_{IF}}{\Delta \nu_s}\right)^2} \tag{3.1.30}$$

The frequency shift is typically on the order of a few hundred megahertz, which is much higher than the typical linewidth of a semiconductor laser and is sufficient to move the IF spectrum away from the noisy low-frequency region [8, 9].

The most popular frequency shifter for this purpose is an acousto-optic frequency modulator (AOFM). Frequency shift of an AOFM is introduced by the interaction between the lightwave signal and a traveling wave RF signal. An oversimplified explaination is that the frequency of the lightwave signal is shifted by the moving grating created by the RF traveling wave through the Doppler effect, as illustrated in Figure 3.1.10. In contrast to a conventional frequency modulator, which often creates two modulation sidebends, an AOFM shifts the signal optical frequency to only one direction. Another unique advantage of an AOFM is its polarization independency, which is desired for practical linewidth measurement based on a fiber-optical system.

It is important to note that in the linewidth measurement using delayed homodyne or heterodyne techniques, the differential delay between the two interferometer arms has to be much longer than the coherence time of the laser source under test. This ensures the incoherent mixing between the optical signal and its delayed version. On the other hand, if the differential delay is much shorter than the source coherence time, the setup would become a classic Mach-zehnder interferometer in which the optical signal coherently mixes with its delayed version. In this later case, the interferometer can be used as an optical frequency discriminator, which will be discussed in the next section.

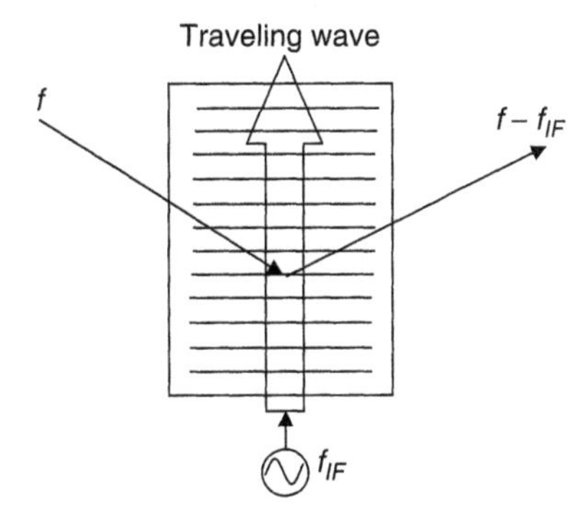

Figure 3.1.10 Illustration of the operating principle of an acousto-optical frequency shifter.

3.2 MEASUREMENT OF ELECTRO-OPTIC MODULATION RESPONSE

Electro-optic modulation is an important functionality of an optical transmitter, which translates electrical signals into optical domain. Electro-optic modulation can be accomplished via either direct modulation of semiconductor lasers or external modulation using external modulators. In terms of modulation parameters, the optical signal can be modulated in its intensity, frequency, or optical phase. The quality of an electro-optic modulator can be measured by parameters such as frequency response, bandwidth limitation, linearity, dynamic range, on/off ratio, level of distortion, and modulation efficiency. In general, direct modulation of laser diodes is simple and low cost, but it usually suffers from high modulation chirp and low extinction ratio. On the other hand, external modulation significantly improves overall performance but requires a dedicated external electro-optic modulator, which introduces extra attenuation and, of course, an additional cost.

Figure 3.2.1 illustrates the operating principle of direct intensity modulation of a semiconductor laser. To ensure that the laser diode operates above threshold, a DC bias current I_B is usually required. A bias-Tee combines the electrical current signal with the DC bias current to modulate the laser diode. The modulation efficiency is then determined by the slope of the laser diode *P-I* curve. Obviously, if the *P-I* curve is ideally linear, the output optical power is linearly proportional to the modulating current by

$$P_{opt}(t) \approx R_C(I_B - I_{th}) + R_C I_{ele}(t) \tag{3.2.1}$$

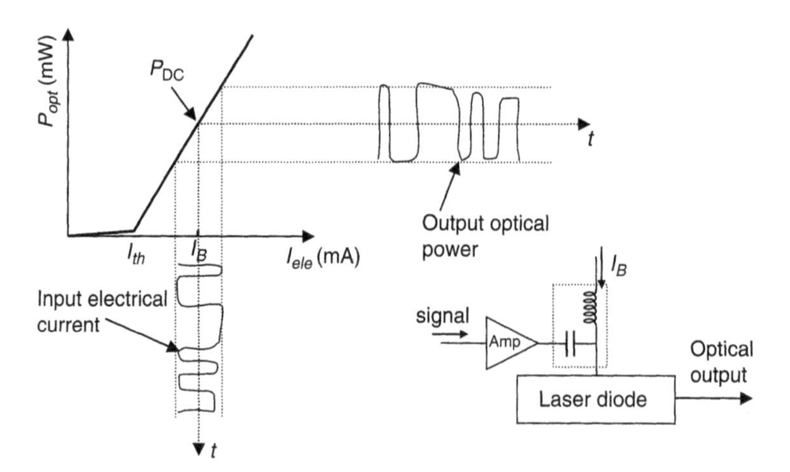

Figure 3.2.1 Direct intensity modulation of a laser diode. *Amp.:* electrical amplifier.

where I_{th} is the threshold current of the laser and $R_C = \Delta P_{opt}/\Delta I_{ele}$ is the slope of the laser diode *P-I* curve. Here we have neglected the optical power level at threshold.

Frequency response of direct modulation mainly depends on the carrier dynamics of the laser diode and >20 GHz modulation bandwidth has been demonstrated. However, further increasing the modulation bandwidth to 40 GHz appears to be quite challenging, mainly limited by the carrier lifetime as well as the parasitic effect of the electrode. Another well-known property of direct modulation in a semiconductor laser is the associated optical frequency modulation, commonly referred to as *frequency chirp*, as discussed in Chapter 1. To improve the performance of optical modulation, various types of external modulators have been developed to perform intensity modulation as well as phase modulation, as discussed in Section 1.5. In this section, we primarily discuss how to characterize modulation response.

3.2.1 Characterization of Intensity Modulation Response

Intensity modulation response of an optical transmitter can be measured either in frequency domain or in time domain. Usually, it is relative easy for a frequency-domain measurement to cover a wide continuous range of modulation frequencies and provide detailed frequency responses of the transmitter. Time-domain measurement, on the other hand, measures the waveform distortion that includes all the nonlinear effects that cannot be obtained by small-signal measurements in frequency domain. In this subsection, we discuss techniques of frequency-domain and time-domain characterizations separately.

3.2.1.1 Frequency-Domain Characterization

Frequency response is a measure of how fast an optical transmitter can be modulated. Typically, the modulation efficiency of a semiconductor laser or an external modulator is a function of modulation frequency Ω. Frequency-domain characterization is a popular way to find the features of device response such as cutoff frequency, uniformity of inband response, and resonance frequencies. Many time-domain waveform features can be predicted by frequency-domain characterizations. A straightforward way to characterize the frequency-domain response of an optical transmitter is to use an RF network analyzer and a calibrated wideband optical receiver as shown in Figure 3.2.2.

In this measurement, the transmitters have to be properly DC biased so that they operate in the desired operating condition. An RF network analyzer is used and is set to the mode of measuring S_{21} parameter. In this mode, port 1 of the

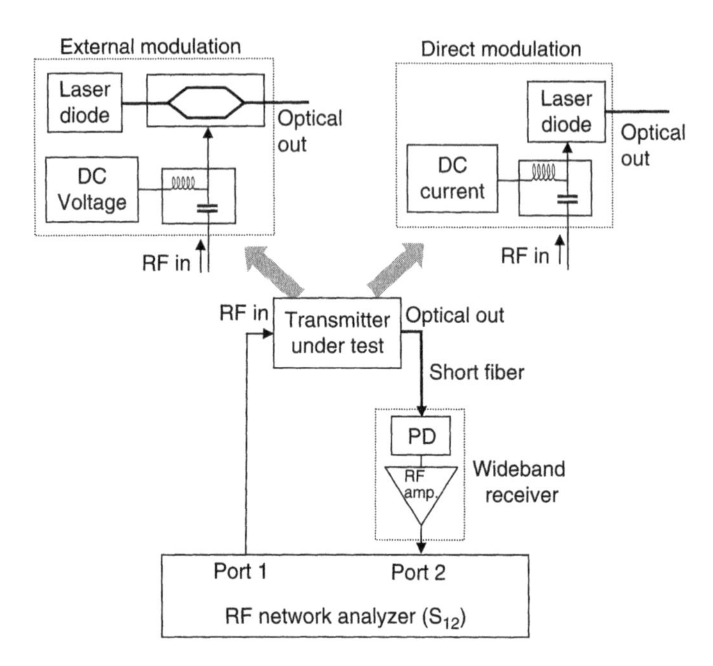

Figure 3.2.2 Frequency domain characterization of an optical transmitter using an RF network analyzer.

network analyzer provides a frequency swept RF signal, which is fed into the transmitter under test. The transmitter converts this frequency swept RF modulation into the optical domain, and the optical receiver then converts this optical modulation back to the RF domain and sent for detection into port 2 of the network analyzer. To characterize the modulation property of the transmitter, the bandwidth of both the network analyzer and the optical receiver should be wide enough and the receiver should be calibrated in advance so that the transfer function of the receiver can be excluded from the measurement.

Figure 3.2.3 shows an example of the measured amplitude modulation response of a DFB laser under direct modulation of its injection current [10]. With the increase of the average output optical power of the laser, the modulation bandwidth increase and the relaxation oscillation peak become strongly damped. This is due to the fact that at high optical power level, the carrier lifetime is short due to strong stimulated recombination.

The advantage of frequency domain measurement is its simplicity and high sensitivity due to the use of an RF network analyzer. Because the frequency of the interrogating RF signal continuously sweeps across the bandwidth of interest, detailed frequency domain characteristics of the optical transmitter can be obtained through a single sweep. However, in the network analyzer,

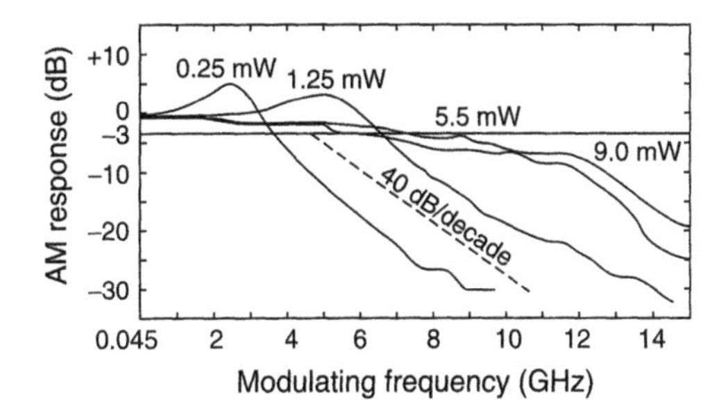

Figure 3.2.3 Example of a direct modulation response of a DFB laser diode at different optical power levels [10]. Used with permission.

the receiver in port 2 only selects the very frequency component sent out from port 1. High-order frequency harmonics generated by the nonlinear transfer characteristics of the electro-optic circuits cannot be measured. In general, as a disadvantage, the frequency domain measurement only provides small-signal linear response, while incapable of characterizing nonlinear effects involved in the electro-optic circuits.

To find nonlinear characteristics of an optical transmitter, the measurement of the strengths of high-order harmonics is necessary especially in analog system applications. This measurement can be performed using an RF signal generator (usually known as a *frequency synthesizer*) and an RF spectrum analyzer, as shown in Figure 3.2.4. To measure the nonlinear response, the transmitter is modulated with a sinusoid signal at frequency Ω generated by the synthesizer. Then if the transmitter response is nonlinear, several discrete frequency components will be measured on the RF spectrum analyzer. In addition to the fundamental frequency at Ω, RF energy will exist at the second order, the third order, and higher orders of harmonic frequencies at 2Ω and 3Ω and so on. The kth order harmonic distortion parameter is defined as

$$HD_k = \frac{P(\Omega_k)}{P(\Omega_1)} \tag{3.2.2}$$

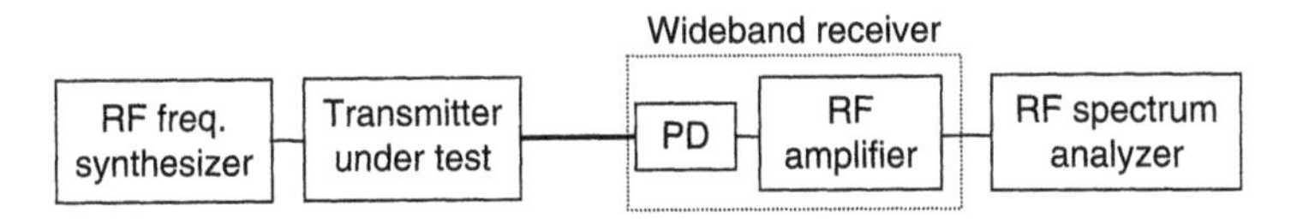

Figure 3.2.4 Frequency domain characterization of an optical transmitter use an RF spectrum analyzer.

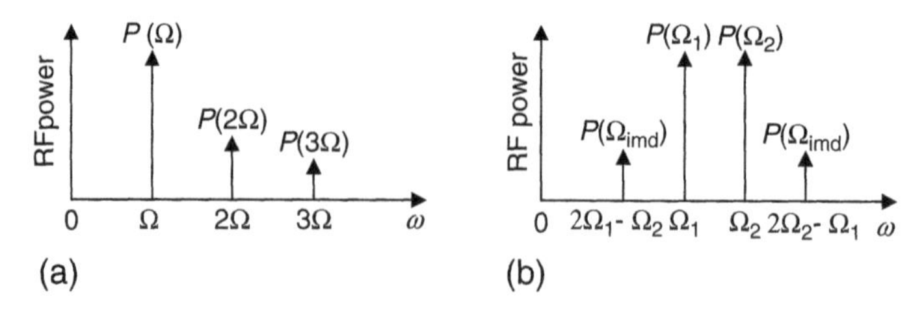

Figure 3.2.5 Illustration of (a) the measurements of second and third orders of harmonic distortions and (b) intermodulation distortion on an RF spectrum analyzer.

where $\Omega_k = k\Omega$, with $k = 2, 3, 4 \ldots,$ $P(\Omega_k)$ is the RF power at the kth order harmonic frequency, and $P(\Omega_1)$ is the power at the fundamental frequency, as shown in Figure 3.2.5(a). From this measurement, the *total harmonic distortion* (THD) can be calculated as

$$THD = \frac{\sum_{k=2}^{\infty} P(\Omega_k)}{P(\Omega_1)} \tag{3.2.3}$$

In general, harmonic distortions in directly modulated laser diodes are functions of both modulation frequency and the modulation index *m*. A larger modulation index implies a wider swing of signal magnitude and therefore often results in higher harmonic distortions, as indicated in Figure 3.2.6(a).

Another type of distortion caused by transmitter nonlinear response is referred to as *intermodulation distortion* (IMD). IMD is created due to the nonlinear mixing between two or more discrete frequency components of the RF

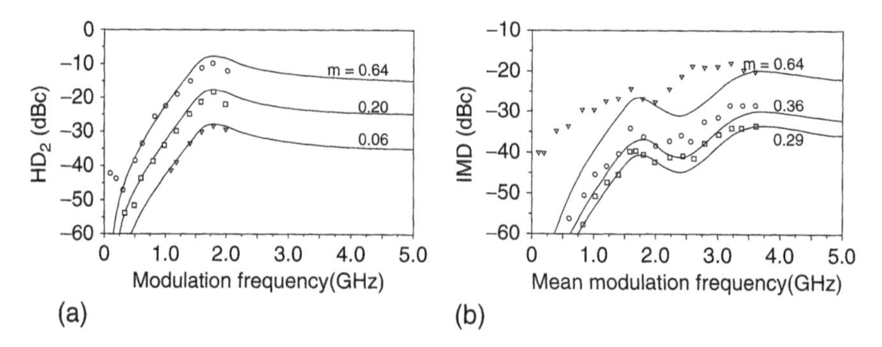

Figure 3.2.6 Examples of (a) measured second-order distortion and (b) intermodulation distortion of a laser diode under direct modulation [11]. Used with permission.

signal in the transmitter. In this nonlinear mixing process, new frequency components are created. If there are originally three frequency components at Ω_i, Ω_j, and Ω_k, new frequency components will be created at $\Omega_{ijk} = \Omega_i \pm \Omega_j \pm \Omega_k$, where i, j, and k are integers. In a simple case, if $i = j$ or $j = k$, only two original frequency components are involved, which is similar to the case of degenerated four-wave mixing in a nonlinear fiber system as discussed in Chapter 1.

For example, two original frequency components at Ω_1 and Ω_2 will generate two new frequency components at $2\Omega_1$–Ω_2 and $2\Omega_2$–Ω_1, as shown in Figure 3.2.5(b). These two new frequency components are created by two closely spaced original frequency components which have the highest power compared to others; these are usually the most damaging terms to the optical system performance. The IMD parameter is defined as

$$IMD(\Omega_{imd}) = \frac{P(\Omega_{imd})}{P(\Omega_i)} \tag{3.2.4}$$

where Ω_{imd} is the frequency of the new components and $P(\Omega_{imd})$ is the power at this frequency, as shown in Figure 3.2.5(b). Figure 3.2.6(b) shows an example of the measured $IMD(\Omega_{imd})$ versus the average modulating frequency $(\Omega_1 + \Omega_2)/2$ [11]. Again, higher modulation index usually results in higher IMD.

In lightwave CATV systems, a large number of subcarrier channels are modulated onto the same transmitter, and even a low level of intermodulation distortion may generate a significant amount of interchannel crosstalk. For CATV applications, both THD and IMD should generally be lower than –55 dBc.

3.2.1.2 Time-Domain Characterization

In binary modulated digital optical systems, waveform distortions represented by eye-closure penalty may be introduced by limited frequency bandwidth as well as various nonlinear effects in the transmitter. Time-domain measurement directly characterizes waveform distortion, which is most relevant to digital optical systems. In addition, frequency-domain response of the transmitter can be indirectly evaluated from the pulse response measurement performed in time domain.

As shown in Figure 3.2.7, time-domain characterization of an optical transmitter requires a waveform generator, a wideband optical receiver, and a high-speed oscilloscope. A unique advantage of time-domain measurement is the ability to characterize the transient effect during the switch of signal level from low to high, and vice versa. In general, the transient effect not only depends on the frequency bandwidth of the transmitter; it also depends on specific patterns of the input signal waveform. Therefore, pseudorandom waveforms are generally used; they contain a variety of combinations of data patterns. We discussed time-

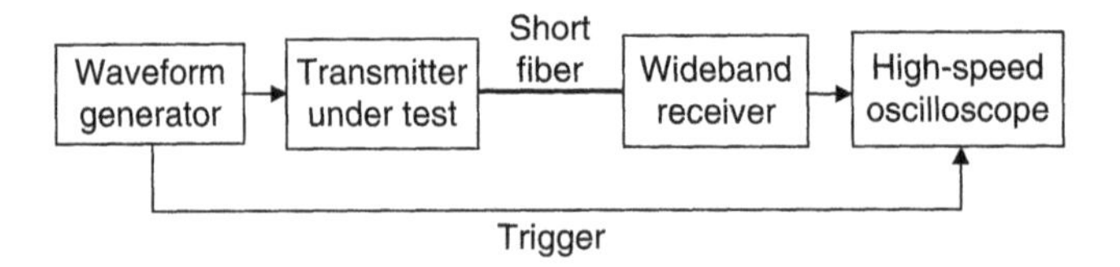

Figure 3.2.7 Block diagram of time-domain measurement of optical transmitter.

domain waveform characterization in Section 2.8, where both electrical domain sampling and optical domain sampling were presented. More detailed descriptions of time-domain waveform measurement and eye diagram evaluation are given in Chapter 5, where we discuss optical transmission systems.

3.2.2 Measurement of Frequency Chirp

Frequency chirp is the associated frequency modulation when an optical transmitter is intensity modulated. Frequency chirp usually broadens the spectral bandwidth of the modulated optical signal and introduces additional performance degradation when the optical signal propagates through a dispersive media. The origins and physical mechanisms of frequency chirp in directly modulated laser diodes and in electro-optic external modulators were discussed in Chapter 1; here we discuss techniques of measuring the frequency chirp.

Let's consider an optical field that has both intensity and phase fluctuations:

$$E(t) = \sqrt{P(t)}e^{j\phi(t)} = \exp\{j\phi(t) + 0.5\ln P(t)\} \tag{3.2.5}$$

where $P(t) = |E(t)|^2$ is the optical power. In this expression, the real part inside the exponent is $0.5\ln P(t)$ and the imaginary part is $\phi(t)$. The ratio between the derivatives of the imaginary part and the real part is equivalent to the ratio between the phase modulation and the intensity modulation, which is defined as the chirp parameter:

$$\alpha_{lw} = 2\frac{d\phi(t)/dt}{d\{\ln P(t)\}/dt} = 2P\frac{d\phi(t)/dt}{dP(t)dt} \tag{3.2.6}$$

The frequency deviation is equal to the time derivative of the phase modulation, $\Delta\omega = d\phi(t)/dt$; therefore the intensity modulation-induced optical frequency shift is

$$\Delta\omega = \frac{\alpha_{lw}}{2}\left(\frac{1}{P}\frac{dP}{dt}\right) \tag{3.2.7}$$

With sinusoidal modulation at frequency Ω and with modulation index m, the signal optical power is $P(t) = P_0(1 + m\sin\Omega t)$ and its derivative is

$$\frac{dP}{dt} = P_0 m\Omega\cos\Omega t$$

where m is the modulation index, P_0 is the average power, and Ω is the RF angular frequency of the modulating signal. Then, according to Equation 3.2.7, the induced frequency modulation is

$$\Delta\omega = \frac{\alpha_{lw}}{2}\frac{m\Omega\cos\Omega t}{1 + m\sin\Omega t} \tag{3.2.8}$$

If the modulation index is small enough ($m \ll 1$), the maximum frequency shift is approximately

$$\Delta\omega_{\max} = \frac{\alpha_{lw}}{2} m\Omega \tag{3.2.9}$$

This indicates that through the modulation chirp, signal optical bandwidth is broadened by the intensity modulation; this broadening is proportional to the linewidth enhancement factor α_{lw} as well as the modulation index m. As we discussed in Chapter 1, frequency chirp may exist in both direct modulation of semiconductor lasers and in external modulation. In both cases, if the linewidth enhancement factor α_{lw} is known, the amount of spectral broadening can be evaluated by Equation 3.2.9.

There are a number of techniques available to characterize modulation-induced frequency chirp in optical transmitters. The often used techniques include modulation spectral measurement, dispersion measurement, and interferometric measurement.

3.2.2.1 Modulation Spectral Measurement

When an optical transmitter is modulated by a sinusoid at frequency Ω with the modulation index m, its output optical power is $P(t) = P_0(1 + m\sin\Omega t)$. Meanwhile, due to the linear chirp, its optical phase is also modulated. If the modulation index is small enough ($m \ll 1$), according to the definition of linewidth enhancement factor in Equation 3.2.6, the phase modulation is

$$\phi(t) = \frac{\alpha_{lw} m}{2}\sin\Omega t \tag{3.2.10}$$

Then the complex optical field can be expressed as

$$E(t) = \sqrt{P_0}\sqrt{1 + m\sin\Omega t}\cdot\exp\left[j\left(\omega t + \frac{m\alpha_{lw}}{2}\sin\Omega t\right)\right] \tag{3.2.11}$$

Since the modulation index is small, Equation 3.2.11 can be linearized to simplify the analysis:

$$E(t) \approx \sqrt{P_0}\left(1 + \frac{m}{2}\sin\Omega t\right) \cdot \exp\left[j\left(\omega t + \frac{m\alpha_{lw}}{2}\sin\Omega t\right)\right] \tag{3.2.12}$$

The right side in Equation 3.2.12 can be expanded into a Bessel series:

$$\begin{aligned} E(t) &= \sqrt{P_0}\sum_k J_k\left(\frac{m\alpha_{lw}}{2}\right)\exp[j(\omega \pm k\Omega)t] \\ &-\frac{jm}{4}\sqrt{P_0}\left[\sum_k J_k\left(\frac{m\alpha_{lw}}{2}\right)\exp[j(\omega \pm (k+1)\Omega)t] - \sum_k J_k\left(\frac{m\alpha_{lw}}{2}\right)\exp[j(\omega \pm (k-1)\Omega)t]\right] \end{aligned} \tag{3.2.13}$$

where $J_k(m\alpha_{lw}/2)$ is the kth order Bessel function. Equation 3.2.13 indicates that if the transmitter is chirp-free ($\alpha_{lw} = 0$), there should be only two modulation sidebands, one at each side of the carrier, which is the nature of a typical intensity modulation. However, in general when $\alpha_{lw} \neq 0$, there will be additional modulation sidebands in the optical spectrum, as indicated by Equation 3.2.13. The amplitude of each modulation sideband, determined by $J_k(m\alpha_{lw}/2)$, is a function of modulation index m as well as the linewidth enhancement factor α_{lw}. Therefore, by measuring the amplitude of each modulation sideband, the linewidth enhancement factor of the transmitter can be determined. In this way, modulation chirp measurement becomes a spectral measurement. However, since the sidebands separation is usually in the multimegahertz range, which is determined by the RF modulation frequency, most optical spectrum analyzers will not be able to resolve the fine spectral features created by modulation. Coherent heterodyne detection probably is the best way to perform this measurement. It translates the modulated optical spectrum into RF domain, which can be measured by an RF spectrum analyzer, as discussed in Chapter 2.

Figure 3.2.8 shows an example of a directly modulated DFB laser diode. In this particular case, the modulation frequency is 1 GHz and the intensity modulation index is $m = 0.1$. When the phase modulation index $\beta = m\alpha_{lw}/2$ is equal to 2.405, the carrier component is suppressed because $J_1(2.405) = 0$, as shown in Figure 3.2.8(a). With the increase of phase modulation index to $\beta = 3.832$, the nearest modulation sidebands at both sides of the carrier are $J_{\pm 2}^2(\beta)$, whereas $J_{\pm 1}^2(\beta)$ components are nulled, which is the solution of $J_1(\beta) = 0$. This measurement allows the precise determination of the linewidth enhancement factor of the transmitter with the known intensity modulation index m, which in turn can be obtained by a time-domain measurement of the modulated intensity waveforms.

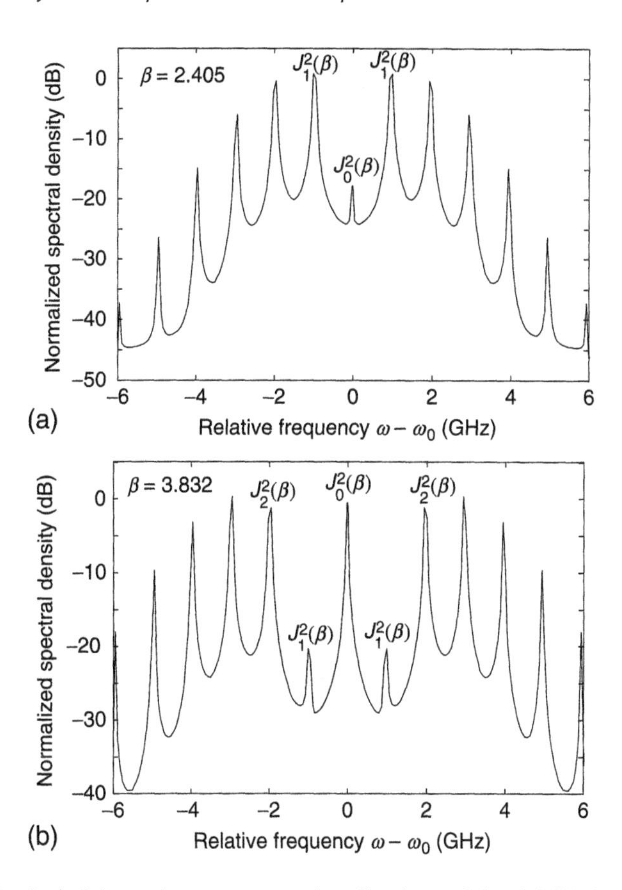

Figure 3.2.8 Optical heterodyne spectrum of a directly modulated DFB laser $m = 0.1$ and $\beta = m\alpha_{lw}/2$ is the phase modulation index.

3.2.2.2 Measurement Utilizing Fiber Dispersion

As mentioned previously, frequency chirp has an effect of broadening the width of the signal optical spectrum and introducing system performance degradation when the transmission fiber is dispersive. For measurement purposes, this dispersive effect in the fiber can be utilized to evaluate the chirp parameter of a transmitter. The experimental setup of this measurement is shown in Figure 3.2.9, where the transmitter under test is modulated by a frequency swept RF signal from a network analyzer (port 1). The optical output from the transmitter passes through an optical fiber, which has a predetermined amount of chromatic dispersion. The transmitted optical signal is detected by a wideband receiver with an optical amplifier, if necessary, to increase the optical power for detection. Then the RF signal detected by the receiver is sent back to the

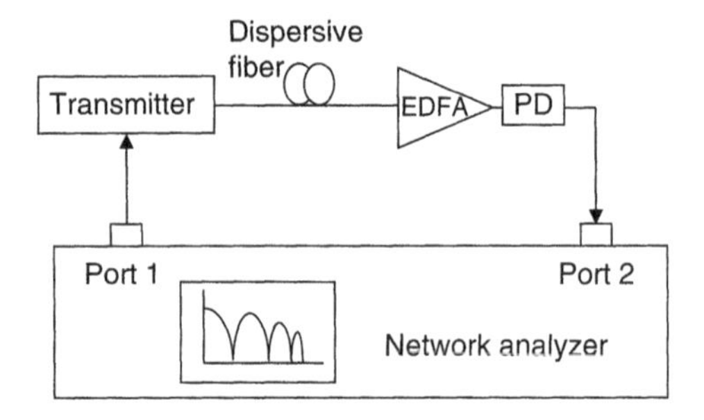

Figure 3.2.9 Experimental setup for frequency chirp measurement using a dispersive fiber and an RF network analyzer.

receiving port (port 2) of the network analyzer. The RF network analyzer measures the S_{21} parameter of the entire optoelectronic system to determine its overall transfer function [12].

To explain the operating principle, we assume that the modulation index m is small enough and the modulation efficiency is linear. Because of the chirp effect, both the optical power and the optical phase are modulated. According to Equation 3.2.11, the modulated complex optical field is

$$E(t) = \sqrt{P_0}\sqrt{1 + m\sin(2\pi ft)} \cdot \exp\left[j\left(\frac{m\alpha_{lw}}{2}\sin(2\pi ft)\right)\right]e^{j2\pi f_0 t} \tag{3.2.14}$$

where f and f_0 are the applied RF modulating frequency and the carrier optical frequency, respectively.

Again, for small-signal modulation ($m \ll 1$), linear approximation can be applied so that

$$E(t) \approx \sqrt{P_0}\left[1 + \frac{1}{2}m\sin(2\pi ft)\right]\left[1 + j\frac{\alpha_{lw}m}{2}\sin(2\pi ft)\right]e^{j2\pi f_0 t} \tag{3.2.15}$$

Neglecting high-order harmonics, Equation 3.2.15 can be simplified as

$$E(t) \approx \sqrt{P_0}\left[1 + m\frac{1 + j\alpha_{lw}}{4}e^{j2\pi ft} + m\frac{1 + j\alpha_{lw}}{4}e^{-j2\pi ft}\right]e^{j2\pi f_0 t} \tag{3.2.16}$$

Obviously this modulated optical spectrum has two discrete sidebands ($f_0 \pm f$), one on each side of the optical carrier f_0. Due to chromatic dispersion of optical fiber, each frequency component will have a slightly different propagation constant β.

$$\beta(\omega) = \beta_0 + 2\pi\beta_1(f - f_0) + (2\pi)^2\frac{\beta_2}{2}(f - f_0)^2 + \ldots \tag{3.2.17}$$

By definition, $\beta_1 = 1/v_g$ is the inverse of the group velocity, $\beta_2 = -\lambda_0^2 D/2\pi c$ is the dispersion coefficient, and $\lambda_0 = c/f_0$ is the center wavelength of the optical carrier. D is the fiber dispersion parameter with the unit of ps/nm/km. Then the propagation constants at carrier, the upper sideband, and the lower sideband can be expressed as,

$$\beta(f_0) = \beta_0 \tag{3.2.18}$$

$$\beta_{+1} = \beta(f_0 + f) = \beta_0 + \frac{2\pi f}{v_g} - \frac{\pi\lambda_0^2 D f^2}{c} \tag{3.2.19}$$

$$\beta_{-1} = \beta(f_0 - f) = \beta_0 - \frac{2\pi f}{v_g} - \frac{\pi\lambda_0^2 D f^2}{c} \tag{3.2.20}$$

With the consideration of chromatic dispersion, after propagating over the fiber with length L, each of these three frequency components at f_0 and $f_0 \pm f$ has a slightly different propagation delay. At the fiber output the optical field is

$$E(t) \approx \sqrt{P_0}\left[e^{-j\beta_0 L} + m\frac{1 + j\alpha_{lw}}{4}e^{j(2\pi ft - \beta_{+1}L)} + m\frac{1 + j\alpha_{lw}}{4}e^{-j(2\pi ft - \beta_{-1}L)}\right]e^{j2\pi f_0 t} \tag{3.2.21}$$

At the photodetector, the photocurrent is proportional to the square of the optical field $I(t) = \Re|E(t)|^2$, where $\Re$ is the responsivity of the photodiode. Collecting the components of the photocurrent only at the modulation frequency f (in fact, this is what an RF network analyzer automatically does), we have

$$\begin{aligned} I(f) &\approx \left|\Re P_0 m\frac{1 + j\alpha_{lw}}{2}\left[\exp\left(j\frac{\pi\lambda_0^2 D f^2 L}{c}\right) + \exp\left(-j\frac{\pi\lambda_0^2 D f^2 L}{c}\right)\right]\right| \\ &= \Re P_0 m\sqrt{1 + \alpha_{lw}^2}\cos\left[\frac{\pi\lambda_0^2 D f^2 L}{c} + \tan^{-1}(\alpha_{lw})\right] \end{aligned} \tag{3.2.22}$$

In this derivation, high-order harmonics, at $2f$, $3f$, and so on, have been neglected. Equation 3.2.22 indicates that the RF transfer function of the system is strongly frequency-dependent and the spectral domain features of this transfer function largely depend on the value of the chirp parameter α_{lw}. One of the important features of this transfer function is the resonance zeros when the following condition is satisfied:

$$\left[\frac{\pi\lambda_0^2 D f^2 L}{c} + \tan^{-1}(\alpha_{lw})\right] = k\pi + \pi/2 \tag{3.2.23}$$

where $k = 0, 1, 2, 3 ..$ is an integer. The solution to Equation 3.2.23 can be found at discrete frequencies $f = f_k$, where

$$f_k = \sqrt{\frac{c}{2DL\lambda_0^2}\left(1 + 2k - \frac{2}{\pi}\tan^{-1}(\alpha_{lw})\right)} \qquad (3.2.24)$$

By measuring the frequencies at which the transfer function reaches zero, the chirp parameter α_{lw} can be evaluated. In practical measurements, the following steps are suggested:

1. Find the accumulated fiber dispersion DL. This can be accomplished by measuring at least two discrete zero-transfer frequencies—for example, f_k and f_{k+1}. From Equation 3.2.24, it can be found that

$$D \cdot L = \frac{c}{(f_{k+1}^2 - f_k^2)\lambda_0^2}, \qquad (3.2.25)$$

   which is independent of the chirp parameter of the transmitter.
2. Once the accumulated fiber dispersion is known, the laser chirp parameter α_{lw} can be found by using Equation 3.2.24 with the knowledge of the frequency of any transfer-zero, f_k. This measurement technique not only provides the absolute value of α_{lw}, it also tells the sign of α_{lw}.

Figure 3.2.10 shows the calculated zero-transfer frequencies versus the chirp parameter in a system with 50 km of standard single-mode fiber with $D = 17$ ps/nm-km. In order to be able to observe high orders of transfer nulls, a wide bandwidth receiver and RF network analyzer have to be used. In general, a

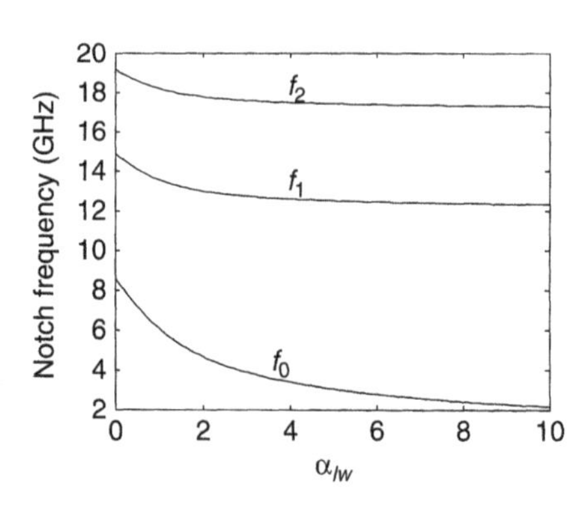

Figure 3.2.10 Transmission notch frequencies versus α_{lw} in a system of 50 km standard single-mode fiber with D = 17 ps/nm-km.

network analyzer with bandwidth of 20 GHz is appropriate for most such measurements.

The following is a practical example of chirp measurement of an electro-absorption (EA) modulator. For this modulator, the chirp parameter is a function of the bias voltage, and the value of the chirp can be both positive and negative depending on the bias level applied on the device. The bias voltage-dependent transfer function of this particular EA modulator is shown in Figure 3.2.11, where the optical power transmission efficiency $T(V)$ decreases with the increase of the negative bias voltage V.

Figure 3.2.12 shows the RF transfer functions, measured with the experimental setup shown in Figure 3.2.9, together with the calculated transfer functions using Equation 3.2.22. Four transfer functions are shown in Figure 3.2.12; each was measured at a different EA modulator bias voltage. In the experimental setup, 50 km standard single mode fiber was used with a dispersion parameter of $D = 17.2$ ps/nm-km; therefore the accumulated dispersion of the fiber system is approximately 860 ps/nm. To fit the measured transfer function at a certain bias voltage using Equation 3.2.22, the only fitting parameter that can be adjusted is the chirp parameter α_{lw}.

By performing the transfer-function measurement at various bias voltage levels, chirp parameter versus bias voltage can be obtained, as shown in Figure 3.13. It is interesting to note that the chirp parameter α_{lw}(V) exhibits a singularity around –1.8 V bias voltage. The reason is that in this particular device, optical transmission efficiency T versus bias voltage reaches a local minimum at the bias level of approximately –1.8 V, where the slope dT/dV changes the sign. Meanwhile, the phase modulation efficiency $d\Phi/dV$ is nonzero in this

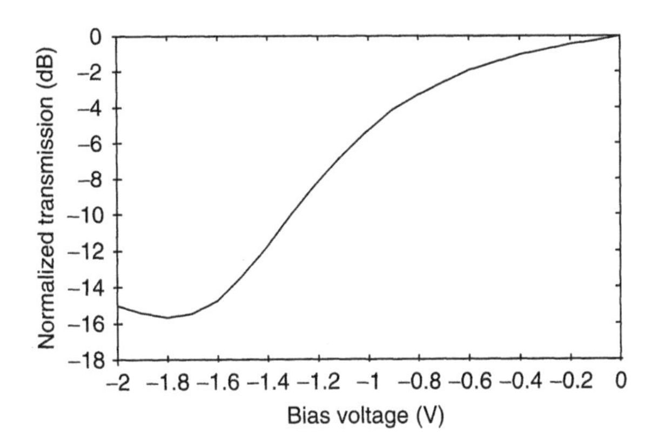

Figure 3.2.11 Voltage-dependent transmission of an electro-absorption modulator.

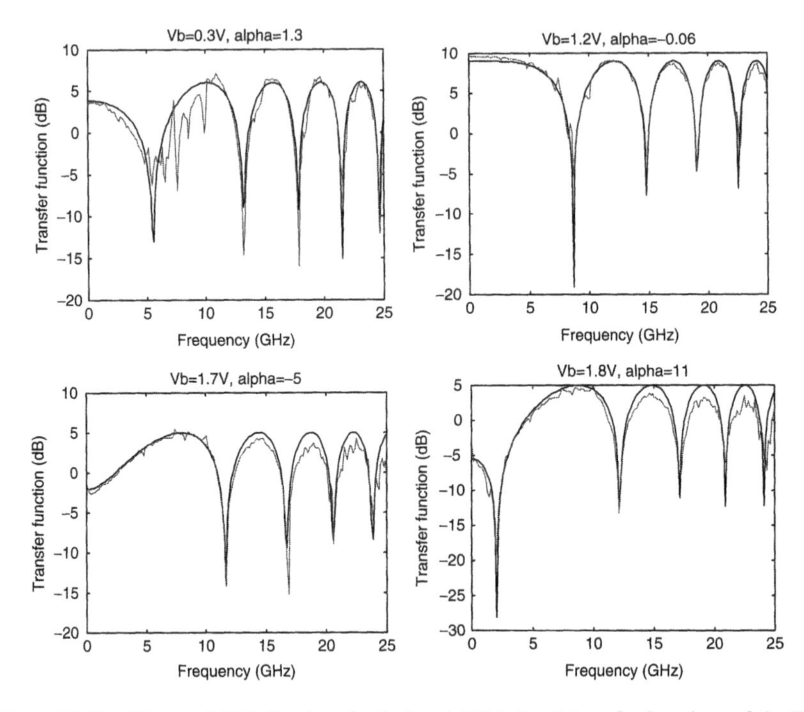

Figure 3.2.12 Measured (thin lines) and calculated (thick lines) transfer functions of the EA modulator at bias voltages of 0.3 V, 1.2 V, 1.7 V, and 1.8 V, respectively.

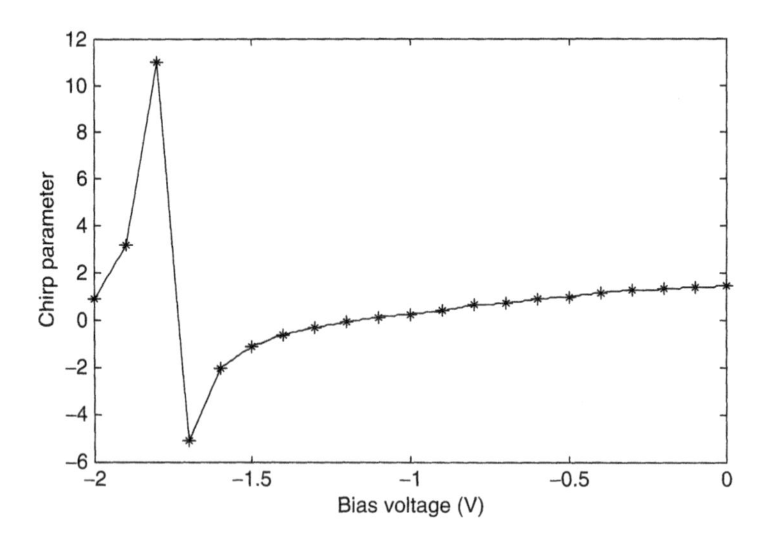

Figure 3.2.13 Measured chirp parameter α_{lw} versus EA bias voltage.

region. Therefore, based on the chirp parameter defined by Equation 3.2.6, the value of α_{lw} should be $\pm\infty$ at this bias level. Based on this measurement, signal optical phase versus bias voltage of the EA modulator can also be obtained by integrating Equation 3.2.6 as

$$\Phi(V) = \int_t \alpha_{lw}(V) \frac{1}{2P(t)} \frac{dP(t)}{dt} dt = \int_P \alpha_{lw}(V) \frac{1}{2P(V)} dP \quad (3.2.26)$$

where $P(V)$ describes the output optical power from the EA modulator. Since the input optical power is constant during the measurement, the output power $P(V)$ should be proportional to the optical transmission efficiency of the device. Based on Equation 3.2.26 and the measured chirp parameter shown in Figure 3.2.13, the bias-dependent signal optical phase can be obtained as shown in Figure 3.2.14. This figure indicates that with the increase of the reverse bias, the signal optical phase first goes negative and then swings back to positive. Even though the chirp parameter exhibits a singularity around –1.8 V bias level, the phase versus voltage curve is continuous. From the practical application point of view, what actually affects the optical system performance is the signal optical phase instead of the chirp parameter.

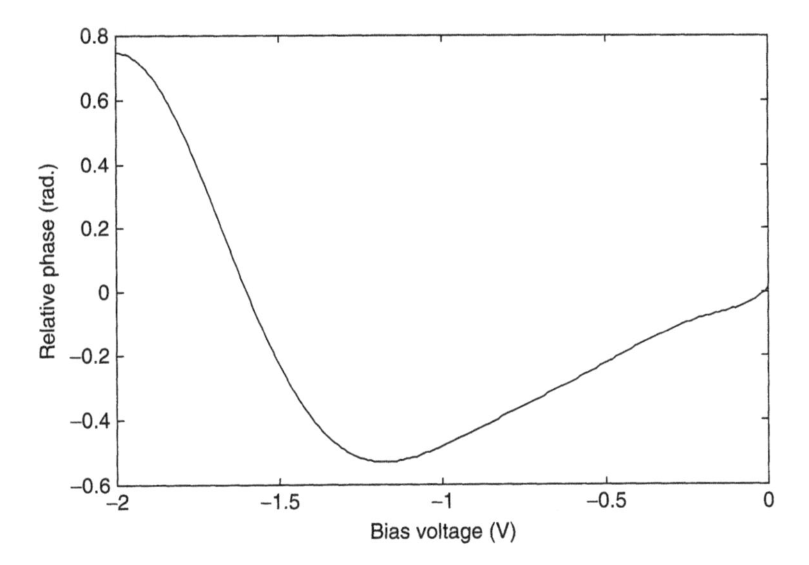

Figure 3.2.14 Measured optical phase versus the EA bias voltage.

3.2.3 Time-Domain Measurement of Modulation-Induced Chirp

Frequency-domain measurement of modulation chirp is a powerful tool that is relatively simple. However, since it is based on the swept-frequency small-signal modulation, it usually provides only linear performance of the transmitter. To measure intensity-dependent waveform distortion, a large-signal characterization in time domain is usually required. Figure 3.2.15 shows an example of time-domain characterization of transmitter chirp using a setup based on a balanced Mach-zehnder interferometer [13]. In this configuration, a waveform generator is used to modulate the transmitter under test. The optical signal is equally split into the two arms of a balanced MZI, which is formed by two optical fibers between two 3-dB fiber couplers. A small portion of the output optical signal from the MZI is detected and used as the feedback to control the relative phase mismatch of the MZI arms, whereas the rest is sent to a wideband optical receiver, and then a high-speed sampling oscillator for the waveform measurement.

To understand the operating principle, let's first consider the transfer function of a MZI, similar to that described by Equation 3.1.28. However, it is important to point out that in the application presented here, the MZI has to operate in the coherent regime where the differential time delay between the MZI arms should be much shorter than the source coherence time ($\Delta\tau \ll t_{coh}$). Under this condition,

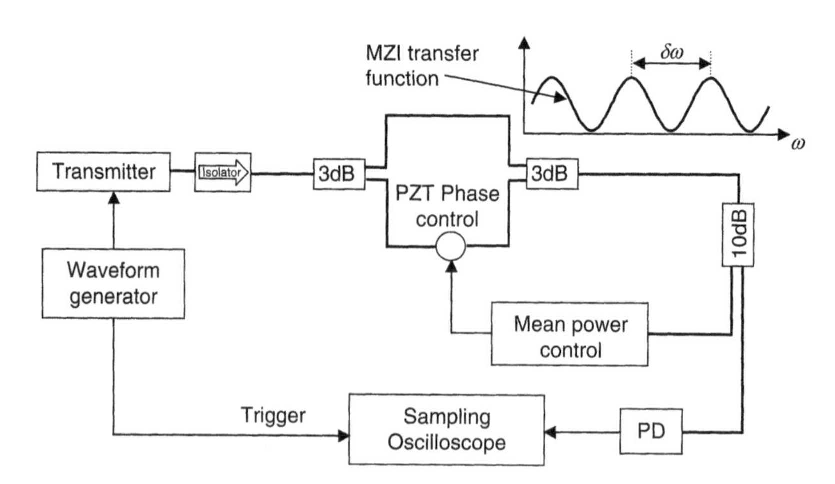

Figure 3.2.15 Time-domain measurement of transmitter chirp using MZI.

$$\phi(t) - \phi(t - \Delta\tau) \approx \frac{d\phi(t)}{dt}\Delta\tau = \Delta\omega(t)\Delta\tau \tag{3.2.27}$$

where $\Delta\omega(t)$ is the modulation-induced frequency deviation and $\Delta\tau$ is the differential time delay between two MZI arms. Therefore, in the coherent regime, the MZI transfer function can be expressed as

$$I(t) = \Re\left\{P_1(t) + P_2(t) + 2\sqrt{P_1(t)P_2(t)}\cos[\omega(t)\Delta\tau + \Delta\phi]\right\} \tag{3.2.28}$$

where $P_1(t)$ and $P_2(t)$ represent the powers of the modulated optical signal at the two MZI arms, $\Re$ is the photodiode responsivity, and $\Delta\phi = \omega_0\Delta\tau$ is the phase constant that is determined by the MZI setting. ω_0 is the central frequency of the source. In the setup shown in Figure 3.1.15, the differential arm length of the MZI can be adjusted by a piezo-electric transducer (PZT) so that the bias phase can be precisely tuned to the quadrature point such that $\Delta\phi = \pi/2 + 2k\pi$, where k is an integer. At this particular bias point, the photocurrent waveform is defined as

$$I(t) = I_Q(t) = \Re\left\{P_1(t) + P_2(t) - 2\sqrt{P_1(t)P_2(t)}\sin[\Delta\omega(t)\Delta\tau]\right\} \tag{3.2.29}$$

Then the frequency chirp $\Delta\omega$(t) can be expressed as

$$\Delta\omega(t) = \frac{1}{\Delta\tau}\sin^{-1}\left\{\frac{P_1(t) + P_2(t) - I_Q(t)/\Re}{2\sqrt{P_1(t)P_2(t)}}\right\} \tag{3.2.30}$$

Based on Equation 3.2.30, to measure the frequency chirp $\Delta\omega$(t), we need to know P_1(t), P_2(t), I_Q(t) and photodiode responsivity $\Re$. The measurement procedure can be summarized as follows:

1. Measure the frequency-domain transfer function of the MZI using an optical spectrum analyzer, which should be a sinusoid. The period $\delta\omega$ of this sinusoidal transfer function is related to the differential delay between the two arms: $\Delta\tau = 1/\delta\omega$. This measurement may be accomplished with a CW wideband light source or with a tunable laser and a power meter.
2. Use the modulated transmitter and block out the first arm of the MZI to measure the photocurrent waveform I_2(t) generated by the second MZI arm: $I_2(t) = \Re P_2(t)$.
3. Use the same modulated transmitter, but while blocking the second arm of the MZI to measure the photocurrent waveform $I_1(t)$ generated by the second MZI arm: $I_1(t) = \Re P_1(t)$.
4. Convert Equation 3.2.30 into the following form:

$$\Delta\omega(t) = \delta\omega \cdot \sin^{-1}\left\{\frac{I_1(t) + I_2(t) - I_Q(t)}{2\sqrt{I_1(t)I_2(t)}}\right\} \tag{3.2.31}$$

Then the time-dependent frequency deviation can be determined with the knowledge of single-arm currents $I_1(t)$, $I_2(t)$, quadrature current waveform $I_Q(t)$, and the period of the MZI $\delta\omega$.

During the measurement, special attention has to be made such that the same modulation data pattern and the same modulation index have to be maintained for step 2 and step 3. Although the measurement procedure we've described looks simple, the requirement of blocking MZI arms may disturb the operating condition of the experimental setup. This is especially problematic if an all-fiber MZI is used; each reconnection of a fiber connector may introduce slightly different loss and therefore create measurement errors. On the other hand, since the phase of the MZI can be easily adjusted with the PZT phase control, the measurement procedure can be modified to avoid the necessity of blocking the MZI arms [14].

Again, referring to the MZI transfer function given by Equation 3.2.28, if the MZI is biased at point A in Figure 3.2.16, where $\Delta\phi = 2m\pi + \pi/2$ and the slope is positive, the photocurrent waveform is

$$I_{Q+}(t) = \Re\left\{P_1(t) + P_2(t) + 2\sqrt{P_1(t)P_2(t)}\sin[\Delta\omega(t)\Delta\tau]\right\} \tag{3.2.32}$$

On the other hand, if the MZI is biased at point B, where $\Delta\phi = 2m\pi - \pi/2$ and the slope is negative, the photocurrent waveform becomes

$$I_{Q-}(t) = \Re\left\{P_1(t) + P_2(t) - 2\sqrt{P_1(t)P_2(t)}\sin[\Delta\omega(t)\Delta\tau]\right\} \tag{3.2.33}$$

Practically, the change of bias point from A to B in Figure 3.2.16 can be achieved by stretching the length of one of the two MZI arms by half a wavelength using the PZT.

We can then define the following two new parameters:

$$I_{AM} = \frac{I_{Q+}(t) + I_{Q-}(t)}{2} = \Re\{P_1(t) + P_2(t)\} \tag{3.2.34}$$

$$I_{FM} = \frac{I_{Q+}(t) - I_{Q-}(t)}{2} = 2\Re\sqrt{P_1(t)P_2(t)}\sin[\Delta\omega(t)\Delta\tau] \tag{3.2.35}$$

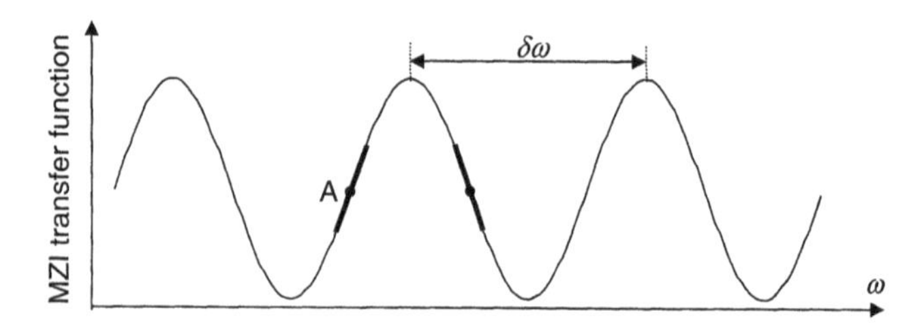

Figure 3.2.16 Biasing a MZI in the positive and negative quadrature points.

where I_{AM} depends only on the intensity modulation, whereas I_{FM} depends on both intensity and frequency modulations. To find the frequency modulation $\Delta\omega(t)$, we can combine Equations 3.2.34 and 3.2.35 so that

$$\Delta\omega(t) = \delta\omega \cdot \sin^{-1}\left[\left(\frac{P_1(t) + P_2(t)}{2\sqrt{P_1(t)P_2(t)}}\right) \cdot \frac{I_{FM}(t)}{I_{AM}(t)}\right] \tag{3.2.36}$$

If both of the two fiber couplers are ideally 3 dB (50 percent), $P_1(t) = P_2(t)$ and Equation 3.2.36 can be simplified to

$$\Delta\omega(t) = \delta\omega \cdot \sin^{-1}\left[\frac{I_{Q+}(t) - I_{Q-}(t)}{I_{Q+}(t) + I_{Q-}(t)}\right] \tag{3.2.37}$$

This measurement is relatively simple without the need to block the MZI arms.

In general, the fiber couplers may not be ideally 3 dB; in that case, we may have to use $P_2(t) = \eta P_1(t)$, where η has a value very close to unity for good couplers which determines the extinction ratio of the MZ interferometer:

$$\Delta\omega(t) = \delta\omega \cdot \sin^{-1}\left[\left(\frac{1+\eta}{2\sqrt{\eta}}\right) \cdot \frac{I_{Q+}(t) - I_{Q-}(t)}{I_{Q+}(t) + I_{Q-}(t)}\right] \tag{3.2.38}$$

Since η is a parameter of fiber coupler, the value of $(1+\eta)/2\sqrt{\eta}$ can usually be calibrated easily.

It is worthwhile to note that in the chirp measurement using an MZI, the differential delay between the two interferometer arms $\Delta\tau = 1/\delta\omega$ is an important parameter. The following are a few considerations in selecting the $\Delta\tau$ value of the MZI:

1. Coherence requirement, which needs the MZI arm length difference to be small enough such that $\Delta\tau \ll t_{coh}$. This is the necessary condition to justify the linear approximation given by Equation 3.2.27. Otherwise, if the MZI arm length difference is too big, the MZI is no longer coherent for an optical signal with wide linewidth, and Equation 3.2.28 will not be accurate.
2. Sensitivity requirement, which desires a longer $\Delta\tau$: Equations 3.2.32 and 3.2.33 show that the FM/AM conversion efficiency at the quadrature point is proportional to $\sin(\Delta\omega\Delta\tau)$. Increase $\Delta\tau$ certainly helps to increase the measured signal level $I_{Q\pm}$ and improve the signal-to-noise ratio.
3. Frequency bandwidth requirement, which desires a short $\Delta\tau$. This frequency bandwidth refers to the maximum frequency deviation of the laser source $\Delta\omega_{max}$ that can be measured by the MZI. Certainly, this maximum frequency deviation has to be smaller than the half free-spectral range $\delta\omega/2$ of the MZI as illustrated in Figure 3.2.16: $\Delta\omega_{max} < \delta\omega/2 = 1/(2\Delta\tau)$. Beyond this limitation, the MZI transfer function would provide multiple solutions. In practice, to maintain a good efficiency of FM/AM conversion, $\Delta\omega_{max} < 1/(8\Delta\tau)$ is usually required.

4. Modulation speed requirement, which requires short $\Delta\tau$. This refers to the requirement that within the differential delay time $\Delta\tau$, there should be negligible intensity variation of the optical signal induced by modulation. Therefore, the differential delay must be much smaller than the inverse of the maximum modulation frequency f_m: $\Delta\tau \ll 1/f_m$. As an example, for a 40 GHz modulation frequency, the differential delay should be much shorter than 25 ps which corresponds to the arm length difference of approximately 0.8 mm in the air.

In the design of the chirp measurement setup, we must consider all four of these requirements and make tradeoffs. Figure 3.2.17 shows an example of the measured waveforms of intensity modulation and the associated frequency modulation of an unbalanced $LiNbO_3$ external modulator [14]. This waveform-dependent frequency chirp usually cannot be precisely characterized by small-signal frequency-domain measurements.

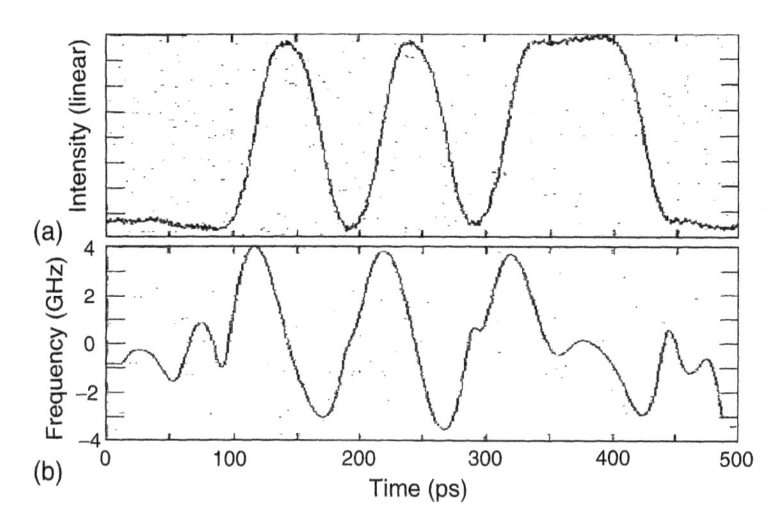

Figure 3.2.17 Example of (a) intensity modulation and (b) frequency modulation of a $LiNbO_3$ intensity modulator [14]. Used with permission.

3.3 WIDEBAND CHARACTERIZATION OF AN OPTICAL RECEIVER

In optical systems, an optical receiver converts the incoming signal from the optical domain to the electrical domain. An optical receiver usually consists of a photodetector and an electrical circuit for transimpedance amplification and signal manipulation. Important parameters of an optical receiver include photodetector responsivity, bandwidth, flatness of frequency response within the bandwidth, noise figure, linearity, and signal wavelength coverage. Optical receiver characterization and calibration are important for both optical communication and instrumentation, which directly affect optical system performance and measurement accuracy. In this section, we discuss techniques to characterize optical receivers, with a focus on the wideband characterization of their frequency response.

3.3.1 Characterization of Photodetector Responsivity and Linearity

Photodetector responsivity (defined by Equation 1.2.2 in Chapter 1) is a measure of optical-to-electrical conversion efficiency of a photodetector and is usually expressed by the value of the photocurrent (mA) generated by each milliwatt of optical signal. Ideally, the value of responsivity $\Re$ should be a constant, but in practice, $\Re$ can be a function of both signal wavelength and signal optical power. Though the wavelength dependency of responsivity in a photodiode is typically weak, the power dependency of $\Re$ may not be neglected, which determines the linearity of the photodetection, that may introduce saturation and high-order harmonics in the optical receiver.

Figure 3.3.1 shows a measurement setup to characterize the responsivity of a photodetector. A tunable laser is used to provide a wavelength tunable light source, and a variable attenuator adjusts the optical power level that is sent to the photodetector under test. The photocurrent signal is usually quite weak, especially if we need to avoid the nonlinear effect of photodetection and keep the signal optical power level low; therefore a transimpedance amplifier (TIA) must be used to linearly convert the photocurrent into a voltage signal. A voltmeter is then used to measure the voltage signal at the output of the TIA which is proportional to the photocurrent generated by the photodetector. A calibrated

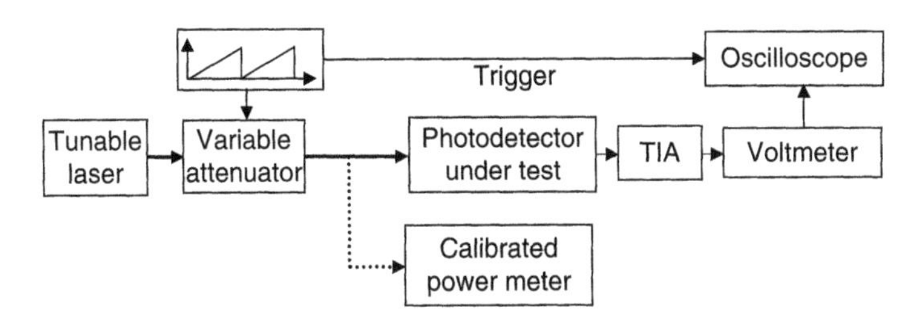

Figure 3.3.1 Experimantal setup to calibrate photodetector responsivity. *TIA:* trans-impedance amplifier.

optical power meter measures the absolute power level of the optical signal, providing the base to calibrate the measurement setup.

In this setup, by scanning the wavelength of the tunable laser, the wavelength-dependent responsivity of the photodetector can be easily characterized. In practice, the responsivity of a photodiode is relatively insensitive to the change of signal wavelength except for specially designed wavelength-selective photodetectors, thus the linewidth of the tunable laser does not have to be very narrow. In fact, in most cases, a wavelength resolution of 1 nm should be fine enough to characterize a photodiode. Since most of the tunable lasers have limited wavelength tuning range (typically <100 nm), for a wide range measurement, a tungsten lamp and a tunable monochromator may be used in place of the tunable laser.

Linearity measurement of photodetector responsivity is accomplished by linearly varying the attenuation of the variable attenuator so that the optical power P that enters the photodetector is linearly varied. Generally, the nonlinearity of the responsivity around an operation power P_0 level is defined as

$$N(I_x) = \frac{I_x/I_0}{P_x/P_0} - 1 \tag{3.3.1}$$

where I_x and I_0 are photocurrents measured at the signal power levels of P_x and P_0, respectively, as illustrated by Figure 3.3.2. From a calibration point of view, the transimpedance gain of the amplifier has to be calibrated using a photodetector with known responsivity. This is essential to obtain the absolute responsivity value of the device under test.

Though the setup shown in Figure 3.3.1 can be used to measure both the wavelength-dependent and the power-dependent responsivities of a photodetector, it is obvious that the system is operated in the continuous wave and provides only the responsivity of the photodetector without optical modulation. In general, the responsivity of a photodetector may strongly depend on the modulation speed of the optical signal; this frequency response is a very important measure of the photodetector property.

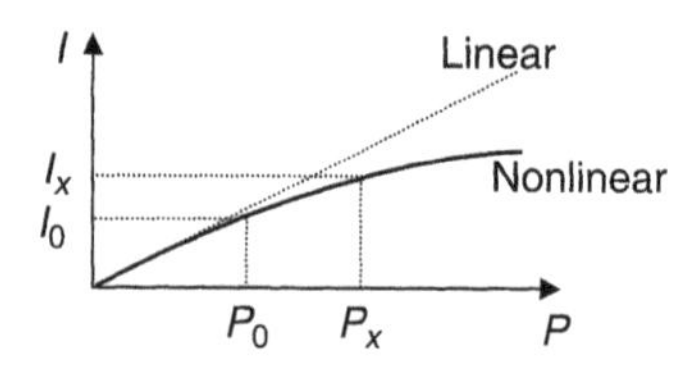

Figure 3.3.2 Illustration of the nonlinear responsivity of a photodetector.

3.3.2 Frequency Domain Characterization of Photodetector Response

A traditional way to characterize the frequency domain response of a photodetector is the use of an RF network analyzer as shown in Figure 3.3.3. This experimental setup is similar to the one used to measure the modulation response of an optical transmitter, as shown in Figure 3.2.2. The only difference

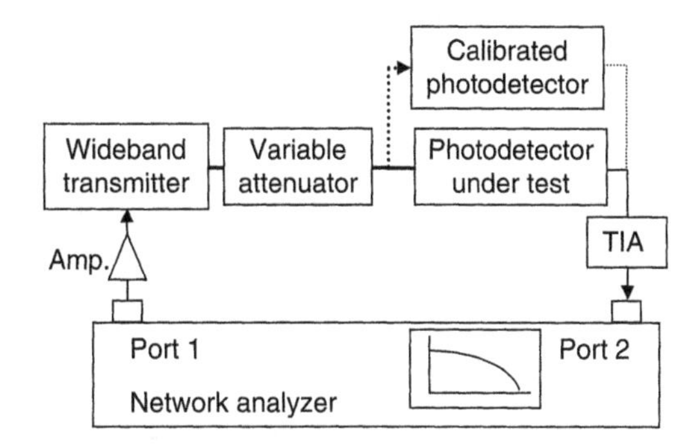

Figure 3.3.3 Measuring frequency response of photodetector using an RF network analyzer and a wideband transmitter.

is that in measuring a transmitter, the receiver bandwidth should be wide enough, or the frequency response of the receiver is known, whereas measuring a photodetector, the transmitter bandwidth should be wide enough, or its frequency response should be known.

The calibration can be done if a calibrated photodetector is available. In this case, we can first measure the overall system response, including the wideband transmitter, the calibrated photodetector, the transimpedance amplifier, and the RF network analyzer, as $F_{ref}(\omega)$. Then, replacing the calibrated photodetector with the photodetector under test, the measured frequency response becomes $F_{test}(\omega)$. Obviously the frequency response of the photodetector under test is

$$F_{detector}(\omega) = \frac{F_{test}(\omega)}{F_{ref}(\omega)} F_{cal}(\omega) \tag{3.3.2}$$

where $F_{cal}(\omega)$ is the frequency response of the calibrated photodetector, which we suppose to be known. Note that nowadays a packaged high-speed photodiode is generally integrated with a transimpedance amplifier, or at least terminated by an impedance-matched 50Ω resistance at the output. In that case, the TIA in Figure 3.3.3 will not be necessary.

Although this measurement technique is simple in principle, it requires a response-calibrated transmitter or a calibrated photodetector as the reference, which are usually difficult to have. In addition, to measure a wideband photodetector, the transmitter should have even wider bandwidth. From a technology standpoint, photodetectors with close to 100 GHz bandwidth have been demonstrated, but an optical transmitter at a comparable speed is rare.

To solve the speed limit of the optical transmitter, coherent heterodyne technique can be used; the block diagram of the measurement setup is shown in Figure 3.3.4 [15]. In this setup, two narrow linewidth tunable lasers are used, and their outputs are combined by a 2 × 2 coupler and sent to the photodetector under test. A polarization controller is used to ensure the polarization state alignment between the two laser outputs. An optical spectrum analyzer monitors the wavelengths of the two tunable lasers through the second output port of the 2 × 2 coupler. The photodetector performs coherent heterodyne detection, which creates a beat note in the RF domain whose frequency is equal to the frequency difference between the two tunable lasers. A wideband RF spectrum analyzer is then used to measure the frequency response of the photodetector.

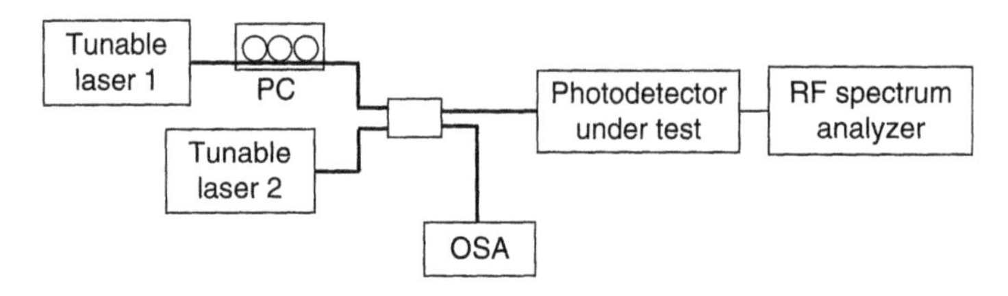

Figure 3.3.4 Photodetector frequency response measurement using the coherent heterodyne technique. *PC:* polarization controller.

In coherent heterodyne detection, as we discussed in Section 2.6, if both of the two lasers are operated in CW mode, the photocurrent signal at the photodetector output can be expressed as

$$I(t) = \Re(0)(P_1 + P_2) + 2\Re(f)\sqrt{P_1 P_2}\cos[2\pi f t + \Delta\phi(t)] \tag{3.3.3}$$

where we have considered that the photodetector responsivity $\Re(f)$ is frequency-dependent. $\Re(0)$ is the responsivity at DC, P_1 and P_2 are the optical powers coming from the two tunable lasers, and $f = |f_2 - f_1|$ is the frequency difference between the two lasers, which are at f_1 and f_2, respectively. $\Delta\phi(t)$ accounts for the phase noise, which can be neglected for the purpose of this measurement because we are only interested in the signal amplitude at the IF frequency. The mean-square photocurrent is

$$\langle I^2(t)\rangle = \Re^2(0)(P_1 + P_2)^2 + 2\Re^2(f)P_1 P_2 = \langle I_{DC}^2\rangle + \langle I_{RF}^2\rangle \tag{3.3.4}$$

where $\langle I_{DC}\rangle^2$ and $\langle I_{RF}\rangle^2$ are the DC component and the heterodyne IF frequency component, respectively, and the ratio between these two components is

$$\frac{\langle I_{RF}^2\rangle}{\langle I_{DC}^2\rangle} = \frac{2\Re^2(f)P_1P_2}{(P_1+P_2)^2\Re^2(0)} \tag{3.3.5}$$

The best measurement condition is obtained when $P_1 = P_2$, which is equivalent to the highest amplitude modulation index in Equation 3.3.3, and ideally this modulation index can be equal to one. In this ideal case, $(P_1+P_2)^2 = 4P_1P_2$ and Equation 3.3.5 can be simplified as

$$\left[\frac{\Re^2(f)}{\Re^2(0)}\right] = 2\frac{\langle I_{RF}^2\rangle}{\langle I_{DC}^2\rangle} \tag{3.3.6}$$

If we define $\Re^2(f)/\Re^2(0)$ as the normalized frequency response of the photodetector, it can be obtained by measuring the ratio between the RF powers at the heterodyne frequency f and at DC.

In this measurement, to maximize the modulation efficiency, not only must we pay attention to equalizing the optical powers from the two tunable lasers, but their polarization states also must be aligned. Although the heterodyne beat frequency between the two tunable lasers can easily be extended to hundreds of gigahertz, the major frequency limitation of this measurement technique is usually due to the limited bandwidth of the RF spectrum analyzer. For a wide bandwidth measurement, high-frequency microwave power sensors may be used, whereas the reading of the IF frequency can be obtained through the OSA in the measurement setup, which is the difference between the optical frequencies of the two CW light sources.

3.3.3 Photodetector Bandwidth Characterization Using Source Spontaneous-Spontaneous Beat Noise

As discussed in the last section, photodetector bandwidth characterization using the optical heterodyne technique provided almost unlimited bandwidth to simulate a modulated optical source. However, the experimental setup is relatively complicated, requiring two tunable lasers with power equalization and polarization alignment between them. Another frequency-domain technique for the characterization of photodetector bandwidth utilizes an optical source with a wideband optical intensity noise. The beating between different frequency components within this wide bandwidth generates different RF frequencies and therefore can be used to test the frequency response of the photodetector [16].

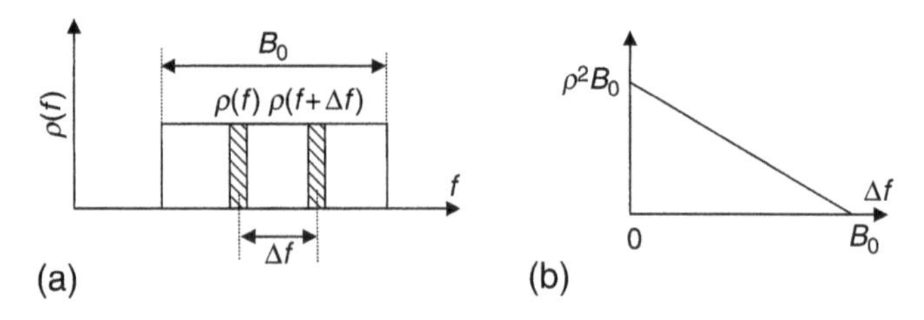

Figure 3.3.5 Illustration of spontaneous-spontaneous beat noise generated by an optical noise source with uniform spectral density: (a) optical spectrum and (b) electrical spectrum.

For this measurement, the light source can be a high-power superluminance LED (SLED) or simply an optical amplifier without an input optical signal. As illustrated in Figure 3.3.5, when we inject a wideband optical noise into a photodetector, the mixing between frequency components f and $f + \Delta f$ in the optical spectrum will produce a frequency component at Δf in the electrical domain through the photodetector. As discussed in the next section on the characterization of optical amplifier noise, this is a self-convolution process and the generated RF noise spectral density can be expressed as

$$S_{sp-sp}(\Delta f) = (B_0 - \Delta f)\Re^2(\Delta f)\rho^2 \tag{3.3.7}$$

where ρ is the input optical signal spectral density which is uniform within the optical bandwidth B_0, Δf is the RF frequency, and $\Re$ is the photodiode responsivity.

Typically the bandwidth of a wideband optical noise source can easily provide a bandwidth of $B_0 > 30$ nm, which is about 4750 GHz in a 1550 nm wavelength window, whereas the interested RF frequency range for characterizing a photodetector is less than 100 GHz. Therefore, $B_0 - \Delta f \approx B_0$ and

$$S_{sp-sp}(\Delta f) \approx \Re^2(\Delta f)\rho^2 B_0 \tag{3.3.8}$$

The frequency dependency of the measured RF noise spectral density is only determined by the frequency response $\Re(\Delta f)$ of the photodetector.

The block diagram of the measurement setup, as shown in Figure 3.3.6, is very simple. The photodetector directly detects the wideband optical noise and sends the converted electrical spectrum into an RF spectrum analyzer.

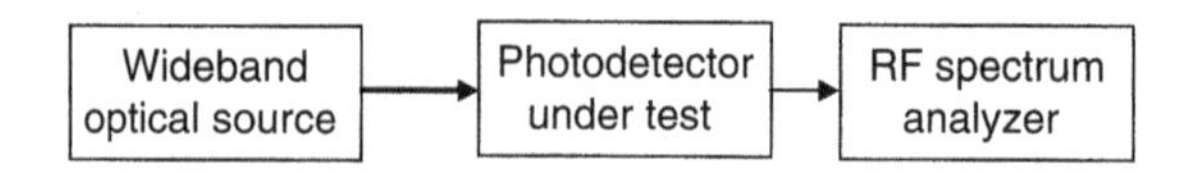

Figure 3.3.6 Block diagram of photodetector bandwidth characterization using a wideband light source and an RF spectrum analyzer.

One major disadvantage of this measurement technique is that the electrical noise spectral density generated by the photodetector, which is related to the frequency response of the photodetector, is usually very low. For example, for a 1 mW optical source with 30 nm bandwidth, if the photodetector responsivity is 1 mA/mW, the spontaneous-spontaneous emission beat noise spectral density generated at the photodetector, which is the useful signal in this measurement, is approximately –157 dBm/Hz. This spectral density level is close to the electrical noise level of a typical wideband RF spectral analyzer, and therefore the amplitude dynamic range of the measured is severely limited.

It is interesting to note from Equation 3.3.8 that if the total power of the optical source is fixed as P_i, the spontaneous-spontaneous emission beat noise is inversely proportional to the optical bandwidth of the source

$$S_{sp-sp}(\Delta f) \approx \frac{\Re^2(\Delta f)P_i^2}{B_0} \tag{3.3.9}$$

For example, if the source optical power is 1 mW but the bandwidth is only 3 nm, for a 1 mA/mW photodetector responsivity, the spontaneous-spontaneous emission beat noise spectral density will be approximately –147 dBm/Hz, which is 10 dB higher than the case when the source bandwidth is 30 nm. Although reducing source optical bandwidth may help increasing the measurement dynamic range, the fundamental requirement that $B_0 \gg f_0$ has to be met for Equation 3.3.9 to be valid, where f_0 is the measurement electrical bandwidth.

In general, the frequency response of a photodiode does not vary dramatically over a narrow wavelength range; therefore discrete sampling in the frequency domain is sufficient. In this case, a Fabry-Perot (FP) filter can be used to select discrete and periodic frequency lines from a wideband ASE noise source. Figure 3.3.7 shows the block diagram of the measurement setup, where a wide spectrum LED is used to generate the spontaneous emission optical noise and a FP filter is used to select a frequency comb. An EDFA is then used to amplify the filtered optical spectrum so that the optical noise spectral density at the transmission peaks of the FP filter can be very high. There are two reasons to use a relatively low total

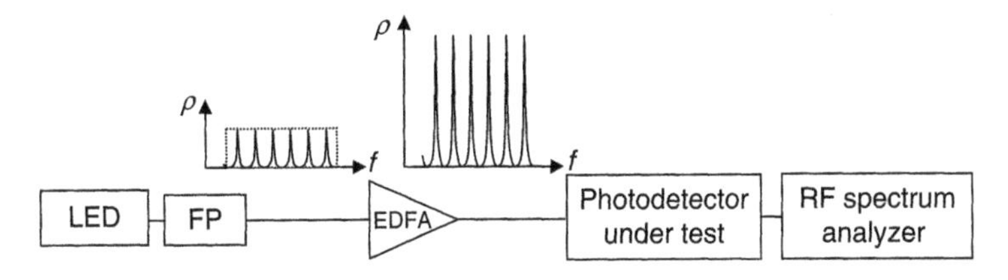

Figure 3.3.7 Block diagram of photodetector bandwidth characterization using an FP filter to increase measurement dynamic range.

optical power while having high peak optical powers at selected frequencies: first, the EDFA is usually limited by its total optical power; if finesse of the FP filter is F, the peak spectral density at peak transmission frequencies can be found as $\rho_{FP} = \rho F/\pi$, where ρ is the uniform power spectral density without FP filter. Second, a photodetector usually has limitations on the total input optical power to avoid saturation and damaging. If the free spectral range of the FP filter is FSR, the mixing between various discrete optical frequency components will generate a series of discrete RF frequency components at the photodetector with the frequency interval of FSR. In this application, the FSR is the actual sampling interval in the frequency domain.

The RF spectral density measured at the spectrum analyzer can be approximated as [17]

$$S_{sp-sp}(f) \approx \frac{\Re^2(f) P_i^2}{B_0} \frac{F}{\pi} \sum_k \frac{1}{1 + \left(\frac{f - k \cdot FSR}{\Delta \nu} \right)^2} \tag{3.3.10}$$

where $\Delta\nu$ is the FWHM width of the FP filter transmission peaks and B_0 is the optical bandwidth of each FP transmission peak. Compare Equation 3.3.10 with Equation 3.3.9; for the same total input optical power P_i to the photodetector, the peak RF spectral density is increased by F/π. For a typical fiber FP filter with a finesse of 500, the improvement in the measurement dynamic range of more than 20 dB can be easily obtained. Using this technique, $B_0 \gg f_0$ can be easily maintained and the signal amplitude can also be significantly increased. One important concern of this technique is the frequency resolution of the measurement. In fact, in the electrical domain, the frequency sampling interval is equal to the FSR of the FP interferometer, which can be easily made into Megahertz regime by increasing the cavity length.

3.3.4 Photodetector Characterization Using Short Optical Pulses

Both the coherent heterodyne technique and the spontaneous-spontaneous emission noise beating technique we've discussed measure linear characteristics of the photodetector because they both characterize small-signal frequency response. Short optical pulses can also be used to characterize photodetector response because of their wide spectral bandwidths. Unlike the frequency-domain measurement techniques we've described, the short-pulse technique measures the time-domain response. With the recent development of ultrafast fiber-optics, picosecond actively mode-locked semiconductor lasers and femtosecond passively mode-locked fiber lasers are readily available.

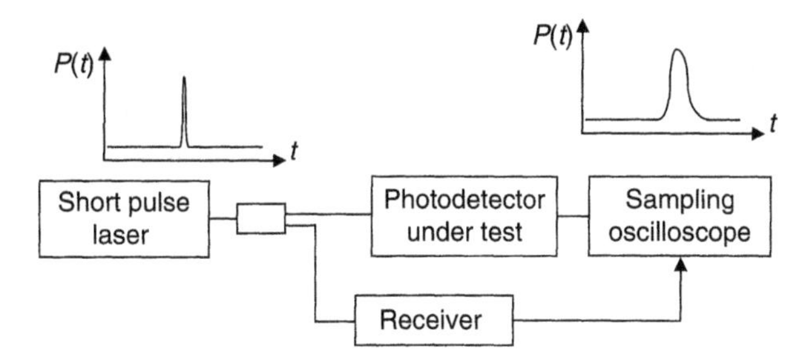

Figure 3.3.8 Block diagram of photodetector bandwidth characterization using an FP filter to increase measurement dynamic range.

Figure 3.3.8 shows the block diagram of an experimental setup to characterize a photodetector in time domain, where the output of a femtosecond laser source is split into two parts by a beam splitter. One part is directly fed to the photodetector under test, which converts the short optical pulses into electrical pulses. A high-speed sampling oscilloscope is then used to measure the electrical signal waveform. The other part of the laser source is sent to an optical receiver, which measures the pulse arrival time and its repetition rate in order to trigger the sampling oscilloscope. In this case the speed of the optical receiver only needs to be much faster than the repetition rate of the optical pulse train for triggering purposes.

Theoretically, the response of the photodetector can be obtained by comparing the shape of the output electrical pulses with that of the input optical pulses. In practice, however, the duration of the femtosecond optical pulses is usually much shorter than the time constant of the photodetector; therefore the width of the input optical pulses can be neglected for most applications. For example, 100 fs optical pulses in a 1550 nm wavelength window have bandwidth of approximately 1 THz, whereas the bandwidth of a photodetector is usually less than 100 GHz.

Generally, the time-domain measurement provides more information than frequency-domain techniques because the frequency response of the photodetector can be easily obtained by a Fourier analysis of the time-domain waveform measured by the oscilloscope. The unique advantage of the time-domain technique is that it measures the actual waveform from the photodetector, which includes transient effect and saturation. For example, a photodiode may have different responses and time constants during the leading edge than the falling edge of a signal pulse. Sometimes, although the input optical pulses are symmetrical, the output electrical pulses are not symmetrical, with a relatively short rise time and a much longer falling tail due to slow carrier transport outside the junction region of the photodiode. In addition, the power-dependent saturation effect of the photodetector can also be evaluated by measuring the electrical waveform change due to the increase in the amplitude of the input optical pulses.

The major limitation of this time-domain measurement is perhaps the limited speed of the sampling oscilloscope. A 50 GHz sampling oscilloscope is commercially available, whereas a sampling oscilloscope with bandwidth higher than 100 GHz is difficult to find.

3.4 CHARACTERIZATION OF OPTICAL AMPLIFIERS

The general characteristics of optical amplifiers have been introduced in Chapter 1. The desired performance of an optical amplifier may depend on the specific application. For example, in optical communication systems, a line amplifier for a multiwavelength WDM system should have high gain over a wide bandwidth to accommodate a large number of wavelength channels. The gain over this bandwidth should be flat to ensure uniform performance for all the channels, and the noise figure should be low so that the accumulated ASE noise along the entire system with multiple line amplifiers is not too high. On the other hand, for a preamplifier used in an optical receiver, the noise figure is the most important concern. For an SOA used for optical signal processing, gain dynamics may be important to enable ultrahigh-speed operation of the optical system. The general characterization techniques and necessary equipment for EDFA and SOA measurements are similar except for some specific characteristics.

In this section, we discuss a number of often used optical amplifier characterization techniques for their important properties such as optical gain, gain dynamics, optical bandwidth, and optical noise.

3.4.1 Measurement of Amplifier Optical Gain

Since the basic purpose of an optical amplifier is to provide optical gain in an optical system, the characterization of optical gain is one of the most important issues that determine the quality of an optical amplifier. In general, optical gain of an optical amplifier is both a function of signal optical wavelength and the signal optical power: $G = G(\lambda, P)$. In addition, when operating in room temperature, both EDFA and SOA can be modeled as homogenously broadened systems, and therefore optical gain at one wavelength can be saturated by optical power of other wavelengths, which is commonly referred to as cross-gain saturation:

$$G(\lambda) = \frac{G_0(\lambda)}{1 + \sum_k P(\lambda_k)/P_{sat}(\lambda_k)} \tag{3.4.1}$$

where $G_0(\lambda)$ is the small-signal optical gain at wavelength λ, $P(\lambda_k)$ is the optical power, and $P_{sat}(\lambda_k)$ is the saturation power at wavelength λ_k.

The optical gain of an amplifier can be measured using either an optical power meter or an optical spectrum analyzer, or even an electrical spectrum analyzer. The important considerations include wide enough optical bandwidth of the source to cover the entire amplifier bandwidth, calibrated signal power variation to accurately measure the saturation effect, and a way to remove the impact of the ASE noise, which might affect the accuracy of optical gain measurement.

A typical experimental setup to measure the gain spectrum of an optical amplifier is to use a tunable laser and an optical power meter as shown in Figure 3.4.1. The tunable laser provides the optical signal that is injected into the optical amplifier through a calibrated variable optical attenuator (VOA) and an optical isolator. After passing through the EDFA under test, the amplified optical signal is measured by an optical power meter. Since the EDFA not only amplifies the optical signal, it also produces wideband amplified spontaneous emission noise, known as ASE noise. A narrowband optical filter is used to select the optical signal and reject the ASE noise. A calibration path consists of an identical ASE noise filter, and the optical power meter. This calibration can usually be performed by simply removing the EDFA from the measurement path. In this setup, the wavelength-dependent nature of the optical amplifier gain can be measured by scanning the wavelength of the tunable laser across the wavelength window of interest. Of course, the ASE noise filter has to be tunable as well in order to synchronize with the wavelength of the tunable laser source.

Assume that the optical signal power at the input of the EDFA is P_s and the EDFA optical power gain is G; then the signal optical power at the EDFA

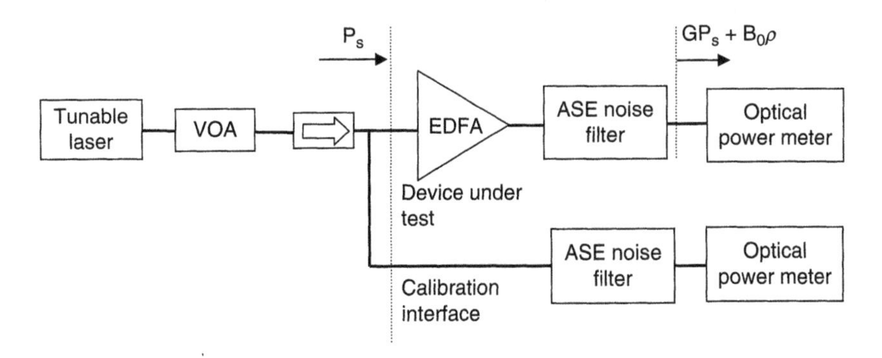

Figure 3.4.1 Block diagram of EDFA gain measurement based on an optical power meter.

output is GP_s. Taking into account the ASE optical noise produced by the EDFA with the power spectral density ρ, and assuming that the ASE noise is wideband, the total ASE noise power after the narrowband ASE optical filter will be

$$P_{ASE} = \int_{-B_0/2}^{B_0/2} \rho(f)df \approx B_0\rho \tag{3.4.2}$$

where B_0 is the ASE filter optical bandwidth. Then the total optical power that enters the power meter is approximately

$$P_{measured} = GP_s + B_0\rho \tag{3.4.3}$$

The measured optical gain is therefore

$$G_{measured} \equiv \frac{P_{measured}}{P_s} = \frac{GP_s + B_0\rho}{P_s} = G + B_0\rho/P_s \tag{3.4.4}$$

In this expression, the second term is the measurement error introduced by the residual ASE noise. This error term is proportional to the output ASE noise divided by the input signal optical power. Since $B_0\rho/P_s \approx G(B_0\rho/P_{out})$, where $P_{out} = GP_s$ is the output signal power, this term is in fact proportional to the optical gain of the amplifier. Nevertheless, because the optical gain G is usually a large number, it is more relevant to evaluate the relative error, which is $(1 + B_0\rho/P_{out})$. Obviously, to ensure the accuracy of the measurement, the input optical signal P_s cannot be too small; otherwise the second term in Equation 3.4.4 can be significant.

On the other hand, because of the gain saturation effect in the optical amplifier, the optical gain G is typically a function of input optical power P_s. Practically, to measure the small-signal optical gain of the amplifier, the input optical signal power has to be small enough to avoid the gain saturation. This usually requires the signal input optical power to be $P_s \ll P_{sat}/G$ or simply $P_{out} = GP_s \ll P_{sat}$, where P_{sat} is the saturation optical power at the signal wavelength.

Combining the ASE noise limitation and the nonlinearity limitation, the input signal optical power should be chosen such that $B_0\rho/G \ll P_s \ll P_{sat}/G$, or $B_0\rho \ll P_{out} \ll P_{sat}$. Small signal gain is the most fundamental property of an optical amplifier, which is typically a function of the wavelength, and by definition, it is not a function of the signal optical power (because the power is assumed to be very small). As an example, for a typical EDFA with a 5 dB noise figure, 13 dBm saturation power, and 30 dB small-signal gain, if the ASE filter has an optical bandwidth of 1 nm, the ASE noise optical power that reaches the power meter is approximately –13 dBm. In this case, the input signal optical power for this measurement should be –43 dBm $\ll P_s \ll$ –17dBm. In this case, a signal optical power on the order of P_s = –30 dBm should be appropriate.

In most practical applications, the signal optical power may not satisfy the small-signal condition. In this situation, the measured optical gain is referred to as large-signal gain, or sometimes *static gain*. The large-signal optical gain spectrum can be measured using the same setup as shown in Figure 3.4.1 where the tunable laser provides a constant but high optical power across the wavelength of interest. In general, large-signal optical gain $G = G(\lambda, P)$ is a function of both wavelength and signal optical power. It is important to note that the wavelength dependency of the large-signal optical gain is usually different from that of the small-signal optical gain. The reason is that the optical gain is wavelength dependent. Although the input signal optical power from the tunable laser is constant, the output optical power varies over the wavelength and the saturation effect is high at wavelengths where the gains are high.

The measurement setup shown in Figure 3.4.1 can also be modified by replacing the optical filter and optical power meter with an optical spectrum analyzer as shown in Figure 3.4.2. In this case the tunable ASE noise filter is no longer required because the spectral resolution of an OSA is determined by the bandwidth of an internal tunable optical filter of the OSA. Therefore, the major advantage of using OSA is the adjustable resolution bandwidth, which allows the measurement of OSNR at different optical signal wavelengths. In addition, a high-quality OSA also provides much higher dynamic range in the measurement than an optical power meter combined with a fixed bandwidth tunable optical filter.

The major advantage of optical gain measurement using an optical power meter or OSA is its simplicity and low bandwidth requirement for electronics. However, the results of measurement may be sensitive to the ASE noise creased

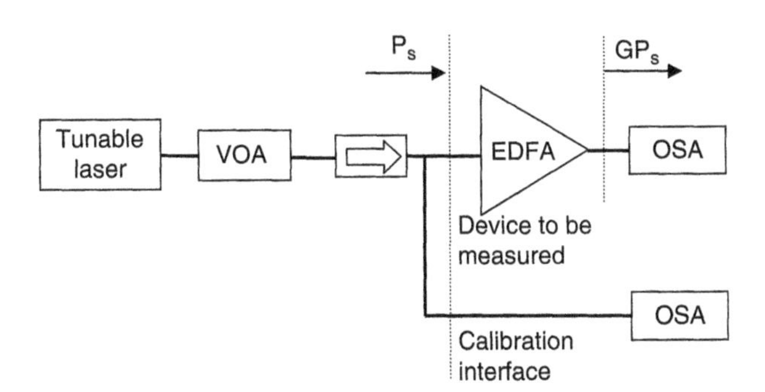

Figure 3.4.2 Block diagram of EDFA gain measurement based on an optical spectrum analyzer (OSA).

by the optical amplifier itself, especially when the signal level is very small. For example, for an OSA with spectral resolution of 0.1 nm, if the ASE noise spectral density of the EDFA is –20 dBm/nm, the total ASE noise power within the resolution bandwidth is –30 dBm. In this case, if the signal optical power at EDFA output is less than –20 dBm for small-signal gain measurement, there will be a significant error (>1 dB) because ASE noise is added to the signal.

Another way to measure EDFA gain is to use an electrical spectrum analyzer (ESA) that is able to provide a much finer spectral resolution and thus less power of ASE noise. The experimental setup using ESA is shown in Figure 3.4.3. In this measurement, the output power of a tunable laser is modulated by a sinusoid at

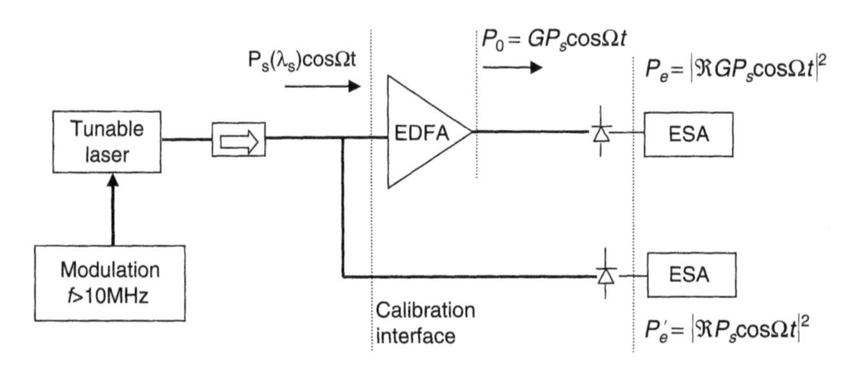

Figure 3.4.3 Block diagram of EDFA gain measurement based on an electrical spectrum analyzer (ESA).

frequency Ω. After the modulated optical signal is amplified, it is detected by an electrical spectrum analyzer. By measuring the amplitude of the RF tune at the modulation frequency Ω and comparing it with the measured value at the reference arm (without EDFA), the optical gain of the EDFA can be evaluated.

In the gain calculation, since the photocurrent measured by the photodetector is proportional to the optical power, the RF power will be proportional to the square of the received optical power. Therefore, one dB of RF power change corresponds to 0.5 dB change in the EDFA optical gain. Using the notation in Figure 3.4.3, $P_e/P_e' = G^2$, where P_e and P_e' are RF powers measured before and after the EDFA.

Since an ESA can have much finer spectral resolution (as low as 30 Hz) compared to an OSA, the contribution from wideband ASE noise can be made negligible. The modulation frequency in this measurement does not have to be very high. It is enough as long as the modulated signal can be easily measured by an ESA, and away from the very low frequency region where the background noise is high in an ESA. A few tens of megahertz should be appropriate.

3.4.2 Measurement of Static and Dynamic Gain Tilt

As we just discussed, the small-signal gain of an optical amplifier can be measured with a very weak optical signal, whereas large-signal optical gain can be evaluated when the output optical signal is comparable to the saturation power of the optical amplifier. In general, both small-signal and large-signal optical gains are functions of signal wavelength. Gain flatness of an optical amplifier is one of the most important parameters, especially when the amplifier is used for WDM operation. Gain tilt is the slope of optical gain across a certain wavelength window, which is an indication of the gain flatness of the amplifier.

3.4.2.1 Static Gain Tilt

The static gain tilt can be subcategorized into small-signal and large-signal gain tilts. By definition, the *small-signal gain tilt* near a central wavelength λ_0 and within a wavelength window $2\Delta\lambda$ *is*

$$m_{sm}(\lambda_0) = \frac{G_{sm}(\lambda_0 + \Delta\lambda) - G_{sm}(\lambda_0 - \Delta\lambda)}{2\Delta\lambda} \quad (3.4.5)$$

where $G_{sm}(\lambda \pm \Delta\lambda)$ is the small-signal gain at wavelengths $\lambda \pm \Delta\lambda$. Obviously, to obtain the small-signal gain tilt, the optical signal power has to be much lower than the saturation power of the amplifier. Likewise, the *large-signal gain tilt* is defined as the gain slope around wavelength λ_0 and within a wavelength window $2\Delta\lambda$ but with the input signal optical power P_{in}, that is,

$$m_{st}(\lambda_0, P_{in}) = \frac{G_{st}(\lambda_0 + \Delta\lambda, P_{in}) - G_{st}(\lambda_0 - \Delta\lambda, P_{in})}{2\Delta\lambda} \quad (3.4.6)$$

where $G_{st}(\lambda \pm \Delta\lambda, P_{in})$ is the large-signal optical gain at wavelengths $\lambda \pm \Delta\lambda$ with input signal optical power P_{in}.

Both the small-signal gain tilt and the large-signal gain tilt are considered *static gain tilt*, which involves only one wavelength-tunable optical signal at the input. Therefore, the static gain tilts can be measured by the simple experimental setups shown in Figure 3.4.1 or Figure 3.4.2.

3.4.2.2 Dynamic Gain Tilt

In practical optical systems, multiple optical signals are usually used, and cross-gain saturation may happen between them. The optical gain spectrum is likely to change if a strong optical signal at a certain fixed wavelength is injected into the optical amplifier. Figure 3.4.4 shows the experimental setup to measure the gain spectrum change of an optical amplifier caused by the existence of a strong optical signal, commonly referred to as the *holding beam*. In this

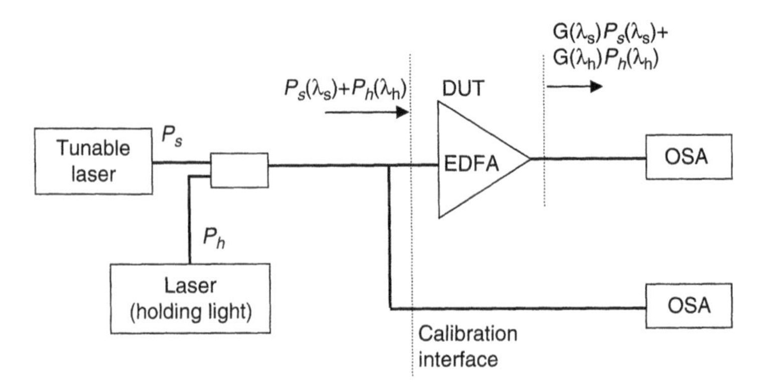

Figure 3.4.4 Block diagram of an optical amplifier dynamic gain tilt meter.

measurement, a strong holding beam with power P_h at a fixed wavelength λ_h is combined with a wavelength-tunable optical probe with a weak power P_s before they are fed into the optical amplifier. By sweeping the wavelength of the small-signal optical probe, gain spectrum can be measured at various power levels of the holding light.

Since the holding beam usually has much higher power than the small-signal tunable probe ($P_h \gg P_s$,), this system basically measures small-signal gain spectrum under a strong holding light. In this case, the total output optical power of the amplifier is essentially determined only by the holding beam: $P_{out} \approx G(\lambda_h)P_h$. Figure 3.4.5 shows an example of the small-signal gain spectrum of an EDFA under the influence of a strong holding beam. In this example, the EDFA is made of 35 m erbium-doped fiber (Lucent HE20P) and the pump power at 980 nm is 75 mW. A holding beam at 1550 nm wavelength is injected into the EDFA together with the small-signal probe. By increasing the power P_h of the holding beam from –30 dBm to –10 dBm, not only is the overall optical gain level decreased, the shape of the gain spectrum is also changed; this effect is commonly known as the *dynamic gain tilt*.

Dynamic gain tilt is defined as the slope of small-signal gain around a central wavelength λ_0, within a wavelength range $2\Delta\lambda$, and in the presence of a strong holding light with power P_h,

$$m_{dy}(\lambda_0, P_h) = \frac{G_{sm}(\lambda_0 + \Delta\lambda, P_h) - G_{sm}(\lambda_0 - \Delta\lambda, P_h)}{2\Delta\lambda} \tag{3.4.7}$$

where $G_{sm}(\lambda \pm \Delta\lambda, P_h)$ is the small-signal gain at wavelengths $\lambda \pm \Delta\lambda$ and with a holding beam of optical power P_h.

The physical mechanism behind the cross-gain saturation effect is the homogenous broadening of the gain medium in the amplifier. Homogenous

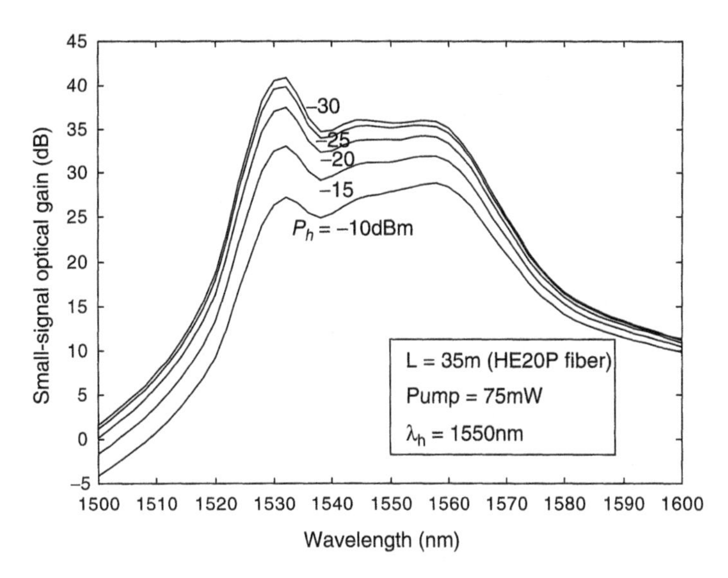

Figure 3.4.5 Small-signal optical gain spectrum of an EDFA with the existence of a holding beam at 1550 nm at various power levels.

broadening is originated from the fact that carrier density in the upper energy band tends to maintain a thermal equilibrium distribution across the energy band. After releasing photons of a certain wavelength due to stimulated emission, carrier density at the corresponding energy level is reduced due to band-to-band transition. As a result, carriers at other energy levels will quickly move in and therefore, the carrier density distribution across the energy band tends to remain unchanged. As a consequence of the homogenous broadening, a large signal at any wavelength within the gain bandwidth can saturate the amplifier optical gain over the entire gain bandwidth.

It is important to note that both EDFA and SOA have homogenously broadening gain medium. The major difference between them is the carrier lifetime, which determines the speed of saturation. The typical carrier lifetime of an EDFA is on the order of 1–10 milliseconds, whereas the carrier lifetime of an SOA is in nanosecond or even picosecond levels. In optical communication systems with data rates higher than megabits per second, dynamic gain tilt in an EDFA is determined by the average power of the optical signal at each wavelength, whereas an SOA may suffer severe intersymbol crosstalk between channels due to the fast carrier response. (We'll discuss the measurement of gain dynamics later.)

Though homogeneous broadening is responsible for cross-gain saturation in an optical amplifier, spectral hole burning is another effect that tends to reduce cross-gain saturation. Spectral hole burning happens when carriers at other energy levels within the energy band cannot move in fast enough to fill the vacancy caused by strong stimulated recombination at a certain energy level. In this case, thermal equilibrium is not reached and carrier density at a particular energy level can be much lower compared to the adjacent energy levels. This can be caused by a strong optical signal at a certain wavelength, which strongly saturates the optical gain at that particular wavelength. In fact, spectral hole burning is a desired property for WDM optical system applications because it will reduce nonlinear crosstalk between different wavelength channels. At room temperature, the effect of spectral hole burning is very weak; however, at a sufficiently low temperature, intraband relaxation can be greatly reduced and spectral hole burning can be significantly increased. EDFA without crosstalk has been demonstrated by cooling the EDF below 70K [18]. The same analogy can be used to explain polarization hole burning, which is related to spin direction of the electrons.

In addition to homogeneous broadening and spectral hole burning, in optical amplifiers, especially in fiber-based amplifiers, carrier density also changes along the amplifier longitudinal direction. In fact, in an EDFA, a strong holding beam not only reduces the upper-level carrier density, it also changes the carrier density distribution along the erbium-doped fiber, which is the major reason for the change in the gain spectral profile.

3.4.3 Optical Amplifier Noise

In an optical amplifier, amplification is supported by stimulated emission in the gain medium with population inversion. At the same time, this inverted carrier population in the gain medium not only supports coherently stimulated emission, which amplifies the incoming optical signal; the carriers also spontaneously recombine to generate spontaneous emission photons. These spontaneously emitted photons are not coherent with the input signal and they constitute optical noise. To make things worse, the spontaneous emission photons are also amplified by the gain medium while they travel through the optical amplifier; thus the optical noise generated by an optical amplifier is commonly known as the *amplified spontaneous emission* (ASE). By its nature, ASE noise is random in wavelength, phase, and the state of polarization.

Generally, the level of ASE noise produced by an optical amplifier depends on both the optical gain and the level of population inversion in the gain medium. ASE noise power spectral density with the unit of $[W/Hz]$ can be expressed as

$$\rho(\lambda) = 2n_{sp}\frac{hc}{\lambda}[G(\lambda) - 1] \tag{3.4.8}$$

In this expression, $G(\lambda)$ is the optical gain of the amplifier at wavelength λ, h is the Planck's constant, c is the speed of light, and the factor 2 indicates two orthogonal polarization states. n_{sp} is the unitless spontaneous emission factor which is a function of population inversion in the gain medium and is defined by

$$n_{sp} = \frac{\sigma_e N_2}{\sigma_e N_2 - \sigma_a N_1} \tag{3.4.9}$$

where σ_e and σ_a are the emission and absorption cross-sections of the erbium-doped fiber and N_1 and N_2 are carrier densities at the lower and upper energy levels, respectively. In general, $0 \leq n_{sp} \leq 1$ and the maximum value, 1, is achieved when the carrier inversion is complete, with $N_1 = 0$ and $N_1 = N_T$, where N_T is the total doping density of erbium ions in the fiber core. This expression is simple, but it is accurate only if both N_1 and N_2 are constant along the fiber length, which is unfortunately not always true. Although $N_1 + N_2 = N_T$ is always a constant, the ratio between N_1 and N_2 usually varies along the fiber. In general, if $N_1(z)$ and $N_2(z)$ are functions of z, the spontaneous emission factor can be generalized in an integral form:

$$n_{sp} = \frac{\sigma_e \int_0^L N_2(z)dz}{\sigma_e \int_0^L N_2(z)dz - \sigma_a \int_0^L N_1(z)dz} \tag{3.4.10}$$

Obviously, since both the optical gain and the cross-section values are functions of wavelength, the ASE noise spectral density is also a function of wavelength. Within a certain optical bandwidth B_0, the total noise optical power can be calculated by integration:

$$P_{ASE} = \int_{-B_0/2}^{B_0/2} \rho(\lambda)d\lambda \tag{3.4.11}$$

If the optical bandwidth is narrow enough and the ASE noise spectral density is flat over this bandwidth, the noise power will be directly proportional to the optical bandwidth and Equation 3.4.11 can be simplified as

$$P_{ASE} \approx 2n_{sp}hf[G(f) - 1]B_0 \tag{3.4.12}$$

For an optical amplifier, the measurement of optical gain spectrum $G(\lambda)$ was discussed in the last section. But this is not enough to accurately calculate ASE noise spectral density, because the spontaneous emission factor n_{sp} is a function of both wavelength and the carrier inversion level. Experimental methods of ASE noise measurement are also very important in the characterization of optical amplifiers.

3.4.4 Optical Domain Characterization of ASE Noise

Since ASE noise is a random process, it is relatively easy to be characterized by its statistic value such as optical power spectral density, which can be measured by an optical spectrum analyzer, as shown in Figure 3.4.6.

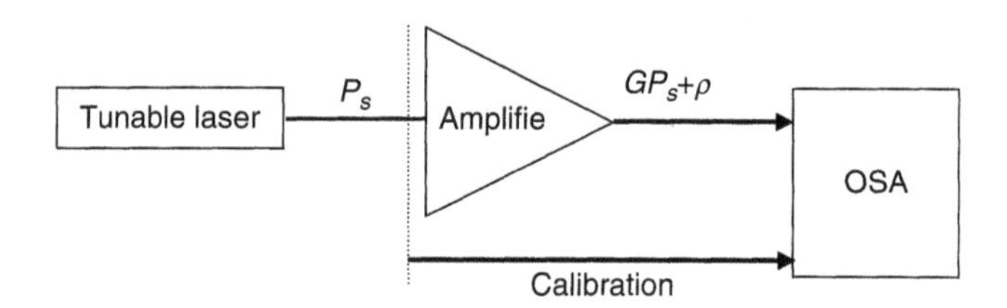

Figure 3.4.6 ASE noise measurement using an OSA.

By comparing the optical spectrum measured before and after the optical amplifier, we can evaluate both optical gain and the optical noise spectral density. Although this experimental setup is simple and the measurement is easy to perform, the accuracy is often affected by the linewidth of the optical signal and the limited dynamic range of the OSA. In practice, an optical signal has a Lorenzian line shape; if the OSNR is high enough, the measured linewidth of the optical signal can be quite wide at the ASE noise level, which is many dB down from the peak, as illustrated by Figure 3.4.7. As a consequence, ASE noise spectral density at the optical signal peak wavelength cannot be directly

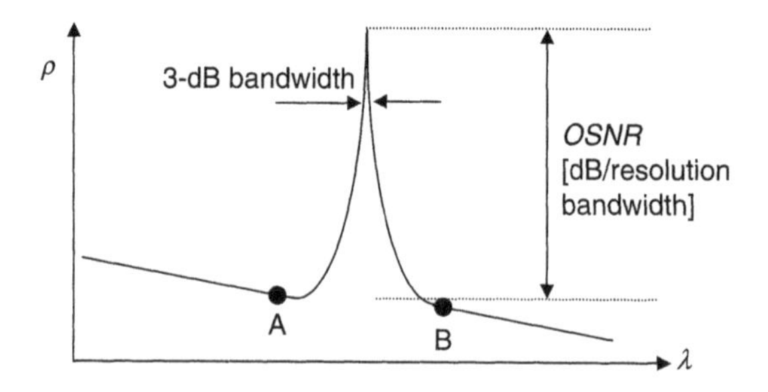

Figure 3.4.7 Illustration of ASE power spectral density measured by an OSA.

measured because it is hidden behind the strong optical signal. It would not be helpful to use an OSA with a finer spectral resolution. The reason is that the ASE noise level is measured by its spectral density, which is the noise power per spectral resolution bandwidth. Since the optical signal from a laser diode is usually narrowband, decreasing OSA resolution bandwidth will only increase the amplitude difference between the signal peak and the measured noise level on the OSA. In general it is more difficult to accurately evaluate the noise level when the OSNR is very high.

If the ASE noise spectral density is known to be flat within the bandwidth hidden behind the optical signal, the measurement should be accurate, but this is not guaranteed. In most cases, the characteristics of the noise are not well known, and the accuracy of the measurement may become a significant concern when the spectral width of the optical signal is too wide. One simple way to overcome this problem is to make interpolation between the measured noise levels at each side of the optical signal, shown as points A and B in Figure 3.4.7. This allows us to estimate the noise spectral density at the signal peak wavelength.

Another way to find the noise spectral density at the wavelength of a strong optical signal is to use a polarizer to eliminate the optical signal, as illustrated by Figure 3.4.8. This measurement is based on the assumption that optical signal is polarized (the degree of polarization is 100 percent), whereas the ASE noise is unpolarized. If this is true, a polarizer always blocks 50 percent of the ASE noise regardless of the orientation of the polarizer. Meanwhile, the polarizer can completely block the single-polarized optical signal if the polarization state of the optical signal is adjusted to be orthogonal to the principle axis of the polarizer. In this measurement, the polarization state of the optical signal is first adjusted to fully pass through the polarizer so that its power GP_s can be measured. The second step is to completely block the optical signal by adjusting the polarization controller, and in this case the ASE noise level $\rho'(\lambda_s)$ at exactly the signal wavelength λ_s can be measured, which is in fact 3 dB lower than its actual value. Then the OSNR can be calculated as $OSNR(\lambda_s) = GP_s/2\rho'(\lambda_s)$.

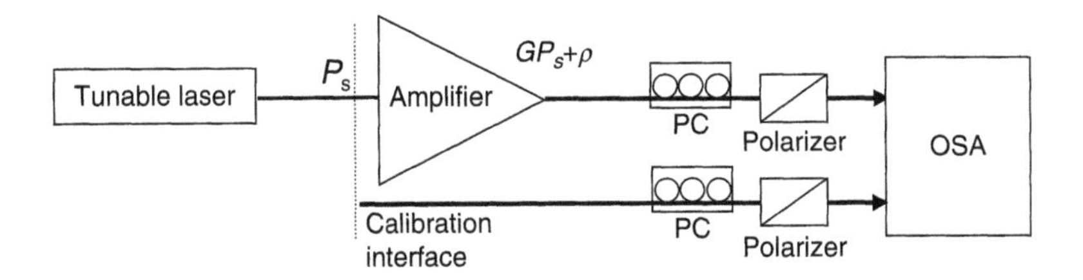

Figure 3.4.8 ASE noise measurement using an OSA and a polarizer to null the strong optical signal. *PC:* Polarization controller.

3.4.5 Impact of ASE Noise in Electrical Domain

Although the noise performance of an optical amplifier can be characterized in the optical domain as described in the last section, it can also be characterized in the electrical domain after the ASE noise is detected by a photodetector, as shown in Figure 3.4.9. In general, both the spectral resolution and the dynamic range of an electrical spectrum analyzer can be several orders of magnitude

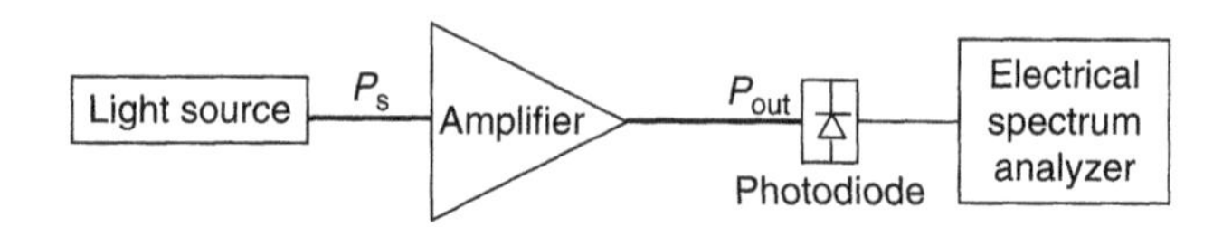

Figure 3.4.9 ASE noise measurement in an electrical domain.

better than an optical spectrum analyzer; thus an electrical-domain measurement should provide better accuracy. In addition, since in an optical system the optical signal will eventually be converted into electrical domain through a photodiode, electrical domain SNR would be a relevant parameter for many applications. Understanding the relationship between optical noise spectral density and the corresponding electrical SNR is essential in the design and performance characterization of optical receivers.

In the system shown in Figure 3.4.9, the photodiode performs square-law detection and the photocurrent is

$$I = \Re P = \Re |E_{sig} + E_{noise}|^2 \tag{3.4.13}$$

where E_{sig} and E_{noise} are the fields of the amplified optical signal and optical noise, respectively. $\Re = \eta q/hf$ is the responsivity, η is the quantum efficiency, and f is the optical frequency. In an electrical domain, in addition to thermal noise and shot noise, the ASE noise created by the optical amplifier will introduce two additional noise terms, often referred to as *signal-spontaneous emission beat noise* and *spontaneous-spontaneous beat noise.*

The origins of thermal noise and short noise were described in Section 1.2.4. In an optical system with amplifiers, the shot noise is generated by both the optical signal and the ASE noise within the receiver optical bandwidth B_0. The shot noise power spectral density (single-side-band in electrical domain) can be expressed by

$$S_i(f) = 2q\Re\langle P_{out} + B_0\rho\rangle \tag{3.4.14}$$

where ρ is the optical spectral density of the ASE noise, $P_{out} = GP_s = |E_{sig}|^2$ is the output optical power of the optical amplifier, G is the amplifier gain, and

P_s is the input signal optical power to the amplifier. The total shot noise power generated by the photodiode within a receiver bandwidth B_e is

$$P_{shot} = \int_0^{B_e} S_i(f)df = 2q\Re\langle GP_s + B_0\rho\rangle B_e \tag{3.4.15}$$

Since the average shot noise electrical power converted from the optical signal and the ASE noise is $\langle P_e\rangle = \Re^2\langle GP_s + B_0\rho\rangle^2$, the relative intensity noise (RIN) measured after the photodiode can be found as

$$RIN_e = \frac{S_i(f)}{\langle P_e\rangle} = \frac{2q\Re\langle GP_s + B_0\rho\rangle}{\Re^2\langle GP_s + B_0\rho\rangle^2} \tag{3.4.16}$$

This is the ratio between the noise power spectral density and the average signal power (converted from the optical signal). The unit of RIN_e is [Hz^{-1}]. When the optical SNR is high enough, $GP_s \gg B_0\rho_{ASE}$, the expression of RIN_e can be simplified to

$$RIN_e \approx \frac{2q}{\Re\langle GP_s\rangle} = \frac{2hf}{\eta\langle GP_s\rangle} \tag{3.4.17}$$

Note that this electrical domain relative intensity noise RIN_e is dependent on the photodiode quantum efficiency η. Sometimes it is useful to define a RIN parameter that is independent of the photodetector characteristics. Thus, it is more convenient to define an optical RIN as $RIN_o = \eta \cdot RIN_e$, and this optical RIN will be independent of the photodiode quantum efficiency η,

$$RIN_o = \frac{2hf}{\langle GP_s\rangle} \tag{3.4.18}$$

This optical RIN can also be explained as the electrical RIN while the photodiode quantum efficiency η is 100 percent.

The noise level measured in an electrical domain after photodetection consists of several components due to different mixing processes in the photodiode. In addition to the noise terms directly generated by the photodiode, as described in Chapter 1, there will be signal-spontaneous emission beat noise and spontaneous-spontaneous emission beat noise, which are unique to receivers in optically amplified systems.

3.4.5.1 Signal-Spontaneous Emission Beat Noise

Signal-spontaneous emission beat noise is generated by the mixing between the amplified optical signal and the ASE noise in the photodiode. Therefore signal-spontaneous beat noise exists only in the electrical domain after photodetection.

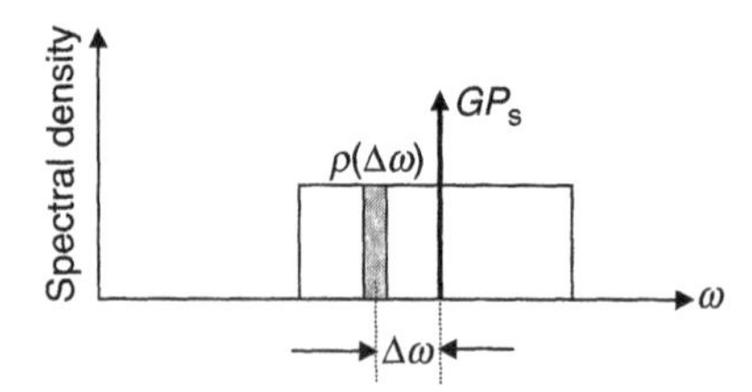

Figure 3.4.10 Illustration of signal-spontaneous beat noise generation.

As illustrated in Figure 3.4.10, at a photodiode, the amplified optical signal GP_s mixes with a bin of ASE noise with spectral density $\rho(\Delta\omega)$, which is separated from the optical signal by $\Delta\omega$ in the optical frequency domain. The electrical field at the input of the photodiode is

$$E_i(\Delta\omega) = \sqrt{GP_s}\exp(-j\omega t) + \sqrt{\rho(\Delta\omega)}\exp(-j(\omega - \Delta\omega)t) \tag{3.4.19}$$

where ω is the optical frequency of the carrier. Similar to coherent detection, the photocurrent is

$$I_i(\Delta\omega) = \Re|E_i|^2 = \Re\left\{GP_s + \rho(\Delta\omega) + 2\sqrt{GP_s\rho(\Delta\omega)}\cos(\Delta\omega t)\right\} \tag{3.4.20}$$

The first and the second terms of Equation 3.4.20 are both DC components, whereas the last term is a time-varying photocurrent at frequency $\Delta\omega$. Therefore the RF power spectral density at frequency $\Delta\omega$ can be found as

$$S_{i,DSB}(\Delta\omega) = 2\Re^2 GP_s\rho(\Delta\omega) \tag{3.4.21}$$

This is a double-sideband, RF spectral density because $\rho(\Delta\omega)$ and $\rho(-\Delta\omega)$ produce the same amount of RF beating noise; therefore the single-sideband spectral density of the signal-spontaneous beat noise is

$$S_{i,SSB}(\Delta\omega) = 4\Re^2 GP_s\rho(\Delta\omega) \tag{3.4.22}$$

Consider that the optical signal is usually polarized while the ASE noise is unpolarized; only half the ASE power can coherently mix with the optical signal. In this case, the actual signal-spontaneous emission beat noise RF spectral density is

$$S_{s-sp}(\Delta\omega) = 2\Re^2 GP_s\rho(\Delta\omega) \tag{3.4.23}$$

Taking into account the contributions of both the amplified optical signal and the ASE noise, the average power in the electrical domain after photodetection is

$$\langle P_e \rangle = \Re^2 \left[GP_s + \int_{-B_0/2}^{B_0/2} \rho(\Delta\omega) d(\Delta\omega) \right]^2 = \Re^2 [GP_s + P_{ASE}]^2 \tag{3.4.24}$$

where $P_{ASE} = \int_{-B_0/2}^{B_0/2} \rho(\Delta\omega) d(\Delta\omega)$ is the total spontaneous emission noise power within the optical bandwidth B_0. Therefore the signal-spontaneous beat noise-limited RIN is

$$RIN_{s-sp}(\Delta\omega) = \frac{S_{s-sp}(\Delta\omega)}{\langle P_e \rangle} = \frac{2GP_s\rho(\Delta\omega)}{[GP_s + P_{ASE}]^2} \tag{3.4.25}$$

3.4.5.2 Spontaneous-Spontaneous Beat Noise Spectral Density

Though signal-spontaneous emission beat noise is generated by the mixing between the optical signal and the ASE noise, spontaneous-spontaneous beat noise is generated by the beating between different frequency components of the ASE noise at the photodiode. Like signal-spontaneous beat noise, spontaneous-spontaneous beat noise exists only in the electrical domain as well. Figure 3.4.11 illustrates that the beating between the two frequency components $\rho(\omega)$ and $\rho(\omega + \Delta\omega)$ in a photodetector generates an RF component at $\Delta\omega$.

Assuming that the optical bandwidth of the ASE noise is B_0 and the ASE noise spectral density is constant within the optical bandwidth as shown in Figure 3.4.12(a), the spontaneous-spontaneous beat noise generation process can be described by a self-convolution of the ASE noise optical spectrum. The size of the shaded area is a function of the frequency separation $\Delta\omega$ and the result of correlation is a triangle function versus $\Delta\omega$, as shown in Figure 3.4.12(b).

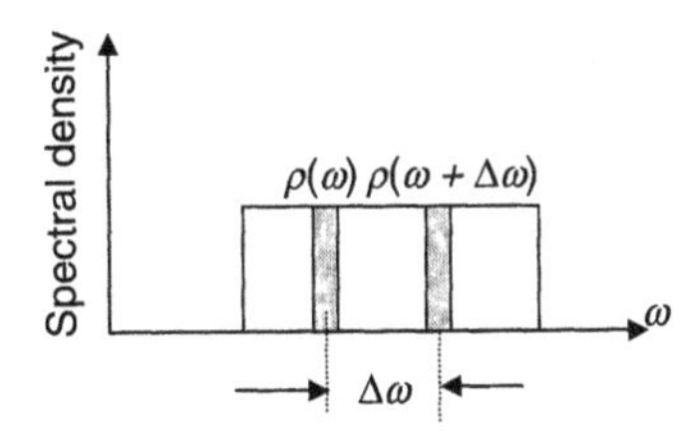

Figure 3.4.11 Illustration of spontaneous-spontaneous beat noise generation.

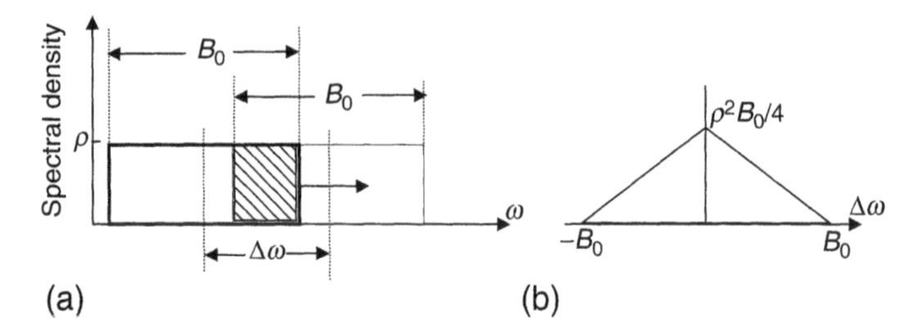

Figure 3.4.12 Illustration of spontaneous-spontaneous beat noise generation.

Another important consideration in calculating spontaneous-spontaneous beat noise is polarization. In general, the ASE noise is unpolarized, which has half energy in the TE mode and half in the TM mode, whereas optical mixing happens only between noise components in the same polarization. Therefore the single-sideband spontaneous-spontaneous beat noise electrical spectral density can be found as

$$S_{sp-sp}(\Delta\omega) = \frac{\Re^2 \rho^2}{2}(B_o - \Delta\omega) \tag{3.4.26}$$

Again, the unit of spontaneous-spontaneous beat noise is in $[A^2/Hz]$. In most cases, the electrical bandwidth is much less than the optical bandwidth ($\Delta\omega \ll B_0$); therefore Equation 3.4.26 is often expressed as

$$S_{sp-sp} \approx \Re^2 \rho^2 B_0/2 \tag{3.4.27}$$

which is a white noise within the RF receiver bandwidth. It is important to notice that in some textbooks [18], the RF spectral density of spontaneous-spontaneous beat noise is expressed as $S_{sp-sp} = 2\Re^2 \rho^2 B_0$, which seems to be four times higher than that predicted by Equation 3.4.27. The reason for this discrepancy is due to their use of single-polarized ASE noise spectral density $\rho = n_{sp}hf[G(f) - 1]$, which is half of the unpolarized ASE noise spectral density given by Equation 3.4.12.

We can also define a spontaneous-spontaneous beat noise-limited relative intensity noise as $RIN_{sp-sp}(\Delta\omega) = S_{sp-sp}(\Delta\omega)/\langle P_e \rangle$, where, $\langle P_e \rangle$ is the average RF power created by the spontaneous emission on the photodetector. For each polarization, the optical power received by the photodiode is

$$\langle P \rangle_{//} = \langle P \rangle_{\perp} = \frac{1}{2}\int_{-B_0/2}^{B_0/2} \rho(\omega) d\omega \tag{3.4.28}$$

After photodetection, the ASE noise-induced average power in the electrical domain is

$$\langle P_e \rangle = \Re^2 \left\{ \langle P \rangle_\perp^2 + \langle P \rangle_{//}^2 \right\} = \frac{1}{2} \Re^2 [\rho B_0]^2 \tag{3.4.29}$$

Therefore the spontaneous-spontaneous beat noise-limited RIN is

$$RIN_{sp-sp} = \frac{S_{sp-sp}}{\langle P_e \rangle} = \frac{1}{B_0} \tag{3.4.30}$$

The unit of RIN is [1/Hz].

3.4.6 Noise Figure Definition and Its Measurement

3.4.6.1 Noise Figure Definition

In electric amplifiers, the noise figure is defined as the ratio between the input electrical SNR and the output electrical SNR. For an optical amplifier, both the input and the output are optical signals. To use the similar noise figure definition as an electrical amplifier, photodetector have to be used both in the input and the output ports and therefore

$$F = \frac{SNR_{in}}{SNR_{out}} \tag{3.4.31}$$

where SNR_{in} and SNR_{out} are the input and the output signal-to-noise power ratios in the electrical domain, which can be measured using the setup shown in Figure 3.4.13.

It is important to note that at the input side of the optical amplifier there is essentially no optical noise. However, since there is a photodetector used to convert the input optical signal into the electrical domain, quantum noise will be introduced even if the photodetector is ideal. Therefore the SNR_{in} in the amplifier input side is determined by the shot noise generated in the photodetector. Assume that the photodetector has an electrical bandwidth B_e; the total shot

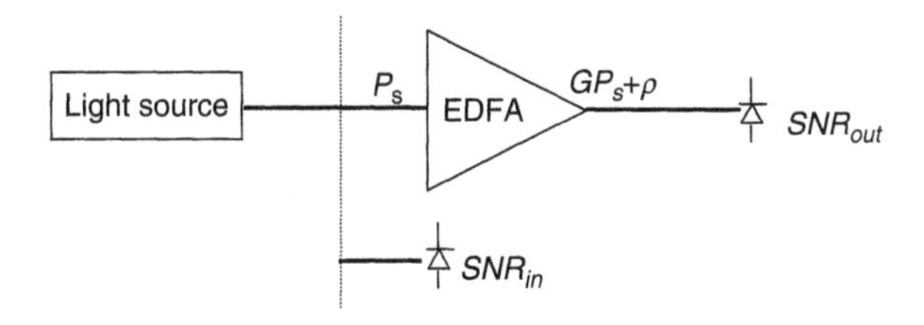

Figure 3.4.13 Measurement of optical amplifier noise figure.

noise power is $P_{shot} = 2q\Re P_s B_e$, where P_s is the input optical signal power. At the same time, the detected signal electrical power at the optical amplifier input side is $(\Re P_s)^2$, so the input SNR is

$$SNR_{in} = \frac{(\Re P_s)^2}{2q\Re P_s B_e} = \frac{\Re P_s}{2qB_e} \tag{3.4.32}$$

At the optical amplifier output, in addition to the amplified optical signal, ASE noise is also generated by the amplifier. After photodetection, the major noise sources are shot noise, signal-spontaneous beat noise, and spontaneous-spontaneous beat noise. Since the spontaneous-spontaneous beat noise can be significantly reduced by using a narrowband optical filter in front of the photodetector, in the calculation of SNR_{out}, only shot noise and signal-spontaneous beat noise are included. Considering that the signal electrical power is $(\Re GP_s)^2$, the signal-spontaneous beat noise electrical power is $S_i(f) = 4\Re^2 GP_s n_{sp} hf(G-1)B_e$, and the shot noise power is $P_{shot} = 2q\Re GP_s B_e$, one can easily find

$$SNR_{out} = \frac{\Re GP_s}{2qB_e + 4\Re n_{sp} hf(G-1)B_e} \tag{3.4.33}$$

Therefore, the noise figure as defined in Equation 3.4.31 is

$$F = \frac{1}{2q}\frac{4\Re n_{sp} hf(G-1)+2q}{G} = \frac{2n_{sp}}{\eta}\frac{(G-1)}{G} + \frac{1}{G}$$

where η is the quantum efficiency of the photodetector. Strictly speaking, the noise figure is the property of the optical amplifier, which should be independent of the property of the photodetector used for the measurement. The precise definition of the optical amplifier noise figure should include a statement that the photodetector used for SNR detection has 100 percent quantum efficiency ($\eta = 1$) and thus

$$F = 2n_{sp}\frac{(G-1)}{G} + \frac{1}{G} \tag{3.4.34}$$

In most practical applications, the amplifier optical gain is large ($G \gg 1$) and therefore the noise figure can be simplified as

$$F \approx 2n_{sp} \tag{3.4.35}$$

From the definition of n_{sp} given by Equation 3.4.10, it is obvious that $n_{sp} \geq 1$, and the minimum value of $n_{sp} = 1$ happens with complete population inversion when $N_1 = 0$ and $N_2 = N_T$. Therefore the minimum noise figure of an optical amplifier is 2, which is 3 dB. Although the definition of the noise figure is straightforward and the expression is simple, there is no simple way to predict the noise factor n_{sp} in an optical amplifier, and therefore actual measurements have to be performed.

3.4.6.2 Optical Domain Measurement of Noise Figure

Although optical amplifier noise figure is originated from an electrical domain definition given by Equation 3.4.31, its value can be evaluated either by optical domain measurements or by electrical domain measurements.

Based on Equation 3.4.8 and 3.4.34, the noise figure of an optical amplifier can be expressed as a function of the ASE noise spectral density ρ and the optical gain G of the amplifier:

$$F = \frac{\lambda\rho}{hcG} + \frac{1}{G} \tag{3.4.36}$$

Using techniques described in Sections 3.4.1 and 3.4.4, ASE noise spectral density and optical gain of an optical amplifier can be precisely characterized; therefore the noise figure of the amplifier can be obtained.

In general, both the ASE noise spectral density and the optical gain of an amplifier are functions of signal wavelength; thus the noise figure should also depend on the wavelength. In addition, the value of the noise figure depends on the signal optical power due to the effect of gain saturation. Therefore if not specially specified, the noise figure should be measured under the small-signal condition where gain saturation effect is negligible. In practice, the accuracy of optical domain measurement of a noise figure is mainly limited by the spectral resolution and the dynamic range of the OSA, especially when the input optical signal level is low.

3.4.6.3 Electrical Domain Characterization of a Noise Figure

Because of the direct relationship between the ASE noise spectral density in the optical domain and the induced electrical noise after photodetection, as we discussed in Section 3.4.5, it is possible to characterize optical amplifier noise in an electrical domain. Thanks to the excellent accuracy and frequency selectivity of RF circuits, electrical domain characterization has unique advantages in many circumstances.

Figure 3.4.14 shows the ASE characterization technique using an electrical spectrum analyzer. This allows the precise measurement of the detected electrical noise spectral densities before and after the optical amplifier. To avoid the measurement in very low RF frequencies where electrical spectrum analyzers are often inaccurate, the optical source is modulated by a sinusoid on the order of tens of Megahertz.

At the calibration interface, if the signal laser source is ideal, the electrical SNR at the photodetector should be given by Equation 3.4.32. However, in practice a signal laser is usually not ideal and its performance can be affected by its relative intensity noise RIN_{source}. This source intensity noise will generate

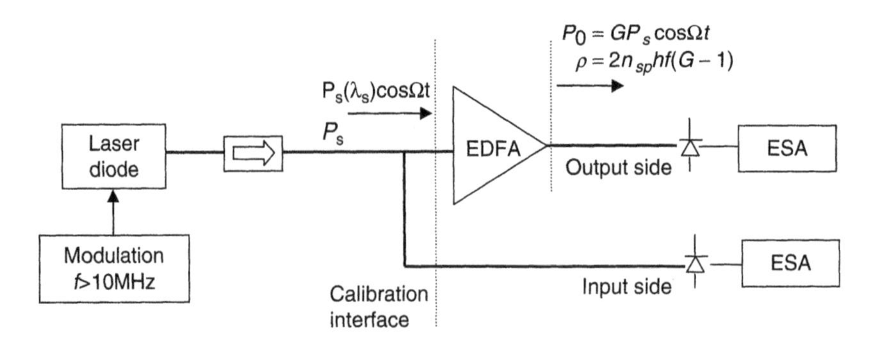

Figure 3.4.14 ASE noise characterization using an electrical spectrum analyzer (ESA).

an extra electrical noise in the photodetector, and by definition the spectral density of this additional electrical noise is

$$\rho_{source} = (\Re P_s)^2 \cdot RIN_{source} \tag{3.4.37}$$

Taking into account this source intensity noise, the *SNR* at the calibration interface becomes

$$SNR_{in} = \frac{\Re P_s}{[2q + RIN_{source}\Re P_s]B_e} \tag{3.4.38}$$

Similarly, at the output of the optical amplifier, the detected signal power in the electrical domain is $G^2(\Re P_s)^2$, the amplified noise spectral density caused by source relative intensity noise is $\rho_{source} = G^2(\Re P_s)^2 RIN_{source}$, and most important, the signal-spontaneous emission beat noise created by the optical amplifier has to be included. Thus the output SNR can be expressed as

$$SNR_{out} = \frac{G\Re P_s}{[2q + RIN_{source}G\Re P_s + 4\Re n_{sp}hf(G-1)]B_e} \tag{3.4.39}$$

Therefore, the measured noise figure is

$$F = \frac{[2q + RIN_{source}G\Re P_s + 4\Re n_{sp}hf(G-1)]}{[2q + RIN_{source}\Re P_s]G} \tag{3.4.40}$$

This equation shows the impact of the source relative intensity noise. Obviously, if $RIN_{source} = 0$, Equation 3.4.40 will be identical to Equation 3.4.34.

3.4.7 Time-Domain Characteristics of EDFA

It is well known that SOA and EDFA have very different time responses. While an SOA has a subnanosecond time constant that is often used for all-optical signal processing, an EDFA has a long time constant in the order of milliseconds, which is ideal for WDM optical systems with negligible crosstalk between channels. The time-domain response of carrier density N_2 in an optical amplifier can be expressed by a simplified first-order differential equation:

$$\frac{dN_2}{dt} = -\frac{N_2}{\tau_{eff}} \tag{3.4.41}$$

where τ_{eff} is the effective carrier lifetime, which depends on both the spontaneous emission carrier lifetime and the operation condition such as photon density within the amplifier. To switch the carrier density from the initial value N_{int} to the final value N_{fin}, the transient process can be described by the following equation:

$$N_2(t) = N_{fin} + (N_{int} - N_{fin})\exp\left(\frac{-t}{\tau_{eff}}\right) \tag{3.4.42}$$

and this transition is illustrated in Figure 3.4.15.

Instantaneous carrier density is an important parameter in an optical amplifier, which, to a large extent, determines the time-dependent optical gain of the amplifier. Although time-invariant carrier density is desired for the stable operation of an optical amplifier, the value of carrier density may be changed by variations in the pump power, signal optical power, and sudden channel add/drops in a WDM system. Figure 3.4.16 illustrates the general response of an EDFA to a square-wave optical signal. If the repetition time of the square wave is much longer than the carrier lifetime, the EDFA provides the small-signal gain only immediately after the optical signal is turned on, and at these instances the amplified output optical power is high. After that, the carrier density inside

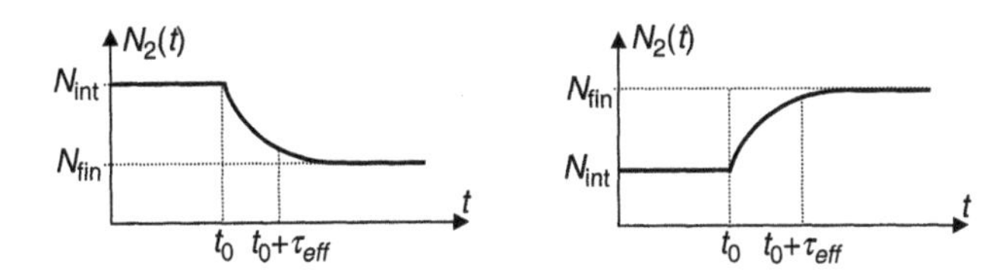

Figure 3.4.15 Carrier density transient between two values.

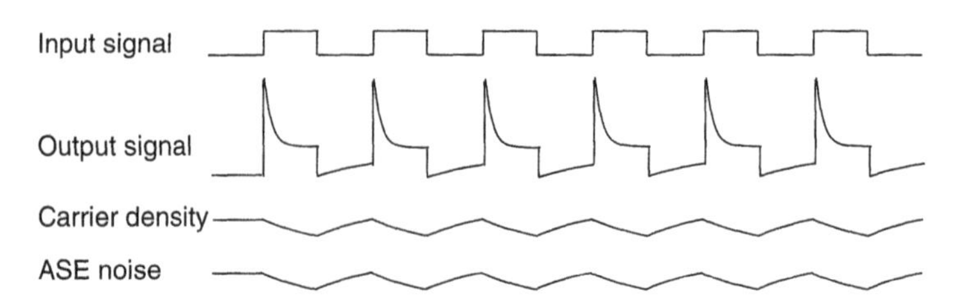

Figure 3.4.16 Illustration of EDFA response to a square-wave optical signal.

the amplifier is reduced exponentially over time due to strong stimulated recommendation, as indicated by Equation 3.4.42, and the optical gain of the amplifier is decreased. The carrier density and the optical gain vary between the small-signal values and the saturated values within each cycle. Since the saturation effect is determined by the total optical power of the signal, crosstalk may happen in WDM optical systems during the add/drop of wavelength channels.

On the other hand, the relatively slow carrier dynamics in an EDFA can be utilized to measure its noise performance using the time gating technique [20, 21]. As we discussed in Section 3.4.4, that the accuracy of optical domain characterization of EDFA is often limited by the spectral resolution and dynamic range of the OSA. It is usually difficult to measure the ASE noise spectral density at wavelengths very close to a strong optical signal, as illustrated by Figure 3.4.7.

Figure 3.4.17(a) shows the measurement setup of the time-gated technique to characterize ASE noise performance of an EDFA. The input and the output sides of the EDFA are each gated by an optical shutter with complementary on-off status, as shown in Figure 3.4.16(b). The repetition period of the gating waveform is chosen to be much shorter than the carrier lifetime of the EDFA; therefore the carrier density is determined by the average optical power that enters the EDFA. Though the optical signal from the source laser provides a useful saturation effect in the EDFA, it never reaches the OSA, because the two gating switches are always complementary. In this way, the OSA is able to measure the ASE noise spectral density of the EDFA under various operation conditions and optical signal levels, whereas the measurement accuracy is not hampered by the strong amplified optical signal on the OSA.

In practice, however, even though the carrier density fluctuation over time is small when the repetition period of the gating signal is much shorter than the carrier lifetime, this fluctuation may still impact the accuracy of the measurement to some extent. As an example, assume the carrier lifetime of the EDFA is 1 *ms* and the gating pulse repetition period is 10 μs with a 50 percent duty cycle. The maximum fluctuation of the carrier density is approximately 0.5 percent.

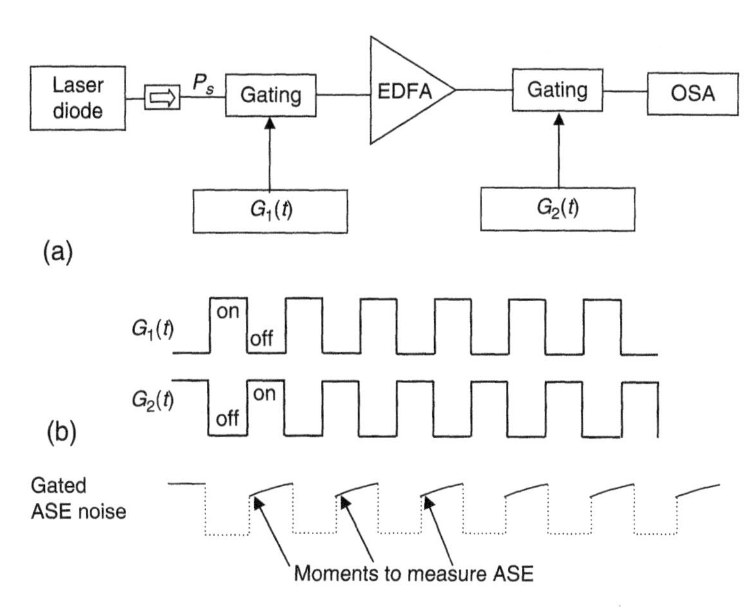

Figure 3.4.17 EDFA noise characterization using a time-gating technique: (a) test setup and (b) gating waveforms.

In terms of the accuracy requirement of optical amplifier measurement, since large numbers of in-line optical amplifiers may be used in an optical system, even a small amount of uncertainty of each amplifier parameter may mean a significant error to estimate the system performance.

3.5 CHARACTERIZATION OF PASSIVE OPTICAL COMPONENTS

Passive optical components are essential in optical systems for the manipulation and processing of optical signals. In this section, we discuss the basic properties and techniques of characterizing several often used passive optical components such as fiber-optic couplers, optical filters, WDM multiplexers and demultiplexers, and optical isolators and circulators. From an optical measurement point of view, optical components are not only the subjects of measurements, they are also fundamental building blocks of experimental setups of the measurements. A good understanding of optical components is very important in the design of optical measurement setups and the assessment of measurement accuracy.

3.5.1 Fiber-Optic Couplers

An optical fiber directional coupler is one of the most important inline fiber-optic components, often used to split and combine optical signals. For example, a fiber coupler is a key component of a fiber-based Michelson interferometer; it is also required in coherent detection optical receivers to combine the received optical signal with the local oscillator. A Mach-zehnder interferometer can be made by simply combining two fiber directional couplers. Since the power-coupling efficiency of a fused fiber directional coupler is generally wavelength dependent, it can also be used to make wavelength selective devices such as wavelength division multiplexers. On the other hand, if the coupler is used for broadband applications, the wavelength dependency has to be minimized through proper design of the coupler structure.

A fiber directional coupler is made by fusing two parallel fibers together. When the cores of the two fibers are brought together sufficiently close to each other laterally, their propagation mode fields start to overlap and the optical power can be transferred periodically between the two fibers, as illustrated in Figure 3.5.1.

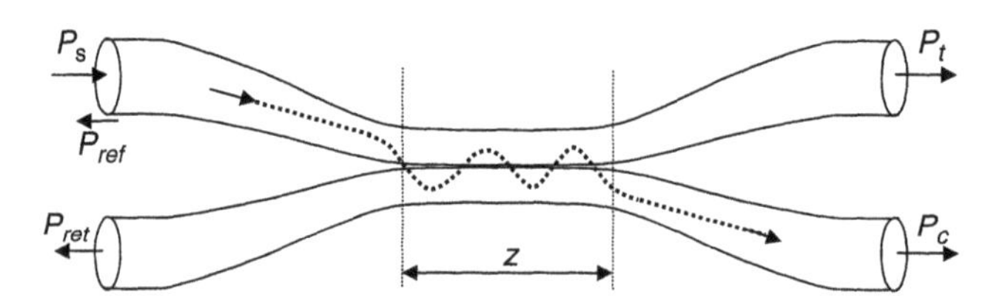

Figure 3.5.1 Illustration of a fused fiber directional coupler. z is the length of the coupling region.

Fused fiber directional couplers are easier to fabricate compared to many other optical devices, and their fabrication can be automated by online monitoring of the output optical powers. It is important to note that by modifying the fabrication process, the same fabrication equipment can also be used to make other wavelength selective optical devices, such as wavelength-division multiplexers/demultiplexers and WDM channel interleavers.

As shown in Figure 3.5.1, the power-splitting ratio of the coupler is defined as

$$\alpha = \frac{P_c}{P_t + P_c} \tag{3.5.1}$$

This is the ratio between the output of the opposite fiber P_c and the total output power. In an ideal fiber coupler, the total output power is equal to the input power P_s; therefore, $\alpha \approx P_c/P_s$. In practical fiber couplers, absorption and scattering loss always exist and add up to an excess loss:

$$\eta = \frac{P_c + P_t}{P_s} \tag{3.5.2}$$

This is another important parameter of the fiber coupler, which is a quality measure of the device. In high-quality fiber couplers, the excess loss is generally lower than 0.3 dB. From an application point of view, *insertion loss* of a fiber coupler is often used as a system design parameter. The insertion loss is defined as the transmission loss between the input and the designated output:

$$T_{c,t} = \frac{P_{c,t}}{P_s} \tag{3.5.3}$$

The insertion loss is, in fact, affected by both the splitting ratio and the excess loss of the fiber coupler, that is, $T_c = P_c/P_s = \alpha\eta$ and $T_t = P_t/P_s = (1 - \alpha)\eta$. In a 3 dB coupler, although the intrinsic loss due to power splitting is 3 dB, the actual insertion loss is generally in the vicinity of 3.3 dB.

In practical fiber couplers, due to the back scattering caused by imperfections in the fused region, there might be optical reflections back to the input ports. Directionality is defined as

$$D_r = \frac{P_{ret}}{P_s} \tag{3.5.4}$$

and reflection is defined as

$$R_{ref} = \frac{P_{ref}}{P_s} \tag{3.5.5}$$

In high-quality fiber couplers, both the directionality and the reflection are typically lower than –55 dB.

From a coupler design point of view, the coupling ratio is a periodic function of the coupling region length z as indicated in Figure 3.5.1:

$$\alpha = F^2 \sin^2\left(\frac{K}{F}(z - z_0)\right) \tag{3.5.6}$$

where $F \leq 1$ is the maximum coupling ratio, which depends on the core separation between the two fibers and the uniformity of the core diameter in the coupling region; z_0 is the length of the fused region of the fiber before stretching, and K is the parameter that determines the periodicity of the coupling ratio. An experience-based formula of K is often used for design purposes:

$$K \approx \frac{21\lambda^{5/2}}{a^{7/2}} \tag{3.5.7}$$

where a is the radius of the fiber within the fused coupling region and λ is the wavelength of the optical signal. Since the radius of the fiber reduces when the fiber is fused and stretched, the parameter K is a function of z. For example, if the radius of the fiber is a_0 before stretching, the relationship between a and z should be approximately $a \approx \sqrt{a_0^2 z_0/z}$.

Figure 3.5.2 shows the calculated splitting ratio α based on Equations 3.5.6 and 3.5.7, where $z_0 = 6$ mm, $a_0 = 62.5$ μm, $\lambda = 1550$ nm, and $F = 1$ were assumed. Obviously, any splitting ratio can be obtained by gradually stretching the length of the fused fiber section. Therefore, in the fabrication process, an *in situ* monitoring of the splitting ratio is usually used. This can be accomplished by simply launching an optical signal P_s at the input port while measuring the output powers P_t and P_c during the fusing and stretching process.

Equations 3.5.6 and 3.5.7 also indicate that for a certain fused section length z, the splitting ratio is also a function of the signal wavelength λ. This means that the splitting ratio of a fiber directional coupler is, in general, wavelength-dependent. The design of a wideband fused fiber coupler is challenging. For a commercial 3-dB fiber directional coupler, the variation of the splitting ratio is typically 0.5 dB across the telecommunication wavelength C-band. Figure 3.5.3(a) shows the calculated splitting ratio of a 3 dB fiber coupler with the same parameters as for Figure 3.5.2 except that the fused section length was chosen as $z = 12.59$ mm. In this case the variation of splitting ratio within the 1530 nm and 1565 nm wavelength band is less than ±0.2 dB.

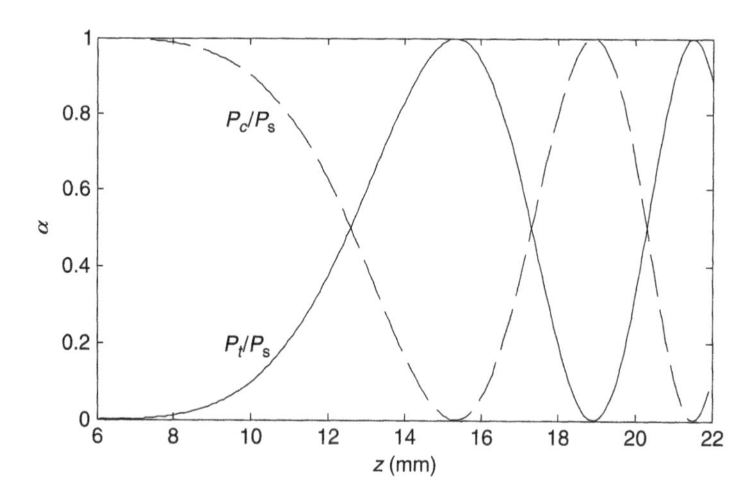

Figure 3.5.2 Calculated splitting ratio as a function of the fused section length, with $z_0 = 6$ mm, $a_0 = 62.5$ μm, and $\lambda = 1550$ nm.

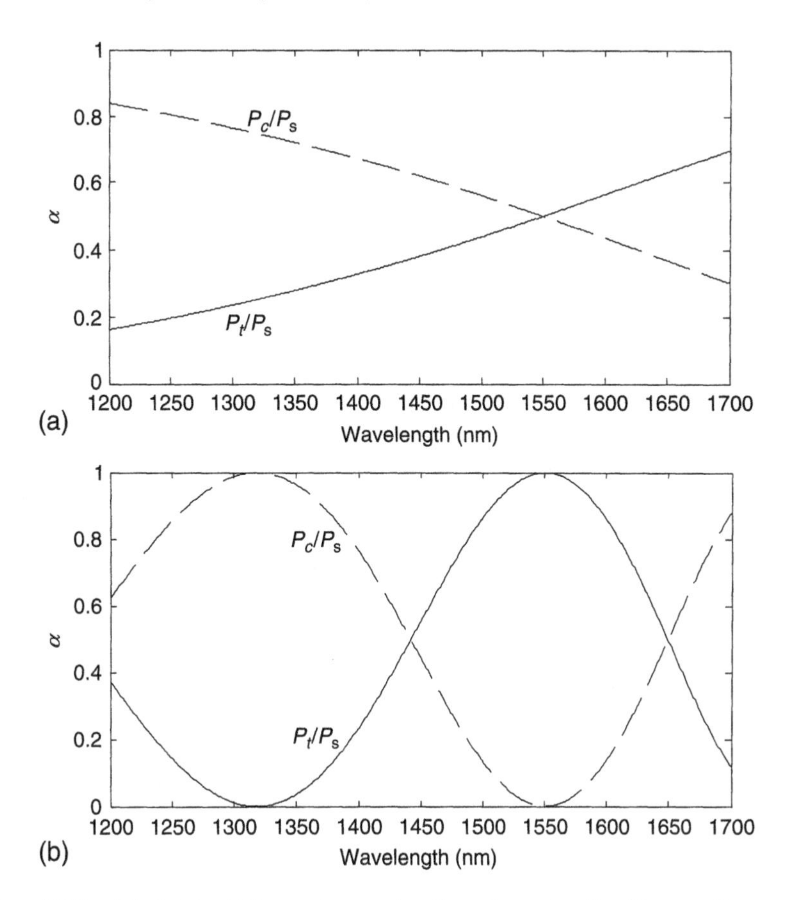

Figure 3.5.3 Calculated splitting ratio as a function of wavelength with $z_0 = 6$ mm and $a_0 = 62.5$ μm. The fused section lengths are $z =$ (a) 12.59 mm and (b) 21.5 mm.

On the other hand, the wavelength-dependent power-splitting ratio of a fiber coupler can be utilized to make a WDM multiplexer. Figure 3.5.3(b) shows the splitting ratio versus wavelength in a fiber coupler, again with the same parameters as in Figure 3.5.2. But here the fused section length was chosen as $z = 21.5$ mm, where $P_t/P_s = 100$ percent, as can be seen from Figure 3.5.2. It is interesting that in this case, the coupling ratio for 1320 nm is $P_t/P_s = 0$ percent, which means $P_c/P_s = 100$ percent. As illustrated in Figure 3.5.4, this fiber coupler can be used as a multiplexer to combine optical signals of 1550 nm and 1320 nm wavelengths or as a demultiplexer to split optical signals at 1550 nm and 1320 nm. Compared to using wavelength-independent 2×2 couplers, the wavelength-division multiplexing and demultiplexing do not suffer from intrinsic combining and splitting losses. From the measurement point of view, the

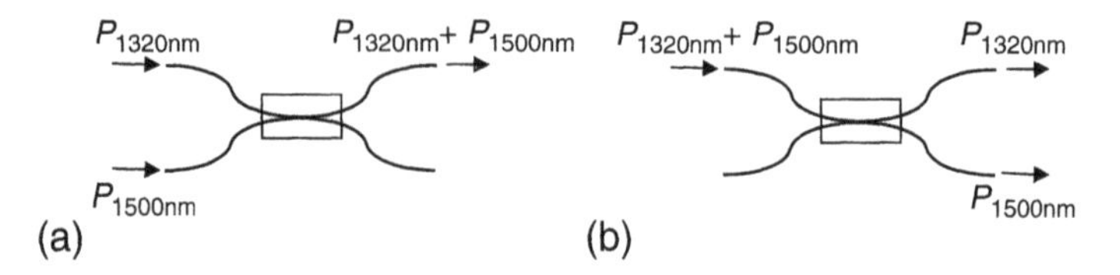

Figure 3.5.4 A wavelength-dependent fiber coupler used as (a) a mux and (b) a demux for optical signals at 1320 nm and 1550 nm wavelengths.

wavelength-dependency of power splitting ratio and excess loss of fiber directional couplers are critical for many applications; therefore these parameters have to be measured carefully before being used in the systems.

The characterization of a fiber directional coupler requires a tunable light source and optical power meters, as shown in Figure 3.5.5. In this setup, a wideband light source such as a super-luminescent LED or a tungsten light bulb can be used with a tunable optical filter to select the wavelength. A polarizer and a polarization controller select the state of polarization of the optical signal that enters the fiber coupler, which allows the measurement of polarization-dependent characteristics of the coupler. Four optical power meters, PM1, PM2, PM3, and PM4, are used to measure the coupled, the transmitted, the returned, and the reflected optical powers, respectively; therefore we can determine the splitting ratio, the insertion loss, the directionality, and the reflectivity as defined by Equations 3.5.1–3.5.5. It is important to note that the measurement setup, including the wideband light source, the polarizer, and the polarization controller are possibly wavelength-dependent, and therefore careful calibration is necessary to guarantee the measurement accuracy. Assuming that the measurement is performed in the linear regime, a simple calibration can be done by measuring the wavelength dependency and the polarization dependency of the source power P_s at the calibration interface, which is then used to correct the measured results.

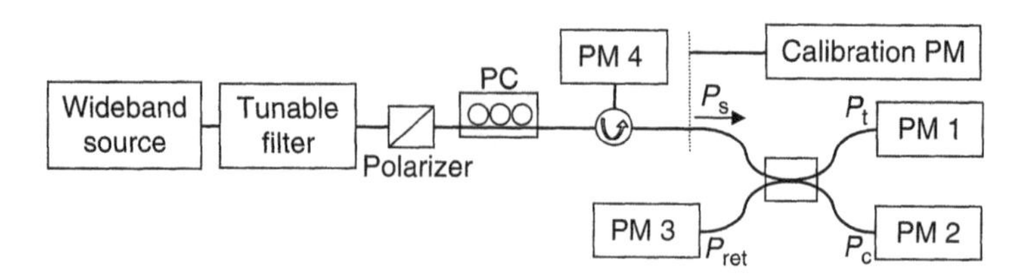

Figure 3.5.5 Measuring wavelength-dependent characteristics of a fiberdirectional coupler using a tunable wavelength source and optical power meters (PM).

3.5.2 Fiber Bragg Grating Filters

Fiber Bragg grating (FBG) is an all-fiber device which can be used to make low-cost, low-loss, and compact optical filters and demultiplexers. In an FBG, the Bragg grating is written into the fiber core to create a periodic refractive index perturbation along the axial direction of the fiber, as illustrated in Figure 3.5.6. The periodic grating can be made in various shapes, such as sinusoid, square, or triangle; however, the most important parameters are the grating period Λ, the length of the grating region L, and the strength of the index perturbation δn. Although the details of the grating shape may contribute to the higher-order harmonics, the characteristic of the fiber grating is mainly determined by the fundamental periodicity of the grating. Therefore, in the simplest case, the index profile along the longitudinal direction z, which is most relevant to the performance of the fiber grating, is

$$n(z) = n_{core} + \delta n\left\{1 + \cos\left(\frac{2\pi}{\Lambda}z\right)\right\} \tag{3.5.8}$$

where n_{core} is the refractive index of the fiber core.

The frequency selectivity of FBG is originated from the multiple reflections from the index perturbations and their coherent interference. Obviously, the highest reflection of an FBG happens when the signal wavelength matches the spatial period of the grating, $\lambda_{Bragg} = 2n_{core}\Lambda$, which is defined as the Bragg wavelength. The actual transfer function around the Bragg wavelength can be calculated using coupled mode equations. Assume a forward-propagating wave $A(z) = |A(z)|\exp(-j\Delta\beta z)$ and a backward-propagating wave $B(z) = |B(z)|\exp(j\Delta\beta z)$, where $\Delta\beta = \beta - \pi/(\Lambda n_{core})$ is the wave number detune around the Bragg wavelength. These two waves couple with each other due to the index perturbation of the grating, and the coupled-wave equations are

$$\frac{dA(z)}{dz} = -j\Delta\beta A(z) - j\kappa B(z) \tag{3.5.9a}$$

$$\frac{dB(z)}{dz} = j\Delta\beta B(z) + j\kappa A(z) \tag{3.5.9b}$$

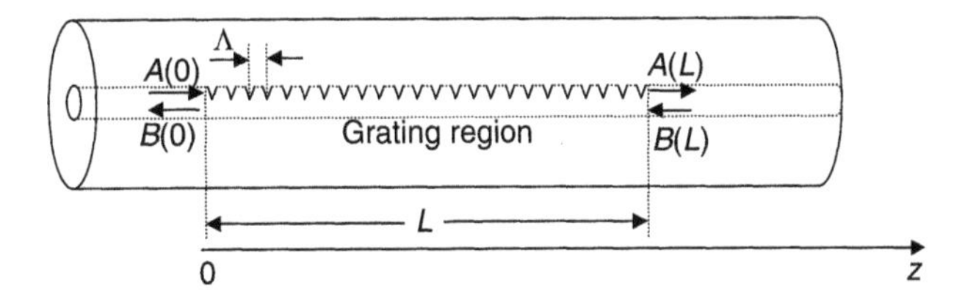

Figure 3.5.6 Illustration of fiber Bragg grating. Λ: grating period; L: grating length.

where κ is the coupling coefficient between the two waves, which is proportional to the strength of the index perturbation δn of the grating. To solve the coupled-wave Equation 3.5.9, we assume

$$A(z) = a_1 e^{-\gamma z} + a_2 e^{\gamma z} \tag{3.5.10a}$$

$$B(z) = b_1 e^{-\gamma z} + b_2 e^{\gamma z} \tag{3.5.10b}$$

where a_1, a_2, b_1, and b_2 are constants. Substituting Equation 5.10 into Equation 3.5.9, the coupled-wave equations become

$$(\Delta\beta - \gamma)a_1 = -j\kappa b_1 \tag{3.5.11a}$$

$$(\Delta\beta + \gamma)b_1 = -j\kappa a_1 \tag{3.5.11b}$$

$$(\Delta\beta + \gamma)a_2 = -j\kappa b_2 \tag{3.5.11c}$$

$$(\Delta\beta - \gamma)b_2 = -j\kappa a_1 \tag{3.5.11d}$$

To have nontrivial solutions for a_1, a_2, b_1, and b_2, we must have

$$\begin{vmatrix} \Delta\beta - \gamma & 0 & j\kappa & 0 \\ 0 & \Delta\beta + \gamma & 0 & j\kappa \\ j\kappa & 0 & \Delta\beta + \gamma & 0 \\ 0 & j\kappa & 0 & \Delta\beta - \gamma \end{vmatrix} = 0$$

This leads to $\Delta\beta^2 - \gamma^2 + \kappa^2 = 0$ and therefore

$$\gamma = \pm\sqrt{\Delta\beta^2 + \kappa^2} \tag{3.5.12}$$

Now we define a new parameter:

$$\rho = \frac{j(\Delta\beta + \gamma)}{\kappa} \equiv \frac{\kappa}{j(\Delta\beta - \gamma)} \tag{3.5.13}$$

The relationships between a_1, b_1, a_2 and b_2 in Equation 3.5.11 become $b_1 = a_1/\rho$ and $b_2 = a_2\rho$. If we know the input and the reflected fields A and B, the boundary conditions at $z = 0$ can be found as $A(0) = a_1 + a_2$ and $B(0) = a_1/\rho + a_2\rho$. Equivalently, we can rewrite the coefficients a_1 and a_2 in terms of $A(0)$ and $B(0)$ as

$$a_1 = \frac{\rho A(0) - B(0)}{\rho - 1/\rho} \tag{3.5.14a}$$

$$a_2 = \frac{A(0) - \rho B(0)}{1 - \rho^2} \tag{3.5.14b}$$

Therefore, Equation 3.5.10 can be written as

$$A(L) = \frac{\rho[B(0) - \rho A(0)]}{1 - \rho^2} e^{-\gamma L} + \frac{[A(0) + \rho B(0)]}{1 - \rho^2} e^{\gamma L}$$

$$B(L) = \frac{[B(0) - \rho A(0)]}{1 - \rho^2} e^{-\gamma L} + \frac{\rho[A(0) + \rho B(0)]}{1 - \rho^2} e^{\gamma L}$$

where L is the length of the grating region. This is equivalent to a transfer matrix expression,

$$\begin{bmatrix} A(L) \\ B(L) \end{bmatrix} = \begin{bmatrix} S_{11} & S_{12} \\ S_{21} & S_{22} \end{bmatrix} \begin{bmatrix} A(0) \\ B(0) \end{bmatrix} \tag{3.5.15}$$

where the matrix elements are

$$S_{11} = \frac{e^{\gamma L} - \rho^2 e^{-\gamma L}}{1 - \rho^2} \tag{3.5.16a}$$

$$S_{22} = \frac{e^{-\gamma L} - \rho^2 e^{\gamma L}}{1 - \rho^2} \tag{3.5.16b}$$

$$S_{12} = -S_{21} = \frac{\rho(e^{-\gamma L} - e^{\gamma L})}{1 - \rho^2} \tag{3.5.16c}$$

Since there is no backward-propagating optical signal at the fiber grating output, $B(L) = 0$, the reflection of the fiber grating can be easily found as

$$R = \frac{B(0)}{A(0)} = -\frac{S_{21}}{S_{22}} = \rho\left(\frac{e^{-2\gamma L} - 1}{e^{-2\gamma L} - \rho^2}\right) \tag{3.5.17}$$

Figure 3.5.7 shows the calculated grating reflectivity versus the frequency detune from the Bragg wavelength. Since the reflectivity is a complex function of the frequency detune, Figure 3.5.7 shows both the power reflectivity $|R|^2$ and the phase angle of R. The power reflectivity clearly shows a bandpass characteristic and the phase shift is quasi-linear near the center of the passband. The dashed line in Figure 3.5.7 shows the reflectivity in a grating with uniform coupling coefficient $\kappa = 1.5\ \mathrm{m}^{-1}$, and in this case the outband rejection ratio is only approximately 15 dB. This poor outband rejection ratio is mainly caused by the edge effect because the grating abruptly starts at $z = 0$ and suddenly stops at $z = L$. To minimize the edge effect, apodization can be used in which the coupling coefficient κ is nonuniform and is a function of z. The solid line in Figure 3.5.7 shows the reflectivity of a fiber

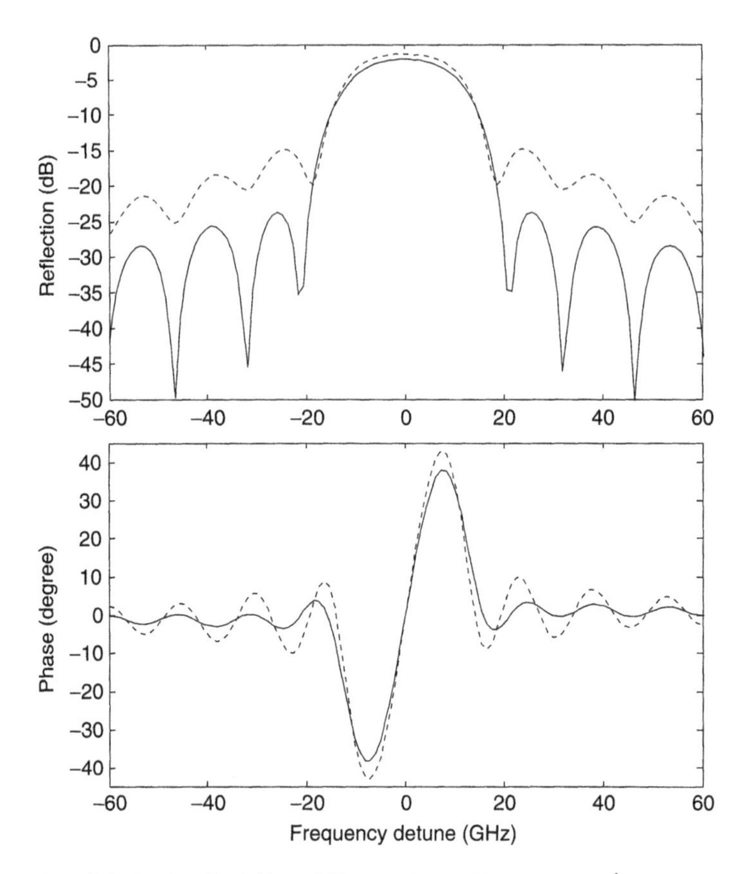

Figure 3.5.7 Calculated reflectivities of fiber gratings with $\kappa = 1.5\ \mathrm{m}^{-1}$, $L = 10$ mm, and $\lambda_{bragg} = 1557$ nm. Dashed lines: uniform grating; solid line: apodized grating.

grating with $\kappa(z) = \kappa_0 \exp\left\{-5(z - L/2)^2\right\}$, as shown in Figure 3.5.8(a), where $\kappa_0 = 1.5\ \mathrm{m}^{-1}$ is the peak coupling coefficient of the grating and $L = 10$ mm is the grating length. The apodizing obviously increases the outband rejection ratio by an additional 10 dB compared to the uniform grating. In general, more sophisticated apodization techniques utilize both z-dependent coupling coefficient $\kappa(z)$ and z-dependent grating period $\Lambda(z)$, which help to improve the FBG performance including the outband rejection, the flatness of the passband, and the phase of the transfer function [22, 23].

For nonuniform coupling coefficient $\kappa = \kappa(z)$ and grating period $\Lambda(z)$, calculation can be performed by dividing the grating region into a large number of short sections, as illustrated in Figure 3.5.8(b), and assuming both κ and Λ are constant within each short section. Therefore, the input/output relation of each section can

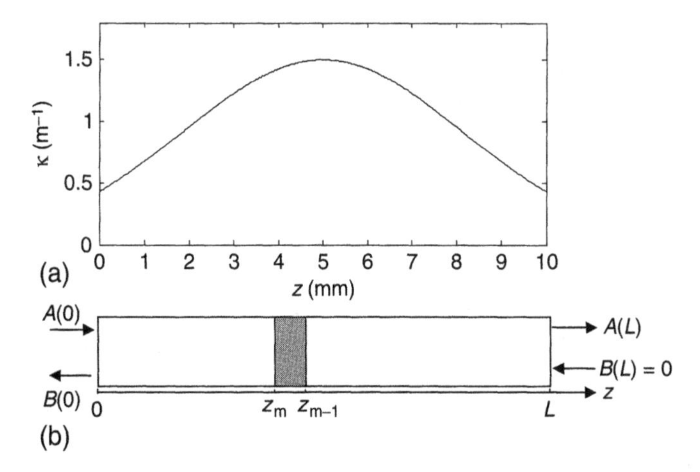

Figure 3.5.8 (a) Nonuniform coupling coefficient $\kappa(z)$ and (b) dividing FBG into short sections for calculation using transfer matrices.

be described by Equation 3.5.15. The overall grating transfer function can be found by multiplying the transfer matrices of all short sections:

$$\begin{bmatrix} A(L) \\ B(L) \end{bmatrix} = \left\{ \prod_{m=1}^{N} \begin{bmatrix} S_{11}^{(m)} & S_{12}^{(m)} \\ S_{21}^{(m)} & S_{22}^{(m)} \end{bmatrix} \right\} \begin{bmatrix} A(0) \\ B(0) \end{bmatrix} \tag{3.5.18}$$

where N is the total number of sections and $S_{i,j}^{(m)}$, $(i = 1, 2, j = 1, 2)$ are transfer matrix elements of the m-th short section that use the κ and Λ values of that particular section.

From an application point of view, the bandpass characteristic of the FBG reflection is often used for optical signal demultiplexing. Since FBG attenuation away from the Bragg wavelength is very small, many FBGs, each having a different Bragg wavelength, can be concatenated, as illustrated in Figure 3.5.9, to make multiwavelength demuxes. With special design, the phase shift introduced by an FBG can also be used for chromatic dispersion compensation in optical transmission systems. In recent years, FBGs are often used to make distributed sensors, which utilize the temperature or mechanical sensitivities of FBG

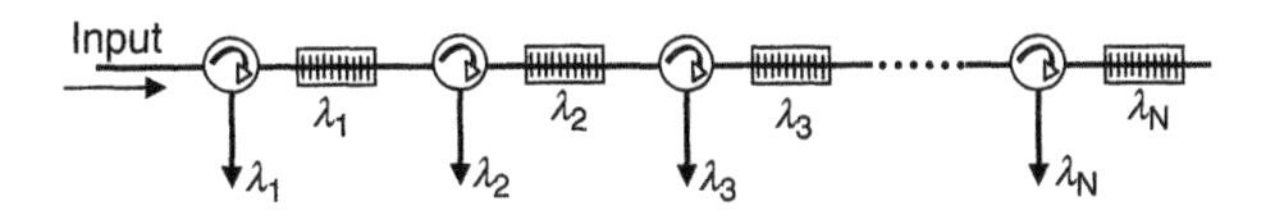

Figure 3.5.9 Configuration of a WDM demux based on FBGs.

transfer functions. Another note is that although an FBG can be made low-cost, an optical circulator has to be used to redirect the reflection from an FBG, which significantly increases the cost. Although a 3-dB fiber directional coupler can be used to replace the circulator, it will introduce a 6-dB intrinsic loss for the roundtrip of the optical signal.

3.5.3 WDM Multiplexers and Demultiplexers

In WDM optical systems, multiple wavelengths are used to carry wideband optical signals; therefore precisely filtering, multiplexing, and demultiplexing these optical channels are very important tasks. Similar to RF filters, the specification of optical filters includes bandwidth, flatness of passband, stopband rejection ratio, transition slope from passband to stopband, and phase distortion. Traditional multilayer thin film optical filters are able to provide excellent wavelength selectivity, especially with advanced thin film deposition technology; hundreds of thin film layers with precisely controlled index profiles can be deposited on a substrate. Multichannel WDM multiplexers (muxes), demultiplexers (demuxes), and wavelength add/drop couplers have been built based on the thin film technology. In terms of disadvantages, since thin film filters are free-space devices, precise collimation of optical beams is required and the difficulty becomes significant when the channel count is high. On the other hand, arrayed waveguide grating (AWG) is another configuration to make mux, demux, and add/drop couplers.

AWG is based on planar lightwave circuit (PLC) technology, in which multipath interference is utilized through multiple waveguide delay lines. PLC is an integrated optics technology that uses photolithography and etching; very complex optical circuit configuration can be made with submicrometer-level precision. WDM mux and demux with very large channel counts have been demonstrated by AWG.

3.5.3.1 Thin Film-Based Interference Filters

Figure 3.5.10(a) shows the simplest thin film fiber-optic filter in which the optical signal is collimated and passes through the optical interference film. At the output, another collimator couples the signal light beam into the output fiber. With a large number of layers of thin films deposited on transparent substrates, high-quality edge filters and bandpass filters can be made with very sharp transition edges, as illustrated in Figure 3.5.10(c). If the transmitted and reflected optical signals are separately collected by two output fibers, a simple two-port WDM coupler can be made as shown in Figure 3.5.10(b).

Figure 3.5.11(a) shows the configuration of a four-port WDM demux that uses four interference films, each having a different edge frequency of the

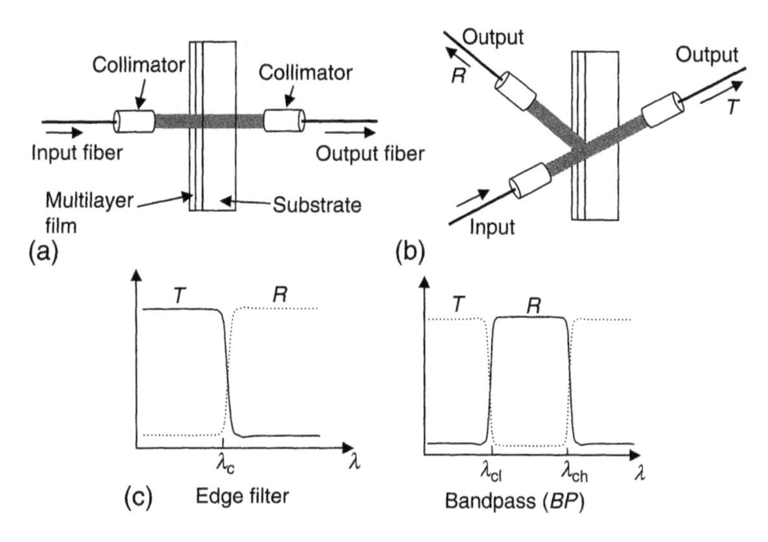

Figure 3.5.10 Thin film-based optical filters. *T*: transmission, *R*: reflection.

transmission spectrum. Figure 3.5.11(b) shows an alternative arrangement of the thin film filters to construct an eight-channel WDM demux in which only edge filter characteristics are required for the thin films. Figure 3.5.11(c) is an example of the typical transfer function of a four-port WDM demux, which shows the maximum insertion loss of about 1.5 dB, very flat response within the passband, and more than 40 dB attenuation in the stopband. It is also evident that the insertion loss increases at longer wavelengths because long wavelength channels pass through larger numbers of thin film filters.

In addition to the intensity transfer function, the optical phase of a thin-film filter is also a function of the wavelength, which creates chromatic dispersion as illustrated in Figure 3.5.12 [24]. The characterization of an optical filter has to include both intensity and phase transfer functions.

Thin-film technology has been used for many years; sophisticated structural design, material selection, and fabrication techniques have been well studied. However, thin film-based WDM muxes and demuxes are based on discrete thin film units. The required number of cascaded thin film units is equal to the number of wavelength channels, as shown in Figure 3.5.11; therefore insertion loss linearly increases with the number of ports. In addition, the optical alignment accuracy requirement becomes more stringent when the number of channels is high and thus the fabrication yield becomes low. For this reason, the channel counts of commercial thin film filter-based muxes and demuxes rarely go beyond 16. For applications involving large numbers of channels, such as 64 and 128, AWGs are more appropriate.

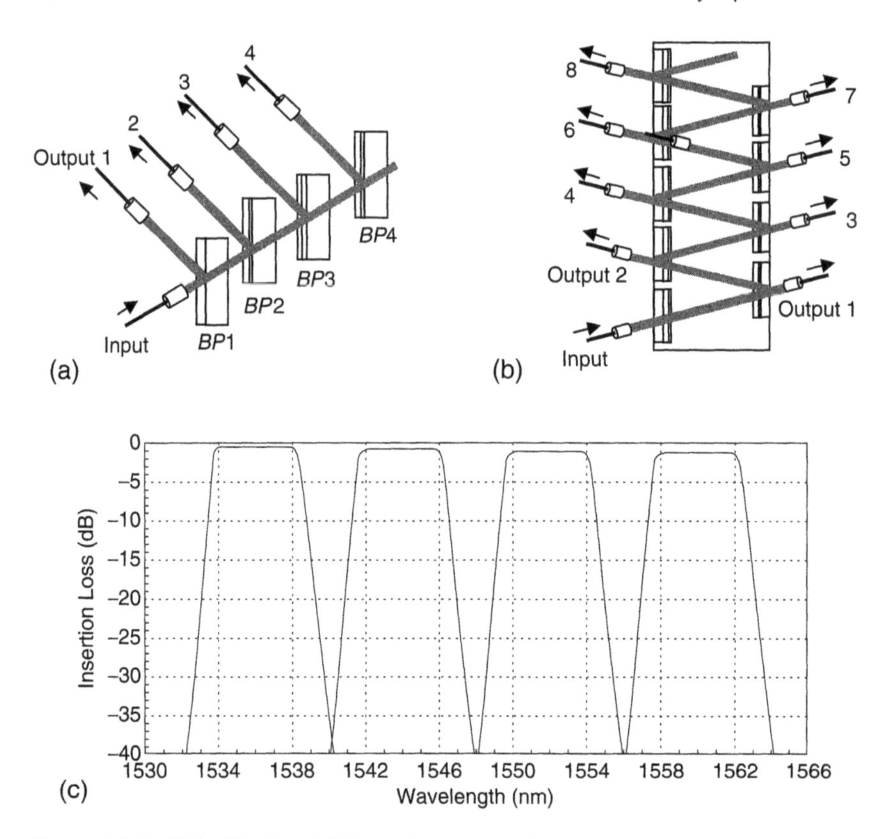

Figure 3.5.11 Thin film-based WDM demux. (a), (b) optical circuit configuration and (c) transfer function of a four-channel demux.

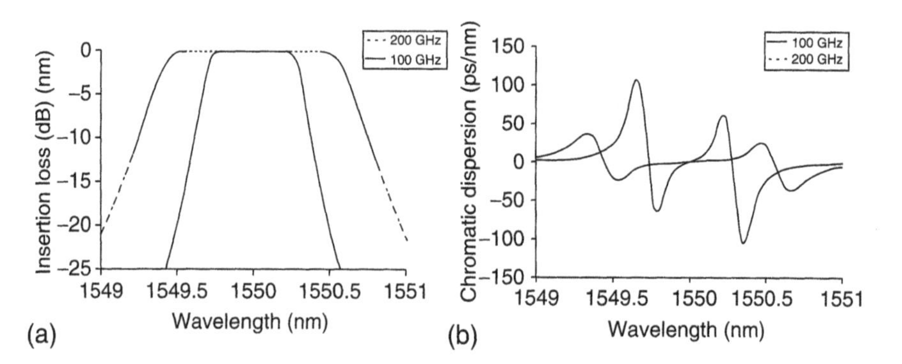

Figure 3.5.12 (a) Intensity transfer function and (b) chromatic dispersion of thin film bandpass filters [24]. Used with permission.

3.5.3.2 Arrayed Waveguide Gratings

The wavelength selectivity of an AWG is based on multipath optical interference. Unlike transmission or reflection gratings or thin film filters, an AWG is composed of integrated waveguides deposited on a planar substrate, commonly referred to as *planar lightwave circuits*. As shown in Figure 3.5.13, the basic design of an AWG consists of input and output waveguides, two star couplers, and an array of waveguides bridging the two star couplers. Within the array,

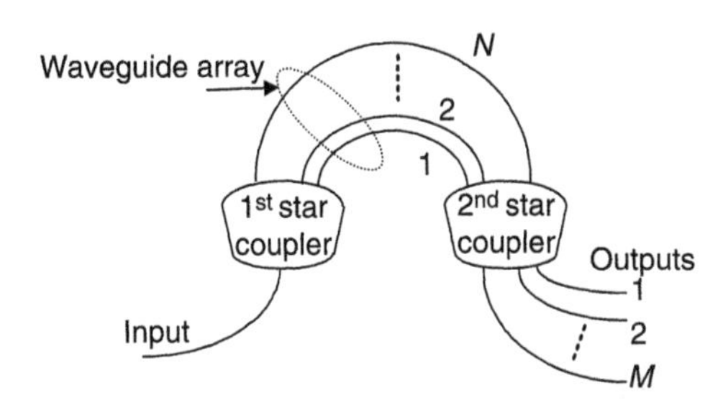

Figure 3.5.13 Illustration of an arrayed waveguide grating structure.

each waveguide has a slightly different optical length, and therefore interference happens when the signals combine at the output. The wavelength-dependent interference condition also depends on the design of the star couplers. For the second star coupler, as detailed in Figure 3.5.14, the input and the output waveguides are positioned on the opposite sides of a Roland sphere with a radius of $L_f/2$, where L_f is the focus length of the sphere. In an AWG operation, the optical signal is first distributed into all the arrayed waveguides by the input star coupler, and then at the output star coupler each wavelength component of the optical field emerging from the waveguide array is constructively added up at the entrance of an appropriate output waveguide. The phase condition of this constructive interference is determined by the following equation [25]:

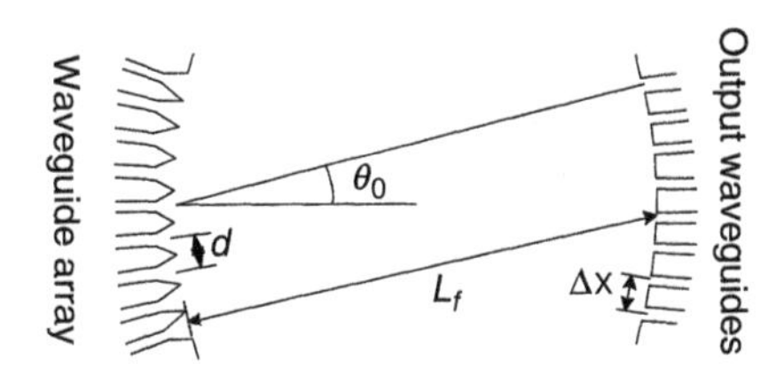

Figure 3.5.14 Configuration of the star coupler used in AWG.

$$n_c \cdot \Delta L + n_s d \sin\theta_o = m\lambda \tag{3.5.19}$$

where $\theta_0 = j \cdot \Delta x / L_f$ is the diffraction angles in the output star coupler, Δx is the separation between adjacent output waveguides, j indicates the particular output waveguide number, ΔL is the length difference between two adjacent waveguides in the waveguide array, n_s and n_c are the effective refractive indices in the star coupler and waveguides, m is the diffraction order of the grating which is an integer, and λ is the wavelength.

Obviously, the condition for a constructive interference at the center output waveguide at wavelength λ_0 is determined by the differential length of the waveguide array:

$$\lambda_0 = \frac{n_c \cdot \Delta L}{m} \tag{3.5.20}$$

On the other hand, the wavelength separation between output waveguides of an AWG depends on the angular dispersion of the star coupler. This can be easily found from Equation 3.5.19:

$$\frac{d\theta}{d\lambda} = -\frac{m}{n_s d}\frac{1}{\cos\theta_0} \tag{3.5.21}$$

When d is a constant, the dispersion is slightly higher for waveguides further away from the center. Based on this expression, the wavelength separation between adjacent output waveguides can be found as

$$\Delta\lambda = \frac{\Delta x}{L_f}\left(\frac{d\theta_0}{d\lambda}\right)^{-1} = \frac{\Delta x}{L_f}\frac{n_s d}{m}\cos\theta_0 \tag{3.5.22}$$

Because AWG is based on multi-beam interference, spectral resolution is primarily determined by the number of waveguides in the array between the two star couplers. Larger number of waveguides will provide better spectral resolution.

As an example, Figure 3.5.15 shows the calculated insertion loss spectrum of a 32-channel AWG, in which 100 grating waveguides are used between the two star couplers. In this example, the pass-band of each channel has a Gaussian-shaped transfer function. With advanced design and apodizing, flat-top transfer function can also be designed, and this type of AWGs are also commercially available.

Both thin-film filters and AWGs are often used to make mux, demux, and wavelength add/drop devices for WDM applications. To characterize these devices, important parameters include: width of the passband, adjacent channel isolation, non-adjacent channel isolation, passband amplitude and phase ripple, return loss and polarization-dependent loss.

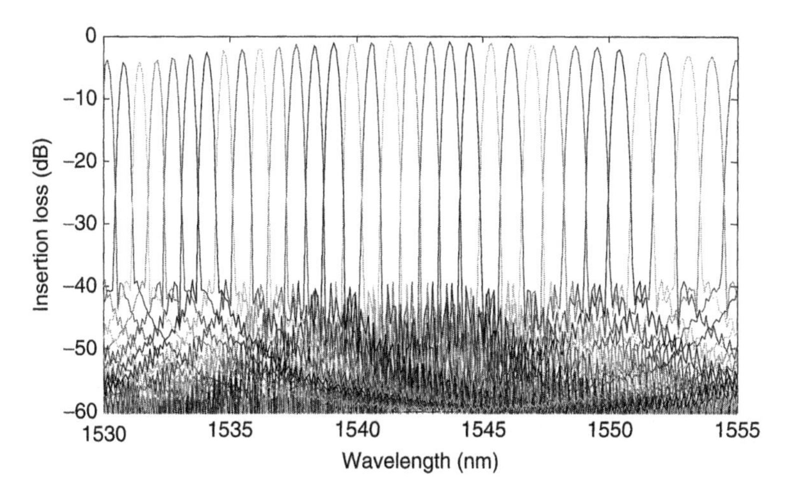

Figure 3.5.15 Calculated transfer function of an AWG with 32 wavelength channels.

3.5.4 Characterization of Optical Filter Transfer Functions

From the measurement point of view, for wavelength-dependent optical devices such as FBGs, optical filters, wavelength-division muxes and demuxes, insertion loss, intensity transfer function, wavelength-dependent phase delay, temperature coefficient as well as the sensitivity to mechanical stresses are all important parameters which have to be characterized.

Figure 3.5.16 shows a simple measurement setup which uses a wideband light source and an optical spectrum analyzer. Both the transmitted and the reflected optical spectra are measured to determine the filtering characteristics of an optical device. The source optical spectrum is also measured for calibration purpose. A polarizer and a polarization controller are used to control the state of polarization of the optical signal, and therefore to characterize the polarization dependency of the optical device performance. Although this measurement setup is

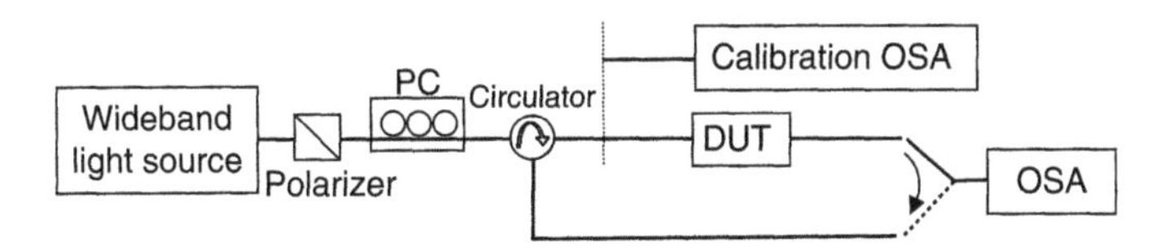

Figure 3.5.16 Measurement of wavelength-dependent characteristics of an optical filter using a wideband wavelength source and OSA.

simple, it only provides the intensity transfer function of the device under test while the differential phase delay information cannot be obtained. In addition, since most of the grating-based OSAs have spectral resolution in the order of 0.05 nm, which is about 6.25 GHz in 1550 nm wavelength window, narrowband optical filters cannot be precisely measured, which require better spectral resolution.

3.5.4.1 Modulation Phase-Shift Technique

Figure 3.5.17 shows the experimental setup known as RF phase shift technique. In this setup, a tunable laser source is used which provides a continuous wavelength scan over the interested wavelength region. The signal optical intensity is modulated by a sinusoid waveform at frequency f_m through an electro-optic modulator (EOM). The optical signal passing through the device under test is detected by a wideband photodiode and then amplified for signal processing. The received RF waveform can be measured and compared with the original modulating waveform either by using a vector voltmeter or by a RF network analyzer as shown in Figure 3.5.17(a) and (b), respectively. If you have a scalar lightwave network analyzer (which is also known as optical component analyzer) as described in Chapter 2, the measurement can be made easier because the tunable laser, the EO-modulator and the photodetection unit are already built-in and the response is calibrated.

Assume the waveform of the received RF signal is

$$I_{RF}(t) = I(\lambda_l) \cdot \cos(2\pi f_m t + \phi(\lambda_l)) \tag{3.5.23}$$

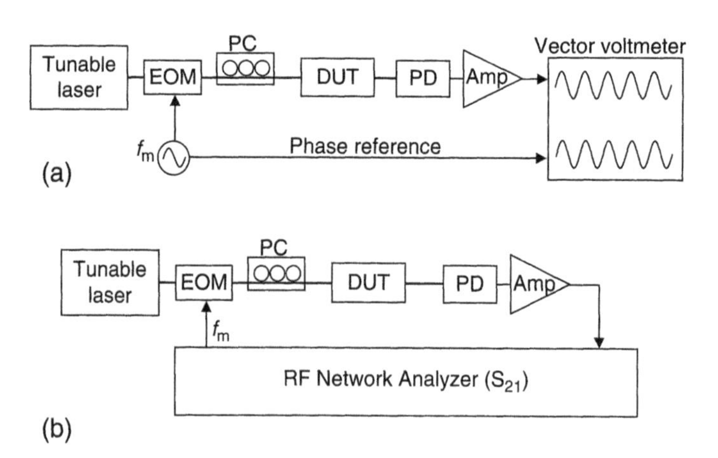

Figure 3.5.17 Experimental setup of measuring optical device characteristics with (a) modulation phase-shift technique using a vector voltmeter and (b) an RF network analyzer.

where, λ_l is the wavelength of the tunable laser, the intensity transfer function of the optical device can be obtained directly from $I(\lambda_l)$ and the group delay can be calculated from the wavelength-dependent RF phase $\phi(\lambda_l)$.

In fact, when an optical carrier at frequency $f_l = c/\lambda_l$ is intensity modulated by an RF frequency f_m, the separation between the two optical sidebands will be $2f_m$. Assume the relative optical phase shift between these two modulation sidebands is $\theta(f_l + f_m) - \theta(f_l - f_m)$, and assume the group delay does not significantly change within the modulation bandwidth $f_l \pm f_m$, the group delay in the vicinity of optical frequency f_l can be found as,

$$\tau_g(f_l) = \frac{d\theta}{d\omega} \approx \frac{\theta(f_l + f_m) - \theta(f_l - f_m)}{2(2\pi f_m)} \tag{3.5.24}$$

Then, the relative RF phase of the recovered intensity modulation waveform is proportional to the group delay and the RF frequency,

$$\phi(f_l) = 2\pi f_m \tau_g(f_l) = [\theta(f_l + f_m) - \theta(f_l - f_m)]/2 \tag{3.5.25}$$

Therefore, the wavelength-dependent group delay can be expressed as the function of the relative RF phase shift,

$$\tau_g(\lambda_l) = \frac{\phi(\lambda_l)}{2\pi f_m} \tag{3.5.26}$$

The experimental setups shown in Figure 3.5.17 are able to measure both the intensity transfer function $I(\lambda_l)$ and the RF phase delay $\phi(\lambda_l)$ as functions of wavelength set by the tunable laser. Smaller wavelength tuning step-size ensures good frequency resolution of the measurement and commercially available tunable lasers are able to provide ~125 MHz tuning step size which is approximately 1pm in 1550 nm wavelength window. In terms of measurement accuracy of RF phase shift, since a 2π RF phase shift corresponds to a group delay difference of $1/f_m$, obviously a high modulation frequency would create a larger relative phase shift for a certain group delay difference. Therefore high modulation frequency generally helps to ensure phase measurement accuracy. However, since the measured group delay $\tau_g(f_l)$ using modulation phase shift technique is an averaged value across the bandwidth of $f_l - f_m < f < f_l + f_m$, if the group delay varies significantly within this frequency window, the measured results will not be accurate.

As illustrated in Figure 3.5.18, assume that the optical phase has a sinusoidal variation [26],

$$\theta(f_l) = \Delta\theta_p \sin\left(\frac{2\pi f_l}{F_p}\right) \tag{3.5.27}$$

where $\Delta\theta_p$ is the magnitude and F_p is the period of the phase ripple. Based on Equation 3.5.24, the measured group delay would be

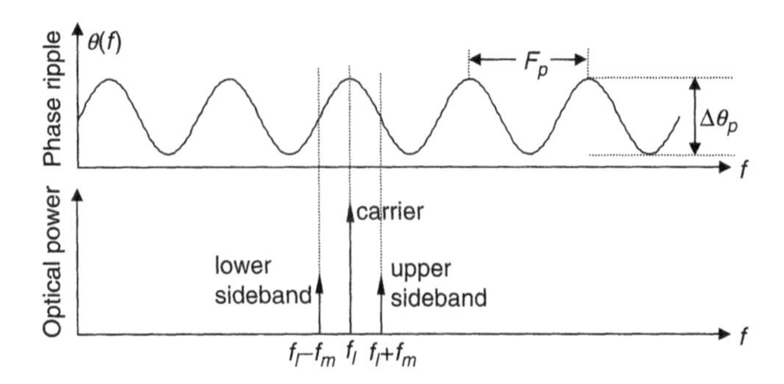

Figure 3.5.18 Illustration of phase ripple and the signal modulation sidebands.

$$\tau_g(f_l) = \Delta\theta_p \frac{\sin\lfloor 2\pi(f_l + f_m)/F_p \rfloor - \sin\lfloor 2\pi(f_l - f_m)/F_p \rfloor}{2(2\pi f_m)} = \Delta\tau_p \cdot \text{sinc}(2\pi f_m/F_p) \tag{3.5.28}$$

where $\Delta\tau_p = \Delta\theta_p F_p \cos(2\pi f_l/F_p)$ is the actual ripple magnitude of the group delay. Figure 3.5.19 shows the *sinc* function, which represents the ratio between the measured and the actual group delay $\tau_g(f_l)/\Delta\tau_p$. Obviously, the measurement is accurate only when the modulation frequency is much lower than the ripple period, $f_m \ll F_p$, so that $sinc(2\pi f_m/F_p) \approx 1$. When the modulation frequency is too high, the measured results may strongly depend on the modulation frequency.

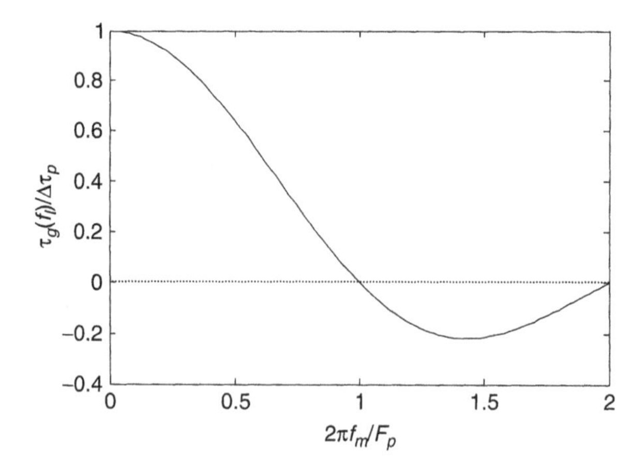

Figure 3.5.19 Measurement error introduced by modulation phase-shift technique, which is a *sinc* function, as indicated by Equation 3.5.28.

In general, if the group delay is frequency-dependent, any ripple function can be decomposed into discrete periodical functions using Fourier transform:

$$T(s) = F[\tau_g(f)] \tag{3.5.29}$$

which represents the magnitude of the group delay ripple as a function of s. Obviously s is the inverse of the ripple period F_p in optical frequency [27]. On the other hand, for each Fourier component of the group-delay ripple s, the modulation phase-shift technique introduces a measurement error according to Equation 3.5.28, which is equivalent to a filter transfer function,

$$H(s) = \text{sinc}\left[\frac{\pi s}{s_m}\right] \tag{3.5.30}$$

where the parameter $s_m = 1/(2f_m)$ is determined by the RF modulation frequency f_m. For each Fourier component of the group-delay ripple, when the modulation frequency f_m is known, the filter transfer function $H(s)$ is deterministic. Using standard signal processing algorithm, this filtering error can be corrected by using an inverse Fourier transform to obtain the correct group-delay function:

$$\tau_g(f) = F^{-1}\left[\frac{T(s)}{H(s)}\right] \tag{3.5.31}$$

The procedure is (1) measure $\tau_g(f)$ using the setup shown in Figure 3.5.17 with a certain modulation frequency f_m; (2) take the Fourier transform of $\tau_g(f)$ to find the magnitude of the group delay ripple $T(s)$, and (3) correct the measurement error caused by $H(s)$ using inverse Fourier transform shown in Equation 3.5.31. However, one difficulty of this signal processing algorithm is that for a certain ripple period s, if $\pi s/s_m$ is close to unity, the value of $H(s)$ is close to zero and Equation 3.5.31 will have a singularity that makes the signal processing invalid. To solve this problem, two modulation frequencies f_{m1} and f_{m2} can be used to measure the same device [27]. Suppose the group-delay functions measured with these two modulation frequencies are $\tau_{g1}(f)$ and $\tau_{g2}(f)$; the following equation can be used for correcting the results:

$$\tau_g(f) = F^{-1}\left\{\frac{H_1(s)T_1(s)}{H_1^2(s) + H_2^2(s)} + \frac{H_2(s)T_2(s)}{H_1^2(s) + H_2^2(s)}\right\} \tag{3.5.32}$$

where

$$T_1(s) = F(\tau_{g1}(f))$$

$$T_2(s) = F(\tau_{g2}(f))$$

$$H_1(s) = \text{sinc}(\pi s/s_{m1})$$

$$H_2(s) = \text{sinc}(\pi s/s_{m2})$$

with $s_{m1} = 1/(2f_{m1})$ and $s_{m2} = 1/(2f_{m2})$.

The weighting method given in Equation 3.5.32 for the two measurements was chosen so that the resulting summation would yield the original signal while avoiding zeros in the denominator that would cause the expression to diverge. As long as $H_1(s)$ and $H_2(s)$ do not get to zero at the same time, this correction algorithm is effective.

3.5.4.2 Interferometer Technique

The interferometer is another technique to measure the complex transfer function of optical devices. Because of the optical coherence nature of interferometers, this technique can potentially provide better phase resolution compared to noncoherent techniques. On the other hand, because the signal optical phase is used in the measurement, special attention needs to be paid to maintain the stability of the system.

Figure 3.5.20 shows a measurement setup in which the device under test (DUT) is placed in one of the two arms of a Mach-zehnder interferometer. A piezoelectric transducer (PZT) is used in the other (reference) arm to stretch the fiber and thus to control the relative optical phase delay in the arm. A wavelength-swept laser is used as the light source; a fixed-wavelength laser is also used to provide a phase reference to stabilize the measurement system. At the output the signal and the reference wavelengths are separated by a wavelength demux and detected independently.

Assume that both fiber couplers equally split the optical power (3-dB couplers); at signal wavelength, the photocurrent at the photodiode is,

$$I = I_0\left\{1 + T + 2\sqrt{T}\cos[A\sin(\omega_m t) + \varphi]\right\} \tag{3.5.33}$$

where T is the optical power attenuation of the DUT, which is generally wavelength-dependent, and I_0 is proportional to the signal optical power. Because the length of the reference arm is modulated at a frequency of ω_m by the PZT stretcher, the phase mismatch between the two arms is modulated with $A\sin(\omega_m t)$, where A is the magnitude of this phase modulation. φ is a constant

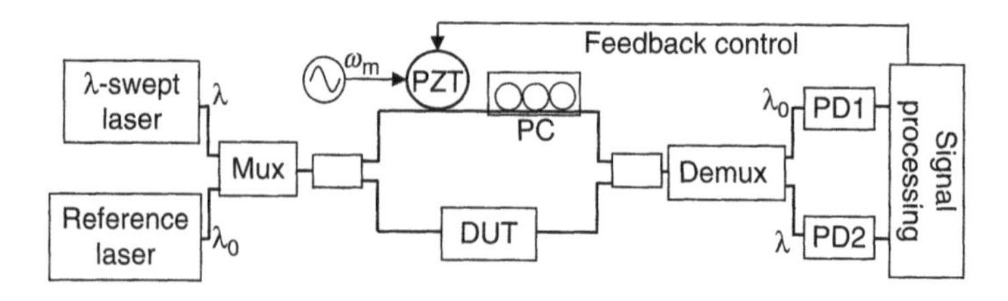

Figure 3.5.20 Measuring optical device phase shift using the interferometer technique. *PZT:* Piezo-electric transducer, *PC:* polarization controller.

phase mismatch between the two arms, which can be adjusted by a CW voltage on the PZT.

The term $\cos[A\sin(\omega_m t)]$ can be expanded in a Bessel series, and the time-varying part of Equation 3.5.33 may be written as

$$\begin{aligned} i(t) &= I_0 \cos(\varphi)[J_0(A) + 2J_2(A)\cos(2\omega_m t) + \ldots.] \\ &-I_0 \sin(\varphi)[2J_1(A)\sin(\omega_m t) + 2J_3(A)\sin(3\omega_m t) + \ldots.] \end{aligned} \tag{3.5.34}$$

At the receiver, if we only select the frequency components at ω_m and $2\omega_m$, their amplitudes are $i(\omega_m) = 2I_0 \sin(\varphi)J_1(A)$ and $i(2\omega_m) = 2I_0 \cos(\varphi)J_2(A)$, respectively. Since the phase modulation index A is a constant, one can precisely determine the constant phase mismatch φ through the measured values of $i(\omega_m)$ and $i(2\omega_m)$ [28].

Since the phase delay of the DUT is wavelength-dependent, the constant phase mismatch between the two arms is

$$\varphi(\lambda) = \frac{2\pi}{\lambda}[\Delta l + \delta l(\lambda)] \tag{3.5.35}$$

where Δl is a fixed optical length mismatch between the two arms and δl is the wavelength-dependent optical path length through the DUT. In practical systems, the optical path length difference Δl is environmentally sensitive and varies with temperature. This may be caused by both thermal expansion and the temperature-dependent refractive index of the silica material. For example, if the thermal sensitivity of the refractive index is $dn/dT = 10^{-6}$, for a one-meter fiber, each degree of temperature change will generate $\sim 1\,\mu m$ optical path length change, which is comparable to the optical wavelength and will change the optical phase mismatch significantly.

The purpose of the fixed-wavelength reference laser in Figure 3.5.20 is to help stabilize the initial phase mismatch through the feedback to control the PZT. In the reference receiver PD1, we can detect the constant phase $\varphi(\lambda_0)$ through the measurements of $i(\omega_m)$ and $i(2\omega_m)$. With the active feedback control, one can minimize the $i(\omega_m)$ component by adjusting the DC bias voltage on the PZT and therefore ensuring that $\varphi(\lambda_0) = m\pi$, where m is an integer. Then, in the signal channel, the phase shift can also be obtained through the measurements of $i(\omega_m)$ and $i(2\omega_m)$. Because of the active feedback, the phase shift is no longer sensitive to the environmental variation:

$$\varphi(\lambda) = \frac{2\pi}{\lambda}\left[\left(\frac{m\pi}{\lambda_0} - \delta l(\lambda_0)\right) + \delta l(\lambda)\right] \tag{3.5.36}$$

where the first term in the bracket is a constant, and the measured phase shift changes only with the optical path length of the DUT. The group delay caused by DUT can then be calculated as

$$\tau_g(\lambda) = -\frac{\lambda^2}{2\pi c}\frac{d\varphi(\lambda)}{d\lambda} = -\frac{\lambda}{c}\frac{d}{d\lambda}(\delta l(\lambda)) \tag{3.5.37}$$

Because this technique is based on optical interferometry, very high phase measurement accuracy can be expected. Experimentally, group delay measurement accuracy on the order of 0.1 fs has been reported in a 600–640 nm wavelength window [28].

There are a number of variations in terms of practical implementation. The biggest disadvantage of using a PZT fiber stretcher or other mechanical methods to modulate the optical path length difference is the low modulation speed and possibly the lack of long-term stability. Figure 3.5.21 shows an alternative implementation of interferometry using an acousto-optic frequency shifter (AOFS). In this implementation, the upper output port of the AOFS collects the 0th-order beam in which the optical frequency is not changed, while the lower output port of the AOFS collects the first-order Bragg diffraction and the optical frequency is shifted by ω_m, which is the RF driving frequency of the AOFS. Therefore the optical signals carried by the lower and the upper arms of the interferometer have slightly different frequencies. Each optical receiver performs a coherent heterodyne detection with the intermediate frequency ω_{IF} equal to the frequency shift ω_m of the AOFS. Assuming equal splitting of the optical power by both the AOFS and the fiber directional coupler, at each optical signal wavelength the photocurrent at the receiver is

$$I = I_0\left(1 + T + 2\sqrt{T}\cos(\omega_m t + \varphi)\right) \tag{3.5.38}$$

where T is the optical power attenuation of the DUT, which is generally wavelength-dependent, and I_0 is proportional to the signal optical power. In the measurement, the phase $\varphi = \varphi(\lambda_0)$ of the intermediate frequency resulted from

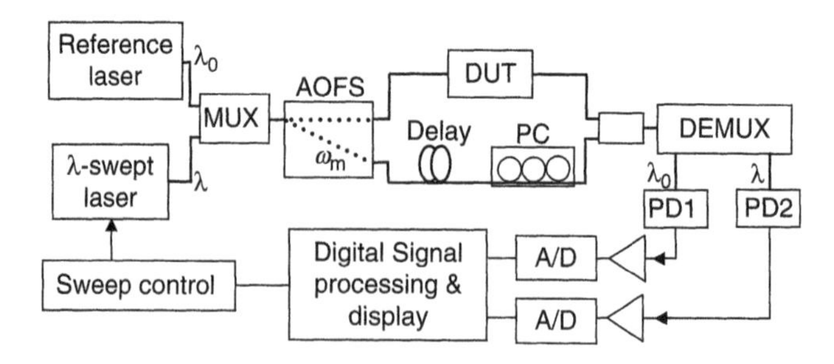

Figure 3.5.21 Real-time interferometric measurement of optical device transfer function using an acousto-optic frequency shift (AOFS).

heterodyne detection at the reference wavelength provides a phase reference. The wavelength-dependent phase $\varphi(\lambda)$ measured with the tunable laser needs to be corrected by $[\varphi(\lambda) - \varphi(\lambda_0)]$ to remove the random phase fluctuations due to environmental and temperature variations [29]. To accurately track the phase of the intermediate frequency, the linewidth of the laser source δv has to be narrow enough such that $\delta v \ll \omega_m$.

From an instrumentation point of view, the measurement accuracy can be further improved by digitizing the recovered RF signal and utilizing advanced digital signal processing (DSP) techniques for frequency locking and phase tracking. Figure 3.5.22 shows an example of the measured phase delay and intensity transmission spectrum of a narrowband fiber Bragg grating. This was measured with an interferometer technique using AOFS. The modulation frequency on the acousto-optic modulator was $\omega_m = 100$ MHz. The linewidth of the laser source was less than 1 MHz.

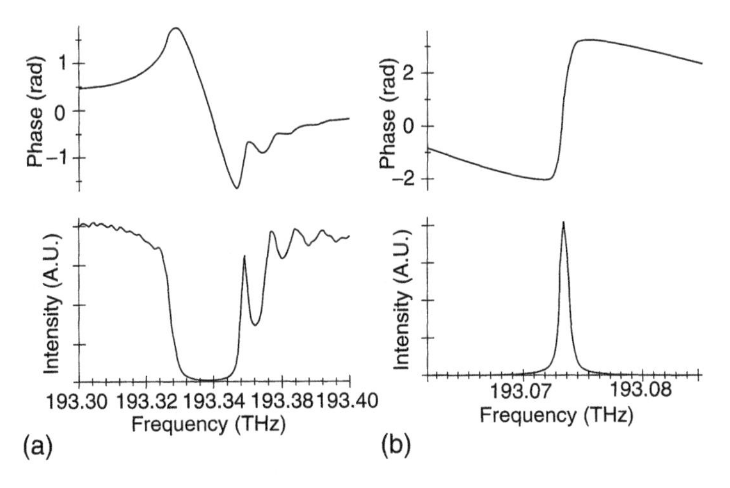

Figure 3.5.22 Examples of the measured phase and intensity transfer functions of (a) a fiber Bragg grating and (b) a thin film filter [29]. Used with permission.

3.5.5 Optical Isolators and Circulators

3.5.5.1 Optical Isolators

An optical isolator is a device that only allows unidirectional transmission of the optical signal. It is often used in optical systems to avoid unwanted optical reflections. For example, a single-frequency semiconductor laser is very sensitive to external optical feedback. Even a very low level of optical reflection from an external optical circuit, on the order of −50 dB, is sufficient to cause a significant increase in laser phase noise, intensity noise, and wavelength instability. Therefore an optical isolator is usually required at the output of each laser diode in applications that require low optical noise and stable optical frequency. Another example is in optical amplifiers, where unidirectional optical amplification is required. In this case, bidirectional optical amplification provided by the optical gain medium would cause self-oscillation if the external optical reflections from, for example, connectors and other optical components are strong enough.

The traditional optical isolator is based on a Faraday rotator sandwiched between two polarizers, as shown in Figure 3.5.23. In this configuration, the optical signal coming from the left side passes through the first polarizer whose optical axis is in the vertical direction, which matches the polarization of the input optical signal. Then a Faraday rotator rotates the polarization of the optical signal by 45° in a clockwise direction. The optical axis of the second polarizer is oriented 45° in respect to the first polarizer, which allows the optical signal to pass through with little attenuation. If there is a reflection from the optical circuit on the right side, the reflected optical signal has to pass through the Faraday rotator from right to left. Since the Faraday rotator is a nonreciprocal device, the polarization state of the reflected optical signal will rotate for an additional 45° in the same direction as the input signal, thus becoming perpendicular to the optical axis of the first polarizer. In this way the first polarizer effectively blocks the reflected optical signal and assures the unidirectional transmission of the optical isolator. For the Bi-YIG-based Faraday rotator of a certain thickness, the angle of polarization rotation versus the intensity of the applied magnetic field is generally not linear. It saturates at around 1000 Gauss. Therefore a magnetic field of >1500 Gauss will guarantee a stable rotation angle when the Bi-YIG thickness is defined. The value of isolation is, to a large extent, determined by the accuracy of the polarization rotation angle and thus by the thickness of the YIG Faraday rotator.

The Faraday rotator is the key component of an optical isolator. In long wavelength applications in 1.3 μm and 1.5 μm windows, Bismuth-substitute yttrium iron garnet (Bi-YIG) crystals are often used; they have high Verdet constant and

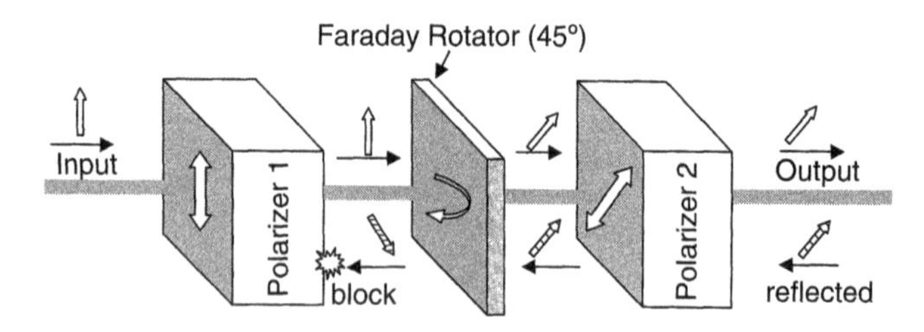

Figure 3.5.23 Optical configuration of a polarization-sensitive optical isolator.

relatively low attenuation. Packaged optical isolators with <1 dB insertion loss and >30 dB isolation are commercially available. If a higher isolation level is required, multistage isolators can be used to provide optical isolation of >50 dB.

Although this isolator configuration is simple, the polarization state of the optical signal has to match the orientation of the polarizer in the isolator. Otherwise significant optical attenuation will occur. Although this type of isolator can be used in the same package with a semiconductor laser, and therefore the SOP of the optical signal is deterministic; its polarization sensitivity is a major concern for many inline fiber-optic applications for which the SOP of the optical signal may vary. For these applications, polarization-insensitive optical isolators are required.

Figure 3.5.24 shows the configuration of a polarization-independent optical isolator made by two birefringence beam displacers and a Faraday rotator. In the 1550 nm wavelength window, a YVO_4 crystal is often used to make the birefringence beam displacers due to its high birefringence, low loss, and relatively low cost. The operation principle of this polarization-independent optical isolator can be explained using Figure 3.5.25. The incoming light beam is first split into vertical and horizontal polarized components (*o* beam and *e* beam) by the first YVO_4 beam splitter; they are shown as solid and dashed lines, respectively, in Figure 3.5.25. These two beams are separated after passing through the YVO_4 crystal. The Bi-YIG Faraday rotator rotates the polarization states of both these two beams by 45° without changing their spatial beam positions. The second YVO_4 crystal has the same thickness as the first one. By arranging the orientation of the birefringence axis of the second YVO_4 crystal, the *o* beam and *e* beam in the first YVO_4 beam displacer become the *e* beam and *o* beam, respectively, in the second YVO_4 beam displacer. Therefore the two beams converge at the end of the second YVO_4 crystal.

In the backward direction, the light beams pass through the second YVO_4 crystal along the same routes as for the forward direction. However, due to its

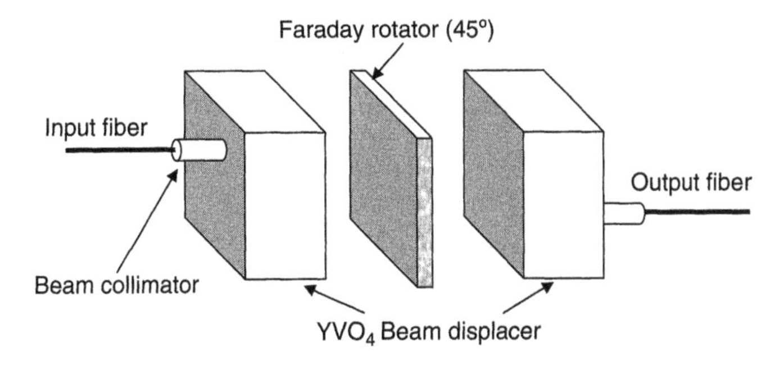

Figure 3.5.24 Configuration of polarization independent optical isolator.

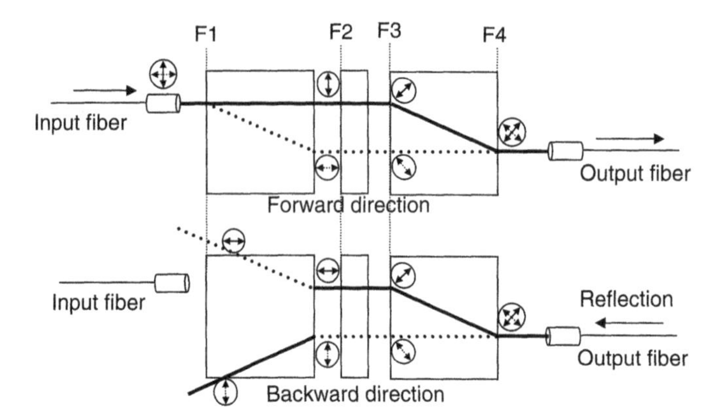

Figure 3.5.25 Illustration of the operating principle of a polarization-independent optical isolator.

nonreciprocal characteristic, the Faraday rotator rotates the polarization of the backward-propagated light beams by an additional 45° (in the same rotation direction as the forward-propagated light beams). The total polarization rotation is 90° after a roundtrip through the Faraday rotator. Therefore the initially $\perp(//)$ polarized light beam in the forward direction becomes $\perp(//)$ polarized in the backward propagation. The spatial separation between these two beams will be further increased when they pass through the first YVO_4 crystal in the backward direction and will not be captured by the optical collimator in the input side. Figure 3.5.25 also illustrates the polarization orientation of each light beam at various interfaces. In the design of an optical isolator, the thickness of YVO_4 has to be chosen such that the separation of o and e beams is larger than the beam cross-section size.

3.5.5.2 Optical Circulators

An optical circulator is another device that is based on the nonreciprocal polarization of an optical signal by Faraday effect. A basic optical circulator is a three-terminal device as illustrated in Figure 3.5.26, where terminal 1 is

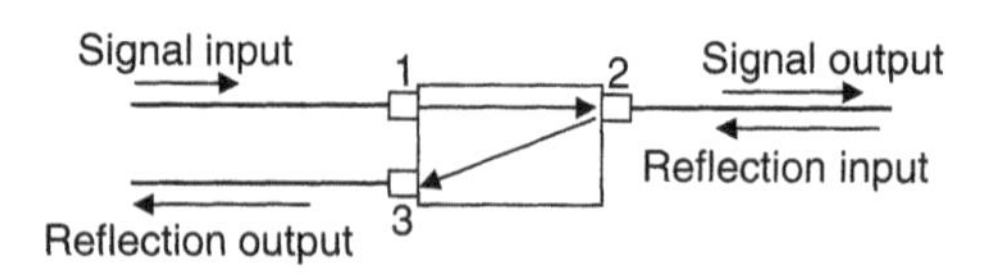

Figure 3.5.26 Basic function of a three-terminal optical circulator.

the input port and terminal 2 is the output port, while the reflected signal back into terminal 2 will be redirected to terminal 3 instead of terminal 1.

Optical circulators have many applications in optical communication systems and optical instrumentations for redirecting optical signals. One example is the use with fiber Bragg gratings, as shown in Figure 3.5.27(a). Since the reflection

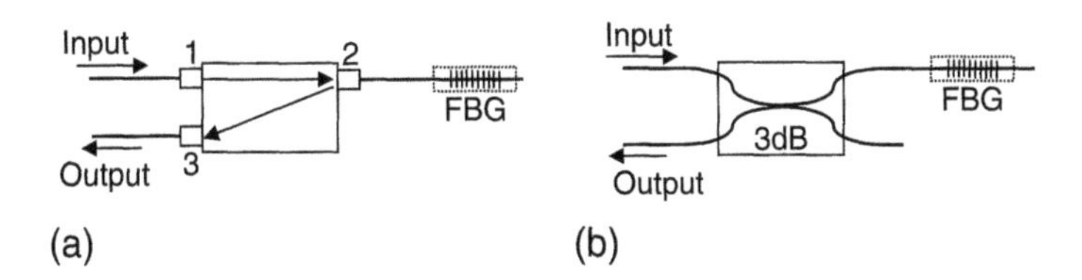

Figure 3.5.27 Redirect FBG reflection using (a) a circulator and (b) a 3-dB fiber directional coupler.

characteristic of a fiber Bragg grating can be used either as a bandpass optical filter or as a dispersion compensator, an optical circulator has to be used to redirect the reflected optical signal into the output. Although a 3-dB fiber directional coupler can also be used to accomplish this job, as shown in Figure 3.5.27(b), there will be a 6 dB intrinsic insertion loss for the optical signal going through a roundtrip in the fiber coupler.

Figure 3.5.28 illustrates the configuration of a polarization-independent optical circulator. Similar to a polarization-independent optical isolator discussed previously, an optical circulator also uses YVO_4 birefringence material as beam displacers and Bi-YIG for Faraday rotators. However, configuration of an optical circulator is obviously much more complex than an isolator because the backward-propagated light has to be collected at port 3.

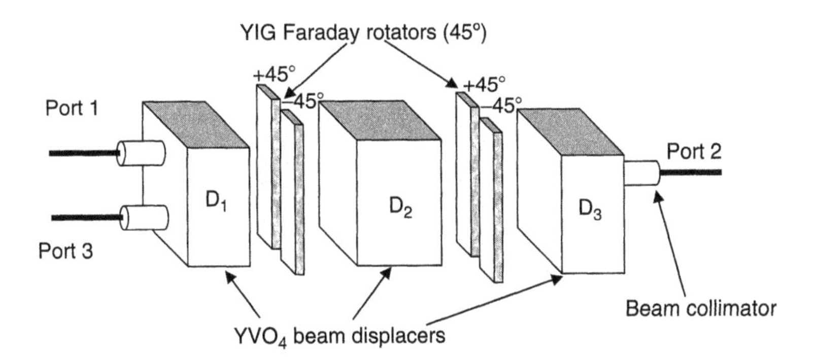

Figure 3.5.28 Configuration of a polarization-insensitive optical circulator.

The operating principle of the optical circulator can be explained using Figure 3.5.29. In the forward-propagation direction, the incoming light beam in port 1 is first split into *o* and *e* beams by the first YVO_4 beam displacer, which are shown as solid and dashed lines, respectively. The polarization state is also labeled near each light beam. These two beams are separated in the horizontal direction after passing through the first YVO_4 displacer (D_1) and then they pass through a pair of separate Bi-YIG Faraday rotators. The left Faraday rotator (a_1) rotates the *o* beam by +45° and the right Faraday rotator (b_1) rotates the *e* beam by –45° without shifting their beam spatial positions. In fact, after passing through the first pair of Faraday rotators, the two beams become copolarized and they are both *o* beams in the second YVO_4 displacer (D_2). Since these two separate beams now have the same polarization state, they will pass the second displacer D_2 without further divergence. At the second set of Faraday rotators, the left beam will rotate an additional +45° at (a_2) and the right beam will rotate an additional –45° at (b_2), then their polarization states become orthogonal with each other. The third beam displacer D_3 will then combine these two separate beams into one at the output, which reconstructs the input optical signal but with a 90° polarization rotation.

For the reflected optical signal into port 2, which propagates in the backward direction, the light beams pass through D_3, creating a beam separation. However, because of the nonreciprocal characteristic, the Faraday rotator a_2 rotates the reflected beam by +45° and b_2 rotates the reflected beam by –45°, all in the same

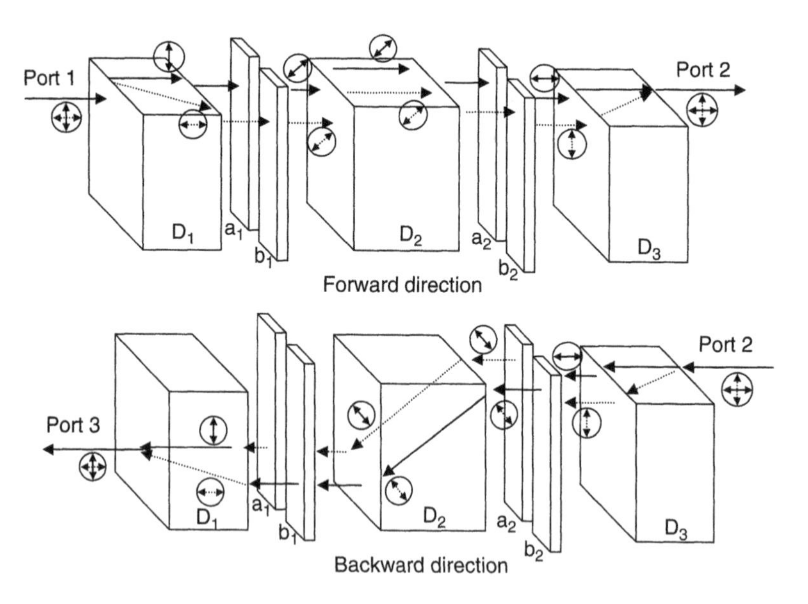

Figure 3.5.29 Illustration of the operating principle of an optical circulator.

direction as rotating the forward-propagated light beams. The total polarization rotation is then 90° after a roundtrip through the second pair of Faraday rotators. Therefore, in the second beam displacer D_2, the backward-propagated beams are again copolarized, but their polarization orientations are both *e* beams. (Recall that the two beams were both *o* beams in the forward-propagating direction.) Because of this polarization rotation, the backward-propagated beams will not follow the same routes as the forward-propagated beams in the second beam displacer D_2, as shown in Figure 3.5.29. After passing through the first set of Faraday rotators and the first beam displacer D_1, the backward-propagated beams are eventually recombined at port 3 at the input side, which is in a different spatial location than port 1.

From a measurement point of view, since both optical isolators and circulators are wideband optical devices, the measurements do not usually require high spectral resolution. However, these devices have very stringent specifications on optical attenuation, isolation, and polarization-dependent loss, so accurate measurements of these parameters are especially important.

For an optical isolator, the important parameters include isolation, insertion loss, polarization-dependent loss (PDL), polarization-mode dispersion (PMD), and return loss. The insertion loss is defined as the output power divided by the input power, whereas isolation is defined as the reflected power divided by the input power when the output is connected to a total reflector. Return loss is defined as the reflected power divided by the input when the output side of the isolator has no reflection; thus return loss is the measure of the reflection caused by the isolator itself. For a single stage isolator within a ±15 nm window around 1550 nm wavelength, the insertion loss is around 0.5 dB, the isolation should be about –30 dB, and the PDL should be less than 0.05 dB. With good antireflection coating, the return loss of an isolator is on the order of –60 dB.

Figure 3.5.30 shows a measurement setup for characterizing optical isolator insertion loss and isolation. In this setup, a tunable laser is used as the light source; its output is intensity modulated by an electro-optic modulator (EOM). The purpose of this intensity modulation is to facilitate the use of a lock-in amplifier in the receiver, so the modulation frequency f_m is only in the kilohertz range. Some commercially available tunable lasers have the built-in function of intensity modulation up to Megahertz frequency. In that case the external EOM is no longer necessary. A polarization controller at the laser output makes it possible to measure the polarization-dependent effects of insertion loss and isolation.

Figure 3.5.30(a) shows the setup to measure insertion loss in which the photodiode and the lock-in amplifier are placed at the output of the isolator. The insertion loss is defined as the lowest value of P_2/P_1 over all input signal state of polarization (SOP) and wavelengths within the specified wavelength window. To measure the isolation, a total reflector is connected to the output side of the

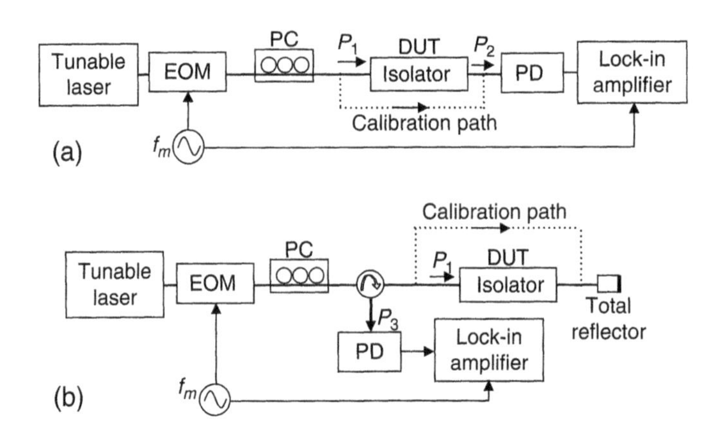

Figure 3.5.30 Measuring (a) optical isolator insertion loss and (b) isolation.

isolator and an optical circulator is used to redirect the reflected optical power as shown in Figure 3.5.30(b). Then the isolation is obtained as the highest value of P_3/P_1 over all signal SOP and wavelength. Polarization-dependent and wavelength-dependent parameters can also be evaluated with this setup. The measurement can be simplified by replacing the tunable laser with a wideband light source and replacing the photodiode and lock-in amplifier with an OSA. However, the measurement accuracy would not be as good as using a tunable laser, because the power spectral density of a wideband light source is generally low and the detection sensitivity of an OSA is limited.

For an optical circulator, important specifications also include insertion loss, isolation, PDL, and return loss. In addition, since a circulator has more than two terminals, directionality is also an important measure. For a three-terminal circulator, as illustrated in Figure 3.5.26, the insertion loss includes the losses from port 1 to port 2 and from port 2 to port 3. Likewise, the isolation also includes the isolation from port 2 to port 1 and from port 3 to port 2. The directionality is defined by the loss from port 1 to port 3 when port 2 is terminated without reflection.

Figure 3.5.31 shows the measurement setups to characterize a three-port isolator. In Figure 3.5.31(a), the insertion losses are measured separately for $P_{2,out}/P_{1,in}$ and $P_{3,out}/P_{2,in}$. During the measurement, the polarization controller adjusts the SOP of the input optical signal and the tunable laser adjusts the wavelength to find the worst-case insertion loss values within the specified wavelength window. Figure 3.5.31(b) shows how to measure isolation. To measure isolation between ports 1 and 2, the input signal is connected to port 1 and a total reflector is linked to port 2. Then the isolation is $I_{1,2} = P_r/P_{1,in}$. Similarly, to measure isolation between ports 2 and 3, the input signal is connected

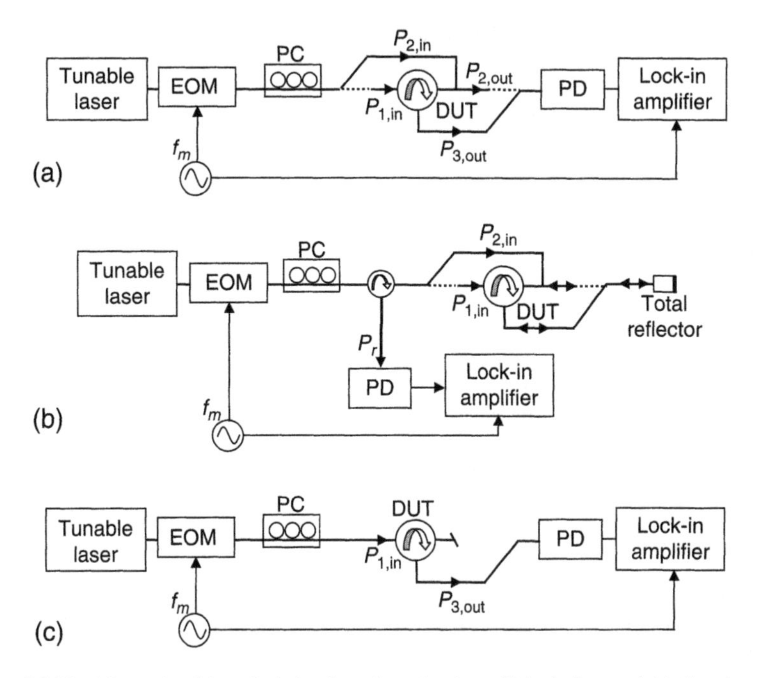

Figure 3.5.31 Measuring (a) optical circulator insertion loss, (b) isolation, and (c) directionality.

to port 2 and a total reflector is linked to port 3 and $I_{2,3} = P_r/P_{2,in}$. To measure the directionality of an optical circulator from port 1 to port 3, port 2 has to be terminated without reflection. This can be easily done by connecting port 2 to an open-ended angled connector such as an angle-polished connector (APC). A typical circulator operating in a 1550 nm wavelength window usually has insertion loss of about 0.8 dB, which is slightly higher than an isolator due to its more complicated optical structure. The isolation value is on the order of –40 dB and the directionality is usually better than –50 dB.

REFERENCES

[1] G. P. Agrawal, N. K. Dutta, *Long wavelength semiconductor lasers*, Van Nostrand Reinhold Co. Inc., New York, NY, 1986.

[2] J. L. Gimlett and N. K. Cheung, "Effects of phase-to-intensity noise conversion by multiple reflections in Gigabit-per-second DFB laser transmission systems," *J. Lightwave Technol.*, Vol. 7, pp. 888–895, 1989.

[3] R. W. Tkach and A. R. Chraplyvy, "Regimes of feedback effects in 1.5-μm distributed feedback lasers," *J. Lightwave Technol.* LT-4, pp. 1655–1661 (1986).

[4] J. Wang and K. Petermann, "Small signal analysis for dispersive optical fiber communication systems," *J. Lightwave Technol.*, Vol. 10, No. 1, pp. 96–100, 1992.
[5] C. Henry, "Phase noise in semiconductor lasers," *J. Lightwave Technol.* LT-4, No. 3, pp. 298–311 (1986).
[6] C. Henry, "Theory of the phase noise and power spectrum of a single mode injection laser," *IEEE Journal of Quantum Electronics,* Vol. 19, No. 9, pp. 1391–1397, 1983.
[7] M. Nazarathy, W. V. Sorin, D. M. Baney, and S. Newton, "Spectral Analysis of Optical Mixing Measurements," *J. Lightwave Technology*, Vol. 7, No. 7, pp. 1083–1096, 1989.
[8] T. Okoshi, K. Kikuchi, and A. Nakayama, "Novel method for high-resolution measurement of laser output spectrum," *Electron. Lett.* 16, 630–631 (1980).
[9] P. Horak and W. H. Loh, "On the delayed self-heterodyne interferometric technique for determining the linewidth of fiber lasers," *Opt. Express* 14, pp. 3923–3928, 2006.
[10] R. S. Vodhanel, A. F. Elrefaie, R. E. Wagner, M. Z. Iqbal, J. L. Gimlett, and S. Tsuji, "Ten-to-Twenty Gigabit-per-Second Modulation Performance of 1.5-pm Distributed Feedback Lasers for Frequency-Shift-Keying Systems," *J. Lightwave Technology*, Vol. 7, No. 10, pp. 1454–1460, 1989.
[11] J. Helms, "Intermodulation and harmonic distortions of laser diodes with optical feedback," *J. Lightw. Technol.,* Vol. 9, No. 11, pp. 1567–1575, Nov. 1991.
[12] F. Devaux, Y. Sorel, and J. F. Kerdiles, "Simple Measurement of Fiber Dispersion and of Chirp Parameter of Intensity Modulated Light Emitter," *J. Lightwave Tech.*, Vol. 11, No. 12, pp. 1937–1940 (1993).
[13] U. Kruger and K. Kruger, "Simultaneous Measurement of the Linewidth, Linewidth Enhancement Factor *a,* and FM and AM Response of a Semiconductor Laser," *J. Lightw. Technol.,* Vol. 13, No. 4, pp. 592–597, Nov. 1995.
[14] R. A. Saunders, J. P. King, and O. Hardcastle, "Wideband chirp measurement technique for high bit rate sources," *Electronics Letters*, Vol. 30, No. 16, pp. 1336–1338, 1994.
[15] S. Kawanishi, A. Takada, and M. Salruwatari, "Wideband frequency-response measurement of optical receivers using optical heterodyne detection," *J. Lightwave Technol.*, Vol. 7, pp. 92–98, 1989.
[16] E. Eichen, J. Schlafer, W. Rideout, and J. McCabe, "Wide-bandwidth receiver/photodetector frequency response measurements using amplified spontaneous emission from a semiconductor optical amplifier," *J. Lightwave Technol,* Vol. 8, pp. 912–916, 1990.
[17] D. M. Baney, W. V. Sorin, and S. A. Newton, "High-Frequency Photodiode Characterization Using a Filtered Intensity Noise Technique," *IEEE Photonics Technology Letters*, Vol. 6, pp. 1258–1260, 1994.
[18] E. L. Goldstein, V. da Silva, L. Eskildsen, M. Andrejco, and Y. Silberberg, "Inhomogeneously broadened fiber-amplifier cascade for wavelength-multiplexed systems," *IEEE Photonics Technology Letters*, Vol. 5, pp. 543–545, May 1993.
[19] Gerd Keiser, *Optical Fiber Communications*, 3rd ed., McGraw-Hill, 2000.
[20] D. M. Baney and J. Dupre, "Pulsed-source technique for optical amplifier noise figure measurement," European Conference on Optical Communications, ECOC 92, pp. 509–512, 1992.
[21] D. M. Baney, "Characterization of Erbium-doped Fiber Amplifiers," in *Fiber-optic Test and Measurement*, Dennis Derickson, ed., Prentice Hall, 1998.
[22] S. J. Mihailov, F. Bilodeau, K. O. Hill, D. C. Johnson, J. Albert, and A. S. Holmes, "Apodization Technique for Fiber Grating Fabrication with a Halftone Transmission Amplitude Mask," *Appl. Opt.* 39, pp. 3670–3677 (2000).
[23] Turan Erdogan, "Fiber Grating Spectra," *Journal of Lightwave Technology*, Vol. 15, No. 8, pp. 1277–1294, 1997.

[24] Keqi Zhang, Jue Wang, Erik Schwendeman, David Dawson-Elli, Ralf Faber, and Robert Sharps, "Group delay and chromatic dispersion of thin-film-based, narrow bandpass filters used in dense wavelength-division-multiplexed systems," *Applied Optics*, Vol. 41, No. 16, pp. 3172–3175, 2002.

[25] H. Takahashi, K. Oda, H. Toba, Y. Inoue, "Transmission Characteristics of Arrayed Waveguide $N \times N$ Wavelength Multiplexer," *J. of Lightwave, Technol.*, Vol. 13, pp. 447–455, March 1995.

[26] Tapio Niemi, Maria Uusimaa, and Hanne Ludvigsen, "Limitations of Phase-Shift Method in Measuring Dense Group Delay Ripple of Fiber Bragg Gratings," *IEEE Photonics Technology Letters*, Vol. 13, No. 12, pp. 1334–1336, 2001.

[27] Rance Fortenberry, Wayne V. Sorin, and Paul Hernday, "Improvement of Group Delay Measurement Accuracy Using a Two-Frequency Modulation Phase-Shift Method," *IEEE Photonics Technology Letters*, Vol. 15, No. 5, pp. 736–738, 2003.

[28] M. Beck and I. A. Walmsley, "Measurement of group delay with high temporal and spectral resolution," *Optics Letters*, Vol. 15, No. 9, pp. 492–494, 1990.

[29] Kensuke Ogawa, "Characterization of chromatic dispersion of optical filters by high-stability real-time spectral interferometry," *Applied Optics*, Vol. 45, No. 26, pp. 6718–6722, 2006.

Chapter 4

Optical Fiber Measurement

4.0. Introduction
4.1. Classification of Fiber Types
4.1.1. Standard Optical Fibers for Transmission
4.1.2. Specialty Optical Fibers
4.2. Measurement of Fiber Mode-Field Distribution
4.2.1. Near-Field, Far-Field, and Mode-Field Diameter
4.2.2. Far-Field Measurement Techniques
4.2.3. Near-Field Measurement Techniques
4.3. Fiber Attenuation Measurement and OTDR
4.3.1. Cutback Technique
4.3.2. Optical Time-Domain Reflectometers
4.3.3. Improvement Considerations of OTDR
4.4. Fiber Dispersion Measurements
4.4.1. Intermodal Dispersion and Its Measurement
4.4.2. Chromatic Dispersion and Its Measurement
4.5. Polarization Mode Dispersion (PMD) Measurement
4.5.1. Representation of Fiber Birefringence and PMD Parameter
4.5.2. Pulse Delay Method
4.5.3. The Interferometric Method
4.5.4. Poincare Arc Method
4.5.5. Fixed Analyzer Method
4.5.6. The Jones Matrix Method
4.5.7. The Mueller Matrix Method
4.6. Determination of Polarization-Dependent Loss
4.7. PMD Sources and Emulators
4.8. Measurement of Fiber Nonlinearity
4.8.1. Measurement of Stimulated Brillouin Scattering Coefficient
4.8.2. Measurement of the Stimulated Raman Scattering Coefficient
4.8.3. Measurement of Kerr effect nonlinearity

4.0 INTRODUCTION

Optical fiber is an indispensable part of fiber-optic communication systems; it provides a low-loss and wideband transmission medium. The performance of an optical fiber system depends, to a large extent, on the characteristics of optical fibers. The realization of low-loss optical fibers in the early 1970s made optical fiber communication a viable technology, and the research and development of optical fibers have since become among the central focuses of telecommunications industry.

In addition to standard multimode fiber and standard single-mode fiber, many different types of optical fibers have been developed to provide modified chromatic dispersion properties, engineered nonlinear properties, and enlarged low-loss windows. Effects such as chromatic dispersion, polarization mode dispersion (PMD), polarization-dependent loss (PDL), and nonlinearities, which may be negligible in low-speed optical systems, have become extremely critical in modern optical communications using high-speed time-division multiplexing (TDM) and wavelength-division multiplexing (WDM). The improved fiber parameters and the introduction of new types of optical fibers enabled high-speed and long-distance optical transmission systems as well as various applications of fiber optics such as optical sensing and imaging. Precise measurement and characterization of optical fiber properties are important in the research, development, and fabrication of fibers as well as in the performance evaluation of optical systems.

This chapter reviews various techniques to characterize the properties of optical fibers, their operating principles, and the comparison among techniques. Section 4.1 briefly reviews various types of optical fibers, including standardized fibers for telecommunications and specialty fibers that are developed for various other applications ranging from optical signal processing to optical sensing. Section 4.2 discusses mode field distributions, the definition of mode field diameter, and techniques to measure near-field and far-field profiles. Section 4.3 introduces fiber attenuation measurement techniques. The primary focus of this section is to discuss the operating principle, accuracy considerations, and application of optical time-domain reflectometers (OTDRs).

In Section 4.4 we discuss the measurement of fiber dispersions, including modal dispersion in multimode fibers and chromatic dispersion in single-mode fibers. Both time-domain and frequency-domain measurement techniques are discussed and compared. Section 4.5 reviews a number of techniques for the characterization of PMD in optical fibers, including pulse delay, interferometric methods, Poincare-arc length, fixed analyzer, Jones Matrix, and Mueller Matrix methods. Understanding these measurement techniques, their pros and cons, and the comparison between them is critical in experimental system design and accuracy assessment. Section 4.6 discusses the measurement of PDL in optical fibers based on the Mueller Matrix technique. Since both the PMD and the PDL in optical systems are randomly varying, the measured results have to be presented to reflect the statistic nature of the process. Section 4.7 discusses implementations of PMD sources and emulators, which are useful instruments in optical transmission equipment testing and qualification. Section 4.8, the last section of this chapter, reviews various fiber nonlinear effects and techniques to characterization them.

4.1 CLASSIFICATION OF FIBER TYPES

As we all know, optical fiber is a cylindrical waveguide that supports low-loss propagation of optical signals. The general properties of optical fibers have been discussed in Chapter 1. In recent years, numerous fiber types have been developed and optimized to meet the demand of various applications. Some popular fiber types that are often used in optical communication systems have been standardized by the International Telecommunication Union (ITU-T). The list includes graded index multimode fiber (G.651), nondispersion-shifted single-mode fiber (G.652), dispersion-shifted fiber (G.653), and nonzero dispersion shifted fiber (G.655). In addition to fibers designed for optical transmission, there are also various specialty fibers for optical signal processing, such as dispersion compensating fibers (DCF), polarization maintaining (PM) fibers, photonic crystal fibers (PCF), and rare-earth doped active fibers for optical amplification. Unlike transmission fibers, these specialty fibers are less standardized.

4.1.1 Standard Optical Fibers for Transmission

The ITU-T G.651 multimode fiber (MMF) has a 50-μm core diameter and a 125-μm cladding diameter. The attenuation parameter for this fiber is on the order of 0.8 dB/km at 1310 nm wavelength. Because of its large core size, MMF is relatively easy to handle with large misalignment tolerance for optical coupling and connection. However, due to its large modal dispersion, MMF is often used for short-reach and low data-rate optical communication systems. Although this fiber is optimized for use in the 1300-nm band, it can also operate in the 850-nm and 1550-nm wavelength bands [1, 2].

The ITU-T G.652 fiber, also known as *standard single-mode fiber*, is the most commonly deployed fiber in long distance optical communication systems. This fiber has a simple step-index structure with a 9-μm core diameter and a 125-μm cladding diameter. It is single-mode with a zero-dispersion wavelength around $\lambda_0 = 1310$ nm. The typical chromatic dispersion value at 1550 nm is about

17 ps/nm-km. The attenuation parameter for G.652 fiber is typically 0.5 dB/km at 1310 nm and 0.2 dB/km at 1550 nm. An example of this type of fiber is Corning SMF-28 [1].

Although standard SMF has low loss in the 1550 nm wavelength window, which makes it suitable for long-distance optical communications, it shows relatively high chromatic dispersion values in this wavelength window. In high-speed optical transmission systems, chromatic dispersion introduces significant waveform distortion, which may significantly degrade system performance. The trend of shifting the transmission wavelength window from 1310 nm to 1550 nm in the early 1990s initiated the development of dispersion-shifted fiber (DSF). Through proper cross-section design, DSF shifts the zero-dispersion wavelength λ_0 from 1310 nm to approximately 1550 nm, making the 1550 nm wavelength window have both the lowest loss and the lowest dispersion. The core diameter of the DSF is about 7 μm, which is slightly smaller than the standard SMF. DSF is able to significantly extend the dispersion-limited transmission distance if there is only a single optical channel propagating in the fiber, but people soon realized that DSF is not suitable for multiwavelength WDM systems due to the high-level nonlinear crosstalk between optical channels through four-wave mixing and cross-phase modulation. For this reason, the deployment of DSF did not last very long.

To reduce chromatic dispersion while maintaining reasonably low nonlinear crosstalk in WDM systems, nonzero dispersion-shifted fibers (NZDSF) were developed. NZDSF moves the zero-dispersion wavelength near to but outside the 1550 nm window so that the chromatic dispersion for optical signals in the 1550 nm wavelength is less than that in standard SMF but higher than that of DSF. The basic idea of this approach is to keep a reasonable level of chromatic dispersion at 1550 nm, which prevents high nonlinear crosstalk from happening but without the need of dispersion compensation in the system. There are several types of NZDSF depending on the selected value of zero-dispersion wavelength λ_0. In addition, since λ_0 can be either longer or shorter than 1550 nm, the dispersion at 1550 nm can be either negative (normal) or positive (anomalous). The typical chromatic dispersion for G.655 fiber at 1550 nm is 4.5 ps/nm-km. Although NZDSF usually has core sizes smaller than standard SMF, which increases the nonlinear effect, some designs such as Corning LEAF (large effective area fiber) have the same effective core area as standard SMF, which is approximately 80 μm^2.

Figure 4.1.1 shows typical dispersion vs. wavelength characteristics for several major fiber types that have been offered for long-distance links [3]. The detailed specifications of these fibers are listed in Table 4.1.1, where M is the product of the chromatic dispersion and the effective cross section area. In optical communication system applications, the debate on which fiber has the best performance can never settle down, because data rate, optical modulation formats, channel spacing, the number of WDM channels, and the optical powers used in each channel may affect the conclusion. In general, standard SMF has

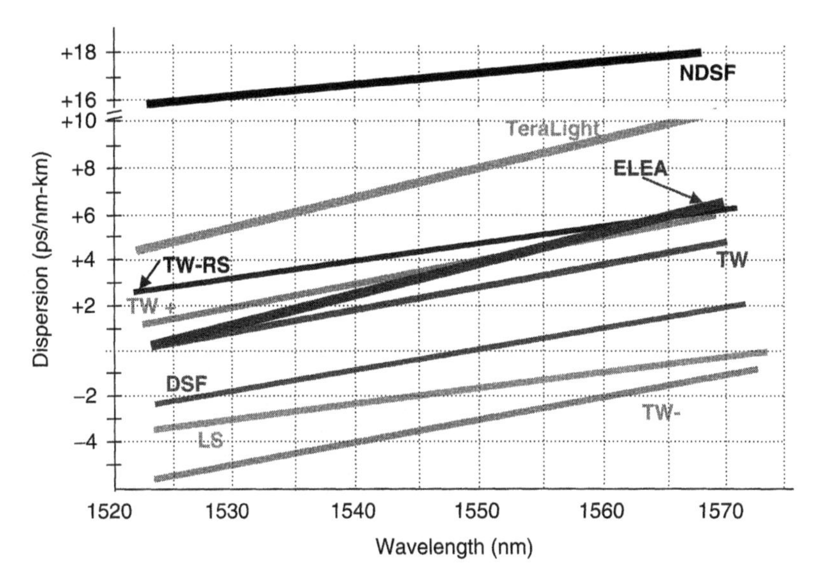

Figure 4.1.1 Chromatic dispersions of non-dispersion-shifted fiber (NDSF) and various different non-zero-dispersion-shifted fibers (NZDSF)

Table 4.1.1

Important Parameters for Standard SMF (NDSF), Long-Span (LS) Fiber, Truewave (TW) Fiber, Truewave Reduced-Slope (TW-RS) Fiber, Large Effective Area Fiber (LEAF), and Teralight Fiber [3]

Fiber Type	ITU	Dispersion @ 1550nm (ps/nm/km)	Dispersion Slope (ps/km/nm^2)	Effective Area (A_{eff}) (μm^2)	M
NDSF(SMF-28)	NDSF/G.652	16.70	0.06	86.6	1446
LS	NZNDSF/G.655	−1.60	0.075	50	80
Truewave (TW)	NZNDSF/G.655	2.90	0.07	55.42	161
TW – RS	NZNDSF/G.655	4.40	0.042	55.42	244
LEAF	NZNDSF/G.655	3.67	0.105	72.36	266
TERALIGHT	NZNDSF/G.655	8.0	0.058	63	504

high chromatic dispersion so that the effect of nonlinear crosstalk between channels is generally small; however, a large amount of dispersion compensation must be used, introducing excess loss and requiring higher optical gain in the amplifiers. This in turn will degrade OSNR at the receiver. On the other hand, low dispersion fibers may reduce the requirement of dispersion compensation but at the risk of increasing nonlinear crosstalk.

4.1.2 Specialty Optical Fibers

In addition to the fibers designed for optical transmission, a number of specialty fibers have also been developed for various purposes ranging from linear and nonlinear optical signal processing to interconnection between equipment. The following are a few examples of specialty optical fibers that have been widely used.

Dispersion compensating fiber (DCF) is a widely used specialty fiber that usually provides a large value of negative (normal) dispersion in a 1550 nm wavelength window. It is developed to compensate for chromatic dispersion in optical transmission systems that are based primarily on standard SMF. The dispersion coefficient of DCF is typically on the order of $D = -95$ ps/nm-km at a 1550 nm wavelength window. Therefore approximately 14 km DCF is required to compensate for the chromatic dispersion of 80 km standard SMF in an amplified optical span. For practical system applications, DCFs can be packaged into modules, which are commonly referred to as dispersion compensating module (DCM).

Compared to other types of dispersion compensation techniques such as fiber Bragg gratings, the distinct advantage of DCF is its wide wavelength window, which is critical for WDM applications, and its high reliability and negligibly small dispersion ripple over the operating wavelength. In addition, DCF can be designed to compensate the slope of chromatic dispersion, thus making it an ideal candidate for WDM applications involving wide wavelength windows. However, due to the limited dispersion value per unit length, DSF usually has relatively higher attenuation compared to fiber gratings, especially when the required total dispersion is high. In addition, because a large value of waveguide dispersion is needed to achieve normal dispersion in 1550 nm wavelength window, the effective core area of DCF can be as small as $A_{eff} \approx 15\mu m^2$, which is less than $1/5^{th}$ that of a standard SMF. Therefore the nonlinear effect in DCF is expected to be significant, which has to be taken into account in designing a measurement setup involving DCF.

Polarization maintaining (PM) fiber is another important category of specialty fibers. It is well known that in an ideal single-mode fiber with circular cross-section geometry, two degenerate modes coexist, with mutually orthogonal polarization states and identical propagation constants. The effect of external stress may cause the fiber to become birefringent and the propagation constants of these two degenerate modes will become different. The partitioning of the propagating optical signal into the two polarization modes not only depends on the coupling condition from the source to the fiber but also on the energy coupling between the two modes while propagating in the fiber, which is usually random. As a consequence, the polarization state of the output optical signal is usually random, even after only a few meters of propagation length in the fiber; the mode coupling and the output polarization state are very sensitive to external perturbations such as temperature variation, mechanical stress change, and micro and macro bending.

It is also known that energy coupling between the two orthogonal polarization modes can be minimized if the difference between the propagation constants of these two modes is large enough. This can be accomplished by incorporating extra elements in the fiber cladding that apply asymmetric stress to the fiber core. Due to the difference in the thermal expansion coefficients of different materials, asymmetrical stress in the fiber core can be achieved from the manufacturing process when the fiber is drawn from a preform. Depending on the shape of the stress-applying parts (SAP), PM fibers can be classified as "Panda" and "Bowtie," as illustrated in Fig. 4.1.2. The bowtie structure is depicted in which the SAPs are arranged in an arced manner around the fiber core; the Panda structure is named based on its similarity to the face of a panda bear. In the direction of the SAPs, which is parallel to the field of tension (horizontal in Figure 4.1.2), the fiber core has a slightly higher refractive index so that this axis is also referred to as the *slow axis* because the horizontally polarized mode propagates slower than the vertically polarized mode. The principle axis perpendicular to the slow axis is thus called the *fast axis*.

It is important to note that a PM fiber is simply a highly birefringent fiber, in which the coupling between the two orthogonally polarized propagation modes is minimized. However, for a PM fiber to maintain the polarization state of an optical signal, the polarization state of input optical signal has to be aligned to either the slow or the fast axis of the PM fiber. Otherwise, both of the two degenerate modes will be excited, although there is minimum energy coupling between them; their relative optical phases will still be affected by the fiber perturbation, and the output polarization state will not be maintained because of the vectorial summation of these two mode fields. To further explain, assume

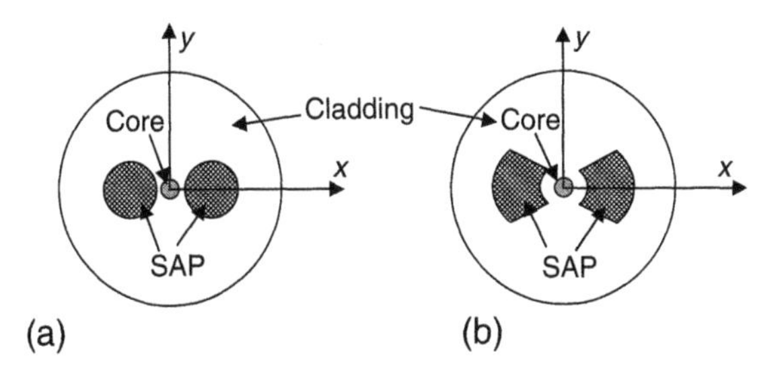

Figure 4.1.2 Cross-sections of polarization fibers: (a) Panda fiber and (b) Bowtie fiber. *SAP:* stress-applying parts.

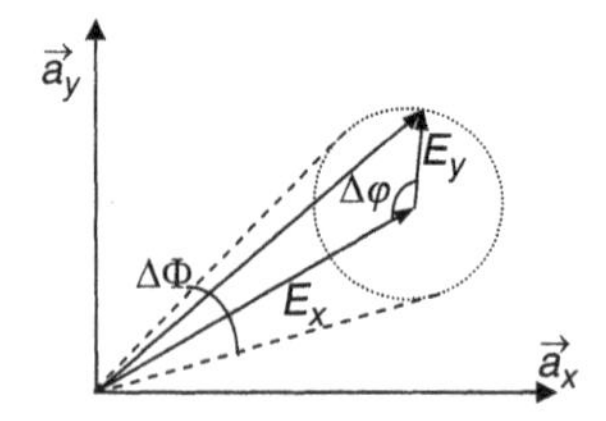

Figure 4.1.3 Illustration of the output field vector of a PM fiber when the input field polarization state is not aligned to the principle axis of the fiber.

that the input optical field is linearly polarized with an orientation angle θ with respect to the birefringence axis of the PM fiber. The optical field E_0 will be split into the two polarization modes such that $E_x = E_0 \cos\theta$ and $E_x = E_0 \sin\theta$. At the ouptut of the fiber, the composite optical field vector $\vec{E} = E_0 e^{-\alpha L}(\vec{a}_x \cos\theta + \vec{a}_y \sin\theta \cdot e^{j\Delta\varphi})$, where $\vec{a}_x$ and $\vec{a}_y$ are unit vectors, L is the fiber length, and α is the attenuation parameter. $\Delta\varphi = (\beta_x - \beta_y)L$ with β_x and β_y the propagation constants of the two modes. It is important to note that the differential phase $\Delta\varphi$ is very sensitive to external perturbations of the fiber. When $\Delta\varphi$ varies, the polarization state of the output optical field also varies if $\theta \neq 0$ (or 90°), as illustrated in Figure 4.1.3. In this case, since the phase difference $\Delta\varphi$ between the two mode fields E_x and E_y are considered random, the variation of the output polarization orientation can be as high as $\Delta\Phi$, as shown in Figure 4.1.3.

Therefore, in an optical system, if a PM fiber is used, one has to be very careful in the alignment of the signal polarization state at the fiber input. Otherwise the PM fiber could be even worse than a standard single-mode fiber in terms of the output polarization stability. Another issue regarding the use of PM fibers is the difficulty of connecting and splicing. When connecting between two PM fibers, we must make sure that their birefringence axes are perfectly aligned. Misalignment between the axes would cause the same problem as the misalignment of input polarization state, as previously discussed. To provide the functionality of precisely controlled axes rotation and alignment, a PM fiber splicer can be five times more expensive than a conventional fiber splicer due to its complexity.

Photonic crystal fiber (PCF), also known as photonic bandgap fiber, is an entirely new category of optical fibers because of its different wave-guiding mechanisms. As shown in Figure 4.1.4, a PCF usually has large number of air holes periodically distributed in its cross-section; for that reason it is also known as the "holey" fiber. The guiding mechanism of PCF is based on the Bragg resonance effect in the transversal direction of the fiber; therefore the low-loss transmission window heavily depends on the bandgap structure design.

Figure 4.1.4(a) shows the cross-section view of a hollow core PCF in which the optical signal is guided by the air core. In contrast to the conventional optical wave-guiding mechanism where high refractive index solid dielectric material

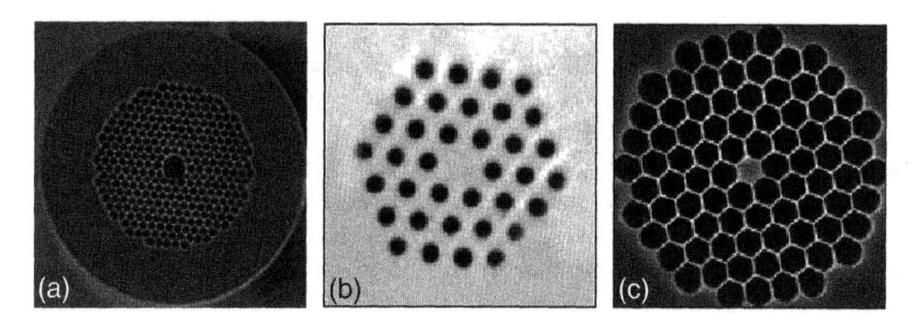

Figure 4.1.4 Cross-sectional view of photonic crystal fibers: (a) hollow-core PCF, (b) large core area PCF, and (c) highly nonlinear PCF [4].

is required for the core, the photonic bandgap structure in the cladding of PCF acts as a virtual mirror that confines the propagating lightwave to the hollow core. In general, the periodical photonic bandgaps structure of PCF is formed by the index contrast between silica and air by incorporating the holes into a silica matrix. In most of the hollow core PCFs, more than 95 percent of the optical power exists in the air; therefore the interaction between signal optical power and the glass material is very small. Because the nonlinearity of the air is approximately three orders or magnitude lower than in the silica, hollow core PCF can have extremely low nonlinearity and can be used to deliver optical signals with very high optical power. In addition, because of the design flexibility of the photonic bandgap structures, by varying the size and the pitch of the holes, zero dispersion can be achieved at virtually any wavelength. One drawback, though, for the hollow core PCF is its relatively narrow transmission window, which is typically on the order of 200 nm. This is the consequence of the very strong resonance effect of the periodic structure that confines the signal energy in the hollow core.

Another category of PCF as shown in Figure 4.1.4(b) is referred to as *large core area PCF.* This type of fiber has a solid silica core to guide the optical signal, whereas the periodic holes in the cladding are used to facilitate optical wave guiding and help confine signal optical power in the core area. Generally, standard single-mode fiber with a 10 μm core diameter has a cutoff wavelength at approximately 1100 nm, below which should be micro meter (um) the fiber will have multiple propagation modes. Large core PCF, on the other hand, allows for single-mode operation in a very wide wavelength window—for example, from 750 to 1700 nm—while maintaining a very large core area. Compared to hollow core PCF, large core area PCF has much wider low-loss window. Though a large core area PCF has a lower nonlinear parameter than standard single-mode fiber, its nonlinear parameter is typically much higher than a hollow core PCF.

Highly nonlinear PCF, as shown in Figure 4.1.4(c), is also a very useful fiber for optical signal processing which has very small solid core cross-section; therefore the power density in the core is extremely high. For example, for a highly nonlinear PCF with zero-dispersion wavelength at $\lambda_0 = 710$ nm, the core diameter is as small as 1.8 μm and the nonlinear parameter is $\gamma > 100\text{W}^{-1}\ \text{km}^{-1}$, which is 40 times higher than that of a standard single-mode fiber. Highly nonlinear PCFs are usually used in nonlinear optical signal processing, such as parametric amplification and supercontinuum generation.

Photonic crystal fibers are considered high-end fiber types that are usually expensive due to their complex fabrication processes. They are generally sold in meters instead of kilometers because relatively short PCF fibers are enough for most of the applications for which PCFs are designed. They are also delicate and difficult to handle; for example, end surface treatment, termination, connection, and fusion splicing are not straightforward because of the existence of air holes in the fiber.

Plastic optical fiber (POF) is a low-cost type of optical fiber that is easy to handle. The core material of POF is typically made of PMMA (polymethyl methacrylate), which is a general-purpose resin, whereas the cladding is usually made of fluorinated polymer, which has a lower refractive index than the core. The cross-section design of POF is more flexible than silica fibers, and various core sizes and core/cladding ratio can be easily obtained. For example, in a large-diameter POF, 95 percent of the cross-section is the core that allows the transmission of light. The fabrication of POF does not need an expensive MOCVD process as usually required for the fabrication of silica-based optical fibers; this is one of the reasons for the low cost of the POFs. Although silica-based fiber is widely used for telecommunications, POF has found more and more applications due to its low cost and high flexibility. The costs associated with POF connection and installation are also low, which are especially suitable for fiber to the home application. On the other hand, POFs have transmission losses on the order of 0.25 dB/m, which is almost three orders of magnitude higher than silica fibers. This excluded POFs from being used in long-distance optical transmission. In addition, the majority of POFs are multi-mode, and therefore they are often used in low-speed, short-distance applications such as fiber-to-the-home networks, optical interconnections, networks inside automobile, and flexible illumination and instrumentation.

4.2 MEASUREMENT OF FIBER MODE-FIELD DISTRIBUTION

Mode-field distribution is an important parameter in the specification of an optical fiber. Many practical characteristics of the optical fiber, such as the mode-field diameter, the coupling efficiency between fibers, and the effective cross-section area, are all determined by mode-field distribution. Although the core diameter can be easily determined by the geometry of the fiber, the determination of its mode-field distribution is more complex, depending on the refractive index profile as well as the wavelength of the optical signal propagating in the fiber. Mode-field distribution can be described by near-field, far-field, or a specially defined mode-field diameter [5, 6, 7]. In this section, we discuss various mode-field definitions and how these parameters can be measured.

4.2.1 Near-Field, Far-Field, and Mode-Field Diameter

In practice, because of the circular geometry of an optical fiber, the field distribution of the fundamental mode in a single-mode optical fiber is circularly symmetrical. This simplifies the problem, and the fiber mode field can be specified by a single parameter known as the *mode-field diameter* (MFD), and the electrical field distribution can often be assumed as Gaussian:

$$E(r) = E_0 \exp\left(-\frac{r^2}{W_0^2}\right) \tag{4.2.1}$$

where r is the radius, E_0 is the optical field at $r = 0$, and W_0 is the width of the field distribution. Specifically, the MFD is defined as $2W_0$, which is

$$2W_0 = 2\left(\frac{2\int_0^\infty r^3 E^2(r)dr}{\int_0^\infty rE^2(r)dr}\right)^{1/2} \tag{4.2.2}$$

Figure 4.2.1 illustrates the mode-field distribution of a single mode fiber in which Gaussian approximation is used. The physical meaning of the MFD definition given by Equation 4.2.2 can be explained as follows: The denominator in

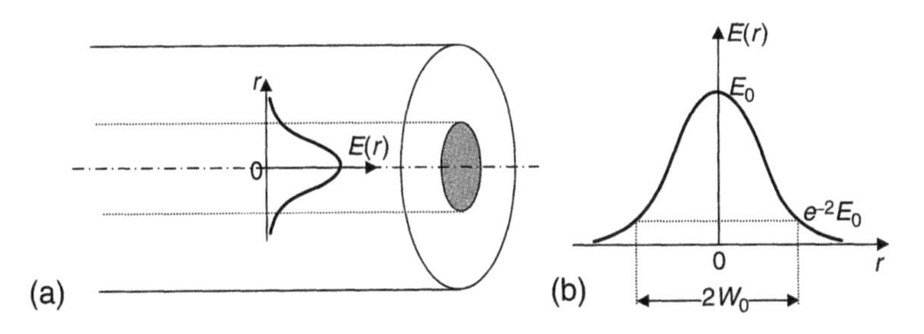

Figure 4.2.1 (a) Illustration of mode distribution in a single-mode fiber and (b) definition of mode-field diameter under Gaussian approximation.

Equation 4.2.2 is proportional to the integration of the power density across the entire fiber cross-section, which is the total power of the fundamental mode, whereas the numerator is the integration of the square of the radial distance (r^2) weighted by the power density over the fiber cross-section. Therefore MFD defined by Equation 4.2.2 represents a root mean square (*rms*) value of the mode distribution of the optical field on the fiber cross-section.

It is important to point out that mode-field distribution $E(r)$ in Equation 4.2.1 represents the field distribution inside the fiber; thus it is equivalent to the optical field distribution exactly on the output end surface of the fiber. It is commonly referred to as the *near-field* (NF) distribution. Near-field distribution is a very important parameter of the fiber that determines the effective cross-section area of the fiber as

$$A_{eff} = \frac{2\pi \left[\int_0^\infty r|E(r)|^2 r dr \right]^2}{\int_0^\infty r|E(r)|^4 r dr} \tag{4.2.3}$$

As introduced in Equation 1.3.95, if the total optical power P carried by the fiber is known, the power density in the fiber core can be determined using the effective cross-section area as $I_{density} = P/A_{eff}$.

On the other hand, the field radiation pattern at a large distance from the fiber output facet is usually defined as the *far-field* (FF) distribution, which is in general different from the NF distribution; their relationship is [Artiglia 1989]

$$\Psi(R, \theta) = \frac{jk}{R} e^{ikR} E_H(\theta) \tag{4.2.4}$$

where

$$E_H(\theta) = \int_0^\infty E(r) J_0(r \cdot k \sin\theta) r dr \tag{4.2.5}$$

represents the angular distribution of the field, which is the 0th-order Hankel transform of the near-field distribution $E(r)$ on the fiber exit facet. As illustrated in Figure 4.2.2, R is the distance between the observation point in the far-field region and the center of the fiber exit facet, θ is the angle of the observation point with respect to the fiber axis, and r is the distance between the emission point on the fiber facet and the center of the fiber core.

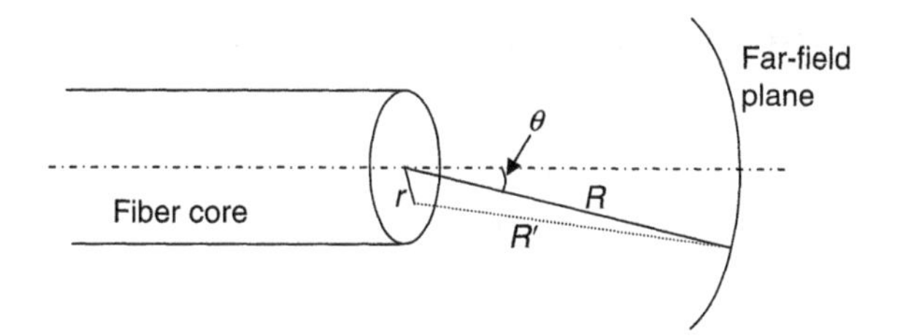

Figure 4.2.2 Coordinates of fiber end surface and far-field plane.

$k = 2\pi/\lambda$ is the wave number and λ is the free-space wavelength. J_0 is the 0th-order Bessel function. As defined previously, the far-field plane has to be far away from the fiber exit facet, and this in general requires $R \approx R'$ and $R \gg W_0^2/\lambda$. In general, if the far-field angular radiation pattern $E_H(\theta)$ can be measured, according to the property of the Hankel transform, the near-field distribution can be easily obtained by an inverse Hankel transform,

$$E(r) = \int_0^\infty E_H(q) J_0(2\pi q r) q dq \tag{4.2.6}$$

where $q = (\sin\theta)/\lambda$ is commonly referred to as the *spatial frequency*.

Similar to the mode-field diameter for the near field as defined by Equation 4.2.2, the mode-field diameter D_0 for the far field can also be defined as

$$2D_0 = 2\left(\frac{2\int_0^\infty \rho^3 E_H^2(\rho) d\rho}{\int_0^\infty \rho E_H^2(\rho) d\rho} \right)^{1/2} \tag{4.2.7}$$

where $\rho = R\sin\theta$ is the radial distance between the observation point and the center on the observation plane. Both the near-field and the far-field mode-field diameters are determined by the distribution of the optical power density on the fiber cross-section; thus the most important task is the precise measurement of the actual distribution of the mode field. There are a number of techniques to measure mode-field distribution, such as the far-field scanning technique and the near-field scanning technique. Again, since the relationship between the near-field and the far-field diameters is deterministic through Equations 4.2.2 to 4.2.7, the measurement of either one of them is considered to be sufficient and the other one can be derived accordingly.

4.2.2 Far-Field Measurement Techniques

Figure 4.2.3 shows the block diagram of the far-field measurement setup. The lightwave signal from a laser diode is coupled into the input end of the optical fiber through a focusing lens. A cladding mode filter is usually necessary to remove the optical power carried in the cladding of the fiber and to accelerate the process of achieving equilibrium mode-field distribution. The aperture of the optical detection head determines the spatial resolution of the measurement, which can be a pinhole in front of a photodiode, or simply using an optical fiber as the probe.

The normalized far-field power distribution $P_{angle}(\theta) = u|E_H(\theta)|^2$ defined by Equation 4.2.5 can be obtained through scanning the angular position of the detector head, where u is a constant representing the detector efficiency and the aperture size. Figure 4.2.4 shows examples of measured angular distributions of far-field power density for three different types of fibers. To make

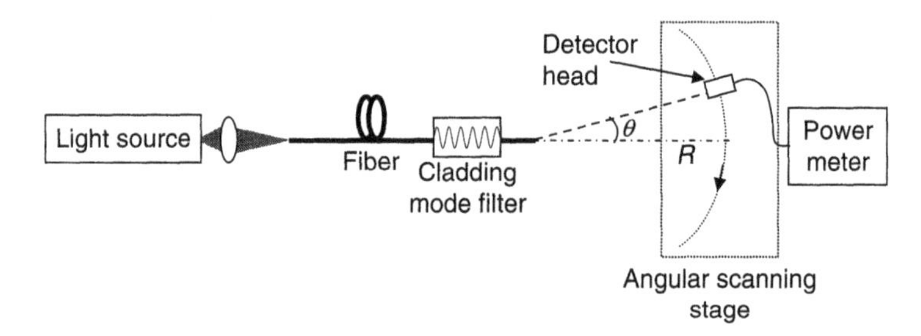

Figure 4.2.3 Far-field measurement setup using angular scanning.

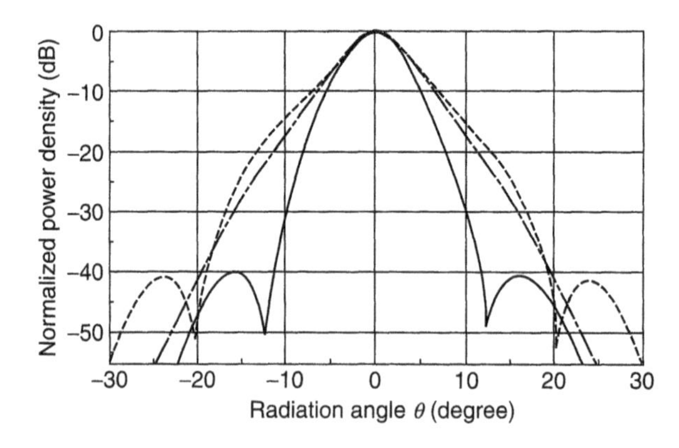

Figure 4.2.4 Examples of measured angular distribution of far-field power density of conventional single-mode step-index fiber at $\lambda = 1300$ nm (solid line), dispersion shifted fiber at $\lambda = 1300$ nm (dot-dashed line), and dispersion flattened fiber at $\lambda = 1550$ nm (dashed line) [8]. Used with permission.

precise estimation of the mode diameter, the far-field power distribution pattern has to be measured over a wide angular range to cover at least the entire main lobe. One practical problem is that the absolute power density $|\Psi(R, \theta)|^2$ of the far field is inversely proportional to the distance R, as indicated by Equation 4.2.4, and thus the measured signal power levels will be very weak if the detector is placed far away from the fiber facet. But the definition of "far field" does require the detector to be far away from the fiber facet such that $R \gg W_0^2/\lambda$. Meanwhile, the acceptable spatial resolution of the measurement requires the small size of the pinhole in front of the detector, which further reduces the optical power at the detector. To make things worse, the power level is even weaker when the detector is placed at large angles with respect to the fiber axis. Therefore very sensitive detectors have to be used in these measurements, with extremely low noise levels. Lock-in amplifiers are usually used in most of the practical implementations. The detection and amplification circuitry also must have reasonably large dynamic ranges to ensure the measurement accuracy. As shown in Figure 4.2.4, the dynamic range has to be higher than 55 dB to clearly show the side lobes of mode distribution.

Fortunately, the mode-field distribution is stationary and the measurement does not require high-speed detection. Therefore, the far-field detection can also be accomplished using a two-dimensional charge-coupled detection (CCD) array, which is basically a video camera. The mode-field distribution at various emission angles can be simultaneously detected by sensing pixels in the camera and can be processed in parallel. This avoids the requirement of mechanically scanning the position of the detector head, which could easily introduce uncertainties. As shown in Figure 4.2.5, since the CCD is a planar array, the far-field power density $P_{plane}(z,\rho)$ measured on the plane is slightly different from the one measured by the angular scanning, and their relationship is

$$P_{plane}(z, \rho) \propto P_{angle}(\theta)\cos^3\theta \tag{4.2.8}$$

where $\rho = z\tan\theta$ is the radial distance from the center of the array plan. The factor $\cos^2\theta$ in Equation 4.2.8 comes from the fact that $R = z/\cos\theta$ and the far-field power density is proportional to R^2, as shown in Equation 4.2.4. The other $\cos\theta$ factor comes from the angle of the detector surface normal, which has an angle θ in respect to the incident light ray.

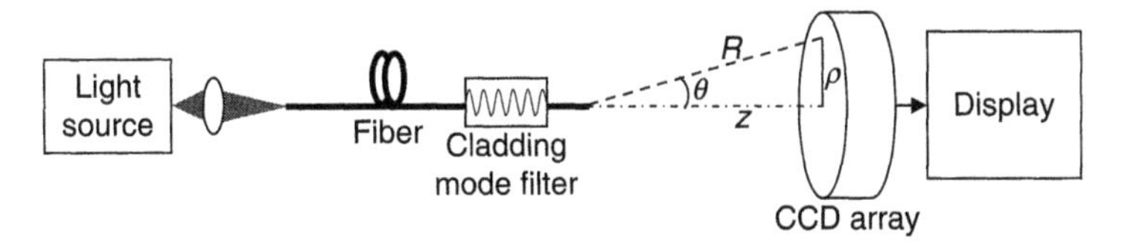

Figure 4.2.5 Far-field measurement setup using a planar detection array.

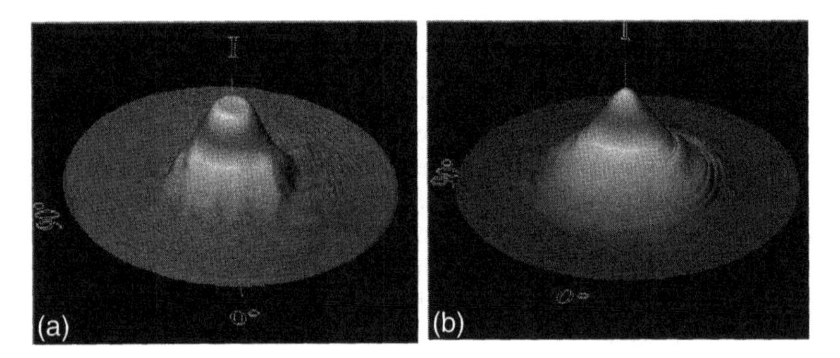

Figure 4.2.6 Examples of the measured far-field profiles of two different fibers [9]. Used with permission.

Instrumentations are commercially available now for the measurement of mode-field distributions [10]. Figure 4.2.6 shows examples of the measured far-field profiles of two different fibers using the far-field imaging technique. Obviously the two mode-field profiles are different, determined by the difference of the index profiles of the two fibers.

4.2.3 Near-Field Measurement Techniques

As signified by its definition, near field is the radial distribution of the electrical field on the fiber end facet. Since the transversal dimension of a fiber facet is very small, it is not feasible to directly measure the field distribution using a probe. It is usually necessary to magnify the image on the fiber exit facet with a microscope objective lens. A block diagram of a near-field measurement setup is shown in Figure 4.2.7, where an objective lens magnifies the image of the fiber exit facet and an optical probe mounted on a translation stage scans across the image plane. A photodiode converts the optical signal sampled by the optical probe into a voltage signal. In this measurement, the scanning probe can simply be a piece of single-mode fiber with the core diameter of approximately 9 μm.

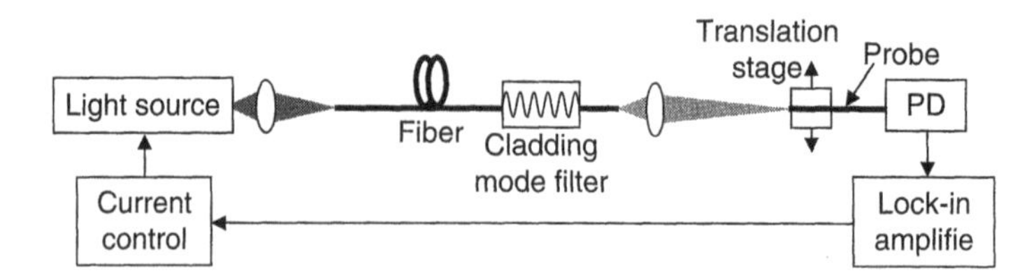

Figure 4.2.7 Near-field measurement setup using a transversal translation stage and a lock-in amplifier.

To make accurate measurements with enough sampling points, the magnification of the optical system has to be large enough, say 100×, with good linearity. A careful calibration has to be made because the absolute value of image magnification is critical in determining the mode-field diameter of the fiber under test. Similar power budget concerns as in the case of far-field measurement also exist in the near-field measurement: For a 100× magnification of the optical system, the power density on the image plane is 10^4 times (or equivalently 40 dB) lower than that on the fiber exit facet. In addition to having a large dynamic range, the noise level in the detection system has to be extremely low; therefore a lock-in amplifier is usually required.

Another consideration of the near-field measurement setup is the numerical aperture of the magnification system. Due to diffraction, the detailed feature of the near-field distribution on the imaging plane may be smoothed out. Based on the Rayleigh criterion, to be able to resolve two adjacent points separated by d on the object plane, the numerical aperture of the optical system has to be $NA_{opt} \geq 0.61\lambda/d$. As an example, if the signal wavelength carried in the fiber is 1550 nm, to resolve 10 sampling points along a fiber core diameter of 9 μm ($d = 1$ μm), the numerical aperture of the optical system has to be $NA_{opt} \approx 0.95$.

Again, similar to the far-field measurement setup shown in Figure 4.2.5, the scanning optical probe, the translation stage, and the photodiode can be replaced by a two-dimensional CCD array. This eliminates the repeatability concern and position uncertainties due to translation stage, and the measurement is also much faster due to parallel sampling and signal processing. As an example, Figure 4.2.8 shows the measured near-field image on the fiber facet together with the near-field intensity profiles in the horizontal and vertical directions. In recent years, the

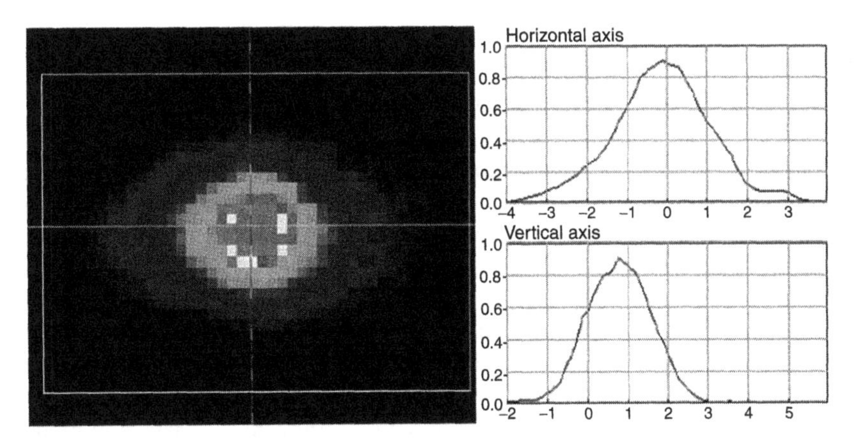

Figure 4.2.8 Examples of the measured near-field profile with the numerical values in the horizontal and vertical directions (scale: 1μm/div) [9]. Used with permission.

advances in digital signal processing and high-sensitivity CCD technology greatly simplified the processing and analysis of digital images, allowing real-time observation of the far-field and near-field images [10].

In principle, the far-field and the near-field are related and they are all determined by the optical field distribution $E(r)$ on the fiber cross section. In practice, near-field measurement requires a lens with large magnification, which may introduce image distortion for the measurement. In comparison, far-field measurement is more straightforward which only requires a scanning probe. If mechanical scanning stage is stable enough far-field measurement can potentially be more accurate with better signal to noise ratio and dynamic range, which can be observed by comparing Figure 4.2.6 with Figure 4.2.8 [9].

There are a number of alternative techniques to measure fiber mode-field diameter, such as the transverse offset method, the variable aperture method, and the mask method. The details of these methods can be found in [11].

4.3 FIBER ATTENUATION MEASUREMENT AND OTDR

Optical attenuation in an optical fiber is one of the most important issues affecting all applications that use optical fibers. A number of factors may contribute to fiber attenuation, such as material absorption, optical scattering, micro or macro bending, and interface reflection and connection. Some of these factors are uniform, whereas some of them vary along the fiber, especially if different spools of fibers are fusion-spliced together. Characterization of fiber attenuation is fundamental to optical system design, implementation, and performance estimation.

4.3.1 Cutback Technique

In early days, the cutback technique was often used to measure fiber attenuation. As illustrated in Figure 4.3.1, the cutback technique measures fiber transmission losses at different lengths. Suppose the attenuation coefficient α is uniform; the power distribution along the fiber is $P(z) = P_0 e^{-\alpha z}$. With the same amount of optical power coupled into the fiber, if the output power measured as P_1, P_2, and P_3 at the fiber lengths of L_1, L_2, and L_3, respectively, the fiber attenuation coefficient can be calculated as $\alpha = [\ln(P_2/P_1)]/(L_1 - L_2)$ or $\alpha = [\ln(P_3/P_2)]/(L_2 - L_3)$.

For a single-mode fiber, there are only two orthogonal fundamental modes and the difference in their attenuations is generally negligible. For a multimode fiber, on the other hand, there are literally hundreds of propagation modes and

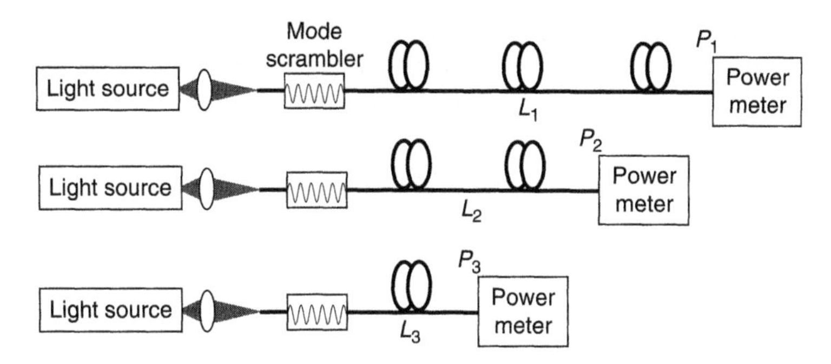

Figure 4.3.1 Illustration of the cutback technique to measure fiber loss.

different modes may have different attenuation coefficients. Therefore, launching condition from the source to the fiber is an important consideration in the loss measurement, especially in comparing results measured by different laboratories. A mode scrambler can be used to stabilize the power distribution of the guided modes and stripe out the cladding mode.

It is well known that a single-mode fiber will turn into multimode if the signal wavelength is shorter than the cutoff wavelength. Immediately below the cutoff wavelength, the second-order LP_{11} mode starts to exist in addition to the fundamental mode. Cutback technique can be used to evaluate the single-mode condition in a fiber by measuring the attenuation coefficient of the second-order LP_{11} mode. Figure 4.3.2 illustrates a technique of measuring the attenuation coefficients of both the fundamental LP_{01} mode and the second-order LP_{11} mode based on the fact that the bending loss of LP_{11} mode is much higher than LP_{01} mode [12]. The measurement procedure is as follows: First, bend the fiber with small radius (~1–2cm) near the input end of the fiber and measure the power P_1 and P_2 at fiber lengths of $L_1 + L_2$, and L_1, respectively.

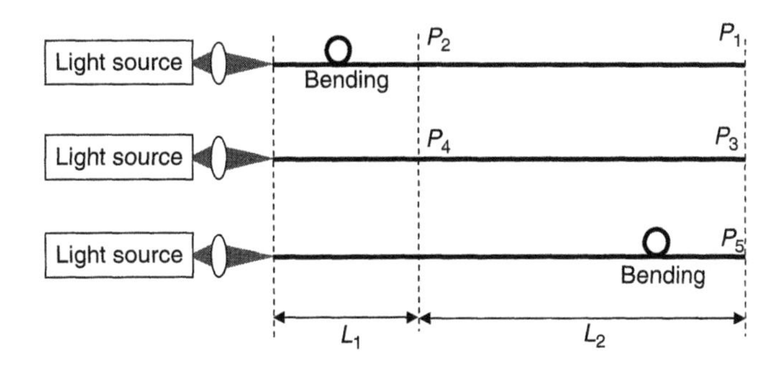

Figure 4.3.2 Technique to selectively measure the attenuation coefficient of the fundamental mode LP_{01} and the second-order mode LP_{11}.

In this case, the bending is equivalent to a mode filter that removes the power carried by the LP_{11} mode while introducing negligible loss for the fundamental LP_{01} mode. The attenuation coefficient of LP_{01} mode can be obtained as

$$\alpha_{01} = \frac{1}{L_2} \ln\left(\frac{P_2}{P_1}\right) \tag{4.3.1}$$

Next, release the bending and measure the powers for the straight fiber at the same locations. In this case, the measured powers P_3 and P_4 include both the LP_{01} mode and the LP_{11} mode. Then bend the fiber near the output terminal so that the measured power P_5 only includes the LP_{01} mode. Through these measurements, the attenuation coefficient of LP_{11} mode can be found as

$$\alpha_{11} = \frac{1}{L_2} \ln\left(\frac{P_1 P_4 - P_2 P_5}{P_1 P_3 - P_1 P_5}\right) \tag{4.3.2}$$

The cutback measurement technique is simple, direct, and it can be very accurate, however, it has several disadvantages. First, cutting the fiber step by step will certainly make the fiber less useable. Especially if the fiber loss is low; a very long fiber has to be used to have measurable losses. For example, if the attenuation coefficient of the fiber is $\alpha = 0.25$ dB/km, the output optical power only increases by 0.025 dB by cutting off 100 m of fiber, and obviously the accuracy of the measurement is limited by the accuracy and the resolvable digits of the power meter. In addition, it would be difficult to use the cutback technique to evaluate fibers after cabling; it is even impossible to measure in-service fibers that are already buried under ground. Therefore nondestructive techniques, such as optical time-domain reflectometer, would be more practical.

4.3.2 Optical Time-Domain Reflectometers

Rayleigh backscattering is one of the most important linear effects in a single-mode optical fiber; it sets a fundamental limit of fiber loss and is responsible for the major part of the attenuation in modern optical fibers in which other losses are already minimized. Meanwhile, Rayleigh backscattering can also be utilized to measure the fiber attenuation distribution along the fiber [13]. In this backscattering measurement technique, a short and high-peak power optical pulse train is launched into the fiber and the waveform of the backscattered optical signal from the fiber is detected, providing the detailed local loss information throughout the fiber. An important advantage of this technique is that it only needs to access a single end of the fiber because both the source and the detector are colocated, so the method is nondestructive. Since Rayleigh

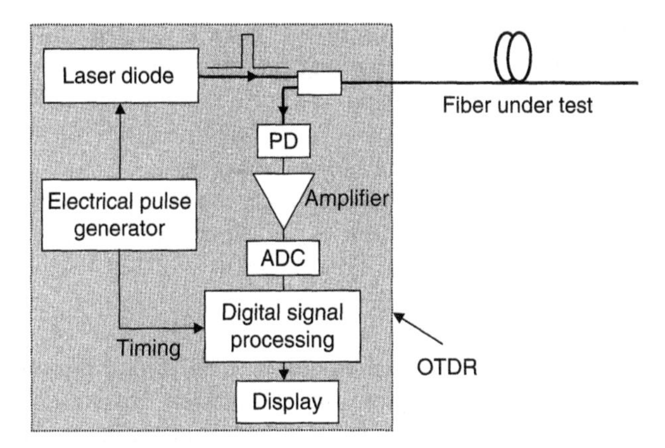

Figure 4.3.3 Block diagram of an OTDR. *PD:* photodetector, *ADC:* analog-to-digital converter.

backscattering is a linear process, at any point along the fiber, the magnitude of the backscattered optical power is linearly proportional to the optical power at that location. Due to the effect of fiber loss, both the transmitted and the backscattered powers are gradually attenuated along the fiber. The measurement of the time-dependent waveform of the backscattered power at the fiber input terminal provides the information about the loss distribution along the fiber; this information can be used to precisely calculate the attenuation coefficient.

Figure 4.3.3 shows the block diagram of an optical time-domain reflectometer (OTDR). A laser diode is driven by an electrical pulse generator that produces a train of short optical pulses. A photodiode is used to detect the backscattered optical power from the fiber through a directional coupler. The detected waveform of the optical signal is amplified and digitized through an ADC and analyzed by a digital-signal processing (DSP) unit. The timing of the DSP is synchronized with the source optical pulses so that the propagation delay of each backscattered pulse can be precisely calculated.

In general, the loss in an optical fiber may be caused by absorption, scattering, bending, and connecting. Most of these effects can be nonuniform along the fiber, especially if different fiber spools are connected together. Therefore, the fiber attenuation coefficient is a function of the location along the fiber. The power distribution along the fiber can be expressed as

$$P(z) = P(0)\exp\left\{-\int_0^z [\alpha_0(x) + \alpha_{SC}(x)]dx\right\} \tag{4.3.3}$$

where $P(0)$ is the input optical power, $\alpha_{SC}(x)$ is the attenuation coefficient due to Rayleigh scattering, and $\alpha_0(x)$ is that caused by attenuation effects other than scattering along the fiber. Suppose an optical pulse is injected into the fiber at time t_0 with the pulse width τ; neglecting the effect of chromatic dispersion and fiber nonlinearity, the locations of the pulse leading edge and the trailing edge along the fiber at time t are $z_{le} = v_g(t - t_0)$ and $z_{tr} = v_g(t - t_0 - \tau)$, respectively, where v_g is the group velocity of the optical pulse.

Then we consider the Rayleigh backscattering caused by a short fiber section of length dz. According to Equation 4.3.3, the optical power loss due to Rayleigh scattering within this short fiber section is

$$dP_{SC}(z) = P(z)\alpha_{SC}(z)dz \tag{4.3.4}$$

Only a small fraction of this scattered energy is coupled to the guided mode, which can propagate back to the input side of the fiber. Therefore the reflected optical power that is originated from the location z and reaches to the input end of the fiber can be calculated as

$$dP_{BS}(z) = P(0) \cdot \eta \cdot \alpha_{SC}(z) \cdot dz \cdot \exp\left\{ -2 \int_0^z \alpha(x)dx \right\} \tag{4.3.5}$$

where $\alpha(x) = \alpha_0(x) + \alpha_{SC}(x)$ is the composite attenuation coefficient of the fiber, which includes both the scattering and other attenuation effects. $\eta = (1 - \cos\theta_1)/2$ is the conversion efficiency from the scattered light to that captured by the fiber, as explained by Equation 1.3.59, and θ_1 is the maximum trace angle of the guided mode in the fiber, which is proportional to the numerical aperture, as illustrated in Figure 1.3.12. More precisely, if the normalized frequency, V, is between $1.5 < V < 2.4$, this conversion efficiency can be estimated as [Nakazawa, 1983]

$$\eta \approx \frac{1}{4.55}\left(\frac{NA}{n_1}\right) \tag{4.3.6}$$

where n_1 is the refractive index of the fiber core. As an example, in a silica-based single-mode fiber with $NA = 0.11$ and $n_1 = 1.45$, the conversion efficiency is approximately $\eta \approx 0.13\%$.

As illustrated in Figure 4.3.4, at time t_1, the backscattered optical signal that reaches the input terminal of the fiber is originated from a short fiber section of length $\Delta z/2 = (z_{le} - z_{tr})/2$, and the amplitude is therefore

$$P_{BS}(z) = \frac{v_g\tau}{2} P(0) \cdot \eta \cdot \alpha_{SC}(z) \cdot \exp\left\{ -2 \int_0^z \alpha(x)dx \right\} \tag{4.3.7}$$

where $\Delta z = v_g\tau$ has been used. Assume that both the scattering loss α_{sc} and the capturing coefficient η are constants along the fiber; the attenuation

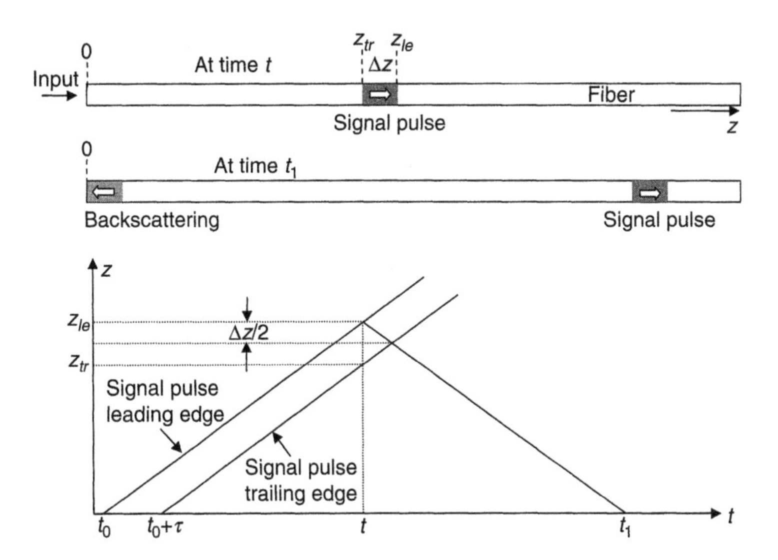

Figure 4.3.4 Illustration of the locations of the signal optical pulse, the scattering section, and the backscattered pulse in the fiber.

coefficient at location z can be evaluated from the differential return loss from Equation 4.3.7:

$$\alpha_{dB}(z) = -\frac{4.343}{2} \cdot \frac{d}{dz} \ln\left(\frac{P_{BS}(z)}{P(0)}\right) = -\frac{1}{2}\frac{d}{dz}\left(\frac{P_{BS}(z)}{P(0)}\right)_{dB} \tag{4.3.8}$$

where the unit of the attenuation coefficient α_{dB} is in dB and $\left(P_{BS}(z)/P(0)\right)_{dB}$ is the normalized return loss, also measured in dB. It is worthwhile to note that the slope of the return loss as the function of z alone is sufficient to determine the fiber attenuation coefficient. However, the source power $P(0)$, the pulse duration τ, and the efficiency of capturing the backscattered power into guided mode $\eta\alpha_{SC}$ are important to determine the actual scattering optical power that reaches the receiver, and thus to evaluate the signal-to-noise ratio (SNR). Although the Rayleigh backscattering power P_{BS} is usually measured as the function of time, it is straightforward to convert the time scale into the fiber length scale z based on the well-known group velocity of the optical pulse. The following is an example of OTDR measurement.

A standard single-mode fiber has the attenuation coefficient $\alpha_{dB} = 0.25\ dB/km$ in a 1550 nm wavelength window. Assume that the attenuation is uniform along the fiber and a large part of this attenuation is caused by Rayleigh backscattering such that $\alpha_{SC,dB} = 0.2\ dB/km$ and $\alpha_{0,dB} = 0.05\ dB/km$. We also assume that the efficiency of capturing the backscattered power into guided mode is

$\eta = 1.19 \times 10^{-3}$ and the signal optical pulse launched into the optical fiber has the peak power of $P(0) = 10\ mW$. The backscattered optical power that reaches the receiver at the fiber input terminal is

$$P_{BS}(z) = \frac{v_g \tau}{2} P(0) \cdot \eta \cdot \frac{\alpha_{SC,dB}(z)}{4.343} \cdot \exp\left\{ -2\int_0^z \alpha(x)dx \right\}$$
$$= 5.59 \times 10^{-3} \tau \cdot \exp\left\{ -2 \times \frac{0.25}{4.343} z \right\} \tag{4.3.9}$$

where τ is the pulse width and $v_g = 2.04 \times 10^{-8} m/s$ has been used considering that the refractive index of silica is approximately 1.47.

Figure 4.3.5 shows the backscattered optical power that reaches the receiver at the fiber input terminal as the function of the pulse location z along the fiber or as the function of time t. Due to the low capturing efficiency of the backscattered optical power by the fiber propagating mode, the level of the received signal power is at least 60 dB lower than the original optical signal peak power when the pulse width is 1 μs. This power level is linearly reduced with further decreasing the pulse width. Therefore, in practical implementations, there are very stringent requirements on the receiver sensitivity, the dynamic range, and the linearity.

If two fibers are connected together either by a connector or by fusion splicing, there will be a loss at the location of connection. In addition, because of the discontinuity possibly caused by the air gap if a connector is used, there is usually a discrete reflection peak in the OTDR trace. The connection loss can be

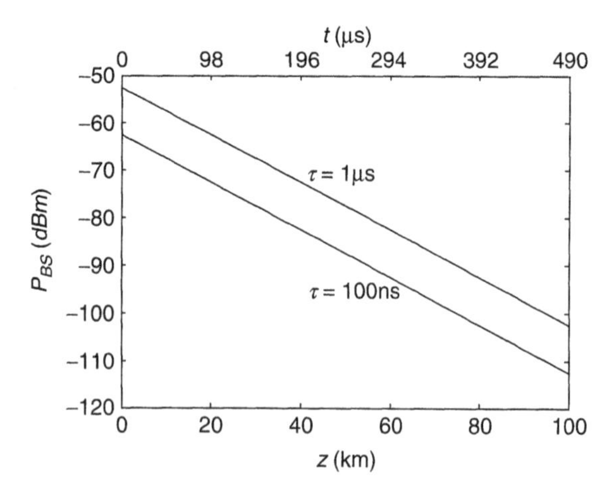

Figure 4.3.5 The backscattered power that reaches the receiver as the function of the location of the signal optical pulse. The input pulse peak power is 10 dBm.

evaluated straightforwardly by OTDR measurements. Assuming that the two fibers are identical with a uniform attenuation coefficient, the backscattered optical power measured by the OTDR is

$$P_{BS}(z) = \begin{cases} \frac{v_g\tau}{2} P(0) \cdot \eta \cdot \frac{\alpha_{SC,dB}}{4.343} \cdot \exp^{-2\alpha z} & (z < L_s) \\ A_s^2 \frac{v_g\tau}{2} P(0) \cdot \eta \cdot \frac{\alpha_{SC,dB}}{4.343} \cdot \exp^{-2\alpha z} & (z > L_s) \end{cases} \quad (4.3.10)$$

where L_s is the location and A_s is the fractional loss of the connection. A_s is squared in Equation 4.3.10 because the roundtrip pass of the optical signal across the connection point. Therefore the connection loss can be obtained from the measured OTDR trace; simply compare the backscattered optical power immediately before and immediately after the connection point, as illustrated in Figure 4.3.6:

$$A_s = \sqrt{\frac{P_{BS}(L_s + \delta)}{P_{BS}(L_s - \delta)}}\Bigg|_{\delta \to 0} \quad (4.3.11)$$

where δ is a very short fiber length.

However, this measurement may not be accurate if the two fibers are not of the same type. In fact, even if they are of the same type, different fiber spools may still have slightly different numerical apertures or different attenuation coefficients. In this more practical case, the backscattered optical powers measured immediately before and after the splicing point are

$$P_{BS}(L_s - \delta) = \frac{v_g\tau}{2} P(0) \cdot \eta_1 \cdot \frac{\alpha_{SC1,dB}}{4.343} \cdot \exp^{-2\alpha_1 L_s} \quad (4.3.12a)$$

$$P_{BS}(L_s + \delta) = A_s^2 \frac{v_g\tau}{2} P(0) \cdot \eta_2 \cdot \frac{\alpha_{SC2,dB}}{4.343} \cdot \exp^{-2\alpha_1 L_s} \quad (4.3.12b)$$

where η_1 and η_2 are the conversion efficiencies defined by Equation 4.3.6 for the first and second fiber sections and α_{SC1} and α_{SC2} are the scattering

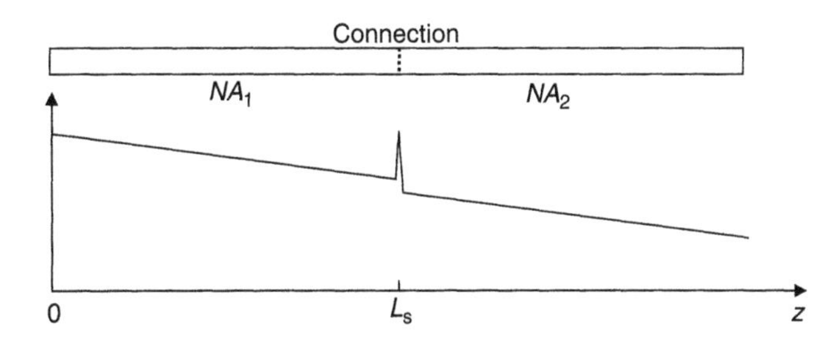

Figure 4.3.6 Illustration of an OTDR trace for a fiber with splicing.

coefficients for the two fiber sections, respectively. Since $\delta \to 0$, the fiber attenuation in the second fiber section is negligible; therefore only α_1 is used in Equation 4.3.12, which is the fiber loss coefficient in the first fiber section. Obviously,

$$A_{s,L} = \sqrt{\frac{P_{BS}(L_s+\delta)}{P_{BS}(L_s-\delta)}}\Bigg|_{\delta\to 0} = A_s\sqrt{\frac{\eta_2\cdot\alpha_{SC2,dB}}{\eta_1\cdot\alpha_{SC1,dB}}} \tag{4.3.13}$$

may not accurately predict the splicing loss. The measurement error can be caused by the difference in the scattering and capturing efficiencies between the two fiber sections.

To overcome this measurement error, a useful technique often used in fiber testing is to perform OTDR measurement from both sides of the fiber. First, if the OTDR is used from the left side of the fiber shown in Figure 4.3.6, the measured OTDR trace in the vicinity of the splicing point, $z = L_s$, is given by Equation 4.3.12. As the second step, the OTDR measurement is launched from the right side of the fiber. The measured backscattering values immediately after and before the splicing point are,

$$P'_{BS}(L_s-\delta) = A_s^2\frac{v_g\tau}{2}P(0)\cdot\eta_1\cdot\frac{\alpha_{SC1,dB}}{4.343}\cdot\exp^{-2\alpha_2 L'_s} \tag{4.3.14a}$$

$$P'_{BS}(L_s+\delta) = \frac{v_g\tau}{2}P(0)\cdot\eta_2\cdot\frac{\alpha_{SC2,dB}}{4.343}\cdot\exp^{-2\alpha_2 L'_s} \tag{4.3.14b}$$

where α_2 is the loss coefficient and L'_s is the length of the second fiber section, respectively (from the splicing point to the right end). From this second measurement, one can obtain

$$A_{s,R} = \sqrt{\frac{P'_{BS}(L_s-\delta)}{P'_{BS}(L_s+\delta)}}\Bigg|_{\delta\to 0} = A_s\sqrt{\frac{\eta_1\cdot\alpha_{SC1,dB}}{\eta_2\cdot\alpha_{SC2,dB}}} \tag{4.3.15}$$

Because the actual connection loss is reciprocal and should be independent of the direction in which the OTDR signal is launched, the value of A_s in Equations 4.3.13 and 4.3.15 should be identical. Combining the results of these two measurements, the correct connection loss can be obtained as

$$A_s = \sqrt{A_{s,L}A_{s,R}} \tag{4.3.16}$$

This automatically removes the measurement error introduced by fiber type mismatch [15].

In general, an optical fiber may consist of many sections that are concatenated together by splicing and connecting. Figure 4.3.7 illustrates a typical OTDR trace for a fiber which is composed of multiple sections. In this case each fiber section may have different attenuation coefficient which can be seen from the slope of the section on the OTDR trace. The sharp peaks in the OTDR trace

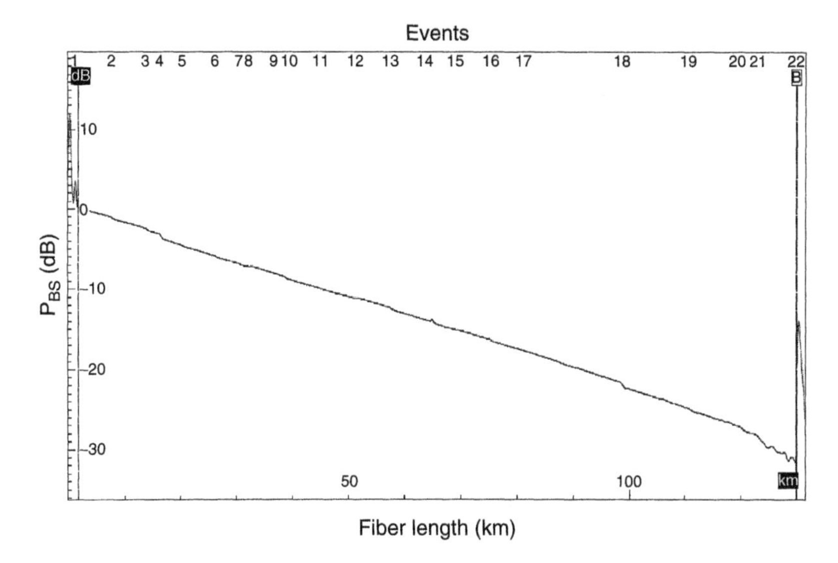

Figure 4.3.7 Illustration of an OTDR trace for a fiber with multiple sections.

indicate the locations and the strengths of reflections from the connectors, while the splices simply introduce attenuation and the reflection is usually small. In addition, negative attenuations can be found in a number of fiber splices (events 8, 12, 15 and 17 in Tab. 4.3.1). This is the result of the change in the numerical aperture of the fiber sections. Therefore, more accurate evaluation of splicing loss has to use the measurements from both directions as indicated in Equation 4.3.16.

For accurate fiber attenuation measurement using OTDR, receiver SNR is a very important concern; it limits the total fiber length the OTDR can measure. For example, for a high-quality receiver with noise-equivalent power of $NEP = 0.5\ pW/\sqrt{Hz}$, if the bandwidth of the receiver is 1 MHz, the required signal optical power that reaches the receiver is $P_{BS} = 0.5\ nW = -63\ dBm$ to achieve $SNR = 1$. Because the noise is random while the signal is deterministic, the SNR can be improved by averaging. In general, the average over N pulses will increase the SNR by simply a factor of $\sqrt{N}$. Therefore, an OTDR measurement usually takes time, from a few seconds to a few minutes, depending on the number of averages required.

4.3.3 Improvement Considerations of OTDR

One possibility to improve the sensitivity of an OTDR is to use coherent detection. Unlike direct detection where the SNR is mainly limited by the receiver thermal noise and the dark current noise of the photodiode, the SNR

Table 4.3.1

Explanation of the OTDR trace in Fig. 4.3.6

	Event (23)	Distance (km)	Attenuation (dB)	Reflectance (dB)	Slope (dB/km)	Rel. Dist. (km)	Link Budget (dB)
1		0.98226	0.545	−41.09		0.98226	0.191
2		7.53062	0.168		0.195	6.54837	2.013
3		13.75157	0.219		0.202	6.22095	3.446
4		16.20721	0.532		0.202	2.45564	4.152
5		20.13623	0.100		0.216	3.92902	5.541
6		26.19347	0.131		0.204	6.05724	6.868
7		30.44991	0.098		0.195	4.25644	7.837
8		32.33257	−0.159		0.187	1.88266	8.280
9		37.32570	0.037		0.209	4.99313	9.174
10		38.63537	0.187		0.214	1.30967	9.492
11		44.93817	0.076		0.200	6.30280	10.937
12		51.07727	−0.181		0.194	6.13909	12.195
13		57.38007	0.234		0.202	6.30280	13.295
14		63.60102	0.153		0.211	6.22095	14.843
15		68.92157	−0.066		0.199	5.32055	16.047
16		75.14252	0.188		0.208	6.22095	17.285
17		81.11791	−0.012		0.200	5.97539	18.665
18		98.47108	0.407		0.232	17.35317	22.682
19		110.74927	0.165		0.227	12.27819	25.862
20		119.83513	0.372		0.228	9.08586	28.093
21		122.94560	0.692		0.229	3.11047	29.174
22		129.98510		−9.81	0.360	7.03950	32.422
23				<31.07			

in coherent detection is primarily limited by the shot noise due to the strong local oscillator. As shown in Figure 2.7.4, the SNR decreases with signal optical power at the rate of –20 dB/decade for direct detection, whereas for coherent detection this slope is only –10 dB/decade. That is why coherent detection is especially advantageous at very low signal optical power levels in comparison with direct detection.

Figure 4.3.8 shows the block diagram of a coherent OTDR, where the output from a single-frequency laser diode is split into two parts. One of them is used as

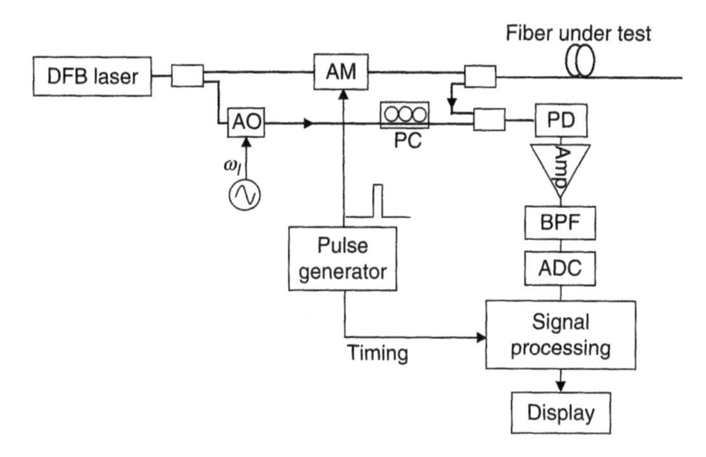

Figure 4.3.8 Block diagram of an OTDR with coherent detection. *AM:* amplitude modulator, *AO:* acousto-optic frequency modulator, *PC:* polarization controller, *PD:* photodetector, *BPF:* RF bandpass filter, *ADC:* analog-to-digital converter.

the stimulating signal that is modulated into a pulse train by an external modulator and sent into the fiber under test. The other part is used as the local oscillator for coherent detection. An acousto-optic (AO) modulator is used to shift the optical frequency of the local oscillator by ω_{IF}, which determines the intermediate frequency (IF) of the coherent heterodyne detection. The backscattered optical signal from the fiber is combined with the frequency-shifted local oscillator through a fiber coupler and detected by a wideband photodiode. A polarization controller is used in the local oscillator branch to match the state of polarization and maximize the coherent detection efficiency. An RF bandpass filter with the central frequency at ω_{IF} is used to select the intermediate frequency component. Generally, in coherent detection the power of the local oscillator is strong enough such that the receiver SNR is mainly limited by the shot noise of the local oscillator.

As an example, assume that the photodiode responsivity is $\Re = 1A/W$ and the receiver bandwidth is 1 MHz. Based on Equation 2.7.16, the required signal optical power is approximately –95 dBm to achieve SNR = 1. Compared to the direct detection case discussed at the end of Section 4.3.2 in which the minimum detectable power level was –63 dBm, the OTDR with coherent detection has more than three orders of magnitude sensitivity improvement over its direct detection counterpart. However, the price paid for this sensitivity improvement are increased complexity as well as the requirements of a high-quality single-frequency laser diode and the frequency shifter for the local oscillator.

Spatial resolution is another concern for an OTDR. A wide optical pulse carries high optical energy for a fixed-peak power level, and at the same time the required receiver electrical bandwidth is also relatively low. Both of these factors may contribute to increasing the receiver SNR. However, wide pulse width corresponds to lower spatial resolution in the OTDR measurement. In fact, the spatial resolution of an OTDR is proportional to the pulse width τ by $R_{resolution} = 0.5v_g\tau$. For example, a pulse width of 100 ns in an OTDR will provide a spatial resolution of approximately 10 m in the fiber. Therefore trade-offs have to be made between the resolution and the SNR.

Dynamic range of the optical receiver in an OTDR is also an important issue. In the example shown in Figure 4.3.5, there is 50 dB reduction in the received optical power level over 100 km single-mode fiber. Theoretically the dynamic range requirement is determined by the roundtrip fiber attenuation over the total length L, which is $\Delta P_{dB} = 2\alpha_{dB}L$. Commercial OTDRs usually provide >90 dB dynamic range to allow the measurement of long fibers with high losses.

Another practical issue is that one has to consider the Fresnel reflection at the fiber input terminal. For a well-cleaved fiber end facet, considering the refractive indices between air ($n = 1$) and silica ($n = 1.47$), the Fresnel reflection is approximately 3.6 percent, or –14.4 dB, which is several orders of magnitude higher than the Rayleigh backscattering from inside the fiber. Therefore most OTDR designs have to block this front-end reflection using proper time gating. This usually introduces difficulty in the measurement of fiber characteristics near the input, which is commonly referred to as the *dead zone*. Nevertheless, OTDR is by far the most popular equipment for fiber attenuation measurement, fiber length measurement, and troubleshooting in fiber-optical systems.

4.4 FIBER DISPERSION MEASUREMENTS

The phenomenon of dispersions in optical fibers, their creation mechanisms, and their consequences in fiber-optic systems were discussed briefly in Chapter 1. Fiber dispersion can be categorized into intermodal dispersion and chromatic dispersion. Intermodal dispersion is caused by the fact that different propagation modes in a fiber travel at different speeds. Usually a large number of modes coexist in a multimode fiber; therefore intermodal dispersion is the major source of dispersion in multimode fibers. Chromatic dispersion is originated from the fact that different frequency components in each propagation mode may travel at slightly different speeds and therefore it is the dominant dispersion source in single-mode fibers. Both intermodal dispersion and chromatic dispersion may cause optical signal pulse broadening and waveform distortion in fiber-optic systems.

4.4.1 Intermodal Dispersion and Its Measurement

In a multimode optical fiber, the optical signal is carried by many different propagating modes. Each mode has its own spatial field distribution and its propagation constant in the longitudinal direction. Because different modes have different propagation speeds, an optical pulse has different arrival times after traveling through the fiber.

Figure 4.4.1 shows a ray trace of propagation modes in a multimode fiber, where each mode travels in a different trace angle. According to Fresnel refraction theory, the largest incidence angle at the core/cladding interface is determined by the critical angle, $\theta \geq \theta_c = \sin^{-1}(n_2/n_1)$. Obviously the slowest propagation mode is the one with the incidence angle $\theta = \theta_c$ at the core/cladding interface, whereas the fastest propagation mode is the one that propagates exactly in the direction of the fiber axis ($\theta = \pi/2$). Thus to the first approximation, the maximum intermodal dispersion can be estimated by the propagation delay difference between these two extreme ray traces. The maximum differential delay for a fiber of unit length is

$$D_{\text{mod}} = \frac{n_1}{c}\left(\frac{1}{\sin\theta_c} - 1\right) = \frac{n_1}{c}\left(\frac{n_1 - n_2}{n_2}\right) \tag{4.4.1}$$

The unit of intermodal dispersion is in [ns/km]. Equation 4.4.1 indicates that intermodal dispersion is proportional to the index difference between fiber core and cladding. For example, for a multimode fiber with a numerical aperture of $NA = 0.2$ and $n_1 = 1.5$, the absolute maximum intermodal dispersion is approximately 45 ns/km. However, in practice most of the optical signal energy in a multimode fiber is carried by lower-order modes and their intermodal dispersion values are much lower than that predicted by Equation 4.4.1. The actual intermodal dispersion value in a multimode fiber is on the order of 200 ps/km.

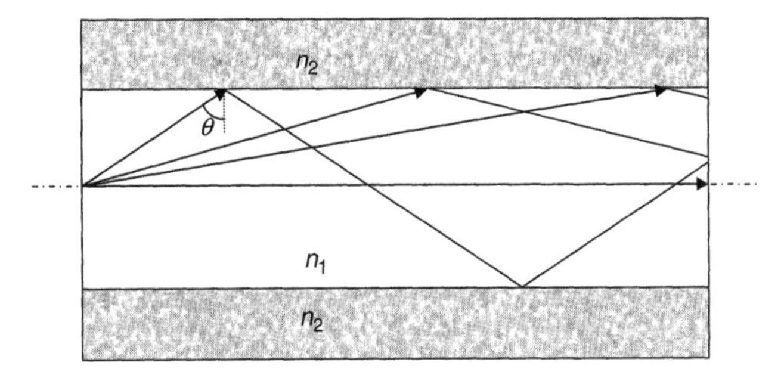

Figure 4.4.1 Geometric representation of propagation modes in a multimode fiber.

Generally, with an arbitrary launching condition, the lightwave signal in a multimode fiber excites only a certain number of modes instead of all the possible guided modes that can be supported by the fiber. Therefore the optical signal energy distribution among various modes is somewhat arbitrary near the input of the fiber which is quite sensitive to the launching condition. This may cause uncertainties in the intermodal dispersion measurements. In practice, an optical fiber is not ideally uniform in geometry. Signal optical energy may randomly couple between modes when there are bending, twisting and other intrinsic and extrinsic perturbations. Thus after a certain transmission distance, energy will be eventually coupled into most of the modes that the fiber can support and reaches to the equilibrium distribution of modes. To speed up this mode redistribution process, mode scrambler is often used in the measurement setups to make the results more stable and reliable. This can be accomplished by twisting or winding the fiber on an array of mechanical bars; therefore the measured intermodal dispersion is less sensitive to other mechanical disturbances.

Intermodal dispersion of a multimode fiber can be measured both in time domain and in frequency domain [16]. Since the dispersion value in a multimode fiber is typically >100 ps/km, the measurement accuracy requirement is generally not as stringent as it is for the chromatic dispersion measurement in single-mode fibers, where the dispersion values are much smaller. However, due to the randomness of mode coupling in a multimode fiber, the measurement only provides a statistic value of intermodal dispersion and a mode scrambler has to be used to ensure the repeatability of the measurement, especially when the fiber is short.

4.4.1.1 Pulse Distortion Method

Figure 4.4.2 shows the block diagram of an intermodal dispersion measurement setup using the pulse distortion method, which is also referred to as the pulse-broadening technique. In this setup, a short optical pulse is generated by a laser diode that is directly modulated by an electrical short pulse train or by active mode locking. The optical pulses are injected into the multimode fiber through a mode scrambler to ensure the equilibrium modal distribution in the fiber. A high-speed photodetector is used at the receiver to convert the received optical pulses into the electrical domain; therefore they can be amplified and the waveforms can be displayed on a sampling oscilloscope. Because of the intermodal dispersion of the fiber, the output pulses will be broader than the input pulses, and the amount of pulse broadening is linearly

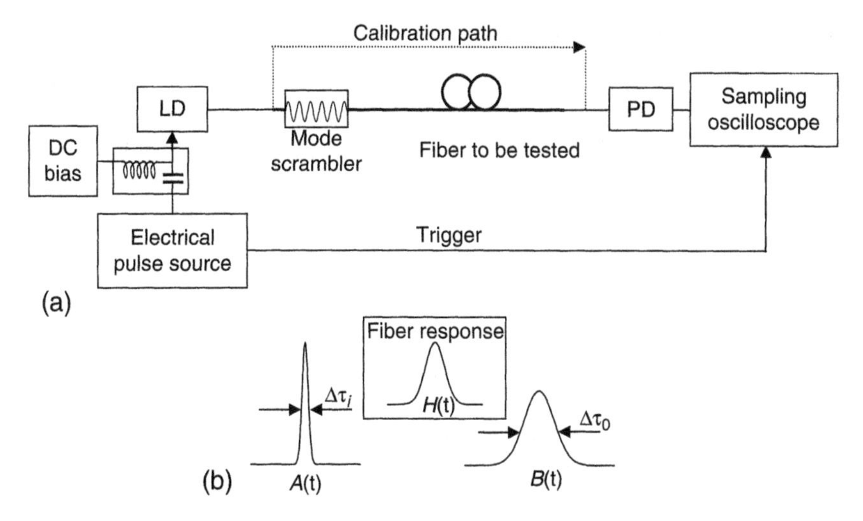

Figure 4.4.2 Pulse-broadening method to measure intermodal dispersion: (a) measurement setup and (b) input and output waveforms.

proportional to the intermodal dispersion of the fiber. In this way, the fiber dispersion characteristics can be directly measured in the time domain.

Assuming that the optical signal power is low enough so that the system operates in the linear regime, the relationship between the input waveform A(t) and the output waveform B(t) is related by

$$B(t) = A(t) \otimes H(t) \tag{4.4.2}$$

where $\otimes$ stands for convolution and H(t) is the time-domain transfer function of the fiber. Consider in a multimode fiber, intermodal dispersion is the dominant source of dispersion effect; thus H(t) essentially represents intermodal dispersion of the fiber. To take into account the finite bandwidth of the measurement system other than the fiber, we need a system calibration that directly connects the transmitter to the receiver, as illustrated by the calibration path in Figure 4.4.2. Since this is a linear system in which superposition is valid, the waveform measured in the calibration process can be regarded as the input waveform A(t) in the calculation of the fiber transfer function H(t). If we use Gaussian approximations for both the input and output signal waveforms and the fiber transfer function, the bandwidth of the fiber can be simply evaluated by the pulse width difference between the input A(t) and the output B(t). If the input pulse width is $\Delta\tau_i$ and the output pulse width is $\Delta\tau_0$, the response time of the fiber will be $\Delta\tau_H = \Delta\tau_0 - \Delta\tau_i$.

4.4.1.2 Frequency Domain Measurement

Although the time-domain pulse-broadening technique is easily understandable, a more straightforward way to find the fiber transfer function is in frequency domain. By converting the time-domain waveforms $A(t)$ and $B(t)$ into frequency-domain $A(\omega)$ and $B(\omega)$ through Fourier transform, the frequency domain transfer function of the fiber can be directly obtained by

$$H(\omega) = \frac{B(\omega)}{A(\omega)} \tag{4.4.3}$$

Figure 4.4.3 shows the block diagram of intermodal dispersion measurement in the frequency domain using an electrical network analyzer, where the network analyzer is set in the S_{21} parameter mode. A laser diode is intensity-modulated by the output of the network analyzer, which is a frequency-swept RF source. The DC bias and the RF modulation to the laser are combined through a bias-tee. The optical signal is launched into the fiber under test through a mode scrambler. A wideband photodiode is used to convert the received optical signal into electrical domain and the RF signal is then fed to the receiving terminal of the network analyzer. Because the effect of fiber dispersion is equivalent to a lowpass filter, by comparing the transfer functions

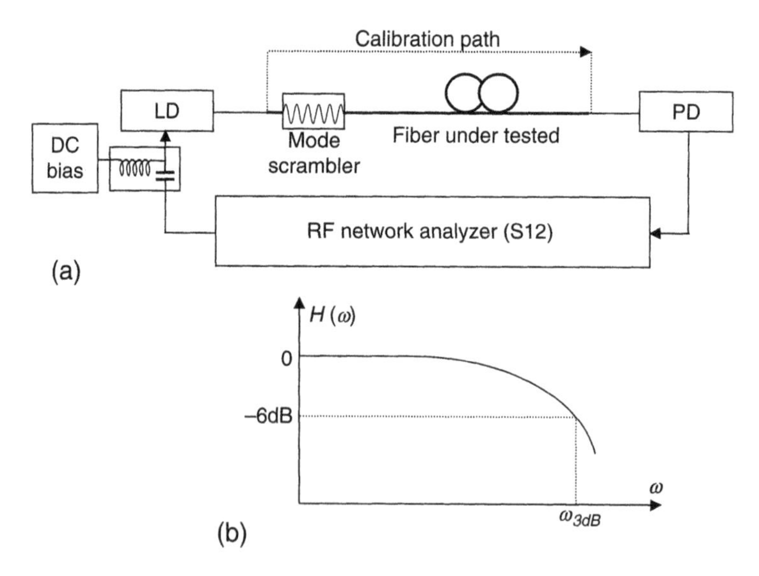

Figure 4.4.3 Frequency-domain method to measure intermodal dispersion: (a) measurement setup and (b) electrical power transfer function.

with and without the fiber in the system, the net effect of bandwidth limitation caused by the fiber can be obtained. The use of an RF network analyzer makes this measurement very simple and accurate. Compared to the time-domain measurement discussed previously, this frequency-domain technique does not require a short pulse generator and a high-speed oscilloscope. The fiber transfer function $H(t)$ is measured directly without the need for additional calculation using FFT.

The measurement procedure can be summarized as follows:

1. Remove the fiber and use the calibration path to measure the frequency response of the measurement system (transmitter/receiver pair and other electronic circuitry), which is recorded as $H_{cal}(\omega)$.
2. Insert the optical fiber and measure the frequency response again, which is recorded as $H_{sys}(\omega)$. Then the net frequency response of the fiber is

$$H_{fib}(\omega) = \frac{H_{sys}(\omega)}{H_{cal}(\omega)} \tag{4.4.4}$$

It is important to note that in this setup, the S_{21} parameter of the electrical power spectral transfer function is measured by the RF network analyzer. We know that at the output of the photodiode, signal electrical power is proportional to the square of the received optical power; therefore 3 dB optical bandwidth is equivalent to 6 dB bandwidth of the electrical transfer function of $H_{fib}(\omega)$. Or simply, a 6 dB change in the measured $H_{fib}(\omega)$ is equivalent to a 3 dB change in the optical power level.

Obviously, both the pulse-broadening method and the frequency-domain method are very effective in characterizing fiber dispersion by measuring the dispersion-limited information bandwidth. In general, the pulse-broadening method is often used to characterize large-signal response in the time domain, whereas the frequency-domain method is only valid for small-signal characterization. However, in fiber dispersion characterization, especially in the measurement of multimode fibers, the nonlinear effect is largely negligible, and thus the time-domain and frequency-domain methods should result in the same transfer function.

Another important note is that the unit of intermodal dispersion is in [ps/km], whereas the unit of chromatic dispersion is in [ps/nm/km]. The reason is that pulse broadening caused by intermodal dispersion is independent of the spectral bandwidth of the optical source and it is only proportional to the fiber length. Because the value of intermodal dispersion in multimode fibers is on the order of 100 ps/km, the RF bandwidth requirement of the measurement equipment is not very stringent. Typically a measurement setup with 15 GHz RF bandwidth is sufficient to characterize a multimode fiber of >2 km in length.

4.4.2 Chromatic Dispersion and Its Measurement

Chromatic dispersion is introduced because different frequency components within an optical signal may travel in different speed, which is the most important dispersion effect in single-mode fibers. But this does not mean a multimode fiber would not have chromatic dispersion. Since chromatic dispersion is usually much smaller than intermodal dispersion for the majority of multimode fibers, its effect is often negligible. From a measurement point of view, because of the relatively small chromatic dispersion values, the dispersion-limited information bandwidth is usually very wide; it is rather difficult to directly measure this equivalent bandwidth. For example, for a standard single-mode fiber operating in the 1550 nm wavelength window, the chromatic dispersion is approximately 17 ps/nm/km. It would require a picosecond level of pulse width for the optical source and >100 GHz optical bandwidth in the receiver to directly measure the pulse broadening in a 1 km fiber, which would be extremely challenging. Therefore alternative techniques have to be used to characterize chromatic dispersion in single-mode fibers. There are a number of different methods for this purpose, but in this section we will discuss the two most popular techniques: (1) the phase shift method, which operates in the time domain, and (2) the AM response method, which operates in the frequency domain.

4.4.2.1 Modulation Phase Shift Method

Figure 4.4.4 shows the measurement setup of the modulation phase shift method to characterize fiber chromatic dispersion. A tunable laser is used as the light source whose wavelength can be adjusted to cover the wavelength window in which the chromatic dispersion needs to be measured. An electro-optic intensity modulator converts a sinusoidal electrical signal at frequency f_m, into

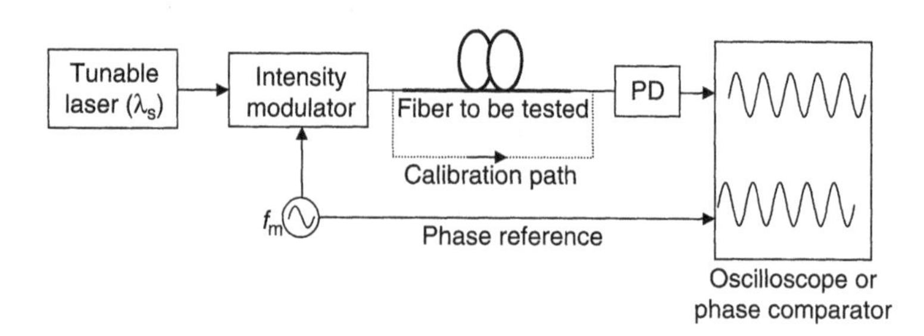

Figure 4.4.4 Experimental setup of the modulation phase shift method to measure fiber chromatic dispersion.

optical domain. After the modulated optical signal passes through the optical fiber, it is detected by a photodiode, and a sinusoidal electrical signal is recovered that is then amplified and sent to an oscilloscope. The waveform of the same sinusoid source at frequency f_m is also directly measured by the same oscilloscope for phase comparison.

Because of the chromatic dispersion in the fiber, the propagation delay of the optical signal passing through the fiber is different at different wavelengths: $\tau_g = \tau_g(\lambda)$. To measure this wavelength-dependent propagation delay, the optical source is modulated by a sinusoidal wave and the propagation delay can be evaluated by the relative phase retardation of the received RF signal. By varying the wavelength of the tunable laser, the RF phase delay as a function of the source wavelength, $\phi = \phi(\lambda)$, can be obtained [17, 18].

If the modulating frequency of the sinusoidal wave is f_m, a relative phase delay of $\Delta\phi = 360°$ between two wavelength components corresponds to a group delay difference of $\Delta\tau_g = 1/f_m$ between them. Therefore the group delay versus optical signal wavelength λ can be expressed as

$$\Delta\tau_g(\lambda) = \frac{\phi(\lambda) - \phi(\lambda_r)}{360°}\frac{1}{f_m} \tag{4.4.5}$$

where $\phi(\lambda) - \phi(\lambda_r)$ is the RF phase difference measured between wavelength λ and a reference wavelength λ_r. With the knowledge of group delay as a function of wavelength, $\phi(\lambda)$, the chromatic dispersion coefficient $D(\lambda)$ can be easily derived as

$$D(\lambda) = \frac{1}{L}\frac{d(\Delta\tau_g(\lambda))}{d\lambda} = \frac{1}{360° L f_m}\frac{d\phi(\lambda)}{d\lambda} \tag{4.4.6}$$

where L is the fiber length. Since the value of chromatic dispersion is determined by the derivative of the group delay instead of its absolute value, the reference phase delay $\phi(\lambda_r)$ and the choice of reference wavelength are not important. In comparison, the selection of the RF modulation frequency may have a much bigger impact on the measurement accuracy. In general, a high modulation frequency helps increase phase sensitivity in the measurement. For a given fiber dispersion, the phase delay per unit wavelength change is linearly proportional to the modulation frequency: $d\phi(\lambda)/d\lambda = 360° D(\lambda) L f_m$.

However, if the modulation frequency is too high, the measured phase variation may easily exceed 360° within the wavelength of interest; therefore the measurement system has to track the number of full cycles of 2π phase shift. In addition, the wavelength tuning step of the source needs to be small enough such that there is enough number of phase measurements within each 2π cycle. Therefore, the selection of the modulation frequency depends on the chromatic dispersion value of the fiber under test and the wavelength window the measurement needs to cover.

The tunable laser in Figure 4.4.4 can also be replaced by the combination of a wideband light source, such as superluminescence LED (SLED), and a tunable bandpass optical filter. The SLED can be directly modulated by injection current, and the bandpass optical filter can be placed either immediately after the SLED (before the fiber) or at the receiver side before the photodiode to select the signal wavelength.

As an example, a spool of 40 km standard single-mode fiber needs to be characterized in the wavelength window between 1530 nm and 1560 nm. We know roughly that the zero-dispersion wavelength of the fiber is around $\lambda_0 = 1310$ nm and the dispersion slope is approximately $S_0 = 0.09$ ps/nm^2-km. Using the Sellmeier equation shown in Equation 1.3.84, the wavelength-dependent group delay is expected to be

$$\tau_g(\lambda) = L\int_{\lambda_r}^{\lambda} D(\lambda)d\lambda = \frac{S_0 L}{8}\left(\lambda^2 + \frac{\lambda_0^4}{\lambda^2}\right) + \tau_g(\lambda_r) \tag{4.4.7}$$

Figure 4.4.5 shows the expected relative group delay $\Delta\tau_g(\lambda) = \tau_g(\lambda) - \tau_g(\lambda_r)$ and the number of full 2π cycles of RF phase shifts for the modulation frequencies of 1 GHz, 5 GHz, and 10 GHz. To obtain this plot, the reference wavelength was set at 1530 nm. In terms of the measurement setup, the speed requirement of the electronic and opto-electronic devices and instrument is less than 10 GHz, which is usually easy to obtain. The measurement accuracy may be limited by the wavelength accuracy of the tunable laser and the step size of wavelength tuning. It may also be limited by noises in the measurement system when the signal level is low. Since this is a narrowband measurement, an RF bandpass filter at the modulation frequency can be used at the receiver to significantly reduce the noise level and improve the measurement accuracy.

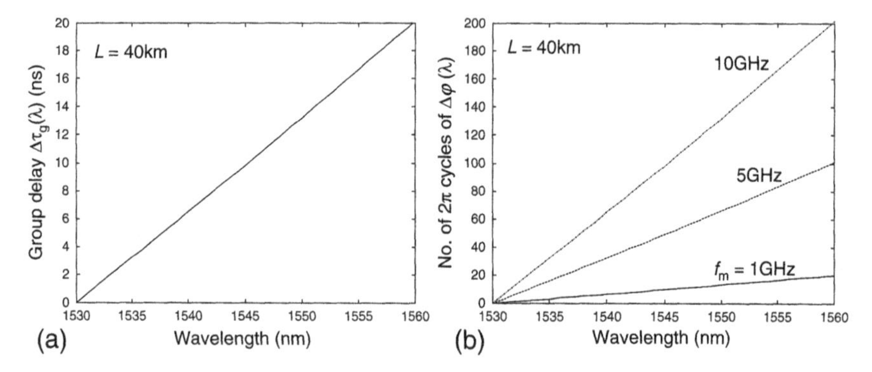

Figure 4.4.5 (a) Relative group delay and (b) the number of full 2π cycles of phase shift for the modulation frequencies of 1 GHz, 5 GHz, and 10 GHz. Fiber length is 40 km.

4.4.2.2 Baseband AM Response Method

The modulation phase shift method previously discussed relies on the RF phase comparison between the reference path and the signal passing through the fiber under test. An oscilloscope or a digitizing system may be used to perform waveform comparison. In this section we discuss the baseband AM response method, which measures chromatic dispersion based on the interference between modulation sidebands [19, 20]. Because of the chromatic dispersion, different modulation sidebands may experience different phase delays, and their interferences at the receiver can be used to predict the dispersion value.

Figure 4.4.6 shows the experimental block diagram of the AM response method to measure fiber chromatic dispersion. A tunable laser is used as the optical source whose output is intensity modulated by an external electro-optic modulator. The modulator is driven by the output from an RF network analyzer operating in the S_{21} mode. The network analyzer provides a frequency-swept RF signal to modulate the optical signal. After passing through the optical fiber under test, the optical signal is detected by a wideband photodiode. Then the weak RF signal is amplified and fed to the receiving port of the network analyzer. In fact, this measurement technique is similar to that shown in Figure 3.2.9 in Chapter 3, where the modulation chirp of an optical transmitter was characterized.

Through the analysis in Section 3.2 we know that the AM response of the fiber system shown in Figure 4.4.6 is not only determined by the transmitter modulation chirp; it also depends on the chromatic dispersion of the fiber. According to the AM modulation response given by Equation 3.2.22 the photocurrent in the receiver is,

$$I(f) \propto \cos\left[\frac{\pi\lambda_s^2 D(\lambda_s) f^2 L}{c} + \tan^{-1}(\alpha_{lw})\right] \tag{4.4.8}$$

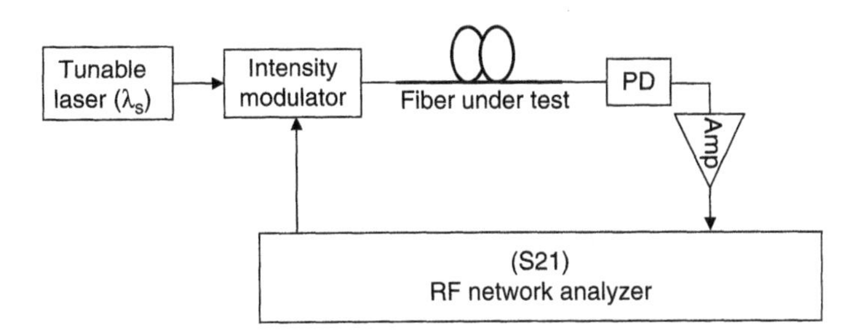

Figure 4.4.6 Experimental block diagram of AM response method to measure fiber chromatic dispersion.

where α_{lw} is the modulator chirp parameter and f is the modulation frequency. An important feature of this AM modulation response is the resonance zeroes at frequencies determined by the following equation:

$$f_k = \sqrt{\frac{c}{2D(\lambda_s)L\lambda_s^2}\left(1 + 2k - \frac{2}{\pi}\tan^{-1}(\alpha_{lw})\right)} \tag{4.4.9}$$

In Chapter 3, the purpose of the experiment was to measure the modulator chirp parameter; we assumed that the fiber dispersion parameter D and the fiber length L were known. In this section, the purpose is to measure fiber chromatic dispersion, so we need to eliminate the impact of modulator chirp of the transmitter. From Equation 4.4.9, we can find that two consecutive zeroes in the AM response, f_k and f_{k+1}, are related only to the accumulated chromatic dispersion:

$$D(\lambda_s) \cdot L = \frac{c}{\left(f_{k+1}^2 - f_k^2\right)\lambda_s^2}, \tag{4.4.10}$$

This allows the accurate determination of fiber dispersion by measuring the spectrum of AM modulation response at each wavelength of the laser source. By adjusting the wavelength λ_s of the laser source, the wavelength-dependent chromatic dispersion parameter $D(\lambda_s)$ can be obtained. Compared to the modulation phase shift method, this AM response technique directly measures dispersion parameter $D(\lambda_s)$ at each source wavelength λ_s without the need to measure group delay around this wavelength. Since the zeroes in the spectrum of the AM response are sharp enough, as indicated by Figure 3.2.12, the measurement can be quite accurate. However, one practical concern might be the required RF bandwidth of the electro-optic devices as well as the RF network analyzer. In general, short fibers and low chromatic dispersions in the fiber require wide bandwidth of the measurement system. As an example, Figure 4.4.7(a) shows the normalized AM response for 100 km standard single-mode fiber with a 17 ps/nm/km dispersion parameter and 100 km True-wave fiber (TWF) with a 2 ps/nm/km dispersion parameter. The source modulation chirp is assumed to be zero. Figure 4.4.7(b) shows the first three frequencies of resonance zero in AM response as the function of the accumulated fiber dispersion $D \cdot L$. Using a network analyzer with 20 GHz bandwidth, the minimum accumulated dispersion the system can measure is about 500 ps/nm; this is because at least two resonance zeroes f_0 and f_1 need to be observed within the 20 GHz network analyzer bandwidth. 500 ps/nm corresponds to approximately 29 km of standard single-mode fiber with 17 ps/nm/km dispersion parameter.

From our discussion on chromatic dispersion measurement so far, we've found both modulation phase shift and AM modulation response techniques suitable for measuring fibers with relatively large accumulated dispersion. If

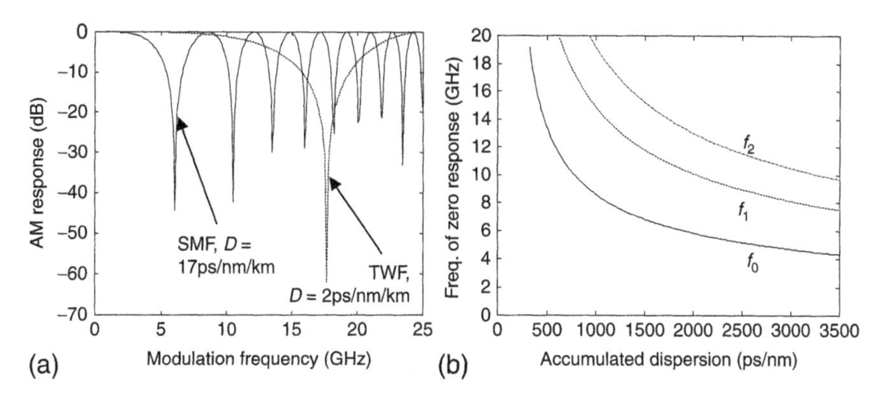

Figure 4.4.7 (a) Normalized AM responses of SMF with $D = 17$ ps/nm/km and TWF with $D = 2$ ps/nm/km. $L = 100$ km and $\alpha_{lw} = 0$ are assumed. (b) Frequencies of resonance zeroes in the AM response versus the accumulated chromatic dispersion DL.

a fiber sample is very short, for example, only a few meters, the required RF bandwidth for both of these techniques would have to be extremely wide. In this case, alternative techniques have to be used such as interferometric method.

4.4.2.3 Interferometric Method

The basic block diagram of the interferometric method for the measurement of chromatic dispersion is shown in Figure 4.4.8 [21,22]. A tungsten lamp is usually used as a wideband light source that can cover a very wide wavelength window from 500 nm to 1700 nm. The signal wavelength is selected by a tunable optical bandpass filter with the FWHM bandwidth of $\Delta\lambda$. The wavelength selection can also be accomplished with a monochromator before the photodetector. In that case, the bandwidth can be adjusted by changing the resolution of the monochromator. The optical signal is then fed into a Mach-zehnder interferometer with the fiber-under-test put in one arm while the length of the other arm is continuously variable. A photodiode at the output of the interferometer detects the optical signal and the waveform is displayed on an oscilloscope.

For the measurement using a Mach-zehnder interferometer, as discussed in Section 3.2.3, if the length difference between the two arms is much shorter than the source coherence length, coherent interference happens, and the photocurrent generated in the photodiode is

$$i(t) = \Re\left\{P_1 + P_2 + 2\sqrt{P_1 P_2}\cos[\Delta\phi(\Delta l)]\right\} \tag{4.4.11}$$

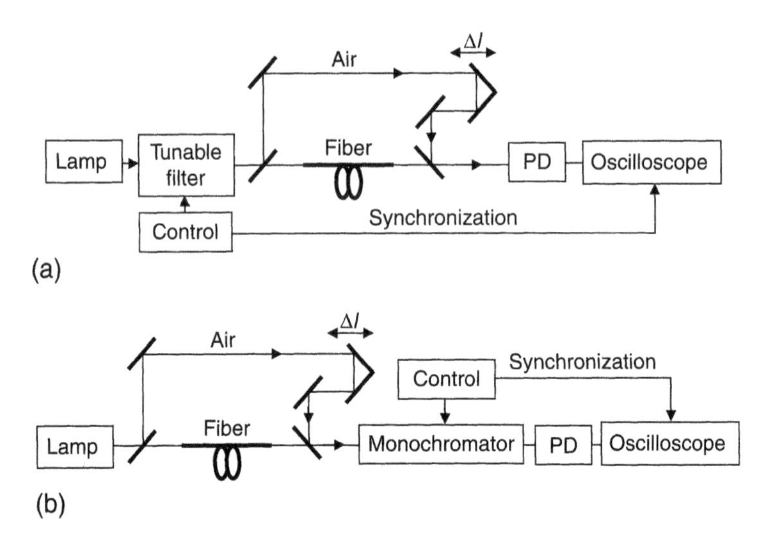

Figure 4.4.8 Interferometric technique to measure fiber chromatic dispersion: (a) using a tunable filter and (b) using a monochromator.

where Δl is the arm length difference, P_1 and P_2 are the optical powers in the two arms, $\Re$ is the photodiode responsivity, and $\Delta\varphi(\Delta l)$ is the relative optical phase difference between the two arms. On the other hand, if the length difference between the two interferometer arms is much longer than the source coherence length, the interference is incoherent and the photocurrent generated in the photodiode is simply proportional to the power addition of the two arms, that is,

$$i(t) = \Re\{P_1 + P_2\} \tag{4.4.12}$$

For the tunable optical filter with the FWHM bandwidth of $\Delta\lambda$, the coherence length of the selected optical signal is approximately

$$L_c = \frac{\lambda^2}{n_g \Delta\lambda} \tag{4.4.13}$$

where n_g is the group index of the material, and λ is the central wavelength selected by the optical filter. For example, for a signal optical bandwidth of $\Delta\lambda = 1$ nm in a 1550 nm wavelength region, the coherence length is approximately 2.4 mm in air.

Obviously, when the two interferometer arms have the same group delay, a slight change in the differential length Δl will introduce the maximum variation of photocurrent. Though outside the coherence length, the photocurrent is independent of the differential arm length Δl, as illustrated in Figure 4.4.9.

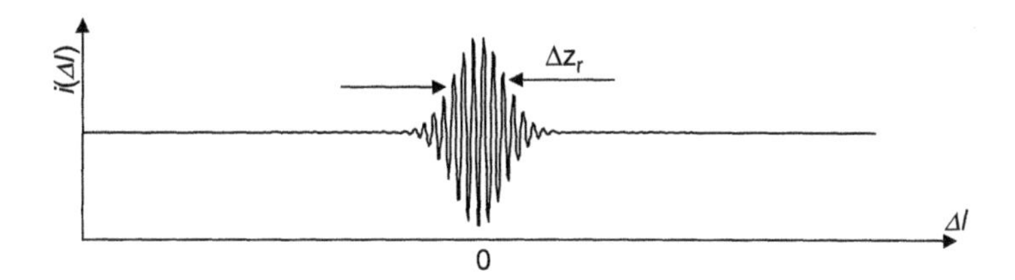

Figure 4.4.9 Illustration of the photocurrent versus the arm length difference.

The width of the envelope of the coherent interference pattern is approximately

$$\Delta z_r = \frac{\lambda^2}{n_g \Delta\lambda} \tag{4.4.14}$$

A narrower width of Δz_r is desired to help accurately determine the differential arm delay of the interferometer. According to the definition, chromatic dispersion is proportional to the derivative of the group delay:

$$D(\lambda) = \frac{1}{L}\frac{d(\tau_g(\lambda))}{d\lambda} \tag{4.4.15}$$

where L is the fiber length and $\tau_g(\lambda)$ is the wavelength-dependent group delay. To measure the chromatic dispersion in a fiber, its wavelength dependency of the group delay is the key parameter that needs to be evaluated. This can be accomplished by sweeping the wavelength of the tunable optical filter and measuring the relative delay as a function of the wavelength.

In the interferometer configuration shown in Figure 4.4.8, the reference arm for which group delay is wavelength-independent is in the air, and therefore the important parameter to evaluate is the differential arm length difference as a function of wavelength $\Delta l(\lambda)$, which indicates the chromatic dispersion of the fiber under test.

$$D(\lambda) = \frac{1}{Lc}\frac{d(\Delta l(\lambda))}{d\lambda} \tag{4.4.16}$$

As illustrated in Figure 4.4.10, when the tunable optical filter sweeps across the wavelength, the peak of the interference envelope will move, providing $\Delta l(\lambda)$. In practical measurements, the resolution of $\Delta l(\lambda)$ measurement is determined by the sharpness of the interference pattern Δz_r, as indicated by Equation 4.4.14, while the minimum wavelength step is limited by the bandwidth of the tunable filter $\Delta\lambda$. Although it is desirable for both of them to be small, Equation 4.4.14 indicated that their product is only determined by the square of the wavelength. As an example, if the filter bandwidth is 2 nm in a 1550 nm wavelength

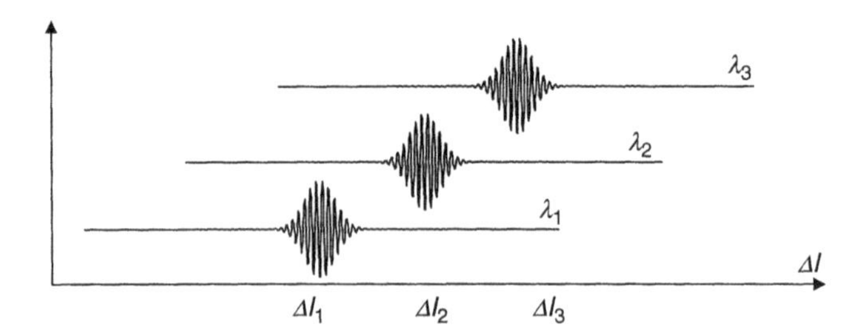

Figure 4.4.10 Illustration of measured arm length difference as a function of the filter wavelength.

window, the spatial resolution Δz_r is approximately 0.8 mm in the fiber, which corresponds to a relative time delay of 2.7 ps in air. The light source used in this technique can have a very wide wavelength window in which the measurement can take place. This fiber dispersion characterization technique covers wide bandwidth. In practice, the wavelength-dependent group delay and thus the chromatic dispersion can be measured in the wavelength window between 500 nm and 1700 nm by fitting the measured dispersion characteristic with the well-known three-term Sellmeier equation for standard single-mode fibers. It allows the precise determination of chromatic dispersion, even for short pieces of fiber samples.

However, the major disadvantage of this technique is that the fiber length cannot be too long. The major reason is that the reference arm is in free space so that a long reference arm may introduce mechanical instability and thus excessive coupling loss. Because the reference arm length has to be roughly equal to the length of the fiber sample, the typical fiber length in the interferometric measurement setup should not be longer than one meter. For fibers with very small dispersion parameters, or if the dispersion is a strong function of the wavelength that cannot be fitted by the Sellmeier equation, such as dispersion shifted or dispersion flattened fibers, it would be quite difficult to ensure the accuracy of the measurement.

To overcome the reference arm length limitation, an all-fiber setup can be useful for measuring the characteristics of long fibers. In the all-fiber setup, a foldback interferometer using a 3-dB fiber coupler may be employed with total reflection at the end of the reference fiber arm, as illustrated in Figure 4.4.11. This all-fiber setup is simple and easy to align; however, we must know precisely the dispersion characteristic of the fiber in the reference arm before the measurement. Another difficulty is that the bandwidth of a 3-dB fiber coupler is usually narrow compared to a thin film-based free-space beam splitter, and therefore the measurement may be restricted to a certain wavelength subband.

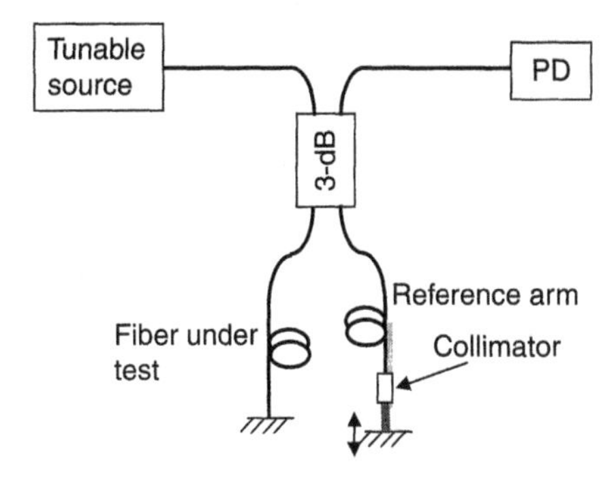

Figure 4.4.11 Dispersion measurement using a foldback all-fiber interferometer.

4.5 POLARIZATION MODE DISPERSION (PMD) MEASUREMENT

Unlike chromatic dispersion, which represents wavelength-dependent differential group delay, polarization mode dispersion is a special type of intermodal dispersion in single-mode optical fibers. It is well known that in a "single-mode" fiber there are actually two propagation modes co-exists, and they are orthogonally polarized; they're commonly referred to as HE_{11}^{x} and HE_{11}^{y}. In an ideal cylindrically symmetrical optical fiber, these two modes are degenerate and they have identical propagation constant. However, in practical optical fibers, the cylindrical symmetry may not be precise; as a result, these two orthogonal propagation modes may travel in different group velocities due to the birefringence in the fiber.

4.5.1 Representation of Fiber Birefringence and PMD Parameter

The birefringence in an optical fiber is generally introduced through intrinsic or extrinsic perturbations. *Intrinsic perturbation* is related to permanent features in the fiber due to errors in the manufacturing process. Manufacturing errors during fiber production may cause noncircular fiber core, which is usually referred to as *geometric birefringence*, and nonsymmetric stress due to the material in the fiber perform, which is known as *stress birefringence.* On the other hand, *extrinsic perturbation* is often introduced due to external forces in the cabling and installation process. Both of these two types of perturbations contribute to the effect of birefringence; as a consequence, the two orthogonally polarized propagation modes exhibit different propagation constants [Ulrich 1980].

Assume that the propagation constant of the vertically polarized propagation mode is $\beta_\perp = \omega n_\perp / c$, the propagation constant of the horizontally polarized propagation mode is $\beta_{//} = \omega n_{//}/c$ where ω is the optical frequency, $n_\perp$ and $n_{//}$ are the effective refractive indices of the two polarization modes, and c is the speed of light. Because of the birefringence in the fiber, $n_\perp$ and $n_{//}$ are not equal; therefore $\beta_\perp \neq \beta_{//}$. Their difference is

$$\Delta\beta = \beta_{//} - \beta_\perp = \frac{\omega(n_{//} - n_\perp)}{c} = \frac{\omega}{c}\Delta n_{eff} \tag{4.5.1}$$

where $\Delta n_{eff} = n_{//} - n_\perp$ is the effective differential refractive index. Typically Δn_{eff} is on the order of 10^{-5} to 10^{-7} in coiled standard single-mode fibers [24]. Then for a fiber of length L, the relative propagation delay between the two orthogonally polarized modes is

$$\Delta\tau = \frac{(n_{//} - n_\perp)}{c}L = \frac{L\Delta n_{eff}}{c} \tag{4.5.2}$$

This is commonly referred to as *differential group delay* (DGD) of the fiber.

Because of fiber birefringence, the state of polarization (SOP) of the optical signal will rotate when propagating in the fiber. For a total phase walk-off $\Delta\beta L = 2\pi$ between the two orthogonal polarization modes, the SOP of the signal at the fiber output completes a full 2π rotation. According to Equations 4.5.1 and 4.5.2, this polarization rotation can be induced by the changes of the fiber length L, the differential refractive index Δn_{eff}, or the signal optical frequency ω.

As a simple example, consider an optical signal with two frequency components ω_1 and ω_2. If these two frequency components are copolarized when the optical signal enters an optical fiber, then at the fiber output their polarization states will walk off from each other by an angle:

$$\Delta\theta = \frac{\Delta\omega \cdot \Delta n_{eff}}{c}L = \Delta\omega \cdot \Delta\tau \tag{4.5.3}$$

where $\Delta\omega = \omega_2 - \omega_1$. In other words, the fiber DGD value $\Delta\tau$ can be measured through the evaluation of differential polarization walk-off between different frequency components of the optical signal. In fact, this is the basic concept of the frequency domain PMD measurement technique, discussed later in this section.

The concept of birefringence is relatively straightforward in a short fiber where refractive indices are slightly different in the two orthogonal axes of the transversal cross-section. However, if the fiber is long enough, the birefringence axes may rotate along the fiber due to banding, twisting, and nonuniformity. Meanwhile, there is also energy coupling between the two orthogonally polarized propagation modes in the fiber. In general, both the birefringence axis rotation and the mode coupling are random and unpredictable, which make polarization mode dispersion a complex problem to understand and to solve [25].

The following are a few important parameters related to fiber polarization mode dispersion:

1. *Principal state of polarization* (PSP) indicates two orthogonal polarization states corresponding to the fast and slow axes of the fiber. Under this definition, if the polarization state of the input optical signal is aligned with one of the two PSPs of the fiber, the output optical signal will keep the same SOP. In this case, the PMD has no impact in the optical signal, and the fiber only provides a single propagation delay. It is important to note that PSPs exist not only in "short" fibers but also in "long" fibers. In a long fiber, if 2nd order PMD is negligible, although the birefringence along the fiber is random and there is energy coupling between the two polarization modes, an equivalent set of PSPs can always be found. Again, PMD has no impact on the optical signal if its polarization state is aligned to one of the two PSPs of the fiber. However, in practical fibers the orientation of the PSPs usually change with time, especially when the fiber is long. The change of PSP orientation over time is originated from the random changes in temperature and mechanical perturbations along the fiber.

Using Stokes parameter representation on the Poincare sphere, the two PSPs are represented as two vectors, which start from the origin and point to the two opposite extremes on the Poincare sphere as shown Figure 4.5.1. In a short fiber without energy coupling between the two polarization modes, the PSP of the fiber is fixed and is independent of optical signal frequency. In Figure 4.5.1, different circles represent different input polarization states of the optical signal. For each input polarization state, the output polarization vector rotates on the Poincare sphere in a regular circle when varying the optical signal frequency. If the fiber is long enough, the PSP will be largely dependent on the signal

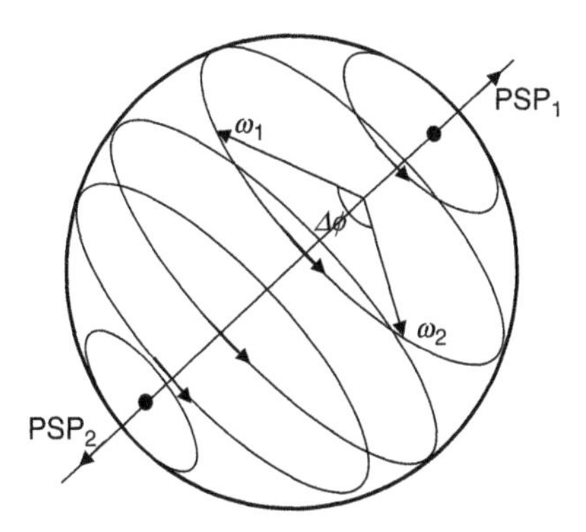

Figure 4.5.1 Definition of PSPs on Poincare sphere representation.

optical frequency, and the regular circles shown in Figure 4.5.1 will no longer exist.

2. *Differential group delay* (DGD) indicates the propagation delay difference between the signals carried by the two PSPs. In general, DGD is a random process due to the random nature of the birefringence in the fiber. The probability density function (PDF) of DGD in an optical fiber follows a Maxwellian distribution [25]:

$$P(\Delta\tau) = \sqrt{\frac{2}{\pi}}\frac{\Delta\tau^2}{\alpha^3}\exp\left(-\frac{\Delta\tau^2}{2\alpha^2}\right) \tag{4.5.4}$$

where $P(\Delta\tau)$ is the probability that the DGD of the fiber is $\Delta\tau$ and α is a parameter related to the mean DGD. In fact, according to the Maxwellian distribution given in Equation 4.5.4, the average value of DGD is

$$\langle\Delta\tau\rangle = \alpha\sqrt{\frac{8}{\pi}} \tag{4.5.5}$$

This is usually referred to as the mean DGD, while the root mean square (RMS) value of DGD is $\alpha\sqrt{3}$. A Maxwellian distribution with $\alpha = 0.5$ ps is shown in Figure 4.5.2, which is plotted both in the linear scale and in the logarithm scale. It is important to notice that even though the average DGD is only 0.8 ps in this case, low-probability events may still happen; for example, the instantaneous DGD may be 2 ps but at a low probability of 10^{-5}. This is an important consideration in calculating the outage probability of optical communication systems.

3. *Mean DGD*, which is the average DGD, $\langle\Delta\tau\rangle$ as defined in Equation 4.5.5. The unit of mean DGD is in picoseconds, mean DGD in a fiber is usually counted as the average value of DGD over certain ranges of wavelength,

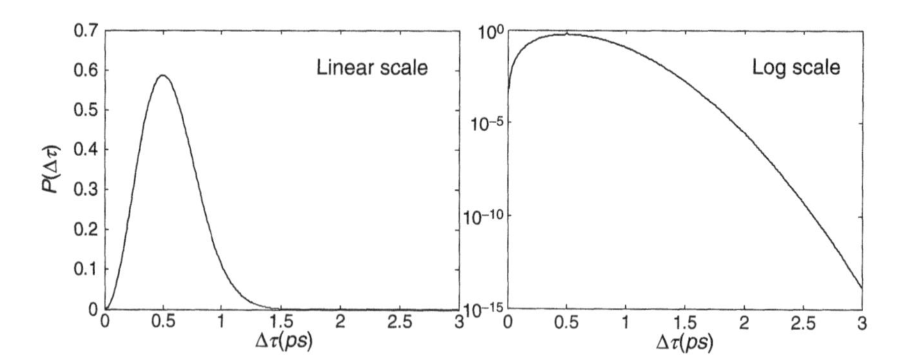

Figure 4.5.2 Probability of fiber DGD at the value of Δt, which has a Maxwellian distribution with $\alpha = 0.5$ ps and a mean DGD of 0.8 ps.

time, and temperature. For short fibers, mean DGD is proportional to the fiber length L: $\langle \Delta\tau \rangle \propto L$, whereas for long fibers, the mean DGD is proportional to the square root of fiber length $\sqrt{L}$: $\langle \Delta\tau \rangle \propto \sqrt{L}$. This is due to the random mode coupling and the rotation of birefringence axes along the fiber. The mean DGD is an accumulated effect over fiber; therefore random mode coupling makes the average effect of PMD smaller. However, due to the random nature of mode coupling, the instantaneous DGD can still be quite high at the event when the coupling is weak.

4. *PMD parameter* is defined as the mean DGD over a unit length of fiber. For the reasons we've discussed, for short fibers, the unit of PMD parameter is [ps/km], whereas for long fibers, the unit of PMD parameter becomes $[ps/\sqrt{km}]$. Because of the time-varying nature of DGD, the measurement of PMD is not trivial. In the following sections, we discuss several techniques that are often used to measure PMD in fibers.

4.5.2 Pulse Delay Method

The simplest technique to measure DGD in an optical fiber is the measurement of differential delay of short pulses that are simultaneously carried by the two polarization modes [26]. Figure 4.5.3(a) shows a block diagram of the measurement setup. Typically a mode-locked laser source is used to provide short optical pulses and is injected into the fiber under test through a polarization controller. A high-speed photodiode converts the output optical signal

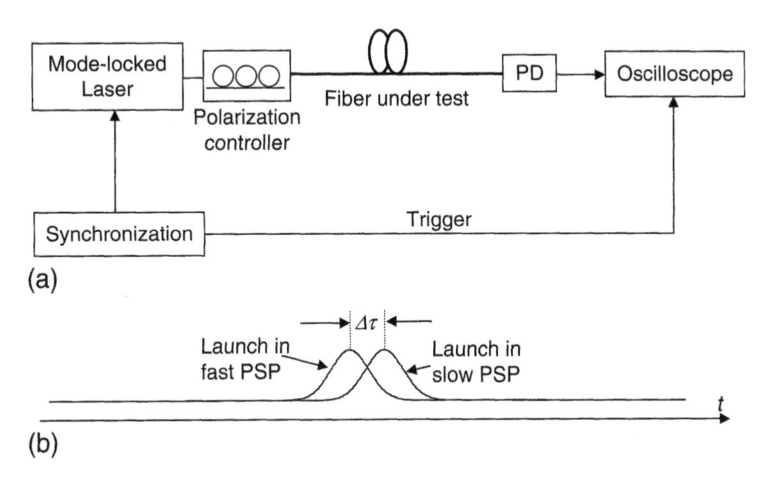

Figure 4.5.3 PMD measurement using differential pulse delay technique: (a) experimental setup and (b) illustration of differential pulse delay.

from the fiber into electrical domain, and the pulse waveform is displayed by an oscilloscope.

The propagation delay of the optical pulse depends on the group velocity of the optical signal that propagates in the fiber. Since the DGD is defined as the group delay difference between the fast axis and the slow axis of the fiber, it can be evaluated by directly measuring the fastest and the slowest arrival times of the optical pulses at the receiver, as shown in Figure 4.5.3(b). Since the optical signal from the source laser is polarized, the search for the fastest and the slowest group velocities in the fiber is accomplished by adjusting the polarization controller before the optical signal is launched into the fiber.

The underlying principle of the differential pulse-delay technique is straightforward and easy to understand. But the accuracy of this measurement, to a large extent, depends on the temporal width of the optical pulses used. Narrower optical pulses are obviously desired to measure fibers with low levels of DGD. However, chromatic dispersion in the fiber will broaden and distort the optical pulses. For example, to measure a 10 km standard single-mode fiber with 17 ps/nm/km dispersion parameter in a 1550 nm wavelength window, if the input optical pulse width is 10 ps, the optical spectral bandwidth is approximately 0.8 nm. The pulse width at the output of the fiber will be about 136 ps only due to chromatic dispersion. Therefore it would not be helpful to use very narrow pulses for DGD measurement when the fiber has significant chromatic dispersion. In fact, if the width of the input optical pulse is Δt_{in}, the output pulse width can be calculated by

$$\Delta t_{out} = \Delta t_{in} + \frac{D\lambda^2}{c\Delta t_{in}} L \tag{4.5.6}$$

where D is the fiber chromatic dispersion parameter, λ is the operation wavelength, and L is the fiber length. If we assume that $D = 17$ ps/nm/km and $L = 10$ km, the relationship between the widths of the output and the input pulses is shown in Figure 4.5.4. The minimum width of the output pulse is about 75 ps; obviously, it is not feasible to measure DGD values of less than a few picoseconds without proper dispersion compensation. On the other hand, if dispersion compensation is applied, the DGD of the dispersion compensator has to be taken into account and its impact in the overall result of measurement may become complicated.

To summarize, the differential pulse delay method is a time-domain technique that directly measures the DGD in a fiber, and the explanation of the results is rather straightforward. By changing the wavelength of the optical source, the wavelength dependency of DGD can be evaluated; thus the PMD parameter can also be obtained. Although the operating principle of this technique is simple; its accuracy may be limited by the achievable pulse width. This technique is usually used to measure fibers with low chromatic dispersion and relatively high levels of DGD.

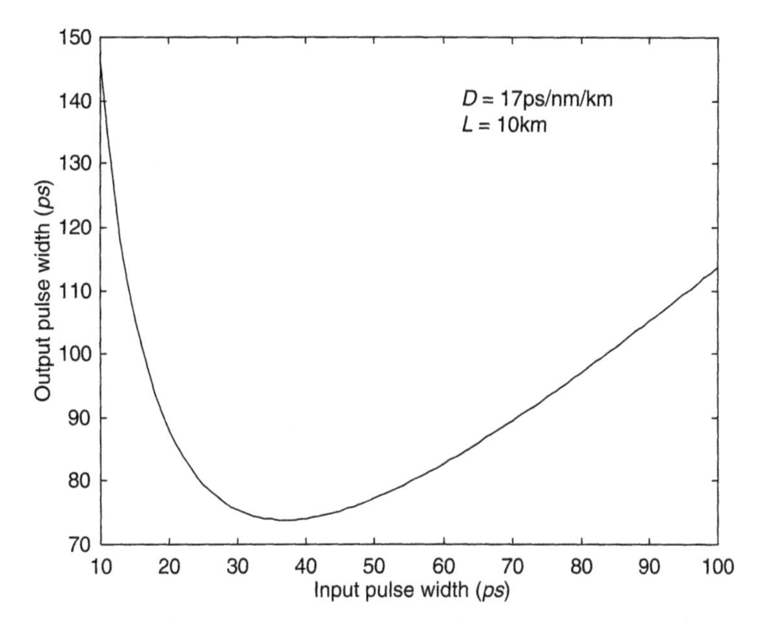

Figure 4.5.4 Output pulse width versus input pulse width in a 10 km fiber with 17 ps/nm/km chromatic dispersion parameter.

4.5.3 The Interferometric Method

As we just discussed in the last section, the major difficulty of the differential pulse delay method is the requirement of short optical pulses and ultrafast detection. These can be avoided using the optical interferometric technique. The principle of the interferometric method is based on the measurement of the differential delay between the signals carried by two PSPs using the low-coherent interferometer technique [27, 28].

Figure 4.5.5 shows the schematic diagram of a DGD measurement setup using the interferometric technique. In this system, a wideband lightwave source, such as a halogen lamp, is used. A monochromator selects the wavelength of the optical signal and the spectral bandwidth. If the spectral bandwidth of the light source is wide enough, it has a very short coherence length, and the optical signals reflected from the two interferometer arms of the Michelson interferometer are coherent only when the two arms have almost the same length. Assume that the spectral bandwidth selected from the light source is $\Delta\lambda$ and the center wavelength is λ; its coherence length is approximately

$$\Delta l = \frac{1}{2n}\frac{\lambda^2}{\Delta\lambda} \tag{4.5.7}$$

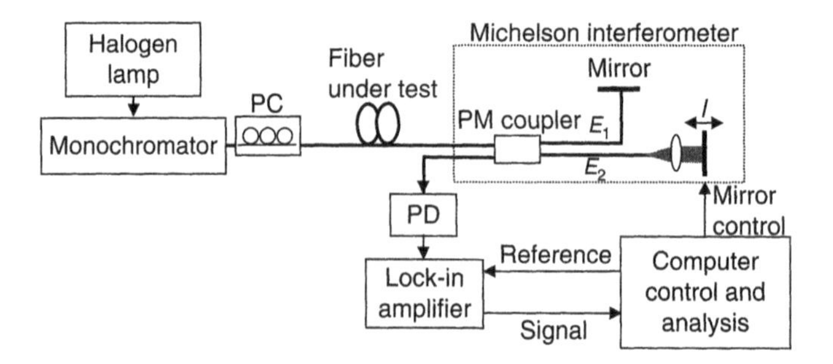

Figure 4.5.5 Optical setup for DGD measurement using interferometry method.

where n is the refractive index. In the Michelson interferometer, if the arm length difference is shorter than the source coherence length, the mixing between the optical fields coming from the two arms is coherent in the photodetector. On the other hand, if the arm length difference is longer than the coherence length, there is no phase relation between optical fields coming from the two arms, and therefore the optical powers add up at the photodetector incoherently. To demonstrate the operation principle, we assume that the polarization-maintaining fiber coupler has a 50 percent splitting ratio and the optical signal amplitudes in the two arms are equal, i.e., $|E_1| = |E_2|$. Within coherence length the mixed optical power at the photodiode will oscillate between zero (destructive interference) and $2|E_1|^2$ (constructive interference), depending on the phase relation between the two arms. Outside the coherence length, the total optical power will simply be a constant $|E_1|^2$, as illustrated in Figure 4.5.6.

When a birefringent optical fiber is inserted between the light source and the interferometer, the optical signal will be partitioned into the fast axis and the slow axis. At the fiber output, the optical field is

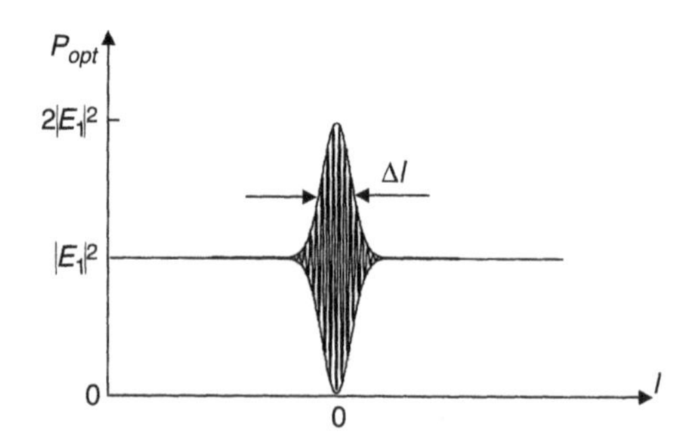

Figure 4.5.6 Illustration of interference pattern in a low-coherence interferometer. $|E_1|$ is the optical field in each of the two interferometer arms and l is the arm length difference.

$$E = \left(E_x e^{j\beta_x L} + E_y e^{j\beta_y L}\right) \tag{4.5.8}$$

where x and y are the two orthogonal PSPs of the fiber, β_x and β_y are the propagation constants of the fast and the slow axis, E_x and E_y are the amplitude of the optical signals carried by these two orthogonal modes, and L is the fiber length. Then, after a roundtrip through the Michelson interferometer, the optical signal at the photodiode is

$$E_0 = \frac{1}{2}\left(E_x e^{j\beta_x L} + E_y e^{j\beta_y L}\right)\left(e^{j\beta l} + 1\right) \tag{4.5.9}$$

where l is the differential delay between the two interferometer arms and $\beta = 2\pi n/\lambda$ is the propagation constant in the interferometer arm. We have assumed that the power-splitting ratio of the optical coupler is 50 percent. Then the square-law detection of the photodiode generates a photocurrent, which is

$$I = \eta\left|E_0\right|^2 = \frac{\eta}{4}\left|E_x\left(e^{j\beta_x L} + e^{j(\beta_x L + \beta l)}\right) + E_y\left(e^{j\beta_y L} + e^{j(\beta_y L + \beta l)}\right)\right|^2 \tag{4.5.10}$$

As illustrated in Figure 4.5.7(a), a strong coherent interference peak is generated when $l = 0$, which is mainly caused by the self-mixing terms of $|E_x|^2$ and $|E_y|^2$ in Equation 4.5.10. In addition, two satellite peaks are generated when the differential arm length l satisfies

$$(\beta_y - \beta_x)L \pm \beta l = 0 \tag{4.5.11}$$

or approximately

$$l = \pm\Delta\tau \cdot c \tag{4.5.12}$$

where $\Delta\tau$ is the DGD of the fiber under test. These two coherent interference peaks represent the contribution of the cross-mixing between E_x and E_y terms

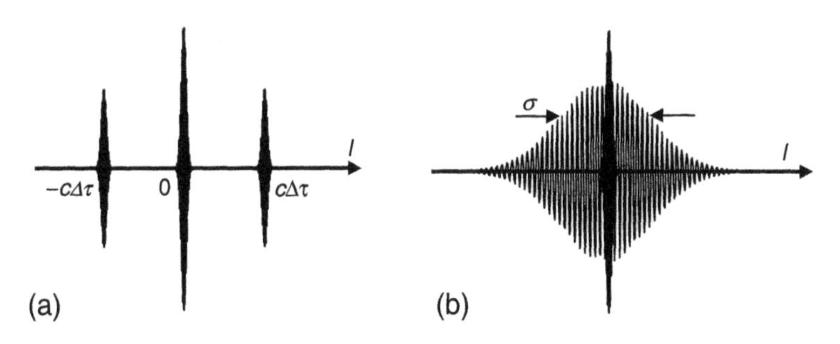

Figure 4.5.7 Interference pattern as the function of differential delay l of the interferometer: (a) short fiber and (b) long fiber with random mode coupling.

as given by Equation 4.5.10, and the width of each interference peak is determined by the spectral width of the source. By measuring the location of these two secondary interference peaks, the fiber DGD can be estimated with Equation 5.4.12. A good measurement resolution (and thus good accuracy) requires narrow interference peaks, which corresponds to a wide optical bandwidth used.

On the other hand, if the fiber is long enough, random-mode coupling exist between the two polarization modes and the PSPs vary along the fiber. In this case, the interference pattern consists of a stable central peak and randomly scattered sidebands due to random-mode coupling. The overall envelope of the energy distribution is generally Gaussian, as shown in Figure 4.5.7(b). Suppose the standard deviation of this Gaussian fitting is σ; then the mean DGD can be calculated as $\langle \Delta\tau \rangle = \sigma\sqrt{2/\pi}$, whereas the RMS value of DGD is $\langle \Delta\tau^2 \rangle^{1/2} = \sigma\sqrt{3/4}$ [29].

In the long fiber situation, due to mode coupling in the fiber, the DGD value is generally a function of signal wavelength. This measurement essentially provides an averaged DGD over the signal spectral bandwidth $\Delta\lambda$. Referring to Figure 4.5.5, the wavelength resolution of the DGD measurement can be selected by choosing a proper width of the exit slit of the monochromator.

4.5.4 Poincare Arc Method

As discussed in Section 4.5.1, PMD is originated from the birefringence of the fiber, which can be measured by the frequency dependency of polarization rotation. A more sophisticated representation of the SOP is the use of Stokes parameters, in which a polarized optical signal is represented as a vector $\vec{S}$ on the Poincare sphere, as discussed in Section 2.6.2. After propagating through a birefringent fiber, the $\vec{S}$ vector will rotate around the PSP on the Poincare sphere when the wavelength (or the frequency) is changed.

In a *short fiber* without mode coupling, the PSP of the fiber is stable and independent of optical signal frequency. In this case the SOP vector of the optical signal rotates on the Poincare sphere in a regular circle around the PSP when the signal optical frequency is swept as shown in Figure 4.5.8(a). On the other hand, in a *long fiber,* where random mode coupling is significant, the PSP is no longer stable and is a function of signal optical frequency. In this case each frequency change of the optical signal has its corresponding PSP and the SOP vector wonders on the Poincare sphere irregularly with the change of signal optical frequency, as shown in Figure 4.5.8(b). By definition, the PMD vector Ω originates from the center of the Poincare sphere and points toward the PSP.

In either case in Figure 4.5.8, with the change of optical signal frequency ω, the signal polarization state at fiber output will change. This is represented by

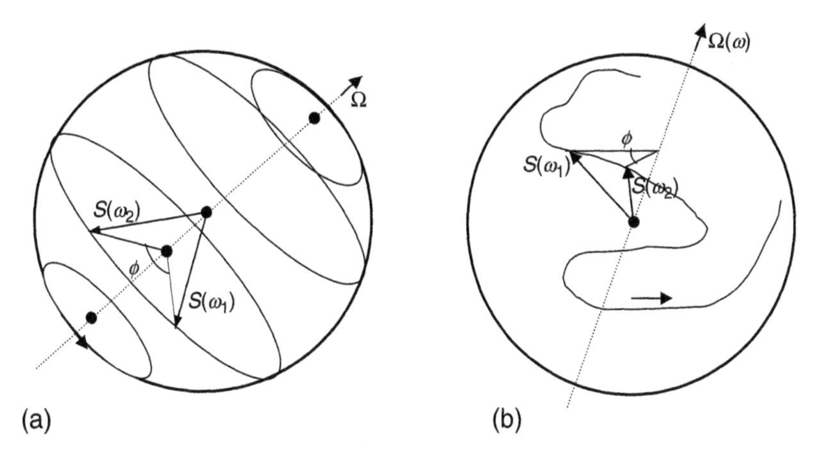

Figure 4.5.8 Illustration of traces of signal polarization vector on Poincare spheres when the optical frequency is varied (a) for a short fiber and (b) for a long fiber.

the change of the SOP vector $\vec{S}$ versus ω. Obviously, the amount of this change is directly proportional to fiber birefringence (or PMD). For an infinitesimal change of the signal optical frequency dω, this vectorial relationship is

$$\frac{d\vec{S}}{d\omega} = \Omega \times \vec{S} \tag{4.5.13}$$

For a step change in the optical frequency, $\Delta\omega = \omega_2 - \omega_1$, it is also convenient to use the scalar relationship [29]:

$$\phi = \Delta\tau \cdot \Delta\omega \tag{4.5.14}$$

where ϕ is the angular change of the polarization vector $\vec{S}$ in radians on the plan perpendicular to the PSP, as shown in Figure 4.5.8, and $\Delta\tau$ is the DGD between the two PSP components. Obviously, the simple relation given in Equation 4.5.14 can be used to measure fiber DGD if a polarimeter is available and the variation of Stokes parameters versus signal frequency can be measured.

Figure 4.5.9 shows the block diagram of the measurement setup using Poincare arc technique, where a tunable laser is used which provides a source whose optical frequency can be swept. A polarization controller is placed before the optical fiber

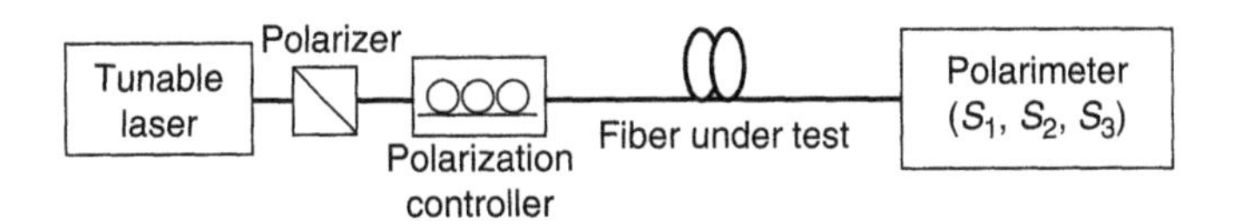

Figure 4.5.9 Block diagram of DGD measurement using a polarimeter.

under test to explore various launching SOPs of the optical signal into the fiber. A polarimeter is used to measure the Stokes parameters corresponding to each frequency of the optical source, $\vec{S}(\omega) = \vec{a}_x S_1(\omega) + \vec{a}_y S_2(\omega) + \vec{a}_z S_3(\omega)$, where $\vec{a}_x$, $\vec{a}_y$, and $\vec{a}_z$ are unit vectors. For a small wavelength increment $\Delta\omega = \omega_2 - \omega_1$, the angular rotation of the polarization vector on the plan perpendicular to the PSP vector can be evaluated by

$$\phi = \cos^{-1}\left\{ \frac{\left(\vec{S}(\omega_1) \times \vec{\Omega}(\omega)\right) \cdot \left(\vec{S}(\omega_2) \times \vec{\Omega}(\omega)\right)}{\left|\vec{S}(\omega_1) \times \vec{\Omega}(\omega)\right| \left|\vec{S}(\omega_2) \times \vec{\Omega}(\omega)\right|} \right\} \tag{4.5.15}$$

Then the DGD value at this frequency can be obtained by

$$\Delta\tau(\omega) = \frac{\phi}{\Delta\omega} \tag{4.5.16}$$

where $\omega = (\omega_2 + \omega_1)/2$ is the average frequency. Overall, the Poincare arc measurement technique is easy to understand and the measurement setup is relatively simple. However, it requires a polarimeter, which is a specialized instrument. In addition, the frequency tuning of the tunable laser has to be continuous to provide the accurate trace of the polarization rotation, as illustrated in Figure 4.5.8(b).

4.5.5 Fixed Analyzer Method

Compared to the Poincare arc measurement technique, the fixed analyzer method replaces the polarimeter with a fixed polarizer; thus it is relatively easy to implement because of the reduced requirement of specialized instrumentation. Figure 4.5.10 shows the schematic diagrams of two equivalent versions of the fixed

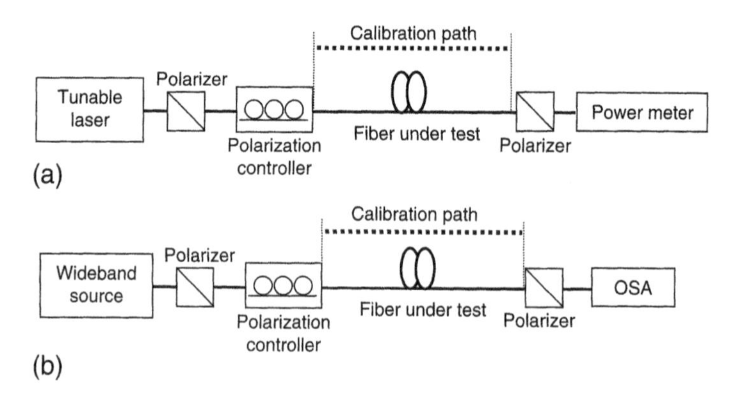

Figure 4.5.10 Block diagram of PMD measurement using a fixed analyzer method: (a) tunable laser + power meter combination and (b) wideband source + OSA combination.

analyzer measurement apparatus [30]. Figure 4.5.10(a) uses a tunable laser as the optical source, while at the receiver side a fixed polarizer is followed by an optical power meter. By varying the frequency of the tunable optical source, the signal polarization state at the fiber output changes, and the polarizer converts this signal SOP change into an optical power change, which is then detected by the power meter. The system shown in Figure 4.5.10(b) uses a wideband source that provides a signal with a broad optical spectrum. Because of the birefringence in the fiber, different signal frequency components will exhibit different polarization states at the output of the fiber, and the polarizer converts this frequency-dependent polarization rotation into a frequency-dependent optical power spectral density, which can be accurately measured by an optical spectrum analyzer. In both implementations, a polarizer immediately after the source is used to make sure the source has a fixed polarization state. The polarization controller enables the change of the polarization state of the optical signal that is injected into the fiber.

In the Poincare sphere representation, the power transfer function through a perfect polarizer can be expressed as

$$T(\omega) = \frac{P_{out}}{P_{in}} = \frac{1}{2}[1 + \hat{s}(\omega) \cdot \hat{p}] \tag{4.5.17}$$

where $\hat{s}(\omega)$ is the unit vector representing the polarization state of the input optical signal into the polarizer and $\hat{p}$ is the unit vector representing the high transmission state of the polarizer.

Because of the birefringence in the fiber, the SOP of the optical signal at fiber output will rotate around the PSP on the Poincare sphere when the optical signal frequency is varied. In a long optical fiber, the birefringence axis is randomly oriented along the fiber and the mode coupling may also be significant; therefore the polarization vector may essentially walk all over the Poincare sphere with the change of the signal optical frequency. As a result, the power transmission efficiency $T(\omega)$ through the fixed polarization analyzer at the fiber output can have all possible values between 0 and 1 as illustrated Figure 4.5.11. If the birefringence orientation along the fiber is truly random, the transfer function $T(\omega)$ will have a uniform probability distribution between 0 and 1 with the average value of 0.5.

If we define

$$T' = \frac{dT(\omega)}{d\omega} \tag{4.5.18}$$

as the frequency derivative of the power transfer function of the fixed polarizer, then statistically, the expected value of $|T'|$ is $E\{|T'|, T = \langle T \rangle\}$, which is obtained under the condition that the transfer function is at its mean level $T(\omega) = \langle T(\omega) \rangle$. According to the fundamental rules of statistics, the mean-level

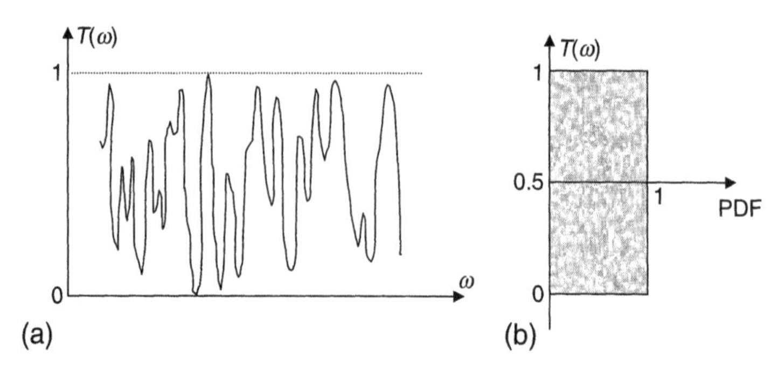

Figure 4.5.11 (a) Illustration of fixed polarizer transfer function $T(\omega)$ with random input polarization state of the signal and (b) probability distribution of $T(\omega)$.

crossing density, which specifies how often the random variable $T(\omega)$ crosses through its average value per unit frequency interval, is

$$\gamma_m = f_T(\langle T\rangle)(E\{|T'|, T = \langle T\rangle\} \tag{4.5.19}$$

where $f_T(\langle T\rangle)$ is the probability density function (PDF) of the transfer function $T(\omega)$ evaluated at its average level $\langle T(\omega)\rangle$. In this specific case of uniform distribution, since $\langle T(\omega)\rangle = 0.5$ and $f_T(\langle T\rangle) = f_T(0.5) = 1$, Equation 4.5.19 can be simplified into

$$\gamma_m = E\{|T'|, T = \langle T\rangle\} \tag{4.5.20}$$

Use the definition of $T(\omega)$ shown in Equation 4.5.17, its derivative shown in Equation 4.5.18 can be expressed as

$$T'(\omega) = \frac{1}{2}\left[\frac{d\hat{s}(\omega)}{d\omega} \cdot \hat{p}\right] \tag{4.5.21}$$

Meanwhile, Equation 4.5.13 in the last section indicated that

$$\frac{d\hat{s}(\omega)}{d\omega} = \Omega \times \hat{s}(\omega) \tag{4.5.22}$$

where Ω is the PMD vector. Then one can find:

$$T'(\omega) = \frac{1}{2}[(\Omega \times \hat{s}) \cdot \hat{p}] = \frac{1}{2}[(\hat{s} \times \hat{p}) \cdot \Omega] \tag{4.5.23}$$

According to Equation 4.5.17, at the mean value of $T(\omega) = 0.5$, $\hat{s}(\omega) \cdot \hat{p}$ has to be equal to zero, which is equivalent to $\hat{s}(\omega) \times \hat{p} = \hat{1}$, where $\hat{1}$ is a unit vector and therefore, $T'(\omega) = 0.5[\hat{1} \cdot \Omega]$. Then Equation 4.5.20 can be expressed as

$$\gamma_m = \frac{1}{2}\langle|\hat{1} \cdot \Omega|\rangle = \frac{1}{2}\langle|\Omega|\rangle\langle|\cos\theta|\rangle \tag{4.5.24}$$

In a long fiber with truly random mode coupling, $\cos\theta$ is uniformly distributed between $-1 \le \cos\theta \le 1$ so that $\langle|\cos\theta|\rangle = 0.5$. Also, by definition, the magnitude of the PMD vector Ω is equal to the DGD of the fiber, $\Delta\tau = |\Omega|$; therefore

$$\gamma_m = \frac{\langle\Delta\tau\rangle}{4} \tag{4.5.25}$$

Recall that the physical meaning of γ_m can be explained by how often the transfer function $T(\omega)$ crosses through its average value per unit frequency interval. Then within a frequency interval $\Delta\omega$, if the average number of cross-overs is $\langle N_m\rangle$, we will have $\gamma_m = \langle N_m\rangle/\Delta\omega$, that is,

$$\langle\Delta\tau\rangle = 4\frac{\langle N_m\rangle}{\Delta\omega} \tag{4.5.26}$$

In this way, by scanning the signal frequency and counting the number of transmission crossovers through its average value within a certain frequency interval, the average DGD can be evaluated.

Similarly, the measurement can also be performed by counting the average number of extrema (maximum + minimum) within a frequency interval to decide the average DGD value. Without further derivation, the equation to calculate the DGD can be found as [30]

$$\langle\Delta\tau\rangle = 0.824\pi\frac{\langle N_e\rangle}{\Delta\omega} \tag{4.5.27}$$

where $\langle N_e\rangle$ is the number of extrema within a frequency interval $\Delta\omega$.

If we choose to use a wavelength interval instead of a frequency interval, Equation 4.5.27 can be expressed as

$$\langle\Delta\tau\rangle = \frac{0.412\langle N_e\rangle\lambda_{start}\lambda_{stop}}{c(\lambda_{stop} - \lambda_{start})} \tag{4.5.28}$$

where λ_{start} and λ_{stop} are the start and stop wavelengths of the measurement.

One practical consideration using the fixed polarizer method is how small the frequency step size should be used for the tunable laser if the setup shown in Figure 4.5.10(a) is implemented. This question is equivalent to the selection of the spectral resolution for the OSA if the setup shown in Figure 4.5.10(b) is used. Assuming that we need at least three data points between adjacent transmission extrema, according to Equation 4.5.28 the wavelength step size has to be

$$\delta\lambda < \frac{0.412\lambda^2}{4c\langle\Delta\tau\rangle} \tag{4.5.29}$$

Approximately $\delta\lambda < \lambda^2/(8c\langle\Delta\tau\rangle)$ can be used as a rule of thumb to determine the size of the wavelength step. For example, in the 1550 nm wavelength window,

$\lambda^2/(8c) \approx 10^{-21}$; therefore the wavelength step size of the source (or the spectral resolution of the OSA) has to be $\delta\lambda < 10^{-21}/\langle\Delta\tau\rangle$, where $\langle\Delta\tau\rangle$ is the expected DGD in seconds and $\delta\lambda$ is the wavelength step size in meters. For convenience, if we use $\langle\Delta\tau\rangle$ in the unit of picoseconds and $\delta\lambda$ is in the unit of nanometers, this requirement becomes very simple:

$$\delta\lambda < \frac{1}{\langle\Delta\tau\rangle} \tag{4.5.30}$$

For instance, for a fiber with the mean DGD of $\langle\Delta\tau\rangle = 20\ ps$, the source wavelength step size (or the OSA spectral resolution) must be better than 0.05 nm.

4.5.6 The Jones Matrix Method

The Jones Matrix method is a comprehensive PMD measurement technique, although its measurement setup, as shown in Figure 4.5.12, is similar to the Poincare arc technique discussed previously, the sequence of the input SOP of the optical signal is sequentially adjusted and the response at the fiber output is systematically measured [31, 32].

To explain the measurement principle of the Jones Matrix method, it is useful to review what the Jones Matrix is about. In general, a Jones Matrix has been widely used to represent the polarization state of a lightwave signal or the transfer matrix of a passive optical device. For example, the transversal electrical field of a polarized light can be represented by a two-element complex vector in the Cartesian coordinate system. These two vectors specify the magnitude and the phase of the x and y components of the electrical field at a particular point in space [33]:

$$\vec{E} = \begin{pmatrix} E_x \\ E_y \end{pmatrix} = \begin{pmatrix} E_{x0}e^{j\varphi_x} \\ E_{y0}e^{j\varphi_y} \end{pmatrix} \tag{4.5.31}$$

where φ_x and φ_y represent the phase of the two elements.

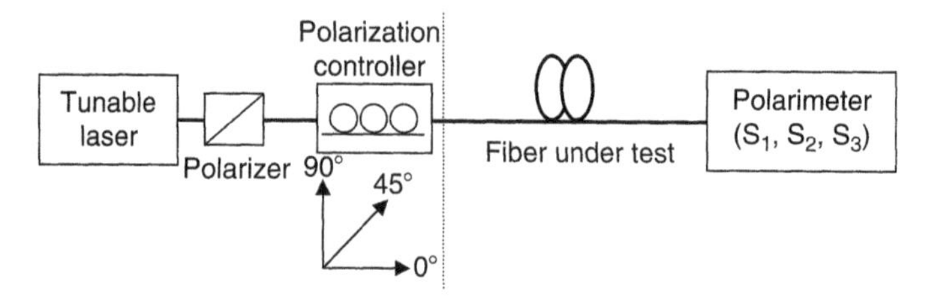

Figure 4.5.12 Block diagram of the Jones Matrix technique to measure fiber PMD.

On the other hand, the transfer function of a passive optical device can be written as a four-element matrix in the form:

$$M = \begin{bmatrix} M_{11} & M_{12} \\ M_{21} & M_{22} \end{bmatrix} \tag{4.5.32}$$

The transfer matrix of an optical device specifies the relationship between the input and output Jones vectors of the lightwave signal. This transfer matrix can be characterized by measuring three output Jones vectors in response to three known input Jones vectors. It is worthwhile to note that although it is measured with a few specific polarization states of the optical signal, a Jones Matrix describes a passive optical device such as an optical fiber and is independent of the input launching condition of the optical signal.

Suppose the input Jones vector of the optical signal is $\vec{E}_{in} = \begin{pmatrix} E_{x,in} \\ E_{y,in} \end{pmatrix}$ and the corresponding output Jones vector is $\vec{E}_{out} = \begin{pmatrix} E_x \\ E_y \end{pmatrix}$. The Jones Matrix of the optical device must be

$$\begin{bmatrix} E_x \\ E_y \end{bmatrix} = \begin{bmatrix} M_{11} & M_{12} \\ M_{21} & M_{22} \end{bmatrix} \begin{bmatrix} E_{x,in} \\ E_{y,in} \end{bmatrix} \tag{4.5.33}$$

To obtain the transfer function of a passive optical device, it is convenient to use three input stimuli with linear polarization states oriented at 0°, 45°, and 90°. In the following, we explain how the four matrix elements can be obtained through the three independent measurements.

1. The input optical signal is linearly polarized at 0° (linear horizontal). That is,

$$\vec{E}_{in} = \begin{pmatrix} E_{x,in} \\ 0 \end{pmatrix} = E_{x,in} \begin{pmatrix} 1 \\ 0 \end{pmatrix} \tag{4.5.34}$$

If the measured output Jones vector is

$$\vec{E}_{out1} = \begin{pmatrix} E_{x1} \\ E_{y1} \end{pmatrix} \tag{4.5.35}$$

then the input/output relationship is

$$\begin{bmatrix} E_{x1} \\ E_{y1} \end{bmatrix} = \begin{bmatrix} M_{11} & M_{12} \\ M_{21} & M_{22} \end{bmatrix} \begin{bmatrix} 1 \\ 0 \end{bmatrix} E_{x,in} \tag{4.5.36}$$

which is equivalent to

$$E_{x1} = M_{11} E_{x,in} \tag{4.5.37a}$$

$$E_{y1} = M_{21} E_{x,in} \tag{4.5.37b}$$

Now we can define a new parameter,

$$k_1 = \frac{E_{x1}}{E_{y1}} = \frac{M_{11}}{M_{21}} \tag{4.5.38}$$

Since both E_{x1} and E_{y1} have already been determined by the measurement, k_1 can be easily obtained.

2. The input is linearly polarized at 90° (linear vertical). In this case,

$$\vec{E}_{in2} = \begin{pmatrix} 1 \\ E_{y,in2} \end{pmatrix} = E_{y,in2}\begin{pmatrix} 0 \\ 1 \end{pmatrix} \tag{4.5.39}$$

Suppose the measured output Jones vector corresponding to this input is

$$\vec{E}_{out2} = \begin{pmatrix} E_{x2} \\ E_{y2} \end{pmatrix} \tag{4.5.40}$$

The input/output relation yields

$$\begin{bmatrix} E_{x2} \\ E_{y2} \end{bmatrix} = \begin{bmatrix} M_{11} & M_{12} \\ M_{21} & M_{22} \end{bmatrix} \begin{bmatrix} 0 \\ 1 \end{bmatrix} E_{x,in2} \tag{4.5.41}$$

which is equivalent to

$$E_{x2} = M_{12}E_{x,in2} \tag{4.5.42a}$$

$$E_{y2} = M_{22}E_{x,in2} \tag{4.5.42b}$$

We can then define a new parameter:

$$k_2 = \frac{E_{x2}}{E_{y2}} = \frac{M_{12}}{M_{22}} \tag{4.5.43}$$

Again, k_2 can be determined through the measurement of E_{x2} and E_{y2}.

3. The input is linearly polarized at 45°. In this case $E_{y,in} = E_{x,in}$, so that

$$\vec{E}_{in3} = \frac{E_{y,in3}}{\sqrt{2}}\begin{pmatrix} 1 \\ 1 \end{pmatrix} \tag{4.5.44}$$

Suppose the measured output Jones vector corresponding to this input is

$$\vec{E}_{out3} = \begin{pmatrix} E_{x3} \\ E_{y3} \end{pmatrix} \tag{4.5.45}$$

The input/output relationship is

$$\begin{bmatrix} E_{x3} \\ E_{y3} \end{bmatrix} = \begin{bmatrix} M_{11} & M_{12} \\ M_{21} & M_{22} \end{bmatrix} \begin{bmatrix} 1 \\ 1 \end{bmatrix} \frac{E_{x,in3}}{\sqrt{2}} \tag{4.5.46}$$

which is equivalent to

$$E_{x3} = (M_{11} + M_{12})E_{x,in3}/\sqrt{2} \tag{4.5.47a}$$

$$E_{y3} = (M_{21} + M_{22})E_{x,in3}/\sqrt{2} \tag{4.5.47b}$$

Then the third new parameter can be introduced as

$$k_3 = \frac{E_{x3}}{E_{y3}} = \frac{M_{11} + M_{12}}{M_{21} + M_{22}} \tag{4.5.48}$$

which can be determined through the measurement of E_{x3} and E_{y4}.

If we neglect the absolute propagation delay and the attenuation of the optical device under test, the four elements in the Jones Matrix will not be independent and we can arbitrarily set $M_{22} = 1$. Then the rest of the three matrix elements can be found using the following three equations:

$$k_1 = \frac{M_{11}}{M_{21}} \tag{4.5.49a}$$

$$k_2 = \frac{M_{12}}{M_{22}} = M_{12} \tag{4.5.49b}$$

$$k_3 = \frac{M_{11} + M_{12}}{M_{21} + M_{22}} = \frac{M_{11} + M_{12}}{M_{21} + 1} \tag{4.5.49c}$$

The solutions to this set of equations are

$$M_{11} = k_1 \frac{k_2 - k_3}{k_3 - k_1} = k_1 k_4 \tag{4.5.50a}$$

$$M_{12} = k_2 \tag{4.5.50b}$$

$$M_{21} = \frac{k_2 - k_3}{k_3 - k_1} = k_4 \tag{4.5.50c}$$

where k_4 is defined as

$$k_4 = \frac{k_2 - k_3}{k_3 - k_1} \tag{4.5.51}$$

Finally, the Jones Matrix of the optical device can be represented as

$$[M] = C\begin{bmatrix} k_1 k_4 & k_2 \\ k_4 & 1 \end{bmatrix} \tag{4.5.52}$$

where all the elements can be determined by the three independent measurements described previously. A multiplicative complex constant C is introduced that represents the absolute propagation delay and the attenuation of the optical device. In the following, we discuss how the DGD of an optical fiber can be represented with the Jones Matrix.

As discussed earlier, due to the birefringence in a fiber, the signal SOP at the fiber output is a function of the optical frequency ω. Thus, the Jones Matrix of the fiber also must be a function of the optical frequency, $M = M(\omega)$. For an arbitrary input optical field vector $E_{in}(\omega)$, if the output optical field vector is measured as $E_{out}(\omega)$, the input/output relation is

$$E_{out}(\omega) = M(\omega)E_{in}(\omega) \tag{4.5.53}$$

The output field vector has an amplitude $A(\omega)$ and a phase $\phi(\omega)$

$$E_{out}(\omega) = A(\omega)e^{j\phi(\omega)} \tag{4.5.54}$$

Then the frequency derivative of Equation 4.5.53 yields

$$E'_{out}(\omega) = M'(\omega)E_{in}(\omega) = \left[\frac{A'(\omega)}{A(\omega)} + j\phi'(\omega)\right]E_{out}(\omega) \tag{4.5.55}$$

where the primes denote differentiation with respect to the frequency ω, such as $E'_{out}(\omega) = dE_{out}(\omega)/d\omega$.

If $M^{-1}(\omega)$ represents the inverse Jones Matrix, Equation 4.5.53 is also equivalent to

$$E_{in}(\omega) = M^{-1}(\omega)E_{out}(\omega) \tag{4.5.56}$$

Combining Equations 4.5.55 and 4.5.56, we find

$$M'(\omega)M^{-1}(\omega)E_{out}(\omega) = \left[\frac{A'(\omega)}{A(\omega)} + j\phi'(\omega)\right]E_{out}(\omega) \tag{4.5.57}$$

which is equivalent to a more convenient form,

$$\left\{M'(\omega)M^{-1}(\omega) - \left[\frac{A'(\omega)}{A(\omega)} + j\phi'(\omega)\right]I\right\}E_{out}(\omega) = 0 \tag{4.5.58}$$

where $I = \begin{bmatrix} 1 & 0 \\ 0 & 1 \end{bmatrix}$ is the unit matrix.

Recall that the physical meaning of the differential phase

$$\phi'(\omega) = \frac{d\phi(\omega)}{d\omega} = \tau_g \tag{4.5.59}$$

represents the group velocity of the optical signal in the fiber. For a small frequency interval $\Delta\omega$, a linear approximation can be made such that

$$M'(\omega) = \frac{dM(\omega)}{d\omega} \approx \frac{M(\omega + \Delta\omega) - M(\omega)}{\Delta\omega} \tag{4.5.60}$$

and thus

$$M'(\omega)M^{-1}(\omega) \approx \frac{[M(\omega+\Delta\omega)M^{-1}(\omega) - M(\omega)M^{-1}(\omega)]}{\Delta\omega} = \frac{[M(\omega+\Delta\omega)M^{-1}(\omega) - I]}{\Delta\omega} \quad (4.5.61)$$

If we neglect the frequency dependency of the optical loss, which is often a valid assumption for an optical fiber within a relatively narrow optical band, that is $A'(\omega) = dA(\omega)/d\omega \approx 0$, the combination of Equations 4.5.58 and 4.5.61 becomes

$$\left\{M(\omega+\Delta\omega)M^{-1}(\omega) - (1 + j\tau_g\Delta\omega)I\right\}E_{out}(\omega) = 0 \quad (4.5.62)$$

In addition, assuming that $\Delta\omega$ is small enough such that $\tau_g\Delta\omega \ll 1$, the term $(1 + j\tau_g\Delta\omega)$ in Equation 4.5.62 can be replaced by $\exp(j\tau_g\Delta\omega)$; therefore we have

$$\left\{M(\omega+\Delta\omega)M^{-1}(\omega) - \exp(j\tau_g\Delta\omega)I\right\}E_{out}(\omega) = 0 \quad (4.5.63)$$

In order to satisfy the Eigen-equation 4.5.63, the output optical field $E_{out}(\omega)$ has to be the Eigenvector which is the output PSP. $\exp(j\tau_g\Delta\omega)$ correspond to the Eigen values,

$$\rho_1 = \exp(j\tau_{g1}\Delta\omega) \quad (4.5.64a)$$

$$\rho_2 = \exp(j\tau_{g2}\Delta\omega) \quad (4.5.64b)$$

where, τ_{g1} and τ_{g2} represent the group delays corresponding to the two fiber PSPs, their ratio is,

$$\rho_1/\rho_2 = \exp[j(\tau_{g1} - \tau_{g2})\Delta\omega] \quad (4.5.65)$$

Define a matrix

$$M(\omega+\Delta\omega)M^{-1}(\omega) = \begin{bmatrix} m_{11} & m_{12} \\ m_{21} & m_{22} \end{bmatrix} \quad (4.5.66)$$

for Equation 4.5.63 to be nontrivial, the following relation must be satisfied

$$\begin{vmatrix} m_{11} - \exp(j\tau_g\Delta\omega) & m_{12} \\ m_{21} & m_{22} - \exp(j\tau_g\Delta\omega) \end{vmatrix} = 0 \quad (4.5.67)$$

Thus, the two Eigen-values can be calculated with

$$\rho_{1,2} = \frac{(m_{11} + m_{22}) \pm \sqrt{(m_{11} + m_{22})^2 + 4(m_{12}m_{21} - m_{11}m_{22})}}{2} \quad (4.5.68)$$

Therefore, the fiber DGD can be expressed as the group delay difference as the function of the optical frequency:

$$\Delta\tau(\omega) = \left|\tau_{g1} - \tau_{g2}\right| = \left|\frac{Arg(\rho_1/\rho_2)}{\Delta\omega}\right| \tag{4.5.69}$$

Where, $Arg(\rho_1/\rho_2)$ denotes the phase angle of (ρ_1/ρ_2). Equation 4.5.69 is the major result of this derivation which underlines the principle of Jones Matrix technique for fiber DGD measurement. Based on Equation 4.5.69, the measurement procedure can be outlined in the following steps:

1. Measure Jones Matrix elements at different optical signal frequencies: $M(\omega_1)$, $M(\omega_2)$, $M(\omega_3)$, $M(\omega_4)\ldots$. This is accomplished by using three independent launching conditions at each frequency and measuring their corresponding output SOPs, as described by Equations 4.5.32–4.5.52.
2. Convert the results of step 1 into a series of new matrices shown in Equation 4.5.66 for different wavelengths.
3. Find the fiber DGD value at each optical frequency using Equations 4.5.68 and 4.5.69.

As a practical issue, it is important to decide the frequency step size $\Delta\omega$ for the measurement. On one hand, the step size $\Delta\omega$ should not be too small. If it is too small, not only the measurement would take an unnecessarily long time for the large number of measurements; any fiber path instability, wavelength inaccuracy of laser source, or polarimeter accuracy would have strong impact in the measured results. On the other hand, $\Delta\omega$ should not be too large. As a rule of thumb, the output SOP should not rotate for more than 45° over each frequency step; otherwise the measurement would produce inaccurate results. With these considerations in mind, for a fiber system with an expected DGD of $\Delta\tau$, the maximum frequency step size is limited by $\Delta\omega < \pi/(4\Delta\tau)$. Since $\Delta\omega \approx 2\pi c\Delta\lambda/\lambda^2$, the corresponding wavelength step size is

$$\Delta\lambda < \frac{\lambda^2}{8c\Delta\tau} \tag{4.5.70}$$

In 1550 nm wavelength window, this step size limitation is $\Delta\lambda \cdot \Delta\tau \leq 1$, where $\Delta\lambda$ is in nanometers and $\Delta\tau$ is in picoseconds. Obviously, for fibers with large DGD, the frequency step size must be made small. As a simple example, for a fiber with an expected DGD of $\Delta\tau = 20\ ps$, the source wavelength step size must be smaller than 0.05 nm, which is approximately the same as the requirement of the fixed analyzer technique described in the last section.

Due to the statistical nature of fiber PMD, in practice, each measurement will likely result in a different DGD value, that is, N independent measurements may result in N different DGD results. The RMS value of the N measurements is $\sqrt{\frac{1}{N}\sum_{i=1}^{N}\Delta\tau_i^2}$ since PMD generally has Maxwellian distribution as defined by Equation 4.5.4. This distribution can be specified by a single parameter α:, which is related to its RMS value $\alpha\sqrt{3}$. Therefore the α value can be found through N independent measurements and

$$\alpha^2 = \frac{1}{3N}\sum_{i=1}^{N}\Delta\tau_i^2 \tag{4.5.71}$$

where $\Delta\tau_i$ is the DGD of each measurement.

To summarize, the Jones Matrix technique has a number of advantages compared to other DGD measurement methods. First, it needs only a small wavelength window to perform a measurement. As a comparison, a fixed analyzer method needs a much larger wavelength window to have enough number of zero-crossings to obtain an accurate average. From this point of view, the Jones Matrix method is more suitable to evaluate the detailed wavelength dependency of PMD. Second, Jones Matrix measurement can be made fast using automated procedures of polarization controller and a polarimeter. This allows both time-dependent and wavelength-dependent PMD measurements. Third, the accuracy of the Jones Matrix technique is considered the best compared to other techniques because of the efficiently arranged SOP adjustment procedure and the coordinated polarimeter testing.

4.5.7 The Mueller Matrix Method

The Mueller Matrix method (MMM) is another technique that can be used to characterize the wavelength-dependent polarization rotation of an optical fiber. In general, the Jones Matrix method described in the last section requires small enough wavelength steps, as indicated in Equation 4.5.70, for accurate measurement of DGD. These small wavelength steps, in turn, demand relatively high SNR to ensure the measurement accuracy. On the contrary, the Mueller Matrix method does not have this limitation and relatively large frequency steps are allowed. In addition, the Mueller Matrix method can be used to measure polarization-dependent loss (PDL) in an optical fiber, which is discussed separately in the next section.

A Mueller Matrix represents the polarization rotation characteristics of an optical device such as an optical fiber, which is determined by the relationship between a set of input polarization vectors and their corresponding output polarization vectors,

$$\begin{bmatrix} s_{out1} \\ s_{out2} \\ s_{out3} \end{bmatrix} = \begin{bmatrix} m_{11} & m_{12} & m_{13} \\ m_{21} & m_{22} & m_{23} \\ m_{31} & m_{32} & m_{33} \end{bmatrix} \begin{bmatrix} s_{in1} \\ s_{in2} \\ s_{in3} \end{bmatrix} \tag{4.5.72}$$

where $S_{in} = \begin{bmatrix} s_{in1} & s_{in2} & s_{in3} \end{bmatrix}^T$ is the input polarization vector, $S_{out} = \begin{bmatrix} s_{out1} & s_{out2} & s_{out3} \end{bmatrix}^T$ is the output polarization vector, and $M = [m_{ij}]$ is the Mueller Matrix, which relates the output vector to the input vector. The superscript T stands for matrix transpose. In this section, we assume that the fiber has no PDL and normalized vectors are used with unity magnitude for both the input and the output optical fields.

For three orthogonal input polarization states, $S_{in}^1 = [1 \quad 0 \quad 0]^T$, $S_{in}^2 = [0 \quad 1 \quad 0]^T$, and $S_{in}^3 = [0 \quad 0 \quad 1]^T$, suppose the corresponding polarization vectors of the output fields are

$$\begin{bmatrix} s_{out,11} & s_{out,21} & s_{out,31} \\ s_{out,12} & s_{out,22} & s_{out,32} \\ s_{out,13} & s_{out,23} & s_{out,33} \end{bmatrix} = \begin{bmatrix} m_{11} & m_{12} & m_{13} \\ m_{21} & m_{22} & m_{23} \\ m_{31} & m_{32} & m_{33} \end{bmatrix} \begin{bmatrix} 1 & 0 & 0 \\ 0 & 1 & 0 \\ 0 & 0 & 1 \end{bmatrix} \tag{4.5.73}$$

The Mueller Matrix elements can be determined by measuring the output polarization states so that [34]

$$[m_{ij}] = [s_{out,ji}] \tag{4.5.74}$$

In fact, if there is no PDL, two independent input polarization vectors S_{in}^1, and S_{in}^a are sufficient to determine the Mueller Matrix. In this measurement, S_{in}^a does not even have to be orthogonal to S_{in}^1. If the corresponding output polarization vectors are S_{out}^1 and S_{out}^a, respectively, a third polarization vector can be found through normalization:

$$S_{out}^3 = S_{out}^1 \times S_{out}^a \tag{4.5.75}$$

To complete a three-dimensional Cartesian coordinate systen, another vector that is orthogonal to both S_{out}^1 and S_{out}^3 can be obtained as

$$S_{out}^2 = S_{out}^3 \times S_{out}^1 \tag{4.5.76}$$

where both the input and the output vectors are normalized such that $|S_{in}^1| = |S_{in}^a| = 1$ and $|S_{out}^1| = |S_{out}^2| = |S_{out}^3| = 1$.

The experimental setup of Mueller Matrix measurement, shown in Figure 4.5.13, is similar to that of the Jones Matrix technique. The only difference

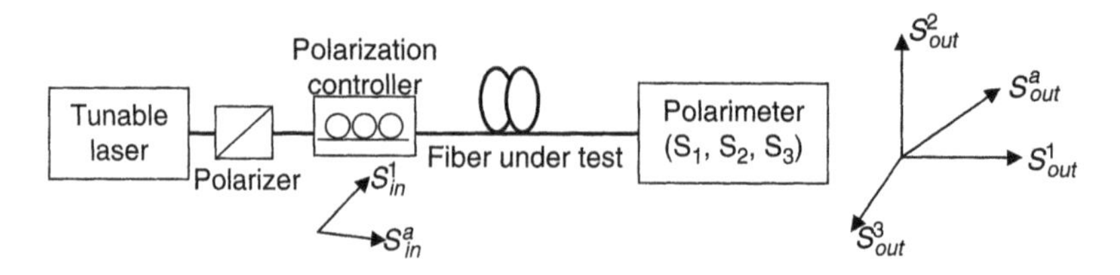

Figure 4.5.13 Block diagram of the Mueller Matrix method to measure fiber PMD.

is that in Mueller Matrix measurement, only two independent linear input polarization states are required, and they do not have to be mutually orthogonal.

Again, since the PMD in a fiber is a frequency-dependent polarization rotation, as indicated in Equation 4.5.13,

$$\frac{dS_{out}}{d\omega} = \Omega \times S_{out} \tag{4.5.77}$$

where Ω is the PMD vector whose magnitude is equal to the DGD of the fiber, which is $\Delta\tau = d\phi/d\omega = |\Omega|$. Performing the measurements at two frequencies ω_1 and $\omega_2 = \omega_1 + \Delta\omega$, we can find two Mueller matrices, $[M(\omega_1)]$ and $[M(\omega_2)]$, through

$$\left[S_{out}(\omega_1)\right] = \left[M(\omega_1)\right]\left[S_{in}(\omega_1)\right] \tag{4.5.78a}$$

$$\left[S_{out}(\omega_2)\right] = \left[M(\omega_2)\right]\left[S_{in}(\omega_2)\right] \tag{4.5.78b}$$

Note that the input polarization vector does not change with the change of the frequency, $[S_{in}(\omega_1)] = [S_{in}(\omega_2)] = [S_{in}]$. Since the Mueller Matrix is normalized, $[M(\omega_1)][M(\omega_1)]^T = I$, where I is the unit matrix, Equations 4.5.78a and 4.5.78b can be combined to eliminate the input $[S_{in}]$,

$$\left[S_{out}(\omega_2)\right] = \left[M_\Delta\right]\left[S_{out}(\omega_1)\right] \tag{4.5.79}$$

where

$$\left[M_\Delta\right] = \left[M(\omega_2)\right]\left[M(\omega_1)\right]^T \tag{4.5.80}$$

The rotational matrix, $[M_\Delta]$, can be determined through the measurement using the setup shown in Figure 4.5.13. However, it is important to understand how to calculate the fiber DGD from the measured rotational matrix.

In fact, the rotational matrix can be defined as a spatial vector, which is described by a rotation angle ϕ around an axis $\vec{r}$ as shown in Figure 4.5.14 [35, 36]. Assume that $S_0 = S_{out}(\omega_0)$ is the polarization vector measured at frequency ω_0 (from the origin to point B as shown in Figure 4.5.14), $S1=S_{out}$ (ω1) is the polarization vector measured at frequency ω_1 (from the origin to point A), and the rotation is around an axis $\vec{r}$ with the rotation angle ϕ. In the following, we shall find the relationship between the vector rotation from S_0 to S_1 and the corresponding angular rotation ϕ, which is directly related to the fiber DGD.

According to the rule of vectoral addition, S_1 can be expressed as

$$S_1 = \overrightarrow{OA} = \overrightarrow{OO'} + \overrightarrow{O'B} = \overrightarrow{OO'} + \overrightarrow{O'C} + \overrightarrow{CA} \tag{4.5.81}$$

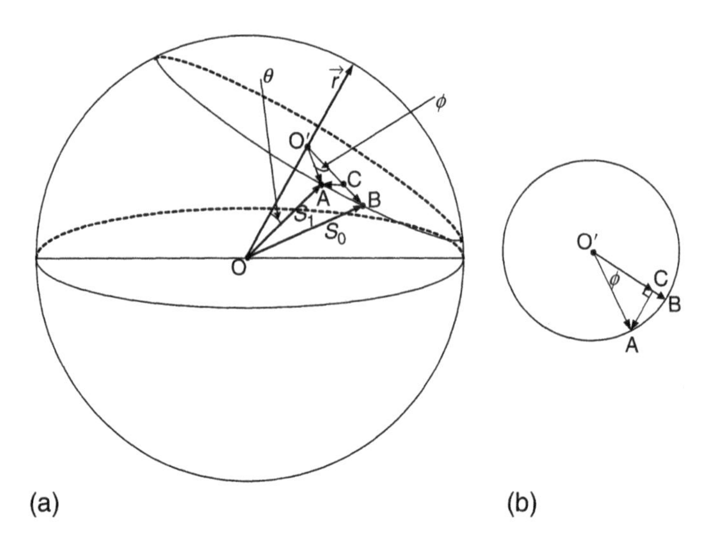

Figure 4.5.14 (a) Geometric expression of rotation matrix; (b) view of a plane perpendicular to the rotation axis.

Without loss of generality, we assume the sphere in Figure 4.5.14(a) has unit radius, and therefore $\vec{r}, S_0$, and S_1 all have the unity length. Then based on the geometric structure shown in Figure 4.5.14, we can find the following relations:

$$\left|\overrightarrow{OO'}\right| = |S_0|\cos\theta = \vec{r}\cdot S_0 \tag{4.5.82}$$

$$\left|\overrightarrow{O'A}\right| = \left|\overrightarrow{O'B}\right| \tag{4.5.83}$$

$$\left|\overrightarrow{O'B}\right| = |S_0|\sin\theta = \left|S_0\times\vec{r}\right| \tag{4.5.84}$$

$$\left|\overrightarrow{CA}\right| = \left|\overrightarrow{O'A}\right|\sin\phi = \left|\overrightarrow{O'B}\right|\sin\phi \tag{4.5.85}$$

$$\left|\overrightarrow{O'C}\right| = \left|\overrightarrow{O'A}\right|\cos\phi = \left|\overrightarrow{O'B}\right|\cos\phi \tag{4.5.86}$$

Since $\overrightarrow{OO'}$ is in the same direction as $\vec{r}$,

$$\overrightarrow{OO'} = \vec{r}\,(\vec{r}\cdot S_0) \tag{4.5.87}$$

Then the vector $\overrightarrow{O'B}$ can be expressed as

$$\overrightarrow{O'B} = \overrightarrow{OB} - \overrightarrow{OO'} = S_0 - \vec{r}(\vec{r} \cdot S_0) \tag{4.5.88}$$

Based on Equation 4.5.86 and take into account that $\overrightarrow{O'C}$ is in the same direction as $\overrightarrow{O'B}$, we can find

$$\overrightarrow{O'C} = [S_0 - \vec{r}(\vec{r} \cdot S_0)]\cos\phi \tag{4.5.89}$$

Using Equations 4.5.84 and 4.5.85 and noting that $\overrightarrow{CA}$ is in the same direction as $(S_0 \times \vec{r})$, we have,

$$\overrightarrow{CA} = \sin\phi(S_0 \times \vec{r}) \tag{4.5.90}$$

Substituting Equations 4.5.87, 4.5.89, and 4.5.90 into Equation 4.5.81, we obtain the following relation:

$$\begin{aligned} S_1 &= \vec{r}\,(\vec{r} \cdot S_0) + [S_0 - \vec{r}\,(\vec{r} \cdot S_0)]\cos\phi + \sin\phi(S_0 \times \vec{r}\,) \\ &= S_0 \cos\phi + (1 - \cos\phi)\vec{r}\,(\vec{r} \cdot S_0) + \sin\phi(S_0 \times \vec{r}) \end{aligned} \tag{4.5.91}$$

Now it is important to review a few matrix concepts and identities. First, a cross-product between two vectors $\vec{a} = [a_1, a_2, a_3]^T$ and $\vec{b} = [b_1, b_2, b_3]^T$ can be expressed in terms of matrix multiplication, which is the product between a skew-symmetric matrix of the first vector and the second vector,

$$\vec{a} \times \vec{b} = \begin{bmatrix} 0 & -a_3 & a_2 \\ a_3 & 0 & -a_1 \\ -a_2 & a_1 & 0 \end{bmatrix} \vec{b} \tag{4.5.92}$$

The cross-product operator of $\vec{a}$ can be written as

$$\vec{a}\times = \begin{bmatrix} 0 & -a_3 & a_2 \\ a_3 & 0 & -a_1 \\ -a_2 & a_1 & 0 \end{bmatrix} \tag{4.5.93}$$

In addition, a dyadic product of the two vectors $\vec{a}$ and $\vec{b}$ is

$$\vec{a}\,\vec{b} = \begin{bmatrix} a_1b_1 & a_1b_2 & a_1b_3 \\ a_2b_1 & a_2b_2 & a_2b_3 \\ a_3b_1 & a_3b_2 & a_3b_3 \end{bmatrix} \tag{4.5.94}$$

Another very useful identity involving matrix cross-product is

$$\vec{a} \times \vec{b} \times \vec{c} = \vec{b}\,(\vec{a} \cdot \vec{c}\,) - \vec{c}(\vec{a} \cdot \vec{b}) \tag{4.5.95}$$

where $\vec{c} = [c_1, c_2, c_3]^T$.

Based on Equation 4.5.93, $(\vec{r}\times)(\vec{r}\times)$ can be written as

$$(\vec{r}\times)(\vec{r}\times) = \begin{bmatrix} 0 & -r_3 & r_2 \\ r_3 & 0 & -r_1 \\ -r_2 & r_1 & 0 \end{bmatrix}\begin{bmatrix} 0 & -r_3 & r_2 \\ r_3 & 0 & -r_1 \\ -r_2 & r_1 & 0 \end{bmatrix}$$
$$= \begin{bmatrix} -r_3^2 - r_2^2 & r_1 r_2 & r_1 r_3 \\ r_2 r_1 & -r_3^2 - r_1^2 & r_2 r_3 \\ r_3 r_1 & r_3 r_2 & -r_1^2 - r_2^2 \end{bmatrix} \quad (4.5.96)$$

Remember, $\vec{r}$ is a unit vector such that $r_1^2 + r_2^2 + r_3^2 = 1$. Based on the relation given in Equation 4.5.94, Equation 4.5.96 can be simplified as

$$(\vec{r}\times)(\vec{r}\times) = \begin{bmatrix} -1 + r_1^2 & r_1 r_2 & r_1 r_3 \\ r_2 r_1 & -1 + r_2^2 & r_2 r_3 \\ r_3 r_1 & r_3 r_2 & -1 + r_3^2 \end{bmatrix}$$
$$= \begin{bmatrix} r_1^2 & r_1 r_2 & r_1 r_3 \\ r_2 r_1 & r_2^2 & r_2 r_3 \\ r_3 r_1 & r_3 r_2 & r_3^2 \end{bmatrix} - \begin{bmatrix} 1 & 0 & 0 \\ 0 & 1 & 0 \\ 0 & 0 & 1 \end{bmatrix} = \vec{r}\,\vec{r} - I \quad (4.5.97)$$

Then use Equation 4.5.95 and 4.5.96, the term $\vec{r}(\vec{r}\cdot S_0)$ can also be simplified as

$$\vec{r}(\vec{r}\cdot S_0) = \vec{r}\times\vec{r}\times S_0 + S_0(\vec{r}\cdot\vec{r}) = [\vec{r}\;\vec{r} - I]\cdot S_0 + S_0 = \vec{r}\,\vec{r}\cdot S_0 \quad (4.5.98)$$

Substituting Equation 4.5.98 to Equation 4.5.91, we find

$$S_1 = S_0\cos\phi + (1-\cos\phi)\vec{r}\,\vec{r}\cdot S_0 + \sin\phi(S_0\times\vec{r})$$
$$= S_0\cos\phi + (1-\cos\phi)\vec{r}\,\vec{r}\cdot S_0 - \sin\phi(\vec{r}\times S_0) \quad (4.5.99)$$
$$= [I\cos\phi + (1-\cos\phi)\vec{r}\,\vec{r} - \sin\phi(\vec{r}\times)]S_0$$

This provides the relationship between the polarization vectors S_0 and S_1 measured at two different frequencies. According to the rotation vector defined in Equations 4.5.79 and 4.5.80, the relationship can be simplified as

$$[M_\Delta] = I\cos\phi + (1-\cos\phi)\vec{r}\,\vec{r} - \sin\phi(\vec{r}\times) \quad (4.5.100)$$

To obtain further explicit relationships between entries of matrix $[M_\Delta]$, the rotation angles and the rotation axis, we can expand Equation 4.5.100 into a matrix form,

$$[M_\Delta] = \begin{bmatrix} m_{\Delta 11} & m_{\Delta 12} & m_{\Delta 13} \\ m_{\Delta 21} & m_{\Delta 22} & m_{\Delta 23} \\ m_{\Delta 31} & m_{\Delta 32} & m_{\Delta 33} \end{bmatrix}$$

$$= \begin{bmatrix} 1 & 0 & 0 \\ 0 & 1 & 0 \\ 0 & 0 & 1 \end{bmatrix} \cos\phi + (1-\cos\phi) \begin{bmatrix} r_1^2 & r_1 r_2 & r_1 r_3 \\ r_2 r_1 & r_2^2 & r_2 r_3 \\ r_3 r_1 & r_3 r_2 & r_3^2 \end{bmatrix} - \sin\phi \begin{bmatrix} 0 & -r_3 & r_2 \\ r_3 & 0 & -r_1 \\ -r_2 & r_1 & 0 \end{bmatrix}$$

$$= \begin{bmatrix} \cos\phi + (1-\cos\phi) r_1^2 & (1-\cos\phi) r_1 r_2 + \sin\phi r_3 & (1-\cos\phi) r_1 r_3 - \sin\phi r_2 \\ (1-\cos\phi) r_2 r_1 - \sin\phi r_3 & \cos\phi + (1-\cos\phi) r_2^2 & (1-\cos\phi) r_2 r_3 + \sin\phi r_1 \\ (1-\cos\phi) r_3 r_1 + \sin\phi r_2 & (1-\cos\phi) r_3 r_2 - \sin\phi r_1 & \cos\phi + (1-\cos\phi) r_3^2 \end{bmatrix} \tag{4.5.101}$$

From Equation 4.5.101, it is straightforward to derive the following relations:

$$Tr([M_\Delta]) = 2\cos\phi + 1 \tag{4.5.102}$$

$$m_{\Delta 12} - m_{\Delta 21} = 2\sin\phi r_3 \tag{4.5.103}$$

$$m_{\Delta 31} - m_{\Delta 13} = 2\sin\phi r_2 \tag{4.5.104}$$

$$m_{\Delta 23} - m_{\Delta 32} = 2\sin\phi r_1 \tag{4.5.105}$$

where $Tr([M_\Delta])$ is the trace of the matrix $[M_\Delta]$. Thus the explicit forms relating $[M_\Delta]$ with the rotation axis $\vec{r}$ and rotation angle ϕ are

$$\cos\phi = \frac{1}{2}(Tr([M_\Delta]) - 1) \tag{4.5.106}$$

$$r_1 = \frac{m_{\Delta 23} - m_{\Delta 32}}{2\sin\phi} \tag{4.5.107}$$

$$r_2 = \frac{m_{\Delta 31} - m_{\Delta 13}}{2\sin\phi} \tag{4.5.108}$$

$$r_3 = \frac{m_{\Delta 12} - m_{\Delta 21}}{2\sin\phi} \tag{4.5.109}$$

The procedure of the Mueller Matrix technique can be summarized as follows:

1. At a certain signal optical frequency ω_1, launch three independent optical signal SOPs into the fiber with their normalized polarization vectors [1 0 0], [0 1 0], and [0 0 1] and measure the corresponding output SOPs

to complete the matrix using Equation 4.5.73 so that the nine elements of a Mueller matrix $[M(\omega_1)]$ can be obtained. (Theoretically, only two independent optical signal SOPs are sufficient to obtain a Mueller Matrix based on the discussion leading to Equations 4.5.75 and 4.5.76 with proper matrix manipulations.)

2. Repeat step 1 but at a different optical frequency $\omega_2 = \omega_1 + \Delta\omega$ and obtain a Mueller Matrix $[M(\omega_2)]$.
3. Derive the rotation matrix $[M_\Delta]$ using Equation 4.5.80, which indicates the Mueller Matrix change due to the change in optical signal frequency.
4. Use the rotation matrix $[M_\Delta]$ to obtain the rotation angle ϕ and the orientation of the PMD vector $\vec{r}$ based on Equations 4.5.106–4.5.109.
5. Calculate DGD based on $\Delta\tau = \phi/\Delta\omega$, which was used in the Poincare arc measurement as indicated in Equation 4.5.16.

An important feature of the Mueller Matrix technique is that it measures the frequency-dependent orientation of the PMD vector, $\vec{r}$, which is not generally provided by the Jones Matrix technique.

4.6 DETERMINATION OF POLARIZATION-DEPENDENT LOSS

In an optical device, the power transfer function may depend on the SOP of the optical signal. An extreme example is a linear polarizer, in which the transmission loss is nearly zero for a specific state of polarization while the transmission loss is very high for the SOP orthogonal to it. For most of the optical devices such as mirrors, beam splitters, and lenses, it is desirable to have low polarization dependency in the transmission loss. In a fiber-optic system, the polarization-dependent loss (PDL) of the optical fiber itself is usually very small, whereas the overall PDL in the system is affected by PDLs of all the individual optical components. From the measurement point of view, even if the PDL of each component may be well defined with a deterministic value, the PDL of the entire system may randomly vary with time. The overall system PDL cannot be obtained by simply adding up the PDL of each component used in the system; this is due to the random variations of the relative orientations between various PDL elements. In addition, the presence of polarization mode dispersion in the fiber further enhances this uncertainty.

A block diagram of PDL measurement of a fiber-optic system is shown in Figure 4.6.1, where the optical system is treated as a two-terminal device under test (DUT); we should have access to both terminals of the device. In this setup, the SOP of the optical signal entering the DUT is adjusted through a polarization controller, and a photodetector is used to measure the signal optical power

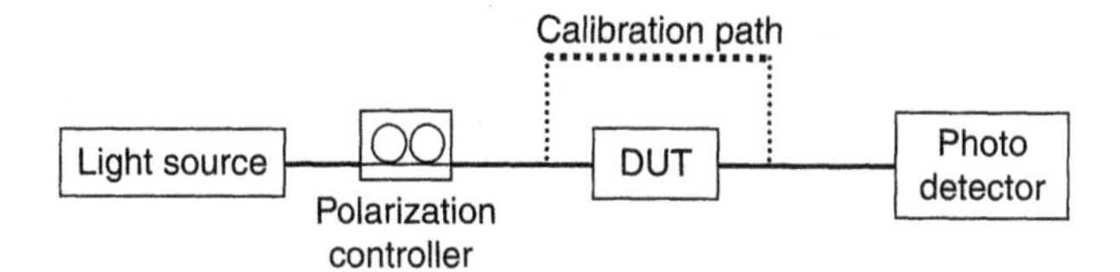

Figure 4.6.1 Measurement setup for polarization-dependent loss.

that passes through the DUT. With the scanning of the signal SOP across the Poincare sphere, a maximum and a minimum value can be detected by the photodetector, and their difference indicates the PDL of the DUT.

In fact, the general definition of PDL is

$$PDL = \frac{T_{\max} - T_{\min}}{T_{max} + T_{\min}} \tag{4.6.1}$$

where $T_{\max}$ and $T_{\min}$ are the maximum and the minimum power transmission efficiencies of the DUT. If the signal optical power, P_{in}, entering the DUT is polarization-independent, Equation 4.6.1 can be written as

$$PDL = \frac{P_{\max}/P_{in} - P_{\min}/P_{in}}{P_{\max}/P_{in} + P_{\min}/P_{in}} = \frac{P_{max} - P_{\min}}{P_{max} + P_{\min}} \tag{4.6.2}$$

where $P_{\max}$ and $P_{\min}$ are the maximum and the minimum optical powers received by the photodetector when the SOP of the optical signal scans to cover the entire Poincare sphere.

PDL in a fiber-optic system is more complicated than that of a simple optical device due to the random mode coupling and birefringence in the optical fiber. It is also impractical to scan the SOP of the optical signal randomly to cover the entire Poincare sphere. A more systematic measurement procedure is necessary, one in which the PDL can be obtained and that requires only a few individual SOP settings of the optical signal. The most commonly used method to characterize PDL in a fiber-optic system is based on the Mueller Matrix method we introduced in the last section. However, when PDL is considered, the 3×3 matrix is no longer sufficient to describe the system, and a 4×4 matrix needs to be used:

$$\begin{bmatrix} S_{out0} \\ S_{out1} \\ S_{out2} \\ S_{out3} \end{bmatrix} = \begin{bmatrix} m_{11} & m_{12} & m_{13} & m_{14} \\ m_{21} & m_{22} & m_{23} & m_{24} \\ m_{31} & m_{32} & m_{33} & m_{34} \\ m_{41} & m_{42} & m_{43} & m_{44} \end{bmatrix} \begin{bmatrix} S_{in0} \\ S_{in1} \\ S_{in2} \\ S_{in3} \end{bmatrix} \tag{4.6.3}$$

where $S_{in} = [s_{in0} \quad s_{in1} \quad s_{in2} \quad s_{in3}]^T$ is the Stokes parameter of the input optical signal, and for a polarized optical signal, $s_{in0} = \sqrt{s_{in1}^2 + s_{in2}^2 + s_{in3}^2}$ represents the input optical power. Similarly, $S_{out} = [s_{out0} \quad s_{out1} \quad s_{out2} \quad s_{out3}]^T$ is the Stokes parameter of the output optical signal and $s_{out0} = \sqrt{s_{out1}^2 + s_{out2}^2 + s_{out3}^2}$ is the output optical power if the signal is polarized.

Although a 4 × 4 Mueller Matrix may be used to characterize both the PMD and PDL, if we are only interested in measuring the PDL, the first row of the Mueller Matrix is sufficient, that is,

$$s_{out0} = m_{11}s_{in0} + m_{12}s_{in1} + m_{13}s_{in2} + m_{14}s_{in3} \tag{4.6.4}$$

The power transmission efficiency through the optical system is defined as the output power versus the input power,

$$T = \frac{s_{out0}}{s_{in0}} = \frac{m_{11}s_{in0} + m_{12}s_{in1} + m_{13}s_{in2} + m_{14}s_{in3}}{s_{in0}} \tag{4.6.5}$$

Since the vector $[m_{12} \quad m_{13} \quad m_{14}]$ describes the optical system property that is independent of the SOP of the input signal $[s_{in1} \quad s_{in2} \quad s_{in3}]$, the following constraint should be valid:

$$-\sqrt{m_{12}^2 + m_{13}^2 + m_{14}^2} \le \frac{m_{12}s_{in1} + m_{13}s_{in2} + m_{14}s_{in3}}{\sqrt{s_{in1}^2 + s_{in2}^2 + s_{in3}^2}} \le \sqrt{m_{12}^2 + m_{13}^2 + m_{14}^2} \tag{4.6.6}$$

Thus, the maximum and the minimum transmission through the fiber should be, respectively,

$$T_{\max} = m_{11} + \sqrt{m_{12}^2 + m_{13}^2 + m_{14}^2} \tag{4.6.7a}$$

$$T_{\min} = m_{11} - \sqrt{m_{12}^2 + m_{13}^2 + m_{14}^2} \tag{4.6.7b}$$

Then, by the definition of PDL as indicated in Equation 4.6.1, we can relate the system PDL with the Mueller Matrix elements:

$$PDL = \frac{\sqrt{m_{12}^2 + m_{13}^2 + m_{14}^2}}{m_{11}} \tag{4.6.8}$$

Now the question is how to measure the Mueller Matrix elements m_{11}, m_{12}, m_{13}, and m_{14}. This can be accomplished using the experimental setup shown in Figure 4.6.1 in which four well-defined SOPs of the input optical signal need to be used. Table 4.6.1 summarizes the input signal SOP and the corresponding output optical power measured by the photodetector [37].

Table 4.6.1

Input Signal SOP Setup for PDL Measurement

Input Signal SOP	Input Stokes Vector	Measured Output Power
Linear horizontal (0°)	$S_{in,a} = (P_a \quad P_a \quad 0 \quad 0)$	$P_1 = m_{11}P_a + m_{12}P_a$
Linear vertical (90°)	$S_{in,b} = (P_b \quad -P_b \quad 0 \quad 0)$	$P_2 = m_{11}P_b - m_{12}P_b$
Linear diagonal (45°)	$S_{in,c} = (P_c \quad 0 \quad P_c \quad 0)$	$P_3 = m_{11}P_c + m_{13}P_c$
Right-hand circular	$S_{in,d} = (P_d \quad 0 \quad 0 \quad P_d)$	$P_4 = m_{11}P_d + m_{14}P_d$

In principle, an ideal polarization controller would not introduce additional signal optical power change when adjusting the signal SOP. However, in practice, there might be internal PDL created by the polarization controller and therefore Table 4.6.1 assumes a slightly different optical power, P_a, P_b, P_c, and P_d, for each of the four input SOP of the optical signal. The output optical power, P_1, P_2, P_3, and P_4, measured corresponding to each input SOP relates to the Mueller Matrix elements through Equation 4.6.3. Solving the four equations shown in the third column of Table 4.6.1, the following expressions can be obtained:

$$m_{11} = \frac{1}{2}\left(\frac{P_1}{P_a} + \frac{P_2}{P_b}\right) \tag{4.6.9a}$$

$$m_{12} = \frac{1}{2}\left(\frac{P_1}{P_a} - \frac{P_2}{P_b}\right) \tag{4.6.9b}$$

$$m_{13} = \frac{P_3}{P_c} - \frac{1}{2}\left(\frac{P_1}{P_a} - \frac{P_2}{P_b}\right) \tag{4.6.9c}$$

$$m_{14} = \frac{P_4}{P_d} - \frac{1}{2}\left(\frac{P_1}{P_a} - \frac{P_2}{P_b}\right) \tag{4.6.9d}$$

The PDL value can finally be obtained by substituting Equation 4.6.9 into 4.6.8. In practical measurements, to create the four well-defined SOPs of the input optical signal, a programmable polarization controller is required with a high level of repeatability. Otherwise, a polarimeter needs to be used to monitor the SOPs of the input optical signal as well.

It is worthwhile to note that both PMD and PDL are system properties that should be independent of the optical signal that is launched into the system. However, the impact of PMD and PDL on the performance of an optical system may depend on the polarization state of the input optical signal. The measurements of PMD and PDL in optical systems that carry live traffic will be discussed in Chapter 5.

4.7 PMD SOURCES AND EMULATORS

Polarization mode dispersion is an important source of performance degradation in fiber-optic transmission systems, especially at high data rates. In system characterization and verification tests, the effect of PMD has to be taking into account to evaluate system margin. It is important to make sure that the transmission equipment can tolerate the PMD level of the fiber system over which the equipment will be deployed. This requires a source to generate a known amount of PMD, either static or statistical, to simulate the PMD characteristic of a fiber system.

In general, a PMD source is referred to as a source that generates a fixed amount of DGD, whereas a PMD emulator is a device that generates a time-varying DGD to emulate the practical PMD characteristics in fiber systems. PMD sources are usually used to test the deterministic PMD impact on systems or subsystems because the added PMD level from the source is precisely known. On the other hand, a PMD emulator is often used to test the statistical nature of the system such as PMD-induced BER outage and the effectiveness of the PMD compensator in the receiver.

The basic configuration of the simplest PMD source is shown in Figure 4.7.1, where two polarization beam splitters (PBS) are used. The input optical signal is first split by PBS1 into horizontal and vertical polarization components. Then one of them passes through an adjustable delay line before the two orthogonally polarized light beams combine in PBS2. The differential group delay (DGD) of the PMD source can be set by the time delay of the adjustable delay line.

To use this type of PMD source correctly, the input optical signal has to be linearly polarized. The SOP of the input signal has to be adjusted midway between the two principle axes of the PBS so that the powers of the horizontal and the vertical polarization components are equal. In fact, the impact of the PMD in an optical receiver depends on the ratio of the signal optical power carried by the two PSPs in the fiber system [25],

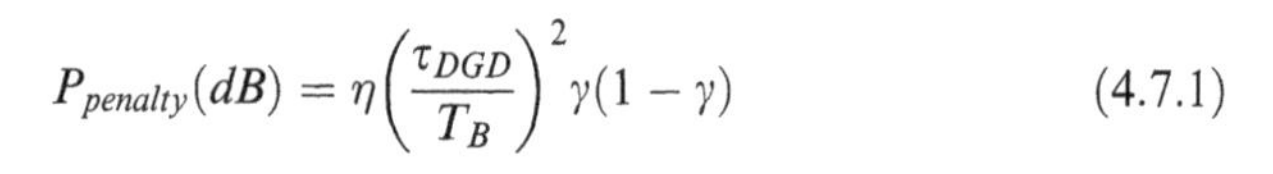

$$P_{penalty}(dB) = \eta\left(\frac{\tau_{DGD}}{T_B}\right)^2 \gamma(1-\gamma) \tag{4.7.1}$$

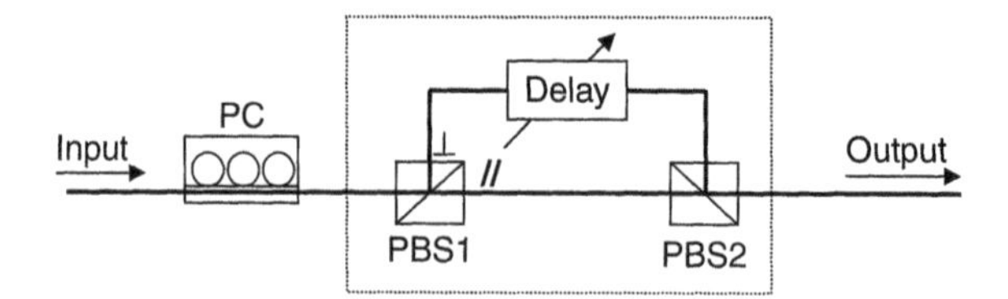

Figure 4.7.1 Block diagram of a simple PMD source.

where $P_{penalty}$ is the power penalty caused by the PMD in the receiver, η is a proportionality parameter that depends on the modulation formats and receiver bandwidth,τ_{DGD} is the first-order DGD of fiber system, and T_B is the bit interval. $0 \leq \gamma \leq 1$ is the power-splitting ratio of the two PSPs, which is defined as $\gamma = P_{fast}/P_{total}$, where P_{fast} is the power carried by the fast axis and $P_{total} = P_{fast}+P_{slow}$ is the total signal power. Obviously for either $\gamma = 0$ or $\gamma = 1$, the signal optical power is carried by one of the two PSPs of the fiber and the PMD-induced penalty should be zero. The principle axes of the PBS in the PMD source are equivalent to the PSPs of a birefringence fiber in a transmission system. When the SOP of the input optical signal is midway between the two PSPs, $\gamma = 0.5$ and $\gamma(1-\gamma) = 0.25$. If we define an angle θ between the SOP of the input optical signal and the PSP of the fiber, Equation 4.7.1 will become

$$P_{penalty}(dB) = \eta\left(\frac{\tau_{DGD}\sin(2\theta)}{2T_B}\right)^2 \tag{4.7.2}$$

where $0 \leq \theta \leq \pi/2$ corresponds to $0 \leq \gamma \leq 1$. For the PMD source using the configuration shown in Figure 4.7.1, if the SOP of the input optical signal is not equal to $\theta = \pi/4$, the equivalent DGD will be $|\tau_{DGD}\sin(2\theta)| \leq \tau_{DGD}$. Therefore, to provide the specified DGD value, we must make sure that the input SOP is correctly aligned at $\theta = \pi/4$. An easy procedure for this polarization alignment is to block one of the two arms and observe a 3 dB drop in the output optical power. This ensures that the optical powers in the two arms are equal.

The PMD source shown in Figure 4.7.1 can be constructed with polarization-maintaining optical fibers between the two PBSs. However, a disadvantage of a fiber-based PMD source is that the output signal SOP always drifts randomly and is hard to stabilize. In fact, if the optical field amplitude of the two arms is equal, $|E_{//}| = |E_{\perp}| = |E_1|$, the output optical field should be $E_0 = |E_1|\left[\vec{a}_x + \vec{a}_y\, e^{j\omega\Delta\tau}e^{j\varphi(t)}\right]$, where $\vec{a}_x$ and $\vec{a}_y$ are unit vectors, $\Delta\tau$ is the differential time delay, and $\varphi(t)$ is the relative optical phase difference between the optical signals passing through the two arms. The SOP of the combined output optical field E_0 is determined directly by the optical phase $\varphi(t)$. For a 10 cm long fiber delay line, for example, if the temperature sensitivity of the refractive index in silica is $\sim 10^{-5}$/°C and the signal wavelength is 1.5 μm, there will be approximately $2\pi/3$ rotation of the output SOP for each degree change in temperature. Temperature-induced mechanical stress will also contribute to phase instability.

A compact PMD source can be made using multiple sections of birefringence crystals, as illustrated in Figure 4.7.2 [38]. An electrically controlled binary polarization switch cell is inserted between adjacent birefringence crystals, which provides a 90° polarization switch when a current is applied. Suppose the SOP of the linearly polarized input optical signal is +45° with respect to the extraordinary axis of the birefringence crystal. The first switch determines

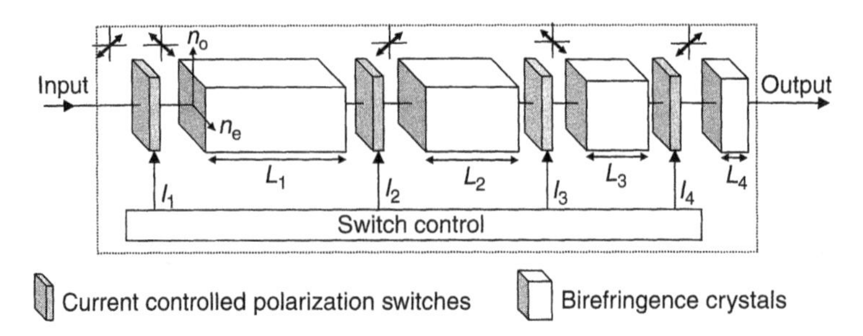

Figure 4.7.2 PMD source using concatenated birefringence crystals and polarization switches.

whether the SOP of the optical signal remains at +45° or it is switched to –45° when it reaches the first crystal section of length L_1. Then, if the second switch cell makes a 90° switch, the (o, e) beams in the first crystal will become the (e, o) beams in the second crystal, otherwise, the signal SOP will remain unchanged. As a consequence, the total DGD for the optical signal will be $(n_0 - n_e)(L_1 \pm L_2)/c$, depending on the status of the polarization switches. In general, for a PMD source with N birefringence crystal sections, the total DGD can be expressed as

$$\tau_{DGD} = \left| \sum_{i=1}^{N} (\pm\Delta\tau_i) \right| \tag{4.7.3}$$

where $\Delta\tau_i = (n_o - n_e)L_i/c$ is the DGD of the i-th crystal section and the $\pm$ sign is determined by the binary status (0 or 90°) of the polarization switch in front of each birefringence crystal. In this implementation, the instability problem for the output signal SOP can be eliminated because the optical signal does not have to be split when passing through the device. The maximum DGD this PMD source can provide depends on the total length of the birefringence crystals, $\tau_{DGD,\max} = (n_0 - n_e)\sum_{i=1}^{N} L_i/c$. As an example, a YVO_4 crystal has the birefringence value of approximately $(n_o - n_e) \approx 0.2$ in the 1550 nm wavelength window. To make a PMD source with the maximum DGD of 50 ps, the aggregated crystal length has to be 75 mm.

Because binary polarization switches are used in this PMD source, it is able to specifically provide first-order DGD at the specified value while maintaining a negligible level of second-order DGD. By dynamically controlling the status of the polarization switches, it is possible to generate time-dependent DGD with any statistical distribution. The correlation time is determined by the speed of the polarization switches. With this dynamic control, the PMD source actually operates as a PMD emulator.

In high-speed optical systems, high orders of PMD may become important, affecting system performance. To create a high-order PMD effect, a large number of birefringence sections can be concatenated with random length and axes

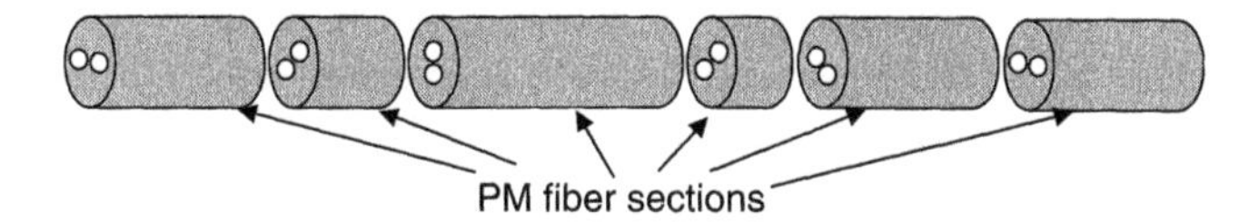

Figure 4.7.3 All-order PMD emulator created by splicing together a large number of PM fiber sections.

alignments. One simple way to make this type of all-order PMD emulator is to splice many pieces of polarization-maintaining (PM) fibers together with random lengths and orientations, as illustrated in Figure 4.7.3. This closely simulates the propagation effect in a practical fiber system. However, since these fiber sections are permanently spliced together, the PMD emulator is no longer adjustable, except for the controlling through environmental perturbations such as temperature and mechanical stress or by sweeping the wavelength of the optical signal. Over a long time in a noncontrolled environment or by sweeping the wavelength across a wide window, a Maxwellian distribution of PMD can be obtained.

However, from system measurement point of view, for a transmission channel with a fixed signal wavelength, it is desirable to accelerate the PMD performance test using an emulator with adequate dynamic control such that the DGD converges rapidly to the expected Maxwellian distribution. This can be achieved by adding a polarization controller or a phase shifter between each pair of birefringence sections, as illustrated in Figure 4.7.4. In this configuration, the birefringence section can be made by either PM fibers or YVO_4 crystals and their birefringence axes can be randomly aligned. The dynamic control of SOP rotation or signal optical phase retardation between birefringence sections accelerates the convergence of PMD statistics toward Maxwellian distribution.

PMD sources and emulators are commercially available with various optical configurations depending on the required maximum DGD level, PMD statistics, and dynamics. Different control mechanisms and algorithms, such as the control of SOP or optical phase retardation between birefringence sections or the

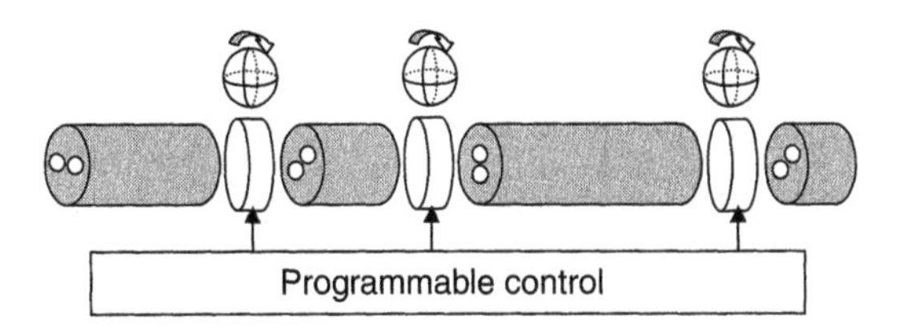

Figure 4.7.4 All-order PMD emulator created by using a polarization controller or a phase shifter between birefringence elements.

control of the DGD provided by each section, may exhibit different PMD statistics. Interested readers on the statistics of DGD generated by different types of PMD emulators can find details at [38, 39, 40, 41].

4.8 MEASUREMENT OF FIBER NONLINEARITY

Nonlinearity in an optical fiber refers to the transmission properties which depend on the optical power carried in the fiber. There are two basic types of nonlinear effects in optical fibers. The first type is due to the interaction of optical signal with phonons, which generates nonlinear scattering, such as stimulated Raman scattering (SRS) and stimulated Brillouin scattering (SBS). As a result, some of the input signal photons are converted into Stokes photons at a longer wavelength. This usually results in the loss of signal optical power, whereas it can also be utilized to perform optical amplification for signals at longer wavelengths.

The second type of nonlinearity is due to the Kerr effect, which produces the intensity-dependent refractive index in the fiber. Several well-known effects such as self-phase modulation (SPM), cross-phase modulation (XPM), four-wave mixing (FWM), and modulation instability are caused by the Kerr effect nonlinearity. In principle, instead of optical power, the power density is the most important quantity that determines the strength of the nonlinear effects in the fiber. Although a single-mode fiber usually carries signal optical power in the level of milliwatts, because the cross-section area of an optical fiber is very small, the power density can be very high. One important parameter to specify the power-handling capability of an optical fiber is its effective cross-section area A_{eff}, which is generally different from the geometric core cross-section area of the fiber.

As illustrated in Figure 4.8.1, the optical power distribution in the fiber core is nonuniform; therefore the power density in the fiber is, strictly speaking, nonuniform. The concept of effective area assumes that a uniform power

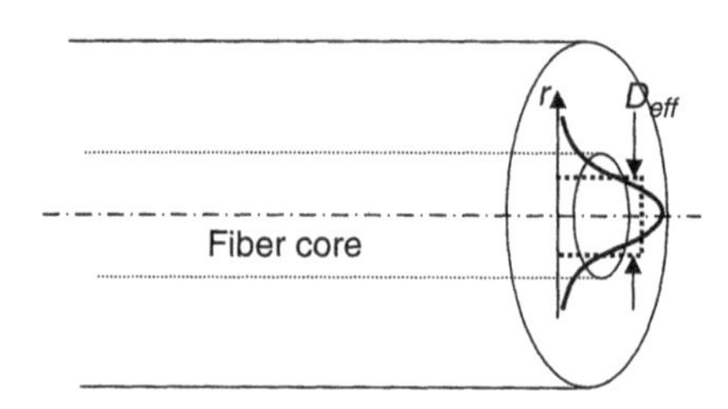

Figure 4.8.1 Illustration of fiber diameter D_{eff}.

distribution in the effective area A_{eff} would produces the correct amount of nonlinear effect, and therefore the definition of fiber effective area is

$$A_{eff} = \frac{\pi}{4} D_{eff}^2 = \frac{2\pi \left[\int_0^\infty P(r) r dr \right]^2}{\int_0^\infty P(r)^2 dr} \tag{4.8.1}$$

where D_{eff} is the effective cross-section diameter and $P(r)$ is the actual power distribution in the fiber. As an example, the effective cross-section area of a standard single-mode fiber is approximately 80 μm^2. If the fiber carries an optical power of 10 mW, the power density is 12.5 kW/cm^2.

In addition, optical signals propagate in the optical fiber for a very long distance, especially for optically amplified multispan fiber systems; the long interaction length between the optical power and the fiber material further magnifies the effect of nonlinearity. For the nonlinear scattering processes such as SBS and SRS, nonlinear gain coefficient and gain spectrum are important parameters, whereas the nonlinear Kerr effect is mainly determined by a nonlinear parameter γ, which is proportional to the nonlinear refractive index of the material and inversely proportional to the fiber core cross-section. All these parameters differ for different types of fibers, and they depend both on the composition of the glass material and on the mode-field distribution of the waveguide. In this section, we discuss measurement techniques of these nonlinear parameters.

4.8.1 Measurement of Stimulated Brillouin Scattering Coefficient

When an intense optical wave, commonly referred to as a *pump wave*, at a frequency f_p propagates in a fiber, it interacts with the mechanical structure of the material and creates an acoustic wave at a frequency f_a, which is about 11.1 GHz for fused silica-based optical fibers. Then a Stokes wave at a frequency of f_B is generated due to the interaction between the original pump wave and the acoustic wave. In this a three-wave mixing process. To satisfy the rule of energy conservation, the frequency of the Stokes wave has to be $f_B = f_p - f_a$. In addition, momentum conservation $\beta_B = \beta_p - \beta_a$ is important for the efficiency of SBS process, where $\beta_{B,p,a} = 2\pi f_{B,p,a}/v_{B,p,a}$ are the wave numbers of the Stokes wave, the pump wave, and the acoustic wave, respectively. Since the propagation speed of the acoustic wave is much slower than both the pump wave and the Stokes wave, $\beta_a \gg \beta_p \approx \beta_B$, SBS in the forward direction does not exist, whereas the backscattering of SBS has the highest efficiency.

When a pump wave with the optical power $P_p(z)$ propagates along an optical fiber, the power of the Stokes wave generated along the fiber $P_B(f, z)$ can be described by the following propagation equation:

$$\frac{dP_B(f,z)}{dz} = -\frac{g_B(f,z)P_p(z)}{A_{eff}} P_B(f,z) + \alpha P_B(f,z) \tag{4.8.2}$$

where the negative sign indicates the backward propagation of the Stokes wave. α is the fiber attenuation coefficient and z is the position in the fiber longitudinal direction.

$$g_B(f,z) = \frac{g_0}{1 + \left[\left(f - f_B(z)\right)/(\delta f_B/2)\right]^2} \tag{4.8.3}$$

is the backscattering Brillouin gain coefficient, which has a Lorentzian spectral shape with g_0 the peak gain. In Equations 4.8.2 and 4.8.3 we have assumed that in general the Brillouin gain coefficient may not be uniform along the fiber, and therefore it is a function of z. The Brillouin gain also has a certain spectral bandwidth δf_B, which is about 35 MHz in the 1550 nm wavelength window for fused silica fibers.

In most practical applications, the power of Brillouin scattering is much smaller than that of the pump and thus the pump depletion is usually negligible. Under this condition, the pump power distribution along the fiber can be simply described by $P_p(z) = P_p(0)e^{-\alpha z}$, and Equation 4.8.2 can be solved analytically so that the backward propagated SBS power at the fiber input can be calculated as

$$P_B(f,0) = P_B(f,L)\exp\left(-\alpha L + \frac{P_p(0)}{A_{eff}} G(f)\right) \tag{4.8.4}$$

where L is the fiber length and $P_s(f, L)$ is the initial backward optical power at the far end of the fiber, which is usually created by spontaneous emission. The nonlinear optical gain introduced by SBS process in the optical fiber is

$$G(f) = \int_0^L \frac{g_0 \exp(-\alpha z)}{1 + \left[\left(f - f_B(z)\right)/(\delta f_B/2)\right]^2} dz \tag{4.8.5}$$

The peak gain coefficient g_0 is mainly determined by the fiber material and can be expressed as [42]

$$g_0 = \frac{2\pi n^7 p_{12} K}{c\lambda^2 \rho v_a \delta f_B} \tag{4.8.6}$$

where n is the refractive index of the material, ρ is the material density, p_{12} is elasto-optic coefficient, λ is the average wavelength, v_a is the acoustic wave velocity, and c is the speed of light in vacuum. K is a polarization mismatch factor, which accounts for the fact that the three-wave mixing process requires the matched polarization states between the pump and the Stokes waves. If the birefringence along the fiber is random, the K value ranges between 1/3 and 2/3.

Because of the Lorentzian line shape of Brillouin gain, as shown in Equation 4.8.5, the FWHM linewidth of the scattered power $P_B(f, 0)$ due to SBS will be

$$\Delta f_{eff} = \frac{\delta f_B \sqrt{\pi}}{2} \sqrt{\frac{\alpha A_{eff}}{g_0 P_p(0)}} \tag{4.8.7}$$

The total backward-propagated Stokes power can be calculated by integrating all the contributions of spontaneous emission along the fiber, which is amplified by the SBS gain of the nonlinear process. It has been shown that for pure spontaneous scattering, this integration is approximately equivalent to the injection of a single Stokes photon at the point along the fiber where the nonlinear gain exactly equals to the attenuation of the fiber. Under this approximation, the effective optical gain of a spontaneous photon at the peak-gain frequency is [43]

$$G_{eff} = \frac{\exp\{P_p(0)g_0/(A_{eff}\alpha)\}}{P_p(0)g_0/(A_{eff}\alpha)} \tag{4.8.8}$$

For the Brillouin scattering process, the phonons are thermally activated and the effective starting Stokes power must be multiplied by $(kT/hf_a + 1) \approx kT/hf_a$, where, f_a is the frequency of the acoustic phonon, k is Boltzmann's constant, T is the absolute temperature, and h is the Planck's constant. Then the total backward-propagated Stokes power that reaches the terminal where the pump is injected can be found simply by

$$P_B(0) = G_{eff} \frac{kT}{hf_a} (hf_B) \Delta f_{eff} \tag{4.8.9}$$

Figure 4.8.2 shows the calculated power of SBS backscattering at the fiber input side using Equation (4.8.9) with the following parameters: $\alpha = 0.25\ dB/km$, $g_0 = 1 \times 10^{-11}\ m/W$, $A_{eff} = 8 \times 10^{-11}\ m^2$, $f_a = 11GHz$, and $f_B = 200\ THz$. Obviously, the backscattered optical power increases sharply when the input pump power reaches a certain level. A parameter that can be used to specify SBS performance in an optical fiber is the *SBS threshold.* A widely adopted definition of the term is the pump power level P_{th} at which the backscattered SBS power equals the launched pump power [43]. This threshold definition is somewhat arbitrary, and in fact from the energy conservation point of view it is even physically impossible for the backscattered Stokes power to be equal to the input pump power. However,

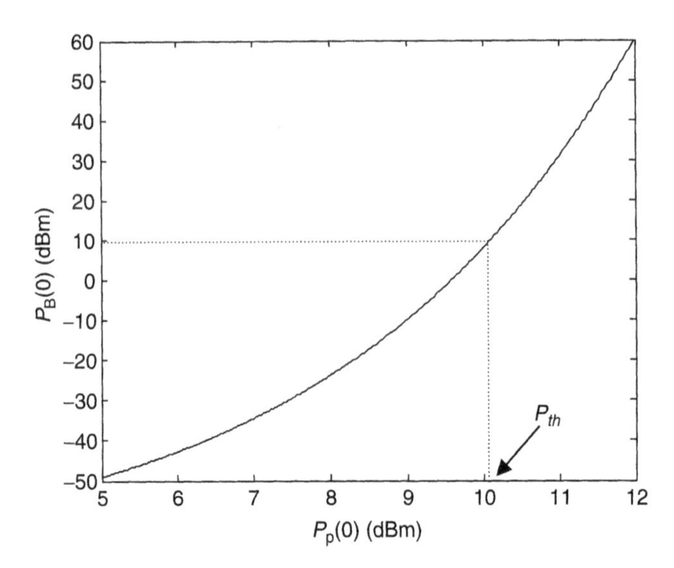

Figure 4.8.2 Backscattered optical power $P_B(0)$ as a function of the input pump power $P_p(0)$ due to SBS calculated with Equation 4.8.9.

since the backscattered power is increased very rapidly in the threshold region, it would not make a significant difference if this threshold was defined at which the backscattered power was equal to, for example, half the input pump power. A good approximation of the SBS threshold is

$$P_{th} \approx \chi_B \frac{A_{eff}\alpha}{g_0} \tag{4.8.10}$$

where the coefficient χ_B ranges from 19 to 21, depending on the characteristics of the fiber [44].

It is worthwhile to point out that the result shown in Figure 4.8.2 was calculated using Equation 4.8.9, where pump depletion was neglected; therefore the backscattered power due to SBS can be even higher than the input pump power in this figure. In this example, the threshold pump power is approximately 10 dBm; according to Equation 4.8.10, this threshold value corresponds to the coefficient of $\chi_B \approx 21$. This χ_B coefficient has a smaller value of 19 for older fibers in which the attenuation is much higher. Although Equation 4.8.9 helps explain and define the SBS threshold, it is valid only when the pump power is much less than the SBS threshold. Practically, when the input pump power increases to a certain value close to the SBS threshold, there will be a rapid increase in the backscattered Stokes power. As a result, the output pump power starts to saturate with further increasing the input pump, and thus pump depletion becomes significant.

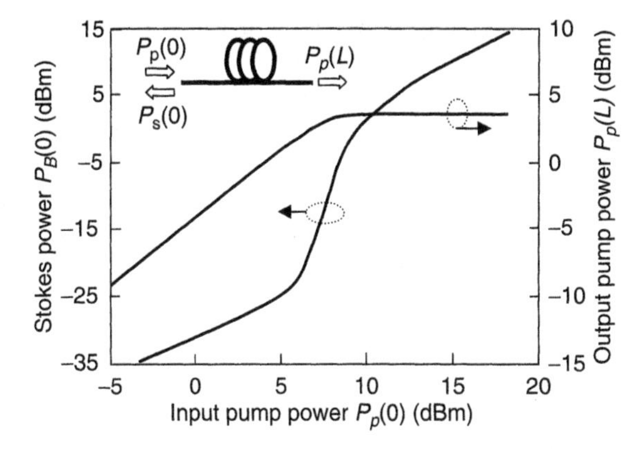

Figure 4.8.3 Illustration of the output optical power saturation in an optical fiber due to the increase of backscattered Stokes power at high input power levels.

Figure 4.8.3 shows the measured backscattering Stokes power at the input side of the fiber together with the pump power at the output side of the fiber as a function of the input pump power in a dispersion-shifted fiber [42]. The fiber parameters are: $\alpha = 0.31\ dB/km$, $A_{eff} = 35.8 \times 10^{-12}\ m^2$ and the fiber length is 13 km. The measurement setup is shown in Figure 4.8.4, where a single-frequency laser is used as the light source whose linewidth is much narrower than the SBS bandwidth of 35 MHz. An optical isolator placed in front of the laser source prevents the optical reflection back to the laser. A 2×2 fiber directional coupler of 10 percent splitting ratio is used to deliver 90 percent of the optical power from the source into the fiber under test, and 10 percent of the power is sent to power meter 2 to monitor the source power. The backscattered optical power from the fiber due to SBS is directed into power meter 1 through the directional coupler. In this configuration the optical powers measured by the three power meters P_1, P_2, and P_3 are related to the parameters in Figure 4.8.3 as $P_1 = 0.1P_s(0)$, $P_2 = 0.1P_p(0)$, and $P_3 = P_p(L)$.

For SBS threshold determination, since the backscattered Stokes power can never be as high as the injected pump power, the SBS threshold can be

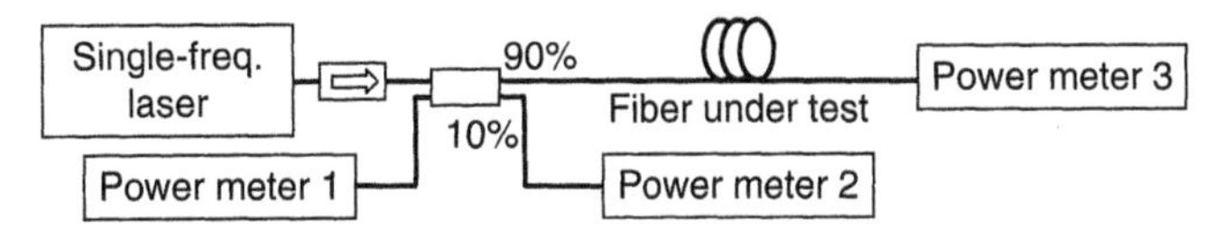

Figure 4.8.4 Experimental setup to measure SBS coefficient in a fiber. Power meters 1, 2, and 3 measure the backscattered power, the source power, and the transmitted power, respectively.

practically defined as the pump power level at which the backscattered power caused by SBS is equal to that due to Rayleigh backscattering. Again, since the power of SBS backscattering increases rapidly near the threshold, this modified definition would provide a very similar SBS threshold value as that defined by Equation 4.8.10.

Another important parameter in the SBS process is the frequency shift, which is in the neighborhood of 11 GHz, depending on the material composition of the fiber. For example, for the fiber doped with F and GeO_2, an empirical equation for SBS frequency shift is

$$f_a(GHz) = 11.045 - 0.277[F] + 0.045[\mathrm{GeO_2}] \quad (4.8.11)$$

where [F] and [GeO2] are the weight percentage of F and GeO_2 concentrations in the glass material.

The measurements of the SBS frequency shift f_a and its bandwidth δf_B are important in fiber selection for optical transmission as well as fiber sensors. The precision of these measurements depends on the spectral resolution of the measurement setup. In general, a tunable laser and an optical spectrum analyzer cannot provide sufficient resolution for these measurements, while an RF spectrum analyzer may be better suited for this purpose.

Figure 4.8.5 shows the experimental setup to measure SBS frequency shift and the SBS bandwidth in a fiber [45]. This setup is based on coherent self-homodyne detection. A single-frequency optical signal from a DFB laser diode operating in CW is amplified by an EDFA and split into two parts through the first fiber coupler. One part is used as the local oscillator for coherent detection; the other part is amplified by another EDFA to boost the optical power

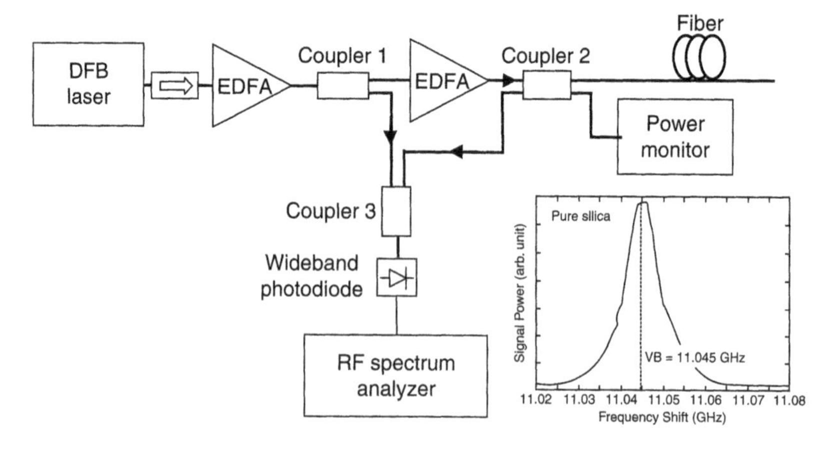

Figure 4.8.5 Experimental setup to measure SBS frequency shift and bandwidth in a fiber.

for SBS measurement. The amplified optical signal is injected into the fiber under test, and the second fiber coupler is used to redirect the backscattered SBS signal into coupler 3 to mix with the original optical signal, which acts as the local oscillator. A photodiode with bandwidth wider than the SBS frequency shift f_a is used for coherent homodyne detection, which brings the SBS optical spectrum into the RF domain. An RF spectrum analyzer is then used to display the SBS spectrum. The inset in Figure 4.8.5 is an example of the measured backscattering power spectral density of a pure silica fiber, where the central frequency shift is 11.045 GHz [45].

4.8.2 Measurement of the Stimulated Raman Scattering Coefficient

In contrast to SBS, in which the frequency shift is approximately 11 GHz and the bandwidth is narrow, SRS has a much higher frequency shift, on the order of 1.3 THz, and much wider bandwidth. In addition to the fact that they have different applications and their impacts on optical systems are very different, the measurement techniques for these two effects are also different.

Stimulated Raman scattering in an optical fiber is caused by the interaction between the optical signal photons and the energy states of the silica molecules. In this scattering process, a photon spends part of its energy to excite a molecule from a lower to a higher vibrational or rotational state. The frequency of the scattered photon f_s, commonly referred to as a *Stokes wave*, is

$$f_s = f_p - \frac{\Delta E_e}{h} \tag{4.8.12}$$

where ΔE_e is the molecular energy level difference of the material, f_p is the input photon frequency, and h is the Planck's constant. If the material structure has a regular crystal lattice, the vibrational or rotational energy levels of the molecules are specific; therefore the spectrum of SRS would be relatively narrow. For amorphous materials, such as fused silica, the SRS spectrum is wide because the electron energy levels are not very well regulated.

Figure 4.8.6 shows the normalized SRS spectrum as a function of the frequency shift $f = f_p - f_s$ in fused silica with $f_p = 1\mu m$ [46]. The peak frequency shift is about 13 THz and the FWHM width of the spectrum is approximately 7 THz. Obviously, the spectral shape of SRS is not Lorizontian. In addition to the much higher frequency shift and the wider spectral bandwidth, SRS differs from SBS by the fact that it propagates in both forward and backward directions in an optical fiber.

When a pump wave with the optical power of $P_p(z)$ propagates along an optical fiber, the power of the Stokes wave $P_s(f, z)$ generated along the fiber can be described by the following propagation equations:

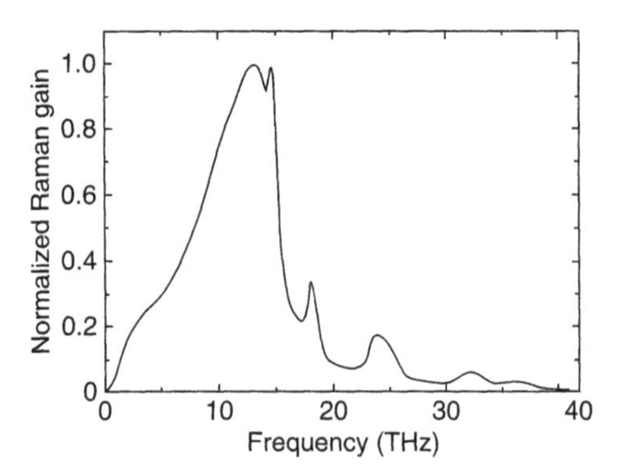

Figure 4.8.6 Raman-gain spectrum of fused silica as a function of the frequency shift. Pump wavelength is 1μm.

$$\left[\frac{d}{dz} \pm \alpha\right] P_s(f,z) = \pm \frac{g_R(f,z)P_p(z)}{A_{eff}} P_s(f,z) \pm \frac{\sigma_{sp}P_p(z)}{A_{eff}} \tag{4.8.13}$$

where the ± sign represents forward (+) and backward (–) propagating Stokes waves, and $g_R(f, z)$ is the SRS gain coefficient, which is generally frequency dependent, as shown in Figure 4.8.6. We have assumed that the SRS gain coefficient may be nonuniform along the fiber and therefore is a function of z. σ_{sp} is the effective spontaneous Raman scattering coefficient. If we neglect pump depletion and assume that the power of the SRS scattering is much smaller than the pump, the pump power along the fiber can be simply described by $P_p(z) = P_p(0)e^{-\alpha z}$ and Equation 4.8.13 can be rewritten as

$$\left[\frac{d}{dz} \pm \alpha\right] P_s(f,z) = \pm \frac{g_R(f,z)P_s(f,z)}{A_{eff}} P_p(0)e^{-\alpha z} \tag{4.8.14}$$

where the spontaneous Raman scattering effect is neglected for simplicity. As a practical application, Raman gain in optical fiber can be used to make optical amplifiers. At the same time, Raman gain also introduces wideband amplified spontaneous emission. If an optical signal $P_s(f, 0)$ in the Stokes wave frequency is injected at the input side of the fiber, it can be amplified by the Raman effect and the output signal power will be

$$P_s(f,L) = P_s(f,0)\exp\left[\frac{g_R(f)}{A_{eff}} P_p(0)\left(\frac{1-e^{-\alpha L}}{\alpha}\right) - \alpha L\right] \tag{4.8.15}$$

where L is the fiber length; for simplicity we have assumed that the Raman gain coefficient $g_R(f)$ is uniform along the fiber. Because pump depletion is neglected, the total optical gain will be the same for both the forward- and the backward-propagated optical signals at the Stokes frequency. The Raman gain coefficient is a material property that is proportional to the resonant contribution associated with molecular vibrations:

$$g_R(f) = \text{Im}\left[\chi_{res}^{(3)}(f)\right]\frac{8\pi^2 f_p}{cn_0} \tag{4.8.16}$$

where $\text{Im}\left[\chi_{res}^{(3)}(f)\right]$ is the imaginary part of the third-order susceptibility and n_0 is the linear refractive index of the material. The peak Raman gain coefficient of $g_R = 1.9 \times 10^{13}\ m/W$ has been measured in fused silica at 526 nm wavelength. This value is about two orders of magnitude smaller than the SBS gain coefficient.

Similar to the SBS, the SRS Stokes power can also be calculated by integrating all the contributions of spontaneous emission along the fiber, which is amplified by the SRS gain process. If we define an *SRS threshold* P_{th}, at which the total SRS power equals the launched pump power [43], this threshold power can still be evaluated by an approximation similar to Equation 4.8.10:

$$P_{th} \approx \chi_R \frac{A_{eff}\alpha}{g_R} \tag{4.8.17}$$

where the value of the coefficient χ_R is approximately 16, which is smaller than that for the SBS threshold. In addition, since the SRS gain coefficient g_R is much smaller than the SBS gain coefficient, the SRS-limited pump threshold can be much higher than that due to SBS. However, it is worthwhile to mention that since SBS is a narrowband process, the SBS threshold can be significantly increased if the optical bandwidth of the pump is much wider than the SBS bandwidth of 35 MHz. In fact, in practical optical fiber communication systems, frequency dithering is usually applied in optical transmitters to widen their spectral linewidths, therefore reducing the nonlinear SBS effect. For SRS, on the other hand, the bandwidth is several terahertzes wide, so there is no simple technique for increasing SRS threshold. In WDM optical systems, the SRS effect converts energy from short wavelength channels to long wavelength channels, not only introducing inter-channel crosstalk but also causing power-level tilt over the wavelength window.

An important application of SRS is performing optical amplification. Fiber-based Raman amplifiers are commercially available and can be designed to operate at different wavelength windows as long as the suitable pump lasers are available for these wavelengths. Distributed Raman amplification can also be used in optical fiber transmission systems, where an intense pump at the wavelength shorter (by $\sim$13 THz) than the optical signal is injected into the optical fiber. In this case the optical signal is amplified due to the Raman gain provided by the pump, and the transmission fiber itself is used as the gain medium.

Raman gain, gain bandwidth, and frequency shift are important parameters for the design of Raman amplifiers. Since these parameters are determined by fiber types, precise measurement and characterization are critically important for amplifier design and system performance evaluation.

Figure 4.8.7 shows the schematic of an experimental setup to measure Raman gain in a fiber. A high power pump laser is used at wavelength λ_p and a low-power wideband light source is used as the probe, which can be a lamp or a superluminescence LED. To minimize the pump SBS effect, the linewidth of the pump laser needs to be controlled, either by frequency dithering through direct current injection or by a phase modulation externally. The pump and the probe are combined in a fiber directional coupler and delivered into the fiber under test. A power meter is used to monitor the pump and the probe optical powers that are actually launched into the fiber. An optical spectrum analyzer is used at the output of the fiber to measure the power spectrum of the Stokes wave.

Based on Equation 4.8.15, the optical gain at the Stokes frequency is

$$G_s(f) = \frac{P_s(f, L)}{P_s(f, 0)} = \exp\left\{\frac{g_R(f)}{A_{eff}} P_p(0) L_{eff}\right\} e^{-\alpha L} \tag{4.8.18}$$

where $L_{eff} = (1 - e^{\alpha L})/\alpha$ is the nonlinear length of the fiber, $P_s(f, 0)$ is the power spectral density of the probe that is injected at the fiber input, and $P_s(f, \mathrm{L})$ is that measured by the optical spectrum analyzer at the fiber output. There are a number of effects that may affect the accuracy of this Raman gain measurement. First, the contribution of spontaneous emission is included in the output optical power spectrum; second, the efficiency of power coupling into the fiber might not be easy to calibrate. However, at each Stokes frequency, if $G_s(f)$ is plotted in logarithm scale, $\log(G_s)$ should be linearly proportional to both the Raman gain coefficient and the injected pump power. In fact, from Equation 4.8.18 we can find

$$\frac{g_R(f)}{A_{eff}} L_{eff} = 0.23 \frac{dG_{s,dB}(f)}{dP_p(0)} \tag{4.8.19}$$

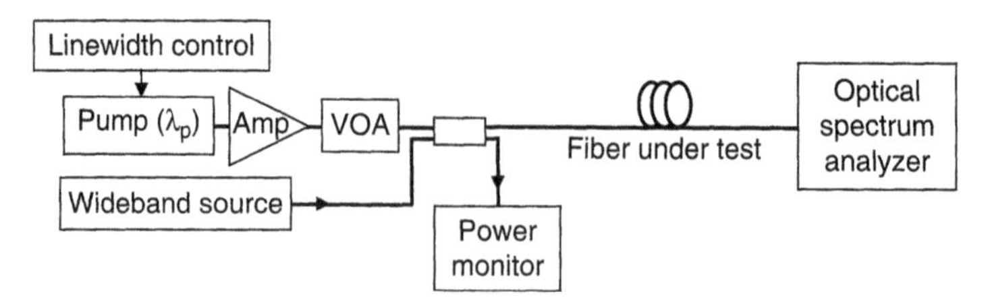

Figure 4.8.7 Experimental setup to measure SRS gain in a fiber. *Amp*: optical amplifier, *VOA*: variable optical attenuator.

where $G_{s,dB}(f) = 10\log G_s(f)$ is the Stokes gain expressed in decibels. By a programmable increase of the pump power through adjusting the variable optical attenuator (VOA), $G_{s,dB}(f)$ as a function of pump power can be measured at each wavelength. Then the composite Raman gain coefficient $g_R L_{eff}/A_{eff}$ can be found through a slope measurement based on Equation 4.8.19.

In the measurement setup shown in Figure 4.8.7, a wideband light source is used as the probe and the spontaneous Raman scattering is neglected because we assumed that the amplified probe is much stronger than the amplified spontaneous emission. In principle, amplified spontaneous Raman scattering can also be used to evaluate the Raman gain coefficient. However, the effective spontaneous Raman scattering coefficient σ_{sp} itself depends on many other effects such as temperature and the efficiency of coupling spontaneous scattering into the guided mode in the fiber.

Figure 4.8.8 shows the block diagram of another measurement technique that uses a square-wave pump for the measurement of fiber Raman gain. In this technique a CW pump generated by a DFB laser diode is phase modulated to minimize the SBS effect in the fiber. The pump wave is then intensity modulated by a square pulse train with duty cycle η and pulse duration τ_0. An optical amplifier is used to boost the pump power, and a bandpass optical filter is applied at the pump wavelength to remove the ASE noise generated by the source and the optical amplifier. Due to the intense optical pump at wavelength λ_p, a Stokes wave is generated in the fiber at wavelength λ_s through the SRS process. Because of chromatic dispersion, the pump and the Stokes waves propagate in different speeds in the fiber; therefore the overlap between the pump and the Stokes waves depends on the pulse width of the pump and their relative delay. Considering the time-varying nature of the pump and the relative walk-off between the pump and the Stokes waves, the propagation Equation 4.8.13 of the forward Stokes power can be written as

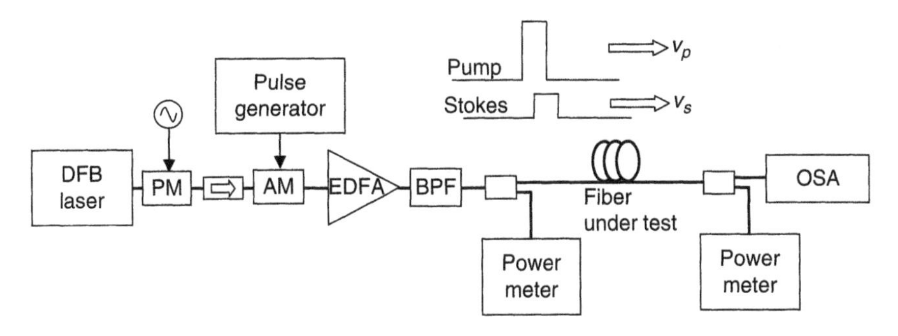

Figure 4.8.8 Experimental setup of SRS measurement using pulsed pump. *PM*: phase modulator, *AM*: amplitude modulator, *BPF*: optical bandpass filter.

$$\frac{\partial P_s(z,t)}{\partial z} = \frac{g_R(z,t)P_p(z,t-z/\Delta v)}{A_{eff}} P_s(z,t) + \frac{\sigma_{sp} P_p(z,t-z/\Delta v)}{A_{eff}} \tag{4.8.20}$$

where $\Delta v = v_p - v_s$ is the group velocity difference between the pump and the forward-propagating Stokes pulses, which is related to the chromatic dispersion parameter D of the fiber by $\Delta v = 1/[D(\lambda_p - \lambda_s)]$. Fiber attenuation has been neglected in Equation 4.8.20 for simplicity.

The effective interaction length between the pump and the Stokes pulses is $L_w = \tau_0 \cdot \Delta v$. For short optical pulses such that $L > L_w$, where L is the total fiber length, the average power density of the Stokes signal at a certain frequency f_s can be found by integrating Equation 4.8.20 as

$$\langle P_s(f_s)\rangle_{L>L_w} = P_{s0}\eta\left\{(2 + P_0 g_s L)\frac{[\exp(P_0 g_s L_w) - 1]}{P_0 g_s L_w} - [\exp(P_0 g_s L_w) + 1]\right\} \tag{4.8.21}$$

Where $P_{s0} \equiv \sigma_{sp}/g_s(f_s)$, $g_s(f_s) \equiv g_R(f_s)/A_{eff}$ and P_0 is the peak power of the square pump waveform $P_p(0)$. For long optical pulses, on the other hand, if the walk-off length is longer than the length of the fiber, the average power density of the Stokes signal is

$$\langle P_s(f_s)\rangle_{L<L_w} = P_{s0}\eta\left\{(2 + P_0 g_s L_w)\frac{[\exp(P_0 g_s L_w) - 1]}{P_0 g_s L_w} - \frac{L}{L_w}[\exp(P_0 g_s L) + 1]\right\} \tag{4.8.22}$$

The effect of fiber loss can be approximately counted for by replacing L and L_w with $[1 - \exp(-\alpha L)]/\alpha$ and $[1 - \exp(-\alpha L_w)]/\alpha$, respectively.

Figure 4.8.9(a) shows the power spectral density of the Stokes signal measured using the setup shown in Figure 4.8.8, where a 10.1 km standard single-mode fiber was used [47]. Different traces in the figure correspond to the pulse widths of 98, 33, 22, 18, 15, 13, 11, 8.3, 6.2, 4, and 1 ns. The Stokes power decreases monotonically with the decrease of pulse width. Figure 4.8.9(b) shows the power spectral density at the peak Stokes gain wavelength of 1680 nm as a function of the pulse width. For the fiber under test, the dispersion varies from $D_1 = 17$ ps/nm/km at 1550 nm to $D_2 = 25$ ps/nm/km near 1680 nm; therefore the pulse width at which the walk-off length equals the fiber length is approximately $\tau_0 \approx 0.5L(D_1 + D_2)\Delta\lambda \approx 26$ *ns*. Figure 4.8.9(b) clearly shows the transition between the two regimes described by Equations 4.8.21 and 4.8.22. The growth rate of the Stokes power as the function of pulse width increase gives the Raman gain coefficient. More precisely, Equation 4.8.20, with the consideration of fiber loss, can be numerically solved to fit the measured results shown in Figure 4.8.9(b) with the Raman gain coefficient as the free parameter for best fitting.

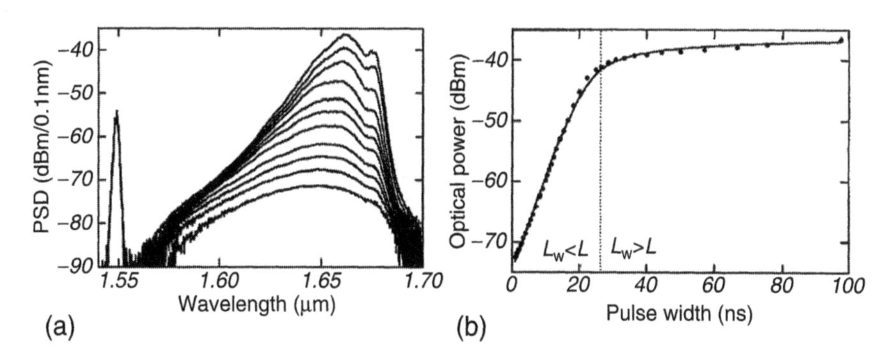

Figure 4.8.9 (a) SRS signal measured from a 10.1 km standard single-mode fiber for 11 different pump pulse widths (from top to bottom): 98, 33, 22, 18, 15, 13, 11, 8.3, 6.2, 4.0, and 1.0 ns, and (b) average Stokes power at the peak of the Raman gain spectrum as a function of pump pulse width [47]. Used with permission.

It is important to note that the Raman gain coefficient is polarization-dependent. The Raman gain for the Stokes wave parallel to the polarization of the pump $g_{R//}$ is much higher than that perpendicular to the pump polarization $g_{R\perp}$. Because of the PMD effect in the fiber, the polarization states of the pump and the Stokes waves walk off from each other rapidly because of their wide wavelength separation. Therefore the Raman gain coefficients measured by both of the discussed techniques are the average effect: $g_R = (g_{R\perp} + g_{R//})/2$.

4.8.3 Measurement of Kerr effect nonlinearity

Kerr effect nonlinearity arises from the intensity dependence of the refractive index in the fiber, which is originated from the third-order susceptibility of the silica material. The refractive index in the fiber can be expressed as

$$n = n_0 + n_2\left(\frac{P}{A_{eff}}\right) \tag{4.8.23}$$

where n_0 is the linear refractive index, which may be wavelength-dependent representing the effect of chromatic dispersion. n_2 is the coefficient of nonlinear refractive index, which represents the dependency of refractive index on the optical power density in the material. For a lightwave signal with power P and wavelength λ_0, the nonlinear phase shift created in a fiber of length L is

$$\Phi_{NL}(P, L) = \left(\frac{2\pi}{\lambda_0}\right)\left(\frac{n_2}{A_{eff}}\right)PL = \gamma PL \tag{4.8.24}$$

where a nonlinear parameter γ is introduced as

$$\gamma = \left(\frac{2\pi}{\lambda_0}\right)\left(\frac{n_2}{A_{eff}}\right) = \frac{n_2\omega_0}{cA_{eff}} \tag{4.8.25}$$

It represents the nonlinear phase change over a unit fiber length with a unit power change. The implication of a nonlinear parameter in the nonlinear Schrödinger equation was discussed in Chapter 1. Several well-studied effects such as self-phase modulation, cross-phase modulation, and four-wave mixing are all originated from the nonlinear refractive index, which are major sources of performance degradation in optical fiber communication networks. On the other hand, Kerr effect nonlinearity has been used to support optical solitons and various other nonlinear optical signal processing techniques such as supercontinuum generation [48] and wavelength conversion.

Kerr effect nonlinearity in an optical fiber depends on the glass composition, material doping, cross-section design, and the core size. In recent years, photonic crystal fibers (PCFs) have been introduced and may provide very high nonlinearity with small solid-core PCFs, or very low nonlinearity with hollow-core PCFs. Precise measurement of the fiber nonlinear parameter is important for both optical system performance evaluation and nonlinear optical signal processing in optical fibers. A number of techniques have been developed to measure nonlinear parameters in fibers based on different effects of the nonlinearity, including SPM, XPM, and modulation instability.

4.8.3.1 Nonlinear Index Measurement Using SPM

When an intense optical pulse $P(t)$ is injected into an optical fiber, a nonlinear phase shift is introduced due to the nonlinear refractive index. The nonlinear phase shift is proportional to the instantaneous power of the optical signal,

$$\Delta\varphi_{NL}(t) = \gamma L_{eff}|A(t)|^2 \tag{4.8.26}$$

where $L_{eff} = (1 - e^{-\alpha L})/\alpha$ is the effective nonlinear length of the fiber, $A(t)$ is the optical field, and $|A(t)|^2 = P(t)$ is the optical power. The time derivative of this nonlinear phase shift is equivalent to a nonlinear frequency shift,

$$\Delta\omega_{NL} = -\frac{\partial\Delta\varphi_{NL}(t)}{\partial t} = -\gamma L_{eff}\frac{\partial}{\partial t}\left|A(t)\right|^2 \tag{4.8.27}$$

At the leading edge of an optical pulse, the nonlinear frequency is downshifted from the original optical frequency, whereas at the trailing edge, the frequency is upshifted. Therefore, after propagating through the fiber, the spectrum of the optical pulse will be broadened and even possibly split. The frequency spectrum of an optical pulse affected by self-phase modulation is given by the Fourier transformation of the pulse optical field $A(t)$,

$$\tilde{A}(\omega) = \int_{-\infty}^{\infty} A(t)\exp\left[i\varphi_{NL}(t)\right]e^{i\omega t}dt \tag{4.8.28}$$

The power spectral density $|\tilde{A}(\omega)|^2$ can be experimentally measured by an optical spectrum analyzer. Nonlinear parameter γ of the fiber can be evaluated by the best fitting between the calculated and the measured optical spectra [49].

Suppose that the pulse shape is Gaussian without frequency chirp at the input of the fiber,

$$A_{in}(t) = A_0 \exp\left\{-\frac{(t-t_0)^2}{2\tau^2}\right\} \tag{4.8.29}$$

where A_0 is the peak amplitude, τ is the FWHM, and t_0 is the center of the pulse. At the fiber output, a time-dependent chirp is introduced by the SPM effect and the optical field becomes

$$A_{out}(t) \propto \exp\left\{-\frac{(t-t_0)^2}{2\tau^2}\right\} \cdot \exp\left\{j\Delta\varphi_{NL,\max}\exp\left[\frac{(t-t_0)^2}{\tau^2}\right]\right\} \tag{4.8.30}$$

where $\Delta\varphi_{NL,\max} = \gamma L_{eff}|A_0|^2$ is the maximum SPM phase shift that happens at the center of the pulse. The chromatic dispersion effect is neglected for simplicity, which is valid when the fiber is short enough, and therefore the amplitude waveform will not change over the fiber. The power spectral density of $A_{out}(t)$ can then be easily obtained by a fast-Fourier transform (FFT) as shown in Figure 4.8.10 for several different $\Delta\varphi_{NL,\max}$ values indicated in the figure.

The experimental setup for measuring fiber nonlinear parameters using the SPM effect is shown in Figure 4.8.11. A mode-locked laser is used to generate short optical pulses with the width on the order of 50 ps. Two optical amplifiers are used to boost the optical power. Two bandpass optical filters eliminate the broadband ASE noise produced by the optical amplifiers. A variable optical attenuator (VOA) is used to adjust the optical power injected into the fiber under test. An OSA at the input side of the fiber monitors the optical spectrum without the SMP effect and another OSA at the fiber output measures the broadened optical spectrum caused by the SPM effect. High-speed photodiodes and oscilloscopes are also used at both the input and the output sides to monitor the pulse waveforms in the time domain for comparison.

With the increase of the pulse peak power injected into the fiber, the nonlinear phase shift due to SPM increases. Figure 4.8.12 shows examples of measured optical spectra at two different pulse peak optical powers, corresponding to

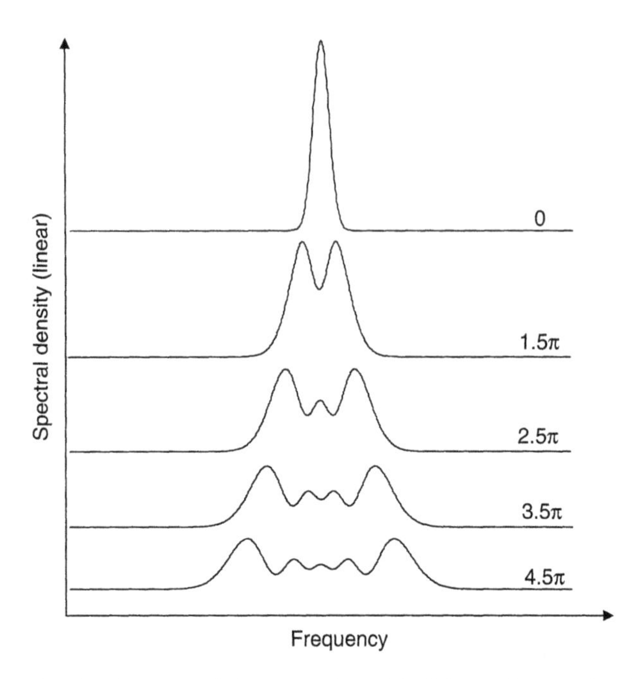

Figure 4.8.10 Spectrum of Gaussian pulses with different phase modulations.

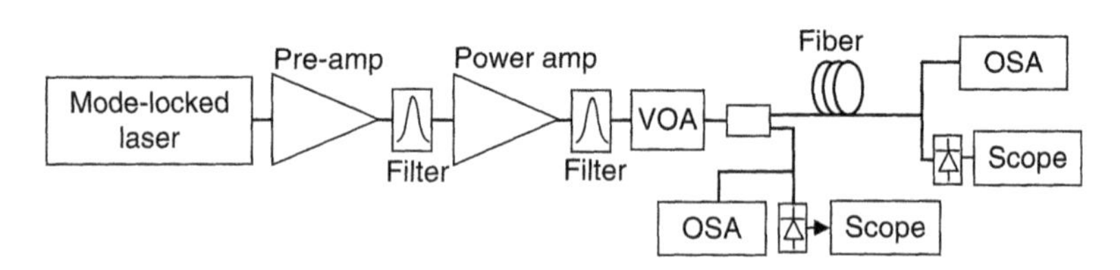

Figure 4.8.11 Block diagram of experimental setup for nonlinear parameter characterization using the SPM technique.

$\Delta\varphi_{NL,\max} = 3.5\pi$ and $\Delta\varphi_{NL,\max} = 7.5\pi$, respectively. If the effect of chromatic dispersion can be neglected, such as in dispersion-shifted fibers, the amplitude of the waveform $|A(t)|^2$ does not change significantly over the fiber and the output optical spectrum can be directly calculated by Equation 4.8.28. However, in general, the effect of chromatic dispersion and its interaction with the nonlinear phase modulation cannot be neglected and the intensity waveform may change dramatically along the fiber. In that case, a numerical simulation is necessary to solve the nonlinear Schrödinger Equation 1.3.96 using the split-step Fourier transforme technique. If fiber parameters such as chromatic dispersion, attenuation and input pulse waveform and chirp are all known, the output waveform

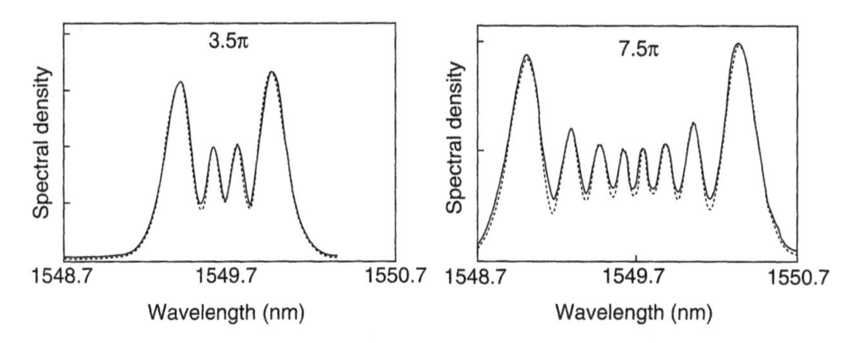

Figure 4.8.12 Measured (solid line) and calculated (dotted line) optical spectra corresponding to the peak phase shift of 3.5π (left) and 7.5π. [50]. Used with permission.

and its optical spectrum can be obtained accurately through numerical analysis. By fitting the calculated optical spectrum using the nonlinear coefficient (n_2/A_{eff}) as the free parameter with the experimentally measured spectrum, the value of n_2/A_{eff} can be precisely determined [50]. The dotted lines in Figure 4.8.12 show the calculated optical spectra that best fit the experimental results, where a 20 m long dispersion-shifted fiber was measured.

This measurement technique using the SPM effect can be very accurate. However, it requires detailed knowledge of the input pulse waveform as well as its frequency chirp. Chromatic dispersion and attenuation of the fiber are also important parameters for the numerical calculation and may change the output spectrum significantly. Asymmetry in the input pulse waveform and its frequency chirp may result in asymmetric broadening of the output optical spectrum. Adequate numerical modeling is also critical to ensure the accuracy of the nonlinear parameter evaluation.

In addition, the Kerr effect nonlinearity in an optical fiber also depends on the state of polarization of the optical signal. The nonlinear refractive index is the maximum for linearly polarized light and drops to 2/3 of its maximum value for circular polarized light. For long fibers, because the SOP of the optical signal varies randomly along the fiber between linear and circular polarization states, the average nonlinear refractive index should be approximately 8/9 of its maximum value [57].

4.8.3.2 Nonlinear Index Measurement Using FWM

When two continuous-wave optical carriers copropagate along an optical fiber, the beating between them through nonlinear Kerr effect modulates the refractive index of the fiber at their frequency difference. Meanwhile, each of these two waves is phase-modulated by this index modulation, creating two extra modulation sidebands. This is commonly known as *degenerate FWM*, as discussed in Chapter 1.

In the simplest case, if we assume that the two carriers at frequencies of ω_1 and ω_2 are coupled into the fiber with identical amplitude A_0, the composite optical field is

$$A_{in}(t) = A_0\left[\exp(j\omega_1 t) + \exp(j\omega_2 t)\right], \tag{4.8.31}$$

and the corresponding optical power at the fiber input is

$$P(t) = \left|A_{in}(t)\right|^2 = 2P_{ave}\left[1 + \cos(\omega_2 - \omega_1)t\right] \tag{4.8.32}$$

where $P_{ave} = |A_0|^2$ is the average optical power of each CW carrier. Due to the SPM effect, the intensity modulation of the optical signal creates a phase modulation, and at the fiber output, the optical field is approximately

$$A(t) = 2A_0 e^{-\alpha L/2}\cos(\Delta\omega t)\cos(\omega_0 t)\exp[i\varphi_{NL,\max}\cos^2(\Delta\omega t)] \tag{4.8.33}$$

where $\omega_0 = (\omega_1 + \omega_2)/2$ is the average frequency of the two CW carriers and $\Delta\omega = (\omega_1 - \omega_2)$ is their frequency difference. $\Delta\varphi_{NL} = \gamma L_{eff} P_{ave}$ is the magnitude of the nonlinear phase shift, and α is the power attenuation coefficient of the fiber. This expression of optical field in Equation 4.8.33 can be expanded into a Bessel series using

$$\exp[ix\cos(\phi)] = \sum_{n=-\infty}^{\infty} i^n J_n(x) e^{in\phi} \tag{4.8.34}$$

where $J_n(x)$ is the Bessel function of the n-th order. Each term of the expanded series represents a discrete frequency component, and there will be a large number of frequency components generated by this degenerate FWM process. However, if we only consider the intensities of the 0th and the first-order harmonics, the ratio between them can be found as [51]

$$\frac{P_0}{P_1} = \frac{J_0^2(\varphi_{NL}/2) + J_1^2(\varphi_{NL}/2)}{J_1^2(\varphi_{NL}/2) + J_2^2(\varphi_{NL}/2)} \tag{4.8.35}$$

where the higher-order terms were neglected because the powers of these terms are much weaker than those at the fundamental frequency and at the first-order sidebands. By experimentally measuring the intensities of P_0 and P_1, the nonlinear phase shift φ_{NL} can be found and the nonlinear parameter γ of the fiber can be determined.

Figure 4.8.13 shows a block diagram of the experimental setup, which measures the fiber nonlinear parameter utilizing the FWM effect. Two separate DFB laser diodes are used and their wavelengths can be adjusted thermally by

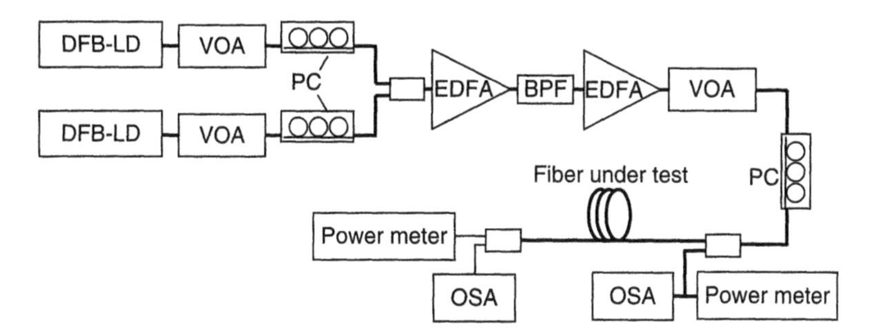

Figure 4.8.13 Block diagram of experimental setup for nonlinear parameter characterization using degenerate FWM technique.

varying the temperature control inside the laser diode package. A variable optical attenuator (VOA) adjusts the optical power of each laser diode, ensuring that the two channels have the same power level. Since FWM requires the polarization alignment between the two signal channels, a polarization controller (PC) is used in each branch to align their polarization states. Two optical amplifiers are used to obtain high optical power levels, and an optical bandpass filter is used between the two amplifiers to avoid the saturation of the second amplifier caused by the wideband ASE noise from the first amplifier. A third VOA at the optical amplifier output allows the control of the signal optical power coupled into the fiber under test. Both the optical power and the optical spectrum are monitored at the fiber input and output with a power meter and an optical spectrum analyzer. If the spacing between the two optical channels is too narrow, the spectral resolution of an OSA may not be sufficient to resolve these channels. In that case, coherent detection may be more appropriate for the spectral measurement.

Figure 4.8.14 shows an example of the measured optical spectrum, where two CW optical carriers separated by approximately 0.25 nm are injected into a fiber and several discrete sidebands are generated in the optical spectrum at the fiber output due to FWM. The ratio between P_0 and P_1 shown in Figure 4.8.14 can be used to calculate the nonlinear phase shift φ_{NL} based on Equation 4.8.35.

Since this technique uses two equal-powered CW optical pumps, the power of the composite optical signal at the fiber input shown in Equation 4.8.32 is equivalent to a single-channel optical signal that is intensity modulated at frequency $\Delta\omega = (\omega_1 - \omega_2)$ and the modulation index is $m = 1$. Therefore this technique is also known as the CW SPM method [51].

It is important to note that the optical field expression given in Equation 4.8.33 neglected chromatic dispersion in the fiber. When dispersion is considered, there is a relative walk-off between the two optical channels, and the FWM efficiency may be reduced, depending on the wavelength separation between

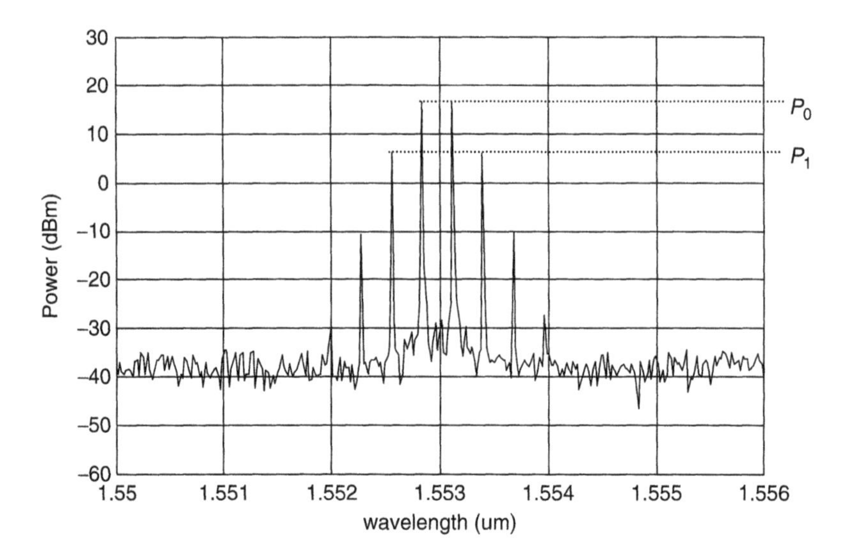

Figure 4.8.14 Example of the measured optical spectrum in FWM measurement.

the two channels, as described by Equation 1.3.119 in Chapter 1. Figure 4.8.15 shows the calculated FWM efficiencies of standard single-mode fiber with dispersion parameter $D = 17$ ps/nm-km and non-zero-dispersion shifted fiber with $D = 5$ ps/nm-km operating in a 1550 nm wavelength window. The wavelength separation between the two channels is assumed to be 0.5 nm and the fiber attenuation is $\alpha = 0.25$ dB/km. Under these conditions, to ensure FWM efficiency of higher than 99 percent, the fiber should not be longer than 110 m for standard single-mode fiber and 360 m for nonzero-dispersion shifted fiber.

The FWM technique may also use two optical signals with very different power levels, also referred to as the *pump-probe technique*, as illustrated in Figure 4.8.16. In this case, because of the large power unbalance between the pump and the probe, the simple relationship given by Equation 4.8.35 does not hold, and the intensity of the generated FWM component at the fiber output is

$$P_{FWM}(L) = \eta_{FWM}\gamma^2 L_{eff} P_{pump}^2(0) P_{probe}(0) \tag{4.8.36}$$

where $P_{pump}(0)$ and $P_{probe}(0)$ are the powers of the pump and the probe at the input of the fiber and

$$\eta_{FWM} = \frac{\alpha^2}{\Delta\beta^2 + \alpha^2}\left[1 + \frac{4e^{-\alpha L}\sin^2(\Delta\beta \cdot L/2)}{(1 - e^{-\alpha L})^2}\right] \tag{4.8.37}$$

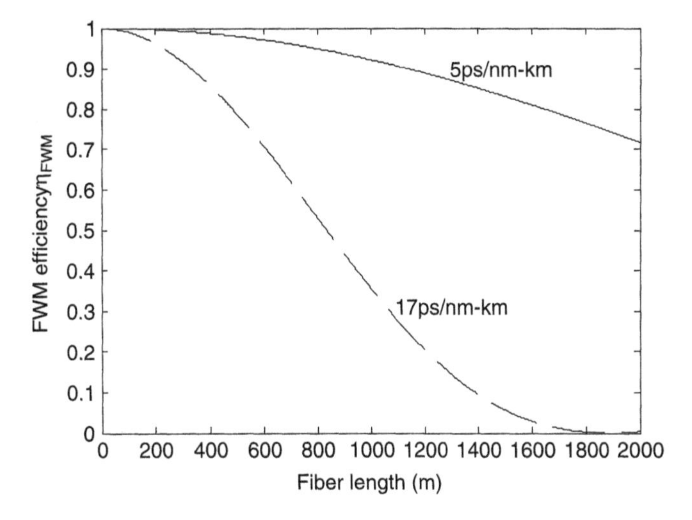

Figure 4.8.15 FWM efficiency for two wavelengths separated by 0.5 nm in a 1550 nm wavelength window. Fiber attenuation: α = 0.25 dB/km, dispersion parameter: 17 ps/nm-km (dashed line) and 5 ps/nm-km (solid line).

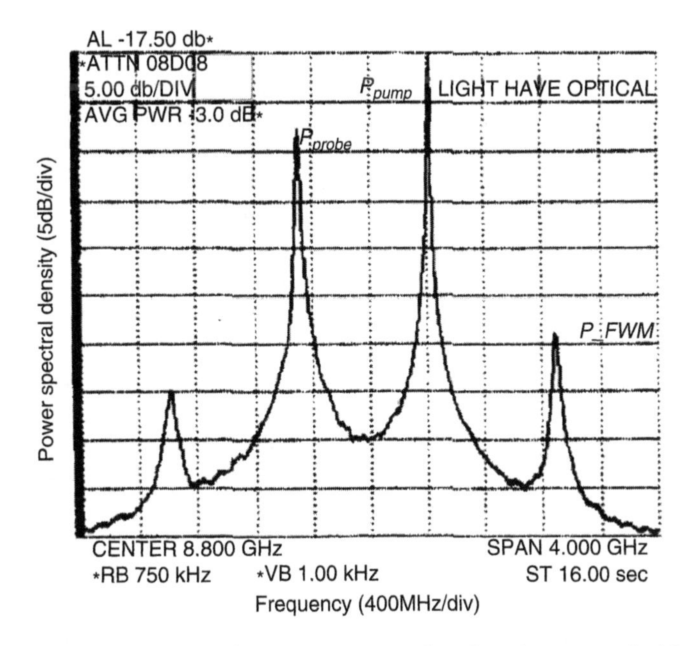

Figure 4.8.16 FWM spectrum with a strong pump and weak probe, measured with coherent detection.

is the efficiency of the degenerate FWM, which is a special case of Equation 1.3.119 for $\lambda_j = \lambda_k$. $\Delta\beta = 2\pi c D(\Delta\lambda/\lambda)^2$ is the phase walk-off and $\Delta\lambda$ is the wavelength separation between the pump and the probe.

In the measurement, the ratio between the output powers of the FWM component and the probe is proportional to the square of the output pump power,

$$R_p = \frac{P_{FWM}(L)}{P_{probe}(L)} = \eta_{FWM}\gamma^2 L_{eff} P_{pump}^2(L)e^{-3\alpha L} \quad (4.8.38)$$

By varying the pump power, the slope of $\log\left(P_{FWM}(L)/P_{probe}(L)\right)$ versus $\log\left(P_{pump}^2(L)\right)$ gives the information $\eta_{FWM}\gamma^2 L_{eff} e^{-3\alpha L}$. In this way the power-coupling efficiency of the pump and the probe at the fiber input does not have to be calibrated, and the relative change in the pump power can easily be controlled by a programmable VOA. In this particular example, as shown in Figure 4.8.16, the spectrum was measured using coherent heterodyne detection, which involves the use of a tunable laser as the local oscillator. In this case, the frequency spacing between the pump and the probe is only 900 MHz, which is approximately 0.008 nm in terms of the wavelength separation; a grating-based OSA would not be able to provide a good measurement.

4.8.3.3 Nonlinear Index Measurement Using Cross-Phase Modulation

Cross-phase modulation (XPM) is another effect originated from the nonlinear refractive index. The process of XPM can be explained briefly as follows: An intensity-modulated strong optical carrier propagating along an optical fiber modulates the refractive index of the fiber. As a consequence, another optical signal propagating in the same fiber will be phase modulated by this index modulation in the fiber. XPM is often a notorious source of performance degradation in dense WDM optical systems due to the interchannel crosstalk it introduces. On the other hand, the effect of XPM can also be utilized to measure the nonlinear parameter of the fiber.

Figure 4.8.17 shows a typical experimental setup for the measurement of a fiber nonlinear index using XPM effect. In this setup, two lasers are used with wavelengths of λ_1 and λ_2, respectively. The output of the laser at wavelength λ_1, commonly referred to as the pump, is intensity modulated by an external optical modulator and amplified by a high-power EDFA. The other laser known as the probe at wavelength λ_2 is used, which operates in continuous wave with a low power level. The pump and the probe are combined by a fiber coupler and sent to the fiber under test. At the output side of the fiber, an optical bandpass filter is used to reject the pump and only allows the probe to pass through. Due to the effect of XPM, the optical phase of the CW probe is modulated by the intensity waveform of the pump. This phase modulation can be

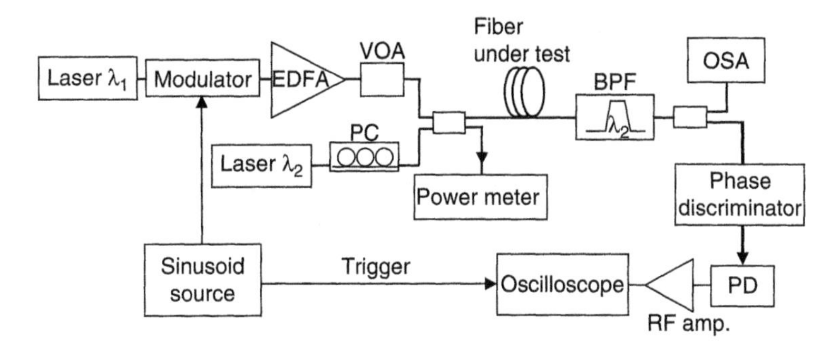

Figure 4.8.17 Block diagram of experimental setup for nonlinear parameter characterization using cross-phase modulation.

monitored either by an optical spectrum analyzer or by an optical phase discriminator that converts the phase modulation into an intensity modulation. The waveform of this converted intensity modulation is then detected by a photodiode and displayed on an oscilloscope.

Theoretically, at the fiber output the nonlinear phase shift of the probe channel at wavelength λ_2 can be expressed as

$$\Delta\varphi_{NL,\lambda_2}(t) = \gamma L_{eff}\left\{\left|A_{\lambda_2}(t)\right|^2 + 2b\left|A_{\lambda_1}(t)\right|^2\right\} \tag{4.8.39}$$

where $\left|A_{\lambda_1}(t)\right|^2 = P_{\lambda_1}(t)$ and $\left|A_{\lambda_2}(t)\right|^2 = P_{\lambda_2}(t)$ are the powers of the pump and the probe channels, and $1/3 < b < 1$ is a parameter representing the effect of polarization mismatch between the pump and the probe along the fiber. The first term represents the SPM effect of the probe channel itself; the second term represents the effect of XPM. The highest value of b is 1, which happens when the pump and the probe are both linearly polarized and parallel to each other; the minimum value $b = 1/3$ also happens when the pump and the probe are both linearly polarized but are perpendicular with each other. In practice, if the fiber is not polarization maintained and the SOP of the optical signal is randomly evolving over the fiber, an average value of $b = 2/3$ should be used [52].

To measure the effect of XPM, we can simply use a low-power probe such that the SPM effect in the probe is negligible. Assume that the intensity of the pump is modulated by a sinusoid, $P_{\lambda_1}(t) = P_{ave,\lambda_1}[1 + \cos\Omega t]$, where Ω is the modulation frequency and P_{ave,λ_1} is the average power of the pump at the fiber input. The phase modulation in the probe channel created by XPM is

$$\Delta\varphi_{XPM,\lambda_2}(t) = \Delta\varphi_{XPM,\max}\left(\frac{1+\cos\Omega t}{2}\right) \tag{4.8.40}$$

where $\Delta\varphi_{XPM,\max} = 4b\gamma L_{eff} P_{ave,\lambda_1}$ is the peak phase shift. Therefore at fiber output the optical field of the probe is

$$A_{\lambda_2}(t) = A_{0,\lambda_2} \exp(i\varphi_{XPM,\max}) \cdot \exp\left[i\frac{\varphi_{XPM,\max}}{2}\cos\Omega t\right] \tag{4.8.41}$$

where A_{0,λ_2} is the probe field amplitude at the fiber output. This expression of the phase-modulated probe can be expanded in a Bessel series based on Equation 4.8.34,

$$A_{\lambda_2}(t) = A_{0,\lambda_2} \exp(i\varphi_{XPM,\max}) \cdot \sum_{n=-\infty}^{\infty} i^n J_n\left(\frac{\varphi_{XPM,\max}}{2}\right) e^{in\Omega t} \tag{4.8.42}$$

A Fourier transform of Equation 4.8.41 yields the optical spectrum of the output probe. In addition to the central optical carrier, discrete sidebands exist with equal frequency separation by Ω between each other. The amplitude of each sideband is proportional to the nth order Bessel function $J_n(\varphi_{XPM,\max}/2)$.

The nonlinear index of the fiber can be evaluated by measuring the relative amplitudes of the carrier and the sidebands of the probe optical spectrum as a function of the pump power. By fitting the measured amplitude with the theoretical prediction as shown in Figure 4.8.18, the XPM-induced peak nonlinear phase shift $\varphi_{XPM,\max}/2$ can be precisely determined. From the practical instrumentation point of view, a conventional grating-based OSA usually has the frequency resolution in the order of 0.05–0.1 nm, or equivalently, 6.25–12.5 GHz. To clearly resolve the sidebands of different orders, the modulation frequency Ω has to be higher than 15 GHz; this imposes the requirement for high-speed electronics and wideband electro-optic modulators. Otherwise, a scanning FP interferometer or coherent detection may be used to provide higher spectral resolution.

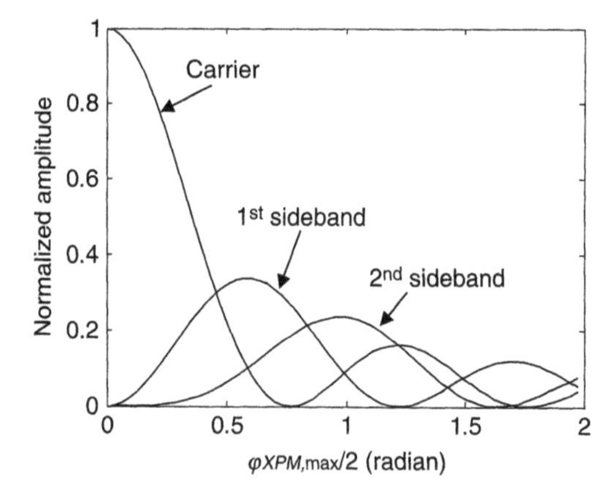

Figure 4.8.18 Relative amplitude of sidebands due to phase modulation.

The phase modulation in the probe can be characterized using either a phase discriminator or a frequency discriminator. Several different techniques can be used for this purpose, such as delayed self-heterodyne detection [53], a Michelson interferometer, or a Mach-zehnder interferometer [54]. The phase discriminator converts the XPM-induced phase modulation into intensity modulation, which can be easily detected by a photodiode, and the waveform can be displayed on an oscilloscope. In this case, the calibration of phase-to-intensity modulation efficiency is extremely important to ensure the accuracy of the measurement.

It is important to note that in the previous analysis, the effect of chromatic dispersion in the fiber was neglected. With the consideration of chromatic dispersion, the pump and the probe waves may propagate in different group velocities in the fiber, and the relative walk-off between them will reduce the efficiency of XPM. At the same time, chromatic dispersion may also convert phase modulation into intensity modulation in the probe channel [55]. This introduces additional complications in the measurement of the optical spectrum and its comparison to the theory. For the dispersion effect to be negligible, the dispersion-induced walk-off between the pump and the probe along the fiber has to be much smaller than the time period of the modulating signal, that is,

$$f < \frac{1}{100} \cdot \frac{1}{D \cdot L \cdot \Delta\lambda} \tag{4.8.43}$$

where $f = \Omega/2\pi$ is the RF modulation frequency, D is the dispersion parameter of the fiber, L is the fiber length and $\Delta\lambda$ is the pump-probe wavelength separation. For example, to measure a 10-km standard single mode fiber in a 1550 nm wavelength window with $D = 17$ ps/nm-km and $\Delta\lambda = 0.2$ nm, the modulation frequency f should not exceed 300 MHz.

In the XPM measurement, a strong pump and a weak probe copropagate along the same fiber. In the experimental setup shown in Figure 4.8.17, only the phase modulation induced in the probe channel is measured, whereas the pump channel is rejected at the fiber output by a bandpass optical filter. In fact, instead of rejecting the pump channel at the fiber output, if we use an OSA to measure the spectral characteristics of both the pump and the probe, the similar experimental setup shown in Figure 4.8.17 can be expanded to measure the effects of both XPM and SPM so that the precise relationship between XPM and SPM can be characterized. Figure 4.8.19 shows an experimental setup that is useful to measure both XPM and SPM in the same fiber, where three lasers are used. The two lasers at wavelengths λ_{11} and λ_{12} are combined in the first directional coupler and amplified by a high-power EDFA to provide the pump waves. The degenerate FWM between these two equal-powered pumps creates sidebands in the output optical spectrum, as already discussed in Equations

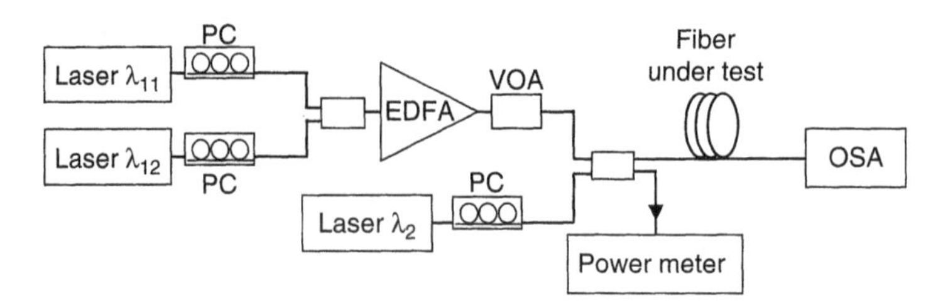

Figure 4.8.19 Experimental setup to measure SPM and XPM simultaneously.

4.8.31–4.8.35. This technique is also known as *CW self-phase modulation* because the combination of these two pump lasers is equivalent to a single laser that is intensity modulated at the difference frequency $\Delta\omega$ between them.

The third laser at wavelength λ_2 is used as the probe and is combined with the pump before coupled into the testing fiber through the second directional coupler. Due to XPM, the probe is phase modulated along the fiber by the intensity modulation created by the pumps. The OSA at the fiber output simultaneously measures the spectra of both the pump and the probe. Figure 4.8.20 shows an example of the measured optical spectrum where the pump and the probe are separated by approximately 2 nm [56]. The relative amplitudes of P_0 and P_1, in the pump output can be used to determine the nonlinear phase shift responsible for SPM with Equation 4.8.35. Meanwhile, by measuring the amplitudes of the carrier and the first-order sideband in the probe, the nonlinear phase shift due to XPM can also be estimated.

It has been observed that if the modulation frequency is lower than 1 GHz, the measurement using the XPM effect often observed a larger nonlinear

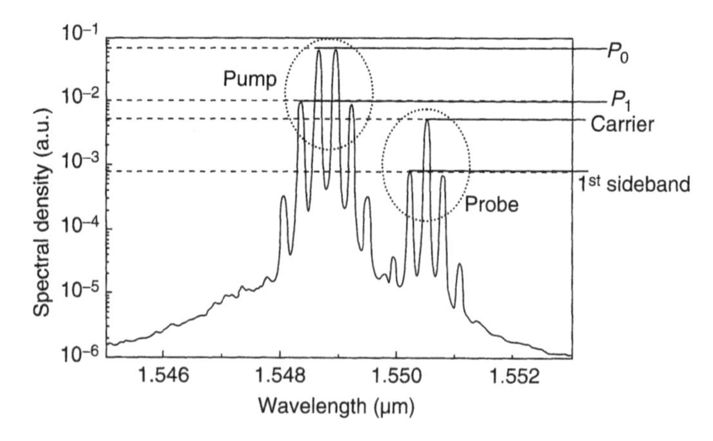

Figure 4.8.20 Experimental setup to measure SPM and XPM simultaneously [56]. Used with permission.

parameter compared to SPM-based techniques. An explanation for this discrepancy is related to electrostrictive contribution to the nonlinear refractive index that occurs at relatively low modulation frequencies [57]. By varying the frequency difference between the two pump lasers, the experimental setup shown in Figure 4.8.19 may be used to verify this discrepancy. However, for the modulation frequency to be lower than 1 GHz, it might be easier to use an intensity-modulated single-pump laser to replace the beating between the two pumps. For the spectrum measurement, a high-resolution optical spectrum analyzer such as coherent detection has to be used.

4.8.3.4 Nonlinear Index Measurement Using Modulation Instability

Modulation instability (MI) is another phenomenon originated from the nonlinear refractive index in an optical fiber. When an intensive optical pump propagates along an optical fiber with anomalous dispersion, optical gain is created in two spectral regions symmetrically located on each side of the pump wavelength. This nonlinear parametric gain process can also be explained as the FWM between the pump and the wideband ASE noise.

Based on the nonlinear Schrödinger Equation 1.3.96, assuming that the steady-state solution of the optical field envelope is $A = \sqrt{P_0}$, where P_0 is the signal optical power. Considering small perturbations in the amplitude and the phase, the general solution can be assumed to be

$$A = (\sqrt{P_0} + a)\exp(i\Phi_{NL}) \tag{4.8.44}$$

where a is the amplitude perturbation and ϕ_{NL} is the nonlinear phase shift as defined in Equation 1.3.106. Substituting Equation 4.8.44 back into the nonlinear Schrödinger equation, the amplitude perturbation can be found as [57]

$$a(z,t) = a_0 \exp\left[i(Kz - \Omega t + \varphi_0)\right] \tag{4.8.45}$$

where a_0 and φ_0 are the complex amplitude and the initial phase, and the wave number is

$$K = \pm\frac{1}{2}\left|\beta_2\right|\Omega\sqrt{\Omega^2 + \mathrm{sgn}(\beta_2)\Omega_C^2} \tag{4.8.46}$$

where $sgn(\beta_2)$ indicates the sign of β_2: $sgn(\beta_2) = 1$ for normal dispersion and $sgn(\beta_2) = -1$ for anomalous dispersion. Ω_C is related to the nonlinear parameter γ, the dispersion parameter β_2, and the signal optical power P_0 by

$$\Omega_C^2 = \frac{4\gamma P_0}{|\beta_2|} \tag{4.8.47}$$

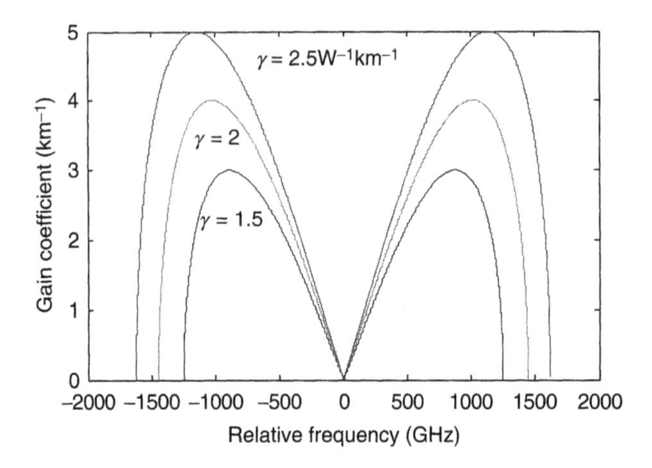

Figure 4.8.21 Parametric gain coefficient $g(\Omega)$ calculated using Equation 4.8.48 with $\beta_2 = 3.82$ ps^2/km, $P_0 = 1$ W and $\gamma = 1.5, 2, 2.5$ W^{-1}km^{-1}.

For a fiber with normal dispersion, the wave number K is real and the amplitude of the perturbation does not increase over the fiber length z. On the other hand, if the fiber dispersion is anomalous, the wave number becomes imaginary for $\Omega < \Omega_C$; therefore the amplitude of the perturbation will grow exponentially along the fiber. This is the origin of modulation instability, which is also referred to as *parametric gain*. From Equations 4.4.46 and 4.4.47, it is evident that the power gain coefficient in the fiber caused by modulation instability is

$$g(\Omega) = 2\mathrm{Im}(K) = |\beta_2|\Omega\sqrt{\frac{4\gamma P_0}{|\beta_2|} - \Omega^2} \tag{4.8.48}$$

The frequency of the maximum gain can be found where $dg(\Omega)/d\Omega = 0$, that is at, $\Omega_{max} = \sqrt{2\gamma P_0/|\beta_2|}$. Therefore, if the signal optical power P_0 and the fiber dispersion parameter β_2 are known, the nonlinear parameter of the fiber can be determined by measuring the peak frequency of the parametric gain.

Figure 4.8.21 shows the calculated parametric gain coefficient using Equation 4.8.48 with the optical power level $P_0 = 1$ W at 1550 nm wavelength and the fiber chromatic dispersion parameter is $D = 1$ ps/nm/km (or equivalently, $\beta_2 = 3.82$ps^2/km). Three values of fiber nonlinear parameters are used $\gamma = 1.5$ W^{-1}km^{-1}, $\gamma = 2$W^{-1}km^{-1}, and $\gamma = 2.5$ W^{-1}km^{-1}, respectively. Obviously, with a higher nonlinear parameter, the nonlinear gain is increased and the gain peak shifts to a higher frequency.

It is worthwhile to note that in our analysis of modulation instability so far, fiber attenuation was not included. Optical attenuation in the fiber will decrease the nonlinear effect because the optical power level is reduced along

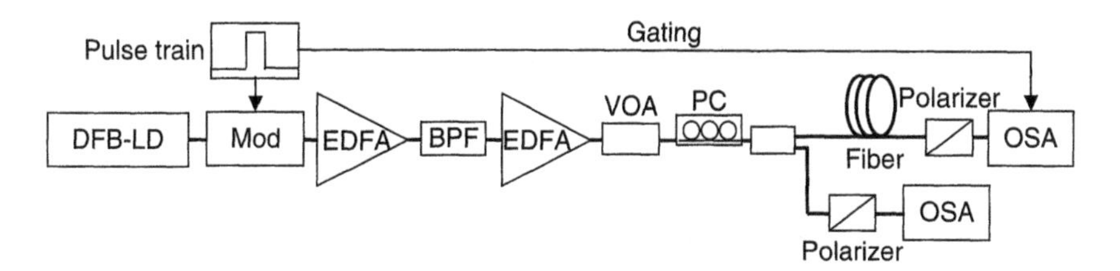

Figure 4.8.22 Nonlinear parameter characterization based on modulation instability. *BPF:* optical bandpass filter, *Mod:* electro-optic modulator, *PBS:* polarization beam splitter, *VOA:* variable optical attenuator, *PC:* polarization controller.

the fiber. However, with the consideration of fiber attenuation, it is not possible to obtain a simple analytic solution for the perturbation term; therefore a numerical simulation becomes necessary. Or a semi-analytic method can be used based on the transfer matrix analysis [58], which will be further discussed in Chapter 5.

Figure 4.8.22 shows the experimental setup for the nonlinear parameter characterization based on MI effect. The output of a laser diode is intensity modulated by a pulse train with a low duty cycle. Two EDFAs are used to produce high power in the optical signal. The purpose of the bandpass optical filter between the two EDFAs is to avoid the gain saturation of the second EDFA due to wideband ASE noise produced by the first EDFA. A variable optical attenuator allows the adjustment of optical power coupled into the fiber, and a polarization controller is used to adjust the SOP of the optical pump before injecting into the fiber. At the fiber output, an optical spectrum analyzer measures the wideband optical noise spectrum which is generated and amplified by the nonlinear parametric gain in the fiber due to MI. Since the MI effect only amplifies the ASE noise of the same polarization as the pump, a polarization beam splitter at the fiber output is adjusted to reject the perpendicular polarization component of the ASE noise. The purpose of using low duty cycle intensity modulation for the pump is to further increase the peak power, since the MI measurement especially requires a high power level for the pump. Generally, if the signal pulse width is shorter than the carrier lifetime of an EDFA, the output power of the EDFA is only limited by its average value. For example, suppose that an EDFA output is limited to a maximum of 10 dBm average power and the EDFA carrier lifetime is 10 ms; if the optical signal is a pulse train with 100 ns pulse width and 10 kHz repetition rate, the pulse peak power at the output of the EDFA could be as high as 40 dBm, or 10 W.

Due to the fact that the pump is pulsed, MI only exists during the peaks of the pulses and therefore the spectral measurement at the fiber output should only be limited to these time windows. Most of the commercial OSAs provide the gating function, in which the OSA operation is enabled by a gating pulse

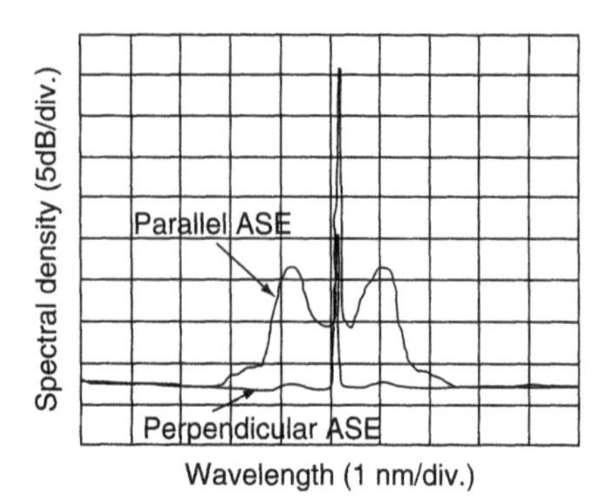

Figure 4.8.23 Measured ASE spectrum created due to modulation instability in the parallel and perpendicular polarization directions with respect to the pump [59]. Used with permission.

train. When the gating signal is at its low level, signals measured by the OSA will be discarded. By comparing the ASE noise spectra measured before and after the fiber, the nonlinear parametric gain spectrum caused by MI can be obtained and the nonlinear parameter of the fiber can be determined.

Figure 4.8.23 shows an example of the measured optical spectrum at the output of 10.1 km fiber. The zero-dispersion wavelength of the fiber is at 1539 nm and the optical signal wavelength is 1553 nm. For the pump optical pulse train, the pulse width is 25 ns, the pulse repetition period is 250 ns, and the pulse peak power is 180 mW [59]. With an intense optical signal propagating in the fiber, the parametric gain generated by MI amplifies the spontaneous emission noise and therefore a very pronounced side peak is generated on each side of the optical carrier, as predicted by Figure 4.8.21. The measurement of the amplitude and the peak frequencies of these amplified spontaneous emission noises shown in Figure 4.8.23 can be used to evaluate the fiber nonlinear parameter. Taking into account fiber chromatic dispersion, dispersion slope, attenuation, and Kerr effect nonlinearity, the measured optical spectra have to be compared with numerical simulations so that the actual fiber nonlinear parameter can be determined accurately.

Figure 4.8.23 shows the measured output ASE spectra at the same polarization as the pump and perpendicular to it. The two traces in the figure were obtained by adjusting the polarizer at the fiber output, as indicated in Figure 4.8.22, to maximize and minimize the pump. Since the modulation instability process only amplifies the ASE noise in the same polarization as the pump, the difference between the parallel and the perpendicular ASE shown in Figure 4.8.23 represents the actual parametric gain caused by the modulation instability.

REFERENCES

[1] www.corning.com
[2] www.ofsoptics.com
[3] K. Demarest, D. Richards, C. Allen, and R. Hui, "Is standard single-mode fiber the fiber to fulfill the needs of tomorrow's long-haul networks?" National Fiber Optic Engineers Conference (NFOEC), pp. 939–946, September 15–19, 2002.
[4] www.crystal-fibre.com
[5] K.Petermann, "Microbending loss in monomode fibres," *Electronics Letters*, Vol. 20, pp. 107–109, 1976.
[6] K. Petermann, "Constraints for fundamental-mode spot size for broadband dispersion-compensated single-mode fibers," *Electronics Letters*, Vol. 19, pp. 712–714, 1983.
[7] C. Pask, "Physical interpretation of Petermann's strange spot size for single-mode fibers," *Electronics Letters*, Vol. 20, pp. 144–145, 1984.
[8] M. Artiglia, G. Coppa, P. D. Vita, M. Potenza, and A. Sharma, "Mode-field diameter measurements in single-mode optical fibers," *J. Lightw. Technol.*, Vol. 7, No. 8, pp. 1139–1152, Aug. 1989.
[9] Jeffrey L. Guttman , "Mode-Field Diameter and 'Spot Size' Measurements of Lensed and Tapered Specialty Fibers," National Institute of Standards and Technology Symposium on Optical Fiber Measurements, September 24–26, 2002.
[10] www.photon-inc.com
[11] E. G. Neumann, *Single-Mode Fibers Fundamentals*, Springer Series in Optical Science, Vol. 57, Springer-Verlag, 1988, ISBN 3-54018745-6.
[12] M. Ohashi, et al., "LP_{11}-mode loss measurement in the two-mode propagation region of optical fibers," *Optics Letters*, Vol. 9, pp. 303–305, 1984.
[13] J. K. Barnoski and S. M. Jensen, "Fiber waveguides: A novel technique for investigation attenuation characteristics," *Appl. Opt.*, Vol. 15, pp. 2112–2115, 1976.
[14] M. Nakazawa, "Rayleigh backscattering theory for single-mode optical fibers," *Journal of Optical Society of America*, Vol. 73, No. 9, pp. 1175–1180, 1983.
[15] M. S. O'Sullivan and R. S. Lowe, "Interpretation of SM fiber OTDR signatures," *Proceedings of SPIE*, International Society for Optical Engineering Optical Testing and Metrology, Vol. 661, pp. 171–176, 1986.
[16] Paul Hernday, "Dispersion Measurements," in *Fiber-optic Test and Measurement*, Dennis Derickson, ed., Prentice Hall, 1998.
[17] B. Costa, M. Puleo, and E. Vezzoni, "Phase-shift technique for the measurement of chromatic dispersion in single-mode optical fibres using LEDs," *Electronics Letters*, Vol. 19, pp. 1074–1076, 1983.
[18] L. G. Cohen, "Comparison of Single-Mode Fiber Dispersion Measurement Techniques," *Journal of Lightwave Technology*, Vol. 3, No. 5, pp. 958–966, 1985.
[19] F. Devaux, Y. Sorel, and J. F. Kerdiles, "Simple Measurement of Fiber Dispersion and of Chirp Parameter of Intensity Modulated Light Emitter," *J. Lightwave Tech.*, Vol. 11, No. 12, pp. 1937–1940, 1993.
[20] B. Christensen, J. Mark, G. Jacobsen, and E. Bodtker, "Simple dispersion measurement technique with high resolution," *Electronics Letters*, Vol. 29, pp. 132–133, 1993.
[21] M. J. Saunders and W. B. Gardner, "Precision interferometric measurement of dispersion in short single-mode fibers," in *Tech Dig. NBS Symp. on Opt. Fib. Meas.* (Boulder, CO), pp. 123–126, October 2–3, 1984.
[22] H. T. Shang, "Chromatic dispersion measurement by white-light interferometry on metre-length single-mode optical fibers," *Electron. Lett.*, Vol. 17, pp. 603–605, 1981.

[23] R. Ulrich, S. C. Rashleigh, and W. Eickhoff, "Bending-induced birefringence in single-mode fibers," *Opt. Lett.* 5, 273–275, 1980.
[24] A. Galtarossa, L. Palmieri, A. Pizzinat, M. Schiano, and T. Tambosso, "Measurement of local beat length and differential group delay in installed single-mode fibers," *Journal of Lightwave Technology*, Vol. 18, No. 10, pp. 1389–1394, 2000.
[25] C. D. Poole and J. Nagel, "Polarization effects in lightwave systems," in *Optical Fiber Telecommunications* IIIA, I. P. Kaminow and T. Koch, eds., Academic Press (1997).
[26] C. D. Poole and C. R. Giles, "Polarization-dependent pulse compression and broadening due to polarization dispersion in dispersion-shifted fiber," *Optics Letters*, Vol. 13, No. 2, pp. 155–157, 1988.
[27] N. Gisin, J.-P. Von der Weid, J.-P. Pellaux, "Polarization mode dispersion of short and long single-mode fibers," *Journal of Lightwave Technology*, Vol. 9, No. 7, pp. 821–827, 1991.
[28] J.-P. Von der Weid, L. Thenenaz, and J.-P. Pellaux, "Interferometer measurements of chromatic dispersion and polarization mode dispersion in highly birefringent single mode fibers," *Electron. Lett.* Vol. 23, pp. 151–152, 1987.
[29] Paul Hernday, "Dispersion Measurements," in *Fiber-optic Test and Measurement*, Dennis Derickson, ed., Prentice Hall, 1998.
[30] C. D. Poole, and D. L. Favin, "Polarization-Mode Dispersion Measurements Based on Transmission Spectra Through a Polarizer," *Journal of Lightwave Technology*, Vol. 12, No. 6, pp. 917–929, 1994.
[31] B. L. Heffner, "Automatic measurement of polarization mode dispersion using Jones Matrix Eigenanalysis," *IEEE Ph. Tech. Let.* 4, 1066–1069, 1992.
[32] B. L. Heffner, "Accurate, automated measurement of differential group delay dispersion and principal state variation using Jones Matrix Eigenanalysis," *IEEE Ph. Tech. Let.* 5, 814–817, 1993.
[33] R. C. Jones, "A new calculus for the treatment of optical systems I," *Journal of Optics Society of America*, Vol. 31, pp. 488–503, 1941.
[34] R. M. Jopson, L. E. Nelson, and H. Kogelnik, "Measurement of second-order polarization-mode dispersion vectors in optical fibers," *IEEE Ph. Tech. Let.*, Vol. 11, No. 9, pp. 1153–1155, 1999.
[35] H. Goldstein, "Finite Rotations," §4–7 in *Classical Mechanics*, 2nd ed., Reading, MA: Addison-Wesley, pp. 164–166, 1980.
[36] D. A. Varshalovich, A. N. Moskalev, and V. K. Khersonskii, "Description of Rotations in Terms of Rotation Axis and Rotation Angle," §1.4.2 in *Quantum Theory of Angular Momentum*, Singapore: World Scientific, pp. 23–24, 1988.
[37] C. Hentschel and S. Schmidt, "PDL Measurements Using the Agilent 8169A Polarization Controller," Product Note, Agilent Technologies.
[38] L. Yan, X. S. Yao, M. C. Hauer, and A. E. Willner, "Practical Solutions to Polarization-Mode-Dispersion Emulation and Compensation," *J. Lightwave Technology*, Vol. 24, pp. 3992–4005, 2006.
[39] J. N. Damask, "Methods to Construct Programmable PMD Sources—Part I: Technology and Theory," *J. Lightwave Technology*, Vol. 22, pp. 997–1005, 2004.
[40] J. N. Damask, P. R. Myers, G. J. Simer, and A. Boschi, "Methods to Construct Programmable PMD Sources—Part II: Instrument Demonstrations," *J. Lightwave Technology*, Vol. 22, pp. 1006–1013, 2004.
[41] A. E.Willner and M. C. Hauer, "PMD emulation," *Journal of Optical Fiber Communication Reports*, Vol. 1, pp. 181–200, 2004.

[42] X. P. Mao, R. W. Tkach, A. R. Chraplyvy, R. M. Jopson, and R. M. Derosier, "Stimulated Brillouin Threshold Dependence on Fiber Type and Uniformity," *IEEE Photonics Technology Letters*, Vol. 4, No. 1, pp. 66–68, 1992.
[43] R. G. Smith, "Optical Power Handling Capacity of Low Loss Optical Fibers as Determined by Stimulated Raman and Brillouin Scattering," *Applied Optics*, Vol. 11, No. 11, pp. 2489–2494, 1972.
[44] C. C. Lee and S. Chi, "Measurement of Stimulated-Brillouin-Scattering Threshold for Various Types of Fibers Using Brillouin Optical-Time-Domain Reflectometer," *IEEE Photonics Technology Letters*, Vol. 12, No. 6, pp. 672–674, 2000.
[45] K. Shiraki, M. Ohashi, and M. Tateda, "SBS threshold of a fiber with a Brillouin frequency shift distribution," *J. Lightwave Technology*, Vol. 14, No. 1, pp. 50–57, 1996.
[46] R. H. Stolen and E. P. Ippen, "Raman gain in glass optical waveguides," *Applied Physics Letters*, Vol. 22, No. 6, p. 276–278, 1973.
[47] D. Mahgerefteh, D. L. Butler, J. Goldhar, B. Rosenberg, and G. L. Burdge, "Technique for measurement of the Raman gain coefficient in optical fibers," *Optics Letters*, Vol. 21, No. 24, pp. 2026–2028, 1996.
[48] C. Lin and R. Stolen, "New nanosecond continuum for excited-state spectroscopy," *Appl. Phys. Lett.* 28, 216–218, 1976.
[49] R. H. Stolen and C. Lin, "Self-phase modulation in silica optical fibers," *Physical Review* A. 17, 1448–1453, 1978.
[50] Roger H. Stolen, William A. Reed, Kwang S. Kim, and G. T. Harvey, "Measurement of the Nonlinear Refractive Index of Long Dispersion-Shifted Fibers by Self-Phase Modulation at 1.55 μm," *Journal of Lightwave Technology*, Vol. 16, No. 6, pp. 1006–1012, 1998.
[51] A. Boskovic, S. V. Chernikov, J. R. Taylor, L. Gruner-Nielsen, and O. A. Levring, "Direct continuous-wave measurement of *n*2 in various types of telecommunication fiber at 1.55 μm," *Optics Letters*, Vol. 21, No. 24, pp.1966–1968, 1996.
[52] S. G. Evangelides Jr., L. F. Mollenauer, J. P. Gordon, and N. S. Bergano, "Polarization Multiplexing with Solitons," *Journal of Lightwave Technology*, Vol. 10, No. 1, p. 28–35, 1992.
[53] T. Kato, Y. Suetsugu, M. Takagi, E. Sasaoka, and M. Nishimura, "Measurement of the nonlinear refractive index in optical fiber by the cross-phase-modulation method with depolarized pump light" *Optics Letters*, Vol. 20, No. 9, pp. 988–990, 1995.
[54] A. Melloni, et al., "Polarization independent nonlinear refractive index measurement in optical fiber," in *Proc. Symposium on Optical Fiber Measurements*, October 1996, pp. 67–70.
[55] R. Hui, K. Demarest and C. Allen, "Cross-phase modulation in multi-span WDM optical fiber systems," *IEEE J. Lightwave Technology*, Vol. 17, No. 7, pp. 1018–1026, June 1999.
[56] S. V. Chernikov and J. R. Taylor, "Measurement of normalization factor of n_2 for random polarization in optical fibers," *Optics Letters*, Vol. 21, No. 9, pp. 1559–1561, 1996.
[57] G. P. Agrawal, *Nonlinear Fiber Optics*, 3rd ed., Academic Press, 2001.
[58] R.Hui, M.O'Sullivan, A. Robinson and M. Taylor, "Modulation instability and its impact in multi-span optical amplified systems: Theory and experiments," *IEEE J. Lightwave Technol.*, Vol. 15, No. 7, pp. 1071-1082, 1997.
[59] M. Artiglia, E. Ciaramella, and B. Sordo, "Using modulation instability to determine Kerr coefficient in optical fibers," *Electronics Letters*, Vol. 31, No. 12, pp. 1012–1013, 1995.

Chapter 5

Optical System Performance Measurements

5.0. Introduction
5.1. Overview of Fiber-Optic Transmission Systems
5.1.1. Optical System Performance Considerations
5.1.2. Receiver BER *and* Q
5.1.3. System Q *Estimation Based on Eye Diagram Parameterization*
5.1.4. Bit Error Rate Testing
5.2. Receiver Sensitivity Measurement and OSNR Tolerance
5.2.1. Receiver Sensitivity and Power Margin
5.2.2. OSNR Margin and Required OSNR (R-OSNR)
5.2.3. BER vs. Decision Threshold Measurement
5.3. Waveform Distortion Measurements
5.4. Jitter Measurement
5.4.1. Basic Jitter Parameters and Definitions
5.4.2. Jitter Detection Techniques
5.5. In-Situ Monitoring of Linear Propagation Impairments
5.5.1. In Situ *Monitoring of Chromatic Dispersion*
5.5.2. In Situ *PMD Monitoring*
5.5.3. In Situ *PDL Monitoring*
5.6. Measurement of Nonlinear Crosstalk in Multi-Span WDM Systems
5.6.1. XPM-Induced Intensity Modulation in IMDD Optical Systems
5.6.2. XPM-Induced Phase Modulation
5.6.3. FWM-Induced Crosstalk in IMDD Optical Systems
5.6.4. Characterization of Raman Crosstalk with Wide Channel Separation
5.7. Modulation Instability and Its Impact in WDM Optical Systems
5.7.1. Modulation-Instability and Transfer Matrix Formulation
5.7.2. Impact of Modulation Instability in Amplified Multispan Fiber Systems
5.7.3. Characterization of Modulation Instability in Fiber-Optic Systems
5.8. Optical System Performance Evalution Based on Required OSNR
5.8.1. Measurement of R-SNR Due to Chromatic Dispersion
5.8.2. Measurement of R-SNR Due to Fiber Nonlinearity
5.8.3. Measurement of R-OSNR Due to Optical Filter Misalignment
5.9. Fiber-Optic Recirculating Loop
5.9.1. Operation Principle of a Recirculating Loop
5.9.2. Measurement Procedure and Time Control
5.9.3. Optical Gain Adjustment in the Loop

5.0 INTRODUCTION

One of the most important applications of fiber-optic technology is optical communication. The introduction of optical fiber into communications revolutionized the entire telecommunications industry. The wide transmission bandwidth and low propagation loss make optical fiber an ideal media for transmission. Nowadays, more than 99 percent of long-distance communication traffic is carried by optical fibers all over the world. Although fundamental communication protocols, modulation formats, and performance evaluation criteria for traditional communications systems are still applicable, optical fiber communication has unique characteristics due to its high data rate and the special properties of optical fibers. The measurement and characterization techniques are very important in the design, development, and installation of optical transmission systems and networks.

This chapter discusses techniques to measure performance of digital fiber-optic systems and subsystems. Measurement objectives in long-distance, high-speed optically amplified systems differ substantially from those passive systems such as PON or LAN. Accordingly, we review performance measurements in the light of the intended system applications.

Section 5.1 provides an overview of digital optical transmission systems and their performance specifications, such as bit error rate (BER), the quality factor (Q), and techniques to measure them. Section 5.2 introduces the definitions of receiver sensitivity and the required optical signal-to-noise ratio (R-OSNR). In practical applications, receiver sensitivity is useful for receiver noise-dominated optical systems, whereas R-OSNR is often used in systems with inline optical amplifiers in which performance is mainly determined by the OSNR of the optical signal.

Section 5.3 briefly discusses the measurement of high-speed optical waveforms and the importance of time-based calibration in a sampling oscilloscope. Section 5.4 introduces a time jitter definition and basic techniques to characterize time jitter in digital optical signals. Section 5.5 presents a number of *in situ* performance-monitoring techniques for optical transmission systems, where data traffic carried in the system is used as probing signal for the measurement. Physical parameters of the fiber system, such as chromatic dispersion, PMD, and PDL, can be extracted from these measurements without interrupting system operation.

Section 5.6 reviews nonlinear crosstalk between wavelength channels in WDM systems such as XPM, FWM, and the Raman effect. Section 5.7 discusses modulation instability and its impact in long-distance optical transmission systems. Modulation instability is a parametric amplification process caused by the interaction between the optical carrier and the broadband ASE noise in the system. Section 5.8 provides several examples of transmission performance measurements based on R-OSNR. The impact on R-OSNR due to chromatic dispersion, SPM, and optical filter misalignment are presented.

The last section, Section 5.9, discusses fiber-optic recirculating loops, a very useful instrument in long-distance optical transmission system equipment measurement.

5.1 OVERVIEW OF FIBER-OPTIC TRANSMISSION SYSTEMS

The physical layer of a digital optical transmission system comprises a transmitter, a line system, and a receiver. The transmitter provides a means of encoding digital data onto an optical carrier. The line system delivers the encoded optical carrier to the receiver, where it is converted back to the electrical domain and detected as data.

The line system can be as simple as a length of transmission fiber or as complex as a multispan, optically amplified and switched wavelength-division multiplexed (WDM) network. The fidelity of data transmission is quantified by the bit error rate (BER). The BER is the fraction of transmitted data that is mistakenly decoded by the receiver. BER is a function of the system quality factor, Q. The quality factor is an electrical domain measure of the ratio of the separation between digital states to the noise associated with the states. Both the numerator and the denominator of Q can be partitioned into contributions whose sources are objects of system design. Examples include accumulated optical noise generated by optical amplifiers, signal optical power, polarization-dependent loss (PDL) and polarization mode dispersion (PMD), receiver and transmitter transfer function, net dispersion, and nonlinear propagation noise and distortion.

Commercial optical systems are designed to operate with some minimum fidelity over their lifetime. For example, a maximum BER of 10^{-15} is commonly required for links spanning cities and continental distances. Such links are often designed with forward error correction (FEC) wherein overhead bits are encoded with the data payload in such a way as to allow limited correction of errors upon decoding at the receiver [1]. There is a wide range of FEC implementations offering a variety of correction capabilities, and these can deliver corrected error rates as low as 10^{-15} given uncorrected (raw) error rates as high as 10^{-3}. In the absence of FEC, a maximum BER of 10^{-12} is often required.

In this chapter we describe methods and procedures designed to measure the performance of optical transmission systems. Though the description is focused on intensity-modulated direct detection (IMDD) systems with binary coding, most of the techniques and their fundamental principles are applicable to other types of modulation formats. Measurements are presented in the context of the link application because this motivates not only the measurement but also the measurement conditions. For example, optical noise loading is an important measurement technique for the characterization of receivers that are designed for operation in amplified links, but is of limited use for receivers operating in the absence of optical amplifiers. For the latter receivers, sensitivity is instead a more useful measure of performance.

5.1.1 Optical System Performance Considerations

The majority of digital optical transmission systems are based on binary modulation. In the intensity modulation and direct detection (IMDD) mode, data is encoded on the optical power emitted from the transmitter, and the transmitter output has two digital states that are usually chosen to be light-pass (mark) and light-block (space). At the receiver, the optical power is converted into a photocurrent by means of a photodiode, in which the photocurrent is linearly proportional to the optical power received. This conversion eliminates channel wavelength information as well as the phase noise of the optical signal at the receiver. The digital data can also be encoded as frequency or phase of the optical carrier, such as frequency shift key (FSK) and phase shift key (PSK). In these systems, frequency or phase decoders have to be included in the optical receivers before the photodiodes so that the data embedded in the optical signal can be recovered.

A diagram of data flow through the components of an IMDD link is shown in Figure 5.1.1. Data is usually encapsulated in a digital wrapper that can be used to record content partitioning, source and destination, enable synchronization, time-domain partitioning performance monitoring, fault isolation, internodal communication, and FEC, to name a few. Additional overhead may be added to simplify clock recovery. These often essential network functionalities increase the line transmission rate (equivalently bandwidth) for a given data rate. Depending on transmission standard and FEC used, such overhead can possibly add 25 percent

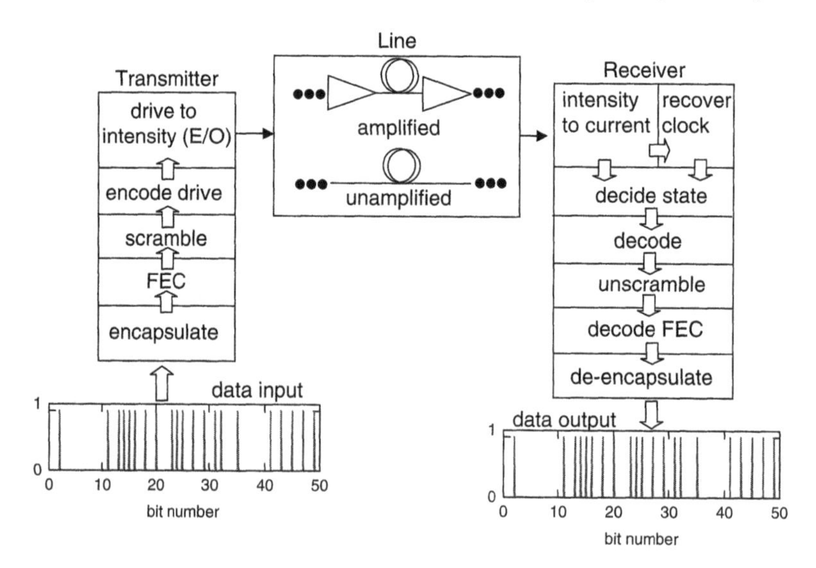

Figure 5.1.1 Schematic diagram of data flow through the components of an IMDD link.

to the line rate, but it is typically 3 percent to 7 percent in practice. Even though the data and overhead are often scrambled to regulate pattern length, in some cases the framing structure (which is not scrambled) can contain long patterns, which place demands on the receiver low-frequency response and can have a bearing on measurement conditions and modulation technology as it relates to performance. For example, long pseudorandom bit sequences (PRBS) such as 2^{31}-1 are used to properly exercise the pattern dependence of a link.

In short-distance and low-speed optical systems without inline optical amplifiers, the system performance is often limited by the signal optical power level that reaches the receiver. Receiver sensitivity is defined as the signal optical power required at the receiver to achieve the targeted bit error rate. If the signal optical power is too low at the receiver, the noise generated by the receiver would make the SNR unacceptable. For high-speed long-distance optical systems employing multiple inline optical amplifiers, signal waveform distortion and accumulated ASE noise throughout the transmission system may become major limitations in the transmission distance. In this case, receiver sensitivity is no longer the most relevant parameter to specify an optical receiver. Instead, the receiver should be qualified by its ability to resist the influences of waveform distortion and optical noise.

One of the most important sources of linear performance impairments in high-speed optical transmission is chromatic dispersion. For a standard SMF, the chromatic dispersion at a 1550 nm wavelength is approximately 17 ps/nm-km. This limits the transmission distance of a 10 Gb/s IMDD optical system to about 100 km. Beyond which the intersymbol interference (ISI) due to chromatic dispersion will introduce significant waveform distortion. Dispersion shifted fibers have been developed to minimize this problem, but they were later found unsuitable for WDM systems due to the increased nonlinear crosstalk between different wavelength channels.

Dispersion compensation has emerged as an effective way to overcome the dispersion limit. This is usually accomplished using dispersion-compensating modules (DCM) that have the opposite sign of dispersion to the transmission fiber. DCM can be made by dispersion-compensating fibers (DCF) or by passive optical devices such as fiber Bragg gratings. The overall accumulated dispersion in a transmission system can be reduced to an acceptable level with proper system design of dispersion compensation. In WDM systems with large numbers of channels, different wavelengths may experience different levels of dispersion due to the dispersion slope in optical fibers. In this case, slope compensation has also to be applied in high-speed optical systems to equalize the performance of WDM channels. Adding a dispersion compensator in each fiber span has become a standard industrial practice for long-distance optical systems. In fact, a dispersion compensator can often be packaged into an inline optical amplifier module to simplify optical system implementation. The disadvantage of dispersion compensation in the optical domain is the increased optical attenuation due to DCM, which requires a higher level of optical amplification in the system to compensate for this additional loss. This increased gain requirement of optical amplifiers will in turn generate excess ASE noise, which tends to degrade optical SNR when the signal reaches the receiver.

In an amplified multispan WDM optical system with a large number of wavelength channels, interchannel crosstalk is another important concern, especially when the system has a large number of spans and the signal optical power level is high. In addition to linear crosstalk that might be caused by leakage from optical filters and switches, nonlinear crosstalk is especially notorious because it cannot be eliminated by improving the qualities of optical components. The major sources of nonlinear crosstalk in high-speed optical transmission systems include cross-phase modulation (XPM), four-wave mixing (FWM), and Raman crosstalk. The characterization of these crosstalks helps in understanding the mechanisms of system performance degradation, which is essential for the system design and performance specification.

An important way to reduce performance degradation due to linear and nonlinear impairments in fiber-optic systems is to use advanced modulation formats. In general, an optical signal with longer pulse duration and (or) narrower spectral width would suffer less from chromatic dispersion. Multilevel modulation [2], phase-shaped binary (PSB) modulation [3], and digital subcarrier multiplexing [4] have been used to reduce the impact of chromatic dispersion because of their reduced spectral width. More recently, electrical domain digital signal processing (DPS) was applied to reduce transmission impairments, which has the potential to completely eliminate the requirement of optical domain dispersion compensation [5]. Without going into great detail on these techniques, it is useful to discuss the fundamental parameters that specify the quality of an optical transmission system.

5.1.2 Receiver *BER* and *Q*

Bit error rate (BER) is a fundamental measure of digital transmission quality. BER is essentially an error probability of digital bits in the received signal; it is also known as *bit error probability*. By definition, BER is

$$BER = \frac{Bit_{Error}}{Bit_{Total}} \tag{5.1.1}$$

where Bit_{Error} is the number of misinterpreted bits by the receiver and Bit_{Total} is the total number of received bits. Both the misinterpreted bits and the total received bits are measured within a certain time window ΔT, which is referred to as *gating time*.

A useful approximation for the estimation of BER is the system Q function. It is a quality factor that relates the BER to the ratio of separation between implemented digital states and the approximate Gaussian noise associated with those states at the receiver. The received signal voltage is presented to a decision circuit. This latter is typically a gated threshold device synchronized to the recovered clock. The decision circuit reports a logical one for signal voltage above a reference, threshold, value and a logical zero otherwise. A decision is made at each clock cycle. This scheme is shown in Figure 5.1.2. An eye diagram, formed by overlapping consecutive segments of the received electrical waveform, shows the site in phase (horizontal axis) and in voltage (vertical axis) of the decision instant and threshold, respectively. Depending on receiver design, the decision instant and threshold might be optimized once at start of life or in a continuous and automatic manner dictated by a performance cost function.

The eye diagram shown in Figure 5.1.2 is noiseless and the spread of the voltage values above and below threshold at the sampling instant is attributable to intersymbol interference (ISI). Sources of ISI include channel memory stemming from receiver and transmitter transfer functions, linear and nonlinear propagation effects such as residual chromatic dispersion, PMD, SPM, XPM, FWM, and optical filter transfer functions. Phase delay variations within the information bandwidth contribute to spreading at eye crossings, usually located ½ a clock cycle from the decision instant. This spreading is a constituent of timing jitter.

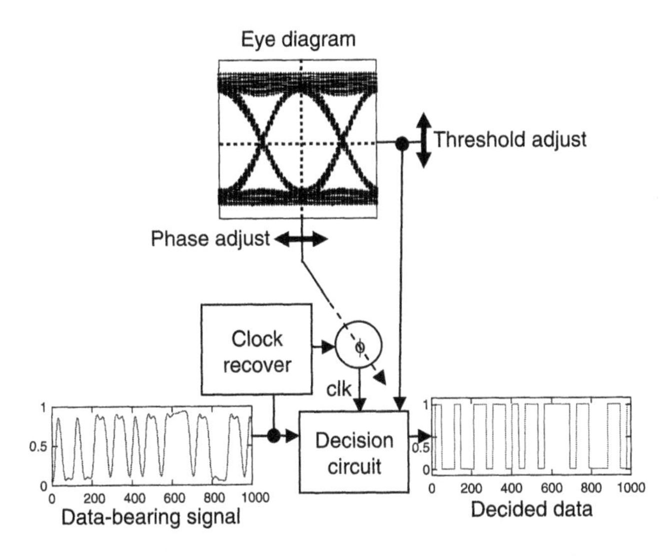

Figure 5.1.2 Illustration of bit decision in a binary receiver.

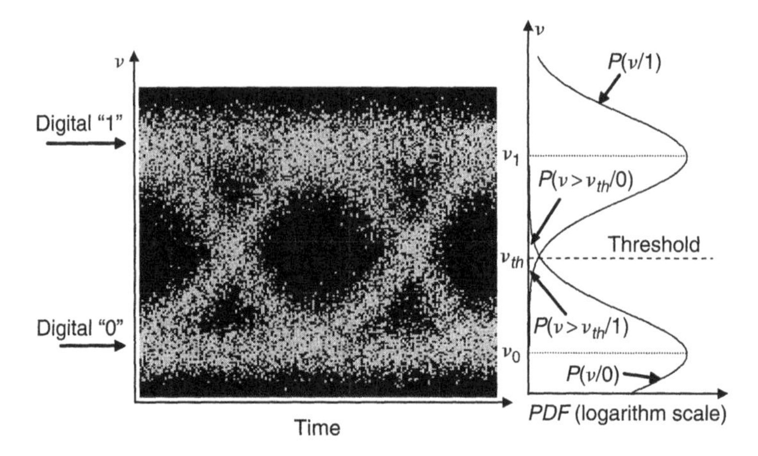

Figure 5.1.3 Probability distribution function (PDF) of the eye diagram.

Figure 5.1.3 shows the probability distribution function (PDF) of the eye diagram so that we can derive the fundamentals of BER calculation. In fact, BER is a conditional probability of receiving signal y while the transmitted signal is x, $P(y/x)$, where x and y can each be digital 0 or 1. Since the transmitted signal digital states can be either 0 or 1, we can define $P(y/0)$ and $P(y/1)$ as the PDFs of the received signal at state y while the transmitted signals are 0 and 1, respectively. Suppose that the probability of sending digital 0 and 1 are $P(0)$ and $P(1)$ and the decision threshold is v_{th}; the BER of the receiver should be

$$BER = P(0)P(v > v_{th}/0) + P(1)P(v < v_{th}/1) \tag{5.1.2}$$

where v is the received signal level. In most of the binary transmitters, the probabilities of sending 0 and 1 are the same, $P(0) = P(1) = 0.5$. Also, Gaussian statistics can be applied to most of the noise sources in the receiver as a first-order approximation,

$$P_{Gaussian}(v) = \frac{1}{\sigma\sqrt{2\pi}}\exp\left(-\frac{(v - v_m)^2}{2\sigma^2}\right) \tag{5.1.3}$$

where σ is the standard deviation and v_m is the mean value of the Gaussian probability distribution. Then the probability for the receiver to declare 1 while the transmitter actually sends a 0 is

$$P(v > v_{th}/0) = \frac{1}{\sigma_0\sqrt{2\pi}}\int_{v_{th}}^{\infty}\exp\left(-\frac{(v - v_0)^2}{2\sigma_0^2}\right)dy = \frac{1}{\sqrt{2\pi}}\int_{Q_0}^{\infty}\exp\left(-\frac{\xi^2}{2}\right)d\xi \tag{5.1.4}$$

where σ_0 and v_0 are the standard deviation and the mean value of the received signal photocurrent at digital 0, $\xi = (v - v_0)/\sigma_0$ and

$$Q_0 = \frac{v_{th} - v_0}{\sigma_0} \tag{5.1.5}$$

Similarly, the probability for the receiver to declare 0 while the transmitter actually sends 1 is

$$P(v < v_{th}/1) = \frac{1}{\sigma_1\sqrt{2\pi}}\int_{-\infty}^{v_{th}} \exp\left(-\frac{(v_1 - v)^2}{2\sigma_1^2}\right)dy = \frac{1}{\sqrt{2\pi}}\int_{Q_1}^{\infty} \exp\left(-\frac{\xi^2}{2}\right)d\xi \tag{5.1.6}$$

where σ_1 and v_1 are the standard deviation and the mean value of the received signal photocurrent at digital 1, $\xi = (v_1 - v)/\sigma_1$ and

$$Q_1 = \frac{v_1 - v_{th}}{\sigma_1} \tag{5.1.7}$$

According to Equation 5.1.2, the overall error probability is

$$\begin{aligned} BER &= \frac{1}{2}P(v > v_{th}/0) + \frac{1}{2}P(v < v_{th}/1) \\ &= \frac{1}{2\sqrt{2\pi}}\left\{\int_{Q_0}^{\infty} \exp\left(-\frac{\xi^2}{2}\right)d\xi + \int_{Q_1}^{\infty} \exp\left(-\frac{\xi^2}{2}\right)d\xi\right\} \end{aligned} \tag{5.1.8}$$

where $P(0) = P(1) = 0.5$ is assumed.

Mathematically, a widely used special function, the error function, is defined as

$$erf(x) = \frac{2}{\sqrt{\pi}}\int_0^x \exp(-y^2)dy \tag{5.1.9}$$

and a complementary error function is defined as

$$erfc(x) = 1 - erf(x) = \frac{2}{\sqrt{\pi}}\int_x^{\infty} \exp(-y^2)dy \tag{5.1.10}$$

Therefore Equation 5.1.8 can be expressed as complementary error functions:

$$BER = \frac{1}{4}\left\{erfc\left(\frac{Q_0}{\sqrt{2}}\right) + erfc\left(\frac{Q_1}{\sqrt{2}}\right)\right\} \tag{5.1.11}$$

Since both Q_0 and Q_1 in Equation 5.1.11 are functions of the decision threshold v_{th}, and usually the lowest BER can be obtained when $P(0)P(v > v_{th}/0) = P(1)P(v < v_{th}/1)$, we can simply set $Q_0 = Q_1$, which is

$$\frac{v_{th} - v_0}{\sigma_0} = \frac{v_1 - v_{th}}{\sigma_1}$$

Or equivalently

$$v_{th} = \frac{v_0\sigma_1 + v_1\sigma_0}{\sigma_0 + \sigma_1} \tag{5.1.12}$$

Under this "optimum" decision threshold, Equations 5.1.5 and 5.1.7 are equal and

$$Q = Q_1 + Q_2 = \frac{v_1 - v_0}{\sigma_1 + \sigma_0} \tag{5.1.13}$$

The BER function in Equation 5.1.11 becomes

$$BER = \frac{1}{2}erfc\left(\frac{Q}{\sqrt{2}}\right) \tag{5.1.14}$$

This is a simple yet important equation that establishes the relationship between BER and the receiver Q-value, as shown in Figure 5.1.4. As a rule of thumb, $Q = 6$, 7, and 8 correspond to the BER of approximately 10^{-9}, 10^{-12}, and 10^{-15}, respectively.

In practical systems, the statistics of noise sources are not always Gaussian. For example, shot noise is a Poisson process whose PDF follows a Poisson distribution [6]. Another example is that the received photocurrent at digital 0 should never be negative because the received optical power is always positive. Therefore the tail of the PDF should be limited to the positive territory and a

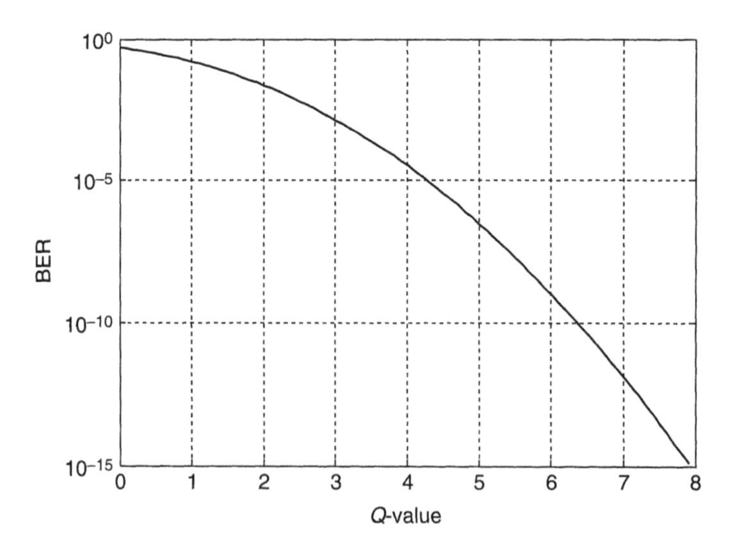

Figure 5.1.4 BER as a function of receiver Q-value.

Rayleigh distribution may be more appropriate to describe the noise statistics. Nevertheless, Gaussian approximation is widely adopted because of its simplicity.

In the Gaussian approximation discussed so far, we have assumed that at the decision instant, the eye diagram only has a single line at the digital 1 level and another single line at the digital 0 level. In practice, the eye diagram may have many lines at each digital level due to pattern-dependent waveform distortion. An eye diagram is recast in Figure 5.1.5. This diagram plots the signal voltage referred (as relative equivalent optical power) to the input connector of the receiver. Associated with each of the noise-free lines of the eye diagram is an approximately Gaussian noise distribution that is generated from a handful of independent processes. The aggregate noise power distributions at the sampling instant are drawn on the right side of the figure. The noise distribution on transmitted 1 s is typically wider than the distribution on transmitted 0 s due to the signal dependence of some of the noise processes.

Noise processes also depend on receiver optical-to-electrical conversion technology. For example, a receiver based on a PIN photodiode has different noise properties than an avalanche photodiode (APD)-based optical receiver. In optically amplified systems, the presence of ASE noise generated from optical amplifiers makes the receiver performance specification and analysis quite different from unamplified optical systems.

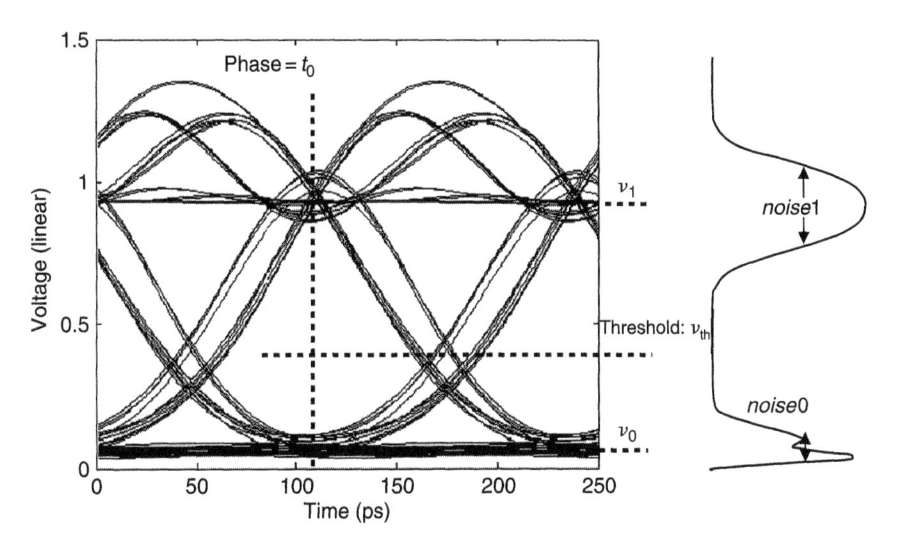

Figure 5.1.5 Statistic distribution of noise when the eye diagram is distorted.

In the absence of waveform distortion and optical nonlinearity, the system Q for a PIN based receiver is

$$Q = \frac{\Re(P_1 - P_0)}{\sqrt{\left(2q(\Re P_1 + I_d) + \frac{4kT}{R_L} + 2\rho_{ASE}\Re^2 P_1 + \rho_{ASE}^2\Re^2(2B_o - B_e) + 2\Re P_1 \times 10^{-\frac{RIN}{10}}\right)B_e} + \sqrt{\left(2q(\Re P_0 + I_d) + \frac{4kT}{R_L} + 2\rho_{ASE}\Re^2 P_0 + \rho_{ASE}^2\Re^2(2B_o - B_e) + 2\Re P_0 \times 10^{-\frac{RIN}{10}}\right)B_e}}$$

where P_0 and P_1 are the instantaneous signal optical powers at digital 0 and digital 1, respectively, at the decision time. For an APD-based optical receiver, the APD multiplication gain M and the excess noise factor F_A have to be considered. The equivalent Q function is then

$$Q = \frac{M\Re(P_1 - P_0)}{\sqrt{\left\{M^2F_A\left(2q(\Re P_1 + I_d) + 2\rho_{ASE}\Re^2 P_1 + \rho_{ASE}^2\Re^2(2B_o - B_e) + 2\Re P_1 \times 10^{-\frac{RIN}{10}}\right) + \frac{4kT}{R_L}\right\}B_e} + \sqrt{\left\{M^2F_A\left(2q(\Re P_0 + I_d) + 2\rho_{ASE}\Re^2 P_0 + \rho_{ASE}^2\Re^2(2B_o - B_e) + 2\Re P_0 \times 10^{-\frac{RIN}{10}}\right) + \frac{4kT}{R_L}\right\}B_e}}$$

Definition of the parameters used in these two Q-functions can be found in Table 5.1.1.

In addition, when wavelength-division multiplexing is used, linear and nonlinear crosstalk will be introduced, adding another layer of complexity to system analysis and characterization. The impact of these noises and waveform distortions depends on the type of optical detection device used in the receiver and the configuration of the system. The single-sided spectral densities of the noises and the sources of crosstalk in WDM systems are listed in Table 5.1.1. The frequency dependence of noise spectral densities in this table has been averaged over the receiver bandwidth. The units of parameters defined in Table 5.1.1 are shown in Table 5.1.2.

Some of the noise spectral densities in Table 5.1.1 depend on the value of the instantaneous signal power P, commonly referred to as signal-dependent noise. It is a reasonable approximation to calculate such contributions based on the instantaneous power value associated with each line of a measured eye diagram for a given sampling instant.

We assume a Gaussian aggregate noise on any digital 1 or 0 level with standard deviations, σ_{1k}, and σ_{0k}, respectively. The received waveform is sampled at times $t_{sk} = t_0 + kT$, where T is the bit period, t_0 is the sample phase, and k is an integer. The BER associated with the kth sampling instant is

Table 5.1.1

Noise Spectral Densities and Sources of Waveform Distortion

	PIN	APD	EDFA	WDM	Single-Sided Noise Current Variance Spectral Density $A^2\text{-}Hz^{-1}$
PIN signal shot noise	X				$2\Re qP$
PIN dark current noise	X				$2qI_d$
Multipath interference (MPI)	X				
PIN thermal noise	X				$4kT/R_L$
APD dark current noise		X			$2MF_AqI_d$
APD signal shot noise		X			$2MF_A\Re qP$
Signal-ASE beat noise			X		$2\rho_{\text{ASE}}P\Re^2$
ASE-ASE beat noise			X		$\rho_{\text{ASE}}^2\Re^2(2B_o - B_e)$
Relative intensity noise (RIN)					$2P\cdot 10^{-\frac{\text{RIN}}{10}}$
Linear crosstalk				X	$\sum_m \Re^2\overline{P}_m^2\eta_m$
Raman crosstalk				X	
Modulation instability (MI)			X		
Four-wave mixing (FWM)				X	
Cross-phase modulation (XPM)				X	

Table 5.1.2

Parameters Used in Table 5.1.1

Variable	Constant	Name	Value	Unit
	k	Boltzmann's constant	1.28×10^{-23}	J-K^{-1}
	h	Planck's constant	6.626×10^{-34}	J-s
	q	Electron charge	1.6023×10^{-19}	Coulomb
P		Instantaneous optical power		W
ζ		Coupling efficiency x quantum Efficiency		—
ρ_{ASE}		ASE noise optical spectral Density		W-Hz^{-1}
B_o		Optical bandwidth		Hz
B_e		Electrical bandwidth		Hz
I_d		Dark current		A
T		Absolute temperature		K
RIN		Relative intensity noise coefficient		—
F_A		APD excess noise factor		—
M		APD multiplication gain		—
R_L		Load resistance		Ohm
$\Re$		Responsivity	$q\zeta\lambda/(\text{hc})$	A-W^{-1}
λ		Signal wavelength		nm

$$BER_{1k} = \frac{1}{2} erfc\left(\frac{v_{1k} - v_{th}}{\sqrt{2}\sigma_{1k}}\right) \tag{5.1.15a}$$

for a transmitted digital 1 or

$$BER_{0k} = \frac{1}{2} erfc\left(\frac{v_{th} - v_{0k}}{\sqrt{2}\sigma_{0k}}\right) \tag{5.1.15b}$$

for a transmitted digital 0, respectively.

v_{0k} and v_{1k}, are the signal voltages corresponding to the digital 1 and 0 of the kth sampling instant, respectively. Then the BER reported from an observation spanning M samples, i.e., over the time interval between t_0 and $t_0 + (M\text{-}1)T$, is

$$BER = \frac{\sum_{k=1}^{M/2} BER_{1k} + \sum_{k=1}^{M/2} BER_{0k}}{M} \tag{5.1.16}$$

The threshold value is chosen to minimize the BER such that

$$\sum_{k=1}^{M/2} BER_{1k} = \sum_{k=1}^{M/2} BER_{0k}$$

For a special case, if there is no ISI at the sample phase, $v_{1k} = v_1$, $v_{0k} = v_0$, $\sigma_{1k} = \sigma_1$, and $\sigma_{0k} = \sigma_0$, the minimum BER can be obtained with the Q-value given in Equation 5.1.13.

In general, when the waveform distortion is considered, the Q-value measured within the time interval $[t_0, t_0 + (M - 1)T]$ can be found as

$$Q = \sqrt{2} \cdot erfc^{-1}\left\{ 2 \times \frac{\sum_{k=1}^{M/2} BER_{1k} + \sum_{k=1}^{M/2} BER_{0k}}{M} \right\} \tag{5.1.17}$$

Equation 5.1.17 can be used to convert a measured BER to a system Q under the assumption of Gaussian noise statistics.

5.1.3 System Q Estimation Based on Eye Diagram Parameterization

The Q-value and the BER of an optical system are affected by random noise and waveform distortion. Both these effects can be observed in the eye diagram. Eye-mask parameterization can be used to estimate system performance. This technique is based on the separation of distortion (eye closure) and noise for the evaluation of system Q and is suitable for systems designed to operate at high Q values ($Q > 7$ or BER $< 10^{-12}$). An optical eye distortion mask parameterization can be made at any reference interface of the system to define the distortion-related link performance contributions, independent of the particular noise characteristic [7].

Figure 5.1.6 is an example of an eye diagram measured at the output of a dispersive fiber link where averaging has been used to remove the random noise. An eye distortion mask is defined by a four-level feature (P_1/P_{av}, P_0/P_{av}, A, B) over a timing window, W, which represents the worst-case phase uncertainty at the sampling instant. W comprises the sum of all bounded uncertainties plus seven times the standard deviation of statistical uncertainty of the decision phase. The factor 7 was chosen to guarantee a bit error rate of 10^{-12} in the absence of forward error correction. P_1 and P_0 are the power levels associated with signal long 1 s and long 0 s in the pseudo-random NRZ bit pattern. The dimensionless parameters A and B are the lowest inner upper eye and the highest inner lower eye measured within the phase window W, and they are independent of the noise. According to definitions in Figure 5.1.6, the average signal optical power is $P_{av} = (P_1+P_0)/2$, given that signal 1 s and 0 s have the same probability.

In a practical system, waveform distortion, signal-dependent noise, and signal-independent noise are all mixed together at the receiver. The receiver Q factor can be written as

$$Q = \frac{(A - B)2\Re P_{av}}{\sqrt{\sigma_{ind}^2 + \eta A 2 P_{av}} + \sqrt{\sigma_{ind}^2 + \eta B 2 P_{av}}} \tag{5.1.18}$$

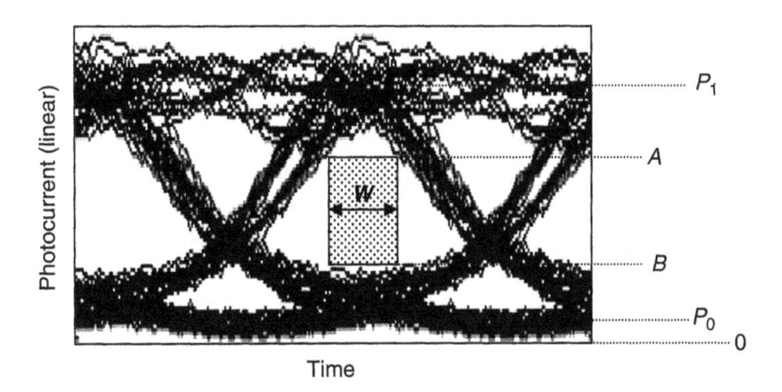

Figure 5.1.6 Schematic of an optical eye distortion mask mapped onto a measured eye diagram.

where σ^2_{ind} is the signal independent noise variance, and η is a system-dependent multiplication factor originated from the signal-dependent noise.

In the absence of distortion, $B = 0$ and $A = 1$, the system Q is determined only by the noise contribution. In this case,

$$Q = Q_0 = \frac{2\Re P_{av}}{\sqrt{\sigma^2_{ind} + \eta 2P_{av}} + \sigma_{ind}} \tag{5.1.19}$$

By this definition of Q_0, the system Q degradation caused only by eye distortion can be written in a general form as

$$D(A, B, x) = \frac{Q}{Q_0} = \frac{A - B}{Y_e} \tag{5.1.20}$$

In this expression, Y_e is an important factor that shows the effect of interaction between distortion and noise:

$$Y_e(A, B, x) = \frac{\sqrt{1 + xA} + \sqrt{1 + xB}}{1 + \sqrt{1 + x}} \tag{5.1.21}$$

where $x = 2\eta P_{av}/\sigma^2_{int}$ is the ratio of signal-dependent noise to signal-independent noise.

In a case where signal-independent noise dominates, $x = 0$ and $Y_e = 1$, so that $D = A - B$. On the other hand, if signal-dependent noise dominates, $x = \infty$ and $Y_e = \sqrt{A} + \sqrt{B}$, therefore $D = \sqrt{A} - \sqrt{B}$. In general, within $x \in (0, \infty)$, the maximum value of Y_e that corresponds to the worst-case distortion can be expressed as

$$Y_0 = \begin{cases} \sqrt{A} + \sqrt{B} & (\sqrt{A} + \sqrt{B}) \geq 1 \\ 1 & (\sqrt{A} + \sqrt{B}) < 1 \end{cases} \tag{5.1.22}$$

The two possible maximums $Y_{max} = \sqrt{A} + \sqrt{B}$ and $Y_{max} = 1$ correspond to $x = \infty$ and $x = 0$, respectively. Using Equation 5.1.20, the worst-case distortion factor, defined as D_{wc}, can be written as a function of Y_{max}:

$$D_{wc} = (A - B)/Y_{max} \tag{5.1.23}$$

Obviously, D_{wc} is a global worst-case distortion effect, which is independent of the nature of the noise. To demonstrate the impact of noise characteristic on the system distortion penalty, D, defined by Equation 5.1.20, is plotted in Figure 5.1.7 as a function of 10log(x). In this plot, two sets of eye-closure parameters were used, corresponding to the conditions for the two solutions of Equation 5.1.22. In one case, $A = 0.7$, $B = 0.15$, and $\sqrt{A} + \sqrt{B} > 1$ so that the worst-case distortion happens at $x = \infty$. In the other case, $A = 0.4$, $B = 0.05$, and $\sqrt{A} + \sqrt{B} < 1$, the worst-case distortion happens at $x = 0$. The dashed lines in Figure 5.1.7 are $10\log(\sqrt{A} - \sqrt{B})$, which represents the case where signal-dependent noise dominates ($x = \infty$), whereas the dash-dotted lines are

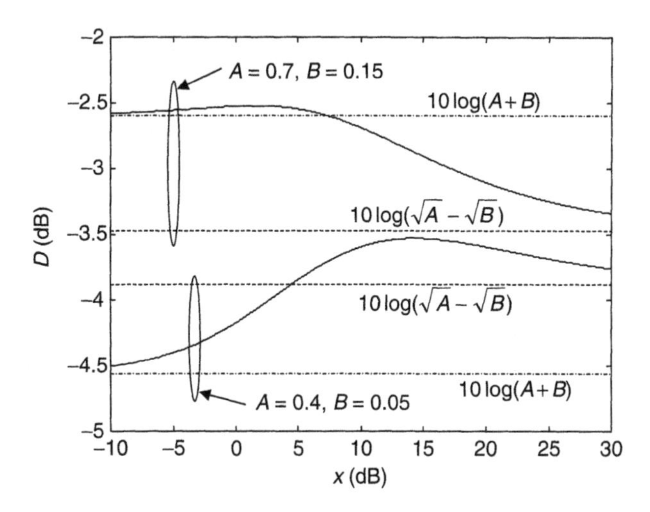

Figure 5.1.7 Q-degradation parameter D as a function of x (solid line). Dashed line: 10log(A − B); dashed-dotted line: 10log(A+B).

10log(A+B), which represents the case of $x = 0$. Shown as the solid lines in Figure 5.1.7, $D(x)$ vs. x characteristics are not monotonic; however, D symptomatically approaches its worst-case D_{wc} with either $x = 0$ or $x \to \infty$, depending on the value of $\sqrt{A} + \sqrt{B}$. It is worthwhile to note that generally, Equations 5.1.22 and 5.1.23 overestimate the distortion penalty because $x \in (0, \infty)$ was used to search for the worst case, but in real systems x value can never be infinity. The existence of a worst-case distortion factor D_{wc} implies a possibility to separate distortion from noise in the system link budgeting. Equation 5.1.23 clearly demonstrates a simple linear relationship between system Q and the worst-case distortion factor D_{wc}. Regardless of the fundamental difference in the origins of noise and distortion, isolation of these two gives a clear picture of system budget allocations.

The linear relationship between Q and D_{wc}, can be easily verified experimentally. One way to accomplish this experimental verification is to add a predetermined polarization-mode dispersion (PMD) to simulate the distortion in a fiber-optic system. In fact, PMD introduces system eye distortion without adding noise, so Q_0 does not change with the added PMD.

The experimental setup is shown in Figure 5.1.8. A 10 Gb/s commercial transmitter was used to generate optical signals. A polarization controller was placed between the amplified transmitter output and a PMD source. The polarization state was adjusted such that 50 percent of the optical power travels through each arm of the PMD source, as explained in Section 4.7. Two EDFAs were used to increase optical signal power, and a tunable bandpass optical filter

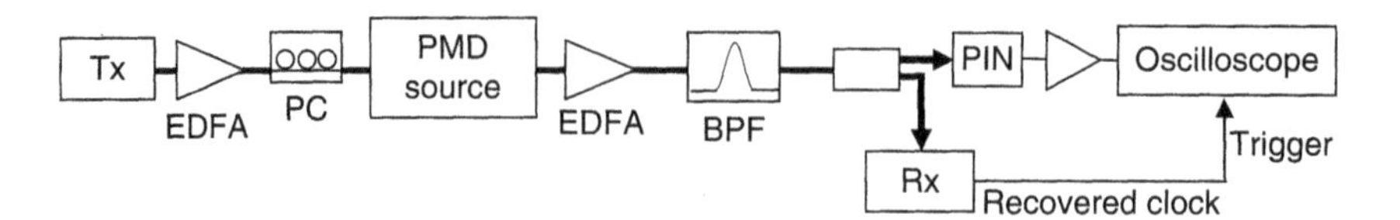

Figure 5.1.8 Experimental setup of PMD-induced eye distortion measurement. Tx: 10 Gb/s transmitter, PC: polarization control, BPF: bandpass optical filter.

is used to remove wideband ASE noise from EDFAs. At the receiver, the system Q value was obtained by a BER-vs.-decision voltage (BER-V) measurement, which will be discussed in Section 5.2.3. The noise-free signal waveform at the oscilloscope was measured through the average of repeated 2^7-1 bit patterns. Then the corresponding system distortion level D_{wc} was evaluated through the measured signal waveforms using Equations 5.1.22 and 5.1.23. The maximum x value of 30 was assumed, which is relevant for most of the practical systems.

Figure 5.1.9 shows the measured $10\log(Q)$ versus $10\log[D_{wc}(A,B)]$ as defined in Equation 5.1.23. The corresponding A and B values are also given in the inset of Fig.5.1.9. A linear relationship between dBQ and dBD was obvious. The slope of the best linear fit to these measured points is 0.99. Since optical power at the receiver was kept constant during the measurements (–13 dBm in this case) and thus Q_0 was constant, this verifies the theoretical prediction in Equation 5.1.23.

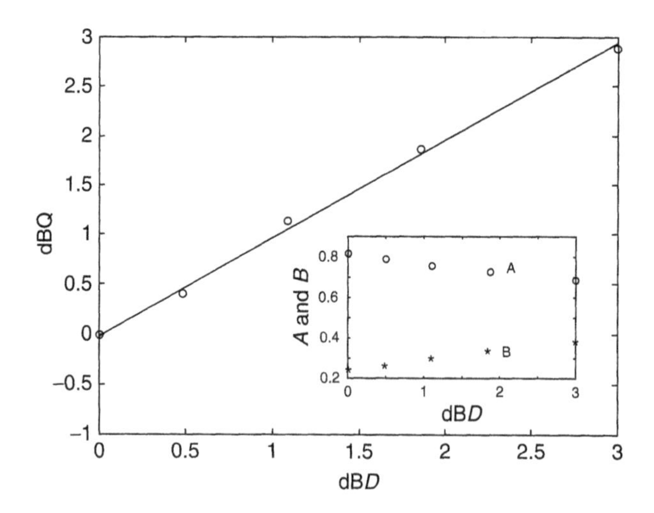

Figure 5.1.9 Measured system penalty vs. eye distortion introduced by a variable PMD. Q and D have been normalized by their values without PMD. Circles: measured data; solid line: best linear fit. The slope of the fitting line is $\Delta Dwc(\text{dB})/\Delta Q(\text{dB}) = 0.99$. The jitter window W used in the eye-mask measurement was 7 ps.

5.1.4 Bit Error Rate Testing

BER is the ultimate performance measure of a digital communication system. A number of important factors may contribute to the degradation of signal BER; these include signal waveform distortion, accumulated excessive noise, and time jitter. For example, if the bit errors are caused primarily by Gaussian noise, the measured BER is relatively stable versus time and in this case BER is relatively easy to measure. On the other hand, as illustrated in Figure 5.1.10, if BER is caused by random waveform distortion, such as due to the effects of interchannel crosstalk and polarization mode dispersion in the optical fiber, bust errors may happen and the measurement requires a much longer time to be statistically relevant.

Because of the random nature of the signal impairments, BER is typically a function of time. Meanwhile, since BER is an average value within each measurement time interval Δt, which is known as gating time, varying the gating time may in principle result in different BER values. To minimize this gating time-related uncertainty, we usually need to collect >100 errors within each gating time window to evaluate the average effect of BER performance. In practical measurements, the time required to acquire 100 errors determines the time of each measurement. This time window is inversely proportional to the system data rate and also inversely proportional to the bit error rate

$$\Delta T = \frac{1}{BER \cdot B} \tag{5.1.24}$$

where B is the bit rate of the digital system in [bit/sec]. As an example, for a 10 Gb/s digital system in which $B = 10^{10}\ \mathrm{s}^{-1}$, if the average BER of the signal

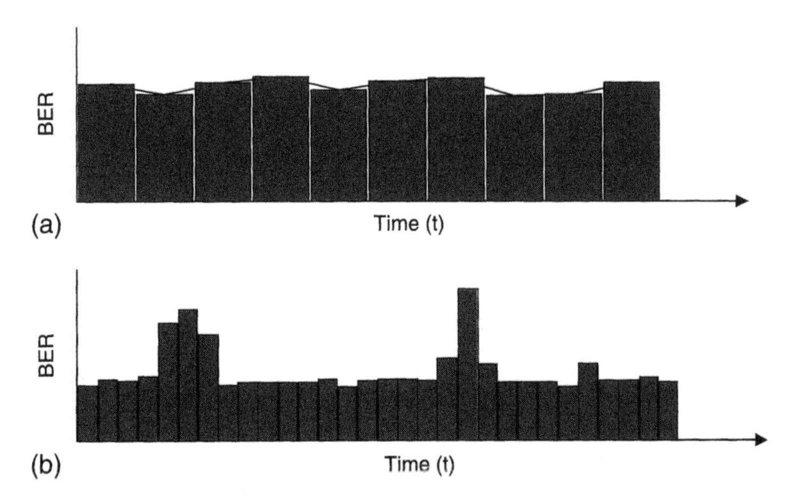

Figure 5.1.10 (a) Stationary BER and (b) bust BER.

is on the order of 10^{-12}, the gating time window has to be approximately 10,000 seconds to accumulate for about 100 errors. This is almost three hours. In commercial system qualification tests, BER measurements may take several days to obtain stable trustworthy BER statistics. For a system with a lower data rate, the measurement time will even be longer. That is why in some measurements, an extra amount of noise is added into the system, which accelerates the error occurrence and significantly speeds up the measurement. This concept is discussed in Section 5.2.

According to its general definition, BER measurement can be reported either in error ratio, error-free seconds or a percentage of unavailable time. *Error ratio* is simply the basic BER, which is the counted error bits divided by the total number of bits received within the gating period ΔT; *error-free seconds* indicates the time interval within which no error occurs; and *percent of unavailability* is the percentage of the extended error period within the gating period, as illustrated in Figure 5.1.11. The reason for this later definition is that the system may be unavailable during a period of excessive errors. In addition, the system needs a guard band (for example, 10 seconds before and 10 seconds after this bad period) to guarantee that the performance beyond this period is acceptable. Therefore, the entire extended error period is treated as a time window within which the channel is unavailable.

Sometimes the system performance can also be measured by the percentage of error seconds, which is defined by the time with measurable errors divided by the total gating time, or similarly by the percentage of severely errored seconds.

Figure 5.1.12 shows the block diagram of a BER test set. The two major components include (1) a pattern generator that generates signal to be transmitted

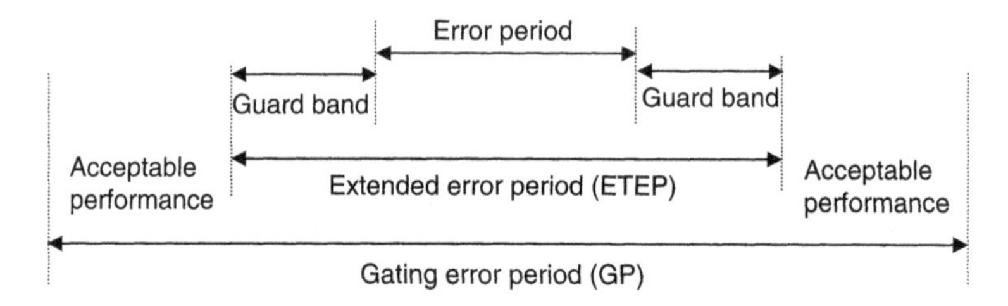

Figure 5.1.11 Illustration of percentage of unavailability.

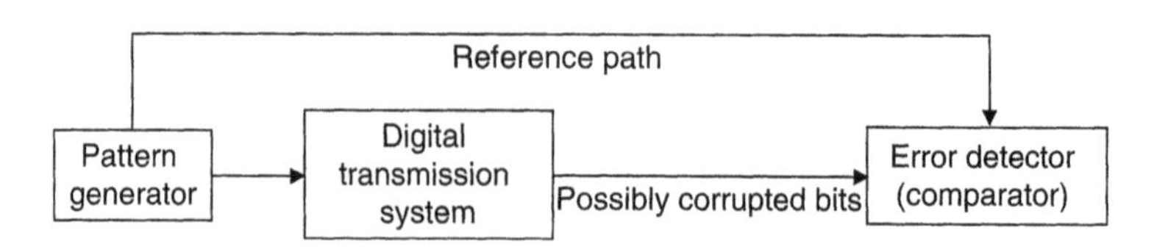

Figure 5.1.12 Basic configuration of a BER test set.

and (2) an error detector that compares bits received from the system and those from the reference path. It predicts an error if these two signals are different, and tells the number of errors that have occurred during the transmission within a gating time window. In the following, we discuss these two fundamental building blocks in more details.

5.1.4.1 Pattern Generator

To simulate the actual digital data traffic in transmission systems, the pattern generator in a BER test set usually produces pseudo-random binary sequence (PRBS) as the test bit pattern. The PRBS is a periodical bit pattern, but there are a large number of bit combinations within each period; they compromise between the randomness of the data signal carried in practical systems and the repetitiveness that simplifies the measurements. Major parameters to specify a PRBS bit pattern include sequence length in bits, maximum continuous digital 1 s, and maximum continuous digital 0 s.

By definition, *pseudo*-random implies that the pattern is not really "random" and in fact is only quasi-random. The PRBS pattern generator generates a random pattern with a certain length and the pattern repeats itself after every pattern length, as illustrated in Figure 5.1.13. Obviously, within each bit pattern, the combination of bits should be as random as possible to simulate actual digital data traffic. This requires the length of the pattern to be long enough. Generally, a long pattern length allows the use longer continuous 1 s and continuous 0 s. This helps stretch the test to the worst case of the system.

For example, a clock recovery circuit usually works well for a pattern with alternative 0 s and 1 s but not as well for long continuous 0 s or 1 s because there is much less clock frequency components. In addition, the frequency spectrum of a PRBS consists of an envelope that is a *sinc* function determined by the time-domain waveform of each individual bit and bit duration (for example, 100 ps for a 10 Gb/s datarate), whereas the frequency spacing between adjacent spectral lines is determined by the inverse of the pattern length, $\Delta f = f_b/N_b$, as shown in Figure 5.1.14, where f_b is the bit rate and N_b is the pattern length (the number of bits per pattern). Obviously, a longer pattern length results in a narrower spacing between spectral lines, which is equivalent to a more closely spaced sampling in the frequency domain.

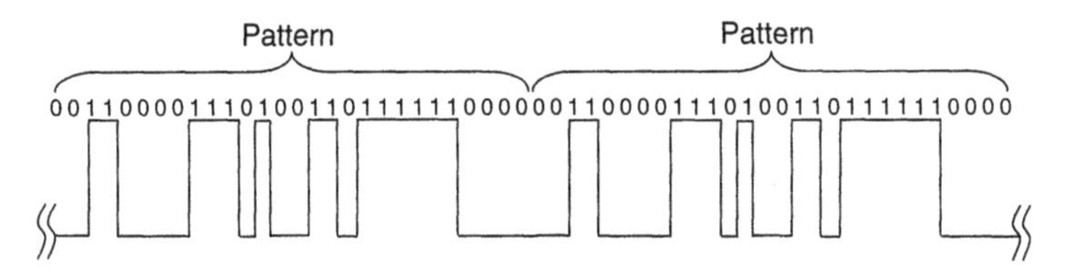

Figure 5.1.13 Nonreturn-to-zero PRBS pattern.

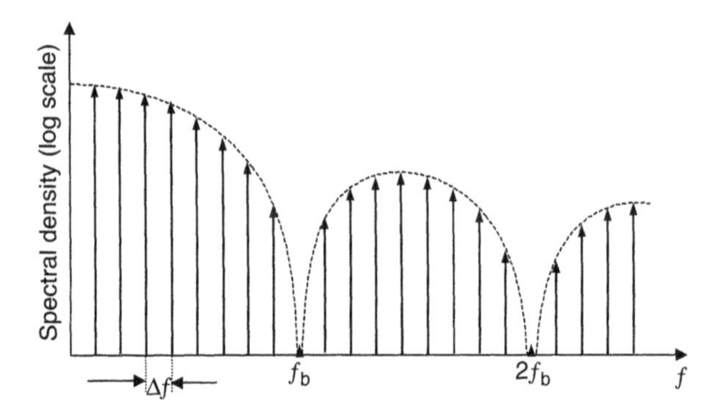

Figure 5.1.14 Illustration of the spectrum of a PRBS signal with NRZ modulation.

From the system transfer function point of view, a probe signal with densely spaced frequency components would be desired for system performance evaluation. For example, if the data rate is 10 Gb/s, to test system transfer function for frequencies as low as 100 kHz the number of bits in each PRBS pattern has to be at least 10^5.

PRBS patterns have been standardized by the ITU for testing digital transmission systems. The most commonly used patterns in digital transmission system testing are $(2^N - 1)$ with $N = 7$, 10, 15, 20, 23 and 31. The corresponding pattern length (sometimes referred to as *word length*) is 127, 1023, 32767, 1048575, 8388607, and 2.1475×10^9 bits, respectively, per pattern. In a typical implementation, PRBS patterns are generated using shift registers with feedback as shown in Figure 5.1.15, where $D_1, D_2, \ldots D_N$ are shift registers.

To explain the operation of PRBS sequence generation, Figure 5.1.16 shows the simplest case, with $N = 3$ and $m = 1$, where m is the number of shifting shown in Fig, 5.1.15. In this example, suppose that the initial states of the shift registers are $D_1 = 1$, $D_2 = 1$, and $D_3 = 1$. According to the truth table shown in Figure 5.1.16(b), the output of the exclusive OR gate should be $C = 0$. Then in the next time slot, the states become $D_1 = 0$, $D_2 = 1$, and $D_3 = 1$ and the output of the exclusive OR gate becomes $C = 1$. As shown in Table 5.1.3, this process continues until the 7th bit slot, where $D_1 = 1$, $D_2 = 1$, and $D_3 = 0$; this makes the exclusive OR gate output $C = 1$. After that the states of the shift registers repeat themselves.

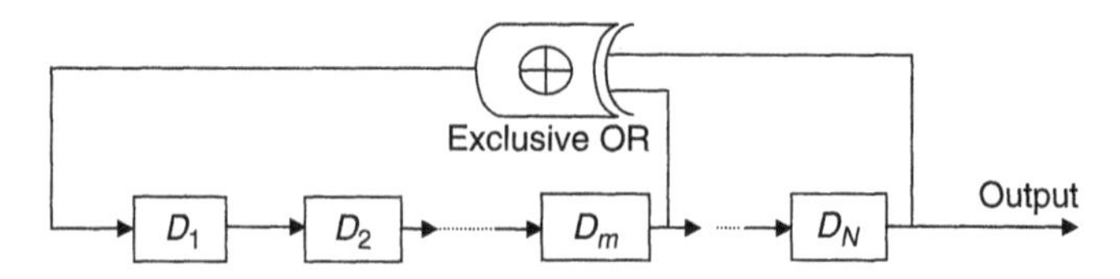

Figure 5.1.15 PRBS generation using shift registers with feedback.

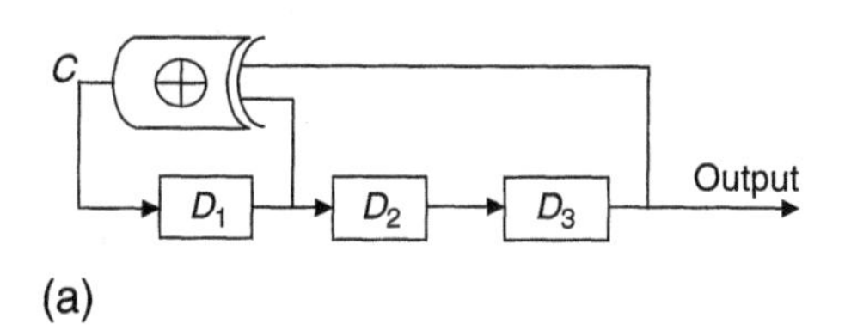

Exclusive OR

D_1	D_3	C
0	0	0
1	0	1
0	1	1
1	1	0

(b)

Figure 5.1.16 (a) Simplest PRBS generator with $N = 3$ and (b) truth table of the exclusive OR gate.

In this example, since $N = 3$, the total number of bits per pattern is $N_b = 2^3 - 1 = 7$. This is the main reason that $2^N - 1$ is used as the word length for PRBS instead of simply 2^N. Another reason for using $2^N - 1$ as the standard pattern length is that the pattern repetition rate is not harmonically related to the data rate. In a $2^N - 1$ PRBS bit pattern, the lowest frequency component is

$$f_{start} = \frac{f_b}{N_b} = \frac{f_b}{2^N - 1} \tag{5.1.25}$$

Table 5.1.3

Logical States of Shift Registers for a PRBS Generator with N 3

Bit	C	D_1	D_2	D_3	Output = D_3
1	1 + 1 = 0	1	1	1	1
2	0 + 1 = 1	0	1	1	1
3	1 + 1 = 0	1	0	1	1
4	0 + 0 = 0	0	1	0	0
5	0 + 1 = 1	0	0	1	1
6	1 + 0 = 1	1	0	0	0
7	1 + 0 = 1	1	1	0	0
8	0	1	1	1	1
9	1	0	1	1	1
10	0	1	0	1	1
11	0	0	1	0	0
12	1	0	0	1	1
13	1	1	0	0	0

For example, for a $2^7 - 1$ PRBS bit pattern at 10 Gb/s data rate, the lowest frequency component is about 78 MHz. If $2^{31} - 1$ PRBS is used at the same data rate, the lowest frequency will be as low as 4.65 Hz. This gives a guideline for choosing RF amplifiers of digital transmission equipment.

In most commercial BER test sets, in addition to the selection of pattern length, there are usually a number of other choices to further specify the PRBS pattern, such as mark-density pattern, zero-substitution pattern, and manual input pattern. Mark-density pattern allows the user to vary the ratio of the numbers of 1 s (marks) and 0 s (spaces) in the PRBS pattern. It is usually used to test system response to unequal mark and space distributions. Zero substitution pattern allows a certain length of sections within the pattern to be filled with spaces. Usually this technique is used to test clock recovery circuitry in the receiver. An exceptionally large percentage of spaces in the digital signal usually makes it more difficult to recover the clock. Manual-input pattern allows the user to manually key in the desired patterns. This is useful if we want to test a specific sequence of marks and spaces.

5.1.4.2 Error Detection

In a BER test set, the PRBS signal from the pattern generator is split into two parts; one of them passes through the system under test while the other part is used as the reference. The basic error detection process involves the comparison between the data stream that passes through the system under test and the reference signal in an exclusive OR gate, as illustrated in Figure 5.1.17. When both data patterns are identical and timely synchronized, the exclusive OR gate gives a zero output. Otherwise, the output from the exclusive OR gate is a binary 1, which indicates an error. It is also often desirable for a BER test set to be able to identify the nature of the errors between transmitting 0 but receiving 1 and transmitting 1 but receiving 0. This functionality in a BER test set may aid in troubleshooting and finding the sources of errors.

To correctly compare the signal received from the transmission system and the reference pattern, they have to be perfectly synchronized in time; otherwise

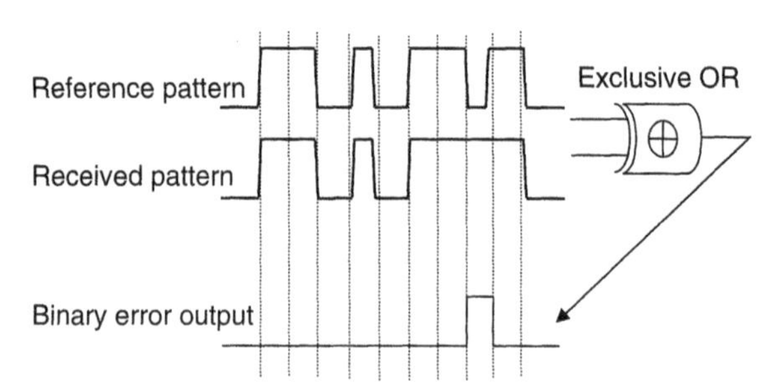

Figure 5.1.17 Bit error testing.

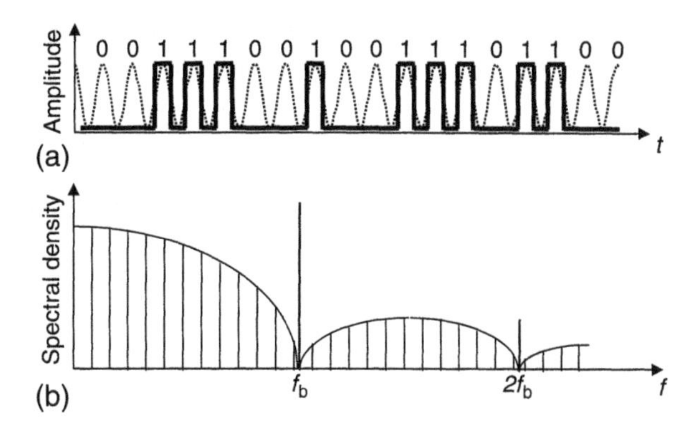

Figure 5.1.18 (a) Time-domain waveform and (b) its spectrum of an RZ modulated signal. Dashed line in (a) is the clock waveform.

the comparison would be invalid. Clock recovery from the received PRBS signal involves the selection of the clock frequency component using a narrowband filter and stabilizes the phase with a phase locked loop. As illustrated in Figure 5.1.18, for a signal with return-to-zero (RZ) modulation format, the signal waveform returns to 0 in each bit and the continuous 1 pattern has exactly the same periodicity as the clock. Therefore the signal spectrum contains a strong clock frequency component. In this case, a simple narrowband filter at the clock frequency can pick up the clock component for clock recovery.

On the other hand, for signals with non-return-to-zero (NRZ) modulation format, as illustrated in Figure 5.1.19, the signal waveform maintains a constant level over an entire bit length; therefore the fundamental frequency of an NRZ

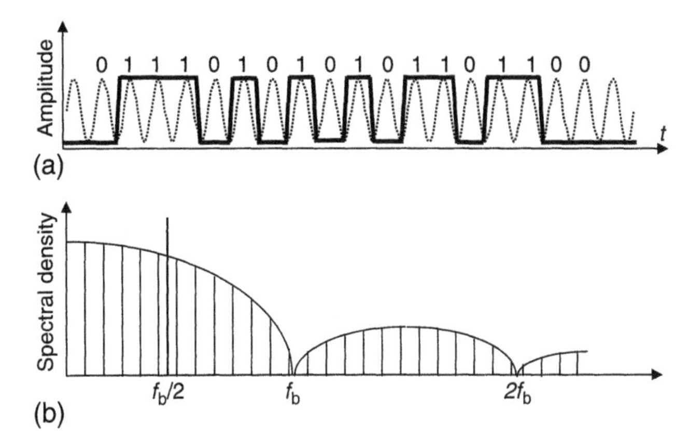

Figure 5.1.19 (a) Time-domain waveform and (b) its spectrum of an NRZ modulated signal. Dashed line in (a) is the clock waveform.

signal is equal to one half the clock frequency. This corresponds to the 010101. . .bit pattern. The spectrum of an NRZ signal has no energy at the clock frequency. Therefore clock extraction from an NRZ signal is more complicated than that from RZ. Typically, to recover the clock from an NRZ data pattern, the narrowband filter has to select the frequency component at half data rate, which is the fundamental frequency of the NRZ signal. Then a nonlinear circuit has to be used to perform frequency doubling to recover the actual clock.

Figure 5.1.20 shows the block diagram of a clock recovery circuit for an NRZ signal, where a narrowband filter selects the fundamental frequency component at the half bit rate, $f_b/2$, and a nonlinear circuit doubles this frequency to f_b. Then this single-frequency signal is compared with a sinusoid signal generated by a voltage-controlled oscillator (VCO) and their frequency difference is used as the error signal to control the VCO. This ensures that the sinusoid generated by the VCO has exactly the same frequency and phase as the clock extracted from the incoming signal. This also allows the clock to be recovered from PRBS signals with long zeroes.

In practice, although the clock recovery functionality may be available in a BER test set, clock recovery is usually not necessary for BER testing when the pattern generation and the error detector are at the same location and the clock can be obtained directly from the master clock generator inside the BER test set. As shown in Figure 5.1.21(a), in this case the common clock used to synchronize the received signal and the reference signal is regarded as the "cheat' clock because it is not actually recovered from the signal transmitted through the testing system. On the other hand, clock recovery is usually an important part of an optical receiver. If the system performance testing requires the inclusion of the clock recovery circuitry performance, the clock recovery circuit inside the optical receiver is regarded as part of the system to be tested, as illustrated in Figure 5.1.21(b).

In addition to clock recovery, precise synchronization between the received signal and the reference signal is also very important in bit error detection. Obviously, without synchronizing the received signal with the reference, it is impossible to make a comparison between them and to decide the rate of errors. Figure 5.1.22 illustrates the block diagram of the circuitry for synchronization in an error detector. The relative time delay between the received signal and the reference is actively adjusted by a synchronization controller, which is

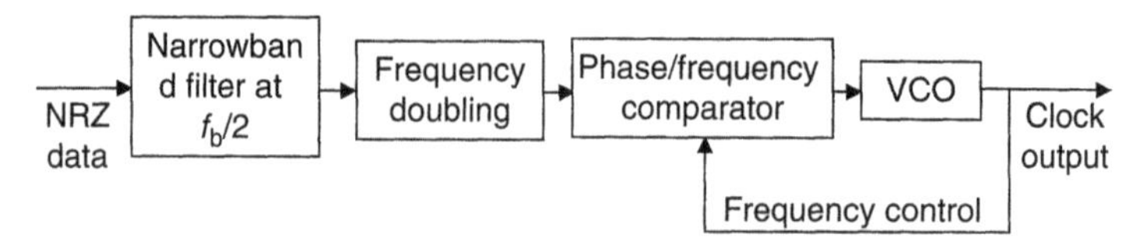

Figure 5.1.20 Block diagram of a clock recovery circuit for NRZ signal. *VCO*: voltage controlled oscillator.

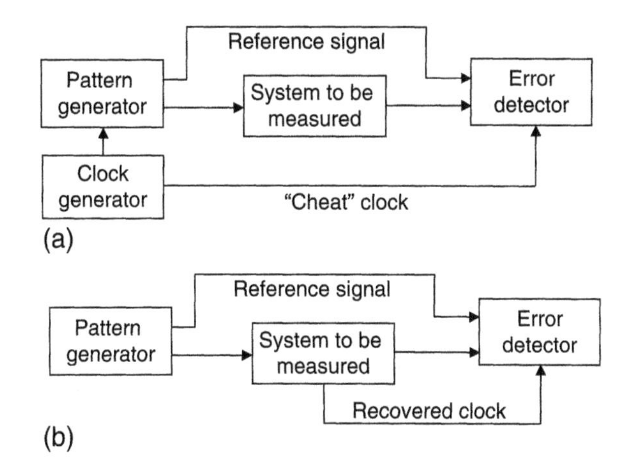

Figure 5.1.21 BER testing (a) using a cheat clock or (b) recovered clock from the system.

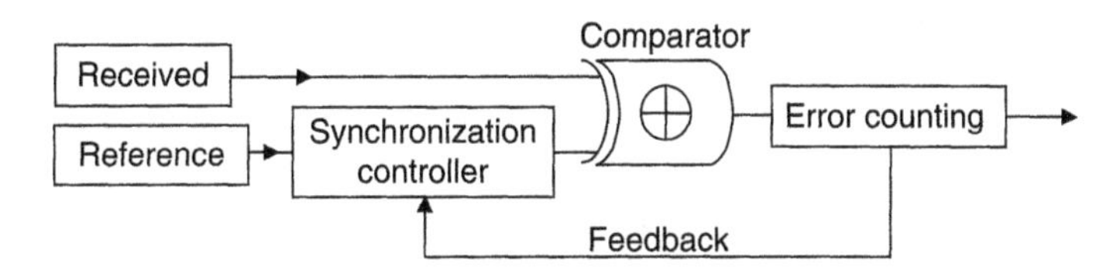

Figure 5.1.22 Synchronization between the received signal and the reference.

usually made by a tunable digital delay line. After proper synchronization, the two signals are compared in an exclusive OR gate so that their difference can be found, which determines the bit errors. If the received signal and the reference are not perfectly synchronized, significant errors will certainly occur. The measured error counts are then used as feedback to optimize the synchronization through minimizing the errors. In practice, at the beginning of each BER test, the synchronization has to be performed; after this process is done, the optimum value of the relative delay is determined and the BER testing can begin.

In most commercial BER test sets, the synchronization can be manually controlled if for some reason one does not trust the optimization process embedded in the test set. Generally the maximum time delay the synchronization controller has to provide should not be shorter than the word length of the PRBS. In fact, beyond the PRBS word length, the patterns will repeat themselves. In practical measurements of long-distance optical systems, the transmission delay may slowly vary. In this case, synchronization may have to be performed periodically. Usually, if the synchronization is not optimum, the degradation in the measured BER will be significant. Theoretically, for a PRBS signal with equal probability of marks and spaces, the measured BER will be equal to 0.5 if the reference is out of synchronization. Usually a threshold can be set for the BER—for example, at BER = 0.1, beyond which "out of synchronization" is declared. This is referred to as *synch loss*.

Another important issue in error detection of a digital signal is the optimization of the decision threshold and the decision timing. Due to various linear and nonlinear impairments in the transmission channel, the eye diagram is not symmetric in the vertical direction; it might be asymmetric in the horizontal direction as well. We must also choose the decision time (also referred to as the *decision phase*) at which the decision is made for whether the received signal is a mark or a space. Obviously, this decision instant should correspond to the time at which the eye diagram has the widest opening, as illustrated in Figure 5.1.5. The decision threshold is a voltage value. If the received signal is higher than this voltage it is declared as a mark; otherwise it is interpreted as a space. If the system under test is a complete system that includes clock recovery and a decision circuit, the setting of the decision time and decision threshold is performed automatically inside the receiver. But if the system does not include a decision circuit in the receiver, the decision time and threshold have to be optimized carefully in the adjustment of BER testing. In a typical BER test set, the decision threshold is usually adjustable either manually or automatically to optimize the BER performance. The selection of a decision threshold depends on the nature of the received signal. As illustrated in Figure 5.1.5, the optimum decision threshold may not be at the exact middle between the mark and the space levels.

5.2 RECEIVER SENSITIVITY MEASUREMENT AND OSNR TOLERANCE

In a fiber-optic transmission system, the receiver has to correctly recover the data carried by the optical carrier. Because the optical signal that reaches the receiver is usually very weak after experiencing attenuation of the optical fiber, a high-quality receiver has to be sensitive enough. In addition, high-speed and long-distance optical transmission systems suffer from waveform distortion, linear and nonlinear crosstalk during transmission along the fiber, and the accumulated ASE noise due to the use of inline optical amplifiers. A high-quality receiver also has to tolerate these additional quality degradations of the optical signal. The criterion of optical receiver qualification depends on the configuration of the optical system and the dominant degradation sources. In this section we discuss receiver sensitivity and receiver OSNR tolerance, two useful receiver specifications, and the techniques to measure them.

5.2.1 Receiver Sensitivity and Power Margin

Receiver sensitivity is one of the most widely used specifications of optical receivers in fiber-optic systems. It is defined as the minimum signal optical power level required at the receiver to achieve a certain BER performance. For example, in a specific optical system, for the BER to be less than 10^{-12}, the minimum signal optical power reaching the receiver has to be no less than -35 dBm; therefore the receiver sensitivity is -35 dBm. Obviously the definition of receiver sensitivity depends on the targeted BER level and the signal data rate. However, signal waveform distortion and optical SNR are, in general, not clearly specified in the receiver sensitivity definition. In fact, the receiver sensitivity measurement assumes that the noise generated in the receiver is the major limiting factor of receiver performance.

Neglecting waveform distortion and the impact of crosstalk in the optical signal for a moment, the Q-value will only depend on the signal power and the noise generated by the receiver. For example, in an intensity-modulated system with direct detection, if no optical amplifier is used in the system, the receiver sensitivity is mainly limited by receiver thermal noise, shot noise, and photodiode dark current noise. In this case, Equation 5.1.19 can be written as

$$Q = \frac{2\Re P_{ave}}{\sqrt{(4kT/R_L + 4q\Re P_{ave} + 2qI_D)B_e} + \sqrt{(4kT/R_L + 2qI_D)B_e}} \tag{5.2.1}$$

Figure 5.2.1 shows the calculated receiver Q-value as a function of the received average signal optical power P_{ave}. This is a 10 Gb/s binary system with direct detection, and the electrical bandwidth of the receiver is $B_e = 7.5$ GHz. Other parameters used are $\Re = 0.85\ mA/mW$, $R_L = 50\Omega$, $I_d = 5$nA, and

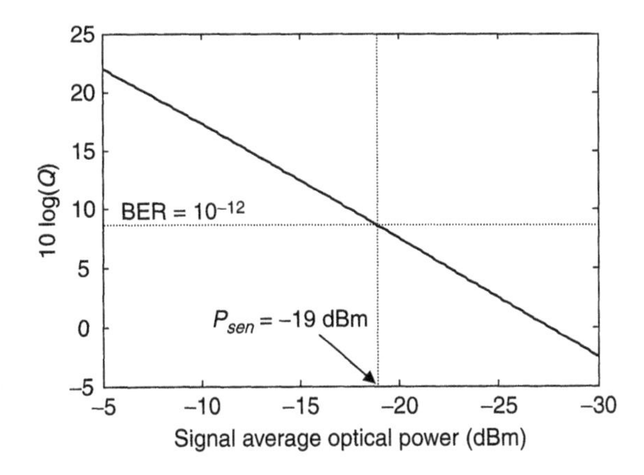

Figure 5.2.1 Receiver sensitivity plot for a 10 Gb/s system using a PIN photodiode.

$T = 300K$. Figure 5.2.1 indicates that to achieve a targeted BER of 10^{-12}, or equivalently $Q = 7$ ($10\log(Q) = 8.45$ dB), the received average signal optical power has to be no less than -19 dBm. Therefore the sensitivity of this 10 Gb/s receiver is -19 dBm. Every dB decrease in signal optical power will result in a dB decrease of the Q-value, as indicated in Figure 5.2.1.

Another type of optical receiver employs an optical preamplifier before the PIN photodiode as illustrated in Figure 5.2.2(b). The optical preamplifier increases the signal optical power before it reaches the photodiode, while at the same time introducing ASE noise. In this case, the level of the ASE noise is proportional to the gain of the optical preamplifier; therefore the receiver Q-value still decreases with the decrease of the input signal optical power. In the preamplified optical receiver, the Q-value can be calculated by

$$Q = \frac{2\Re P_r}{\sqrt{\left(2q(2\Re P_r + I_d) + \frac{4kT}{R_L} + 4\rho_{ASE}\Re^2 P_r + \rho_{ASE}^2\Re^2(2B_o - B_e)\right)B_e} + \sqrt{\left(2qI_d + \frac{4kT}{R_L} + \rho_{ASE}^2\Re^2(2B_o - B_e)\right)B_e}} \tag{5.2.2}$$

where the parameters are defined by Table 5.1.2. Again, the waveform distortion is neglected and the average signal power is equal to ½ of the instantaneous power at signal digital 1. Suppose the amplified signal optical power that reaches the PIN photodiode is fixed at $P_r = -3$ dBm; the gain of the optical preamplifier becomes a function of the input signal optical power, as does the ASE noise level. For an optical amplifier with an $F = 5$ dB noise figure which corresponds to $n_{sp} = 1.58$, the ASE noise spectral density is

$$\rho_{ASE}(f) = 2n_{sp}\text{hf}[G(\lambda) - 1] = 4 \times 10^{-19}\left(\frac{P_r}{P_{ave}} - 1\right) \tag{5.2.3}$$

in the unit of Watt per Hertz. Figure 5.2.3 shows the calculated receiver Q-value as a function of the received average signal optical power P_{ave} at the input of

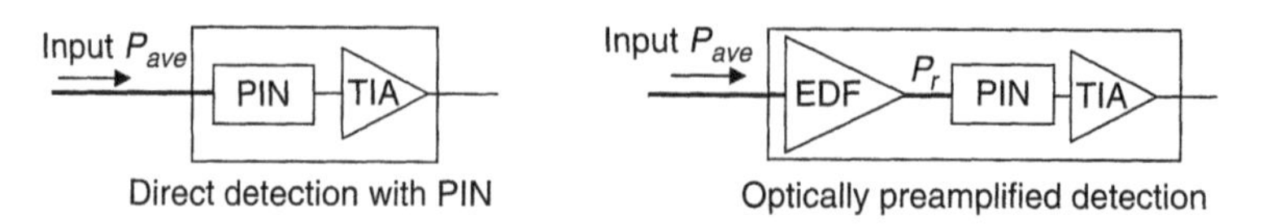

Figure 5.2.2 Direct detection receivers with and without optical preamplifier.

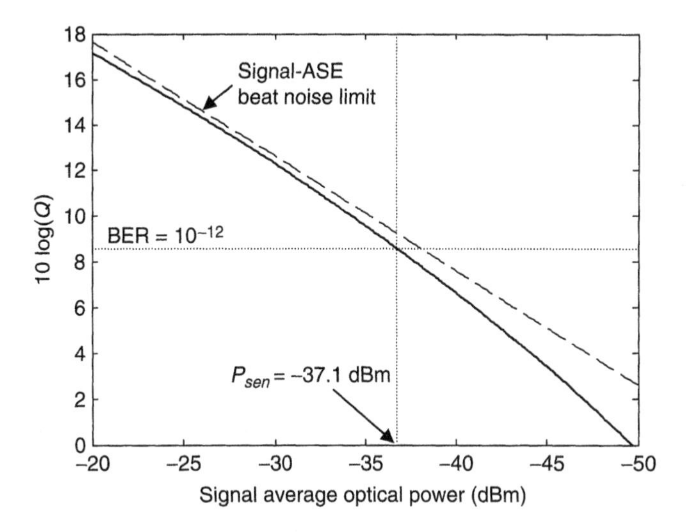

Figure 5.2.3 Receiver sensitivity plot for a 10 Gb/s system with an optically preamplified PIN receiver.

the EDFA preamplifier. The parameters used in the calculation are $R_L = 50\Omega$, $I_d = 5$ nA, $T = 300$ *K*, $B_o = 25$ GHz, $B_e = 7.5$ GHz, and $\lambda = 1550$ nm. In this case the receiver sensitivity is $P_{sen} = -37.1$ dBm, which is approximately 18 dB better than the direct detection without the EDFA preamplifier.

For every dB signal optical power decrease, there is only half a dB decrease in 10log(Q). The reason is that the optically preamplified receiver is dominated by the signal-ASE beat noise, which is signal-dependent. In fact, the dashed line in Figure 5.2.3 shows the Q values only considering signal-ASE beat noise, which is obviously a good approximation. Except when the signal optical power is too high or too low, where the signal-ASE beat noise is comparable to or lower than the PIN thermal noise or the ASE-ASE beat noise, then the linear approximation becomes inaccurate.

Figure 5.2.4 shows the measurement setup to characterize the receiver sensitivity in an optical transmission system. A variable optical attenuator (VOA) is used to change the level of signal optical power that reaches the receiver. The optical power is monitored by a power meter through a fiber tap. The optical transmitter is modulated by a digital signal generated from a BER test set and the signal recovered by the optical receiver is analyzed by the error detector of the same BER test set. The Q-value of the receiver can be calculated through the BER measurement using $Q = \sqrt{2} \cdot erfc^{-1}(2 \times BER)$, where $erfc^{-1}$ is the inverse complementary error function. By scanning the value of the VOA and thus the optical power level at the receiver, the Q-value as the function of signal optical power can be obtained systematically, and the receiver sensitivity can be

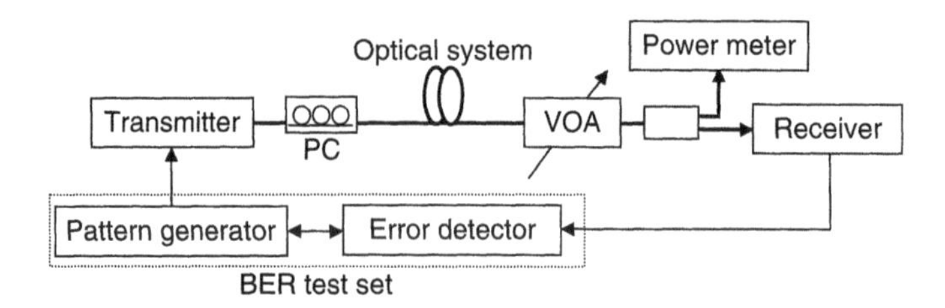

Figure 5.2.4 Schematic diagram of experimental setup to characterize receiver sensitivity.

derived. This is more accurate than only measuring a single power level corresponding to the targeted BER.

Figure 5.2.5 shows a comparison between the calculated and the measured BER as a function of the input signal optical power at the receiver. The 10Gb/s receiver has an optical preamplifier with a 5 dB noise figure. The corresponding Q-value is also indicated along the right-side vertical axis in the figure. It is noticeable that the measured sensitivity at BER $= 10^{-12}$ is approximately 3 dB worse than the calculated results. This is caused by the waveform distortion and eye closure penalty, which are not considered in the calculation. The curve of BER versus signal optical power as shown in Figure 5.2.5 is sometimes referred to as a *waterfall curve* because of its waterfall-like shape.

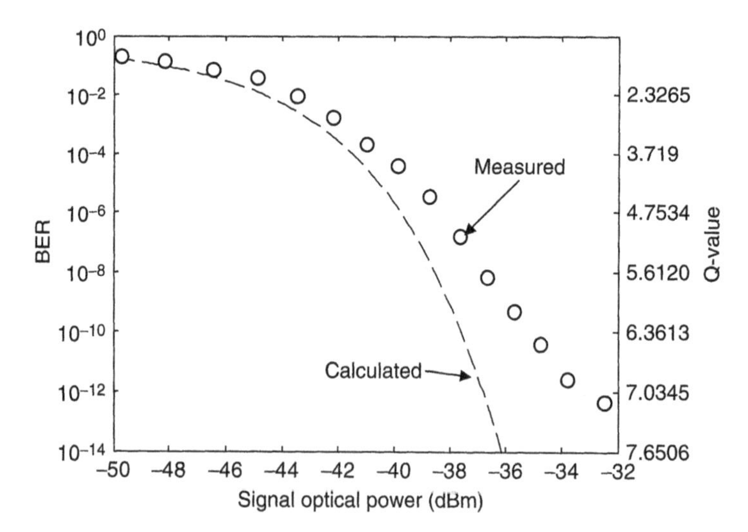

Figure 5.2.5 Illustration of the calculated and measured receiver BER plot.

Accurate prediction of receiver sensitivity must take into account the eye closure penalty caused by waveform distortion, as illustrated in Figure 5.1.6, where $A < 1$ and $B > 0$. More accurately, Equations 5.1.15 and 5.1.16 have to be used to take into account the pattern-dependent waveform distortion effect. In general, the measured receiver sensitivity can be affected by a number of factors. First, different eye openings of the received optical signal may result in different waterfall plots; therefore it has to be precisely specified when reporting the measured sensitivity values. Second, if the optical system has inline optical amplifiers, the accumulated ASE noise in the system will certainly have significant impact in the measured sensitivity at the receiver.

Receiver sensitivity is a useful parameter to find the performance margin of a transmission system. The power margin of a system is defined as the ratio between the available signal optical power and the receiver sensitivity.

$$P_{m\arg in} = P_{actual}/P_{sen} \tag{5.2.4}$$

where P_{actual} is the actual optical power arrives at the receiver, and P_{sen} is the receiver sensitivity. For example, in a transmission system, if the actual signal optical power that reaches the receiver is −25 dBm while the receiver sensitivity is −32 dBm, the power margin is 7 dB. Generally, in optical system design and implementation, a certain power margin has to be budgeted at the time of system installation. This is reserved for system aging and other unexpected degradations during the lifetime of the system. Power margin consideration is an important issue in system design, both technically and economically. If the margin is too small, the integrity of the system transmission performance may not be guaranteed, whereas if the margin is too large, it would be a waste of resources that results in unnecessarily high cost.

Receiver sensitivity and power margin have been widely used to specify the performance of optical receivers and optical transmission systems. In a traditional optical system without inline optical amplifiers, noise generated in the receiver is the dominant source of transmission performance degradation. Therefore, receiver sensitivity and power margins are the perfect measures of the receiver and the system performance with minimum ambiguity. On the other hand, in long-distance and high-speed optical systems using multiple optical amplifiers, signal optical power levels at the photodiode may not be the most important limiting factor in terms of the system performance. For example, the signal optical power can be easily increased by increasing the gain of inline optical amplifiers. In these systems, optical SNR of the signal is more important than the level of signal optical power. Therefore the power margin may no longer be the best performance measure of amplified optical systems. Instead, optical SNR margins may be more relevant.

5.2.2 OSNR Margin and Required OSNR (R-OSNR)

In a multispan optically amplified transmission system, accumulated ASE noise becomes significant when the number of inline optical amplifiers is large. In this case, the system performance is no longer limited by the signal optical power that reaches the receiver; rather, it is limited by the optical signal-to-noise ratio (OSNR). With the use of optical amplifiers, the level of optical power at the receiver can usually be high enough that the thermal noise and dark current noise can be neglected in comparison with signal-ASE beat noise and ASE-ASE beat noise.

If we neglect the thermal noise and dark current noise, considering the effect of waveform distortion, Equation 5.1.18 can be simplified as

$$Q = \frac{2\Re(A-B)P_{ave}}{\sqrt{\left(4\Re(q+\rho_{ASE}\Re)AP_{ave}+\rho_{ASE}^2\Re^2(2B_o-B_e)\right)B_e} + \sqrt{\left(4\Re(q+\rho_{ASE}\Re)BP_{ave}+\rho_{ASE}^2\Re^2(2B_o-B_e)\right)B_e}} \quad (5.2.5)$$

If we further neglect the contribution of ASE-ASE beat noise and the shot noise, Equation 5.2.5 can be further simplified as

$$Q = \frac{(\sqrt{A}-\sqrt{B})\sqrt{P_{ave}}}{\sqrt{\rho_{ASE}B_e}} = \frac{(\sqrt{A}-\sqrt{B})}{\sqrt{B_e}}\sqrt{\text{OSNR}} \quad (5.2.6)$$

Figure 5.2.6 shows the calculated receiver Q-value as the function of the signal OSNR without eye closure penalty ($A = 1$ and $B = 0$) and with eye closure penalty ($A = 0.8$ and $B = 0.25$). The OSNR, based on a resolution bandwidth of 0.1 nm, is the signal optical power divided by the noise power within 0.1 nm optical bandwidth. Other system parameters are P_ave = 3dBm, $\Re = 0.85\ mA/mW$, $B_o = 25$ GHz, $B_e = 7.5$ GHz, and $\lambda = 1550$ nm. The solid lines in Figure 5.2.6 were obtained with Equation 5.2.5, which considered contributions from shot noise, signal-ASE beat noise, and ASE-ASE beat noise; the dashed lines were obtained using Equation 5.2.6, which only considers the contribution from signal-ASE beat noise. In this system, linear approximation with Equation 5.2.6 is reasonably accurate when the OSNR is higher than 15 dB. If the OSNR is too low, ASE-ASE beat noise becomes significant in the Q calculation. The simple relation between receiver Q-value and signal OSNR shown in Equation 5.2.6 suggests that in signal-ASE beat noise limited optical systems, OSNR can be used as an indicator of system performance. From a receiver specification point of view, a high-quality receiver should be able to tolerate low OSNR while still providing acceptable Q-value.

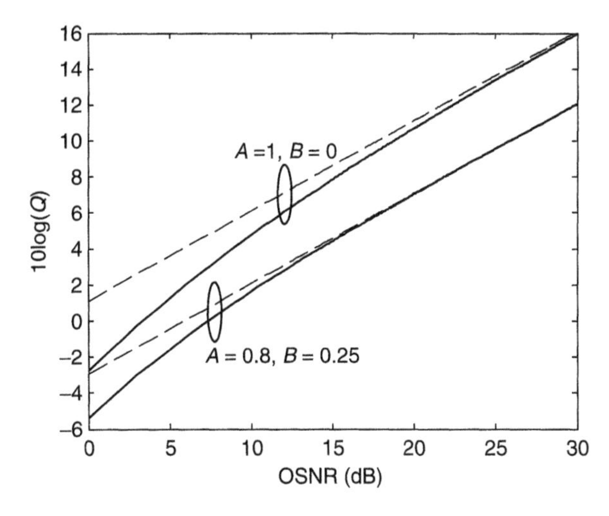

Figure 5.2.6 Receiver Q as the function of signal OSNR with and without eye closure penalty. Solid lines were obtained using Equation 5.2.5 and dashed lines were obtained considering signal-ASE beat noise only using Equation 5.2.6.

Figure 5.2.7 shows the block diagram of an experimental setup that measures the receiver performance against optical signal OSNR. In this measurement, an external ASE source is used, which can simply be an EDFA without input. The ASE noise is added to the optical signal through a fiber directional coupler. The OSNR of the optical signal can be varied by adjusting the attenuation of the VOA which varies the amount of ASE that is added to the system. A band-pass optical filter (BPF) limits the bandwidth of the ASE noise and reduces the impact of ASE-ASE beat noise, which also acts like a WDM demultiplexer in a practical WDM optical system. The OSNR of the optical signal is monitored by an OSA, and the recovered digital signal from the optical receiver is sent back to the BER test set for error detection, which determines the Q-value of the system.

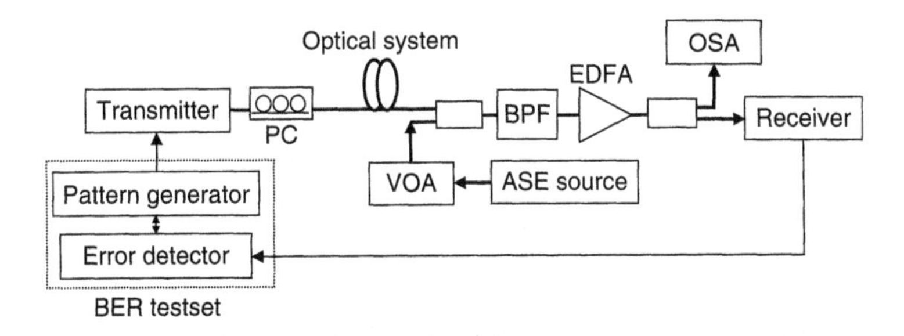

Figure 5.2.7 Schematic diagram of experimental setup to characterize receiver sensitivity.

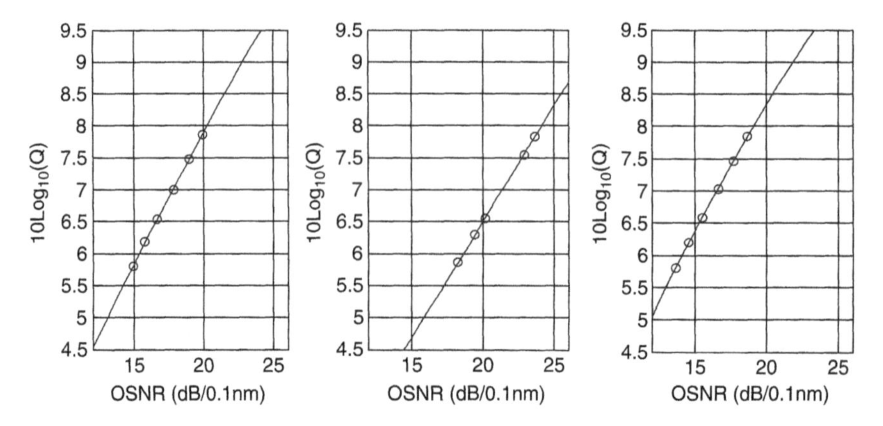

Figure 5.2.8 Examples of measured Q vs. OSNR plots for a 10 Gb/s optical system with different residual chromatic dispersions.

The experimental setup shown in Figure 5.2.7 is commonly referred to as *noise loading technique*; it allows the evaluation of the receiver performance under various optical signal OSNR conditions. Figure 5.2.8 shows examples of the measured receiver dBQ, ($10\log_{10}(Q)$), versus the OSNR of the received optical signal. These were measured in a 10 Gb/s optical transmission system with eight amplified spans of standard SMF. The chromatic dispersion in each span was optically compensated with adjustable dispersion compensation modules. Therefore the accumulated dispersion over the system can be adjusted. For a dBQ of 8.45 ($Q = 7$ or $BER = 10^{-12}$), the required OSNR ranges from 20 dB to 26 dB, depending on the residual chromatic dispersion of the system and thus the effect of eye closure penalty.

Required OSNR (R-OSNR) is a useful parameter to specify an optical receiver. R-OSNR is defined as the OSNR level at which the receiver reaches the targeted Q-value or the targeted BER. An important note is that R-OSNR value depends on the optical spectral resolution in which the OSNR is measured. Practically, a resolution bandwidth of 0.1 nm is usually used for R-OSNR measurement if it is not specifically specified.

OSNR margin is another useful parameter that defines the system performance. OSNR margin is defined as the ratio between the actual OSNR of the optical signal at the receiver and the receiver R-OSNR.

$$OSNR_{margin} = OSNR_{actual}/ROSNR \tag{5.2.7}$$

where $OSNR_{actual}$ is the actual OSNR measured by the receiver. Similar to the power margin in noise-limited optical receivers discussed in the last section, the OSNR margin has to be budgeted in the design of optically amplified systems, taking into account system aging and other unexpected degradation events during the lifetime of the transmission system. OSNR margin specification also directly tells the full potential of the system and how many amplified fiber spans can be added to the system before the limit is reached. The following is a practical example of system OSNR margin measurement in an electrically precompensated ultra-long-distance fiber-optic system [8, 9].

In long-haul optical communication systems, chromatic dispersion is traditionally compensated in the optical domain by placing dispersion compensators in each amplified fiber span. Meanwhile, various electronic means have also been developed to compensate for the effect of chromatic dispersion. These include digital equalization techniques such as feed-forward (FFE) and decision-feedback (DFE) equalizers or maximum-likelihood sequence estimation (MLSE). Electronic dispersion compensation (EDC) can potentially be a low-cost alternative to optical domain compensation thanks to the mass production potential of very large-scale integrated silicon electronic circuits. The example discussed here uses electronic precompensation in the optical transmitter. The transmitter sends out a well-calculated optical signal with predistortion in both amplitude and phase. After transmitting over a long-distance optical fiber system, the signal waveform becomes undistorted at the receiver due to the effect of fiber chromatic dispersion.

Figure 5.2.9 shows the dispersion compensating transmitter in which a tunable laser is used as a CW source. The system is RZ-DPSK modulated at 10 Gb/s. A $LiNbO_3$ dual-parallel Mach-Zehnder modulator (MZM) is used to modulate the optical field (E-field) in the Cartesian plane. To fully control the amplitude and phase of the optical signal, the modulator has two RF inputs for the in-phase (*I*) and quadrature (*Q*) drive signals, respectively. Three bias inputs are also provided, one each for the *I*, *Q*, and quadrature ($\pi/2$) controls. A dual-channel arbitrary waveform generator (AWG), represented by the data encoder and the digital-to-analog converters (DACs) in Figure 5.2.9, is

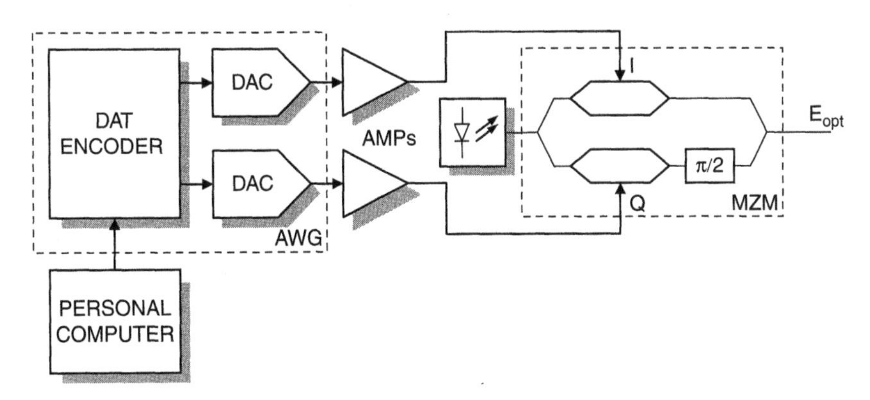

Figure 5.2.9 Block diagram of a dispersion-compensating transmitter. The arbitrary waveform generator (AWG) drives the in-phase (I) and quadrature (Q) ports of the dual-parallel Mach-Zehnder modulator (MZM) [8].

composed of pulse pattern generators, high-speed data multiplexers, wideband RF combiners, phase shifters, and RF attenuators.

Because this system was designed for ultra-long-haul transmission, OSNR margin test is most relevant to evaluate the system transmission performance. Figure 5.2.10 shows the experimental setup for the OSNR margin test using optical noise loading. An optical recirculating loop was used in the experiment and avoided the use of large numbers of optical amplifiers and many spools of optical fibers. The operation of an optical recirculation loop is described in detail in the last section of this chapter. In this particular experiment, the recirculating loop consists of four amplified spans of standard SMF (80 km per span) without dispersion compensation in the optical domain. The total transmission distance of the system can be chosen by selecting the number of circulations in the recirculation loop. The noise figure of the EDFAs used in this experiment was 7 dB. Two average launch power levels, −5 and −7 dBm, were used in the experiment. A Gaussian-shape demultiplexing filter (DMX) with 40 GHz bandwidth placed before the receiver was intended to limit the bandwidth of the ASE noise. A separate ASE noise source provides noise loading at the receiver for the measurement of the BER as a function of the signal OSNR. The average optical power at the input of the receiver was set at −2 dBm.

The DPSK (differential phase-shift-key) receiver is composed of a fiber-based Mach-Zehnder interferometer (MZI) with a free spectral range of

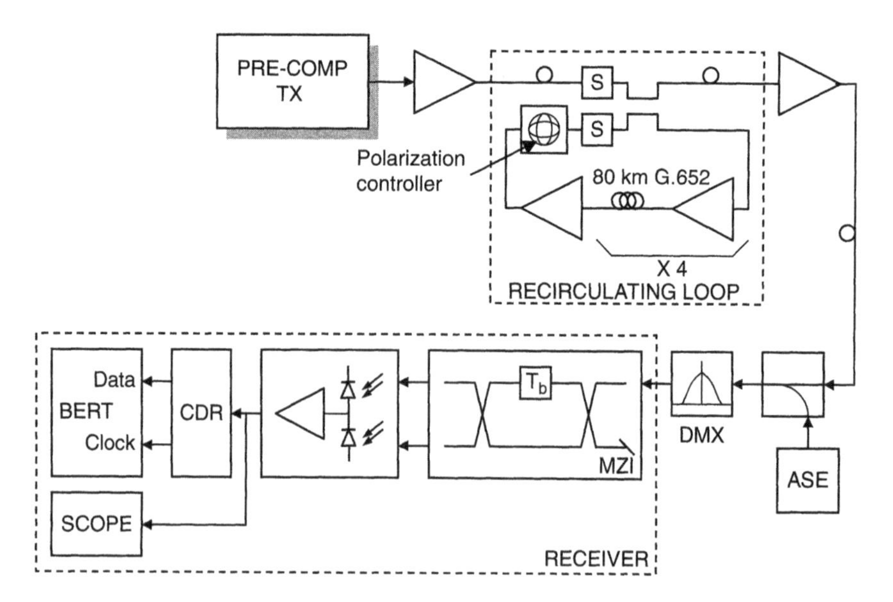

Figure 5.2.10 Experimental setup to measure OSNR margin of the system [8].

10 GHz ($1/T_b$), which converts the received DPSK signal to an intensity-modulated signal. The MZI is followed by balanced PIN detectors integrated with an RF preamplifier and a clock and data recovery (CDR) circuit. The small-signal 3 dB bandwidth of the balanced PIN-preamplifier is 6.5 GHz. The data and clock output of the CDR circuit are then sent to a bit-error rate test set (BERT). Both the sampling phase and threshold level of the CDR circuit are optimized for each measurement of the BER. A sampling oscilloscope is also used to capture eye diagrams at the output of the preamplifier.

Figure 5.2.11 shows the measured eye diagrams at the receiver for back-to-back and after 1600, 3200, and 5120 km, respectively. The measurement was conducted with an average launch power of −7 dBm. For each fiber length, the power of the loading ASE source was varied to adjust the OSNR of the optical signal at the receiver, and thus a BER versus OSNR can be obtained. Figure 5.2.12 shows the measured BER versus OSNR curves for different distances at two average launch powers: −5 and −7 dBm. In this particular system, the lower launch power level (−7 dBm) allows the longest transmission reach, which is 5120 km. This corresponds to an accumulated chromatic dispersion of 82433 ps/nm in the system.

In terms of OSNR margin estimation, traditionally if the targeted BER is 10^{-15}, we must find the OSNR level corresponding to that BER target. It is noticed, however, that in the measured results shown in Figure 5.2.12, the

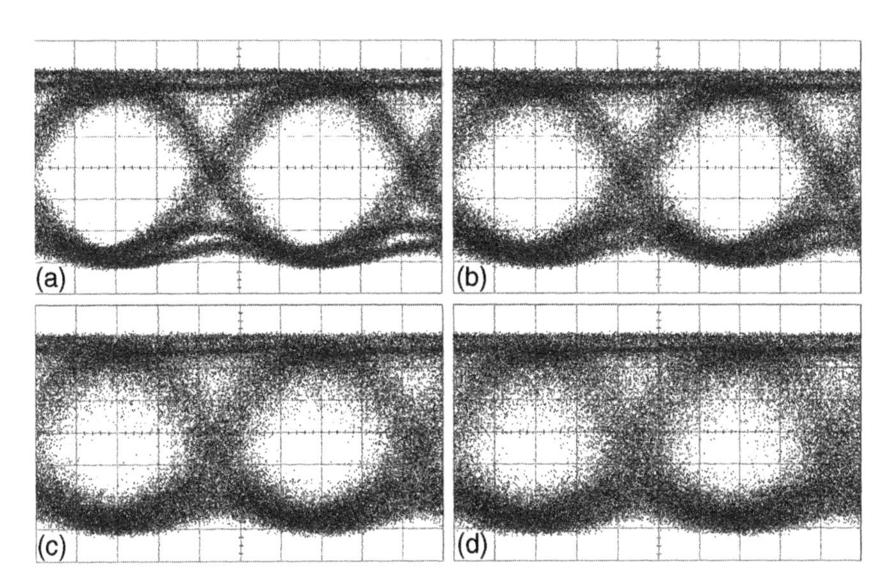

Figure 5.2.11 Eye diagrams at the receiver after propagation through standard SMF. (a) Back-to-back (0 ps/nm), (b) 1600 km (25760 ps/nm), (c) 3200 km (51520 ps/nm), and (d) 5120 km (82433 ps/nm). Launch power is −7 dBm.

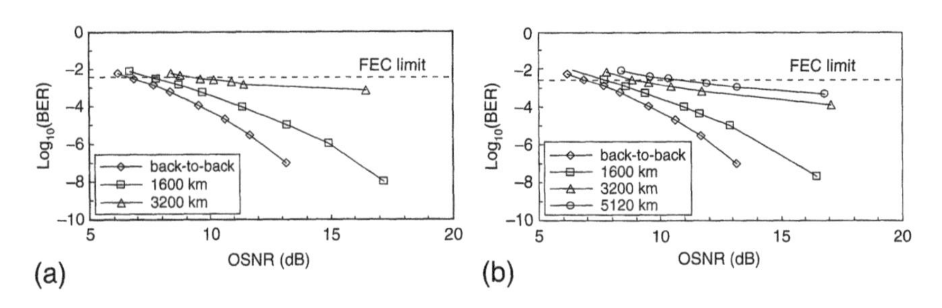

Figure 5.2.12 BER as a function of OSNR in a 0.1 nm resolution bandwidth. Average launch power: (a) −5 dBm and (b) −7 dBm.

minimum BER is only about 10^{-8}. The reason is because the measured BER reported in Figure 5.2.12 did not include the BER improvement due to forward-error correction (FEC). In fact, with the help of FEC, a raw BER of 3×10^{-3} corresponds to 10^{-15} of the corrected BER after FEC. The dashed lines in Figure 5.2.12 indicate the FEC limit of BER $= 3 \times 10^{-3}$.

Based on the measured results shown in Figure 5.2.12, the R-OSNR can be estimated. Figure 5.2.13 shows the R-OSNR as a function of transmission distance. From our previous discussions, the R-OSNR should be independent of the transmission distance if there is no eye closure penalty. The distance-dependent nature of R-OSNR shown in Figure 5.2.13 must be primarily caused by waveform distortion that is not fully corrected by the precompensation in the transmitter. Nevertheless, in this system there is less than 1 dB degradation in the R-OSNR for transmission distances up to 1600 km, whereas at 5120 km, an additional 2 dB R-OSNR is observed.

The maximum OSNR could be achieved in this system was about 17 dB, as shown in Figure 5.2.12, which is determined by the fiber loss and EDFA noise figure. This was obtained with no additional noise from the ASE source. Since

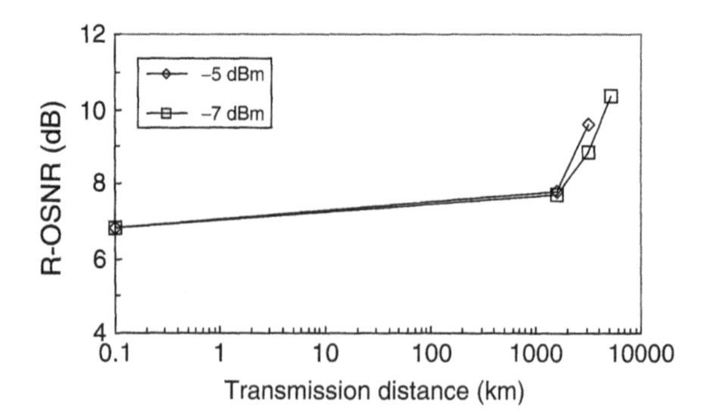

Figure 5.2.13 Required OSNR vs. transmission distance for two average launch powers of −5 dBm and −7 dBm.

the R-OSNR is 10.3 dB for 5120 km transmission distance, the OSNR margin is approximately 6.7 dB.

5.2.3 BER vs. Decision Threshold Measurement

BER is an ultimate performance measure of an optical transmission system. The receiver sensitivity, R-OSNR, and system margins discussed so far were all based on the measurement of BER and the conversion into system *Q*-value. BER measurement is usually time-consuming, especially when low BER levels need to be measured. For a high-quality optical system with a large performance margin, the actual BER could be lower than 10^{-15}. At this low BER level, for a system with a 10 Gb/s data rate, for example, the average time required to observe one error is a million seconds, which is about 11 days. It is simply not feasible to conduct actual BER measurements in these types of systems. Therefore an accelerated testing is necessary for margin evaluation. Noise loading, discussed previously, is one of the techniques for this purpose. Another technique is to measure BER as the function of decision threshold voltage [10], commonly referred to as *BER-V* testing.

Based on Equations 5.1.5 through 5.1.8, the BER can be expressed as the combination of errors from the marks and the spaces:

$$BER = \frac{1}{4}\left\{erfc\left(\frac{v_1 - v_{th}}{\sigma_1\sqrt{2}}\right) + erfc\left(\frac{v_{th} - v_0}{\sigma_0\sqrt{2}}\right)\right\} \tag{5.2.8}$$

where v_{th} is the decision threshold voltage, v_0 and v_1 are mean levels of digital 0 and 1 at the decision phase, and σ_0 and σ_1 are their standard deviations, respectively.

Figure 5.2.14 shows the calculated BER and *Q*-value versus decision threshold for an ideally opened eye ($v_1 = 1$, $v_0 = 0$). The three sets of curves were calculated with ($\sigma_1 = 0.04$, $\sigma_0 = 0.03$), ($\sigma_1 = 0.07$, $\sigma_0 = 0.035$), and ($\sigma_1 = 0.08$, $\sigma_0 = 0.06$), respectively. The BER-V plot not only reveals the noise standard deviations associated with digital 0 and 1; eye closure penalty can also be identified by looking at the BER-V curve near the top and the bottom edges of the eye diagram. Therefore it can help to identify the sources of BER degradation in system design and development.

Figure 5.2.15 shows the experimental setup to measure BER versus decision threshold. Beside a transmitter, a receiver, and an optical transmission system, the only equipment required in this experiment is a BER test set (BERT). In the measurement, one has to disable the automatic threshold searching function of the BERT and use the manual adjustment of the decision threshold. Another important requirement is to find the optimum decision phase where the eye opening is the maximum. By carefully measuring BER at each decision threshold voltage, a BER versus threshold curve like that shown in Figure 5.2.14 can be obtained.

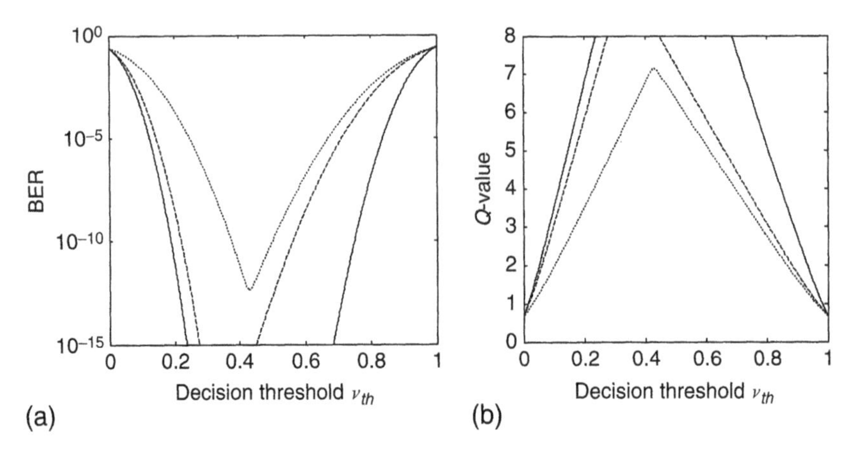

Figure 5.2.14 (a) BER and (b) Q-value versus decision threshold for ($v_1 = 1$, $v_0 = 0$). The three sets of curves were calculated with $\sigma_1 = 0.08$, $\sigma_0 = 0.06$ (dotted line), $\sigma_1 = 0.07$, $\sigma_0 = 0.035$ (dashed line), and $\sigma_1 = 0.04$, $\sigma_0 = 0.03$ (solid line).

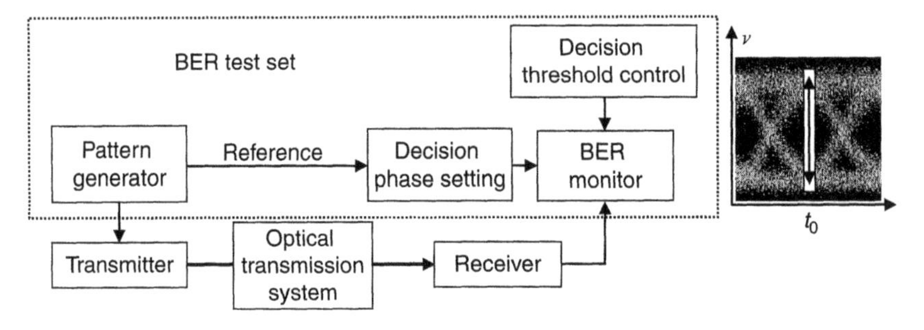

Figure 5.2.15 Block diagram of BER vs. decision threshold testing. t_0 is the decision time.

Practically, to save the measurement time, we only need to measure BER at convenient levels—for example, not lower than 10^{-9}. The signal characteristics such as v_0, v_1, σ_0, and σ_1 can be obtained by best fitting the measured BER-V data with the calculated curves. These parameters provide a good estimation of the minimum BER of the system, which cannot be directly measured with reasonable efforts and time.

Figure 5.2.16 shows an example of a measured Q versus decision threshold plot that was obtained in a WDM system with 40 Gb/s per channel data rate. The horizontal axis is the value of the decision threshold voltage, which was adjusted in the measurement. The Q-value in the vertical axis was converted from the measured BER using Equation 5.1.14. Because of the quasi-linear relationship between the decision threshold voltage and the Q-value, linear extrapolation can help us find the optimum decision threshold and the highest possible Q.

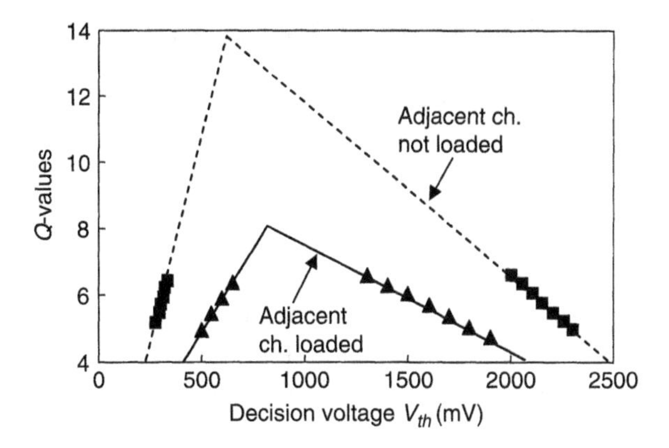

Figure 5.2.16 Example of measured *Q*-*V* plot and linear extrapolation.

In this multichannel dense-WDM system, when adjacent wavelength channels are not loaded (not modulated by data), the eye diagram is wide open, and the extrapolated *Q*-value reaches to approximately 14. With the modulation of the adjacent channels turned on, the crosstalk introduced by interchannel cross-phase modulation (XPM) degrades the eye opening and the extrapolated *Q*-value is significantly reduced to about 8. It is obvious that the slope of the *Q*-*V* line is shallower at digital 1 level in comparison to that at digital 0 level, even without loading the adjacent crosstalk channels. This is due to signal-ASE beat noise in the amplified optical system. The effect of XPM further reduces the slope of the *Q*-*V* line at digital 1 level due to the signal-dependent nature of XPM, which is discussed in Section 5.6.

It is worthwhile to note that the noises in an optical system may not exactly have Gaussian statistics. Therefore Equation 5.2.8 may not perfectly describe the actual system BER performance. Unfortunately, a closed-form expression as simple as Equation 5.2.8 has not been found to predict the exact noise characteristics in the receiver. Since in most practical optical systems, Gaussian approximation provides a reasonably good accuracy in estimating BER performance, it is widely used because of its simplicity.

In optical system specification, an important parameter is the *Q*-margin, which is defined as the difference between the *Q*-value of the system during the measurement and the required minimum *Q*-value to achieve the targeted BER level. *Q*-margin is similar to OSNR margin discussed in the previous section and is used to guarantee the system performance over its specified life span. Because BER-V measurement can predict the estimated system *Q*-value (although it cannot be directly measured), it automatically provides the *Q*-margin information.

5.3 WAVEFORM DISTORTION MEASUREMENTS

In fiber-optic transmission systems, performance degradation can be caused by random noise and deterministic waveform distortion. The ability to identify sources of performance degradation is critical in system design and development. Optical signal waveform measurement at the receiver interface is an important way to estimate system performance. It contains the effects of intersymbol interference (ISI) arising at modulation or as a result of propagation impairments. Combining waveform measurement with separate measurements of noise, system Q can be estimated. For example, eye mask specifications that are based on waveform measurements are commonly used to define minimum acceptable physical layer performance within standard optical formats. Eye mask specification also helps distinguish waveform distortion from random noise in the optical signal.

A generic schematic of a waveform measurement apparatus is shown in Figure 5.3.1. An optical signal under test is amplified, optically filtered and split into two parts for signal detection and clock recovery. The optical filter bandwidth is chosen to exclude broadband ASE noise outside the signal spectrum without introducing ISI and thus limit the ASE-ASE beat noise in optical detection. Part of the optical signal is converted to the electrical domain with a wideband photodetector whose 6 dB bandwidth should be one to two times the bit rate of the optical signal. This converted electrical signal is amplified and filtered with a suitable RF circuit, which imitates a receiver electrical filter of interest, and sent to the input of a digital sampling oscilloscope. A clock signal derived from the other part of the optical splitter provides a synchronous pattern trigger to the oscilloscope after a digital divider.

Alternately, if the transmitter and the receiver are in the same location, the master clock in the transmitter can also be used to trigger the oscilloscope, provided the transmission propagation delay is constant during the full waveform capture. Possible causes of variability in the propagation delay may include link differential group delay and temperature changes in the environment. The trigger pulse rise time

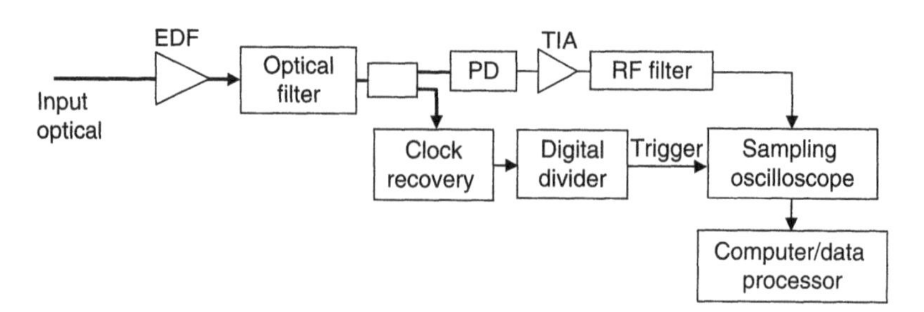

Figure 5.3.1 Experimental setup to measure the waveform of an optical signal using a digital sampling oscilloscope. PD: photodiode, TIA: transimpedance amplifier.

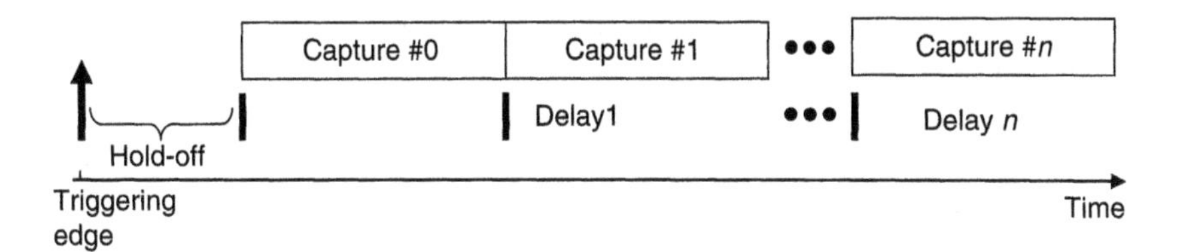

Figure 5.3.2 Timing diagram for segmented waveform capture.

is commensurate with the rise time of the waveform being measured. The number of bits between triggers is controlled by the divider ratio, and the data content of the optical signal is arranged to be repetitive with respect to the trigger. The time between samples is chosen to avoid aliasing of the shortest-lived significant feature in the waveform.

The timing schematic for waveform capture is shown in Figure 5.3.2. The trigger event initiates the acquisition sequence for each datum. The oscilloscope time base is user controllable after a minimum hold-off time designed to ensure adequate settling of circuits such as the ADCs. The window of acquisition is controlled by choice of delay and time-base scale. Long pattern waveforms can be recorded by concatenation of waveform segments captured using a sequence of equally spaced delays. Since the PRBS signal is periodical and repeats itself after each pattern length, noise from ASE and electronic sources can be reduced by averaging over a large number of captures.

In high-speed waveform measurement using a digital sampling oscilloscope, one of the most important issues is time-base calibration. The waveform is often time sampled, but the sampling events may be nonuniformly distributed within a capture. The origin of this nonuniformity may come from the nonlinearity of the oscilloscope time base. Although these nonuniformities are uncorrelated, they are typically reproducible, which contributes to an artificial distortion of the measured waveform. To minimize this artificial distortion on the measured waveform introduced by the oscilloscope, a careful calibration could be very helpful.

A measurement example of time-base calibration is schematically shown in Figure 5.3.3. A stable frequency synthesizer is split into two parts. One part is

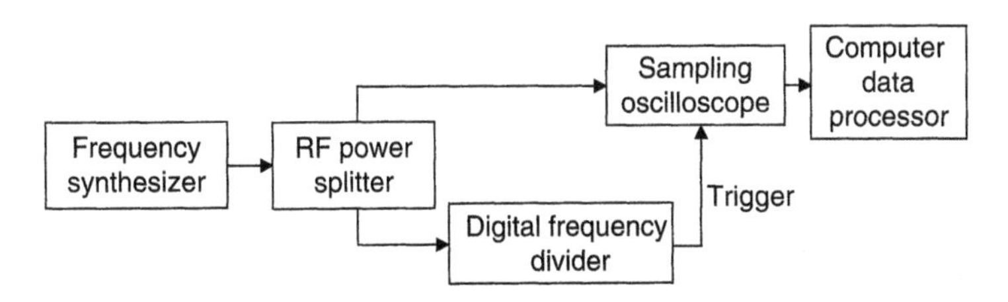

Figure 5.3.3 Experimental setup for oscilloscope time-base calibration.

directly connected to the oscilloscope input port; the other part is sent to a digital divider that generates synchronized low-frequency harmonics of the original signal. This frequency-divided signal is then used as a trigger to synchronize the oscilloscope time base. The time-base error of the oscilloscope can be estimated from the differences between the known sine wave and its measured counterpart on the oscilloscope. Figure 5.3.4 shows an example of a time-base calibration carried out using a 10 GHz sine wave reference. The sampling rate was set at 10 samples per cycle, or equivalently, 10 ps per sample. Each capture window, made up of 4096 samples, is the average of 64 measurements. The expanded view of this figure displays the difference between captured waveform zero crossing times and times dictated by the reference 10 GHz sine wave, which can be regenerated at the computer data processor. Figure 5.3.4 indicates that in this example, the time base can vary by as much as ± 2 ps within each capture window. In addition, there is a slower time-base draft over the longer time span. The calibration processes include the capture of the error waveform, such as the one shown in Figure 5.3.4, and use it to correct the captured waveforms in real measurements.

An eye diagram can be constructed from the captured waveform by overlapping plots, usually two unit intervals (bit length) in duration with a sequential delayed by one unit interval. Therefore the number of lines plotted in the eye

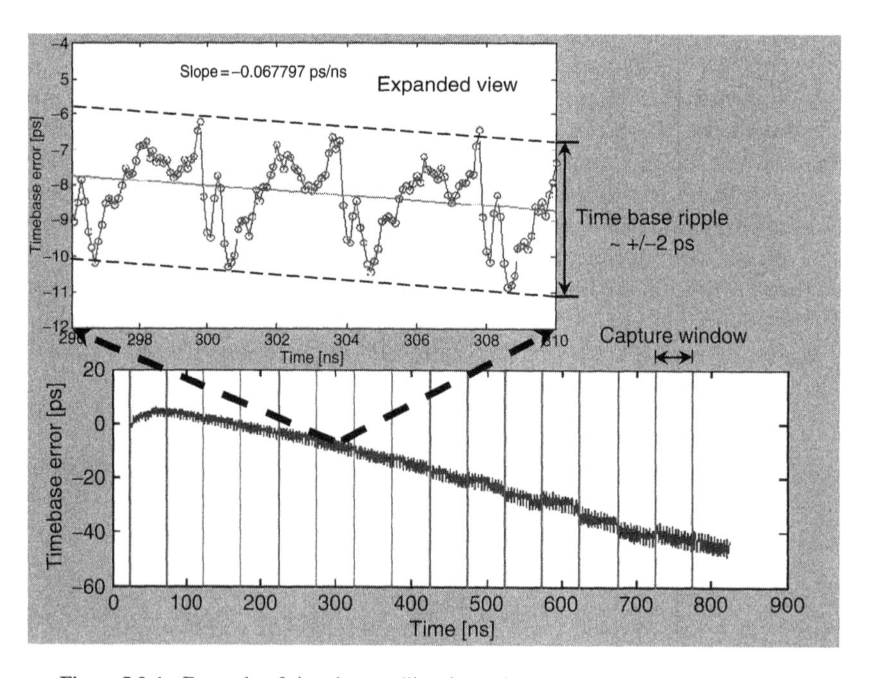

Figure 5.3.4 Example of time-base calibration using a 10 GHz reference sine wave.

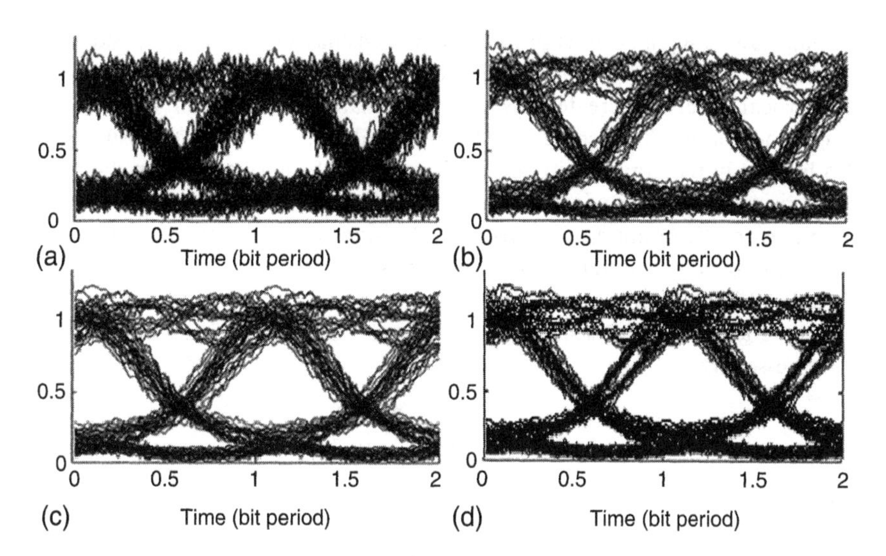

Figure 5.3.5 Example of the measured eye diagram with 2 (a), 16 (b), and 32 averages (c). (d) is the eye diagram with 32 averages and time-base calibration.

diagram equals the number of unit intervals in the waveform. The fidelity of the constructed eye diagram is entirely determined by accuracy of the waveform measurement. A careful calibration of the time base is essential to guarantee the accurate estimation of the eye closure penalty.

As a measurement example, Figure 5.3.5 shows a series of eye diagrams constructed from captured waveforms of a 10 Gb/s transmission system with 1000 ps/nm of accumulated chromatic dispersion. In this measurement, the input signal OSNR was 21 dB, the optical filter bandwidth was 200 GHz, the photodetector electrical bandwidth was 20 GHz, and there was no reference RF filter after photodetection. The oscilloscope was DC coupled. The first three eye diagrams (a), (b), and (c) in Figure 5.3.5 show the effect of noise reduction achieved by averaging, whereas the last diagram (d) was obtained with time-base correction. It can be seen that this last eye diagram puts into evidence a rising double edge in the eye diagram, which was obscured without time-base calibration.

5.4 JITTER MEASUREMENT

5.4.1 Basic Jitter Parameters and Definitions

Time jitter has always been an issue affecting the performance of digital communication systems and networks [11]. This issue is especially important when the data rate is high, such as in fiber-optic transmission systems. Consider the eye diagram of an intensity modulated binary optical signal, the eye-closure in the vertical direction creates system performance degradation because the spaces and the marks become less distinguishable. Similarly, eye closure may also happen in the horizontal direction because of the uncertainties in the times of the rising and falling edges of the signal pulses. This leads to the increase in the bit error rate because of the reduced decision time window at the receiver. This eye closure in the horizontal direction is a signature of time jitter as illustrated in Figure 5.4.1.

Theoretically for a bit sequence with data rate R_b, the jittered waveform can be expressed as

$$P_j(t) = P\left(t + \frac{\Delta\varphi(t)}{2\pi R_b}\right) \tag{5.4.1}$$

where $\Delta\varphi(t)$ is the phase variation introduced by time jitter, which can be in radians or in degrees, and $P(t)$ is the waveform in the absence of time jitter. Naturally, time jitter can also be measured in time, Δt, which is in seconds:

$$\Delta t(t) = \frac{\Delta\varphi(t)}{2\pi R_b} \tag{5.4.2}$$

In terms of the impact of time jitter in system performance, systems with higher data rates are generally more vulnerable to time jitter; therefore a convenient measure of jitter is in the *unit interval* (UI), which is defined as the ratio between the time jitter and the bit period $1/R_b$, that is,

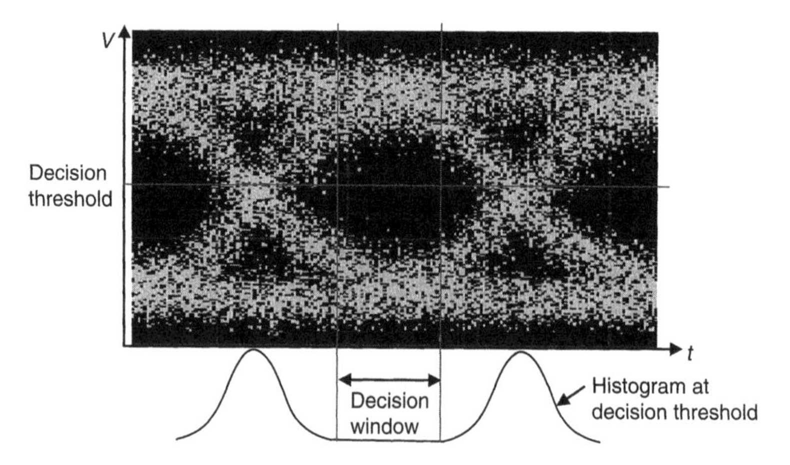

Figure 5.4.1 Illustration of an eye diagram and histograms at the eye crossing edges.

$$\Delta t_{UI} = \frac{\Delta\varphi(t)}{2\pi} \tag{5.4.3}$$

In digital systems, time jitter can be categorized into random jitter and deterministic jitter. Random jitter is caused by noises such as receiver thermal noise, quantum noise, and amplified spontaneous emission noise accumulated throughout the system. To the first-order approximation, the histogram of random jitter can be modeled as a Gaussian process. On the other hand, deterministic jitter is caused by pattern effects mediated by transmission impairments such as chromatic dispersion, SPM, and interchannel crosstalk. The histogram of deterministic jitter is usually bounded and the width of the distribution depends on the specific waveforms as well as on the level of distortion in the waveform. The overall time jitter is the combined effect of the random and the deterministic contributions, and equivalently the histogram should be the convolution of the histograms of these two effects.

As an example, Figure 5.4.2 illustrates eye diagrams with and without noise contributions. Figure 5.4.2(a) shows the NRZ modulated eye diagram without random noise. Because of the waveform distortion introduced by chromatic dispersion, significant time jitter is created, as can be seen at the crossover areas. With NRZ modulation, the signal waveform contains different pulse widths

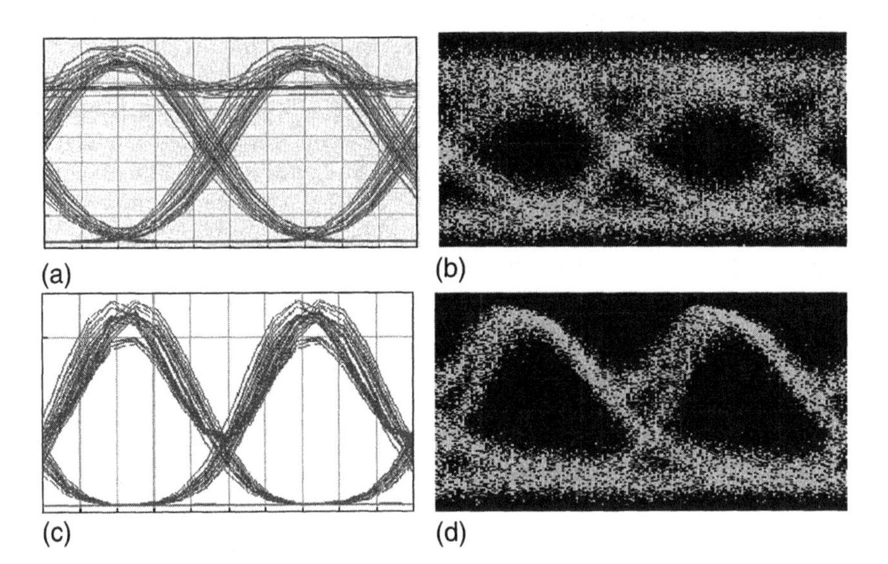

Figure 5.4.2 Eye diagrams without noise (a) and (c) and with noise (b) and (d), for NRZ modulation (a) and (b) and NR modulation (c) and (d).

representing isolated 1 s and continuous 1 s with various lengths. Signals of different pulse widths generally have different responses to chromatic dispersion, therefore creating different amounts of broadening in the time domain. This pattern-dependent effect is shown as the deterministic time jitter. When noise contribution is taken into account, as shown in Figure 5.4.2(b) [not the same waveform as (a)], the waveform is superimposed by the random scattering of the noise. The originally thin waveforms become broadened and the eye opening is thus further reduced. As a comparison, Figure 5.4.2(c) and (d) illustrate eye diagrams of RZ modulated signals. Although pattern-dependent waveform distortion due to chromatic dispersion still exists, it is less significant than that of NRZ. The reason is that the pulse width with RZ modulation is fixed, while only the separations between pulses depend on the data pattern. This reduced deterministic jitter makes RZ a better choice for long-distance fiber-optic transmission.

Time-jitter specifications in a digital system include jitter generation in transmitters, jitter tolerance of the receivers, and jitter transfer of system elements such as transceivers, regenerators, and passive components.

Jitter tolerance is primarily a performance measure of a digital receiver. A high-quality receiver should be able to maintain the targeted BER performance even if the input signal has a significant time jitter. Jitter tolerance is generally a function of jitter frequency; therefore, in jitter tolerance measurement, a digital signal with a variable jitter amplitude and frequency is required. Figure 5.4.3 shows the schematic diagram of an experimental setup for receiver jitter tolerance measurement. In this measurement, a variable jitter source is used whose amplitude and frequency can be adjusted. The time jitter is introduced into a digital data pattern through its clock. This jittered bit pattern is converted into the optical domain through an external electro-optic modulator, and a modulated optical waveform is sent into an optical fiber. Before the optical signal is detected by the receiver under test, its amplitude is adjusted by a variable optical attenuator (VOA). A bit error tester measures the BER at the output of the receiver. First, when the jitter source is switched off, the data

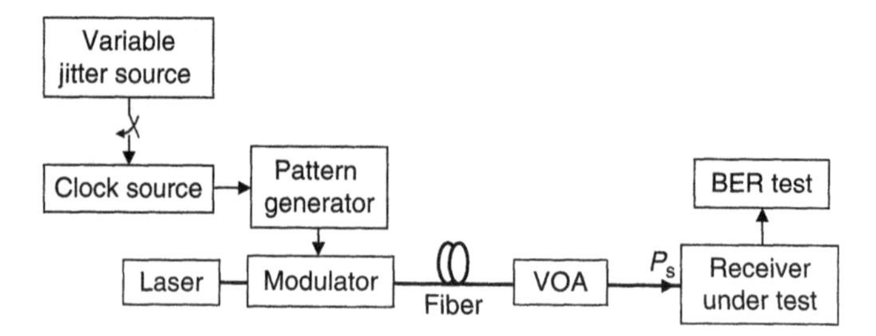

Figure 5.4.3 Experimental setup for receiver jitter tolerance measurement.

pattern is jitter-free. To obtain a targeted BER of, for instance, 10^{-12}, the required optical power at the receiver is P_{s0}. Then when we switch on the jitter source, the BER at the output of the receiver will be degraded with the increase of the jitter amplitude. To achieve the same targeted BER of 10^{-12}, the signal optical power at the receiver has to be increased to, say, P_{sJ}. If the jitter source is sinusoidal, $\Delta\varphi(t) = A_J \sin[\Omega_J(t)]$, where A_J is the jitter amplitude and Ω_J is the jitter frequency, the jitter tolerance of the receiver $\Delta P_{\text{tolrance}}(\Omega_J)$ can be found as the jitter level at which P_{sJ} is 1 dB higher than P_{s0}. This is the jitter level that is equivalent to 1 dB receiver sensitivity degradation.

In receivers for long-distance fiber-optic transmission using inline optical amplifiers, the power penalty is no longer the relevant parameter to specify the performance degradation. As discussed in the previous sections, in these systems the measurement of the R-SNR increase or the reduction in the *Q*-margin may be more appropriate to evaluate the impact of time jitter. This can be done using the ASE noise-loading technique.

Jitter generation usually happens in a transmitter in which the output digital signal is jittered, although the data signal at the input has no jitter. In general, jitter may be generated in any optical and electronic device. Therefore, jitter generation, which is also referred to as *intrinsic jitter*, is defined as the output jitter of a component when there is no jitter at the input. Intrinsic jitter may include both random jitter and deterministic jitter, depending on the nature of the specific component under test. For example, random jitter in an optical transmitter can be caused by the frequency and intensity noises of the laser source, as well as the noise in the modulation circuitry; deterministic jitter in the transmitter may be caused by the nonlinear transfer function of the modulator and the multipath interference in the optoelectronic circuit. Because *rms* value, Δt_{rms}^{RJ}, is the best way to characterize random jitter, whereas deterministic jitter is better characterized by the peak-to-peak value Δt_{pp}^{DJ}, the overall time jitter including these two contributions can be characterized by the combination of Δt_{rms}^{RJ} and Δt_{pp}^{DJ}.

As shown in Figure 5.4.4, the measurement of jitter generation requires a reference clock without jitter, which provides a precise time base. The jittered PRBS pattern generated by the optical transmitter has to be detected by an optical receiver so that it can be compared with the reference clock for jitter detection.

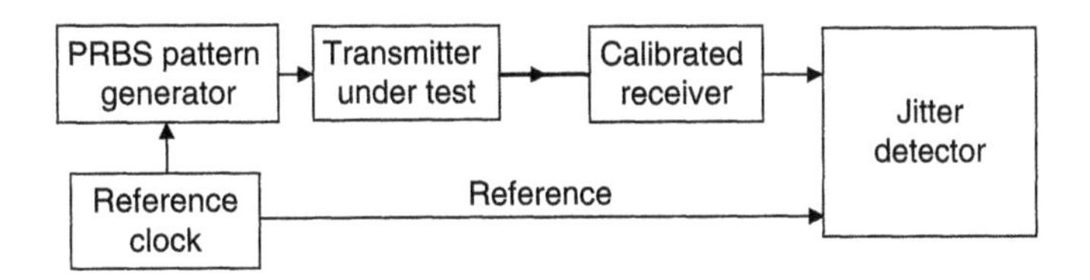

Figure 5.4.4 Measurement of jitter generation.

Jitter transfer is defined as the jitter amplification of a network component from the input to the output. For example, suppose that the input digital signal has a sinusoidal jitter at frequency Ω_J and amplitude $A_{in}(\Omega_J)$. If the jitter amplitude at the output of the network element is $A_{out}(\Omega_J)$ at the same frequency Ω_J, the jitter transfer of this network element is $J_{transfer}(\Omega_J) = A_{out}(\Omega_J)/A_{in}(\Omega_J)$. By sweeping the jitter frequency of the input signal, the frequency response of the jitter transfer can be obtained.

As illustrated in Figure 5.4.5, jitter transfer is a comparison between the input jitter and the output jitter of a network element under test. The jitter source at a certain frequency Ω_J can be introduced to the data clock through a phase modulator so that the PRBS pattern is jittered before been sent to the device under test. The jitter detector has to simultaneously measure the jitter characteristics of both the input and the output waveforms of the device and the jitter transfer function can be calculated. In general, jitter transfer can be measured for electronic devices, optical devices, and optoelectronic devices such as a transceiver or a regenerator. Though Figure 5.4.5 only shows a schematic diagram of jitter transfer measurement of an electronic device, a similar setup can be used for optical device jitter transfer measurement by adding a calibrated (or jitter-free) transmitter/receiver pair.

5.4.2 Jitter Detection Techniques

From the previous discussion, a jitter detector is a key element in all jitter measurement setups. There are a number of techniques to detect time jitter; among them the most widely used techniques are sampling oscilloscope measurement, phase detection measurement, and BER-scan measurement.

5.4.2.1 Jitter Measurement Based on Sampling Oscilloscope

The most straightforward way to characterize time jitter in a PRBS waveform is to use a sampling oscilloscope. As discussed in Chapter 2, taking advantage of the repetitive nature of the PRBS digital signal, sequential sampling and

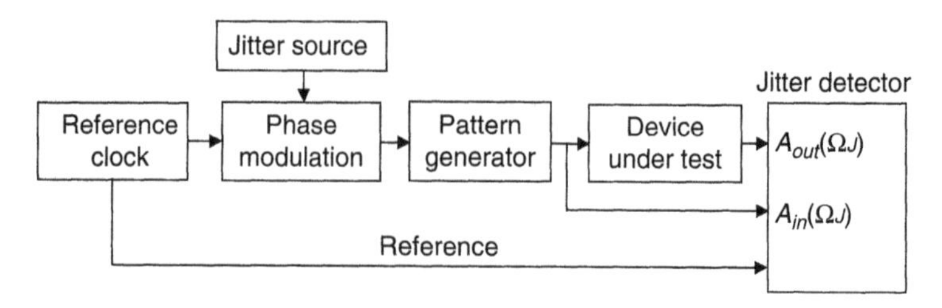

Figure 5.4.5 Measurement of jitter transfer.

random sampling can help a digital sampling oscilloscope measure high-speed waveforms. In these measurements, only one or a small number of data point (s) within each pulse are sampled in each period and the signal eye diagram can be reconstructed after measuring a large number of repetitive periods. Each pixel on the scope represents an independent sampling that provides the amplitude of the sampling as the function of the sampling time. To sample the waveform in a precise time, a reference clock that has no time jitter has to be used to trigger the oscilloscope, as illustrated in Figure 5.4.6.

Since time jitter is a time or phase error in the waveform, the best place to observe the time jitter in an eye diagram is at the cross point, as shown in Figure 5.4.5. Because of the time jitter, the width of the cross point of the rising and falling edges of the pulse is broadened. Choosing a decision threshold voltage that is a horizontal line near the middle of the eye diagram, we can obtain a histogram that is the number of pixels on the threshold line as the function of time. To find reasonably good statistics of time jitter, a large number of pixels are required to construct the histogram. This can be done by increasing the width of the threshold line so that a larger number of pixels are included, but the threshold-level specificity will be reduced. More practically, we can increase the measurement time so that a larger number of samples can be accumulated. However, this is ultimately limited by the memory of the sampling oscilloscope.

The histogram shown in Figure 5.4.6 at the cross point of the eye diagram can be analyzed to find the nature of the time jitter. For example, if random jitter dominates, the measured histogram should be largely Gaussian; on the other hand, if deterministic jitter is the major jitter source, the histogram should be bounded. In practice, the overall jitter is usually the combined effect of random and deterministic jitters, and the statistical distribution is a convolution between them. A proper deconvolution in signal processing may help distinguish the random and deterministic contributions to the time jitter.

Jitter measurement using a sampling oscilloscope is straightforward; it measures jitter characteristics specifically at the decision threshold, which is most relevant to the BER performance of the system. In addition, no specialized equipment is required in this measurement because a sampling oscilloscope is general-purpose equipment that exists in many laboratories. However, since this

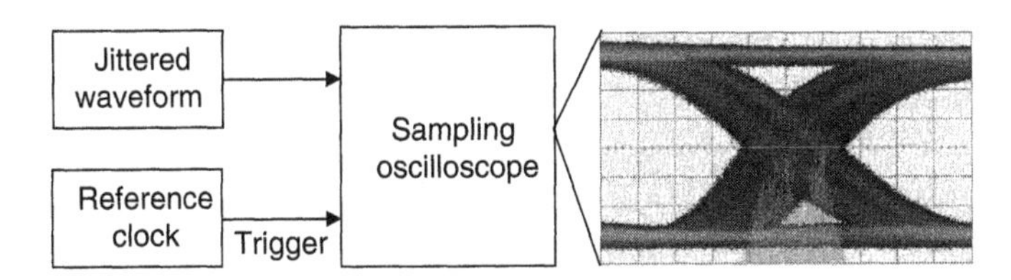

Figure 5.4.6 Time-jitter measurement using a digital sampling oscilloscope [12].

is not a real-time measurement, a disadvantage is that frequency-dependent characteristics of the time jitter cannot be obtained.

5.4.2.2 Jitter Measurement Based on a Phase Detector

A phase detector is basically an RF mixer that multiplies the two input signals and yields their product. Figure 5.4.7 shows the block diagram of jitter measurement setup based on a phase detector. This setup measures the fundamental clock component of the jittered waveform and compares it with a jitter-free reference clock in an RF mixer. Assume that the recovered clock from the jittered waveform is $S_J(t) = A_J \sin[2\pi R_b t + \Delta\varphi(t)]$, where R_b is the data rate and $\Delta\varphi$(t) is the phase jitter. A reference clock at the same fundamental frequency $S_R(t) = A_R \sin(2\pi R_b t + \varphi_0)$ is used for the detection, where φ_0 is a fixed phase delay. The recovered clock and the reference mix at the phase detector, which is an RF mixer, and the output voltage is

$$\begin{aligned} V(t) &= KS_J(t) \cdot S_R(t) \\ &= \frac{1}{2} KA_R A_J \left\{ \sin[4\pi R_b t + \Delta\varphi(t) + \varphi_0] + \sin[\Delta\varphi(t) - \varphi_0] \right\} \end{aligned} \tag{5.4.4}$$

where K is the efficiency of the RF mixer.

The lowpass filter removes the double-frequency component. Suppose the phase noise is small enough such that $|\Delta\varphi(t) - \varphi_0| \ll \pi$; the voltage signal at the filter output is approximately

$$V(t) \approx 0.5\, KA_R A_J [\Delta\varphi(t) - \varphi_0] \tag{5.4.5}$$

This voltage signal is linearly proportional to the phase term $\Delta\varphi(t)$, which is related to the time jitter $\Delta t(t)$ through Equation 5.4.2. This voltage signal is usually much slower than the fundamental clock and can be easily digitized and analyzed with digital signal processing. The distinct advantage of the phase detection technique is the capability of measuring jitter information in real time, which allows the frequency analysis of time jitter.

The jittered waveform $S_J(t) = A_J \sin[2\pi R_b t + \Delta\varphi(t)]$ can also be characterized by a spectrum analyzer. In fact, the spectrum of the jittered signal $S_J(t)$

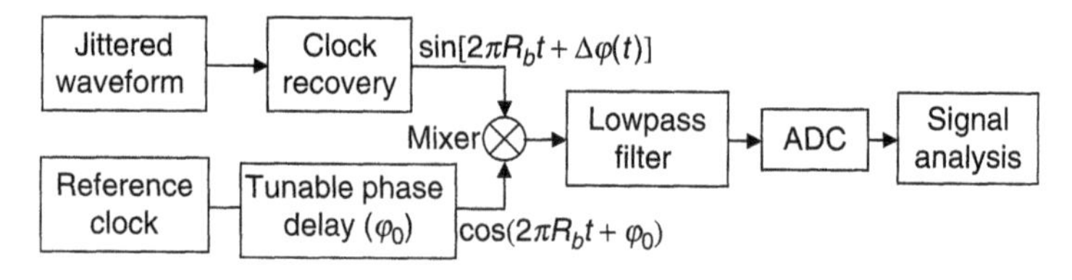

Figure 5.4.7 Time-jitter measurement based on phase detection.

should have a strong fundamental frequency component and phase modulation sidebands. The power density of the phase modulation sideband normalized by the power of the carrier should provide the magnitude as well as the frequency of the jitter. However, due to the limitation of spectral resolution and the dynamic range of a practical spectrum analyzer, the powerful carrier at frequency R_b may influence the measurement of the relatively weak phase modulation sidebands. This is especially problematic for low-frequency jitter components for which the frequency separation between the carrier and the modulation sideband is small.

5.4.2.3 Jitter Measurement Based on a BER-T Scan

BER is an ultimate quality measure of a digital signal. The involvement of time jitter certainly has an impact in the BER, whereas vice versa, a BER measurement can be used to evaluate the time jitter of the digital signal. For this purpose, a technique of measuring BER while scanning the decision time across the eye diagram is known as a *BER-T scan.*

As illustrated in Figure 5.4.8, the measurement uses a general-purpose BER test set that consists of a PRBS pattern generator, a BER monitor, and a variable delay line between them to control the phase of the reference signal. The BER-T scan measures the BER while scanning the decision phase over a bit time T_b across the entire length of the eye diagram, as shown in Figure 5.4.9. For example, it starts at the left eye edge, which is the center of the cross point between the rise and the fall edges of the signal pulse. At this decision time, $t = 0$, the BER is approximately 0.5 if the probability of space and marks are equal in the PRBS pattern. Then shift the decision time toward the middle of the eye; the BER value should decrease because the level difference between the marks and spaces starts to open up. However, due to deterministic jitter, the eye crossing may be wide and the BER does not decrease significantly until $t = T_{DJ,L}$. After that, the BER decreases dramatically, with the decision phase moving toward the middle of the eye diagram. The rate of this BER decrease is determined by the eye opening, the random noise, and the random jitter.

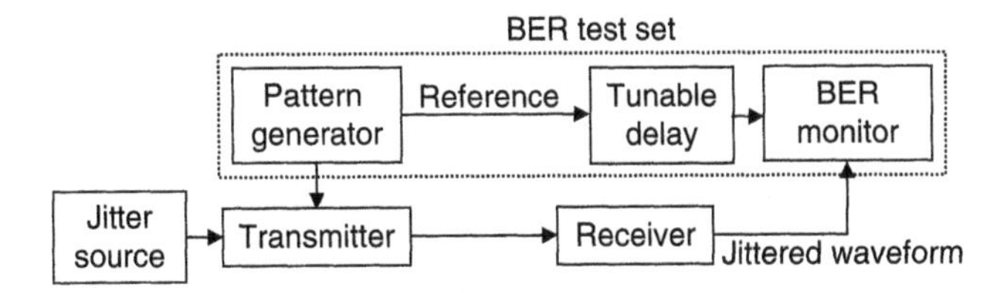

Figure 5.4.8 An example of jitter measurement through BER testing.

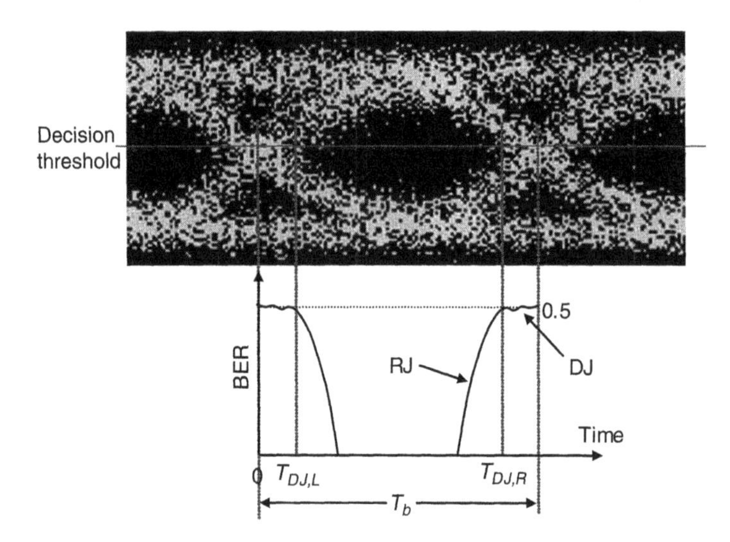

Figure 5.4.9 Illustration of an eye diagram and BER-T scan measurement. *DJ:* deterministic jitter, *RJ:* random jitter.

Assuming that the signal at the receiver is strong enough and the SNR is high, the intensity noise can be neglected. This is equivalent to assume that the bit errors are primarily caused by the time jitter:

$$BER(t) = BER_{DJ,L} \cdot erfc\left(\frac{t - T_{DJ,L}}{\sigma_{RJ,L}\sqrt{2}}\right) \tag{5.4.6}$$

where $BER_{DJ,L} \approx 0.5$ is the BER caused by deterministic jitter at $t = T_{DJ,L}$ and $\sigma_{RJ,L}$ is standard deviation of the Gaussian distribution, which is the *rms* value of the random jitter near the left edge of the eye. The BER expression in Equation 5.4.6 may have underestimated the BER value because it neglected the contribution from the right edge of the eye. In fact, a more accurate expression should be

$$BER(t) = BER_{DJ,L} \cdot erfc\left(\frac{t - T_{DJ,L}}{\sigma_{RJ,L}\sqrt{2}}\right) + BER_{DJ,R} \cdot erfc\left(\frac{T_{DJ,R} - t}{\sigma_{RJ,R}\sqrt{2}}\right) \tag{5.4.7}$$

where $BER_{DJ,R}$, $T_{DJ,R}$, $\sigma_{RJ,L}$ are parameters at the right edge of the eye. In practice, if the SNR of the signal is high enough, the second term in Equation 5.4.7 should be much smaller than the first term when $t < T_b/2$ and the BER impact from the opposite eye edge is negligible.

In optical communication systems, the eye diagrams may not be symmetric in the horizontal direction. The major reason is that the leading edge and the falling edge of a signal pulse may not be distorted in the same way in optical transmission. Therefore the plateaus of the BER-T curve, which are the results of deterministic jitter, may have different widths so that $T_{DJ,L} \neq T_b - T_{DJ,R}$.

In fact, the jitter measurement technique using a BER-T scan is very similar to the histogram technique using a sampling oscilloscope, as illustrated in Figure 5.4.6. Comparing these two techniques, a BER-T scan is able to measure the statistic jitter distribution down to very low probability events—, for example, 10^{-12}. In contrast, it is not feasible for a sampling oscilloscope to memorize $>10^{12}$ pixel points to construct a histogram. In addition, since BER is the ultimate measure of signal quality, BER-T scan measurements probably provide more relevant estimations of the jitter impact in the system performance.

5.5 IN-SITU MONITORING OF LINEAR PROPAGATION IMPAIRMENTS

In the previous chapters we discussed various techniques to characterize the properties of optical fibers and optical components. In the techniques of characterizing fiber chromatic dispersion, PMD and PDL, discussed in Chapter 4, we must have access to both terminals of the fiber system to perform the measurements. In these measurements, an optical probe signal is launched into the fiber from one end and the testing is performed at the other end of the fiber. However, if a fiber is already deployed as part of an optical network, it is not always possible to have access to both terminals of the fiber, because they are usually at distance from each other. In addition, if the optical system carries live traffic, inserting an additional optical probe signal for the measurement may jeopardize the integrity of optical transmission due to possible crosstalk. Therefore, it would be useful to utilize the traffic carried by the fiber to evaluate the overall chromatic dispersion, PMD, and PDL characteristics of the system.

5.5.1 *In Situ* Monitoring of Chromatic Dispersion

It is well known that chromatic dispersion is created due to wavelength-dependent group velocity of lightwave signals, which results in the pulse broadening and waveform distortion of the optical signal when it travels along the fiber. Although the chromatic dispersion parameters of each fiber section can be measured before being deployed in a transmission system, the accumulated dispersion of the overall system may still change if the system is aged or reconfigured and if the fibers are respliced. Chromatic dispersion (CD) may also change with temperature. Since the performance of high-speed optical transmission systems is usually very sensitive to the chromatic dispersion in the fiber, precise measurement and monitoring become important tasks for system design, implementation, and maintenance.

The basic idea behind CD monitoring in digital fiber-optic systems is based on the fact that with digital modulation, the spectrum of an optical signal typically has two redundant clock frequency components, one on each side of the optical carrier. Due to chromatic dispersion, these two clock components propagate at different speeds, creating a differential delay when arrive at the receiver. If we can separately recover the two clocks, their phase difference should tell us the amount of chromatic dispersion the optical signal has experienced. Figure 5.5.1(a) shows the experimental setup for *in situ* CD monitoring using a tunable vestigial sideband (VSB) optical filter. In this setup, the incoming optical signal, which carries digitally modulated data, is split into two parts. One of these parts is received by a photodetector and the clock is recovered by a clock recovery circuit, which can be found in a typical optical receiver. The other part passes through a tunable VSB optical filter that rejects one of the two clock components in the optical spectrum, as illustrated in Figure 5.5.1(b). In the branch with the VSB filter, the recovered clock is produced by mixing the optical carrier with either the upper or the lower clock tone in the optical signal. The relative clock phase shift $\Delta\varphi$ between the upper and the lower clock is proportional to the chromatic dispersion of the fiber [13],

$$\Delta\varphi = \eta\frac{2\pi\lambda^2}{c}R_b\cdot D\cdot L\cdot 2\Delta f \tag{5.5.1}$$

where R_b is the signal data rate, $D\cdot L$ is the accumulated chromatic dispersion of the system, and $2\Delta f$ is the frequency separation of the VSB filter when adjusted to select the upper and lower sidebands. Since the bandwidth of the VSB optical filter is wide enough, it not only selects the discrete clock

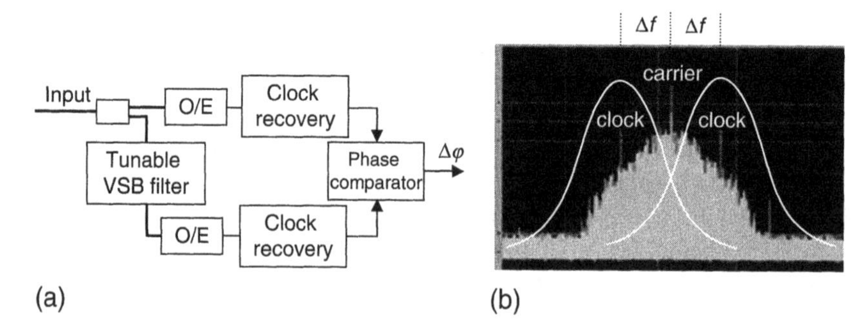

Figure 5.5.1 Monitoring chromatic dispersion using vestigial sideband filtering: (a) block diagram of experimental setup and (b) illustrations of signal optical spectrum and VSB filter transfer functions.

components, it also passes a wideband optical signal. The mixing between optical signal components also contributes to the recovered clock. Therefore a scaling factor η is used in Equation 5.5.1, which is

$$\eta = \frac{1}{2\Delta f}\int f \cdot [S_{USB}(f) - S_{LSB}(f)]df \tag{5.5.2}$$

where $S_{\mathrm{USB}}(f)$ and $S_{\mathrm{LSB}}(f)$ are the power spectral densities of the upper and lower VSB signals, respectively.

In this measurement technique using direct detection in the receiver, the measurement accuracy may depend on the optical filter's parameters such as shape, bandwidth, and detuning as well as the optical signal modulation format as indicated in Equation 5.5.2. In practice, the bandwidth and shape of an optical filter are difficult to control, especially when the required bandwidth is very narrow; therefore a careful calibration is mandatory for this monitoring apparatus before measuring each specific system.

This uncertainty may be reduced using coherent detection [14] because the transfer function of an RF filter can be well controlled and the bandwidth of an RF filter can be made very narrow. Coherent detection is able to down-convert the spectrum of the optical signal into the RF domain and the relative phase-delay information of the optical signal is preserved.

Figure 5.5.2 shows an example of the chromatic dispersion-monitoring experimental setup using coherent detection. A small portion of optical signal is tapped off from a transmission system that carries a 10 Gbit/s RZ PRBS optical signal. At the coherent dispersion monitor, the optical signal is combined with the light emitted from a local oscillator through a 3 dB fiber coupler. The mixed optical signal is detected by a photodiode, which down-converts the spectrum of the 10 Gbit/s optical signal into the RF domain. By adjusting

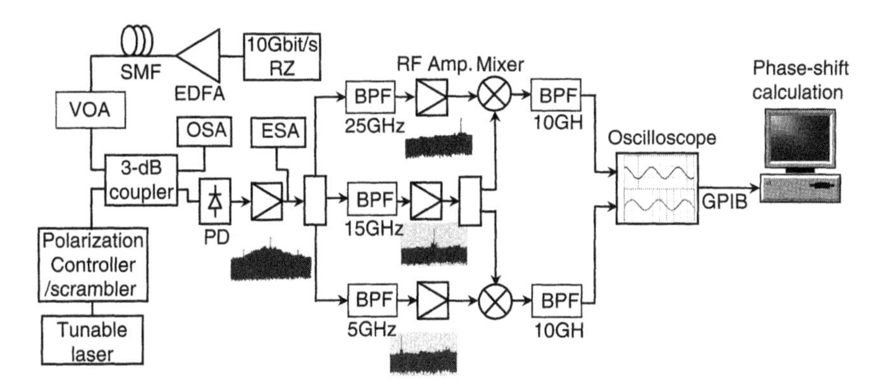

Figure 5.5.2 Block diagram of chromatic dispersion monitoring using coherent detection. *BPF:* bandpass filter, *OSA:* optical spectrum analyzer, *ESA:* electrical spectrum analyzer.

the wavelength of the local oscillator, the carrier frequency of the heterodyne IF signal is tuned to approximately 15 GHz.

Then this IF signal is split into three parts which pass through narrowband RF filters with central frequencies at 5 GHz, 10 GHz, and 15 GHz, respectively to individually select the carrier and the lower and upper clock sidebands. The measured RF spectra at various stages of the circuit are shown in the insets of Figure 5.5.2. The carrier component is further split into two and mix with the upper and the lower clocks independently. This generates two separate clock waveforms after narrowband filtering at the clock frequency of 10GHz. An oscilloscope then displays the two clock waveforms, and a computer collects the data and calculates the relative time delay between the two recovered clocks. The chromatic dispersion can be evaluated from the relative time delay Δt between these two recovered clocks by [15]:

$$\Delta t = DL\lambda^2 R_b / c \tag{5.5.3}$$

Figure 5.5.3 shows examples of the clock waveforms measured from the upper and lower branches of the receiver for standard SMF fiber lengths of 0 km, 29.4 km, and 60.1 km, respectively. These waveforms clearly show the progressive phase shift as the fiber length increases. The results of a systematic measurement on chromatic dispersion versus fiber length are shown in Figure 5.5.4, together with the results calculated with Equation 5.5.3 using the dispersion parameter $D = 17$ ps/nm-km specified on the fiber spool. This figure shows an agreement between the measured and the calculated differential delays between the two clock components.

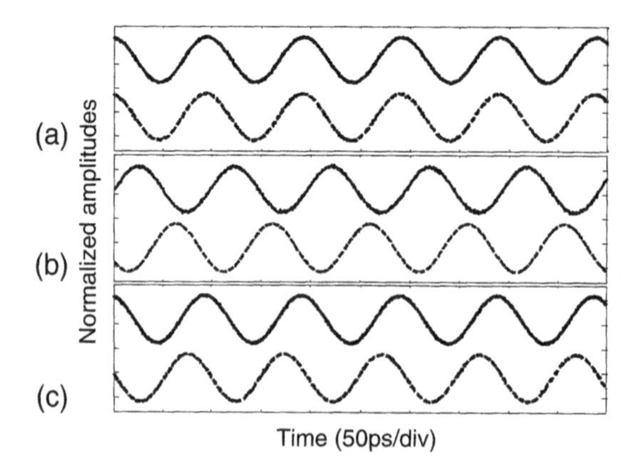

Figure 5.5.3 Measured lower (dashed line) and upper (solid line) clock waveforms after signal transmitted over (a) 0 km, (b) 29.4 km, and (c) 60.1 km of optical fiber.

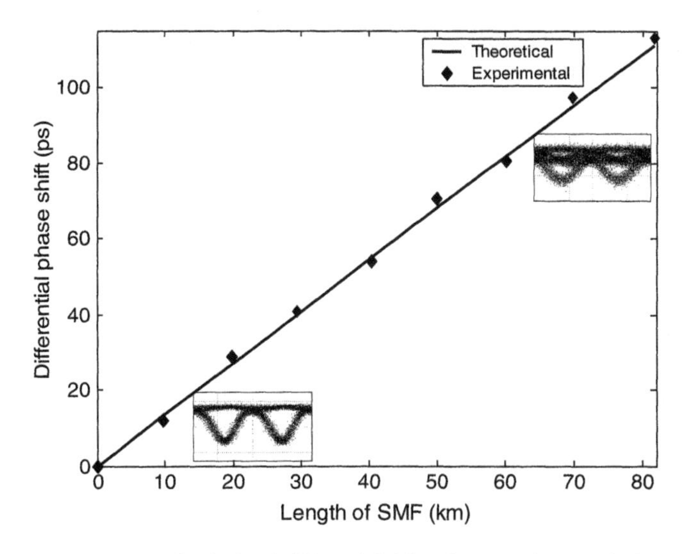

Figure 5.5.4 Measured and calculated differential delays between the two clock components versus fiber length. Solid line: theoretical, solid diamonds: Measured. Inset: eye diagrams measured after 0 km and 81.7 km fiber.

5.5.2 *In Situ* PMD Monitoring

For the PMD and PDL measurement techniques discussed in Chapter 4, a polarization-controlled optical signal is launched into the fiber from one end and the measurement is performed at the other end of a fiber system. If the fiber is already deployed as part of an optical network it is not always possible to have the access to both ends of the fiber because they are usually far away from each other. It would be useful to utilize the WDM optical signal carried in the optical fiber to evaluate the overall PMD and PDL characteristics of the system.

5.5.2.1 Basic Operating Principle

The basic idea behind nonintrusive inline PMD monitoring is the measurement of polarization walk-off, induced by fiber birefringence, between different frequency components within the spectrum of an optical signal. In an optical transmitter, all frequency components within an optical channel usually have the same SOP because the optical signal is originated from one laser diode. After the optical signal passes through the fiber system with PMD, the polarization states of different frequency components will walk off from each other; this SOP walk-off can be used to predict the fiber DGD. A simple relation between the polarization walk-off and the fiber DGD can be easily explained by the Poincare arc representation that was discussed in Section 4.5.5 (Equation 4.5.16) as $\phi = \Delta\omega \cdot \Delta\tau$, where φ is the angle of polarization walk-off between two frequency components separated by $\Delta\omega$ as illustrated in Figure 4.5.8, and $\Delta\tau$ is the DGD of the fiber system.

Figure 5.5.5 shows an experimental setup for PMD monitoring using vestigial sideband (VSB) optical filtering and RF power detection [16]. In this setup, a return-to-zero (RZ) optical transmitter is used and the signal optical spectrum has a carrier and two distinct clock sidebands, one on each side of the carrier. To monitor the DGD this optical signal has experienced after passing through the fiber system, a VSB optical filter selects the carrier and one of the two clock sidebands. At the photodiode (PD), the carrier mixes with the selected clock sideband, producing an RF beating frequency at the clock rate. A bandpass RF filter selects this clock frequency and an RF power meter measures the RF power of the recovered clock. Due to the PMD effect in the fiber system, there is an SOP walk-off between the optical carrier and the selected clock sideband; therefore the efficiency of mixing in the photodiode is reduced. The measured RF power at the clock tone is thus dependent on the DGD of the fiber [16]:

$$P(f_R) = P_0 \cos^2(\pi f_R \Delta\tau_s) \tag{5.5.4}$$

where f_R is the clock frequency of the modulated optical signal, P_0 is the RF power in the absence of PMD, and $\Delta\tau_s$ is the DGD experienced by the optical signal. Because only one of the two clock sidebands is used to mix with the carrier, carrier fading, which is usually associated with chromatic dispersion, will not affect the amplitude of the recovered clock RF power. Thus chromatic dispersion in the fiber system does not have significant impact in the result of DGD measurement.

However, this technique is sensitive to the modulation format of the optical signal because it requires the optical signal to have a discrete clock component in the spectrum. It would be difficult to apply this technique to a system

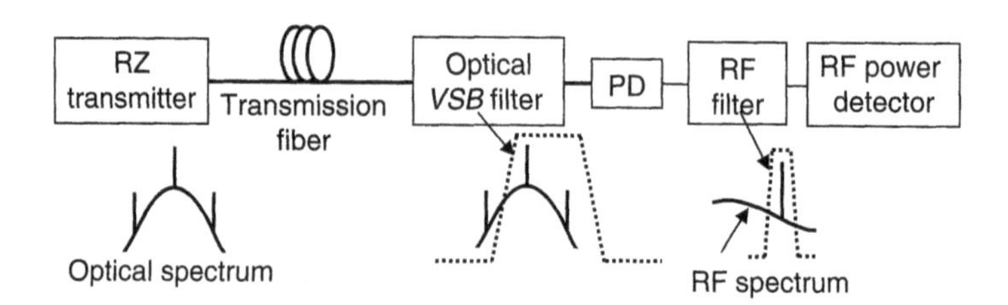

Figure 5.5.5 PMD monitoring using vestigial sideband (VSB) optical filtering and RF power detection.

carrying optical signal with non-return-to-zero (NRZ) modulation format, where the clock component is weak. In addition, vestigial sideband detection has stringent requirements on the optical filter; it has to be narrowband with sharp edges to cut off the unwanted clock sideband. It needs to point out that $\Delta\tau_s$ is the DGD experienced by the optical signal instead of the DGD of the fiber. Therefore $\pi f_R \Delta\tau_s = \varphi/2$, where φ is the center angle between the two SOP vectors instead of the angle ϕ on the plane perpendicular to the fiber PSP, as shown in Figure 4.5.8. This point will be further explained in 5.5.2.3.

5.5.2.2 PMD Monitoring Using Coherent Detection

Another way to monitor system PMD through frequency-dependent polarization walk-off is the use of coherent detection [14]. As described in Chapter 2, coherent heterodyne detection shifts the signal spectrum from the optical domain to the RF domain and therefore it would be much easier to perform signal processing and analysis. Figure 5.5.6 shows the block diagram of a coherent polarimeter, which can be used to characterize the frequency-dependent SOPs within the spectrum of an optical channel that passes through the fiber-optic system [17]. In this setup, the output SOP of the local oscillator (LO) is determined by a polarization controller. Because of the polarization selectivity of coherent detection, the photocurrent at the balanced photodetector is $I \propto \cos^2(\varphi/2)$, where φ is the angle between polarization vectors of the input optical signal and the LO on the Poincare space. From this point of view, the coherent detection receiver can be configured to perform as a polarimeter, where the SOP of the LO is equivalent to the principle axis of the polarizer in a conventional polarimeter with direct detection.

In this coherent detection technique, the central optical frequency difference between the optical signal and the LO determines the intermediate frequency (IF) of the coherent heterodyne detection. Assuming that the linewidth of the LO is negligible, the IF spectrum in the RF domain is identical to the signal optical spectrum. As illustrated in Figure 5.5.7, the bandpass filter (BPF) cuts

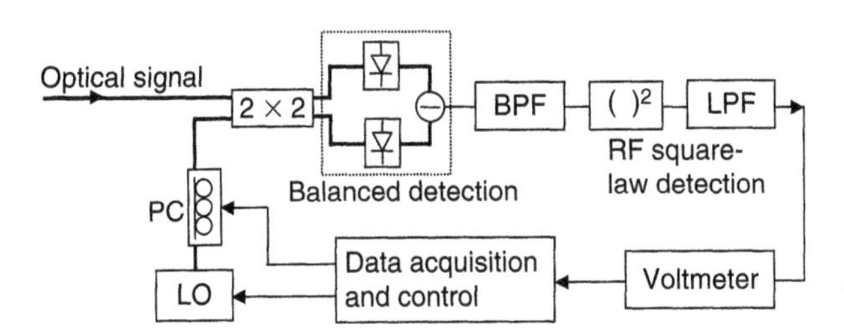

Figure 5.5.6 Block diagram of coherent polarimeter. *BPF:* bandpass RF filter, *LPF:* lowpass RF filter, *LO:* local oscillator.

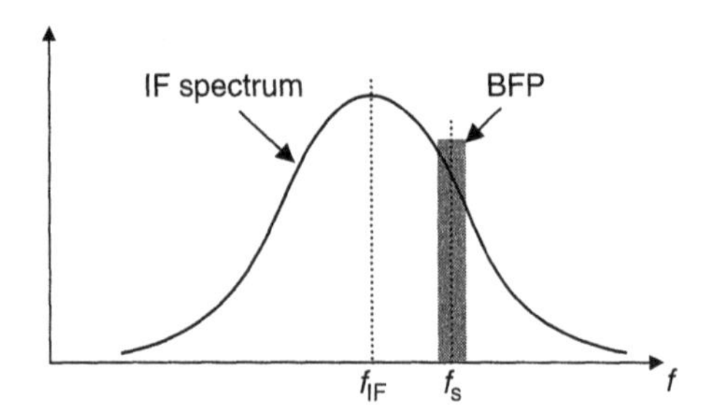

Figure 5.5.7 Illustration of signal optical spectrum and frequency selection with narrowband filters.

a narrow slice of the IF spectrum for processing. The electrical circuit, which is a square-law RF detector, following the BPF is used to measure the RF power of the selected narrowband RF signal. Similar to the procedure of a conventional polarimeter with direct detection as discussed in Chapter 2, by systemically setting the polarization state of the LO at linear 0°, 90°, 45° and circular polarization while measuring the corresponding RP power, the polarization state of the optical signal at frequency f_s selected by the BPF can be fully characterized.

To perform polarization analysis of various frequency components within the optical signal, we can either sweep the LO frequency with a fixed BPF or fix the LO frequency but scan the central frequency of the BPF. In the latter case, the scanning BPF and the RF detection circuit can be replaced by an RF spectrum analyzer, where the bandwidth of the BPF is equivalent to the resolution bandwidth and the lowpass filter (LPF) bandwidth is equivalent to the video bandwidth of the RF spectrum analyzer.

This polarimeter based on coherent heterodyne detection allows the precise determination of frequency-resolved SOP within the spectrum of the optical signal. However, it is not optimized to measure time-dependent DGD of the fiber system because of the requirement for a tunable RF filter or a scanning RF spectrum analyzer.

A simplified but more practical technique to monitor time-dependent fiber DGD using coherent heterodyne detection is to replace the tunable RF bandpass filter with two fixed narrowband RF filters using the setup shown in Figure 5.5.8 [18].

This PMD monitoring technique was based on the measurement of differential polarization walk-off between two fixed frequency components within the optical signal. In general, these two frequencies do not have to be the discrete carrier or clock components; therefore the technique is applicable to most of the practical optical modulation formats. As discussed in the last section, due to first-order DGD ($\Delta\tau_s$) experienced by the optical signal, the differential polarization walk-

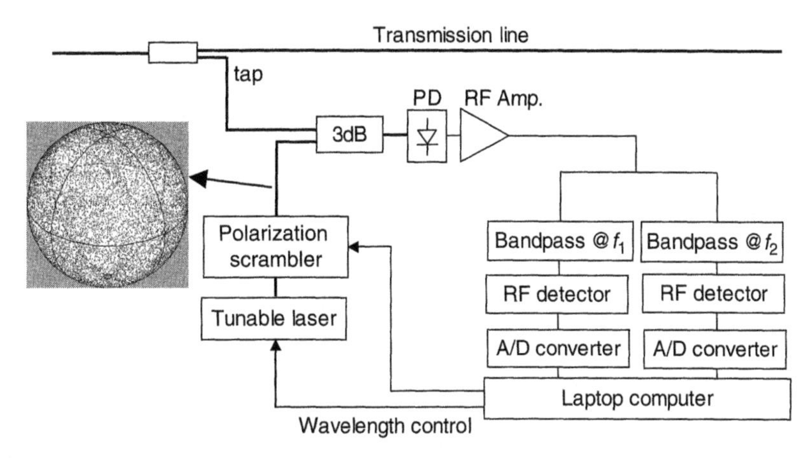

Figure 5.5.8 Block diagram of PMD monitoring using coherent detection and two fixed bandpass RF filters.

off between two frequency components separated by Δf can be represented as their relative angular walk-off on the Poincare sphere $\varphi = 2\pi\Delta f \Delta\tau_s$. Since coherent heterodyne detection shifts the optical spectrum into the RF domain, this measurement can be performed by simultaneously measuring RF powers selected at two frequencies. Thanks to the fact that coherent detection is inherently polarization-selective, the efficiency of coherent detection depends on the SOP mismatch between the signal and the local oscillator, the generated RF powers of the two selected frequency components f_1 and f_2 are, respectively:

$$P_1 = \eta_1 P_{lo} P_{sig}(f_1)\cos^2(\xi/2)\cos^2(\beta/2) \tag{5.5.5}$$

$$P_2 = \eta_2 P_{lo} P_{sig}(f_2)\cos^2(\xi/2)\cos^2(\beta/2 + \varphi/2) \tag{5.5.6}$$

where ξ is angle between the local oscillator and the E(f_1)/E(f_2) plane, β is the angle between the projection of E_{LO} on the E(f_1)/E(f_2) plane and E(f_1), and φ is the SOP angle between the two selected frequency components, as illustrated by Figure 5.5.9. η_1 and η_2 represent the combined effects of the photodetector responsivity and the efficiencies of the RF detectors at the selected two frequency components, P_{lo} and P_{sig}, are the optical powers of the local oscillator and the signal. Although direct measurement of P_1 and P_2 may indicate the angular polarization walk-off φ, uncertainties may be introduced because power spectral densities of the optical signal at the two selected frequencies may be different.

In addition to the difference in the efficiencies of the two RF detectors and frequency-dependent gain of the RF amplifier, this power difference may also change with the signal modulation formats and modulator characteristics in the transmitter. This problem can be solved by randomly scrambling the SOP

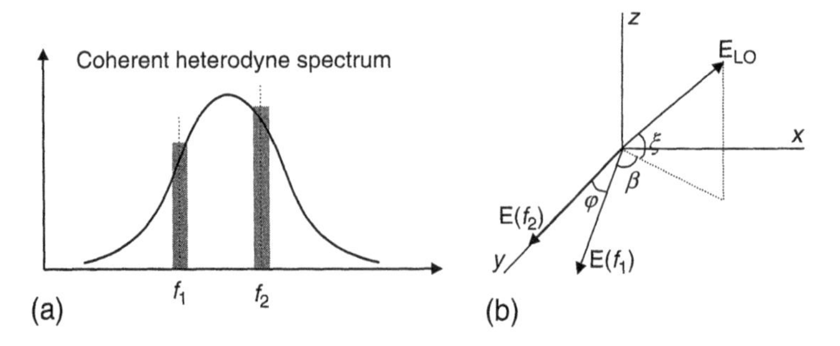

Figure 5.5.9 (a) Slicing of IF spectrum by two RF filters and (b) polarization vectors of the local oscillator (E_{LO}) and the two selected signal frequency components ($E(f_1)$) and ($E(f_2)$).

of the local oscillator, which makes both ξ and β random variables while φ is relatively stable, assuming that the DGD variation in the fiber system is much slower than the polarization scrambling of the local oscillator. After a few cycles of P_1 and P_2 passing through their maxima and minima, the RF power measured at the two branches can easily be normalized, as shown in Figure 5.5.10. The phase difference φ between the two power traces can be obtained and the first-order DGD of the fiber system can be derived. In fact, once one of the two traces reaches the maximum, where $\xi = 0$ and $\beta = 0$, the normalized amplitude of the other trace, x, indicates the DGD value through $\varphi = 2\cos^{-1}(\sqrt{x})$.

As an example, Figure 5.5.11 shows the measured results on a 192 km installed fiber link, which carries a 40 Gb/s live traffic with duobinary modulation. A 7 ps fixed DGD was added at each end of the system using highly

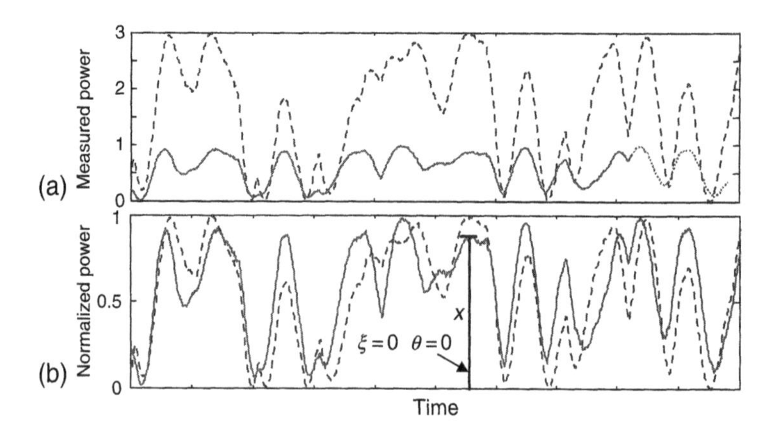

Figure 5.5.10 (a) Example of measured $E(f_1)$ and $E(f_2)$ and (b) after normalization.

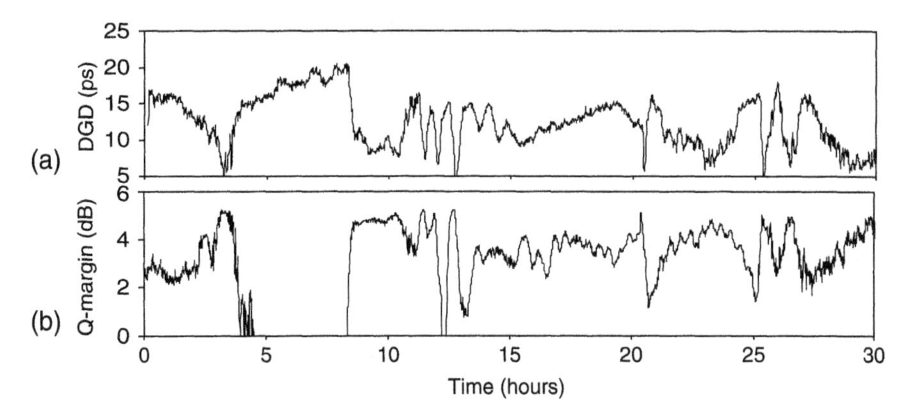

Figure 5.5.11 Comparison between measured DGD values and measured system dBQ margin: (a) DGD measured by PMD monitor and (b) system dBQ margin measured from a separate BER monitoring [18].

birefringence fiber to increase the system DGD level. In this measurement, the IF frequency of heterodyne detection was set at 20 GHz and the central frequencies of the two narrowband RF filters were 15 GHz and 25 GHz, respectively, each with a 1 GHz bandwidth. Figure 5.5.11(a) shows the measured DGD as a function of time within 30 hours' measurement duration. The system Q-margin, shown in Figure 5.5.11(b), was measured through independent bit error testing, where the high Q-margin indicates a better system performance, whereas a 0 dB system margin means that the system BER is at the edge of failing. The results in Figure 5.5.11 clearly show that the system margin is inversely proportional to the instantaneous DGD measured with the setup shown in Figure 5.5.8. The transmission failed whenever the DGD level was higher than 15 ps in this particular system [18].

5.5.2.3 Difference Between Fiber DGD and the DGD Experienced by an Optical Signal

It is important to note that for the in-service PMD monitoring techniques discussed so far using live traffic as the probing signal, the DGD value $\Delta\tau_s$ obtained is an "apparent" DGD, which is in fact the DGD experienced by the probing optical signal rather than the actual PMD parameter of the fiber. In principle, the PMD parameter is a property of the fiber which indicates the differential group delay between the two PSPs; it is independent of the optical signal it carries. If the live traffic carried by the fiber is used as the probe signal, the optical signal at the fiber system input has a relatively stable SOP, which might not be aligned 45 ° between the two PSPs of the fiber. Thus, this probe signal does not necessarily explore the actual DGD of the fiber. It is useful to clarify the relationship between the measured DGD and its statistics using the in-service monitoring technique and the actual fiber PMD characteristics.

To better understand the impact of these differences, let's review the definition of PMD vector $\vec{\Omega}$ in Poincare representation as shown in Figure 5.5.12, which indicates the dependence of output SOP on PMD in an optical system. The output SOP of an optical signal will rotate when its optical frequency changes, with $\vec{\Omega}$ as the rotation axis. The rotation rate $\phi/\Delta\omega$ is defined as the first-order DGD, which equals the modulus of $\vec{\Omega}$ where ϕ is the change in the rotation angle in the plane perpendicular to $\vec{\Omega}$, and $\Delta\omega$ is the corresponding frequency change. The DGD value of the fiber can be evaluated if the rotation angle ϕ is known between two frequency components shown as points A and B in Figure 5.5.12. However, the measurement of ϕ is usually complicated and the evaluation of fiber PMD requires several different SOPs of the input optical signal to complete a Jones or Mueller Matrix. On the other hand, the angle φ shown in Figure 5.5.12 is relatively easy to obtain, and the *in situ* monitoring techniques we've described were based on the measurement of φ. In fact, the relationship between ϕ and φ is simple and can be expressed as

$$\sin(\varphi) = \sin(\phi)\sin\theta \tag{5.5.7}$$

where θ represents the angle between point A and the PMD vector $\vec{\Omega}$ shown in Figure 5.5.12. When ϕ is small enough, which can be ensured by choosing appropriate frequency difference $\Delta\omega$, Equation 5.5.7 can be simplified to

$$\varphi \approx \phi\sin\theta \tag{5.5.8}$$

In the Stokes space, the principle state of polarization model indicates that a long fiber can be regarded as a birefringent wave plate with the differential time

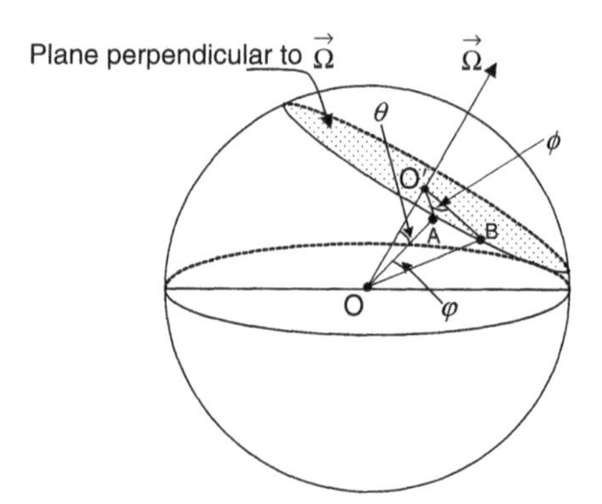

Figure 5.5.12 Poincare sphere representation of polarization dispersion vector and output polarization state rotation with optical frequency change.

retardation equal to the modulus of the PMD vector in the fiber, while the principle axis of the wave plate is aligned with the slow axis of the PMD vector. Thus the angle between the input SOP of the signal $\vec{S}_{in}$ and the fiber PMD vector is also equal to θ, and therefore

$$\cos\theta = \frac{\vec{\Omega}\cdot\vec{S}_{in}}{|\vec{\Omega}||\vec{S}_{in}|} \tag{5.5.9}$$

In a Cartesian coordinate system, the PMD vector can be decomposed into three orthogonal components, $\vec{\Omega} = \vec{a}_x\Omega_1 + \vec{a}_y\Omega_2 + \vec{a}_z\Omega_3$, where $\vec{a}_x$, $\vec{a}_y$, and $\vec{a}_z$ are unit vectors, and thus $|\vec{\Omega}| = \sqrt{\Omega_1^2 + \Omega_2^2 + \Omega_3^2}$. When each of the three orthogonal components Ω_1, Ω_2, and Ω_3 follows an independent Gaussian distribution with zero mean and the same standard deviation q, the statistics of PMD vector will exhibit a Maxwellian distribution:

$$p_3(\tau) = \sqrt{\frac{2}{\pi}\frac{\tau^2}{q^3}}e^{-\frac{\tau^2}{2q^2}} \tag{5.5.10}$$

In general, fiber PMD parameter is regarded as its mean DGD, which is related to the parameter q by

$$\tau_m = q\sqrt{8/\pi} \tag{5.5.11}$$

where the mean DGD τ_m is the average value of the Maxwellian distribution shown in Equation 5.5.10. For *in situ* PMD monitoring as discussed previously, live traffic carried in the fiber was used as the probing signal. Since the SOP of the input optical signal is determined by the laser in the transmitter, its SOP is relatively stable at the fiber input. Without losing generality, we can arbitrarily assume that the SOP of the input optical signal is $\vec{S}_{in} = (1, 0, 0)$; then

$$|\vec{\Omega}|\cos\theta = \Omega_1 \tag{5.5.12}$$

The combination of Equations 5.5.8 and 5.5.12 yields

$$\frac{\varphi}{\Delta\omega} = \frac{\phi}{\Delta\omega}\sin\theta = |\vec{\Omega}|\sin\theta = \sqrt{\Omega_2^2 + \Omega_3^2} \tag{5.5.13}$$

Note that a Maxwellian distribution is referred to as a *Chi* distribution with 3 degrees of freedom because it is related to three independent components Ω_1, Ω_2, and Ω_3. However, in the case of *in situ* monitoring, Equation 5.5.13 indicates that $\phi/\Delta\omega$ is only related to two of the three independent orthogonal components; therefore it should follow a *Chi* distribution with 2 degrees of freedom, which is also known as a Rayleigh distribution, and its probability density function can be expressed as

$$p_2(\tau) = \frac{\tau}{q^2} e^{-\frac{\tau^2}{2q^2}} \tag{5.5.14}$$

where we use the subscript 2 to indicate the 2 degrees of freedom. Instead, a subscript 3 was used in Equation 5.5.10 where the Maxwellian distribution has 3 degrees of freedom. The mean value of this $p_2(\tau)$ distribution is

$$\tau_{s,m} = q\sqrt{\pi/2} \tag{5.5.15}$$

From Equations 5.5.11 and 5.5.15, the relationship between $\tau_{s,m}$ and τ_m can be easily found as

$$\tau_m = 4\tau_{s,m}/\pi \tag{5.5.16}$$

Since the in-service monitoring technique evaluates $\tau_{s,m}$ from the monitoring of φ, we can easily estimate the mean DGD of the fiber, τ_m, from the relationship given by Equation 5.5.16.

For example, Figure 5.5.13 shows the measured DGD on a 750 km fiber link that carries 10 Gb/s live traffic. The *in situ* measurement based on coherent

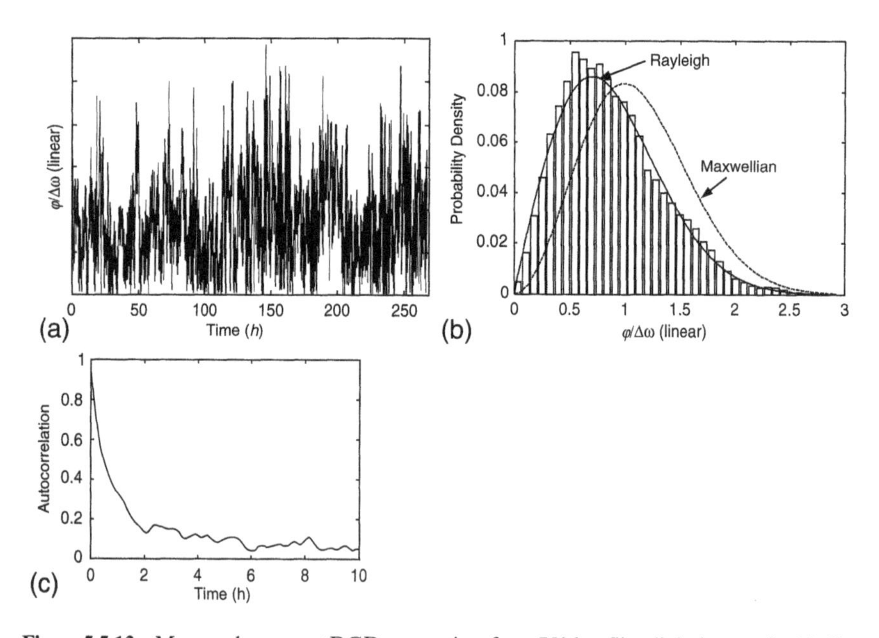

Figure 5.5.13 Measured apparent DGD versus time for a 750 km fiber link that carries 10 Gb/s live traffic: (a) measured DGD as a function of time, and (b) normalized statistic distribution of (a). Solid lines in (b): Rayleigh distribution; dotted lines: Maxwellian distribution with the same mean value. (c) is the autocorrelation function.

detection as discussed previously was performed for more than 250 hours. The measured probability distribution function of DGD shown as vertical bars fits well with a Rayleigh distribution. According to Equation 5.5.16, the mean DGD, τ_m, of the fiber should be $4/\pi$ times the mean DGD, $\tau_{s,m}$, measured by the *in situ* technique and the probability distribution of the fiber DGD should be Maxwellian, which is also plotted as the dashed curve in Figure 5.5.13(b) for comparison.

From the measured DGD as a function of time, we can also find the correlation time of the fiber DGD, which is a measure of PMD time dynamics in the system. Figure 5.5.13(c) shows the autocorrelation of the measured time-dependent DGD shown in Figure 5.5.13(a). The correlation time in this buried 750 km fiber link is about 20 minutes.

To conclude this section, in-service monitoring techniques using live traffic carried by the fiber system as the probe signal are useful to evaluate the PMD performance of installed fiber-optic systems without disturbing the normal operation. The measured DGD is only an "apparent DGD," which is not equal to the traditionally defined PMD parameter of the fiber. However, a simple conversion can be made using Equation 5.5.16 to find the mean DGD of the fiber from the "apparent DGD" measurement.

5.5.3 *In Situ* PDL Monitoring

As we discussed previously, although PDL is a system property independent of the optical signal that is launched into the system, the impact of PDL on the performance of a particular optical system does depend on the SOP of the input optical signal. The traditional PDL measurement method discussed in Section 4.6 requires simultaneous access to both ends of the system as well as the active control of the SOP of the optical signal that couples into the system [19, 20]. This technique is effective for systems in laboratory settings when the system does not carry live traffic. Practically, for installed optical systems and networks that carry live traffic, it would be almost impossible to have simultaneous access to both terminals, which are generally at a distance from each other. Most importantly, the probing optical signal for PDL measurement may disturb the normal operation of the optical system. In this case, the PDL can also be monitored using the live traffic carried in the system as the probe signal [21] similar to the *in situ* PMD measurement discussed in the last section. Again, the SOP of the traffic-carrying input optical signal is relatively stable but usually not controllable.

According to the Mueller Matrix description discussed in Chapter 4 (Equation 4.6.5), the power transmission coefficient through an optical system can be expressed as

$$T = \frac{s_{out0}}{s_{in0}} = m_{11} + \frac{s_{in1}}{s_{in0}} m_{12} + \frac{s_{in2}}{s_{in0}} m_{13} + \frac{s_{in3}}{s_{in0}} m_{14}$$
$$= m_{11} + DOP \cdot [s_1 m_{12} + s_2 m_{13} + s_3 m_{14}] = m_{11} + DOP(\vec{S} \cdot \vec{m}) \tag{5.5.17}$$
$$= m_{11} + DOP\, |\vec{m}| \cos\Psi$$

where matrix elements m_{11} and $\vec{m} = (m_{12}, m_{13}, m_{14})$ are in the first row of the Mueller Matrix and are related to the system PDL. $\vec{S} = (s_1, s_2, s_3)$ is the normalized SOP vector of the input optical signal with $s_j = s_{inj}/[s_{in1}^2 + s_{in2}^2 + s_{in3}^2]^{1/2}$ ($j = 1, 2, 3$). $DOP = [s_{in1}^2 + s_{in2}^2 + s_{in3}^2]^{1/2}/s_{in0}$ is the degree of polarization of input optical signal, and Ψ is the angle between $\vec{m}$ and $\vec{S}$. In general, the input optical signal in a live optical system, which is usually provided by an optical transmitter, has a high degree of polarization and therefore one can simply assumes $DOP = 1$. Thus Equation 5.5.17 can be simplified as $T = m_{11} + |\vec{m}| \cos\Psi$.

The power transmission coefficient T varies with time because of the random nature of the system PDL, as well as the random variation of angle Ψ. However, corresponding to each PDL value, there exists a maximum $T_{max} = m_{11} + |\vec{m}|$ and a minimum $T_{min} = m_{11} - |\vec{m}|$ of the transmission coefficient. The system PDL can be described by a PDL vector $\vec{\Gamma}$, and the amplitude of $\vec{\Gamma}$ is

$$\Gamma = \left| \frac{T_{max} - T_{min}}{T_{max} + T_{min}} \right| = \frac{|\vec{m}|}{|m_{11}|} = \frac{\sqrt{m_{12}^2 + m_{13}^2 + m_{14}^2}}{|m_{11}|} \tag{5.5.18}$$

The direction of the PDL vector is parallel to the SOP vector $\vec{S}$ that corresponds to the direction of the maximum transmission coefficient. Equation 5.5.18 indicates that the orientation of the PDL vector is determined by $\vec{m}$, and thus a general PDL vector can be defined by $\vec{\Gamma} = \vec{m}/m_{11}$. On the other hand, the traditional definition of PDL is $\rho = 10\log_{10}(T_{max}/T_{min})$ and the unit is in dB. The relationship between ρ and $\Gamma = |\vec{\Gamma}|$ can be easily found as $\rho = 10\log_{10}[(1+\Gamma)/(1-\Gamma)]$. When Γ is small enough ($\Gamma \ll 1$), the following approximation is valid:

$$\rho = 10\log\left[1 + \frac{2\Gamma}{1-\Gamma}\right] \approx 8.7\Gamma \tag{5.5.19}$$

which indicates that Γ is linearly proportional to ρ. It is generally accepted that ρ follows a Maxwellian distribution, so that Γ should also follow the same distribution. In the Mueller Matrix, m_{11} represents a constant attenuation of the optical system since its effect on the system is independent of polarization parameters of the input optical signal. Thus each individual component m_{12}, m_{13}, m_{14} in the expression of the PDL vector $\vec{\Gamma}$ should follow a normal distribution with zero mean and equal variance q^2. Then the probability density distribution of $\vec{\Gamma}$ can be expressed as

$$p(\Gamma) = \sqrt{\frac{2}{\pi}} \frac{\Gamma^2}{(q/m_{11})^3} \exp\left(-\frac{\Gamma^2}{2(q/m_{11})^2}\right) \tag{5.5.20}$$

$$\langle \Gamma \rangle = \sqrt{\frac{8}{\pi}} \frac{q}{m_{11}} \tag{5.5.21}$$

where $\langle \Gamma \rangle$ is the average value of Γ. Equation 5.5.20 indicates that system PDL can be fully determined by two parameters, m_{11} and q.

Without loss of generality, we can arbitrarily assume that the SOP of the input optical signal is $\vec{S} = (1, 0, 0)$, and in fact the coordinate rotation does not impact the transmission coefficient. Then Equation 5.5.17 can be simplified as

$$T = m_{11} + m_{12} \tag{5.5.22}$$

Since m_{11} is a constant, Equation 5.5.22 clearly shows that T follows a normal distribution (because m_{12} has normal distribution) with m_{11} as the mean and q^2 as the variance. This indicates that in a practical optical system carrying live traffic, m_{11} and q^2 can be obtained by measuring the statistical distribution of T. It is worth noting that PDL is a system property that should be independent of the optical signal it carries. In a system with fixed input signal SOP, the optical signal may not necessarily explore the worst-case PDL of the system; therefore T defined by Equation 5.5.22 is only related to a *partial* PDL. In this case $|\vec{m}| = |m_{12}|$ and the part of the system PDL vector seen by the optical signal is $\Gamma_{partial} = |m_{12}|/m_{11}$, which follows a half normal distribution.

In terms of PDL measurement, the coherence detection setup similar to that described in the last section for PMD monitoring can also be used to obtain the information of T, including its time-domain characteristics and the statistics. Figure 5.5.14 shows the experimental setup, where a small portion of the optical signal is tapped from the transmission line, which carries WDM traffic, to

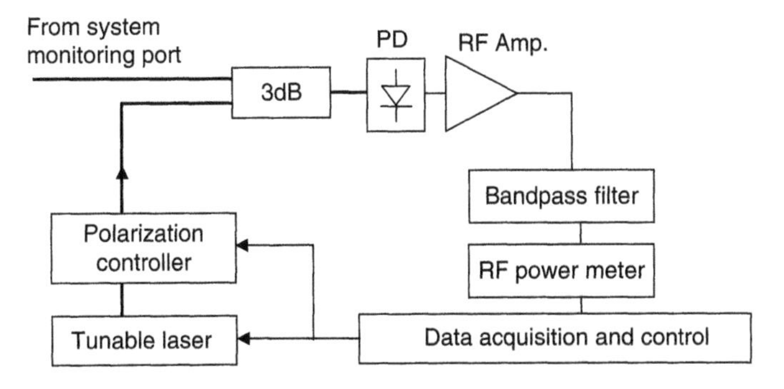

Figure 5.5.14 Block diagram of PDL experiment setup.

perform the measurement. A tunable laser is used as the local oscillator (LO) for coherent heterodyne detection and channel selection. A programmable polarization controller is placed at the output of the LO to periodically scan the SOP of the LO. A 3 dB fiber coupler combines the received optical signal with the LO, and a wideband photodetector is used to perform heterodyne detection. Only one bandpass RF filter with 1 GHz bandwidth is required to select a narrow slice of the amplified heterodyne IF spectrum whose power is then measured by a detector. A data acquisition card reads the measured RF power and converts it into a voltage V_{out} for calculation:

$$V_{out} = \eta_D P_L P_s T \cos^2\frac{\zeta}{2} + V_{offset} = kT\cos^2\frac{\zeta}{2} + V_{offset} \tag{5.5.23}$$

where η_D is a coefficient including the effects of the photodiode responsivity, the coupling coefficient of the fiber coupler, the RF amplifier, the RF filter, the RF detector, and the data acquisition circuitry. P_s is the received signal optical power, P_L is the power of the LO, and $k = \eta_D P_L P_s$. ζ is the angle between the SOP vector of the input optical signal and that of the LO. V_{offset} is the voltage offset resulting from no-ideal electronic circuit.

In the coherent detection measurement, the SOP of the LO is periodically scrambled, which effectively scans the angle ζ in Equation 5.5.23. If the scan is fast enough, we can assume that the system PDL vector remains constant during each period of scanning. The maximum voltage in Equation 5.5.23 is obtained when $\zeta = 0$:

$$V_{out,\max} = kT + V_{offset} \tag{5.5.24}$$

while the minimum voltage corresponds to $\zeta = \pi$:

$$V_{out,\min} = V_{offset} \tag{5.5.25}$$

The difference between $V_{out,\max}$ and $V_{out,\min}$ is then

$$V = V_{out,\max} - V_{out,\min} = kT \tag{5.5.26}$$

The mean, μ, and the variance, σ^2, of V can be directly calculated from the measured data, which are related to the parameters m_{11} and q^2 through T:

$$\mu = km_{11} \tag{5.5.27}$$

$$\sigma^2 = k^2 q^2 \tag{5.5.28}$$

Since $|m_{12}| = |T - m_{11}|$, the partial PDL can be calculated through

$$\Gamma_{partial} = \frac{|m_{12}|}{m_{11}} = \frac{|T - \mu/k|}{\mu/k} = \left|\frac{V}{\mu} - 1\right| \tag{5.5.29}$$

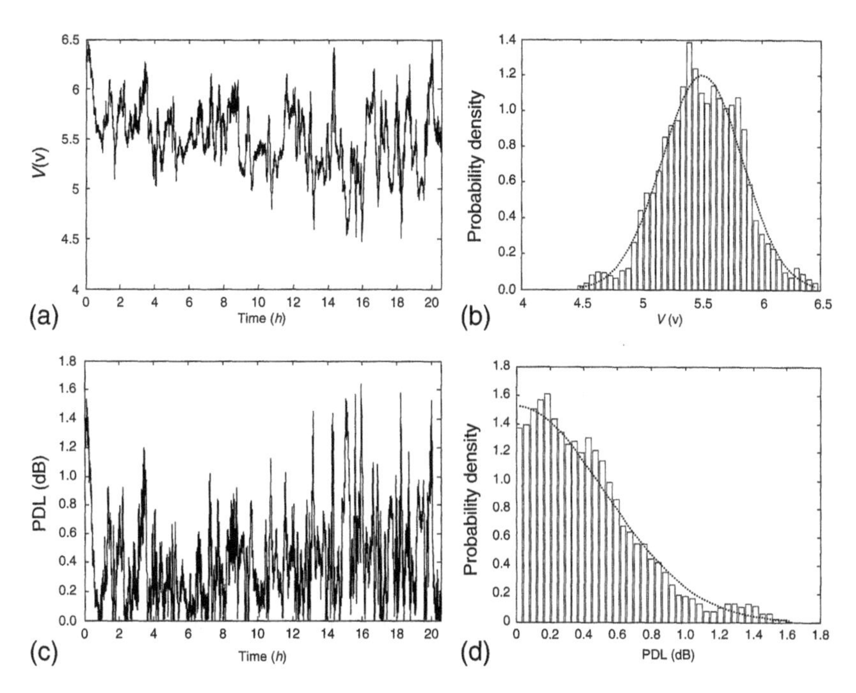

Figure 5.5.15 (a) Variation of V with time; (b) histogram of V; (c) variation of partial PDL in dB unit with time; (d) histogram of partial PDL. The dotted line in (b) is the ideal normal distribution. The dashed line in (d) is the ideal half normal distribution.

This partial PDL randomly varies with time, which is in fact the PDL seen by the optical receiver, and it has direct impact on the performance of the optical system.

As a measurement example, Figure 5.5.15 shows the measured results in a terrestrial fiber-optic system, which carries live WDM traffic at a 10 Gb/s data rate. The system is approximately 626 km long, with 10 inline optical amplifiers. Figure 5.5.15 (a) shows the measured V as a function of time; Figure 5.5.15(b) shows the histogram of V, which clearly follows a normal distribution as predicted by the statistical analysis. Figure 5.5.15(c) shows the partial PDL, $\Gamma_{partial}$, as the function of time; Figure 5.5.15(d) is the histogram of $\Gamma_{partial}$, which fits well with a half-normal distribution.

Although the measured partial PDL reflects its impact on the optical transmission performance, the PDL of the optical system itself is also an important parameter that is often used as part of the system specification. The average PDL of the optical system can be evaluated based on Equation 5.5.21 with the use of the measured parameters μ and σ^2 given by Equations 5.5.27 and 5.5.28:

$$\langle \Gamma \rangle = \sqrt{\frac{8}{\pi}} \frac{\sqrt{\sigma^2}}{\mu} \tag{5.5.30}$$

According to the μ and σ values shown in Figure 5.5.15, the average system PDL value $\langle \Gamma \rangle$ is approximately 0.093, which corresponds to a ρ of approximately 0.8 dB as defined by Equation 5.5.19.

5.6 MEASUREMENT OF NONLINEAR CROSSTALK IN MULTI-SPAN WDM SYSTEMS

In an amplified multispan WDM optical system, interchannel crosstalk is an important concern, especially when the system has a large number of spans and the signal optical power level is high. In addition to linear crosstalk, which might be caused by leakage from optical filters and switches, nonlinear crosstalk is especially notorious because it is difficult to avoid and cannot be eliminated by improving the qualities of optical filters and switches. The major sources of nonlinear crosstalk in high-speed optical transmission systems include cross-phase modulation (XPM), four-wave mixing (FWM), and Raman crosstalk. The characterization of these types of crosstalk helps us understand the mechanisms of system performance degradation that are essential for the system design and performance specification. In this section we discuss causes of these nonlinear crosstalks and introduce techniques to characterize their impact on the performance of multispan WDM systems.

5.6.1 XPM-Induced Intensity Modulation in IMDD Optical Systems

XPM originates from the Kerr effect in optical fibers, in which intensity modulation of one optical carrier can modulate the phases of other copropagating optical signals in the same fiber [22, 23]. Unlike coherent optical systems, intensity-modulation direct-detection (IMDD) optical systems are not particularly sensitive to optical phase fluctuations. Therefore XPM-induced phase modulation alone is not a direct source of performance degradation in IMDD systems. However, due to the chromatic dispersion of optical fibers, phase modulation can be converted into intensity modulation [24] and thus can degrade the IMDD system performance. On one hand, nonlinear phase modulation created by XPM is inversely proportional to the signal baseband modulation frequency [25]; on the other hand, the efficiency of phase noise to intensity noise conversion through chromatic dispersion increases with the modulation frequency [24]. Therefore XPM-induced overall intensity modulation is a complicated function of the signal modulation frequency [26].

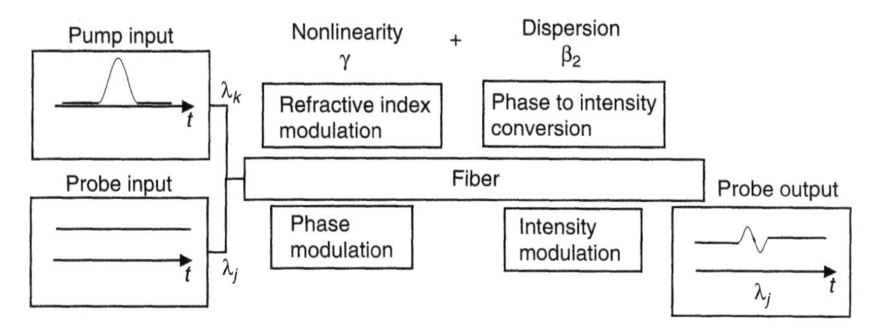

Figure 5.6.1 Illustration of pump-probe interaction through XPM process.

The theoretical analysis of XPM begins with the nonlinear wave propagation equation that was presented in Chapter 1. As illustrated in Figure 5.6.1, assume that there are only two wavelength channels copropagating along the fiber. They are defined as the probe and the pump, and their optical fields are denoted as $A_j(t, z)$ and $A_k(t, z)$, respectively. The evolution of the probe wave (a similar equation can be written for the pump wave) is described by

$$\begin{aligned}\frac{\partial A_j(t,z)}{\partial z} &= -\frac{\alpha}{2}A_j(t,z) - \frac{1}{v_j}\frac{\partial A_j(t,z)}{\partial t} - \frac{i\beta_2}{2}\frac{\partial^2 A_j(t,z)}{\partial t^2} \\ &+ i\gamma_j p_j(t,z)A_j(t,z) + i\gamma_j 2p_k(t - z/v_k, z)A_j(t,z)\end{aligned} \quad (5.6.1)$$

where α is the fiber attenuation coefficient, β_2 is the chromatic dispersion parameter, $\gamma_j = 2\pi n_2/(\lambda_j A_{eff})$ is the nonlinear coefficient, n_2 is the nonlinear refractive index, λ_j and λ_k are the probe and the pump wavelengths, A_{eff} is the fiber effective core area, and $p_k = |A_k|^2$ and $p_j = |A_j|^2$ are optical powers of the pump and the probe, respectively. Due to chromatic dispersion, the pump and the probe waves generally travel at different speeds, and this difference must be taken into account in the calculation of XPM because it introduces the walk-off between the two waves. Now we use v_j and v_k to represent the group velocities of these two channels.

On the right-hand-side of Equation 5.6.1, the first term represents the effect of attenuation, the second term is the linear phase delay, the third term accounts for chromatic dispersion, the fourth term is responsible for self-phase modulation (SPM), and the fifth term is the XPM on the probe signal j induced by the pump signal k. The strength of XPM is proportional to the optical power of the pump and the fiber nonlinear coefficient. To simplify our analysis and focus the investigation on the effect of XPM-induced interchannel crosstalk, the interaction between SPM and XPM can be neglected, assuming that these two effects act independently. To simplify the analysis, we can also assume that the probe is operating in CW, whereas the pump is modulated with a sinusoid at a frequency Ω. Although the effects of SPM for both the probe and the pump channels are neglected in the XPM calculation, a complete system performance evaluation can take into account SPM and other nonlinear effects separately. This approximation is valid as long as the pump signal waveform is not appreciably changed by the SPM-induced distortion within the nonlinear length of the fiber. Under this approximation the fourth term on the right side of Equation 5.6.1 can thus be neglected. Using variable substitutions $T = t - z/v_j$ and $A_j(t,z) = E_j(T,z)\exp(-\alpha z/2)$, Equation 5.6.1 becomes

$$\frac{\partial E_j(T,z)}{\partial z} = -\frac{i\beta_2}{2}\frac{\partial^2 E_j(T,z)}{\partial T^2} + i\gamma_j 2p_k(T - d_{jk}z, 0)\exp(-\alpha z)E_j(T,z) \qquad (5.6.2)$$

where $d_{jk} \equiv (1/v_j) - (1/v_k)$ is the relative pump/probe walk-off, which can be linearized as $d_{jk} = D\Delta\lambda_{jk}$ if the channel separation $\Delta\lambda_{jk}$ is not too wide. $D = -2\pi c\beta_2/\lambda^2$ is the fiber dispersion coefficient, λ is the average wavelength, and c is the light velocity. This linear approximation of d_{jk} neglected high-order dispersion effects.

In general, dispersion and nonlinearity act together along the fiber. However, as illustrated by Figure 5.6.2, in an infinitesimal fiber section dz, we can assume that the dispersive and the nonlinear effects act independently, the same idea as used in the split-step Fourier method [27]. Let $E_j(T,z) = |E_j|\exp[i\phi_j(T,z)]$, where $|E_j|$ and ϕ_j are the amplitude and the phase of the probe channel optical field. Taking into account the effect of XPM alone, at $z = z'$, the nonlinear phase modulation in the probe signal induced by the pump power in the small fiber section dz can be obtained as

$$d\phi_j(T,z') = \gamma_j 2p_k(T - d_{jk}z', 0)\exp(-\alpha z')\mathrm{dz}$$

The Fourier transformation of this phase variation gives

$$d\tilde{\phi}_j(\Omega,z') = 2\gamma_j p_k(\Omega,0)e^{(-\alpha+i\Omega d_{jk})z'}\mathrm{dz} \qquad (5.6.3)$$

Neglecting the intensity fluctuation of the probe channel, this phase change corresponds to a change in the electrical field, $\bar{E}_j\exp[id\phi_j(T,z')] \approx \bar{E}_j[1 + id\phi_j(T,z')]$, or, in the Fourier domain, $\bar{E}_j[1 + id\tilde{\phi}_j(\Omega,z')]$, where $d\tilde{\phi}_j(\Omega,z')$ is the Fourier transform of $d\phi_j(T,z')$, and $\bar{E}_j$ represents the average field amplitude.

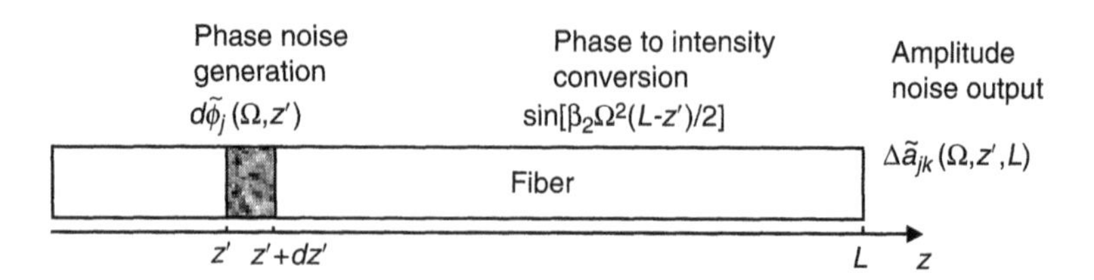

Figure 5.6.2 Illustration of elementary contribution of XPM from a short fiber section.

Due to chromatic dispersion of the fiber, the phase variation generated at location $z = z'$ is converted into an amplitude variation at the end of the fiber $z = L$. Taking into account a source term of nonlinearity-induced phase perturbation at $z = z'$ and the effect of chromatic dispersion, the Fourier transform of Equation 5.6.2 becomes

$$\frac{\partial \tilde{E}_j(\Omega, z)}{\mathrm{dz}} = \frac{i\beta_2\Omega^2}{2} \cdot \tilde{E}_j(\Omega, z) + \bar{E}_j[1 + id\tilde{\phi}(\Omega, z')]\delta(z - z')$$

where the Kronecker delta δ (z−z′) is introduced to take into account the fact that the source term exists only in an infinitesimal fiber section at $z = z'$. Therefore, at the fiber output $z = L$, the probe field is

$$\tilde{E}_j(\Omega, L) = \bar{E}_j + id\tilde{\phi}_j(\Omega, z')\bar{E}_j \exp[i\beta_2\Omega^2(L - z')/2]$$

The optical power variation caused by the nonlinear phase modulation created in the short section dz at $z = z'$ is thus

$$\Delta\tilde{a}_{jk}(\Omega, z', L) = |\tilde{E}_j(\Omega, L)|^2 - \bar{E}_j^2 = -2\bar{E}_j^2 d\tilde{\phi}_j(\Omega, z')\sin[\beta_2\Omega^2(L - z')/2]$$

where a linearization has been made considering that $d\tilde{\phi}_j$ is infinitesimal. Using $E_j(T, z) = A_j(T + z/v_j, z)\exp(\alpha z/2)$ and Equation 5.6.3, integrating all nonlinear phase contributions along the fiber, the accumulated intensity fluctuation at the end of the fiber can be obtained as

$$\Delta\tilde{s}_{jk}(\Omega, L) = -4\gamma_j p_j(0)e^{-(\alpha - i\Omega/v_j)L}\int_0^L p_k(\Omega, 0)\sin[\beta_2\Omega^2(L - z')/2]e^{-(\alpha - i\Omega d_{jk})z'}dz' \tag{5.6.4}$$

where $\Delta\tilde{s}_{jk}(\Omega, L) = \Delta\tilde{a}_{jk}(\Omega, L)e^{-\alpha L}$ represents the fluctuation of A_j at frequency Ω. After integration, we have [26]

$$\Delta\tilde{s}_{jk}(\Omega, L) = 2p_j(L)\gamma_j e^{i\Omega/v_j L}p_k(\Omega, 0)\left\{\frac{\exp(i\beta_2\Omega^2 L/2) - \exp(-\alpha + i\Omega d_{jk})L}{i(\alpha - i\Omega d_{jk} + i\beta_2\Omega^2/2)} - \frac{\exp(-i\beta_2\Omega^2 L/2) - \exp(-\alpha + i\Omega d_{jk})L}{i(\alpha - i\Omega d_{jk} - i\beta_2\Omega^2/2)}\right\} \tag{5.6.5}$$

where $p_j(0)$ and $p_j(L)$ are the probe optical powers at the input and the output of the fiber, respectively. If the fiber length is much longer than the nonlinear length, $\exp(-\alpha L) \ll 1$, and the modulation bandwidth is much narrower than the channel spacing, i.e., $d_{jk} \gg \beta_2\Omega/2$, a much simpler frequency domain description of the XPM-induced intensity fluctuation can be derived for the probe channel:

$$\Delta\tilde{s}_{jk}(\Omega, L) = 4\gamma_j p_j(L) p_k(\Omega, 0) \frac{\sin(\beta_2 \Omega^2 L/2)}{\alpha - i\Omega d_{jk}} e^{i\Omega/v_j L} \tag{5.6.6}$$

Equation 5.6.6 can be further generalized to analyze multispan optically amplified systems, where the total intensity fluctuation at the receiver is the accumulation of XPM contributions created by all fiber spans, as illustrated in Figure 5.6.3. For a system with N amplified fiber spans, the nonlinear phase modulation created in the m-th span produces an intensity modulation $\Delta\tilde{s}_{jk}^{(m)}(\Omega, L_N)$ at the end of the system. Even though the phase modulation creation depends only on the pump power and the pump/probe walk-off within the m-th span, the phase-to-intensity conversion depends on the accumulated dispersion of the fibers from the m-th to the N-th fiber spans, and therefore

$$\Delta\tilde{s}_{jk}^{(m)}(\Omega, L_N) = 4\gamma_j p_j(L_N) p_k^{(m)}(\Omega, 0) \exp\left[i\Omega \sum_{n=1}^{m-1} d_{jk}^{(n)} L^{(n)}\right] \times \frac{\sin\left[\Omega^2 \sum_{n=m}^{N} \beta_2^{(n)} L^{(n)}/2\right]}{\alpha - i\Omega d_{jk}^{(i)}} \exp(i\Omega L_N/v_j) \tag{5.6.7}$$

where $L_N = \sum_{n=1}^{N} L^{(n)}$ is the total fiber length in the system, $L^{(m)}$ and $\beta_2{}^{(m)}$ are fiber length and dispersion of the m-th span (where $L^{(0)} = 0$), $p_k^{(m)}(\Omega, 0)$ is the pump signal input power spectrum in the m-th span, and $d_{jk}^{(m)}$ is the relative walk-off between two channels in the m-th span (where $d_{jk}^{(0)} = 0$). To generalize the single-span XPM expression Equations 5.6.6 to 5.6.7, which represents a multispan system, the term

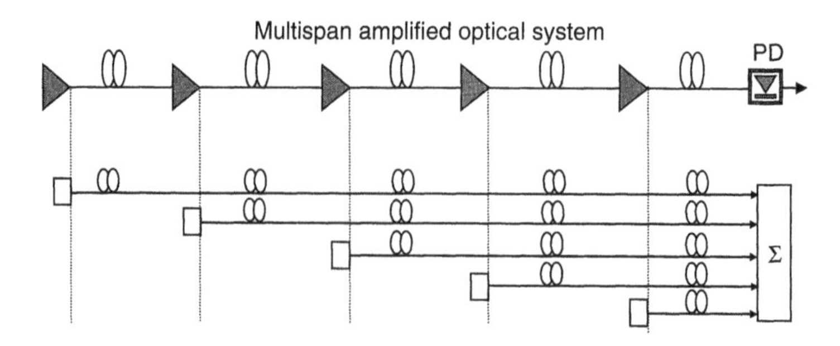

Figure 5.6.3 Linear superposition of XPM contributions from each amplified span.

$\sin(\beta_2\Omega^2 L/2)$ in Equation 5.6.6 has to be replaced by $\sin\left[\Omega^2 \sum_{n=m}^{N} \beta_2^{(n)} L^{(n)}/2\right]$ in Equation 5.6.7 taking into account the linear accumulation of dispersion. Another important effect that has to be taken into account is the different propagation speeds between the pump and the probe wavelengths. The phase mismatch between the pump and the probe waveforms at the input of the m-th span is different from that at the input of the first span. The walk-off-dependent term $\exp\left[i\Omega \sum_{n=1}^{m-1} d_{jk}^{(n)} L^{(n)}\right]$ in Equation 5.6.7 takes into account the walk-off between the probe and the pump channels before they both enter into the m-th fiber span.

Finally, contributions from all fiber spans add up, as illustrated in Figure 5.6.3, and therefore the intensity fluctuation induced by the XPM of the whole system can be expressed as

$$\Delta\tilde{S}_{jk}(\Omega, L_N) = \sum_{m=1}^{N} \Delta\tilde{s}_j^{(m)}(\Omega, L_N) \tag{5.6.8}$$

In the time domain, the output probe optical power with XPM-induced intensity crosstalk is

$$p_{jk}(t, L_N) = p_j(L_N) + \Delta S_{jk}(t, L_N) \tag{5.6.9}$$

where $\Delta S_{jk}(t, L_N)$ is the inverse Fourier transform of $\Delta\tilde{S}_{jk}(\Omega, L_N)$ and $p_j(L_N)$ is the probe output without XPM. $\Delta S_{jk}(t, L_N)$ has a zero mean.

When the probe signal reaches an optical receiver, the electrical power spectral density at the output of the photodiode is the Fourier transform of the autocorrelation of the time domain optical intensity waveform. Therefore we have

$$\rho_j(\Omega, L_N) = \Re^2\left\{p_j^2(L_N)\delta(\Omega) + |\Delta\tilde{S}_{jk}(\Omega, L_N)|^2\right\} \tag{5.6.10}$$

where δ is the Kronecker delta and $\Re$ is the photodiode responsivity. For $\Omega > 0$, the XPM induced electrical domain power spectral density in the probe channel, normalized to its power level without an XPM effect, can be expressed as [26]

$$\begin{aligned}\Delta p_{jk}(\Omega, L_N) &= \frac{\Re^2|\Delta\tilde{S}_{jk}(\Omega, L_N)|^2}{\eta^2 p_j^2(L_N)} \\ &= \left|\sum_{i=1}^{N}\left\{4\gamma_j p_k^{(i)}(\Omega, 0)\exp\left[i\Omega\sum_{n=1}^{i-1} d_{jk}^{(n)} L^{(n)}\right] \frac{\sin\left[\Omega^2 \sum_{n=i}^{N} \beta_2^{(n)} L^{(n)}/2\right]}{\alpha - i\Omega d_{jk}^{(i)}}\right\}\right|^2\end{aligned} \tag{5.6.11}$$

$\Delta p_{jk}(\Omega, L_N)$ can be defined as a normalized XPM power transfer function, which can be directly measured by a microwave network analyzer. It is worth noting that in the derivation of Equation 5.6.11, the waveform distortion of the pump signal has been neglected. This is indeed a small signal approximation, which is valid when the XPM-induced crosstalk is only a weak perturbation to the probe signal [28]. In fact, if this crosstalk level is less than, for example, 20 percent of the signal, the second-order effect caused by the small intensity fluctuation through SPM in the pump is considered negligible.

To characterize the XPM-induced interchannel crosstalk and its impact on optical system performance, it is relatively easy to perform a frequency-domain transfer function measurement. A block diagram of the experimental setup is shown in Figure 5.6.4. Two external-cavity tunable semiconductor lasers (ECL) emitting at λ_j and λ_k, respectively, are used as sources for the probe and the pump. The probe signal is CW and the pump signal is externally modulated by a sinusoidal signal from an RF network analyzer. The two optical signals are combined by a 3 dB coupler and then sent to an EDFA to boost the optical power. A tunable optical filter is used before the receiver to select the probe signal and reject the pump. After passing through an optical preamplifier, the probe signal is detected by a wideband photodiode, amplified by an RF amplifier, and then sent to the receiver port of the network analyzer. The transfer function measured in this experiment is the relationship between the frequency-swept input pump and its crosstalk onto the output probe.

For example, Figure 5.6.5 shows the normalized XPM frequency response measured at the output of a fiber link consisting of a single 114 km span of non-zero dispersion-shifted fiber (NZDSF). The channel spacings used to obtain this figure were 0.8 nm ($\lambda_j = 1559$ nm, $\lambda_k = 1559.8$ nm) and 1.6 nm ($\lambda_j = 1559$ nm, $\lambda_k = 1560.6$ nm). Corresponding theoretical results obtained from Equation 5.6.11 are also plotted in the same figure. To best fit the measured results,

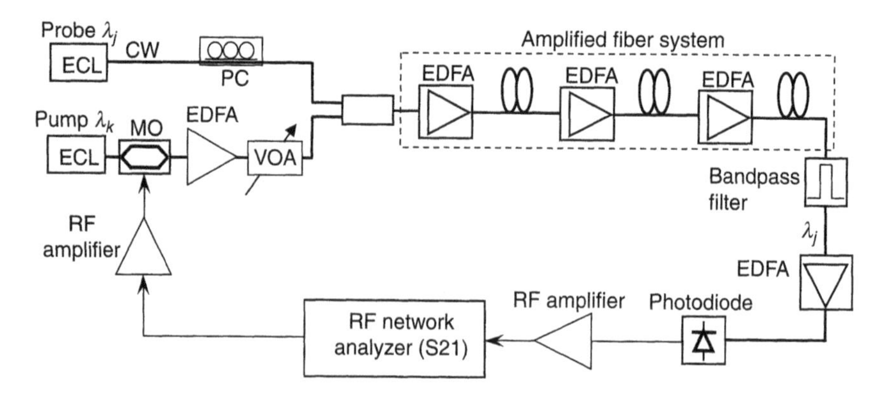

Figure 5.6.4 Experimental setup for frequency domain XPM measurement.

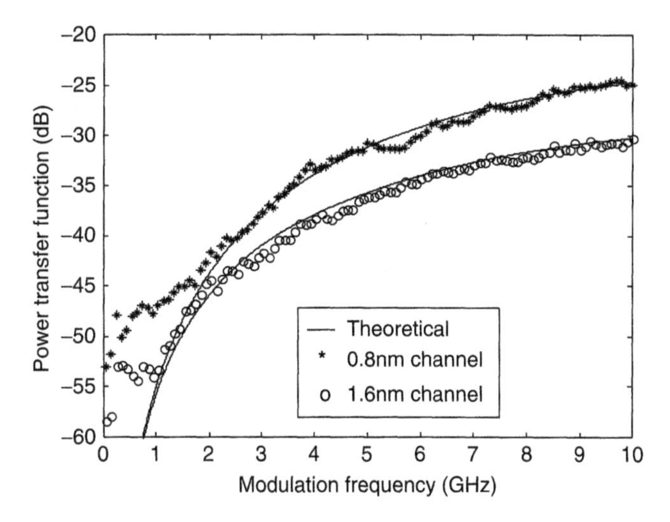

Figure 5.6.5 XPM frequency response in the system with single span (114 km) nonzero dispersion shifted fiber. Stars: 0.8 nm channel spacing (λ_{probe}=1559 nm and λ_{pump}=1559.8 nm), open circles: 1.6 nm channel spacing (λ_{probe}=1559 nm and λ_{pump}=1560.6 nm). Continuous lines are corresponding theoretical results.

parameters used in the calculation were chosen to be $\lambda_0 = 1520.2$ nm for the zero-dispersion wavelength, $S_0 = 0.075$ ps/km/nm^2 for dispersion slope, $n_2 = 2.35 \cdot 10^{-20} \mathrm{m}^2/\mathrm{W}$ for nonlinear index, $A_{eff} = 5.5 \cdot 10^{-11} \mathrm{m}^2$ for the effective area, and $\alpha = 0.25$ dB/km for the fiber loss. These values agree with nominal parameter values of the NZDSF used in the experiment. Both the probe and the pump signal input optical powers were 11.5 dBm, and the pump channel modulation frequency was swept from 50 MHz to 10 GHz. To avoid significant high-order harmonics generated from the $LiNbO_3$ Mach-zehnder intensity modulator, the modulation index is chosen to be approximately 50 percent. Highpass characteristics are clearly demonstrated in both curves in Figure 5.6.5. This is qualitatively different from the frequency dependence of phase-modulation described in the section 5.6.2, where the conversion from phase modulation to intensity modulation through fiber dispersion does not have to be accounted for and the *phase* variation caused by the XPM process has a *lowpass* characteristic [25]. In an ideal IMDD system, the phase modulation of the probe signal by itself does not affect the system performance. However, when a nonideal optical filter is involved, it may convert the phase noise to intensity noise. This is significant in the low frequency part where XPM-induced probe phase modulation is high. The discrepancy between theoretical and experimental results in the low-frequency part of Figure 5.6.5 is most likely caused by the frequency discrimination effect introduced through the narrowband optical filter.

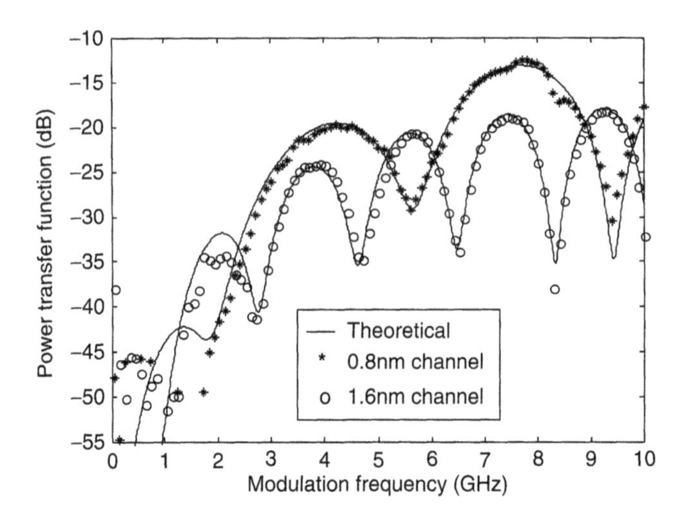

Figure 5.6.6 XPM frequency response in the system with two spans (114 km and 116 km) of NZDSF. Stars: 0.8 nm channel spacing (λ_{probe}=1559 nm and λ_{pump}=1559.8 nm), open circles: 1.6 nm channel spacing (λ_{probe}=1559 nm and λ_{pump}=1560.6 nm). Continuous lines are corresponding theoretical results.

To demonstrate the effect of XPM in multispan systems, Figure 5.6.6 shows the XPM frequency response of a two-span NZDSF system (114 km for the first span and 116 km for the second span), where, again, two channel spacings of 0.8 nm and 1.6 nm were used and the optical power launched into each fiber span was 11.5 dBm. In this multispan case, the detailed shape of the XPM frequency response is strongly dependent on the channel spacing, and the ripples in the XPM spectral shown in Figure 5.6.6 are due to interference between XPM-induced crosstalk created in different fiber spans. For this two-span system, the notch frequencies in the spectrum can be found from Equation 5.6.11 as approximately

$$1 + e^{i\Omega d_{jk} L_1} = 0 \tag{5.6.12}$$

Therefore the frequency difference between adjacent notches in the spectrum is $\Delta f = 1/(d_{jk}L_1)$, where L_1 is the fiber length of the first span. To verify the simple relationship shown in Equation 5.6.12, Figure 5.6.7 shows the measured Δf versus the average signal wavelength at a fixed channel spacing of 1.2 nm. The agreement between theory and experiment suggests that this is also an alternative way to precisely measure fiber dispersion parameters in the system.

Because the resonance structure of the XPM transfer function is caused by interactions between XPM created in various fiber spans, different arrangements of fiber system dispersion maps may cause dramatic differences in the

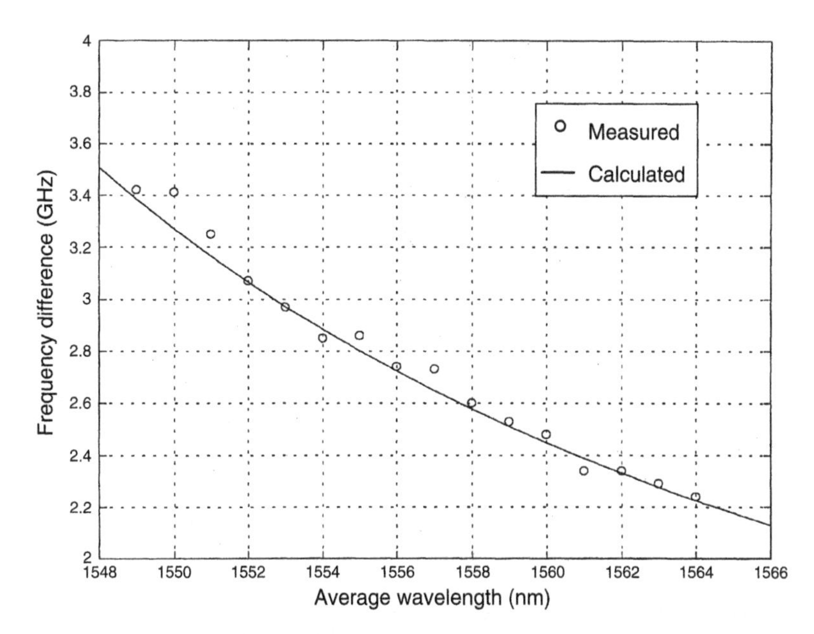

Figure 5.6.7 Frequency differences between adjacent notches in the XPM spectrum versus signal wavelength with channel spacing 1.2 nm. Calculation using fiber parameters $\lambda_0 = 1520.2$ nm, and $S_0 = 0.075$ ps/km/nm^2.

overall XPM transfer function. As another example, the XPM frequency response measured in a three-span system is shown in Figure 5.6.8, where the first two spans are 114 km and 116 km of NZDSF and the third span is 75 km of standard SMF. In this experiment, the EDFAs are adjusted such that the optical power launched into the first two spans of NZDSF is 11.5 dBm and the power launched into the third span is 5 dBm. Taking into account the larger spot size, $A_{eff} = 80\,\mu\text{m}^2$, of the standard SMF (about $55\,\mu\text{m}^2$ for NZDSF) and the lower pump power in the third span, the nonlinear phase modulation generated in the third span is significantly smaller than that generated in the previous two spans.

Comparing Figure 5.6.8 with Figure 5.6.6, it is evident that the level increase in the crosstalk power transfer function in Figure 5.6.8 is mainly due to the high dispersion in the last fiber span. This high dispersion results in a high efficiency of converting the phase modulation, created in the previous two NZDSF spans, into intensity modulation. As a reasonable speculation, if the standard SMF were placed at the first span near the transmitter, the XPM crosstalk level would be much lower.

So far we have discussed the normalized frequency response of XPM-induced intensity crosstalk and the measurement technique. It would also be useful to

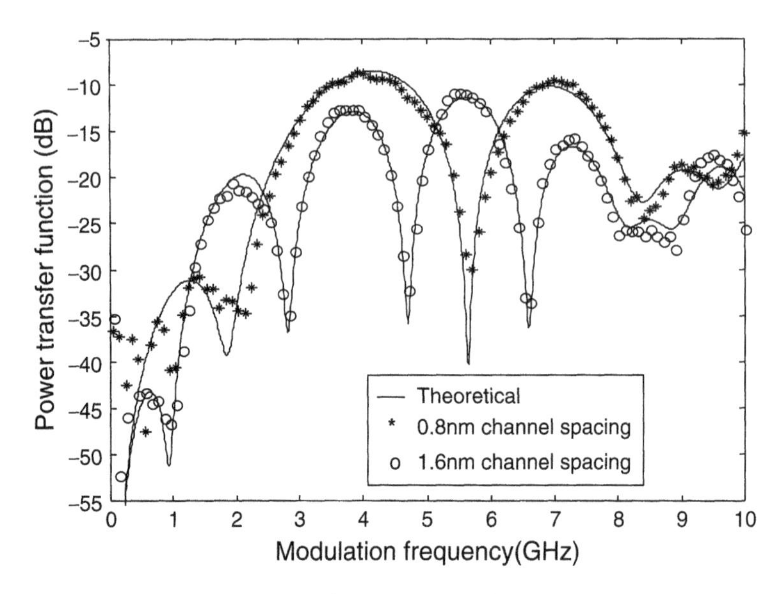

Figure 5.6.8 XPM frequency response in a system with first two spans (114 km and 116 km) of NZDSF and the third span (75 km) of normal SMF. Stars: 0.8 nm channel spacing (λ_{probe} = 1559 nm and λ_{pump} = 1559.8 nm), open circles: 1.6 nm channel spacing (λ_{probe} = 1559 nm and λ_{pump} = 1560.6 nm).

find its impact in the performance of optical transmission systems. Even though the CW waveform of the probe simulates only the continuous 1s in an NRZ bit pattern, the results may be generalized to pseudo-random signal waveforms. It is evident in Equation 5.6.6 that the actual optical power fluctuation of the probe output caused by XPM is directly proportional to the unperturbed optical signal of the probe channel. Taking into account the actual waveforms of both the pump and the probe, XPM-induced crosstalk from the pump to the probe can be obtained as

$$C_{jk}(t) = F^{-1}\left\{F[m_k(t)]\sqrt{\Delta p_{jk}(\Omega, L)}\sqrt{H_j(\Omega)}\right\}m_j(t) \tag{5.6.13}$$

where $m_j(t)$ is the normalized probe waveform at the receiver and $m_k(t)$ is the normalized pump waveform at the transmitter. For pseudo-random bit patterns, $m_{j,k}(t) = u_{j,k}(t)/2P^{av}_{j,k}$ with $u_{j,k}$, the real waveforms, and $P^{av}_{j,k}$, the average optical powers and F and F^{-1} indicate Fourier and inverse Fourier transforms. $H_j(\Omega)$ is the receiver electrical power transfer function for the probe channel.

It is important to mention here that the expression of $\Delta p_{jk}(\Omega, L)$ in Equation 5.6.11 was derived for a CW probe, so Equation 5.6.13 is not accurate during probe signal transitions between 0s and 1s. In fact, XPM during probe signal transitions may introduce an additional time jitter, which is neglected in this analysis. It has been verified experimentally that XPM-induced time jitter due to a probe pattern effect was negligible compared to the XPM-induced eye closure at signal 1 in a system with NRZ coding; therefore the CW probe method might still be an effective approach [29]. Another approximation in this analysis is the omission of pump waveform distortion during transmission. This may affect the details of the XPM crosstalk waveforms calculated by Equation 5.6.13. However, the maximum amplitude of $C_{j,k}(t)$, which indicates the worst-case system penalty, will not be affected as long as there is no significant change in the pump signal optical bandwidth during transmission. In general, the impact of XPM crosstalk on system performance depends on the bit rate of the pump channel, XPM power transfer function of the system, and the baseband filter transfer function of the receiver for the probe channel.

To understand the impact of XPM on the system performance, it is helpful to look at the time-domain waveforms involved in the XPM process. As an example, trace (a) in Figure 5.6.9 shows the normalized waveform (optical power) of the pump channel, which is a 10 Gb/s (2^7-1) pseudo-random bit pattern, band-limited by a 7.5 GHz raised-cosine filter. Suppose that the probe is launched into the same fiber as a CW wave and its amplitude is normalized to 1. Due to XPM, the probe channel is intensity modulated by the pump, and the waveforms created by the XPM process for two different system configurations are shown by traces (b) and (c) in Figure 5.6.9. Trace (b) was obtained in a

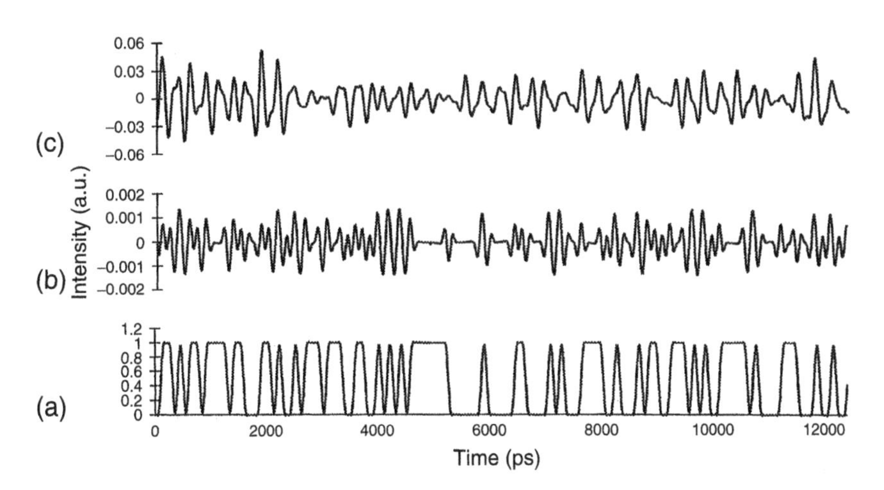

Figure 5.6.9 Time-domain waveforms. Trace a: input pump signal (10 Gb/s (2^7-1) pseudo-random bit pattern). Trace b: XPM crosstalk of the probe channel in a single-span 130 km NZDSF system. Trace c: XPM crosstalk of the probe channel in a three-span system with 130 km NZDSF + 115 km NZDSF + 75 km normal SMF.

single-span system with 130 km NZDSF, whereas trace (c) shows the XPM crosstalk waveform calculated for a three-span system with 130 km NZDSF + 115 km NZDSF + 75 km standard SMF. Looking at these time-domain traces carefully, we can see that trace (b) clearly exhibits a simple highpass characteristic that agrees with the similar waveform measured and reported in [30] in a single-span fiber system. However, in multispan systems, XPM transfer functions are more complicated. Trace (c) in Figure 5.6.9 shows that the amplitude of the crosstalk associated with periodic 0 1 0 1 patterns in the pump waveform has been significantly suppressed.

To help you better understand the features in the time-domain waveforms obtained with various system configurations, Figure 5.6.10 shows the XPM power transfer functions in the frequency-domain corresponding to trace (b) and trace (c) in Figure 5.6.9. In the single-span case, the crosstalk indeed has a simple highpass characteristic. For the three-span system, the XPM power transfer function has a notch at the frequency close to the half bit rate, which suppresses the crosstalk of 0 1 0 1 bit patterns in the time domain.

It is worth mentioning that the crosstalk waveforms shown in Figure 5.6.9 were calculated before an optical receiver. In practice, the transfer function and the frequency bandwidth of the receiver will reshape the crosstalk waveform and may have a strong impact on system performance. After introducing a receiver transfer function, XPM-induced eye closure (*ECL*) in the receiver of a

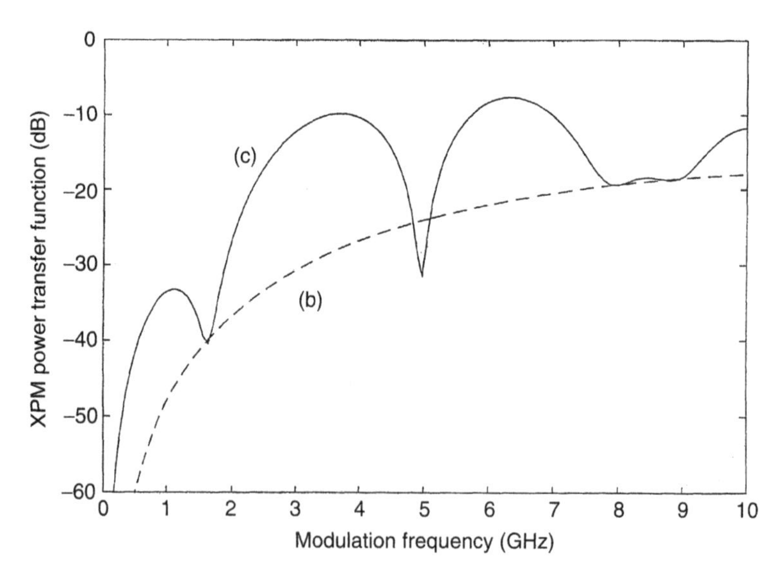

Figure 5.6.10 XPM power transfer functions: (b) corresponds to trace (b) in Figure 5.6.9 and (c) corresponds to trace (c) in Figure 5.6.9.

system can be evaluated from the amplitude of the crosstalk waveform for the probe channel. The worst-case eye closure happens with $m_j(t) = 1$, where $ECL_{(m_j=1)} = \{\max[C_{jk}(t)] - \min[C_{jk}(t)]\}/2$. It is convenient to define this eye closure as a *normalized XPM crosstalk*. In a complete system performance evaluation, this normalized XPM crosstalk penalty should be added on top of other penalties, such as those caused by dispersion and SPM. Considering the waveform distortion due to transmission impairments, the received probe waveform typically has $m_j(t) \leq 1$, especially for isolated 1s. Therefore normalized XPM crosstalk gives a conservative measure of system performance.

In WDM optical networks, different WDM channels may have different data rates. The impact of probe channel bit rate on its sensitivity to XPM-induced crosstalk is affected by the receiver bandwidth. Figure 5.6.11(a) shows the normalized power crosstalk levels versus the probe channel receiver electrical bandwidth for 2.5 Gb/s, 10 Gb/s, and 40 Gb/s bit rates in the pump channel. This figure was obtained for a single-span system of 130 km with a dispersion of 2.9 ps/nm/km, launched optical power of 11.5 dBm, and a channel spacing of 0.8 nm. In this particular system, we see that for a bit rate of higher than 10 Gb/s, the XPM-induced crosstalk is less sensitive to the increase in the pump bit rate. This is because the normalized XPM power transfer function peaks at approximately 15 GHz for this system. When the pump spectrum is wider than 15 GHz, the XPM crosstalk efficiency is greatly reduced. This is the reason that the difference in the XPM-induced crosstalk between 40 Gb/s and 10 Gb/s systems is much smaller than that between 10 Gb/s and 2.5 Gb/s systems.

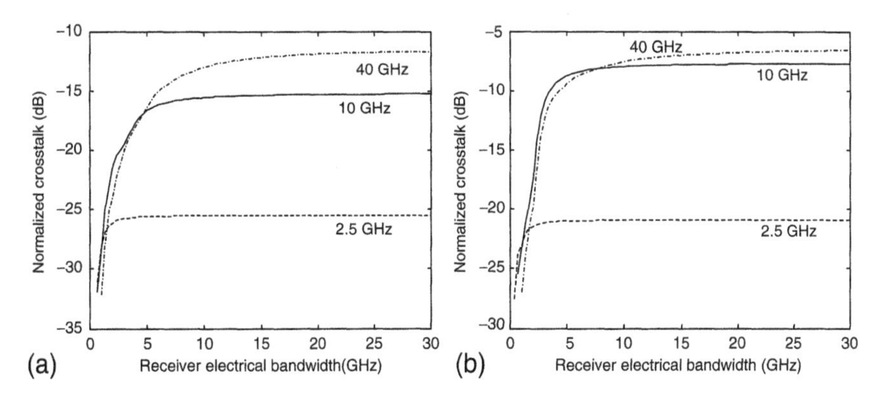

Figure 5.6.11 Normalized power crosstalk levels versus the receiver bandwidth for 2.5 Gb/s, 10 Gb/s, and 40 Gb/s bit rates in the pump channel. (a) The system has a 130 km single fiber span with fiber dispersion of 2.9 ps/nm/km and optical channel spacing of 0.8 nm. Launched pump optical power is 11.5 dBm. (b) The system has five fiber spans (100 km/span) with fiber dispersion of 2.9 ps/nm/km and optical channel spacing of 0.8 nm. Launched pump optical power at each span is 8.5 dBm.

In binary optical transmission, typical receiver bandwidths for 2.5 Gb/s, 10 Gb/s, and 40 Gb/s systems are 1.75 GHz, 7.5 GHz, and 30 GHz, respectively. Figure 5.6.11(a) indicates that when the receiver bandwidth exceeds the bandwidth of the pump channel, there is little increase in the XPM-induced crosstalk level with further increasing of the receiver bandwidth. In principle, the crosstalk between high bit rate and low bit rate channels is comparable to the crosstalk between two low bit rate channels. An important implication of this idea is in hybrid WDM systems with different bit rate *interleaving*; for example, channels 1, 3, and 5 have high bit rates and channels 2, 4, and 6 have low bit rates. The XPM-induced crosstalk levels in both high and low bit rate channels are very similar and are not higher than the crosstalk level in the system with only low bit rate channels. However, when the channel spacing is too low, XPM crosstalk from channel 3 to channel 1 can be bigger than that from channel 2 with a low bit rate channels. Figure 5.6.11(b) shows the normalized crosstalk levels versus receiver electrical bandwidth in a five-span NZDSF system with a 100 km/span. The fiber dispersion is 2.9 ps/nm/km and the launched optical power at each span is 8.5 dBm. There is little difference in the crosstalk levels for the 10 Gb/s system and the 40 Gb/s system. This is because in systems with higher accumulated dispersion, the XPM power transfer function peaks at a lower frequency and the high-frequency components are strongly attenuated.

Figure 5.6.12 shows the normalized crosstalk versus fiber dispersion for the same system used to obtain Figure 5.6.11(b). The fixed receiver bandwidths used for 40 Gb/s, 10 Gb/s, and 2.5 Gb/s systems are 30 GHz, 7.5 GHz, and 1.75 GHz,

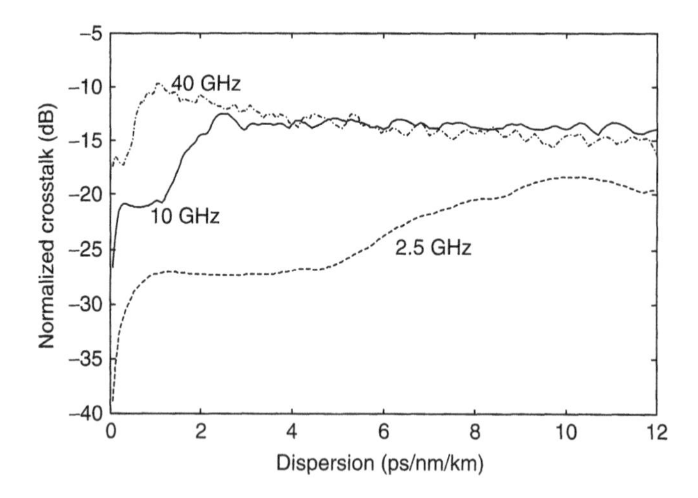

Figure 5.6.12 Normalized power crosstalk levels versus the fiber dispersion for 2.5 Gb/s, 10 Gb/s, and 40 Gb/s bit rates. Five cascaded fiber spans (100 km/span). Optical channel spacing 0.8 nm; 8.5 dBm launched pump optical power at each span.

respectively. The worst-case XPM crosstalk happens at lower dispersion levels with higher signal bit rates. It is worth noting that for the 10 Gb/s system, the worst-case XPM crosstalk happens when the fiber dispersion parameter is 2.5 ps/nm/km, and therefore the total accumulated dispersion of the system is 1250 ps/nm, which is about the same as the dispersion limit for an uncompensated 10 Gb/s system.

It needs to be pointed out that, for simplicity, in both Figure 5.6.11(a) and (b), the signal optical powers were chosen to be the same for systems with different bit rates. A generalization of these results to the case with different signal power levels can be made using the simple linear dependence of XPM crosstalk on the launched power level, as shown in Equation 5.6.6.

Although most people would think that XPM crosstalk was significant only in low dispersion fibers, Figure 5.6.12 clearly indicates that for uncompensated systems, before the system dispersion limit, higher dispersion generally produces more XPM crosstalk. On the other hand, in dispersion compensated optical systems, high local dispersion helps reduce the XPM-induced phase modulation and low accumulated system dispersion will reduce the phase noise to intensity noise conversion.

One important way to reduce the impact of XPM-induced crosstalk in a fiber system is to use dispersion compensation [31]. The position of dispersion compensator in the system is also important. The least amount of dispersion compensation is required if the compensator is placed in front of the receiver. In this position, the dispersion compensator compensates XPM crosstalk created in all fiber spans in the system. The optimum amount of dispersion compensation for the purpose of XPM crosstalk reduction is about 50 percent of the accumulated dispersion in the system [26]. Although this lumped compensation scheme requires the minimum amount of dispersion compensation, it does not give the best overall system performance.

Figure 5.6.13 shows the normalized power crosstalk levels versus the percentage of dispersion compensation in a 10 Gb/s system with six amplified NZDSF fiber spans of 100 km/span. The dispersion of transmission fiber is 2.9 ps/nm/km and the launched optical power into each fiber span is 8.5 dBm. Nonlinear effects in the dispersion-compensating fibers are neglected for simplicity. Various dispersion compensation schemes are compared in this figure. Trace (1) is obtained with compensation in each span. In this scheme XPM-induced crosstalk created from each span can be precisely compensated, so at 100 percent of compensation the XPM crosstalk is effectively eliminated. Trace (2) was obtained with the dispersion compensator placed after every other span. In this case, the value of dispersion compensation can only be optimized for either the first span or the second span but not for both of them. The residual XPM crosstalk level is higher in this case than that with compensation in each span. Similarly, trace (3) in Figure 5.6.13 was obtained with a dispersion compensator placed after every three spans, and trace (4) is with only one lumped compensator placed in

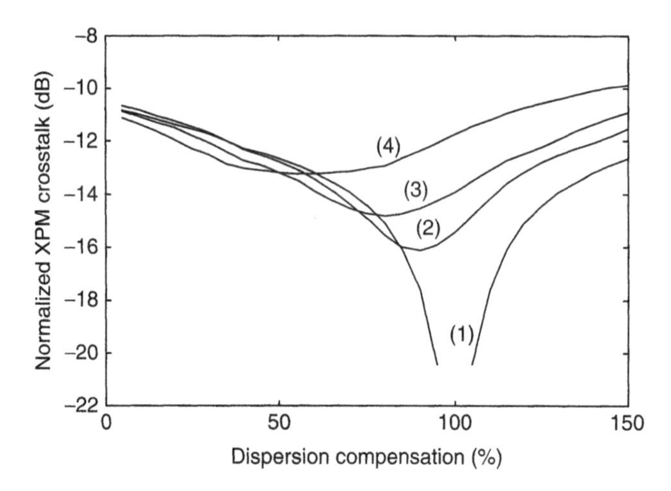

Figure 5.6.13 Normalized power crosstalk levels versus the percentage of dispersion compensation in a 10 Gb/s, six-span system (100 km/span) with fiber dispersion of 2.9 ps/nm/km. An 8.5 dBm launched pump optical power at each fiber span. (1) Dispersion compensation after each span, (2) dispersion compensation after every two spans, (3) dispersion compensation after every three spans, and (4) one lumped dispersion compensation in front of the receiver.

front of the receiver. Obviously, when the number of dispersion compensators is reduced, the level of residual XPM crosstalk is higher and the optimum value of dispersion compensation is closer to 50 percent of the total system dispersion. Therefore, in systems where XPM-induced crosstalk is a significant impairment, per-span dispersion compensation is recommended. However, this will increase the number of dispersion compensators and thus increase the cost.

5.6.2 XPM-Induced Phase Modulation

In the last section, we discussed the intensity modulation introduced by XPM in which the phase modulation is converted into intensity modulation through chromatic dispersion. This XPM-induced intensity crosstalk is a source of performance degradation in IMDD systems and introduces eye closure penalty. On the other hand, if the system is phase modulated, most likely the most relevant impact of XPM is the nonlinear phase modulation itself. In fact, from Equation 5.6.3, if we directly integrate the elementary contribution of nonlinear phase created over the entire fiber, the overall contribution is

$$\tilde{\phi}_j(\Omega) = 2\gamma_j p_k(\Omega, 0) \int_0^L e^{(-\alpha + i\Omega d_{jk})z} dz = 2\gamma_j p_k(\Omega, 0) \sqrt{\eta_{XPM}} L_{eff} e^{j\theta} \quad (5.6.14)$$

where $L_{eff} = (1 - e^{-\alpha L})/\alpha$ is the effective nonlinear length, and

$$\eta_{XPM} = \frac{\alpha^2}{\alpha^2 + \Omega^2 d_{jk}^2}\left[1 + \frac{4\sin^2(\Omega d_{jk} L/2)e^{-\alpha L}}{(1 - e^{-\alpha L})^2}\right] \tag{5.6.15}$$

is the XPM-induced phase modulation efficiency, which is obviously a function of the modulation frequency Ω. The phase term in Equation 5.6.14 is

$$\theta = -\tan^{-1}\left(\frac{\Omega d_{jk}}{\alpha}\right) - \tan^{-1}\left[\frac{e^{-\alpha L}\sin(\Omega d_{jk} L)}{1 - e^{-\alpha L}\cos(\Omega d_{jk} L)}\right] \tag{5.6.16}$$

In time domain, the XPM-induced phase variation of the probe signal can be expressed as [19]

$$\phi_{XPM}(L,t) = \gamma_j[\bar{P}_j(0) + 2\bar{P}_k(0)]L_{eff} + |\tilde{\phi}_j(\Omega)|\cos\left[\Omega\left(t - \frac{L}{v_j}\right) + \theta\right] \tag{5.6.17}$$

where the first term on the right side is a constant nonlinear phase shift with $\bar{P}_j(0)$ and $\bar{P}_k(0)$ the average input powers of the probe and the pump, respectively. In this case the conversion from phase modulation to intensity modulation through chromatic dispersion has been neglected.

If the system has more than one amplified fiber span, the overall effect of cross-phase modulation is the superposition of contributions from all fiber spans,

$$\phi_{XPM}\left(\sum_{l=1}^{N} L^{(l)}, t\right) = \sum_{l=1}^{N}\left|\tilde{\phi}_j^{(l)}(\Omega)\right|\cos\left[\Omega\left(t - \sum_{n=1}^{N}\frac{L^{(n)}}{v_j^{(n)}}\right) + \Omega\sum_{n=1}^{l-1} L^{(n)} d_{jk}^{(n)} + \theta^{(l)}\right] \tag{5.6.18}$$

where N is the total number of amplified fiber spans, and each term on the right hand side of Equation 5.6.18 represents the XPM-induced phase shift created in the corresponding fiber span. The additional phase term $\Omega\sum_{n=1}^{l-1} L^{(n)} d_{jk}^{(n)}$ represents the effect of pump-probe phase walk-off before the l-th span.

The characterization of XPM-induced phase modulation (PM) involves the PM index measurement, which can be accomplished by either coherent heterodyne or self-homodyne measurement. Figure 5.6.14 shows a block diagram of the coherent measurement setup, where two tunable lasers are used as the probe and the pump at the wavelengths of λ_j and λ_k, respectively. The pump laser is intensity modulated by a sinusoidal wave at frequency Ω through an external modulator. The probe and the pump are combined and sent to an optical system with multiple amplified fiber spans. At the output of the optical system, a narrowband optical filter is used to select the probe wave at λ_j while rejecting the pump. Because the probe wave is phase modulated through the XPM process in the fiber system, modulation sidebands will be created in its optical spectrum.

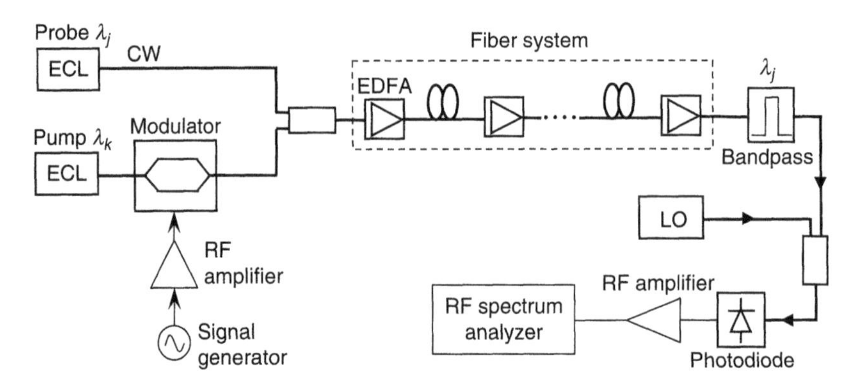

Figure 5.6.14 Experimental setup to measure XPM-induced phase modulation.

The optical spectrum of the output probe wave can be measured by an OSA if its spectral resolution is fine enough to resolve the modulation sidebands. Or it can be measured by a coherent heterodyne detection, as shown in Figure 5.6.14, which provides much better spectral resolution as required if the modulation frequency is on the order of megahertz. Through the measurement of the probe optical spectrum, its phase modulation index can be easily found based on the strength of the modulation sidebands in the spectrum. This is similar to the measurement of modulation chirp in Section 3.2.2. Then, by sweeping the frequency, Ω, of the modulating signal, the frequency response of the XPM-induced phase modulation can be obtained.

Figure 5.6.15 shows examples of the measured XPM-induced PM indices in two-span and three-span amplified fiber-optic systems [25]. In this measurement, the average pump power level $P_k(0)$ at the input of each amplified fiber span was equal; therefore the XPM index was defined as a normalized phase

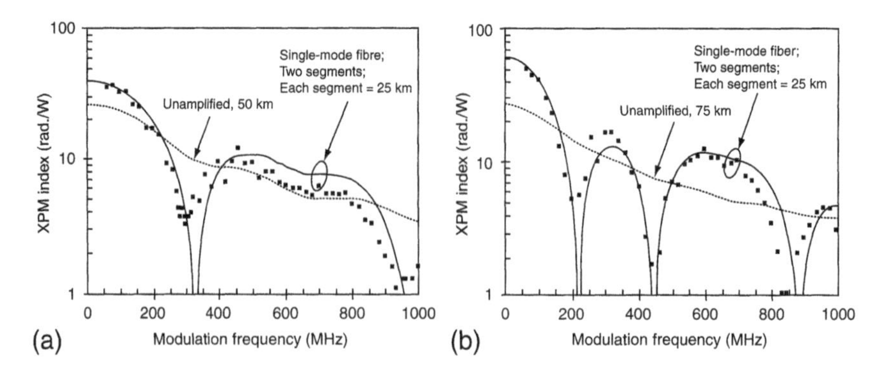

Figure 5.6.15 XPM index for (a) a two-span system and (b) a three-span system [25]. Used with permission.

modulation efficiency $\phi_{XPM}/P_k(0)$. The wavelength spacing between the pump and the probe in this measurement was 3.7 nm.

Figure 5.6.15(a) was obtained in a system with two fiber spans, each having 25 km standard SMF. For Figure 5.6.15(b) there were three fiber spans in the system, again with 25 km standard SMF in each span. In both cases, optical power at the input of each span was set to 7 dBm. It is obvious that if there is no optical amplifier between fiber spans, the XPM index versus modulation frequency decreases monotonically showing a lowpass characteristic. This is shown in Figure 5.6.15 as the dotted lines that were calculated with a single 50 km (a) and 75 km (b) fiber span. When an optical amplifier was added at the beginning of each 25 km fiber span, the XPM indices varied significantly over the modulation frequency. Clearly, this is due to coherent interferences between XPM-induced phase modulations produced in different fiber spans. Compared to the XPM-induced intensity modulation, XPM-induced phase modulation tends to diminish at high modulation frequency, especially when the chromatic dispersion of the fiber is high. From this point of view, XPM-induced intensity modulation discussed in the last section is probably the most damaging effect on high-speed optical systems.

It is important to point out that XPM is the crosstalk originated from the intensity modulation of the pump, which results in the intensity and phase modulations of the probe. In optical systems based on phase modulation on all wavelength channels, XPM will not exist, in principle, because the pump has a constant optical power. However, the coded optical phase modulation carried by the pump wave can be converted into intensity modulation through fiber chromatic dispersion. Then this intensity modulation will be able to produce an XPM effect in the probe channels, thus causing system performance degradation [32].

5.6.3 FWM-Induced Crosstalk in IMDD Optical Systems

Four-wave mixing (FWM) is a parametric process that results in the generation of signals at new optical frequencies:

$$f_{jkl} = f_j + f_k - f_l \tag{5.6.19}$$

where f_j, f_k and f_l are the optical frequencies of the contributing signals. There will be system performance degradation if the newly generated FWM frequency component overlaps with a signal channel in a WDM system, and appreciable FWM power is delivered into the receiver. The penalty will be greatest if the frequency difference between the FWM product and the signal, $f_{jkl} - f_i$, lies within the receiver bandwidth. Unfortunately, for signals on the regular ITU frequency grid in a WDM system, this overlapping is quite probable, especially for long-distance systems where the wavelengths of WDM channels are precisely on the ITU grid.

Over an optical cable span in which the chromatic dispersion is constant, there is a closed form solution for the FWM power to signal power ratio:

$$x_s = \frac{P_{jkl}(L_s)}{P_l(L_s)} = \eta L_{eff}^2 \chi^2 \gamma^2 P_j(0) P_k(0) \tag{5.6.19}$$

where $L_{eff} = (1 - \exp(-\alpha L_s))/\alpha$ is the nonlinear length of the fiber span, L_s is the span length, $P_j(0)$, $P_k(0)$ and $P_l(0)$ are contributing signal optical powers at the fiber input, $\chi = 1, 2$ for nondegenerate and degenerate FWM, respectively, and the efficiency is

$$\eta = \rho \frac{\alpha^2}{\Delta k_{jkl}^2 + \alpha^2} \left[1 + \frac{4e^{-\alpha L_s} \sin(\Delta k_{jkl} L_s/2)}{(1 - e^{-\alpha L_s})^2} \right] \tag{5.6.20}$$

where the detune is

$$\Delta k_{jkl} = -\frac{2\pi c}{f_m^2} D(\lambda_m)[(f_j - f_m)^2 - (f_l - f_m)^2] \tag{5.6.21}$$

$0 < \rho < 1$ is the polarization mismatch factor, and λ_m is the central wavelength, corresponding to the frequency of $f_m = \frac{f_j + f_k}{2} = \frac{f_l + f_{jkl}}{2}$.

In practice, in a multispan WDM system, the dispersion varies not only from span to span but also between cable segments, typically a few kilometers in length, within each span. The contributions from each segment can be calculated using the analytic solution described earlier, with the additional requirement that the relative phases of the signals and FWM product must be taken into account in combining each contribution [33]. The overall FWM is a superposition of all FWM contributions throughout the system:

$$a_F = \sum_{\text{spans}} \sum_{\text{segments}} \exp(i\Delta\phi_{jkl}) \sqrt{\frac{P_{jkl}(z)}{P_l(z)}} \tag{5.6.22}$$

where $\Delta\phi_{jkl}$ is the relative phase of FWM generated at each fiber section. The magnitude of the FWM-to-signal-ratio is quite sensitive, not only to the chromatic dispersions of the cable segments but also to their distribution, the segment lengths, and the exact optical frequencies of the contributing signals. Because of the random nature of the relative phase in each fiber section, statistical analysis may be necessary for system performance evaluation.

Figure 5.6.16 shows the experimental setup to evaluate the impact of FWM in a multispan WDM optical system. In this measurement, a WDM system uses n optical transmitters, and each transmitter is individually adjusted for the optical power and the SOP before being combined in a WDM multiplexer. The PRBS pattern generated by a bit error test set (BERT) is split into n channels to modulate the

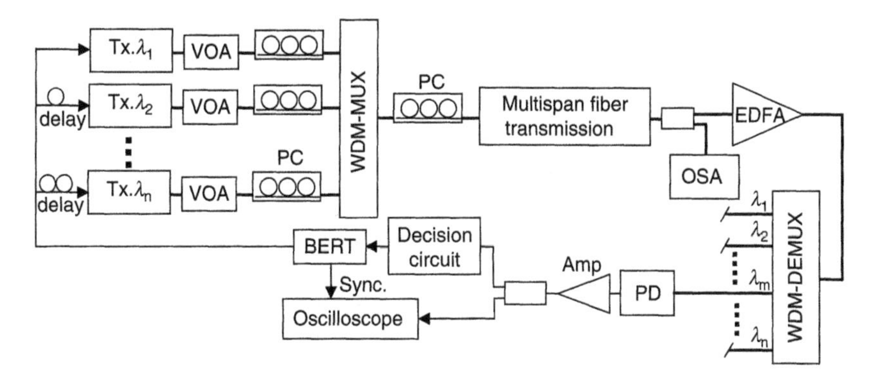

Figure 5.6.16 Measurement of FWM-induced crosstalk in a WDM system.

n optical transmitters. To make sure that the bit sequence of the n channels are not correlated, a relative time delay is introduced for the PRBS pattern before modulating each transmitter. The maximum time delay is determined by the pattern length of the PRBS. The combined multichannel optical signal propagates through a multispan amplified optical system. The optical spectrum of the composite signal at the system output is measured by an OSA which tells the optical power levels of both the WDM optical signal and the FWM crosstalk. Another part of the optical output passes through a WDM demultiplexer, which selects a particular optical channel, and is detected by a wideband photodiode. The signal waveform is displayed on an oscilloscope to measure the time-domain waveform distortion caused by FWM. The signal is also sent for BER testing through a decision circuit.

For optical spectrum measurement using OSA, if the system has only two wavelength channels, degenerate FWM exists and the new frequency components created on both sides of the original optical signal frequencies can be easily measured. However, for a WDM system with multiple equally spaced wavelength channels, the generated FWM components overlap with the original optical signals, and therefore it is not possible to precisely measure the powers of the FWM crosstalk.

For example, in a WDM system with four equally spaced wavelength channels, there are 10 FWM components which overlap with the original signal channels, as illustrated in Figure 5.6.17(a), where f_1, f_2, f_3, and f_4 are the frequencies of the signal optical channels and f_{jkl} (j, k, l = 1, 2, 3, 4) are the FWM components created by the interactions among signals at f_j, f_k, and f_l. One way to overcome this overlapping problem is to use unequal channel spacing in the system for the purpose of evaluating the strength of FWM. We can also deliberately remove one of the signal channels and observe the power of the FWM components generated at that wavelength, as illustrated in

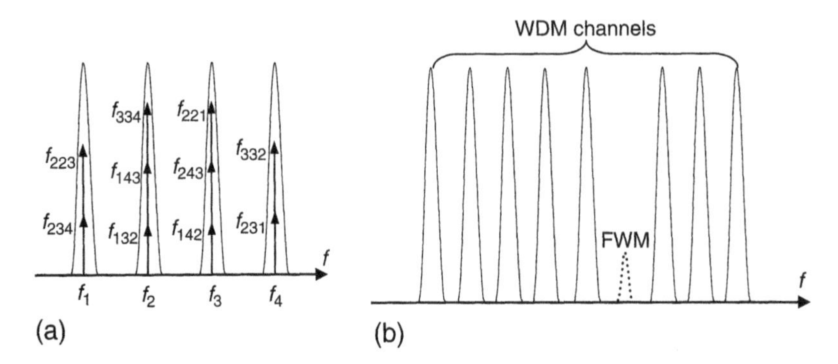

Figure 5.6.17 (a) Illustration of FWM components in a four-channel system and (b) evaluating FWM crosstalk in a WDM system with one empty channel slot.

Figure 5.6.17(b). This obviously underestimates the FWM crosstalk because the FWM contribution that would involve that empty channel is not considered. For example, in the four-channel case, if we remove the signal channel at f_3, FWM components at frequency f_3 would only be f_{221} and f_{142}, whereas f_{243} would not exist.

Figure 5.6.18 shows an example of the measured FWM-to-carrier power ratio defined by Equation 5.6.19. The system consists of five amplified fiber spans with 80 km fiber in each span, and the per-channel optical power level at the beginning of each fiber span ranges from 6 dBm to 8 dBm. The fiber used in the system has zero-dispersion wavelength at $\lambda_0 = 1564$ nm, whereas the average signal wavelength is approximately $\lambda = 1558$ nm. In this system, three channels were used with frequencies at f_1, f_2, and f_3, respectively. The horizontal axis in Figure 5.6.19 is the frequency separation between f_1 and f_3, and the frequency of the FWM component f_{132} is 15 GHz away from f_2, that is, $f_2 = (f_1 + f_3)/2 - 7.5\,GHz$. Because there is no frequency overlap between f_{132} and f_2, the FWM-to-carrier power ratio P_{132}/P_2 can be measured.

However, in practice, both the spectral resolution and the dynamic range of a grating-based OSA are not sufficient to make accurate measurement of this power ratio because the FWM power is typically much lower than the carrier. Coherent heterodyne detection can help us solve this problem, shifting the optical spectrum onto RF domain and utilizing an RF spectrum analyzer. Figure 5.6.18 clearly demonstrates that FWM efficiency is very sensitive to the frequency detune, and there can be more than 10 dB efficiency variation, with the frequency change only on the order of 5 GHz. This is mainly due to the rapid change in the phase match conditions as well as the interference between FWM components created at different fiber spans in the system.

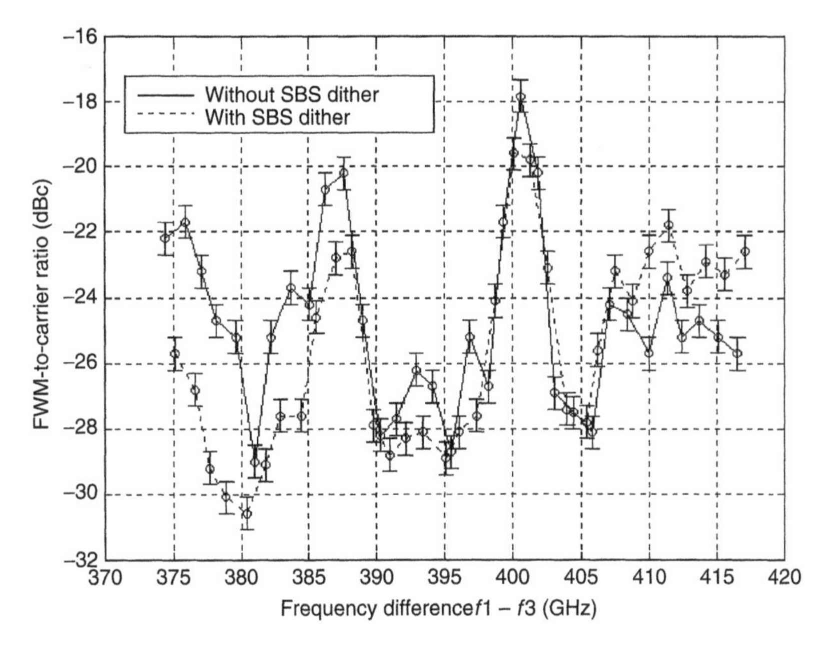

Figure 5.6.18 FWM-to-carrier power ratio measured in a five-span optical system with three wavelength channels. The horizontal axis is the frequency separation of the two outer channels.

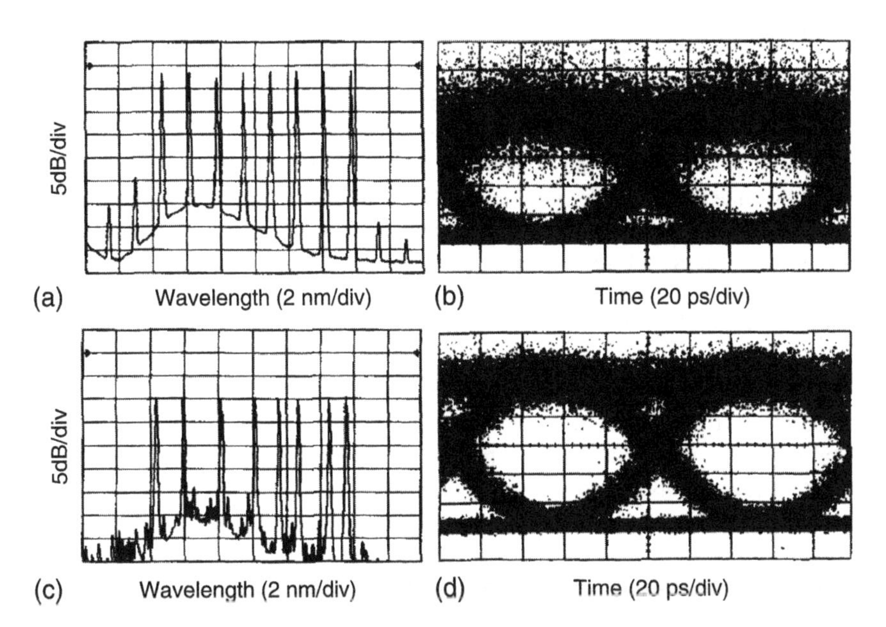

Figure 5.6.19 (a) and (b) WDM optical spectrum with eight equally spaced channels and the measured eye diagram; (c) and (d) WDM optical spectrum with eight unequally spaced channels and the corresponding eye diagram [34]. Used with permission.

The instantaneous four-wave mixing efficiency is also sensitive to the relative polarization states of the contributing signals. In the system measured to obtain Figure 5.6.18, the fiber DGD averaged over the wavelength range 1550–1565 nm is approximately $\tau = 0.056\ ps/\sqrt{km}$. From this DGD value, we can estimate the rate at which signals change their relative polarizations (at the system input they are polarization-aligned). For signal launch states that excite both PSPs of the fiber, the projection of the output SOP onto the Poincaré sphere rotates around the PSP vector at a rate of $d\phi/d\omega = \tau$, where τ is the instantaneous DGD and ω is the (radian) optical frequency. In crude terms, we expect the signals to remain substantially aligned within the nonlinear interaction length of around 20 km. There may be a significant loss of polarization alignment between adjacent spans.

Over the full link, the outer channel polarization alignments between f_1 and f_3 can take more or less arbitrary relative orientations. If transmitters are mutually aligned and the composite signal is aligned with one of the PSPs of the fiber, the output states will be almost aligned. With any residual misalignment arising from second-order PMD (movement of the PSP), there will be misalignment at intermediate spans. So launching on the principal states will not necessarily give worst-case FWM, even ignoring any contributions of fiber birefringence to phase matching. However, for the PMD coefficients given in this system, such effects are not expected to be large. In contrast, if the fiber PMD coefficients exceed $\tau = 0.2\ ps/\sqrt{km}$, the signal polarization states in successive spans will be largely uncorrelated. The worst-case FWM is likely to occur when the signals are aligned in the lowest dispersion spans. If these are intermediate rather than outer spans, then the worst-case FWM will not necessarily occur for signals mutually aligned at the transmitter or for alignment with the principal states at the transmitter or receiver.

In practical long-distance optical transmission systems, frequency dithering is often used to reduce the effect of stimulated Brillouin scattering (SBS). The SBS dither effectively increases the transmission efficiency of the fiber by decreasing the power loss due to SBS while simultaneously smoothing out the phase match peaks. Since the optical powers used in this experiment is less than the SBS threshold, Figure 5.6.18 indicates that SBS dithering only moderately reduces the level of FWM.

Another way to evaluate FWM crosstalk is to measure the closure in the eye diagrams. Figure 5.6.19 shows an example of the measured WDM optical spectra and the corresponding eye diagrams [34]. This figure was obtained in a system with a single-span 137 km dispersion shifted fiber that carries eight optical channels at 10 Gb/s data rate per channel. Since FWM crosstalk is generated from nonlinear mixing between optical signals, it behaves more like a coherent crosstalk than a noise. In an optical system with equally spaced WDM channels, severe degradation on the eye diagram can be observed, as shown in Figure 5.6.19(a) and (b), where the per-channel signal optical power at the input of the fiber was 3 dBm. This eye closure penalty can be significantly reduced if the wavelengths of the WDM channels are unequally spaced as shown in

Figure 5.6.19(c) and (d), although a higher per-channel signal optical power of 5 dBm was used. This is because the FWM components do not overlap with the signal channels and there is no coherent crosstalk.

To explain the coherent crosstalk due to FWM, we can consider that a single FWM product is created at the same wavelength as the signal channel f_i. Assuming that the power of the FWM component is p_{fwm}, which coherently interferes with an amplitude-modulated optical signal whose instantaneous power is p_s, the worst-case eye closure occurs when all the contributing signals are at the high level (digital 1). Due to the mixing between the optical signal and the FWM component, the photocurrent at the receiver is proportional to

$$I' \propto \left| \sqrt{p_s} + \sqrt{p_{fwm}} \cos\left(\Delta\phi + 2\pi(f_i - f_{jkl})\right) \right|^2 \tag{5.6.23}$$

As a result, the normalized signal 1 level becomes

$$A = \frac{p_s + p_{fwm} \pm 2\sqrt{p_s p_{fwm}}}{p_s} \approx 1 \pm 2\sqrt{\frac{p_{fwm}}{p_s}} \tag{5.6.24}$$

This is a "bounded" crosstalk because the worst-case closure caused for the normalized eye diagram is $2\sqrt{p_{fwm}/p_s}$.

In a high-density multichannel WDM system there will be more than one FWM product that overlaps with each signal channel. In general, these FWM products will interfere with the signal at independent beating frequencies and phases. The absolute worst-case eye closure can be found by superpositioning the absolute value of each contributing FWM product. However, if the number of FWM products is too high, the chance of reaching the absolute worst case is very low. The overall crosstalk then approaches Gaussian statistics as the number of contributors increases [35].

5.6.4 Characterization of Raman Crosstalk with Wide Channel Separation

In addition to XPM and FWM, which are originated from Kerr effect nonlinearity, nonlinear Raman scattering is another source that may cause inter-channel crosstalk in fiber-optic WDM systems [22, 36, 37]. Stimulated Raman scattering is capable of transferring optical signal energy from short wavelength channels to longer wavelength channels, especially when the channel spacing is in the vicinity of Raman gain peak. Fiber-optic analog systems such as cable television (CATV) are especially susceptible to the nonlinear crosstalk due to their stringent carrier-to-noise-ratio (CNR) requirement. In a passive optical network (PON), which brings broadband data, voice, and video services to residential and business users, a 1550 nm wavelength video channel travels downstream with a 1490 nm wavelength digital data channel in the same fiber. Although the

1550 nm video channel carries a much stronger optical power compared to the 1490 nm data channel, the performance of the analog CATV channels may still be significantly degraded by the presence of the digital data channel due to nonlinear Raman crosstalk if the system is not designed properly [38].

Suppose two wavelength channels are copropagating along a fiber. It is well known that the nonlinear Raman crosstalk between these two channels can be described by the coupled-wave equations [39]

$$\frac{\partial P_j(z,t)}{\partial z} + \frac{1}{v_j}\frac{\partial P_j(z,t)}{\partial t} = \left[g_s P_j(z,t) - \alpha\right] P_j(z,t) \tag{5.6.25}$$

$$\frac{\partial P_k(z,t)}{\partial z} + \frac{1}{v_k}\frac{\partial P_k(z,t)}{\partial t} = \left[-g_s P_k(z,t) - \alpha\right] P_k(z,t) \tag{5.6.26}$$

where $P_j(z, t)$ and $P_k(z, t)$ are optical signal intensities at wavelengths λ_j and λ_k, v_j and v_k are the group velocities of these two channels, g_s is the stimulated Raman coefficient, α is the loss of the fiber, and z is the propagation direction along the fiber.

This set of coupled-wave equations can be solved using the perturbation method. Similar to the pump-probe techniques for XPM, to obtain the frequency response of the nonlinear Raman crosstalk we can set channel j to be a CW probe with constant optical power at the fiber input $P_j(0,t) = P_{j0}$, whereas channel k is set as the pump that is amplitude modulated by a single RF frequency:

$$P_k(0,t) = P_{k0}[1 + m\cos(\Omega t)] \tag{5.6.27}$$

where Ω is the RF modulation frequency and m is the modulation index.

After transmitting over the fiber with the length L, channel j, which was initially CW, is intensity modulated through the nonlinear Raman process:

$$P_j(L,u) = P_{j0}e^{-\alpha L}(1 + A) \tag{5.6.28}$$

where $u = t - L/v_j$ is the relative time scale, $e^{-\alpha L}$ is the attenuation, and the parameter A represents the Raman crosstalk

$$A = \int_0^L g_s P_k(0, u + d_{jk}z)e^{-\alpha z}dz \tag{5.6.29}$$

and $d_{jk} = (1/v_j - 1/v_k)$ is the relative delay between the two channels.

Combining Equations 5.6.27, 5.6.28, and 5.6.29, the probe channel output is

$$P_j(L,u) = P_{j0}e^{-\alpha L}\left\{1 + g_s P_{k0}L_{eff} + g_s P_{k0}m\frac{\sqrt{1 + e^{-2\alpha L} - 2e^{-\alpha L}\cos(\Omega d_{jk}L)}}{\sqrt{\alpha^2 + (d_{jk}\Omega)^2}}\cos(\Omega u + \Phi_{SRS})\right\} \tag{5.6.30}$$

with the phase angle is

$$\Phi_{SRS} = \tan^{-1}\left(\frac{\Omega d_{jk}}{\alpha}\right) + \tan^{-1}\left(\frac{e^{-\alpha L}\sin(\Omega d_{jk}L)}{e^{-\alpha L}\cos(\Omega d_{jk}L) - 1}\right) \tag{5.6.31}$$

where $L_{eff} = (1 - e^{-\alpha L})/\alpha$ is the nonlinear length of the fiber. In principle, the Raman gain coefficient g_s is affected by the relative polarization states of the two channels involved in the Raman crosstalk process. It is generally accepted that Raman gain is maximum when the two channels have the same polarization states, whereas no Raman crosstalk is expected when the two channels are orthogonally polarized. Usually the SRS-induced crosstalk, C_{SRS}, is defined as the power ratio between the RF signal induced on the CW probe channel and the RF signal on the pump channel, that is,

$$C_{SRS} = 10\log(g_s P_{k0})^2 \frac{1 + e^{-2\alpha L} - 2e^{-\alpha L}\cos(\Omega\, d_{jk}L)}{(\alpha^2 + \Omega^2 d_{jk}^2)(1 + g_s P_{k0} L_{eff})^2} \tag{5.6.32}$$

The characterization of SRS crosstalk can be accomplished using the pump-probe experiment as shown in Figure 5.6.20. Two optical transmitters are used with emission wavelengths at λ_j and λ_k, respectively, for the pump and the probe. An RF network analyzer is set at S_{21} mode, and its swept frequency output is used to modulate the pump while the output of the probe transmitter is CW. The pump and the probe are combined by a WDM multiplexer and their polarization states are individually adjusted by polarization controllers. At the output of the fiber system, a bandpass optical filter selects the probe wavelength and rejects the pump. The probe signal output from the system is then detected by an optical receiver, and the RF signal is sent to the receiving port of the RF network analyzer.

Equivalently, instead of using an RF network analyzer, this measurement can also be accomplished using a separate swept-frequency RF source and an RF spectrum analyzer [40]. Figure 5.6.21 shows the measured SRS crosstalk as a function of the modulation frequency. In this measurement, the power levels at the input of the

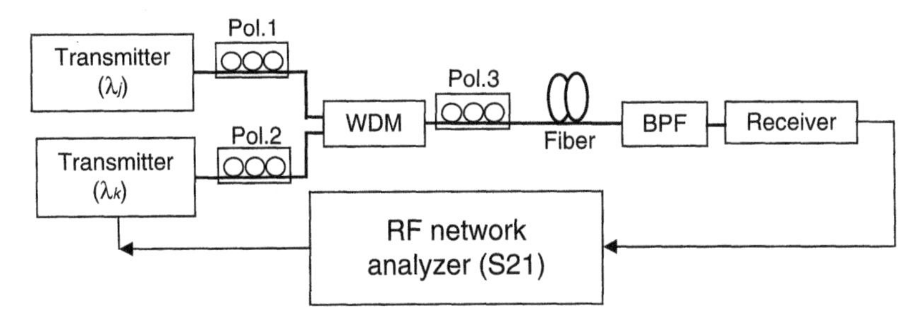

Figure 5.6.20 Characterization of SRS-induced crosstalk using an RF network analyzer.

pump and the probe channels are $P_{k0} = 13$ dBm and $P_{j0} = 6.2$ dBm, respectively, and their wavelengths are $\lambda_j = 1555.6$ nm and $\lambda_k = 1543.5$ nm so that the channel spacing is 12.1 nm. The length of the standard single-mode fiber used for transmission is $L = 25$ km. The dashed line in Figure 5.6.21 is the result calculated from Equation 5.6.32. Other parameters are: fiber loss $\alpha \approx 0.21$ dB/km, and chromatic dispersion $D = 17$ ps/nm/km. The Raman gain coefficient for this 12.1 nm wavelength separation is $g_s \approx 0.5 \times 0.094\ W^{-1}km^{-1}$, where the factor 0.5 accounts for the completely random polarization overlap between the pump and the probe.

It is important to note that in this measurement, uncertainties of SRS crosstalk versus frequency may arise because of the random nature of the PMD in the fiber. In fact, as discussed in Chapter 4, the differential group delay $\Delta\tau$, which is a direct indication of fiber PMD, can be evaluated by the relative polarization walk-off of the two wavelength components:

$$\Delta\tau = \frac{\Delta\phi}{\Delta\omega} \approx -\frac{\lambda_0^2}{2\pi c}\frac{\Delta\phi}{\Delta\lambda} \tag{5.6.33}$$

where $\Delta\phi$ is the angular walk-off between the two wave vectors on the Poincare sphere, which represents the walk-off of their polarization states in radians; c is the speed of light; $\Delta\lambda$ is the wavelength difference between the two channels; and λ_0 is the average wavelength. As an example, for a 10 km of standard single-mode fiber with a typical PMD parameter of $0.1\,\mathrm{ps}/\sqrt{\mathrm{km}}$, the total first-order DGD is approximately $\Delta\tau = 0.1\sqrt{10} = 0.32$ ps. If the wavelength separation between the two wavelength channels is $\Delta\lambda = 12$ nm, the relative polarization walk-off $\Delta\phi$ between the two wavelength components is about π radians.

As mentioned previously, a direct consequence of this polarization walk-off is the reduction of Raman gain coefficient g_s seen by the optical signals propagate

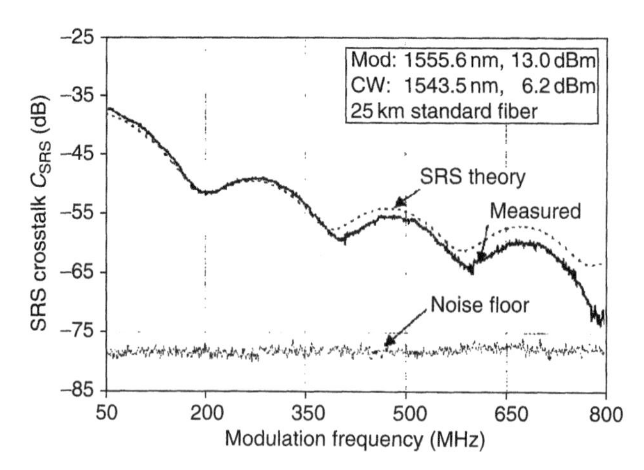

Figure 5.6.21 SRS crosstalk versus modulation frequency for a system with 25 km fiber [40]. Used with permission.

along the fiber. In a typical PON, the wavelength spacing between the digital and the analog channels is about 60 nm; since the polarization walk-off is on the order of multiple π's, the effective Raman gain will be position-dependent along the fiber. Because of the temporal walk-off between the two channels due to chromatic dispersion, nonlinear crosstalk originated from different locations along the fiber will likely contribute to the change of the frequency in the overall transfer function. Although the effect of PMD and thus the angular walk-off of the SOP between the two channels is random in nature, we can describe this phenomenon by representing the Raman gain $g_s(z)$ along the fiber at any given time as a Fourier series:

$$g_s(z) = \frac{g_0}{2}\left[1 + \sum_{n=0}^{N} k_n \cos\left(\frac{2n\pi}{L} z + \phi_n\right)\right] \tag{5.6.34}$$

where g_0 is the maximum Raman gain at the given pump-probe channel spacing that occurs when the two channels are copolarized. n quantifies the rate of rotation of the relative polarization states along the fiber. Higher values of n represent a higher PMD in the fiber. ϕ_n is an initial phase angle, together with coefficient k_n indicating the polarization state mismatch at the fiber input.

With this assumption of location-dependent Raman gain coefficient, the overall Raman crosstalk at the end of a fiber of length L can be expressed in a closed form. The RF power of the Raman crosstalk, normalized by the average signal power at the probe channel, can be found as [41]

$$\delta P_j = \frac{1}{2}(p^2 + q^2)(1 + e^{-2\alpha L} - 2e^{-\alpha L}\cos \Omega d_{jk} L) \tag{5.6.35}$$

where

$$p = \frac{g_0 P_{k0} m}{4}\left\{\frac{-2\Omega d_{jk}}{\alpha^2 + (\Omega d_{jk})^2} - \sum_{n=0}^{N}\left\{k_n \frac{(\Omega d_{jk} + 2n\pi/L)\cos\phi_n + \alpha \sin\phi_n}{\alpha^2 + (\Omega d_{jk} + 2n\pi/L)^2}\right.\right.$$

$$\left.\left. + k_n \frac{(\Omega d_{jk} - 2n\pi/L)\cos\phi_n - \alpha\sin\phi_n}{\alpha^2 + (\Omega d_{jk} - 2n\pi/L)^2}\right\}\right\}$$

$$q = \frac{g_0 P_{k0} m}{4}\left\{\frac{2\alpha}{\alpha^2 + (\Omega d_{jk})^2} - \sum_{n=0}^{N}\left\{k_n \frac{(\Omega d_{jk} + 2n\pi/L)\sin\phi_n - \alpha\cos\phi_n}{\alpha^2 + (\Omega d_{jk} + 2n\pi/L)^2}\right.\right.$$

$$\left.\left. - k_n \frac{(\Omega d_{jk} - 2n\pi/L)\sin\phi_n + \alpha\cos\phi_n}{\alpha^2 + (\Omega d_{jk} - 2n\pi/L)^2}\right\}\right\}$$

Equation 5.6.35 is a general expression of frequency-dependent Raman crosstalk. It takes into account the variation of Raman gain coefficient along the fiber caused by channel polarization walk-off. This normalized crosstalk is proportional to the average input power P_{k0} of the pump channel and peak Raman gain coefficient g_0, which is a function of the wavelength spacing. By adjusting k_n, ϕ_n, and N in Equation 5.6.34, arbitrary variations of Raman gain coefficient along the fiber can be produced.

Figure 5.6.22 shows the normalized transfer functions of nonlinear Raman crosstalk measured using 10 km of standard single-mode fiber (SMF-28), where the normalized crosstalk is defined as $T(f) = \delta P_j(f)/P_{j0}$. In this measurement, the wavelengths of the probe and the pump were set at $\lambda_j = 1550$ nm and $\lambda_k = 1490$ nm, respectively, so that the channel spacing is 60 nm. Their optical power levels were 17 dBm and 0 dBm. Two measured transfer functions in Figure 5.6.22 illustrate the dramatic change in the crosstalk transfer function that was obtained only by changing the signal polarization states and the PSPs of the transmission fiber through the adjustment of the polarization controllers. In addition to the fact that the efficiency of Raman crosstalk depends on the relative polarization states of the two wavelength channels, it is evident that the frequency

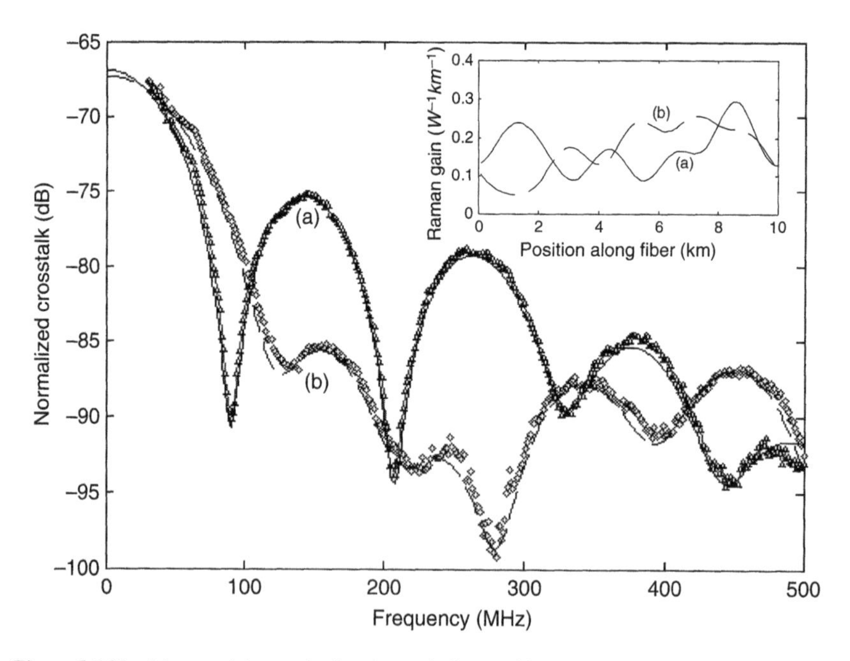

Figure 5.6.22 Measured (opened triangles and diamonds) and calculated (solid and dashed lines) transfer functions of nonlinear Raman crosstalk in the same system but with different polarization conditions. Inset: profiles of Raman gain coefficients along the fiber used in the calculation to fit the measured transfer functions (a) and (b), respectively [41].

dependency of this crosstalk was very sensitive to the polarization states of the input optical signals as well as the fiber PSP.

A direct consequence of this phenomenon is that the system performance will vary over time due to the fluctuation of the fiber PSPs. The solid line (a) and the dashed line (b) in Figure 5.6.22 show the calculated transfer functions. The parameters used in the calculation are $P_{k0} = 1$ mW, $m = 1$, $g_0 = 0.34\ W^{-1}km^{-1}$, and $\Delta\lambda = 60$ nm. To obtain the best match for each curve in Figure 5.6.22, specific profiles of Raman gain coefficient along the fiber had to be assumed, as shown by the inset in Figure 5.6.22. The good agreement between the theoretical calculation and the measurement verifies the hypothesis that the frequency dependency of Raman crosstalk depends on the fiber birefringence. The important conclusion of this measurement is that the frequency transfer function of Raman crosstalk is very sensitive to the signal polarization states, and copolarization at the fiber input does not necessarily imply the worst-case nonlinear crosstalk.

Then we should ask: What is the worst-case Raman crosstalk? It can be obtained using the maximum-hold function on the network analyzer while varying the fiber PSPs and the input signal polarization states. The result is shown in Figure 5.6.23 (by the bold line). This measurement was performed on a system with 20 km of SMF-28. Although the instantaneous transfer functions vary with time, the maximum-hold values are deterministic, which indicates the existence of a worst-case limit.

In practical fiber-optic systems, the PSPs of single-mode transmission fibers change randomly with time due to the effect of PMD. No matter what the condition is at the fiber input, the relative polarization alignment between multi-wavelength channels will always be unpredictable while evolving along the fiber. Therefore, the Raman gain coefficient will vary both with the location along the fiber and with time. For a particular system configuration, a global worst case may be found using Equation 5.6.35. This can be done by searching for the maximum crosstalk level at each RF frequency with all possible distributions of polarization states along the fiber. This worst-case Raman crosstalk transfer function provides a base for potential channel CNR equalization. Higher modulation indices may be used for the RF channels at frequencies with high Raman crosstalk. In Figure 5.6.23(a), the dashed line shows the absolute maximum level of Raman crosstalk calculated versus frequency. The solid line in the same figure indicates the 99 percent worst-case crosstalk level. This was obtained by making 10^5 calculations at each frequency using Equation 5.6.35, assuming a uniform distribution for $k_n \in (0,1)/\sum_{i=1}^{N} k_i$ and $\theta_n \in (0, 2\pi)$, and 10 Fourier terms were used ($N = 10$). This calculated 99 percent worst case agrees well with the experimentally measured transfer function using maximum holding.

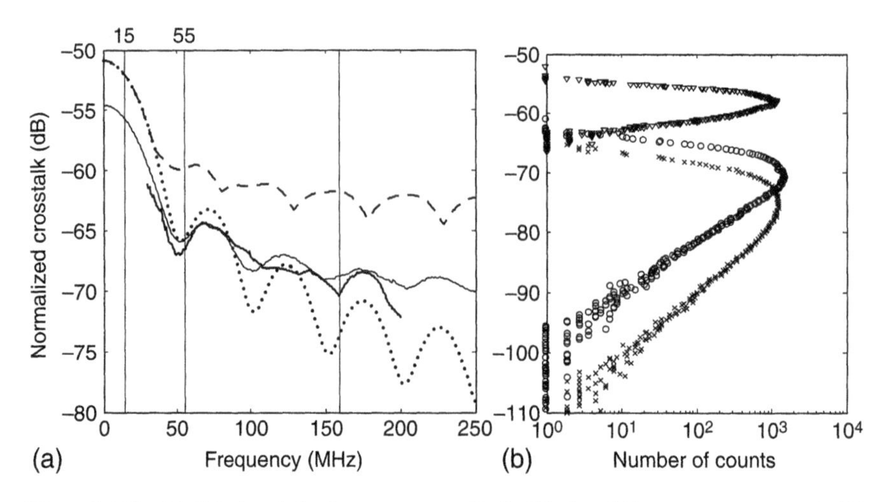

Figure 5.6.23 (a) Calculated absolute worst-case (dashed line) and 99 percent worst-case (solid line) Raman crosstalk in a 20 km fiber system. The bold line represents the measured worst case, which was measured using a maximum-hold on the network analyzer. Dotted line: calculated Raman crosstalk assuming perfect signal polarization alignment along the fiber. (b) Calculated statistical distribution of Raman crosstalk at 15 MHz (triangles), 55 MHz (open circles), and 160 MHz (open triangles) modulation frequencies. 10^5 calculations were made at each wavelength [41].

It was generally accepted that the worst-case Raman crosstalk can be predicted by assuming copolarization of the two wavelength channels all along the fiber and no polarization walk-off. The dotted line in Figure 5.6.23(a) was calculated under this condition. Although in low frequencies (up to 35 MHz) this approximation coincides with the absolute worst case, it underestimates the crosstalk level at high frequencies. This is caused by the interference of Raman crosstalk created at different locations along the fiber. As a reference, Figure 5.6.23(b) shows the statistical distributions of Raman crosstalk calculated at 15 MHz, 55 MHz, and 160 MHz, respectively. At low frequencies, the width of the statistical distribution is relatively narrow; this width becomes widened with the increase of frequency. For 20 km of SMF with dispersion parameter 16.5 ps/nm/km, the accumulated chromatic dispersion is 330 ps/nm. The group delay difference between two wavelength channels 60 nm apart is approximately 19.8 ns. This group delay difference is much shorter than the RF period for a modulation frequency at 15 MHz, and therefore interference effect is not significant. At a modulation frequency of 55 MHz, the RF period is approximately 18 ns, which is on the same order of the group delay difference. Both constructive and destructive interference may occur depending on the polarization walk-off between the two channels, making the crosstalk level less predictable.

It is worth noting that for shorter fibers, the effect of polarization walk-off is not significant. In these cases copolarization between the two wavelength channels all along the fiber does represent the worst-case Raman crosstalk. It was predicted that a global worst-case Raman crosstalk happened with a fiber length of approximately 10 km, assuming aligned polarization states [38], where the worst-case crosstalk was found always at the lower frequency limit of CATV band (55 MHz). To make sure that this global worst case still holds when random polarization walk-off is considered, Figure 5.6.24 shows the calculated Raman crosstalk at 55 MHz RF frequency versus fiber length for several different channel spacings. When the channel spacing is lower than 70 nm, the worst-case crosstalk does happen at a fiber length of approximately 10 km, as predicted by [38]. However, with wider channel spacing [42], the worst-case crosstalk may move to longer fiber lengths, and these worst cases do not correspond to copolarization of the two channels. Figure 5.6.24 was obtained with the typical fiber parameters: dispersion slope $S_0 = 0.086$ ps/(nm^2 km) and zero dispersion wavelength $\lambda_0 = 1313$ nm. It is worthwhile to point out that these worst cases of crosstalk may not easily be found by measurements because of the requirement of precise polarization conditions along the fiber. Although the probability for these worst cases to happen is small, they have to be considered in system design.

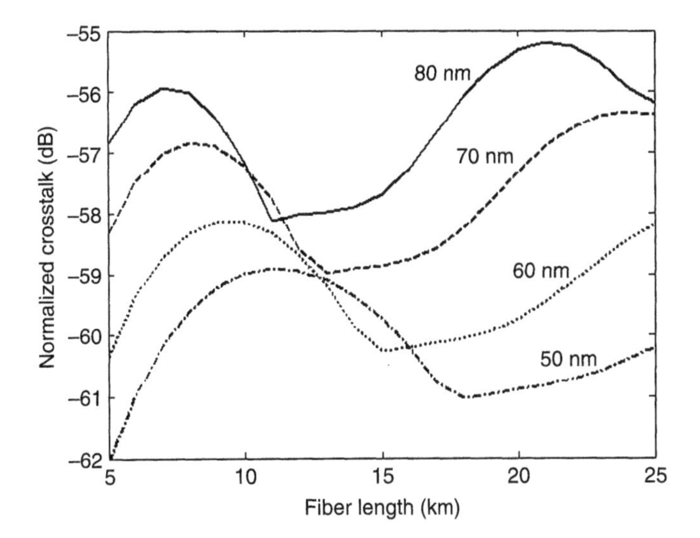

Figure 5.6.24 Calculated worst-case Raman crosstalk at 55 MHz versus fiber length for four different wavelength separations of $\Delta\lambda = 80$ nm (solid line), 70 nm (dashed line), 60 nm (dotted line), and 50 nm (dash-dotted line).

5.7 MODULATION INSTABILITY AND ITS IMPACT IN WDM OPTICAL SYSTEMS

The impact of Kerr effect nonlinearity can be significant in long-distance fiber-optic systems, as discussed in the last section. In addition to nonlinear crosstalk due to XPM and FWM, crosstalk may also happen between the optical signal and the broadband ASE noise in the system to cause performance degradations. This effect is commonly referred to as *modulation instability* (MI) [43]. The mechanism for system performance degradation caused by MI depends on system type. In coherent transmission systems, degradation is mainly caused by the broadening of the optical spectrum [44]. On the other hand, for an intensity-modulated system with direct detection (IMDD), the performance degradation can be introduced by phase noise to intensity noise conversion through chromatic dispersion in the optical fiber. The increased relative intensity noise (RIN) within the receiver baseband is responsible for this degradation.

5.7.1 Modulation-Instability and Transfer Matrix Formulation

Since MI is caused by the Kerr effect nonlinearity, its analysis can be based on the nonlinear Schrodinger equation, which was given in Equation 1.3.96:

$$\frac{\partial A(t,z)}{\partial z}+\frac{i\beta_2}{2}\frac{\partial^2 A(t,z)}{\partial t^2}+\frac{\alpha}{2}A(t,z)-i\gamma\left| A(t,z)\right|^2 A(t,z)=0 \qquad (5.7.1)$$

where $A(z, t)$ is the electrical field, $\gamma = \omega_0 n_2/cA_{eff}$ is the nonlinear coefficient of the fiber, ω_0 is the angular frequency, n_2 is the nonlinear refractive index of the fiber, c is the speed of light, A_{eff} is the effective fiber core area, β_2 is the fiber dispersion parameter, and α is the fiber attenuation. High-order dispersions were ignored here.

In Section 4.8.3, in the discussion of MI and the measurement of fiber nonlinear parameter using MI, fiber attenuation was neglected so that we were able to obtain analytical equations as given by Equation 4.8.44–4.8.48. However, for more accurate analysis for the impact of MI in the performance of optical systems, fiber attenuation has to be taken into account. The steady-state solution of Equation 5.7.1 has to be z-dependent:

$$A_0(z)=\sqrt{P_0}\exp\left(i\gamma\left|A_0(z)\right|^2 z\right)\exp\left(-\frac{\alpha}{2}z\right) \qquad (5.7.2)$$

Since the signal optical power can vary significantly along the fiber in practical optical systems because of the attenuation, a simple mean-field approximation over the transmission fiber is usually not accurate enough. To obtain a semianalytical solution, the fiber can be divided into short sections, as illustrated in Figure 5.7.1, and a mean-field approximation can be applied within each section. For example, in the j-th section with length Δz_j, Equation 5.7.1 becomes:

$$\frac{\partial A_j(z,t)}{\partial z} = \frac{-i}{2}\beta_2 \frac{\partial^2 A_j(z,t)}{\partial t^2} + i\gamma_j \left| A_j(z,t) \right|^2 A_j(z,t) \quad (5.7.3)$$

where

$$\gamma_j = \frac{1 - \exp(-\alpha \Delta z_j)}{\alpha \Delta z_j} \gamma \quad (5.7.4)$$

With the assumption that noise power at fiber input is much weaker than pump power, the solution of Equation 5.7.3 can be written as

$$A_j(z,t) = \left[A_{0j} + \tilde{a}_j(z,t)\right] \exp\left(i\gamma |A_{0j}|^2 z\right) \quad (5.7.5)$$

where A_{0j} is the steady-state solution of Equation 5.7.3, $\tilde{a}_j(z,t)$ is a small perturbation and $\tilde{a}_j(z,t) \ll A_{0j}$ is assumed. With linear approximation of the noise term, the nonlinear Schrodinger Equation 5.7.3 becomes

$$\frac{\partial \tilde{a}_j(z,t)}{\partial z} = \frac{-i}{2}\beta_2 \frac{\partial^2 \tilde{a}_j(z,t)}{\partial t^2} + i\gamma_j \left[|A_{0j}|^2 \tilde{a}_j(z,t) + A_{0j}^2 \tilde{a}_j^*(z,t) \right] \quad (5.7.6)$$

where the symbol * denotes complex conjugate. We need to emphasize that the linearization used to obtain Equation 5.7.6 is valid only when the perturbation is small enough such that the effect of pump depletion can be neglected.

By converting Equation 5.7.6 into frequency domain through Fourier transform [45], the following two equations can be obtained for the real and the imaginary parts of $\tilde{a}_j(z,t)$, respectively:

$$\frac{\partial a_j(\omega,z)}{\partial z} = \frac{i}{2}\omega^2 \beta_2 a_j(\omega,z) + i\gamma_j \left[|A_{0j}|^2 a_j(\omega,z) + A_{0j}^2 a_j^*(-\omega,z) \right] \quad (5.7.7)$$

$$\frac{\partial a_j^*(-\omega,z)}{\partial z} = \frac{-i}{2}\omega^2 \beta_2 a_j^*(-\omega,z) - i\gamma_j \left[|A_{0j}|^2 a_j^*(-\omega,z) + A_{0j}^{*2} a_j(\omega,z) \right] \quad (5.7.8)$$

The formal solution of linear differential Equations 5.7.7 and 5.7.8 can be expressed in a matrix format:

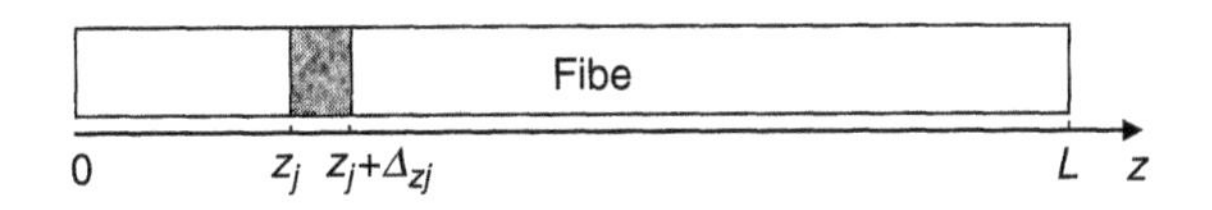

Figure 5.7.1 Illustration of dividing fiber into short sections for transfer matrix analysis.

$$\begin{bmatrix} a_{j+1}(\omega, z_j + \Delta z_j) \\ a_{j+1}^*(-\omega, z_j + \Delta z_j) \end{bmatrix} = \begin{bmatrix} M_{11}^{(j)} & M_{12}^{(j)} \\ M_{21}^{(j)} & M_{22}^{(j)} \end{bmatrix} \begin{bmatrix} a_{j+1}(\omega, z_j) \\ a_{j+1}^*(-\omega, z_j) \end{bmatrix}$$

When we further take into account the linear attenuation of the signal in each section, a factor $\exp(-\alpha\Delta z_j/2)$ has to be added so that

$$\begin{bmatrix} a_{j+1}(\omega, z_j + \Delta z_j) \\ a_{j+1}^*(-\omega, z_j + \Delta z_j) \end{bmatrix} = \begin{bmatrix} M_{11}^{(j)} & M_{12}^{(j)} \\ M_{21}^{(j)} & M_{22}^{(j)} \end{bmatrix} \begin{bmatrix} a_{j+1}(\omega, z_j) \\ a_{j+1}^*(-\omega, z_j) \end{bmatrix} \exp\left(-\frac{\alpha}{2}\Delta z_j\right) \quad (5.7.9)$$

where $\Delta z_j = z_{j+1} - z_j$ and

$$M_{11}^{(j)} = \frac{e^{ik_j z_j} - r_{fj} r_{bj} e^{-ik_j z_j}}{\left|1 - r_{fj} r_{bj}\right|} \quad (5.7.10)$$

$$M_{12}^{(j)} = \frac{r_{bj}\left(e^{-ik_j z_j} - e^{ik_j z_j}\right)}{\left|1 - r_{fj} r_{bj}\right|} \quad (5.7.11)$$

$$M_{21}^{(j)} = \frac{r_{fj}\left(e^{ik_j z_j} - e^{-ik_j z_j}\right)}{\left|1 - r_{fj} r_{bj}\right|} \quad (5.7.12)$$

$$M_{22}^{(j)} = \frac{e^{-ik_j z_j} - r_{fj} r_{bj} e^{ik_j z_j}}{\left|1 - r_{fj} r_{bj}\right|} \quad (5.7.13)$$

$$r_{fj} = \frac{k_j - \beta\omega^2 - \gamma_j |A_{0j}|^2}{\gamma_j A_{0j}^2} = \frac{-\gamma_j A_{0j}^{*2}}{k_j + \beta\omega^2 + \gamma_j |A_{0j}|^2} \quad (5.7.14)$$

$$r_{bj} = \frac{k_j - \beta\omega^2 - \gamma_j |A_{0j}|^2}{\gamma_j A_{0j}^{*2}} = \frac{-\gamma_j A_{0j}^2}{k_j + \beta\omega^2 + \gamma_j |A_{0j}|^2} \quad (5.7.15)$$

Equations 5.7.14 and 5.7.15 have two eigenmodes whose propagation constants are equal in magnitude and opposite in sign, and they are given by

$$k_j = \pm\sqrt{\left(\frac{\beta_2}{2}\omega^2 + \gamma_j |A_{0j}|^2\right)^2 - \left(\gamma_j |A_{0j}|^2\right)^2} \quad (5.7.16)$$

The parameters r_{fj} and r_{bj} can be regarded as effective reflectivities for the two eigenmodes; therefore the sign of k should be chosen such that $| r_{fj} |\leq 1|$ and $|r_{bj}|\leq 1$.

The evolution of the noise along the fiber can then be calculated simply by matrix multiplication:

$$\begin{bmatrix} a(\omega, L) \\ a^*(-\omega, L) \end{bmatrix} = \begin{bmatrix} B_{11} & B_{12} \\ B_{21} & B_{22} \end{bmatrix} \begin{bmatrix} a(\omega, 0) \\ a^*(-\omega, 0) \end{bmatrix} \exp\left(-\frac{\alpha}{2}L\right) \tag{5.7.17}$$

with

$$\begin{bmatrix} B_{11} & B_{12} \\ B_{21} & B_{22} \end{bmatrix} = \prod_{j=1}^{N} \begin{bmatrix} M_{11}^{(j)} & M_{12}^{(j)} \\ M_{21}^{(j)} & M_{22}^{(j)} \end{bmatrix} \tag{5.7.18}$$

where L is the fiber length and N is the total number of sections.

5.7.1.1 Power Spectrum of The Optical Field

According to the Wiener-Khintchine theorem, the power spectral density of the optical field is proportional to the square of the modulus of the Fourier transform of the complex field amplitude. If the field is sampled over a time interval T, this spectral density is

$$S_t(\omega, z) = \left| \left\{ \frac{1}{T} \int_{-T/2}^{T/2} A(z, t) \exp(i\omega t) dt \right\} \right|^2 \Bigg|_{T \to \infty}$$

Separating the field into CW and stochastic components, we have

$$S_t(\omega, z) = \langle | a(\omega, L) |^2 \rangle + |A_0(L)|^2 \delta(\omega)$$

where $< >$ denotes ensemble average, and normalization by the sample interval. $\delta(\omega)$ is the Kronecker delta function. The noise term has zero mean, so $\langle a(\omega, z) \rangle = 0$ and the cross terms vanish. It is convenient to remove the CW contribution from the power spectrum, so we define

$$S(\omega, L) = S_t(\omega, L) - |A_0(L)|^2 \delta(\omega) = \langle | a(\omega, L) |^2 \rangle \tag{5.7.19}$$

Using Equation 5.7.17 we can find

$$S(\omega, L) = \langle |B_{11} \cdot a(\omega, 0) + B_{12} \cdot a^*(-\omega, 0) |^2 \rangle \exp(-\alpha L) \tag{5.7.20}$$

Because $a(\omega,0)$ is a random process, amplitudes at distinct frequencies are uncorrelated, so $<a(\omega,0)a^*(-\omega,0)> = 0$ and

$$S(\omega, L) = [|B_{11}|^2 S(\omega, 0) + |B_{12}|^2 S(-\omega, 0)] \exp(-\alpha L) \tag{5.7.21}$$

where $S(\omega,0) = <a(\omega,0)a^*(\omega,0)>$ is the power spectral density of $\tilde{a}(t, 0)$, which is the input noise. To simplify the analysis, we assume that the input noise spectrum is symmetric around the carrier (e.g., white noise): $S(\omega,0) = S(-\omega,0)$. Equation 5.7.21 becomes

$$S(\omega, L) = \left(|B_{11}|^2 + |B_{12}|^2\right)S(\omega, 0)e^{-\alpha L} \tag{5.7.22}$$

A linear system can be treated as a special case with nonlinear coefficient $\gamma = 0$. In this case, $k = \beta_2\omega^2/2$, $r_f = r_b = 0$, $|B_{11}| = 1$, $B_{12} = 0$, and

$$S_L(\omega, L) = S(\omega, 0)e^{-\alpha L} \tag{5.7.23}$$

Using S_L as a normalization factor so that the normalized optical gain, which is equivalent to an optical noise amplification in the nonlinear system is

$$S_{OG}(\omega, L) = \frac{S(\omega, L)}{S_L(\omega, L)} = |B_{11}|^2 + |B_{12}|^2 \tag{5.7.24}$$

Figure 5.7.2 shows the normalized optical spectra as a function of fiber length in a single-span system using dispersion shifted fibers (DSF) with anomalous (a) and normal (b) dispersions. Fiber parameters used to obtain Figure 5.7.2 are loss coefficient $\alpha = 0.22$ dB/km, input signal optical power $P_{in} =$ 13 dBm, nonlinear coefficient $\gamma = 2.07\ W^{-1}km^{-1}$ and fiber dispersion $D =$ 2 ps/nm/km for Figure 5.7.2(a), and $D = -2$ ps/nm/km for Figure 5.7.2(b) with D defined as $D = 2\pi c\beta_2/\lambda^2$. In both cases of Figure 5.7.2(a) and (b), optical noise is amplified around the carrier. The difference is that in an anomalous dispersion regime, optical spectrums have two peaks at each side of the carrier, whereas in the normal dispersion regime, spectra are single peaked. The amplification of optical spectra near the carrier can be explained as the spectrum broadening of the carrier caused by the nonlinear phase modulation between the signal and the broadband ASE. Figure 5.7.2(c) and (d) show nonlinear amplifications of ASE in a 100 km fiber with positive (c) and negative (d) chromatic dispersions. Input power level used for Figure 5.7.2 (c) and (d) is $P_{\text{in}} =$ 15 dBm. Three different dispersion values are used in each figure, which are solid line: $D = \pm 1$ ps/nm/km, dashed line: $D = \pm 0.5$ ps/nm/km, and dashed-dotted line: $D = \pm 0.05$ ps/nm/km.

In the case of coherent optical transmission, the entire optical spectrum is moved to the intermediate frequency after beating with the local oscillator. The frequency components beyond the range of the baseband filter will be removed, which may cause receiver power reduction. Therefore the broadening of the signal optical spectrum is the major source of degradation in coherent optical transmission systems. For IMDD optical systems, on the other hand, the photodiode detects the total optical power without wavelength discrimination, and the relative intensity noise (RIN) of the optical signal is the major source of degradation related to MI.

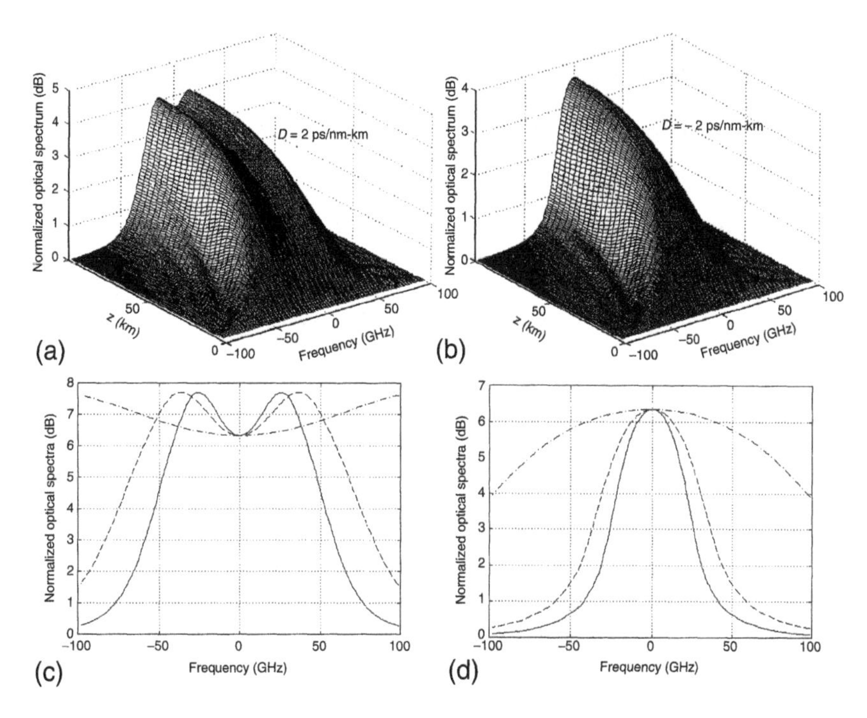

Figure 5.7.2 (a) and (b) Nonlinear amplification of ASE along the longitudinal direction of a single-span fiber. Input optical power P_{in} = 13 dBm, fiber nonlinear coefficient $\gamma = 2.07\ W^{-1}km^{-1}$, and fiber loss α = 0.22 dB/km. (c) and (d) Nonlinear amplification of ASE in a 100 km fiber for positive (c) and negative (d) dispersions. Input power P_{in} = 15 dBm. Solid line: $D = \pm 1$ ps/nm/km, dashed line: $D = \pm 0.5$ ps/nm/km, and dash-dotted line: $D = \pm 0.05$ ps/nm/km [46].

5.7.1.2 Noise Description at a Direct-detection Optical Receiver

After the square-law detection of a photodiode, the photocurrent can be expressed as

$$I(t) = \Re|A_0 + \tilde{a}(t,L)|^2 = \Re[|A_0|^2 + A_0\tilde{a}^*(t,L) + A_0^*\tilde{a}(t,L)] \qquad (5.7.25)$$

where $\Re$ is the photodetector responsivity. For simplicity, second and higher orders of small terms have been omitted in the derivation of Equation 5.7.25.

In an IMDD system, the receiver performance is sensitive only to the amplitude noise of the photocurrent, which can be obtained from Equation 5.7.25 as

$$\delta I(t) = I(t) - I_0 = \Re[A_0\tilde{a}^*(t,L) + A_0^*\tilde{a}(t,L)] \qquad (5.7.26)$$

where $I_0 = \Re|A_0|^2$ is the photocurrent generated by the CW optical signal.

The power spectrum of the noise photocurrent is the Fourier transformation of the autocorrelation of the time-domain noise amplitude:

$$\rho_n(\omega) = \Re^2\left(|A_0^* B_{11} + A_0 B_{12}|^2 S(\omega, 0) + |A_0^* B_{12} + A_0 B_{22}|^2 S(-\omega, 0)\right)e^{-\alpha L} \tag{5.7.27}$$

where $S(\omega, 0)$ is the power spectral density of $\tilde{a}(t, 0)$.

Under the same approximation as we used in the optical spectrum calculation, the input optical noise spectrum is assumed to be symmetric around zero frequency: $S(\omega, 0) = S(-\omega, 0)$, Equation 5.7.27 becomes

$$\rho_n(\omega) = \Re^2\left(|B_{11} + B_{21}|^2 + |B_{12} + B_{22}|^2\right)| A_0(L) |^2 S(\omega, 0)e^{-\alpha L} \tag{5.7.28}$$

Using Equations 5.7.10–5.7.18, it is easy to prove that $|B_{11} + B_{21}|^2 = |B_{12} + B_{22}|^2$, and therefore Equation 5.7.28 can be written as

$$\rho_n(\omega) = 2P_{in}\Re^2|B_{11} + B_{21}|^2 S(\omega, 0)e^{-2\alpha L} \tag{5.7.29}$$

where P_{in} is the input signal power and thus $|A_0(\mathrm{L})|^2 = P_{in}\exp(-\alpha L)$.

Again, a linear system can be treated as a special case of a nonlinear system with the nonlinear coefficient $\gamma = 0$. In this case, $k = \beta_2\omega^2/2$, $r_f = r_b = 0$, $|B_{11}| = 1$, and $B_{21} = 0$. The electrical noise power spectral density in the linear system is then

$$\rho_0(\omega) = 2P_{in}\Re^2 S(\omega, 0)e^{-2\alpha L} \tag{5.7.30}$$

Using $\rho_0(\omega)$ as the normalization factor, the normalized power spectral density of the receiver electrical noise, or the normalized RIN, caused by fiber Kerr effect is therefore

$$R(\omega) = |B_{11} + B_{21}|^2 \tag{5.7.31}$$

Comparing Equation 5.7.31 to Equation 5.7.24, it is evident that these two spectra are fundamentally different: The relative phase difference between B_{11} and B_{21} has no impact on the optical noise amplification of Equation 5.7.24, but this phase difference is important in the RIN spectrum as given in Equation 5.7.31.

Figure 5.7.3 shows the normalized RIN spectra versus fiber length in a single-span system with anomalous dispersion (a) and normal dispersion (b). Fiber parameters used in Figure 5.7.3 are the same as those used for Figure 5.7.2. In the anomalous fiber dispersion regime [Figure 5.7.3(a)], two main side peaks of noise grow along z, the peak frequencies become closer to the carrier and the widths become narrower in the process of propagating along the fiber. On the other hand, if the fiber dispersion is in the normal regime, as shown in Figure 5.7.3(b), noise power density becomes smaller than in the linear case in the vicinity of the carrier frequency. This implies a noise squeezing [47] and a possible system performance improvement. In either regime, the system performance is sensitive to the baseband electrical filter bandwidth. For the purpose of clearer display, Figure 5.7.3(c) and (d) show nonlinear amplifications of RIN over 100 km fiber with positive (c) and negative (d) chromatic dispersions. The input power level is $P_{in} = 15$ dBm. Three different dispersion values are used in both Figure 5.7.3(c) and (d), they are solid line: $D = \pm 1$ ps/nm/km, dashed line: $D = \pm 0.5$ ps/nm/km and dashed-dotted line: $D = \pm 0.05$ ps/nm/km.

It is worthwhile to notice the importance of taking into account fiber loss in the calculation. Without considering fiber loss, the calculated RIN spectra would be qualitatively different. For example, in the case of normal fiber dispersion, the normalized RIN spectra were always less than 0 dB in the mean-field approximation [23], which is in fact not accurate, as can be seen in Figure 5.7.3 (b) and (d).

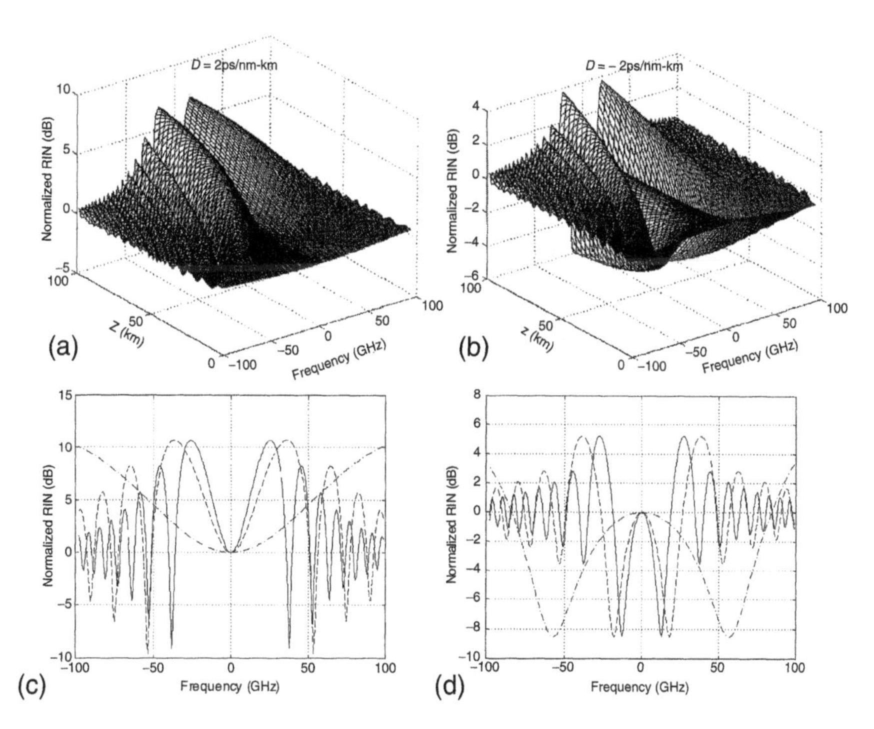

Figure 5.7.3 (a) and (b) Nonlinear amplification of RIN along the longitudinal direction of a single-span fiber. Input optical signal power $P_{in} = 13$ dBm, fiber nonlinear coefficient $\gamma = 2.07\ W^{-1}km^{-1}$ and fiber loss $\alpha = 0.22$ dB/km. (c) and (d) Nonlinear amplification of RIN of 100 km fiber for positive (c) and negative (d) dispersions. Input power $P_{in} = 15$ dBm. Solid line: $D = \pm 1$ ps/nm/km, dashed line: $D = \pm 0.5$ ps/nm/km, and dashed-dotted line: $D = \pm 0.05$ ps/nm/km [46].

5.7.1.3 Effect of Dispersion Compensation

It is well known that dispersion compensation (DC) is an important way to reduce eye closure penalty due to chromatic dispersion in fiber systems. It can also reduce XPM-induced nonlinear crosstalk in IMDD system, as discussed in Section 5.6. Similarly, DC may have important impact on the effect of MI in optical systems, which will be discussed in this section. Neglecting the nonlinear effect of the DC module, its transfer function can be represented by a conventional Jones Matrix:

$$[C] = \begin{bmatrix} \exp[i\Phi(\omega)] & 0 \\ 0 & \exp[-i\Phi(\omega)] \end{bmatrix} \tag{5.7.32}$$

If the DC is made by a piece of optical fiber, the phase term is related to the chromatic dispersion, $\Phi(\omega) = \beta_2\omega^2 z/2$.

The effect of DC on the optical system can be evaluated by simply multiplying the Jones matrix of Equation 5.7.32 to the MI transfer matrix provided by Equation 5.7.17. The RIN spectra at the IMDD optical receiver are sensitive not only to the value of DC but to the position at which the DC module (DCM) is placed in the system. Let's take two examples to explain the reason. First, if the DCM is positioned after the nonlinear transmission fiber (at the receiver side), the combined transfer function becomes:

$$\begin{bmatrix} B_{11} & B_{12} \\ B_{21} & B_{22} \end{bmatrix} = \begin{bmatrix} \exp[i\Phi(\omega)] & 0 \\ 0 & \exp[-i\Phi(\omega)] \end{bmatrix} \begin{bmatrix} B_{11}^f & B_{12}^f \\ B_{21}^f & B_{22}^f \end{bmatrix}$$

$$= \begin{bmatrix} B_{11}^f \exp(i\Phi) & B_{12}^f \exp(i\Phi) \\ B_{21}^f \exp(-i\Phi) & B_{22}^f \exp(-i\Phi) \end{bmatrix}$$

The normalized RIN spectrum in Equation 5.7.31 is then:

$$R(\omega) = \left| B_{11}^f \exp(i\Phi) + B_{21}^f \exp(-i\Phi) \right|^2$$

where $B_{ij}^f (i = 1, 2, j = 1, 2)$ are the transfer function elements of the nonlinear transmission fiber. Obviously, the normalized RIN spectrum is sensitive to the amount of DC represented by the phase shift Φ. Figure 5.7.4 shows an example of normalized RIN spectra without dispersion compensation (solid line), with 50 percent of compensation (dashed-dotted line), and with 100 percent compensation (dashed line). It is interesting to note that 100 percent dispersion compensation does not necessarily bring the RIN spectrum to the linear case.

On the other hand, if the DCM is placed before the nonlinear transmission fiber (at the transmitter side), the total transfer matrix is

$$\begin{bmatrix} B_{11} & B_{12} \\ B_{21} & B_{22} \end{bmatrix} = \begin{bmatrix} B_{11}^f & B_{12}^f \\ B_{21}^f & B_{22}^f \end{bmatrix} \begin{bmatrix} \exp[i\Phi(\omega)] & 0 \\ 0 & \exp[-i\Phi(\omega)] \end{bmatrix}$$
$$= \begin{bmatrix} B_{11}^f \exp(i\Phi) & B_{12}^f \exp(-i\Phi) \\ B_{21}^f \exp(i\Phi) & B_{22}^f \exp(-i\Phi) \end{bmatrix}$$

The normalized RIN spectrum in Equation 5.7.31 then becomes

$$R(\omega) = \left| B_{11}^f \exp(i\Phi) + B_{21}^f \exp(i\Phi) \right|^2 = \left| B_{11}^f + B_{21}^f \right|^2$$

In this case, it is apparent that the frequency-dependent phase term Φ introduced by dispersion compensation does not bring any difference into the normalized RIN spectrum.

Another important observation is that although the RIN spectrum in the electrical domain after photodetection can be affected by the DC, the optical spectrum is not affected by the use of DC, regardless of the position of the DC module (DCM). The reason for this can be found from Equation 5.7.24, where the normalized optical spectrum is related to the absolute value of B_{11}

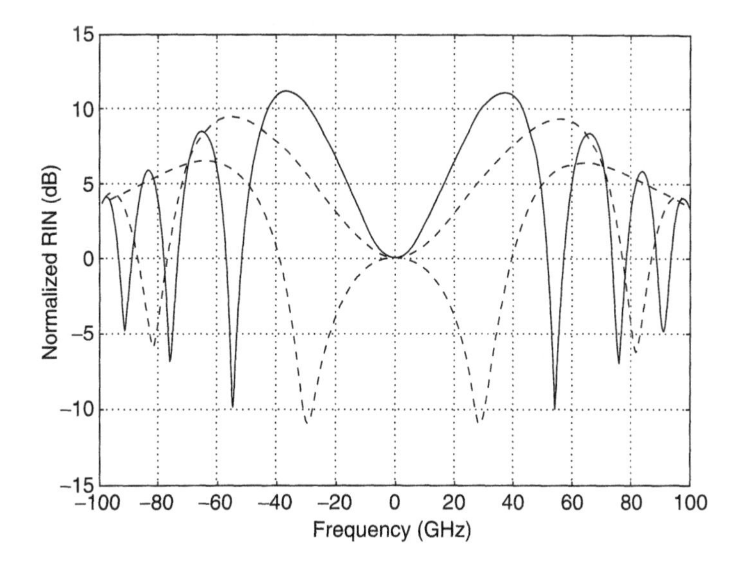

Figure 5.7.4 Normalized RIN spectra for 0 percent (solid line), 50 percent (dash-dotted line), and 100 percent (dashed line) dispersion compensations. Fiber length $L = 100$ km, fiber nonlinear coefficient $\gamma = 2.07\ \mathrm{W^{-1}km^{-1}}$, fiber loss $\alpha = 0.22$ dB/km, $P_{in} = 15$ dBm, and $D = 1$ ps/nm/km [46].

and B_{12}. It does not matter where the DCM is placed in the system; DC has no effect on the normalized optical spectrum.

5.7.2 Impact of Modulation Instability in Amplified Multispan Fiber Systems

In an optical fiber system of N spans with N EDFAs (one post-amplifier and N-1 line amplifiers), the fiber loss per span is compensated by the optical gain of the EDFA. Suppose that all EDFAs have the same noise figure; the ASE noise power spectral density generated by the i-th EDFA is

$$S_i = h\nu(FG_i - 1) \tag{5.7.33}$$

where the ASE noise optical spectrum is supposed to be white within the receiver optical bandwidth. After propagating through fibers, amplified by EDFAs and detected by the photodiode, the power spectrum of the detected RIN can be obtained by the multiplication of the transfer function of each span of optical fiber, supposing that ASE noise generated by different EDFAs are uncorrelated:

$$\rho(\omega) = 2h\nu P_{in}\Re \sum_{m=1}^{N} \left\{ (FG_{N-m+1} - 1) \left| B_{11}^{(N-m+1)} + B_{21}^{(N-m+1)} \right|^2 \right\} \tag{5.7.34}$$

where $B_{ij}^{(k)}$ $(i, j, = 1, 2)$ are matrix elements defined as

$$B^{(k)} = \prod_{m=1}^{k} \begin{bmatrix} B_{11} & B_{12} \\ B_{21} & B_{22} \end{bmatrix}_m \tag{5.7.35}$$

and the matrix

$$\begin{bmatrix} B_{11} & B_{12} \\ B_{21} & B_{22} \end{bmatrix}_m$$

represents the IM transfer function of the m-th fiber span.

Setting $\gamma = 0$ to obtain the normalization denominator, the normalized RIN spectrum is then

$$R(\omega) = \frac{\sum_{m=1}^{N} \left\{ (FG_{N-m+1} - 1) \left| B_{11}^{(N-m+1)} + B_{21}^{(N-m+1)} \right|^2 \right\}}{\sum_{m=1}^{N} (FG_{N-m+1} - 1)} \tag{5.7.36}$$

Assuming Gaussian statistics, the change of the standard deviation of the noise caused by fiber MI can be expressed in a simple way:

$$\delta\sigma = \frac{\sigma^2}{\sigma_0^2} = \int_{-\infty}^{\infty} R(\omega) \left| f(\omega) \right|^2 d\omega \tag{5.7.37}$$

where σ and σ_0 are noise standard deviations in the nonlinear and linear cases, respectively, and $f(\omega)$ is the receiver baseband filter transfer function.
In a direct-detection optical receiver with the effect of MI taken into account, the quality factor Q can be expressed as

$$Q = \frac{\Re(P_1 - P_0)}{\sqrt{\sigma_{sh1}^2 + \sigma_{th1}^2 + \sigma_{sp-sp1}^2 + \sigma_{s-sp1}^2 \delta\sigma_1} + \sqrt{\sigma_{sh0}^2 + \sigma_{th0}^2 + \sigma_{sp-sp0}^2 + \sigma_{s-sp0}^2 \delta\sigma_0}} \tag{5.7.38}$$

where P is the signal level, σ_{sh}, σ_{th}, $\sigma_{sp\text{-}sp}$, and $\sigma_{s\text{-}sp}$ are, respectively, the standard deviations of shot noise, thermal noise, ASE-ASE beat noise, and signal-ASE beat noise in the absence of MI. Subscripts 1 and 0 indicate the symbols at signal logical 1 and logical 0, respectively. In practice, MI introduces a signal-dependent noise that affects more on signal during logical 1 than that during logical 0. The change in the RIN spectrum causes system performance degradation primarily due to the increase of signal-ASE beat noise through the ratio $\delta\sigma_1$. However, considering that if the signal extinction ratio is not infinite, signal power at logical 0 may also introduce MI through $\delta\sigma_0$, although this effect should be relatively small. Meanwhile, optical spectrum change due to MI may also introduce Q degradation through ASE-ASE beat noise, but it is expected to be a second-order small effect in an IMDD receiver.

Equation 5.7.38 indicates that, in general, system performance degradation due to MI depends on the proportionality of signal-ASE beat noise to other noises. Multispan optical amplified fiber systems, where signal-ASE beat noise predominates, are more sensitive to MI in comparison to unamplified optical systems. For signal-ASE beat noise limited optical receiver, to the first-order approximation, the system Q degradation caused by MI can be expressed in a very simple form:

$$10\log(Q) - 10\log(Q_0) = -5\log(\delta\sigma) \tag{5.7.39}$$

where Q_0 is the receiver Q-value in the linear system without considering the impact of MI, and $\delta\sigma$ is the change of noise standard deviations due to MI.

5.7.3 Characterization of Modulation Instability in Fiber-Optic Systems

As has been discussed so far, the major impact of MI in an IMDD optical system is the increase of the RIN at the receiver due to parametric amplification of ASE noise. The enhanced intensity noise spectrum within the receiver electrical bandwidth is the primary cause of system performance degradation. Therefore, it is straightforward to measure MI and its impact in optical transmission systems in frequency domain.

Figure 5.7.5 shows a general block diagram of an experimental setup to characterize the effect of MI. Because a high optical power level is usually required for the MI measurement, especially when the fiber span is short and the fiber nonlinearity is weak, two optical amplifiers are used. The purpose of the bandpass optical filter between the two optical amplifiers is to limit the ASE noise bandwidth from the first amplifier. This avoids the saturation of the second optical amplifier caused by the broadband ASE of the first amplifier. A variable optical attenuator (VOA) adjusts the level of signal optical power that is injected into the fiber system.

Two optical power meters are used to monitor the optical power: PM1 measures the launched optical power into the fiber; PM2 monitors the back-reflected optical power from the fiber due to SBS, which may reduce signal power level in the fiber. To reduce the backscattering due to SBS, a frequency dithering is usually used for the laser source so that the spectral width of the optical signal is much higher than 10 MHz. After the optical signal passes through the optical system, it is measured by an optical spectrum analyzer, which characterizes the optical parametric gain due to MI. The output optical signal is also measured in the electrical domain through a wideband photodiode and an electrical spectrum analyzer (ESA), which characterizes the increase of the RIN spectrum.

For example, Figure 5.7.6 shows the detailed experimental setup for the MI measurement in a dispersion-compensated multispan fiber system with optical amplifiers. The CW laser diode, emitting at 1543 nm, has a sine wave tone superimposed on its injection current to dither the optical frequency so as to suppress Brillouin scattering in the fiber. A separate experiment found that 5 mA p-p at 10 kHz direct injection-current modulation was adequate to ensure that Brillouin

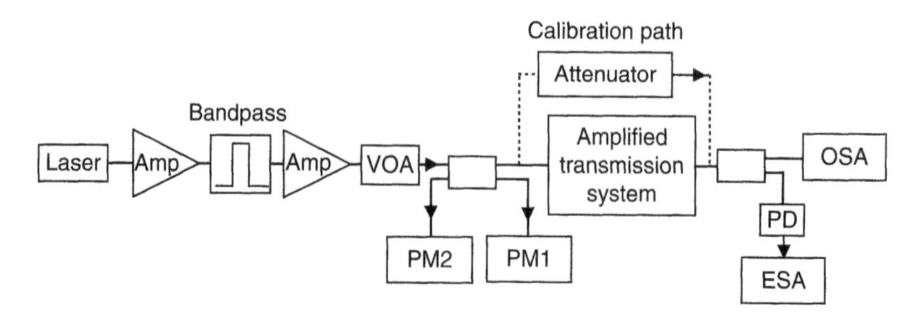

Figure 5.7.5 Block diagram of experimental setup to measure modulation instability in an amplified optical system.

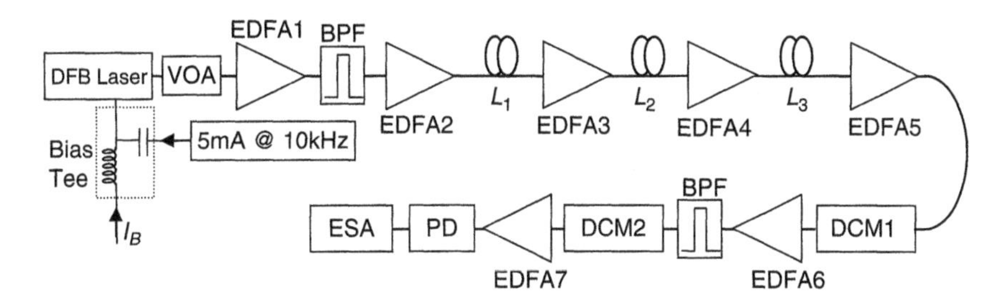

Figure 5.7.6 Experimental setup. L_1: 84.6 km (1307 ps/nm), L_2: 84.6 km (1291 ps/nm), L_3: 90.4 km (1367 ps/nm). Optical powers at the output of EDFA5, EDFA6, and EDFA7 are less than 0 dBm. *DCM:* dispersion compensation module.

scattering was suppressed. In practice, the laser RIN could also corrupt the results in some circumstances if it is not carefully calibrated. When the laser drive current was 35 mA, the RIN had a peak at 6 GHz due to the laser relaxation oscillation 7 dB higher than the DC level. With the insertion of 85 km of nondispersion shifted fiber (NDSF), this peak rose by a further 6 dB due to laser phase noise which was converted to intensity noise by the fiber's dispersion. This converted noise was avoided by increasing the laser current to 70 mA, which moved the relaxation oscillation to higher frequency and increased the damping rate to the relaxation oscillation, thereby reducing the laser-related RIN level.

To make sure that the measured receiver RIN is dominated by signal-ASE beat noise, one can increase the broadband ASE noise from EDFAs intentionally. The role of the first optical amplifier in the link was to inject wideband optical noise. It had a noise figure of 7.5 dB at the input optical power of –32 dBm. The narrowband optical filter that followed kept the total ASE power less than that of the signal so as to decrease the ASE-ASE beat noise. The signal power was typically 0.5 dB less than the total power emerging from the EDFA, and this was taken into account when setting the output power of the line amplifiers. The first EDFA was the dominant source of ASE arriving at the PIN detector. The three inline amplifiers had output power adjustable by computer control. The fiber spans all had loss coefficient measured by OTDR of 0.2 dB/km. The output power of the next two EDFAs was set below 0 dBm to avoid nonlinear effects in the DCM. The narrowband filter after EDFA suppressed ASE in the 1560 nm region, which would have led to excessive ASE-ASE beat noise. The last optical amplifier was controlled to an output power of 4 dBm, just below the overload level of the PIN detector.

First, a calibration run was made, with an attenuator in place of the system. This measurement was subtracted from all subsequent traces so as to take out the effect of the frequency response of the detection system. The calibration trace was >20 dB higher than the noise floor for the whole 0–18 GHz band.

The RIN spectra at the output of the three-span link are shown in Figure 5.7.7 with line amplifier output power (signal power) controlled at 8, 10, 12, and 14 dBm, for inverted triangles, triangles, squares, and circles, respectively. In Figure 5.7.7(a), open points represent the measured spectra with the system configuration described in Figure 5.7.6 except no dispersion compensation was used. Continuous lines in the same figures were calculated using Equation 5.7.31. Similarly, Figure 5.7.7(b) shows the measured and calculated RIN spectra with the dispersion compensation of –4070 ps/nm at the receiver side, as shown in Figure 5.7.6. To obtain the theoretical results shown in Figure 5.7.7, the fiber nonlinear coefficient used in the calculation was $\gamma = 1.19\,\text{W}^{-1}\text{km}^{-1}$, and other fiber parameters such as length and dispersion were chosen according to the values of standard single-mode fiber used in the experiment, as shown in the caption of Figure 5.7.6. Very good agreement between measured and calculated results in the practical power range assures the validity of the two major approximations used in the transfer matrix formulation, namely, the linear approximation to the noise term and the insignificance of pump depletion.

Although the RIN spectra are independent of the signal data rate, the variance of the noise depends on the bandwidth of the baseband filter as explained in Equation 5.7.37. Figure 5.7.8 shows the effect of dispersion compensation on the ratio of noise standard deviation between nonlinear and linear cases for the three-span fiber system described in Figure 5.7.6. The optical power was 12 dBm at the input of each fiber span and a raised-cosine filter was used in the receiver with 8 GHz electrical

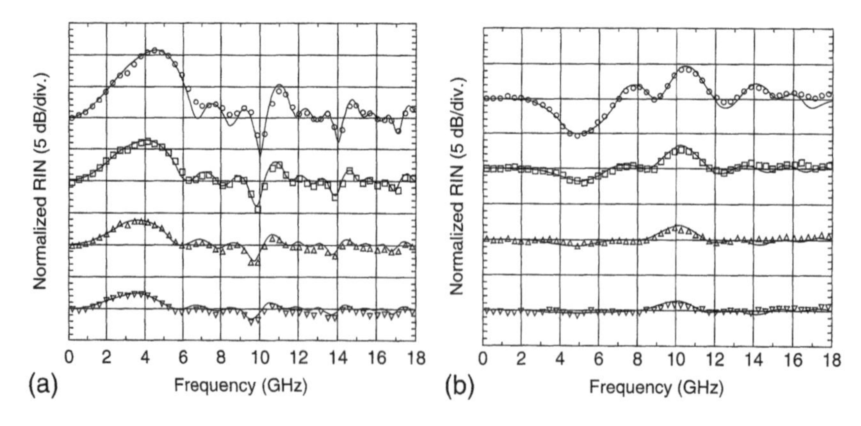

Figure 5.7.7 Measured (open points) and calculated (solid lines) RIN spectra in the three-span standard SMF as described in Figure 5.7.6. The optical power at the output of EDFA2, EDFA3, and EDFA4 is 8 dBm (triangles-down), 10 dBm (triangles-up), 12 dBm (squares), and 14 dBm (circles). Curves are shifted for 10 dB between one and another for better display. (a) without dispersion compensation and (b) with –4070 ps/nm dispersion compensation [46].

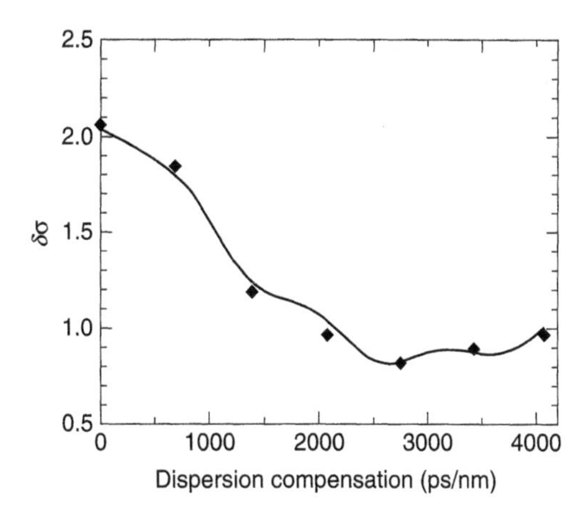

Figure 5.7.8 Comparison of δσ between calculation (solid line) and measurement (diamonds) for the three-span system described in Figure 5.7.6 with optical power $P_{in} = 12$ dBm [46].

bandwidth. Both theoretical and experimental results demonstrate that δσ approaches its minimum when the DC is approximately 70 percent of the total system dispersion. Generally, the optimum level of dispersion compensation depends on the number of spans, electrical filter bandwidth, optical power levels, and the dispersion in each fiber span.

As far as MI is concerned, in the anomalous fiber dispersion regime, system performance always becomes worse with increasing signal power. On the other hand, in the normal dispersion regime, system sensitivity may be improved by the nonlinear process. This is an interesting phenomenon that was explained as a noise squeezing [47]. It can be readily understood from Figure 5.7.7(b), where if the signal optical power is higher than 12 dBm, the noise level is reduced at the low-frequency region. If the receiver bandwidth is less than 7 GHz, the total noise level will be lower compared to the linear system. However, for systems with higher bit rate, sensitivity degradations may also be possible in the normal dispersion regime with high input signal powers. This degradation is caused by the increased noise level at higher frequencies, which happen to be within the receiver baseband.

To conclude this section, MI is in a unique category of fiber nonlinearity. In contrast to nonlinear crosstalk due to XPM and FWM, MI is a single-channel process. The high power of the optical signal amplifies the ASE noise through the parametric gain process. The increased RIN due to MI is originated from the nonlinear interaction between the signal and the broadband ASE noise. This effect is also different from SPM because the consequence of MI is the amplification of intensity noise rather than the deterministic waveform distortion.

5.8 OPTICAL SYSTEM PERFORMANCE EVALUATION BASED ON REQUIRED OSNR

It is well known that transmission performance of a digital optical communication system can be evaluated by the bit error rate or the Q-value at the receiver. In the system design and characterization, another important measure of system performance is the system margin. For example, a 3 dB Q-margin means that the system can tolerate additional 3 dB Q degradation from the current state before it produces unaccepted bit errors. In low data rate or short-distance optical systems without optical amplifiers, the bit error rate degradation is usually caused by receiver thermal noise when the optical signal is too low. In this case, the optical power level at the receiver is the most important measure to determine the system performance.

On the other hand, in high data rate optical systems involving multiple wavelength channels and multiple inline optical amplifiers, signal waveform distortion, and accumulated ASE noise in the system become significant. In amplified optical systems, the signal optical power level at the receiver is no longer the most relevant measure of the system performance. One widely adopted system margin measurement is the required optical signal-to-noise ratio (R-OSNR) as introduced in Section 5.2.2. In this section, we further discuss the impact of R-OSNR due to several specific degradation sources in the system such as chromatic dispersion, SPM, and the limited bandwidth of an optical filter.

The definition of R-OSNR is the required optical SNR of the optical signal at the receiver to achieve the specified bit error rate (BER). Figure 5.8.1 shows a schematic diagram of the experimental setup to measure R-OSNR using optical noise loading. The optical signal from the transmitter is delivered to the receiver through the optical fiber transmission system, which may include multiple amplified optical spans. The output of an independent wideband ASE noise source is combined with the optical signal so that the optical SNR at the receiver can be varied by controlling the level of the inserted ASE noise through adjusting the variable optical attenuator. The BER of the optical signal detected by the receiver is measured at the bit error rate test set (BERT). Since the measured BER is a function of the optical SNR at the receiver, an R-OSNR

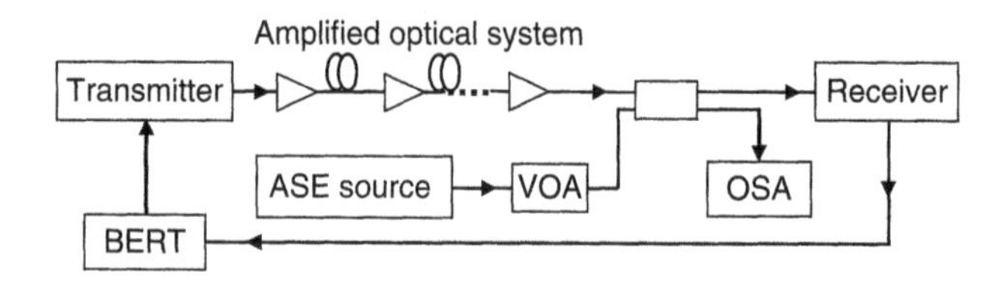

Figure 5.8.1 Block diagram of R-OSNR measurement using optical noise injection.

level can be found at which the BER is at the specified level. A good transmitter/receiver pair should exhibit a low R-OSNR level, whereas obviously a higher R-OSNR implies that the system can tolerate less additional ASE noise, and thus the system margin is smaller.

5.8.1 Measurement of R-SNR Due to Chromatic Dispersion

Chromatic dispersion in a fiber-optic transmission system is one of the biggest sources of performance impairment. It primarily introduces signal waveform distortion and eye closure penalty and deteriorates the receiver BER. Dispersion compensation is widely used in long-distance optical fiber transmission systems to reduce the accumulated chromatic dispersion between the transmitter and the receiver. Since chromatic dispersion in a fiber-optic system is usually wavelength-dependent, different wavelength channels in a WDM system may experience different dispersion. Although dispersion slope compensation may help equalize the dispersion levels for various wavelength channels, it would increase nonlinear crosstalk between channels through XPM and FWM because of the minimized interchannel phase walk-off.

From an optical transmission equipment performance point of view, as one of the most important qualification measures, a high-quality transmitter/receiver pair should be able to tolerate a wide range of chromatic dispersion in the fiber system. A wide dispersion window can be achieved using advanced optical modulation formats, optical, and electronic signal processing. Figure 5.8.2 shows an

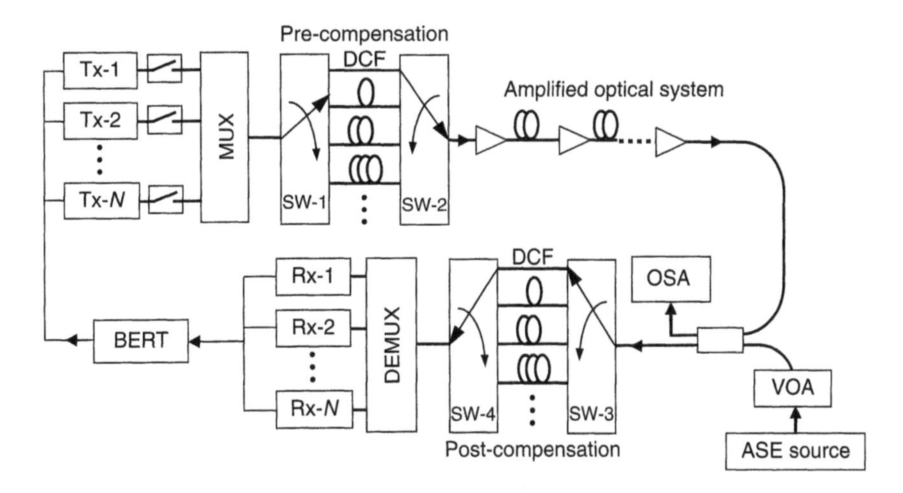

Figure 5.8.2 System test bed for dispersion window measurement.

optical system test bed that is used to perform qualification test for optical transmitters and receivers. Dispersion window and R-OSNR are two critical parameters to qualify the optical transmission equipment.

In Figure 5.8.2, a WDM system with N wavelength channels is tested. The data rate of each channel is 10 Gb/s and the number of channels can be selected by activating the optical switch in front of each transmitter. Both precompensation and post-compensation are included to mitigate the effect of chromatic dispersion. The two switchable dispersion compensation units are made by two groups of dispersion-compensating fibers, each having a different length, sandwiched between two optical switches. Because of the wideband nature of the dispersion-compensating fibers, the switchable compensator is able to support WDM optical system operation while avoiding connection errors compared to manually reconfigurable fiber arrays. An independent ASE noise source with variable power level is added to the optical link so that the OSNR at the receiver can be adjusted. A BERT is used, providing the modulating data pattern for the transmitter and BER testing for signals recovered by the optical receivers. In the OSNR measurement, the spectral resolution of the OSA was set at 0.1 nm.

Figure 5.8.3 shows an example of the measured R-OSNR as the function of the residual chromatic dispersion, which includes precompensation at the transmitter side, post-compensation at the receiver side, and the inline dispersion compensations in

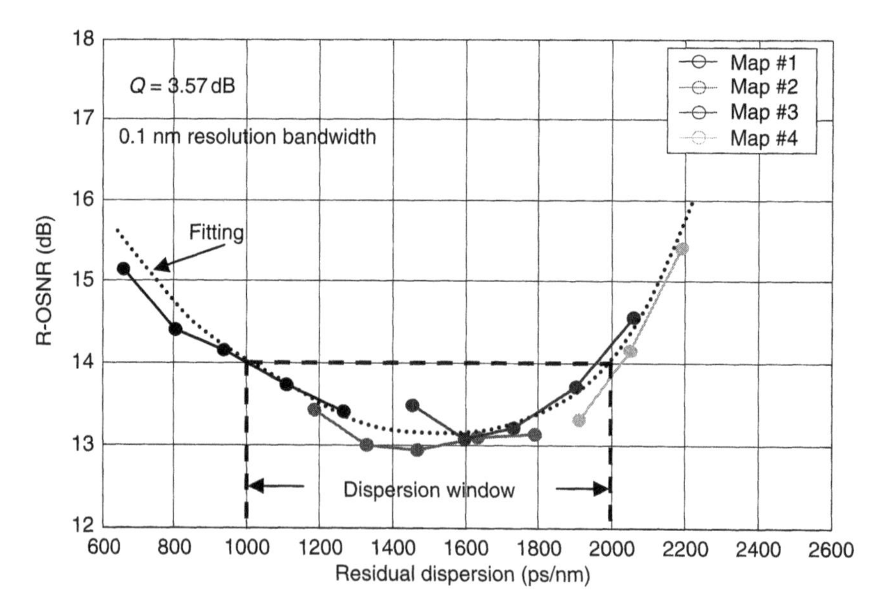

Figure 5.8.3 Example of the measured R-OSNR as a function of residual chromatic dispersion for a signal optical channel. The signal optical power at each fiber span is 3dBm.

front of each EDFA line amplifier along the transmission system. Only one wavelength channel is used in this measurement and thus there is no crosstalk expected. The system has eight amplified fiber spans and the lengths of the eight fiber spans are 92.2 km, 92.4 km, 91.9 km, 103.4 km, 104.4 km, 102.3 km, 99.3 km, and 101.3 km, and at the signal wavelength the chromatic dispersion values of these eight fiber spans are, respectively, 1504 ps/nm, 1500 ps/nm, 1477 ps/nm, 1649 ps/nm, 1697 ps/nm, 1650 ps/nm, 1599 ps/nm, and 627 ps/nm. Figure 5.8.4 shows four different

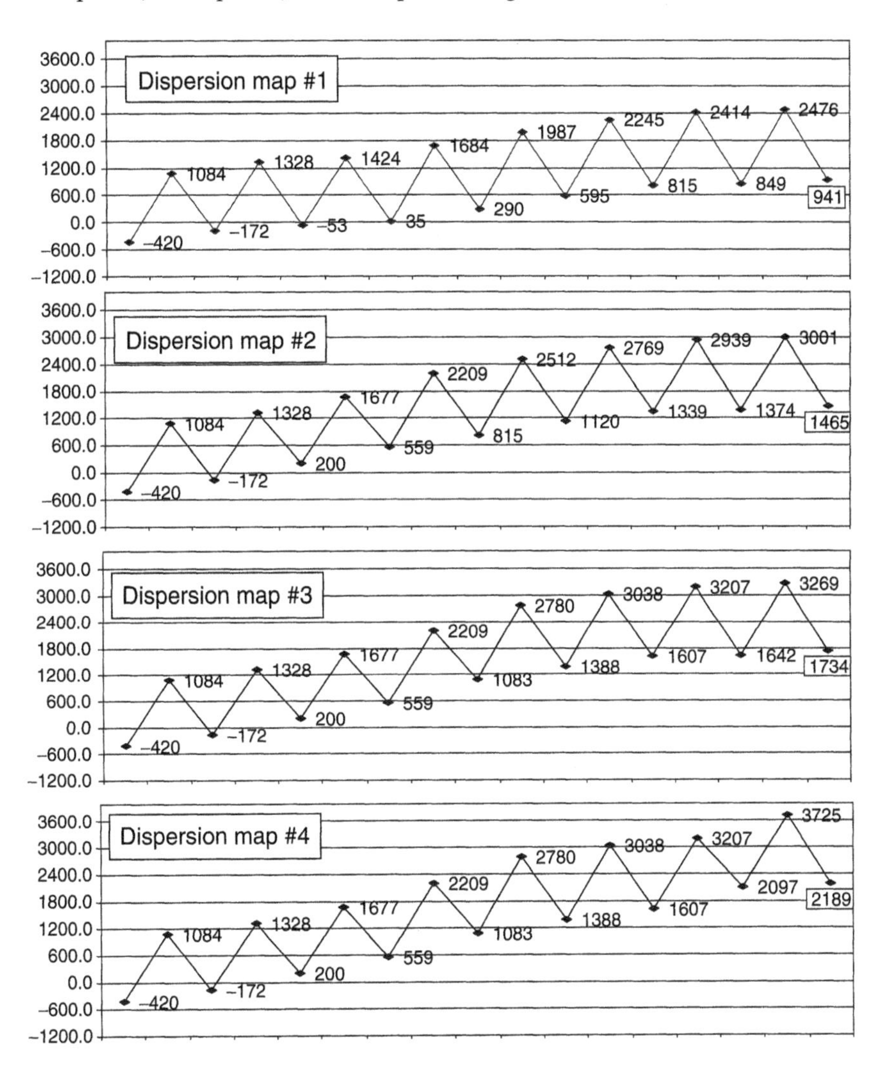

Figure 5.8.4 Four different dispersion maps used to create dispersion window measurement in Figure 5.8.3.

dispersion maps by choosing different values of the inline dispersion compensators in the system. The accumulated chromatic dispersions of the transmission system (excluding pre- and post-compensations) are 941 ps/nm, 1465 ps/nm, 1734 ps/nm, and 2189 ps/nm for the four dispersion maps. Then, by fine adjusting the pre- and post-compensations through switch selection, the residual dispersion of the optical link between the transmitter and the receiver can be varied around the value of each dispersion map. In this particular example, the transmission performance is optimized when the residual dispersion is in the vicinity of 1500 ps/nm, which depends on the transmitter and the receiver design as well as the chirp of the optical signal. If 1-dB OSNR penalty is allowed, the dispersion window of this system is approximately 1000 ps/nm as illustrated in Figure 5.8.3.

5.8.2 Measurement of R-SNR Due to Fiber Nonlinearity

The R-OSNR shown in Figure 5.8.3 was measured with only a single wavelength channel, and the average optical power at the input of each fiber span is approximately +3 dBm, whereas the optical power at each dispersion compensator was at 0 dBm. It is worthwhile to mention that the optimum residual dispersion value may also depend on the signal optical power level used in each fiber span because the effect of SPM is equivalent to an additional chirp on the optical signal. When the optical power level is low enough, the system is linear and the effect of chromatic dispersion can be completely compensated for, no matter where the dispersion compensator is placed along the system. In a nonlinear system, on the other hand, the nonlinear phase modulation created by the SPM process is converted into an intensity modulation through chromatic dispersion in the following fiber spans. Therefore, the location of dispersion compensation is important. Similar to the case of XPM, which was discussed in Section 5.6, per-span dispersion compensation usually works better than lumped compensation either in the transmission side or in the receiver side.

Figure 5.8.5 shows the measured R-OSNR in a 12-span amplified fiber-optic system with 80 km of standard SMF in each span. The chromatic dispersion is also compensated at the end of each fiber span. The experimental setup is similar to that shown in Figure 5.8.2, but the R-OSNR measurements were conducted at the end of each fiber span. The output optical power level, P, at the output of each inline optical amplifier was adjusted to observe the effect due to nonlinear SPM. Figure 5.8.5 indicates that when the signal optical power level is less than –4 dBm, the system can be considered linear. A slight R-OSNR degradation can be observed when the signal optical power increases from –4 dBm to –2 dBm near the end of the system. R-OSNR degradation becomes significant when the signal optical power increases from –2 dBm to 0 dBm.

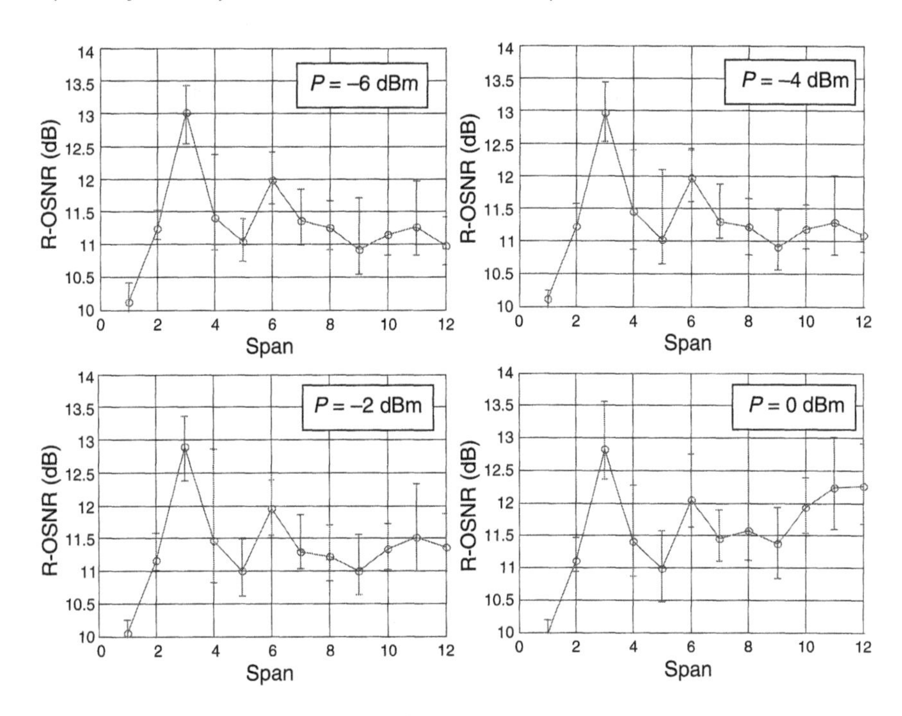

Figure 5.8.5 Measured R-OSNR along the system at four different signal optical power levels.

Figure 5.8.6 shows the R-OSNR measured at the end of the 12th span as the function of the signal optical power, which indicates that an approximately 1.3 dB R-OSNR penalty was introduced due to nonlinear SPM effect when the signal optical power was at 0 dBm.

In general, the optimization of residual chromatic dispersion and the measurement of dispersion window width are affected by the chirp in the optical waveform, the SPM effect, which is determined by the optical power of the signal channel, and the nonlinear crosstalk, which is determined by the optical powers of other channels.

Figure 5.8.7 shows a the schematic diagram of a WDM system over 20 amplified fiber spans, and 80 km of standard single-mode fiber is used in each span [48]. The optical power level of each wavelength channel is equalized by optical power management (OPM) using wavelength selective switch (WSS). In this particular system, the 10 Gb/s transmitters are equipped with electronic precompensations for the chromatic dispersion; therefore no optical dispersion compensation is necessary in the system. This WDM system consists of nine groups of optical channels with eight wavelengths in each group. The optical spectrum is shown in Figure 5.8.8, where the frequency separation between adjacent channels is 50 GHz and the gap between adjacent groups is 100 GHz.

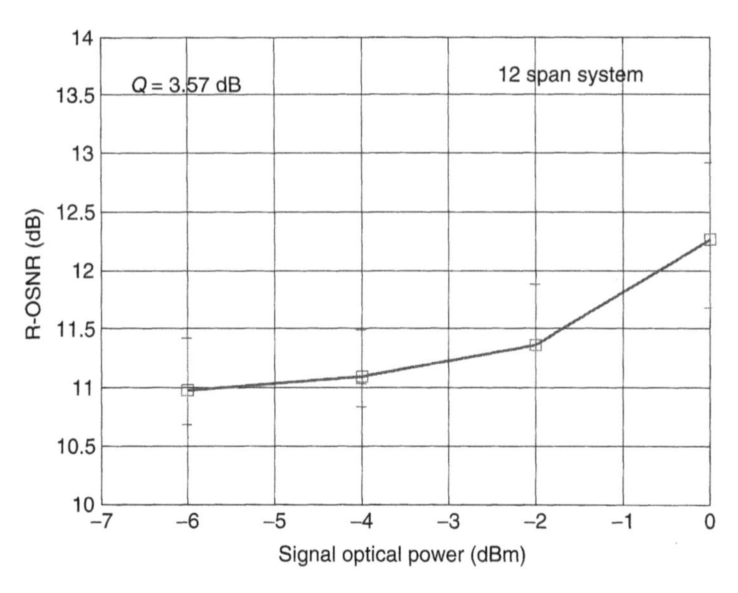

Figure 5.8.6 R-OSNR in a 12-span system as the function of the signal optical power.

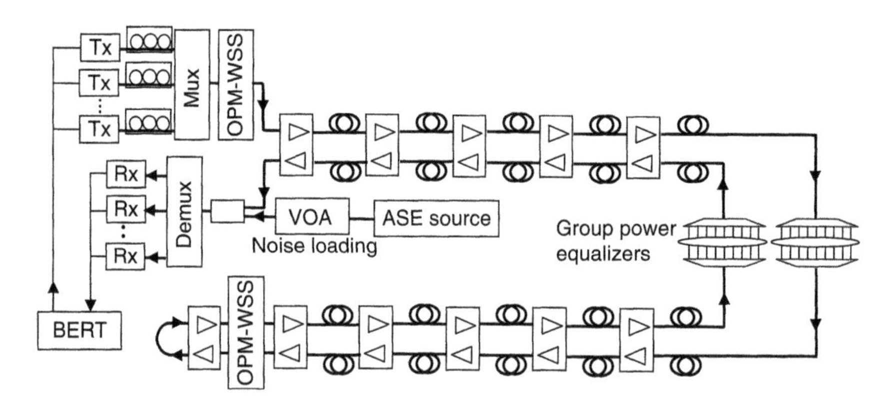

Figure 5.8.7 Simplified schematic configuration of a 20-span fiber transmission system employing bidirectional optical amplifiers and optical power management (OPM).

The polarization controller in front of each transmitter is used to make sure that the SOPs of the transmitters are aligned so that the nonlinear crosstalk in the fiber system reaches the worst case. A BERT supplies 2^{31}-1 PRBS patterns to modulate each transmitter and detect transmission errors at the receivers.

Because there is no optical dispersion compensation in the system, the accumulated chromatic dispersion in the 1600 km transmission system can be as high as

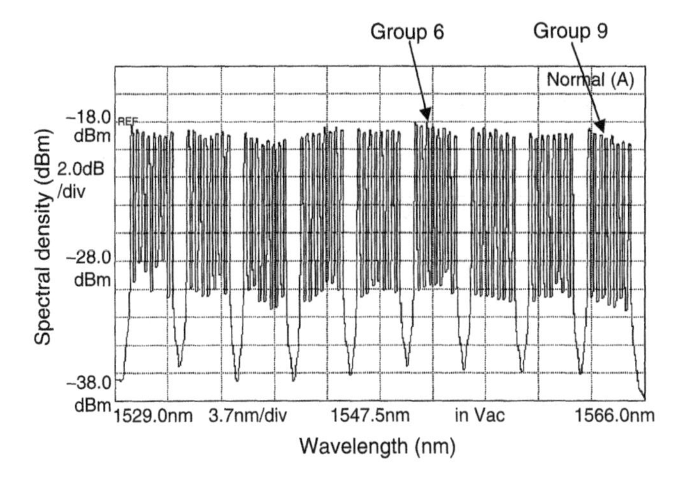

Figure 5.8.8 Optical spectrum at the receiver measured by an OSA with 0.2 nm resolution bandwidth.

27200 ps/nm-km. Crosstalk caused by FWM is expected to be rather weak because of the strong phase mismatch between channels due to rapid walkoff. However, the effect of XPM can still be strong because the efficiency of phase noise to intensity noise conversion is high due to the large uncompensated chromatic dispersion.

The transmission performance was evaluated by ASE noise loading at the receiver side. The required OSNR levels were measured at which the raw BER at the receiver were kept at 3.8×10^{-3} before forward error correction. ASE noise loading measurements were performed in three cases: (a) a single wavelength channel, (b) only one group of eight channels, and (c) all 72 wavelength channels shown in Figure 5.8.8 are turned on. To complete each measurement, the peak power level at each channel was varied from –3.5 dBm to 2 dBm. Figure 5.8.9 shows the measured R-OSNR at the 1551.32 nm channel in group 6 and the 1563.45 channel within group 9, respectively. Figure 5.8.9 indicates that R-OSNR increases with the increase of the per-channel optical power level, which is mainly caused by the nonlinear SPM effect. Meanwhile, with the increase in the number of channels, the R-OSNR also increases, primarily caused by nonlinear crosstalk due to XPM.

From overall system performance point of view, low R-OSNR levels are obtained at low signal optical powers, as shown in Figure 5.8.9. However, if the accumulated ASE noise generated by the inline amplifiers in the system is fixed, OSNR at the receiver (excluding noise loading) should be linearly proportional to the level of signal optical powers. Therefore, low optical power is not necessarily the best choice.

Taking into account the actual ASE noise generated in the transmission system and the signal optical power, the OSNR of each received optical channel

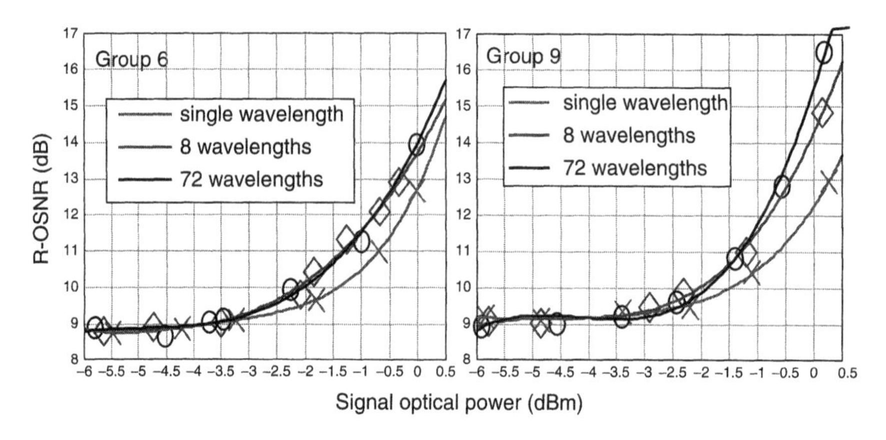

Figure 5.8.9 R-OSNR as the function of the per-channel signal optical power.

can be measured. The difference between this actual OSNR and the R-OSNR is defined as the OSNR margin, which signifies the robustness of the system operation. Figure 5.8.10 shows the measured OSNR margins of group-6 and group-9 in the system with all 72 WDM channels fully loaded [48]. The optimum per-channel signal optical power is approximately –2 dBm.

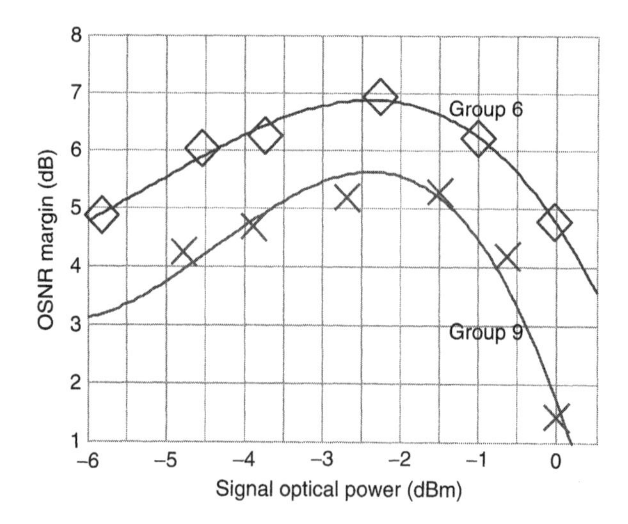

Figure 5.8.10 OSNR margin as a function of the per-channel signal optical power.

5.8.3 Measurement of R-OSNR Due to Optical Filter Misalignment

The complex transfer functions of optical devices such as optical filters, add/drop multiplexers, and narrowband dispersion compensators are important parameters that have significant impact in system performance. Techniques to measure individual optical components were discussed in the previous chapters. However, their impact in the transmission performance depends on system configurations, modulation formats, and the combination with other system building blocks. Therefore a relevent measurement of optical device performance and its impact on an optical system is to place the device into the system. The change of R-OSNR introduced by this particular device can be measured as an indication of the device quality.

Figure 5.8.11 shows the schematic diagram of an experimental setup to characterize the impact of a wavelength-selective switch (WSS) in an optical transmission system. Noise loading technique is used to evaluate the R-OSNR degradation due to the optical device under test. In this measurement, a tunable transmitter is used whose optical wavelength can be continuously tuned. The WSS is placed in a recirculating loop, in which 80 km of standard single-mode fiber is used with two optical amplifiers to compensate for the loss and a dispersion compensator reduces the accumulated dispersion. An optical recirculating loop simulates a multispan optical transmission system; its detailed operation principle and applications are discussed in the next section. Here we simply accept that each circulation in the loop is equivalent to an amplified optical span.

The measurement was performed by sweeping the wavelength of the tunable transmitter across the bandwidth of the optical filter. The R-OSNR at the receiver is measured for each wavelength setting of the transmitter. Figure 5.8.12 shows an

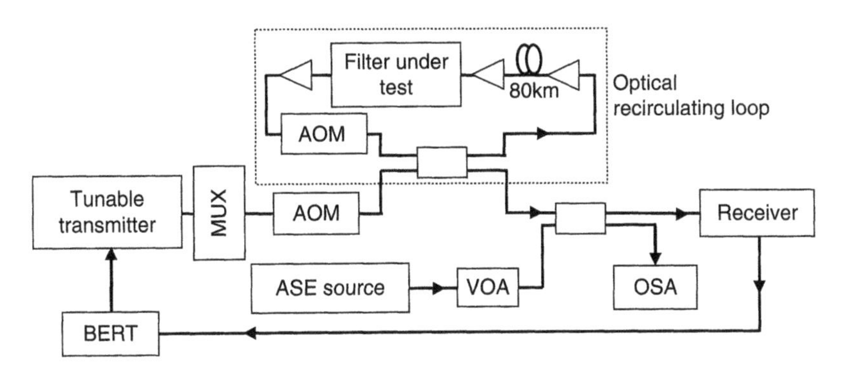

Figure 5.8.11 Measurement of optical filter characteristics using noise loading. AOM: acousto-optic modulator, VOA: variable optical attenuator

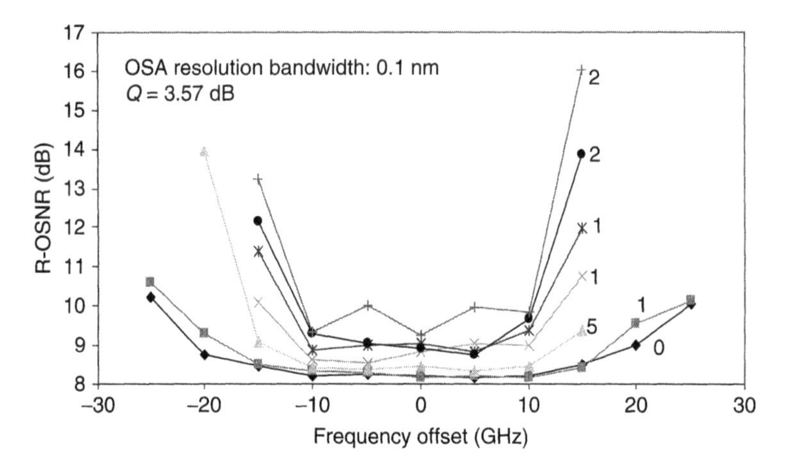

Figure 5.8.12 Measured R-OSNR as a function of the transmitter frequency offset.

example of the measured R-OSNR as a function of the transmitter frequency offset for different numbers of circulations of the optical recirculating loop as indicated beside the corresponding curve. For small numbers of loop circulations, the WSS optical bandwidth extends to approximately ±20 GHz. When the optical signal passes through the WSS more than 10 times in the recirculating loop, the optical bandwidth shrinks down to about ±10 GHz.

Recall that R-OSNR represents the required OSNR at the receiver to guarantee a certain Q-value, which is $Q = 3.57$ dB in this case. Therefore, the sharp R-OSNR degradation at the edges of the WSS passband is not likely caused by the reduced power transmission efficiency at these frequencies, nor the associated ASE noise increase in the system. Rather, it is caused primarily by the waveform distortion due to phase transition and group velocity distortion near the edges of the passband.

5.9 FIBER-OPTIC RECIRCULATING LOOP

To test optical transmission equipment and devices for long-distance fiber-optic systems, the system test-bed may have to include hundreds of kilometers of optical fibers, many optical inline amplifiers, and dispersion compensators. Fiber-optic recirculating loops have been used widely in optical communication laboratories to simplify the system test-bed and enable equipment manufactures to demonstrate their transmission terminal equipment anywhere without bringing with them the entire system [49]. It is a very convenient way to evaluate transmission performance of both ultra-long reach submarine systems and terrestrial fiber-optic systems without actually using many spans of fibers and a large number of optical amplifiers.

5.9.1 Operation Principle of a Recirculating Loop

A fiber-optic recirculating loop is a controlled optical switch which allows the optical signal from a transmitter to pass through an optical system many times to simulate a multi-span optical transmission. A block diagram of an optical recirculating loop is shown in Figure 5.9.1, where two acousto-optic modulators (AOMs) are used as the optical switches. The major reason for choosing AOMs to perform optical switch is due to their superior extinction ratio of typically >50 dB and their polarization insensitivity. Because the optical signal has to pass through the optical switch many times in a transmission experiment, a high on/off ratio is essential to minimize artificial multipath interference caused by the switches.

In Figure 5.9.1, the optical signal from a transmitter passes through the first optical switch AOM-1 and splits into two parts by a 2×2 fiber-optic directional coupler. One output from the coupler is sent to an optical transmission system; the other output is sent to an optical receiver for testing. The timing of the two optical switches are arranged such that when AOM-1 is on, AOM-2 has to be off; this allows the optical signal to fill up the transmission system without

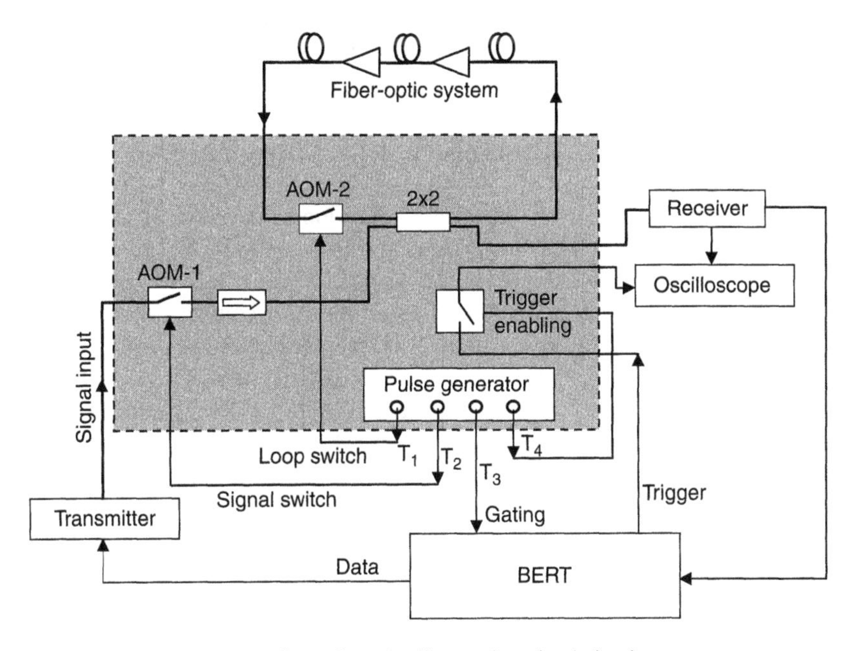

Figure 5.9.1 Configuration of a fiber-optic recirculating loop test set.

multi-path interference. Then AOM-1 is set to off and AOM-2 is set to on so that the optical signal can make roundtrip circulations in the optical transmission system in the loop. Obviously, for every roundtrip, the optical signal suffers an extra power loss due to the fiber coupler. The optical receiver detects the optical signal that passes through the optical system after the desired number of circulations. The waveform is then measured by a digital oscilloscope and the bit error rate is tested by a BERT. To measure only the optical signal that passes through the transmission system for a certain number of circulations, a time-gating pulse enables the BERT to only count the signal digital errors during certain time windows. Similarly, the trigger signal generated by the BERT is enabled only during the same time windows such that the digital oscilloscope is only triggered within these time windows. Outside these time windows, both the BERT and the oscilloscope are disabled.

Usually, a commercial BERT has a gating enabling function that allows a pulse train at the TTL voltage level to control the BER testing. However, the trigger generated by the BERT to synchronize the time sweep of the digital oscilloscope is typically at high frequency, which is a subharmonic of the digital data signal. The trigger enabling is accomplished by a microwave switch that is controlled by a pulse train synchronized with the gating of the BERT. In this measurement, four mutually synchronized switch pulse trains are required, for the two AOMs, the BERT enabling, and the oscilloscope trigger switch, respectively. A multichannel electrical pulse generator can be used which allows the precise control of the relative delays between the four pulse trains.

5.9.2 Measurement Procedure and Time Control

To help the description of the recirculating loop operation, it is useful to define the *loading state*, the *looping state,* and the *loop time*. The loading state is an operation state in which the optical signal is loading into the transmission system through the signal switch AOM-1 and the fiber coupler. In the loading state, the signal switch AOM-1 is on and the loop switch AOM-2 is off. Therefore the optical signal fills up the entire transmission system. The looping state is an operation state when the optical signal loaded in the loop circulates within the loop. In this state, the signal switch AOM-1 is off and the loop switch AOM-2 is on. The loop time is defined as the roundtrip time delay for the optical signal to travel through the fiber system, that is, $\tau = nL/c$, where n is the refractive index of the fiber, L is the length of the fiber, and c is the speed of light. The following are the fundamental steps of operating a recirculating loop:

1. As illustrated in Figure 5.9.2(a), a transmission experiment using a recirculating loop starts with the loading state, with the signal switch on and the loop

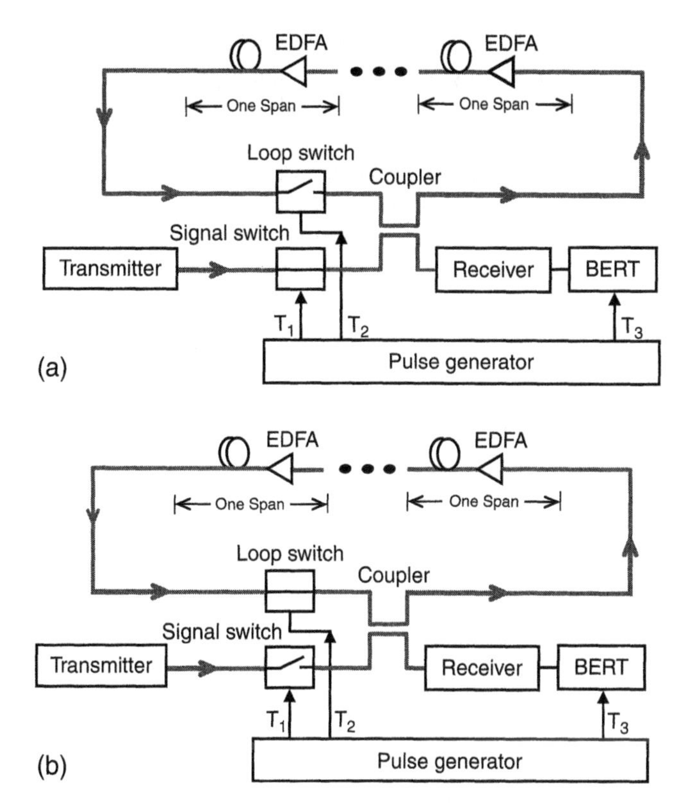

Figure 5.9.2 Two states of switch setting: (a) loading state and (b) looping state.

switch off. The two switches are held in this loading condition for at least one loop time to completely fill up the loop with the optical data signal. In the loading state, the control pulse T_1 is in the high level to turn on the signal switch, and T_2 should be in the low level to turn off the loop switch.

2. Once the transmission system in the loop is fully loaded with data, the switches change to the looping state. In the looping state, T_1 is on the low level to turn off the signal switch, whereas T_2 is on the high level to turn on the loop switch. In this state, the optical signal is allowed to circulate in the loop for a specified number of revolutions. The duration of a looping state can be estimated by $\Delta T = N\tau$, where N is the desired number of revolutions for the signal to travel in the loop, and τ is the roundtrip time of each loop. Even though after each circulation, a portion of the optical signal is coupled to the receiver through the fiber coupler, the eye diagram and the BER are measured only at the end of the looping state after the signal is circulated in the loop for N times. The BERT is gated by a TTL signal T_3; the

gating function, typically available in most of the BER test set, allows the BERT to be activated only when T_3 is on the high level.

3. The measurement continues with the AOMs switching alternatively between the loading state and the looping states so that bit errors can be accumulated over time. The BER is then calculated as the number of errors detected by the BERT error gating period divided by the total number of bits transmitted during the observation period. Since errors are counted only during the error gating period, the effective bit rate for the experiment is diminished by the duty cycle of the error gating signal; thus the real time for demonstrating a particular BER using a recirculating loop will be much longer than conventional system BER measurements.
4. In the eye diagram measurement using a digital oscilloscope, the trigger is usually generated by the BERT which is a subharmonic of the data signal. A microwave switch that is controlled by the synchronized pulse train T_4, as shown in Figure 5.9.1, allows the trigger to be delivered to the oscilloscope only during the BERT gating time window. That is, the trigger enabling switch is in the *on* state only at the end of each looping state, the time window when the eye diagram needs to be measured.

Figure 5.9.3 shows the timing control diagram of the pulse generator in the recirculating loop experiment. Typically, a BERT test set is gated by positive

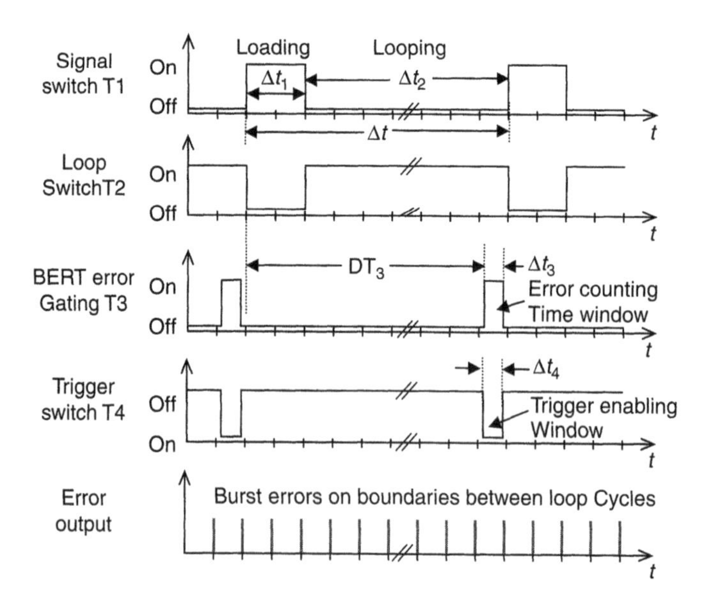

Figure 5.9.3 Timing control diagram for recirculating loop experiment. Time unit is normalized by loop time τ.

pulses of TTL level. To trigger the oscilloscope, the microwave switch embedded in the recirculating loop test set is on the *off* state for positive TTL pulses and in the *on* state at the zero control pulse level. Thus, the oscilloscope is enabled only within the windows when T_4 = 0. It is worth noting that there might be BER burst errors at the beginning and the end of each loop time, as illustrated in Figure 5.9.3, and these errors are artifacts due to the use of recirculating loops. The burst errors happen because of the rise and fall time of the signal and loop switches. Although the switch speed of the AOM is typically less than 1 μs, this time interval corresponds to thousands of corrupted data bits in a multigigabit data stream. By choosing a BERT gating time window to be slightly less than a loop time τ, these burst errors will be excluded from the BER counting, and therefore these burst errors do not affect the emulation of optical fiber transmission with recirculating loop. The following is an example that helps us understand how to establish a practical optical system test bed using a recirculating loop.

Assuming that a two-span fiber-optic system with 80 km of optical fiber in each span is used in the loop test, the loop delay time is

$$\tau = nL/c = \frac{1.5 \times 160 \times 10^3}{3 \times 10^8} = 800\,\mu s \tag{5.9.1}$$

where $n = 1.5$ is assumed for the refractive index in the fiber. Then the time window of loading state should be between 800 μs and 1600 μs ($\tau < \Delta t_1 < 2\tau$). One can simply choose a value in the middle: $\Delta t_1 = 1200$ μs. If one wants to measure the system performance after 10 loops of circulation, that will be equivalently a 20-span system with total transmission distance of 16,000 km. The switch repetition period can be chosen as

$$\Delta t = 800 \times 10 + 1200 = 9200\,\mu s \tag{5.9.2}$$

As shown in Figure 5.9.3, T_1 and T_2 are complementary (T_1 is positive while T_2 is negative) and there is no relative time delay between them. BERT error gating pulse T_3 and oscilloscope trigger switch control T_4 have the same pulse widths but opposite polarities as well, with $\Delta t_3 = \Delta t_4 < 800\ \mu s$. To avoid bust errors caused by AOM leading and trailing edges, we can choose

$$\Delta t_3 = \Delta t_4 = 600\,\mu s \tag{5.9.3}$$

The relative delay between T_2 and T_3 is (the same delay is between T_1 and T_4)

$$DT_3 = 9200 - 800 + 100 = 8500\,\mu s \tag{5.9.4}$$

In this arrangement, the error gating pulse train has the duty cycle of

$$d = \frac{600\,\mu s}{9200\,\mu s} = 0.0652 \tag{5.9.5}$$

Therefore the actual BER of this 20-span system is related to the measured BER by

$$BER_{actual} = \frac{BER_{measured}}{0.0652} \tag{5.9.6}$$

This is because the time used to accumulate the errors (during the time window Δt_3) is only a small fraction of the pulse repetition period. As a consequence, to observe a reasonable number of errors to accurately evaluate BER, a much longer time (about 15 times longer) is required for the measurement in this case because the use of a recirculating loop.

5.9.3 Optical Gain Adjustment in the Loop

In long-distance multispan optical transmission systems, optical amplifiers are used to overcome the loss of optical fibers and other optical components. In a system test bed using an optical recirculating loop, an optical signal travels through the system in the loop for a number of circulations. Therefore optical gain provided by the optical amplifiers in the loop has to be adjusted such that it is equal to the losses in the system, which also includes the losses introduced by the loop switch and the fiber directional couple. This ensures that after each circulation in the loop, the optical power stays at the same level.

An easy way to adjust the loop gain is to use a low-speed photodetector (O/E converter) in place of the optical receiver shown in Figure 5.9.1, which is connected directly to an oscilloscope. A variable optical attenuator (VOA) is added in the loop to adjust the loop gain. Since the O/E converter has low speed, the waveform displayed by the oscilloscope is proportional to the average signal optical power after each loop circulation. Since the trigger to the oscilloscope is synchronized to the loop switch, the measured waveform shows exactly the loop control function and timing. The optical gain of each circulation through the loop can then be easily evaluated by this waveform.

Figure 5.9.4 illustrates the gain adjustment procedure. If the net gain of the loop is smaller than unity, the photocurrent signal displayed on the oscilloscope decreases with each consecutive loop circulation, as shown by Figure 5.9.4(a). On the other hand, if the loop gain is higher than unity, the waveform on the oscilloscope will increase with each consecutive circulation and this is shown in Figure 5.9.4(b). Figure 5.9.4(c) shows the expected waveform when the optical gain exactly compensates for the loss. In this case, there is no optical power change from one loop circulation to the next.

This optical gain adjustment is important for system stability consideration. It is also a necessary procedure to simulate a straight-line optical transmission system in which optical power levels in different spans are usually identical.

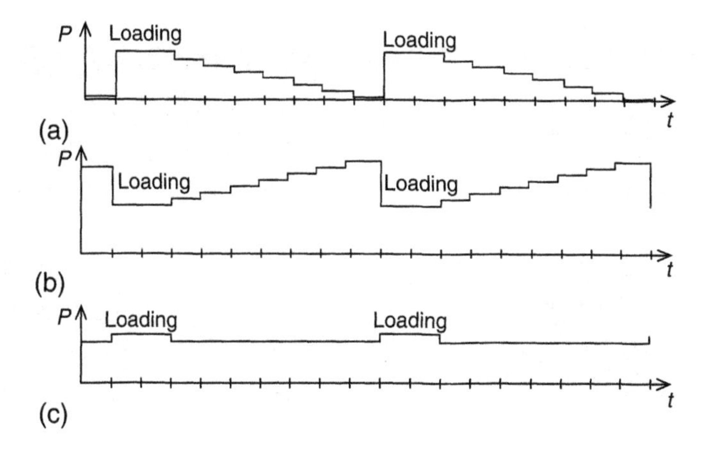

Figure 5.9.4 Illustration of loop optical gain adjustment.

Figure 5.9.5 shows an example of eye diagrams measured after the signal traveled through different loop circulations. These eye diagrams were obtained from a 5 Gb/s system using Truewave fibers with dispersion of about 2.5 ps/nm/km at the 1555 nm optical signal wavelength. Two fiber spans were used in the recirculating loop with two inline fiber amplifiers. The gain of the optical amplifiers was carefully adjusted to compensate for the loss. Since the system was not dispersion compensated, it shows visible waveform distortion after 900 km of equivalent fiber transmission.

Another important note is that a system test bed based on recirculating loop uses a relatively small number of amplified optical spans in the loop. From a system performance point of view it may differ from a long-haul straight-line optical system in a few aspects. First, in a real long-haul optical system with a large number of amplified fiber spans, the ASE noise created by fiber amplifiers accumulates toward the end of the system. As a result, the optical amplifier gain saturation caused by ASE noise is stronger near the end of the system than at the beginning of the system. Obviously, this nonuniform amplifier gain saturation effect along the system does not appear in an optical recirculating loop test bed with a small number of amplified spans in the loop.

Second, a recirculating system with M amplified spans in the loop is equivalent to a straight-line long-haul system in which all the system parameters exactly repeat themselves after each M span. This artificial periodicity would not likely to happen in a practical system. As a consequence the statistics of system performance tested with a recirculating loop may not be the same as a real straight-line system. For example, four-wave mixing (FWM) in a multispan optical transmission system is caused by nonlinear interaction between different

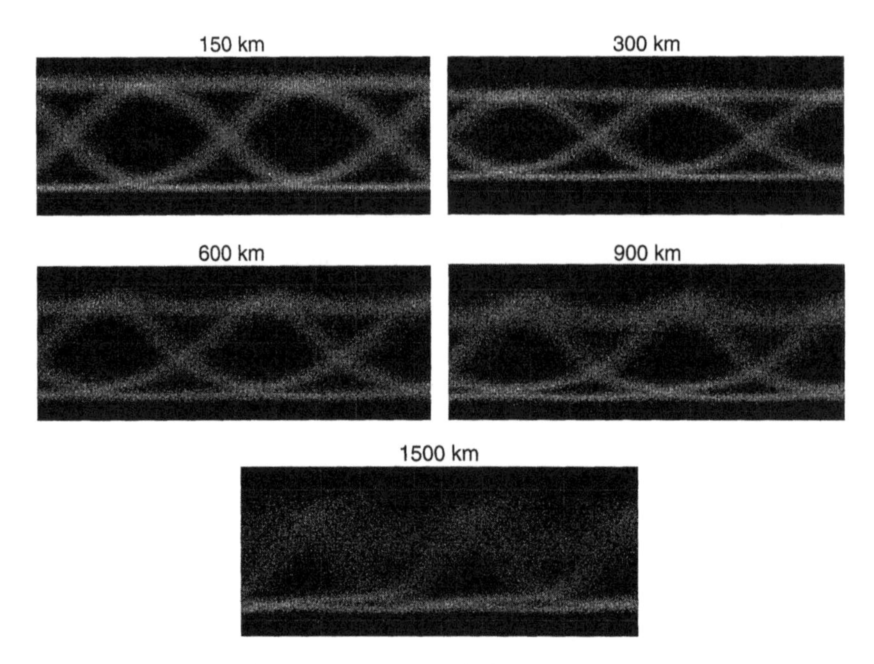

Figure 5.9.5 Examples of eye diagrams measured after signal traveled through different loop circulations; 5 Gb/s system using Truewave fibers.

wavelength components. The efficiency of FWM depends both on the nonlinearity of the fiber and on the phase match condition between participating channels. In a single-span optical system with two wavelength channels, the power of the degenerate FWM component at the fiber output can be expressed as

$$P_{FWM} = (\gamma L_{eff})^2 e^{-\alpha L} P_1(0) P_2^2(0) \left(\frac{\alpha^2}{\alpha^2 + \Delta\beta^2} \right) \left(1 + \frac{4e^{-\alpha L} \sin^2(\Delta\beta L/2)}{(1 - e^{-\alpha L})^2} \right) \quad (5.9.7)$$

where $P_1(0)$ and $P_2(0)$ are the optical power levels of the two optical channels at the input of the fiber, γ is the nonlinear parameter of the fiber, $L_{eff} = (1 - e^{-\alpha L})/\alpha$ is the nonlinear length of the fiber, α is the fiber loss parameter, and L is the fiber length. $\Delta\beta$ is the wave number mismatch between the two wavelength channels, which is

$$\Delta\beta = \beta_{FWM} + \beta_{sig}^{(1)} - 2\beta_{sig}^{(2)} = \frac{2\pi\lambda^2}{c} \Delta f^2 \left(D + \Delta f \frac{\lambda^2}{c} \frac{dD}{d\lambda} \right) \quad (5.9.8)$$

where $\beta_{sig}^{(1)}$, $\beta_{sig}^{(2)}$ and β_{FWM} are wave numbers of the two participating signal channels and the FWM component, Δf is the frequency difference between the two signal channels, and D is the dispersion parameter of the fiber. For an amplified multispan optical system, FWM contributions from different fiber spans will combine coherently at the end of the system. Assuming that the parameters at each fiber span are identical, which is equivalent to use only one fiber span in the recirculating loop, the FWM power at the output of the system will be,

$$P_{FWM\_MS} = P_{FWM}\frac{\sin^2(N_A\Delta\beta L/2)}{\sin^2(\Delta\beta L/2)} \tag{5.9.9}$$

where N_A is the total number of amplified fiber spans, or equivalently, the number of circulations in the single-span recirculating loop.

Figure 5.9.6 shows the block diagram of an experimental setup to measure FWM in a fiber system using recirculating loop test bed. In this experiment the optical spectrum analyzer (OSA) is controlled by an enabling pulse train that is synchronized with the switch of AOMs. The OSA sweeps only during the time window when the enabling pulse is at the high level. Figure 5.9.7(a) shows the measured FWM efficiency as a function of the wavelength separation between the two signal channels. This result was obtained when the optical signal circulated in the loop for five times, which represents a five-span optical system. The optical power at the output of the EDFA was 18 dBm, and the optical fiber used in the experiment was 57 km nonzero dispersion shifted fiber (NZDSF) with the dispersion of $D = 2.5$ ps/nm/km at the signal wavelength. Because of the precise periodical structure of the optical system using recirculating loop, the FWM efficiency versus channel spacing exhibits strong resonance because of perfect phase matching at

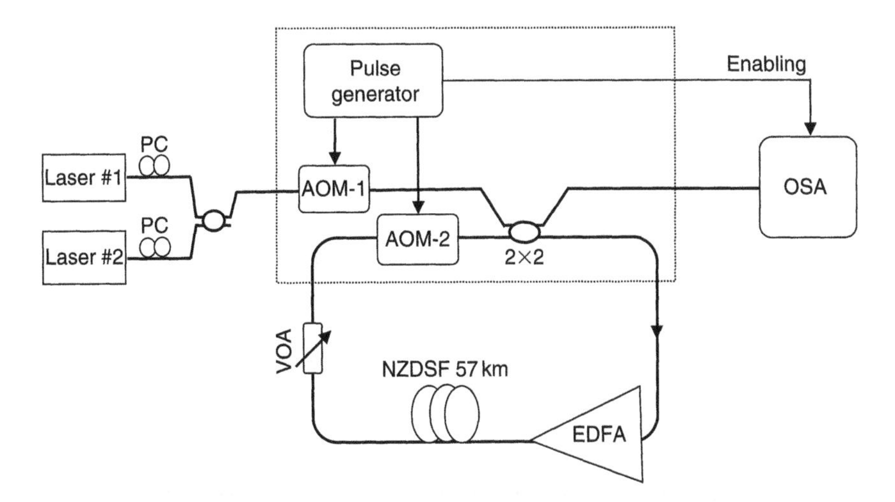

Figure 5.9.6 Measurement of FWM in a fiber-optical system using recirculating loop.

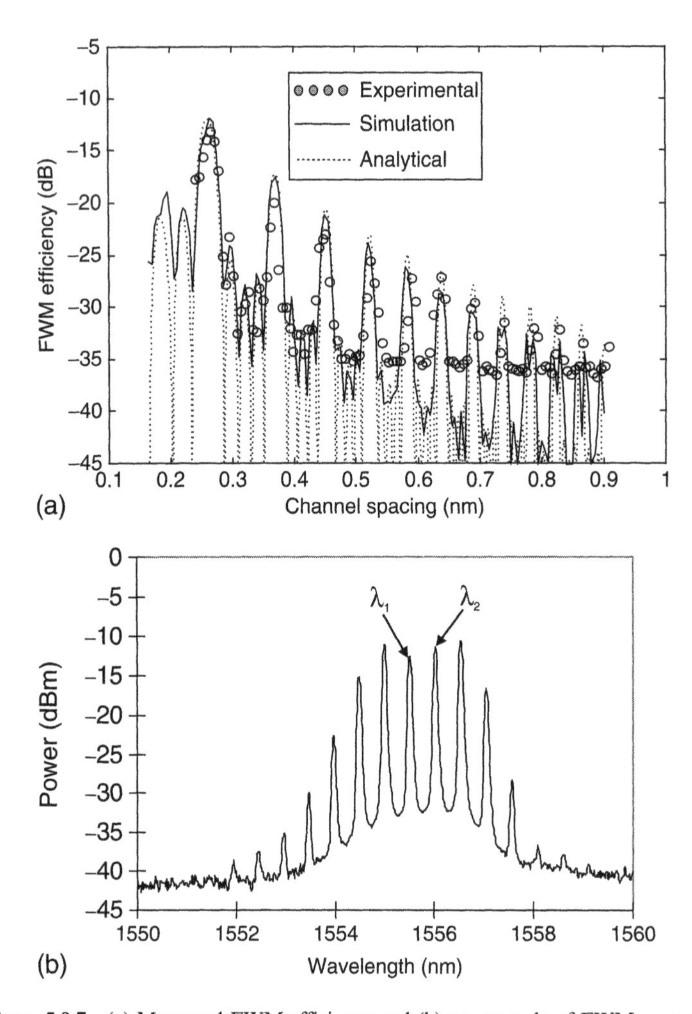

Figure 5.9.7 (a) Measured FWM efficiency and (b) an example of FWM spectrum.

these resonance channel separation. This resonance effect can be easily reproduced by Equation 5.9.9 analytically, shown as the dashed line in Figure 5.9.7(a).

A numerical simulation using split-step Fourier method also predicted the same phenomenon which is shown as the solid line in the same figure. Both of them agree with the measured results. As an example, Figure 5.9.7(b) shows a measured FWM optical spectrum when the channel spacing is at one of the resonance peaks shown in Figure 5.9.7(a). In this case, the FWM sidebands can even have higher powers than that of the original optical signals, which may be due to the effect of parametric gain introduced by modulation instability.

It is worthwhile to note that in practical optical systems with multiple amplified fiber spans, because the fiber length, the dispersion parameters, and the optical power level at different spans are not likely to be identical; this type of perfect phase match would typically not happen. On the positive side, this recirculating loop measurement might be useful to determine the nonlinear and the dispersion parameters of a particular fiber span by putting it in the loop.

Another potential concern of using a recirculating loop is that due to the same reason of perfect periodicity of chromatic dispersion, the PDL and PMD parameters are also precisely periodical. Therefore the statistic natures of PDL and PMD in practical straight-line fiber systems may not be accurately simulated by a recirculating loop measurement. One way to overcome these problems is to add a polarization scrambler in the fiber loop, as illustrated in Figure 5.9.8 [59, 60, 61]. In this system, the polarization scrambler is controlled by a pulse train that is synchronized with the loop switch AOM-2. The polarization scrambler adds a random rotation of the polarization for every loop circulation of the optical signal, which makes the system closer to reality.

To conclude this section, an optical recirculating loop is a very useful instrument for long-distance optical system performance measurements. It significantly reduces the required number of optical fiber spans and optical amplifiers. Time controls and synchronization of AOM switches and enabling pulses for BERT and oscilloscopes are important for recirculating loop measurements. Optical gain in the loop also has to be properly adjusted so that the net gain is zero for each circulation. The periodic nature of the optical system test bed based on a

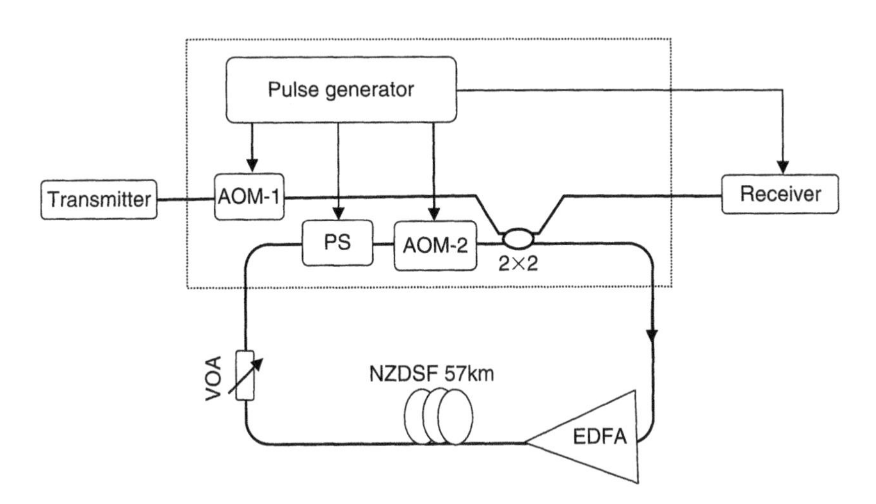

Figure 5.9.8 Adding a synchronized polarization scrambler (PS) to a recirculating loop.

recirculating loop may create artificial phase match and polarization state match over the system, which have to be considered in the measurements.

REFERENCES

[1] P. V. Kumar, M. Z. Win, H. F. Lu, and C. N. Georghiades,"Error–control coding techniques," in *Optical Fiber Telecommunications* IVB, I. Kaminow and T. Li, eds., Academic Press, 2002.
[2] S. Waklin and J. Conradi, "Multilevel signaling for increasing the reach of 10Gb/s lightwave systems," *Journal of Lightwave Technology*, Vol. 17, pp. 2235–2248, 1999.
[3] D. Penninckx, M. Chbat, L. Pierre, and J. P. Thiery, "The Phase-Shaped Binary Transmission (PSBT): a new technique to transmit far beyond the chromatic dispersion limit," *IEEE Photon. Technol. Lett.*Vol. 9, pp. 259–261, 1997.
[4] R. Hui, B. Zhu, R. Huang, C. Allen, K. Demarest, and D. Richards, "Subcarrier multiplexing for high-speed optical transmission," *Journal of Lightwave Technology*, Vol. 20, pp. 417–427, 2002.
[5] J. McNicol, M. O'Sullivan, K. Roberts, A. Comeau, D. McGhan, and L. Strawczynski, "Electrical domain compensation of optical dispersion," in *Proc. OFC*, 2005, paper OThJ3.
[6] S. D. Personick,"Receiver Design for Optical Fiber Systems," *Proc. of the IEEE*, Vol. 65, pp. 1670–1678, 1977.
[7] R. Hui, M. Vaziri, J. Zhou, and M. O'Sullivan, "Separation of noise from distortion for high-speed optical fiber system link budgeting," *IEEE Photonics Technol. Lett.*, Vol. 11, pp. 910–912, 1999.
[8] D. Walker, H. Sun, C. Laperle, A. Comeau, and M. O'Sullivan, "960-km transmission over G.652 fiber at 10 Gb/s with a laser/electro-absorption modulator and no optical dispersion compensation," *IEEE Photonics Technology Letters*, Vol. 17, pp. 2751–2753, 2005.
[9] M. S. O'Sullivan, K. Roberts, and C. Bontum, "Electronic dispersion compensation techniques for optical communication systems," European Conference on Optical Communications (ECOC 2005), Glasgow, Scotland, September 25–29, 2005, pp. 189–190.
[10] N. S. Bergano, F.W. Kerfoot, and C. R. Davidsion, "Margin measurements in optical amplifier system," *IEEE Photonics Technology Letters*, Vol. 5, pp. 304–306, 1993.
[11] R. Stephens, "Analyzing jitter at high data rates," *IEEE Communications Magazine*, Vol. 42, No. 2, pp. S6–S10, 2004.
[12] Agilent Technologies, Application Notes #1432, "Jitter analysis techniques for high data rates."
[13] Qian Yu, Zhongqi Pan, Lian-Shan Yan, and A. E. Willner, "Chromatic dispersion monitoring technique using sideband optical filtering and clock phase-shift detection," *Journal of Lightwave Technology*, Vol. 20, Issue 12, pp. 2267–2271, December 2002.
[14] B. Fu and R. Hui, "Fiber chromatic dispersion and polarization-mode dispersion monitoring using coherent detection," *IEEE Photonics Technology Letters*, Vol. 17, No.7, pp. 1561–1563, 2005.
[15] G. P. Agrawal, *Fiber-optic Communication Systems*, 2nd ed., New York, Wiley, 1992.
[16] T. Luo, A. Pan, S. M. R. M. Nezam, L. S. Yan, A. B. Sahin, and A. E. Willner, "PMD monitoring by tracking the chromatic-dispersion-insensitive RF power of the vestigial sideband," *IEEE Photonics Technology Letters*, Vol. 16, No.9, pp. 2177–2179, 2004.

[17] I. Roudas, G. A. Piech, M. Mlejnek, Y. Mauro, D. Q. Chowdhury, and M. Vasilyev, "Coherent Frequency-Selective Polarimeter for Polarization-Mode Dispersion Monitoring," *Journal of Lightwave Technology*, Vol. 22, No. 4, pp. 953–967, 2004.

[18] R. Hui, R. Saunders, B. Heffner, D. Richards, B. Fu, and P. Adany, "Nonblocking PMD monitoring in live optical systems," *Electron. Lett.*, Vol. 43, No. 1, pp. 53–54, 2007.

[19] R. M. Craig, S. L. Gilbert, and P. D. Hale, "High-Resolution, Nonmechanical Approach to Polarization-Dependent Transmission Measurements," *J. Lightwave Tecnol.*, Vol. 16, No. 7, pp. 1285–1294, July 1998.

[20] B. Heffner, "Deterministic, analytically complete measurement of polarization-dependent transmission through optical devices," *IEEE Photon. Technol. Lett.*, Vol. 4, pp. 451–454, 1992.

[21] J. Jiang, D. Richards, C. Allen, S. Oliva, and R. Hui, "Non-intrusive polarization dependent loss monitoring in fiber optic transmission systems," *Optics Communications*, Vol. 281, pp 4631–4633, 2008.

[22] A. R. Chraplyvy, "Limitations of lightwave communications imposed by optical fiber nonlinearities," *IEEE J. Lightwave Technol.*, Vol. 8, pp. 1548–1557, 1990.

[23] D. Marcuse, A. R. Chraplyvy, and R. W. Tkach, "Dependence of cross-phase modulation on channel number in fiber WDM systems," *IEEE J. Lightwave Technol.*, Vol. 12, pp. 885–890, 1994.

[24] J. Wang and K. Petermann, "Small signal analysis for dispersive optical fiber communication systems," *IEEE Journal of Lightwave Technol.*, Vol. 10, No. 1, pp. 96–100, 1992.

[25] T.-K. Chiang, N. Kagi, M. E. Marhic, and L. Kazovsky, "Cross-phase modulation in fiber links with multiple optical amplifiers and dispersion compensators," *IEEE Journal of Lightwave Technol.*, Vol. 14, No. 3, pp. 249–260, 1996.

[26] R. Hui, K. Demarest and C.Allen, "Cross phase modulation in multi-span WDM optical fiber systems," *IEEE J. Lightwave Technology*, Vol. 17, No. 7, pp. 1018-1026, 1999.

[27] G. P. Agrawal, *Nonlinear Fiber Optics*, Academic Press, San Diego, CA, 1989.

[28] M. Shtaif and M. Eiselt, "Analysis of intensity interference caused by cross-phase modulation in dispersive optical fibers," *IEEE Photonics Technology Letters*, Vol. 9, No. 12, pp. 1592–1594, 1997.

[29] M. Eiselt, M. Shtaif and L. D. Garrett, "Cross-phase modulation distortions in multi-span WDM systems," Optical Fiber Communication Conference OFC '99, paper ThC5, SanDiego, CA, February 1999.

[30] L. Rapp, "Experimental investigation of signal distortion induced by cross-phase modulation combined with distortion," *IEEE Photonics Technology Letters*, Vol. 9, No. 12, pp. 1592–1594, 1997.

[31] R. A. Saunders, B. L. Patel, H. J. Harvey, and A. Robinson, "Impact of cross-phase modulation seeded modulation instability in 10Gb/s WDM systems and methods for its suppression," Proceedings of Optical Fiber Communication Conference OFC'97, paper WC4, pp. 116–117, Dallas TX, February 1997.

[32] K.-P. Ho, "Performance degradation of phase-modulated systems due to nonlinear phase noise," *IEEE Photon. Technol. Lett.* Vol. 15, pp. 1213–1215, 2003.

[33] K. Inoue and H. Toba,"Fiber four-wave mixing in multi-amplifier systems with non-uniform chromatic dispersion," *IEEE J. Lightwave Technology*, Vol. 13, no. 1, pp. 88–93, 1995.

[34] Fabrizio Forghieri, R. W. Tkach, and A. R. Chraplyvy, "WDM Systems with Unequally Spaced Channels," *IEEE J. Lightwave Technology*, Vol. 13, pp. 889–897, 1995.

[35] M. Eiselt, "Limits on WDM Systems Due to Four-Wave Mixing: A Statistical Approach," *J. Lightwave Technol.* Vol. 17, No. 11, pp. 2261–267, 1999.

[36] J. Wang, X. Sun, and M. Zheng, "Effect of group velocity dispersion on stimulated Raman crosstalk in multichannel transmission systems," *IEEE Photonics Technology Letters*, Vol. 10, pp. 540–542, 1998.
[37] K-P. Ho, "Statistical properties of stimulated Raman crosstalk in WDM systems," *J. Lightwave Technology*, Vol. 18, pp. 915–921, 2000.
[38] F. Coppinger, L. Chen, and D. Piehler, "Nonlinear Raman cross-talk in a video overlay passive optical network," Proceedings of Optical Fiber Communication Conference, OFC '2003, March 23–28, 2003, Atlanta, GA, pp. 285–286.
[39] Z. Wang, A. Li, C. J. Mahon, G. Jacobsen, and E. Bodtker, "Performance Limitations imposed by Stimulated Raman scattering in optical WDM SCM video distribution systems," *IEEE Photon. Technol. Lett.*, Vol. 7, pp. 1492–1494, Dec. 1995.
[40] M. R. Phillips and D. M. Ott, "Crosstalk due to optical fiber nonlinearities in WDM CATV lightwave systems," *J. Lightwave Technology*, Vol. 17, pp. 1782–1792, 1999.
[41] F. Tian, R. Hui, B. Colella, and D. Bowler, "Raman crosstalk in fiber-optic hybrid CATV systems with wide channel separations," *IEEE Photonics Technology Letters*, Vol. 16, No. 1, pp. 344–346, 2004.
[42] ITU-T Recommendation G.983.3, Video channel wavelength 1550–1560 nm, data channel wavelength 1480–1500 nm; therefore the widest channel spacing is 80 nm.
[43] M. Yu, G. P. Agrawal, and C. J. McKinstrie, "Pump-wave effects on the propagation of noisy signals in nonlinear dispersive media," *J. Opt. Soc. Am. B*, Vol. 11, pp. 1126–1132, 1995.
[44] A. Mecozzi, "Long-distance transmission at zero dispersion: combined effect of the Kerr nonlinearity and the noise of the inline amplifiers," *J. Opt. Soc. Am. B*, Vol. 12, pp. 462–465, 1994.
[45] D. Marcus, "Noise properties of four-wave mixing of signal and noise," *Electron. Lett.*, Vol. 30, pp. 1175–1177, 1994.
[46] R. Hui, M. O'Sullivan, A. Robinson, and M. Taylor, "Modulation instability and its impact in multispan optical amplified systems: Theory and experiments," *IEEE J. Lightwave Technol.*, Vol. 15, No. 7, pp. 1071–1082, 1997.
[47] R. Hui and M. O'Sullivan, "Noise squeezing due to Kerr effect nonlinearty in optical fibers with negative dispersion," *IEEE Electronics Letters*, Vol. 32, No. 21, pp. 2001–2002, 1996.
[48] M. Birk, X. Zhou, M. Boroditsky, S. H. Foo, D. Bownass, M. Moyer, M. O'Sullivan, "WDM technical trial with complete electronic dispersion compensation," *ECOC* 2006, paper Th2.5.6.
[49] N. S. Bergano and C. R. Davidson, "Circulating loop transmission experiments for the study of long-haul transmission systems using erbium doped fiber amplifiers," *J. Lightwave Technol.*, Vol. 13, pp. 879–888, May 1995.
[50] Q. Yu, L.-S. Yan, S. Lee, Y. Xie, and A. E. Willner, "Loop-Synchronous Polarization Scrambling Technique for Simulating Polarization Effects Using Recirculating Fiber Loops," *J. Lightwave Technol.* Vol. 21, pp. 1593–1600, 2003.
[60] Yu Sun, A.O. Lima, J. Zweck, L. Yan, C.R. Menyuk, and G.M. Carter, "Statistics of the System Performance in a Scrambled Recirculating Loop with PDL and PDG," *IEEE Photonics Technology Letters* Vol. 15, pp. 1067–1069 (2003).
[61] Hai Xu, Hua Jiao, Li Yan, and Gary M. Carter, "Measurement of Distributions of Differential Group Delay in a Recirculating Loop With and Without Loop-Synchronous Scrambling," *IEEE Photonics Technology Letters* Vol. 16, pp. 1691–1693 (2004).

Index

A

aberration in spectrum analyzer lenses, 144
absolute calibration, FPI spectrum analyzers, 157
absolute wavelength accuracy, OSAs, 132
absorption
  cross-section, EDFAs, 102–103
  fiber couplers, 330–331
  optical fibers, 62–63
acceptance angle of fibers. *See* numerical aperture
acoustic phonons, 77
acousto-optic frequency modulators (AOFMs), 275
  in optical recirculating loops, 617–618
acousto-optic frequency shifters (AOFSs), 352
active region, 8
ADCs (analog-to-digital converters), 223–224
AGC. *See* automatic gain control
all-fiber configuration, FPIs, 158
AM response method for measuring chromatic dispersion, 403–405
amplified spontaneous emission (ASE) noise, 107–109, 314
  crosstalk between signal and. *See* modulation instability (MI)
  with forward and backward pumping configurations, 111
  gain measurement and, 307–310
  impact in electrical domain, 318–323
  optical amplifier noise definition, 324–326
  optical domain characterization of, 316–317
  optical preamplification, 510–511
  in optical recirculating loops, 623–624
  OSNR margin and R-OSNR, 514–521
  signal-spontaneous emission beat noise, 318, 319–321
    optical signal-to-noise ratio (OSNR) and, 514–515
  spontaneous-spontaneous emission beat noise, 301–304, 318, 321–323
    optical signal-to-noise ratio (OSNR) and, 514–515
    waveform measurement, 524
amplifiers. *See* optical amplifiers
amplitude stability of OSAs, 133
analog oscilloscopes, 212–214
analog-to-digital converters (ADCs), 223–224
angle-polished fiber surfaces, 60–61
angular spreading of diffracted light, gratings, 136
anomalous dispersions, DSFs, 594
AOFMs (acousto-optic frequency modulators), 275
  in optical recirculating loops, 617–618
AOFSs (acousto-optic frequency shifters), 352
AOMs. *See* AOFMs
APC (angle-polished connector) contactor, 60–61
APC (automatic power control), 111–114
APDs. *See* avalanche photodiodes
aperture for optical spectrum analyzers, 142–143
apparent DGD, 547–551
arrayed waveguide gratings (AWGs), 340, 343–345
ASE. *See* amplified spontaneous emission noise
ASE-ASE beat noise, 301–304, 318, 321–323. *See also* signal-spontaneous emission beat noise
  optical signal-to-noise ratio (OSNR) and, 514–515
  waveform measurement, 524
attenuation coefficient, 64, 385–386
attenuation in optical fibers, 62–67, 368, 382–394
  cutback measurement technique, 382–384
  nonuniformity of, 385
  optical time-domain interferometers (OTDRs), 384–391
    improvement considerations, 391–394

attenuation in optical fibers (*Continued*)
PDL (polarization-dependent loss), 359–360, 431–432, 438–441
in-situ monitoring, 551–556
partial, 558
auto-trigger mode (oscilloscopes), 216
autocorrelators
fundamental operation principle of, 225–226
to measure optical pulses, 224–231
optical receiver characterization, 304–306
automatic gain control (AGC), 111–114
automatic power control (APC), 111–114
avalanche photodiodes (APDs), 41–43
noise and SNR, 43
Q function, 492
average DGD, 412–413. *See also* differential group delay (DGD)
AWGs (arrayed waveguide gratings), 340, 343–345

B

backscattering. *See* scattering
backward pumping (EDFAs), 110–111
balanced coherent detection, 202–204
bandgap
carrier confinement, 7–8
direct and indirect semiconductors, 6–7
bandwidth, photodetectors, 34–36
frequency domain characterization of optical receivers, 299–301
short optical pulse technique, 304–306
spontaneous-spontaneous emission noise beating technique, 301–304
baseband AM response method, 403–405
BBO (Beta-Barium Borate) crystals, 228
BER (bit error rate), 483, 486–494
decision threshold voltage vs., 521–523
testing, 499–508
error detection, 504–508
pattern generators, 501–504
BER-T scan, 535–537
BER-V testing, 521–523
Bessel equation, 55–56
birefringence in optical fibers, 75, 409–413
bit error detection, 504–508. *See also* BER (bit error rate)
bit error rate. *See* BER
bit rate interleaving, 570
Bowtie fibers, 371
Bragg wavelength, 27
Brewster angle, 49
Brillouin scattering. *See* stimulated Brillouin scattering (SBS)

C

calibration (wavelength)
accuracy of, with OSAs, 133
FPI spectrum analyzers, 156–157, 185
optical wavelength meters, 185–186
carrier confinement, 7–8
carrier drift, 6
carrier drift speed, 35–36
carrier dynamics, optical amplifier, 327
carrier lifetime, LED, 12
cathode ray tubes (CRTs), 212
CCDs. *See* charge-coupled devices (CCDs)
CD. *See* chromatic dispersion
charge-coupled devices (CCDs), 145
for far-field measurement, 379
for near-field measurement, 381–382
chirp (phase modulation), 24, 260
from direct current modulation, 114–115
electro-optic intensity modulators, 119–120
measurement of, 282–291
measurement of (time-domain), 292–296
electro-optic phase modulators, 116–117
OLCR (optical low-coherent reflectometry), 234
chirp parameter, external electro-optic modulators, 119–120
chromatic dispersion, 70–73, 394
compensating in FBGs, 339
dispersion compensating fibers (DCFs), 370
in-situ monitoring, 537–540
linewidth measurement, 287–288
measuring in optical fibers, 400–409
baseband AM response method, 403–405
interferometric method, 405–409
modulation phase-shift technique, 400–402
nonlinear phase modulation and, 81
optical fibers, 70–73
optical network performance and, 485
R-OSNR due to, measuring, 607–610
standard single-mode fibers, 367–369
thin-film filters, 341
circular polarization, Stokes parameters for, 192
cladding, propagation in, 53

clock recovery functionality, 505–506
coherence length, optical wavelength meters, 185
coherence requirement for MZI chirp measurement, 295
coherent detection, 196–211
  balanced, polarization diversity and, 202–204
  Fourier-domain reflectometry (FDR), 241
  heterodyne detection, 199–200
    frequency domain characterization of optical receivers, 300–301
    in-situ PMD monitoring, 543–547
    linewidth measurement using, 271, 275, 284–285
  homodyne. *See* homodyne coherent detection
  in-situ monitoring of chromatic dispersion, 539–560
  in-situ PDL monitoring, 543–547
  linear optical sampling, 221–223
  operating principle, 196–199
  with OTDRs (optical time-domain interferometers), 391–393
  phase diversity, 204–207
  phase noise and linewidth measurement, 269–272
  SNR in receivers, 201–202
coherent OSAs, 207–211
collimating lens in optical spectrum analyzers, 142–143
collinear configuration, 149
communication systems. *See* fiber-optic transmission systems
concatenated Michelson interferometers, 175–176
confocal-mirror configuration, FPIs, 157–158
connection loss, 388–390
constants, table of, xvii
contrast (FPIs), 151–152
conversion table, xvii
coupled-wave equations, FBGs, 334–336
coupling ratio, 331–332
critical angle, 48
  phase shift, 48–49
cross-gain saturation effect, 95, 312–314
cross-phase modulation (XPM), 82, 486. *See also* interchannel crosstalk
  FWM-induced crosstalk (IMDD systems), 575–581
  intensity modulation in (IMDD systems), 556–572
  nonlinear index measurement using, 468–473
  phase modulation in (IMDD systems), 572–575
crosstalk, 486, 492
  from four-wave mixing (FWM), 83
  modulation instability (MI), 590–605
    amplified multispan fiber systems, 600–601
    characterization in fiber-optic systems, 601–605
    linear index measurement using, 473–476
    transfer matrix formulation, 590–600
  nonlinear, measuring, 556–589
    FWM-induced crosstalk (IMDD systems), 575–581
    XPM-induced intensity modulation, 556–572
    XPM-induced phase modulation, 572–575
  from stimulated Raman scattering (SRS), 79
    characterizing, with wide channel separation, 581–589
CRTs (cathode ray tubes), 212
current modulation, 115–116
cutback technique to measure fiber attention, 382–384
cutoff wavelength, 34, 383
  attenuation measurement, 382–384
  numerical aperture and, 60
CW self-phase modulation (SPM), 472

D

dark current noise, 38–39
  avalanche photodiodes, 43
data transmission systems. *See* fiber-optic transmission systems
DBR lasers, 242
DC. *See* dispersion compensation
DCFs (dispersion-compensating fibers), 370, 485
DCMs (dispersion-compensating modules), 370, 485
dead zone, 228
decibel (dB), 64–65
decision phase, 508
decision threshold voltage, 521–523
degenerate FWM, 83
degenerate modes, 74
degree of polarization (DOP), 190

delay, network analyzers, 249
delay, propagation, 69. *See also* differential group delay (DGD); dispersion
demultiplexers, WDM, 340–345
depletion region, 6
  carrier confinement and, 7–8
detecting jitter, 532–537
deterministic jitter, 529
DFB laser diodes, 27–28, 278
DGD. *See* differential group delay
differential group delay (DGD), 75, 409, 412
  in autocorrelators, 228
  fiber DGD vs. experienced DGD, 547–551
  measurement of, 431–438
    fixed analyzer method, 420–424
    interferometric method, 415–418
    Jones Matrix method, 424–431
    Mueller Matrix method (MMM), 431–438
    Poincare arc method, 418–420
    pulse delay method, 413–415
  OLCR (optical low-coherent reflectometry), 234
  optical filter transfer functions, 347–349
  PMD sources and emulators, 444
  self-homodyne detection systems, 273, 275
differential pulse-delay technique, 413–415
diffraction angle spreading, gratings, 136
diffraction gratings, fundamentals of, 134–138
digital oscilloscopes, 216–219
  operating principle, 214–216
digital signal processing
  ADCs (analog-to-digital converters), 223–224
  with electro-optical modulators, 121–123
  to improve network performance, 486
  optical filter transfer function characterization, 353
  in optical time-domain interferometers, 385
digital transmission. *See* fiber-optic transmission systems
direct injection current modulation, 115–116
direct intensity modulation. *See* electro-optic modulation response, measuring; electro-optic modulators
direct modulation, 4
direct optical sampling. *See* sampling oscilloscopes
direct semiconductors, 6–7
directionality, fiber couplers, 331
discrete propagation modes in fibers, 54
dispersion
  anomalous and normal in DSFs, 594
  chromatic. *See* chromatic dispersion
  optical circulators, 359
  optical fibers, 69–76, 394–409
    dispersion compensating fibers (DCFs), 370
    group velocity dispersion, 69–70
    modal dispersion, 73–74, 394, 395–399
    polarization mode dispersion (PMD), 74–76
  polarization-mode dispersion (PMD), 359, 409–413
  polarization-mode dispersion (PMD), measurement of, 413–438
    fixed analyzer method, 420–424
    interferometric method, 415–418
    Jones Matrix method, 424–431
    Mueller Matrix method (MMM), 431–438
    Poincare arc method, 418–420
    pulse delay method, 413–415
  waveguides, 344
dispersion-compensating fibers (DCFs), 370, 485
dispersion-compensating modules. *See* DCMs
dispersion compensation (DC), 485–486
  impact on modulation instability (MI), 598–600
dispersion-shifted fibers (DSFs), 368
  anomalous and normal dispersions, 594
  in optical networks, 485
distortion, wavefront. *See* waveform measurement
distributed Bragg reflector (DBR) lasers, 242
distributed feedback (DFB) laser diodes, 27–28, 278
DOP (degree of polarization), 190
double-monochromator OSA configuration, 139–140
double-sideband modulation, 123
DPSK (differential phase-shift-key) receiver, 517–520
DSFs (dispersion-shifted fibers), 368
  anomalous and normal dispersions, 594
  in optical networks, 485
duo-binary optical modulation, 122–123
dynamic gain tilt, 114–115, 311–314
dynamic range of OSAs, 133
  OTDR design and, 394

E

EA modulators, 125–127
  frequency chirp measurement of, 289–291
EDC (electronic dispersion compensation), 517
EDFAs. *See* erbium-doped fiber amplifiers
edge-emitting diodes, 10. *See also* LEDs
edge triggering, 214
efficiency, LED, 11
electric ADCs, optical techniques for, 223–224
electrical bandwidth, LEDs, 12–13
electrical characteristics of photodetectors, 36–37
electrical domain, 264
  ASE noise, optical amplifiers, 317–323
electrical spectrum analyzers (ESAs), 310
electro-absorption (EA) modulation, 125–127
  frequency chirp measurement of, 289–291
electro-optic modulation response, measuring, 276–296
  characterization of intensity modulation response, 277–282
    frequency-domain characterization, 277–281
    time-domain characterization, 281–282
  frequency chirp measurement, 282–291
    fiber dispersion for, 285–291
    modulation spectral measurement, 283–285
    time-domain measurement, 292–296
electro-optic modulators, 115–127
  basic operation principle, 116–121
  frequency doubling and duo-binary modulation, 121–123
  Mach-zender interferometers (MZIs) as, 117–118
    zero-chirp modulators, 120
  optical single-side modulation, 123–125
  using electro-absorption (EA) effect, 125–127
electromagnetic field theory of fiber propagation, 54–58
electronic dispersion compensation (EDC), 517
elliptical polarization, 192
emission cross-section, EDFAs, 102–103
equivalent-time sampling, 217
erbium-doped fiber amplifiers (EDFAs), 100–115. *See also* optical amplifiers; semiconductor optical amplifiers (SOAs)
  absorption and emission cross-sections, 102–103
  coherent detection vs., 196
  design considerations, 110–114
    automatic gain and power control, 111–114
    forward and backward, 110–111
  gain flattening, 114–115
  impact of modulation instability (MI), 600–601
  rate equations, 104–110
  spontaneous-spontaneous emission noise beating technique, 303–304
  time-domain characteristics of, 327–329
error detection, 504–508. *See also* BER (bit error rate)
error-free seconds, 500
error ratio, 500
ESAs (electrical spectrum analyzers), 310
excess loss, fiber-optic couplers, 330–331
extended-cavity lasers, 31
external cavity laser diodes, 28–31
external modulators. *See* electro-optic modulation response, measuring; electro-optic modulators
external optical feedback, 28–29. *See also* external cavity laser diodes
  intensity noise and, 262–263
external quantum efficiency, LED, 11
extrinsic perturbation, 75
eye diagrams, 487
  BER-T curves, 536–537
  estimating system Q with, 494–498
  jitter, 529
  waveform measurement, 526–527

F

Fabry-Perot interferometers. *See* scanning FP interferometers
Fabry-Perot lasers, 26
Fabry-Perot resonators, 26
far-field (FF) distribution, 376
Faraday rotators
  in optical circulators, 357–359
  in optical isolators, 354–355
fast axis, polarization maintaining (PM) fibers, 371
FBG. *See* fiber Bragg grating filters
FDR (Fourier-domain reflectometry), 240–245
FEC (forward error correction), 483
feedback. *See* optical feedback

FF (far-field) distribution, 376–377
FFT with Michelson interferometer–based optical wavelength meters, 182–183
fiber. *See* optical fibers
fiber attenuation. *See* attenuation in optical fibers
fiber-based ring resonators, 166–167
fiber birefringence, 75, 409–413
fiber Bragg grating (FBG) filters, 236–239, 335–340
  for dispersion compensation, 485
  optical circulators with, 357
  power reflectivity, 337
fiber DGD vs. experienced DGD, 547–551. *See also* differential group delay (DGD)
fiber dispersion for linewidth measurement, 285–291
fiber-optic couplers, 330–334
fiber-optic recirculating loops, 616–628
  measurement procedure and time control, 618–622
  operating principle, 617–618
  optical gain adjustment in, 622–628
fiber-optic transmission systems, 483–508
  BER testing, 499–508
    error detection, 504–508
    pattern generators, 501–504
  BER vs. decision threshold voltage, 521–523
  degradation measurements based on R-OSNR, 606–616
    measuring R-OSNR from chromatic dispersion, 607–610
    measuring R-OSNR from fiber nonlinearity, 610–614
    measuring R-OSNR from filter misalignment, 615–616
  in-situ monitoring. *See* in-situ monitoring of linear propagation impairments
  jitter measurement, 527–537
    based on BER-T scan, 535–537
    based on phase detector, 534–535
    basic parameters and definitions, 527–532
    detection techniques, 532–537
  measuring distortion wavefront, 524–527. *See also* waveform measurement
  modulation instability in WDM systems, 590–605
    amplified multispan fiber systems, 600–601
    characterization of, 601–605
    transfer matrix formulation, 590–600
  nonlinear crosstalk measurement, 556–589
    XPM-induced intensity modulation, 556–589
  optical recirculating loops, 616–628
    measurement procedure and time control, 618–622
    operating principle, 617–618
    optical gain adjustment in, 622–628
  performance considerations, 484–486
  receiver BER and Q, 486–494
  receiver OSNR margin and R-OSNR, 514–521
  receiver sensitivity and power margin, 509–513
  system Q estimation, 494–498
field transfer function, FPI, 147–148
filters. *See* optical filters
finesse
  Mach-zender interferometers, 167
  scanning FP interferometers, 151
fixed analyzer method, 420–424
Fizeau wedge interferometer–based wavelength meters, 186–188
focusing optics in optical spectrum analyzers, 142–144
forward-biased pn junctions, 5, 6, 32, 37
forward error correction (FEC), 483
forward pumping (EDFAs), 110–111
four-wave mixing (FWM), 82–85, 486. *See also* interchannel crosstalk
  crosstalk from, IMDD systems, 581–589
  nonlinear index measurement using, 463–468
  nonlinear optical mixer based on, 220
  in optical recirculating loops, 623–625
  wavelength conversion in SOAs, 96–98
Fourier-domain reflectometry (FDR), 240–245
FP lasers, 26
FPIs. *See* scanning FP interferometers
free-space FP interferometers, 158
free-space Michelson interferometers, 168–169
free spectral range (FSR), 150, 155–156
frequency bandwidth requirement for MZI chirp measurement, 295
frequency calibration. *See* wavelength calibration
frequency chirp. *See* chirp
frequency dithering, 580

frequency domain measurement
chromatic dispersion, 403–405
electro-optic modulation response, 277–281
intermodal dispersion, optical fibers, 398–399
optical receivers, 299–301
frequency doubling, 121–123
frequency noise converted to intensity noise, 263
frequency-resolved optical gating (FROG), 230–231
frequency response of APD gain, 42–43
frequency response of direct modulation. *See* electro-optic modulation response, measuring
frequency shift key (FSK), 484
frequency shift, SBS, 452
frequency specificity of Michelson interferometers, 174–179
frequency sweep rate, OSAs, 133, 137
frequency synthesizer, 279
Fresnel reflection coefficients, 45–47
TDR design and, 394
fringe counting, 181
FSK (frequency shift key), 484
FSR (free spectral range), 150, 155–156
FWM. *See* four-wave mixing

G

gain
adjusting in recirculating loops, 622–628
automatic gain control (AGC), 111–114
avalanche photodiodes, 42. *See also* avalanche photodiodes
gain flattening, 114–115
measurement of, 306–310
optical amplifiers, in general, 86–89. *See also* erbium-doped fiber amplifiers (EDFAs); optical amplifiers; semiconductor optical amplifiers (SOAs)
gain bandwidth, 87–89
gain tilt, 311–314
gating time, 486
Gaussian approximations for noise sources, 490–491
geometric birefringence, 75, 409
geometric optics analysis of fiber propagation, 51–54
graded-index fibers, 50
grating-based optical spectrum analyzers (OSAs), 131–145
basic configurations, 138–145
double monochromator, 139–140
focusing optics, 142–144
photodiode arrays, 144–145
polarization sensitivity compensation, 140–142
diffraction grating fundamentals, 134–138
FPI spectrum analyzers combined with, 159–160
FPI spectrum analyzers vs., 153
general specifications, 131–133
gratings, fundamentals of, 134–138
group delay. *See* differential group delay
group velocity, optical fibers, 67–69. *See also* phase velocity, optical fibers
group velocity dispersion, optical fibers, 69–70

H

half-power bandwidth (HPBW), 151
Hankel transform, 377
HE mode, 57
Helmholtz equation, 55
heterodyne coherent detection, 199–200
frequency domain characterization of optical receivers, 300–301
in-situ PMD monitoring, 543–547
linewidth measurement using, 271, 275, 284–285
high-speed electric ADCs, 223–224
high-speed sampling of optical signals, 216–219
linear sampling, 220–221
nonlinear sampling, 220–221
highly nonlinear photonic crystal fibers (PCFs), 373–374
homodyne coherent detection, 200–201
phase diversity in, 204–207
self-homodyne detection, 272–275
homojunctions, 7
HPBW (half-power bandwidth), 151
hybrid propagation mode, 57

I

IMD (intermodulation distortion), 280–281
IMDD. *See* intensity modulation and direct detection (IMDD)
impurities in optical fibers, 62–63
in-situ monitoring of linear propagation impairments, 537–556
chromatic dispersion, 537–540

in-situ monitoring of linear propagation impairments (*Continued*)
  PDL (polarization-dependent loss), 551–556
  PMD (polarization-mode dispersion), 541–551
    fiber DGD vs. experienced DGD, 547–551
    using coherent detection, 543–547
indirect semiconductors, 6–7
InGaAs-based CCDs, 145
injection current modulation, 115–116
input aperture for optical spectrum analyzers, 142–143
insertion loss
  fiber couplers, 331
  optical circulators, 359–360
instantaneous carrier density, optical amplifier, 327
intensity modulation and direct detection (IMDD), 484–485, 556. *See also* fiber-optic transmission systems
  FWM-induced crosstalk in, 575–581
  modulation instability (MI), 590–605
    amplified multispan fiber systems, 600–601
    characterization in fiber-optic systems, 601–605
    linear index measurement using, 473–476
    transfer matrix formulation, 590–600
  XPM-induced intensity modulation, 556–572
  XPM-induced phase modulation, 572–575
intensity modulation response, characterization of, 277–282
intensity noise. *See* relative intensity noise (RIN)
interchannel crosstalk, 486, 492, 556–589. *See also* crosstalk
  FWM-induced (IMDD systems), 575–581
  Raman crosstalk, characterizing, 581–589
  XML-induced, 556–572
  XPM-induced intensity modulation, 556–572
  XPM-induced phase modulation, 572–575
interferometers. *See* Mach-zender interferometers; Michelson interferometers; optical low-coherent interferometry; scanning FP interferometers; vector optical network analyzers
interferometric method
  measuring fiber dispersion, 405–409
  measuring polarization-mode dispersion, 415–418
  measuring transfer functions with, 350–353
interleaving, 570
intermediate frequency (IF), 198
  spectral shape of, 269–271
intermodal dispersion, optical fibers, 73–74, 394, 395–399
  frequency domain measurement, 398–399
  measurement of, 395–399
  pulse distortion method, 396–397
intermodulation distortion (IMD), 280–281
internal quantum efficiency, LED, 11
internal reflection. *See* reflection in optical fibers
intersymbol interference (ISI), 485, 487
intraband relaxation, 101
intrinsic filters, 531
intrinsic layer, semiconductors, 32–34
  preamplification and receiver Q-value, 510–511
  Q function, 492
intrinsic perturbation, 74–75, 409
ionization, 41
ISI. *See* intersymbol interference
isolation, optical circulators, 358, 359–360
ITU-T G.651 multimode fiber, 367
ITU-T G.652 multimode fiber, 367

J

jitter generation, 531
jitter measurement, 527–537
  based on BER-T scan, 535–537
  based on phase detector, 534–535
  basic parameters and definitions, 527–532
  detection techniques, 532–537
jitter tolerance, 530–531
jitter transfer, 532
Jones Matrix method, 424–431
junction capacitance, photodiodes, 36

K

Kerr effect nonlinearity, 79–85
  measurement of, 459–476
    cross-phase modulation (XPM), 468–473
    four-wave mixing (FWM), 463–468
    modulation instability (MI), 473–476
    self-phase modulation (SPM), 460–463, 472

Kerr effect nonlinearity (*Continued*)
  modulation instability (MI), 590–605
    amplified multispan fiber systems, 600–601
    characterization in fiber-optic systems, 601–605
    linear index measurement using, 473–476
    transfer matrix formulation, 590–600

L

Lambertian surfaces, 9
large core photonic crystal fibers (PCFs), 373
large effective area fibers (LEAFs), 369
large-signal gain tilt, 311
laser characterization, 260–275
  phase noise and linewidth, 266–275
    coherent detection, 269–272
    self-homodyne detection, 272–275
  RIN (relative intensity noise), 261–266
laser diodes, 13–26
  direct injection current modulation, 115–116
  photodiodes vs., 32
  rate equations, 16–18
  single-frequency lasers, 26–31
    DFB laser diodes, 27–28, 278
    external cavity laser diodes, 28–31
  threshold carrier density, 18–26
  turn-on delay, 22–23
laser noises, 25–26
  optical isolators and, 354
LDs (laser diodes), 12–13, 13–26
  direct injection current modulation, 115–116
  photodiodes vs., 32
  rate equations, 16–18
  single-frequency lasers, 26–31
    DFB laser diodes, 27–28, 278
    external cavity laser diodes, 28–31
  threshold carrier density, 18–26
  turn-on delay, 22–23
LEAFs (large effective area fibers), 369
LEDs (light-emitting diodes), 9–13
  modulation dynamics, 12–13
  P-I curves, 10–11
light-emitting diodes. *See* LEDs
lightwave polarization. *See entries at* polarization
lightwave propagation. *See entries at* propagation
$LiNbO_3$, 116
line system (transmission systems), 483
linear crosstalk, 486, 492
linear effects in optical fibers. *See* attenuation coefficient; chromatic dispersion; modal dispersion
linear optical sampling, 221–223
linear polarization, Stokes parameters for, 192
linearity measurement of photodetector responsivity, 297–298
linewidth. *See* spectral linewidth
loading state (recirculating loop operations), 618
long fibers, PSP of, 418
long-span (LS) fibers, 369
loop time (recirculating loop operations), 618
looping state (recirculating loop operations), 618
loss
  fiber couplers, 330–331
  optical circulators, 359
  optical fibers. *See* attenuation in optical fibers
low-coherent interferometry, 232–245
  Fourier-domain reflectometry (FDR), 240–245
  OLCR (optical low-coherent reflectometry), 232–240

M

Mach-zender interferometers (MZIs), 160–168
  chromatic dispersion measurement, 405–406
  DPSK (differential phase-shift-key) receiver, 517–520
  as electro-optic modulator, 117–118
    zero-chirp modulators, 120
  as optical filters, 164–168
    Michelson interferometers with, 174–175, 303
  time-domain characterization of transmitter chirp, 292–295
  transfer functions, 162–164
  transfer matrix, 2 x 2 optical couplers, 161–162
macro bending, 64
main mode, 20
manual-input patterns, 504
mark-density patterns, 504
material absorption, optical fibers, 62–63

material dispersion, 71–73
maximum incident angle, fibers, 60
maximum power of OSAs, 132
Maxwell's equations for fiber propagation, 54–58
mean DGD, 412–413. *See also* differential group delay (DGD)
measurement conversion table, xvii
MFD (mode-field diameter), 375–377
MI (modulation instability), 590–605
  amplified multispan fiber systems, 600–601
  characterization of, 601–605
  linear index measurement using, 473–476
  transfer matrix formulation, 590–600
Michelson interferometers, 168–179
  differential group delay (DGD) measurement, 416
  Fourier-domain reflectometry (FDR), 240–245
  measurement and characterization of, 172–174
  operating principle, 169–172
  optical wavelength meters based on, 180–183
  techniques to increase frequency sensitivity, 174–179
micro bending, 64
microwave transition analysis, 219
minimum resolution bandwidth of OSAs, 132, 139
MMFs. *See* multimode fibers (MMFs)
MMM (Mueller Matrix method), 431–438, 440, 551–552
modal dispersion, optical fibers, 73–74, 394, 395–399
mode-field diameter (MFD), 375–377
mode-field distribution, optical fibers, 374–382
  far-field techniques, 378–380
  near-field techniques, 380–382
mode partition noise, 26
modified Bessel functions, 56
modified Scholow-Towns formula, 268
modulation bandwidth, LEDs, 12–13
modulation efficiency, 116
modulation instability (MI), 590–605
  amplified multispan fiber systems, 600–601
  characterization in fiber-optic systems, 601–605
  linear index measurement using, 473–476
  transfer matrix formulation, 590–600
modulation phase-shift technique
  optical fiber chromatic dispersion, 400–402
  optical filter transfer functions, characterizing, 346–350
modulation spectral measurement, 283–285
modulation speed requirement for MZI chirp measurement, 296
modulators, electro-optic. *See* electro-optic modulation response, measuring; electro-optic modulators
monochromator configuration, OSA, 138–139
  double monochromator, 139–140
MSR (mode suppression rate), 20–22
Mueller Matrix method (MMM), 431–438, 440, 551–552
multimodal effects. *See* multiple longitudinal modes
multimode fibers (MMFs), 367
  attenuation measurement, 382–384
  chromatic dispersion, 400
multiple longitudinal modes, 14
  guided modes in optical fibers, 58
  modal dispersion, 73–74, 394, 395–399
  mode partition noise, 26
  MSR (mode suppression rate), 20–22
  WDM multiplexers and demultiplexers, 340–345
multiplexers, WDM, 340–345
MZI. *See* Mach-zender interferometers

N

n-type semiconductors, 5
  APD structure, 41
  pn-junction photodiodes, 32–34
NA. *See* numerical aperture
NDSFs (non-dispersion-shifted fibers), 369
near-field (NF) distribution, 376
negative slope triggering, 215
NEP (noise-equivalent power), 40
net emission rate, EDFAs, 103
network analyzers, 246–256
  S-parameters, 246–249
NF (near-field) distribution, 376
NLS. *See* nonlinear Schrödinger equation
noise. *See also* signal-to-noise ratio (SNR)
  ASE noise. *See* amplified spontaneous emission noise
  charge coupled devices (CCDs), 145
  frequency noise converted to intensity noise, 263
  Gaussian approximations, 490–491
  laser noises, 25–26, 354

noise (*Continued*)
modulation instability and. *See* modulation instability (MI)
observing in eye diagrams, 494–498
optical amplifiers, 314–316
ASE noise, electrical domain, 317–323
ASE noise, optical domain, 316–317
gain measurement and, 307–310
noise figure definition, 323–326
quantum noise, 323
in optical recirculating loops, 623–624
photodetectors, 37–40
avalanche photodiodes, 43
relative intensity noise. *See* RIN
spontaneous-spontaneous emission beat noise, 301–304, 318, 321–323
optical signal-to-noise ratio (OSNR) and, 514–515
waveform measurement, 524
noise-equivalent power (NEP), 40
noise figure definition, 323–326
noise loading technique, 516
non-dispersion-shifted fibers (NDSFs), 369
non-return-to-zero (NRZ) modulation format, 501
error detection, 505–506
nonlinear crosstalk, 486, 492, 556–589. *See also* crosstalk
FWM-induced (IMDD systems), 575–581
Raman crosstalk, characterizing, 581–589
XML-induced, 556–572
XPM-induced intensity modulation, 556–572
XPM-induced phase modulation, 572–575
nonlinear effects in optical fibers, 77–85, 446–476
Kerr effect nonlinearity, 79–85, 459–476
cross-phase modulation (XPM), 468–473
four-wave mixing (FWM), 463–468
modulation instability (MI), 473–476
self-phase modulation (SPM), 460–463, 472
modulation instability (MI), 590–605
amplified multispan fiber systems, 600–601
characterization in fiber-optic systems, 601–605
linear index measurement using, 473–476
transfer matrix formulation, 590–600
R-OSNR due to, measuring, 610–614
stimulated Brillouin scattering (SBS), 77–78, 447–453
stimulated Raman scattering (SRS), 78–79, 453–459
nonlinear optical mixer, 220
nonlinear optical sampling, 220–221
nonlinear parameter, 79
nonlinear phase modulation, XPM-induced, 572–575
nonlinear Schrödinger equation, 79–85, 462–463, 473
modulation instability (MI), 590
nonradiative recombination (LEDs), 11
nonzero dispersion-shifted fibers (NZDSFs), 368
normal dispersions, DSFs, 594
normal incidence, 47
normalized XPM crosstalk, 569
NRZ. *See* non-return-to-zero (NRZ) modulation format
numerical aperture (NA), 58–62
backscattering, 389–390
for near-field measurement, 381
Nyquist criterion, 216
NZDSFs (nonzero dispersion-shifted fibers), 368

O

OCT (optical coherent tomography), 232
OLCR. *See* optical low-coherent reflectometry
one-dimensional charge coupled devices (CCDs), 145
operation wavelength, 14
external cavity laser diodes, 29–30
multiple longitudinal modes, 14
guided modes in optical fibers, 58
modal dispersion, 73–74, 394, 395–399
mode partition noise, 26
MSR (mode suppression rate), 20–22
WDM multiplexers and demultiplexers, 340–345
photodiode responsivity and, 34
small-signal modulation response, 23–24
optical amplifiers, 85–115, 306–329. *See also* semiconductor optical amplifiers (SOAs)
dynamic gain tilt, 114–115, 311–314
erbium-doped fiber amplifiers (EDFAs), 100–115
absorption and emission cross-sections, 102–103

optical amplifiers (*Continued*)
design considerations, 110–114
gain flattening, 114–115
rate equations, 104–110
time-domain characteristics of, 327–329
noise figure definition, 323–326
noise measurement, 314–316
ASE noise, electrical domain, 317–323
ASE noise, optical domain, 316–317
gain measurement and, 307–310
optical gain, 86–89
measurement of, 306–310
in optical recirculating loops, 622–628
OSNR margin and R-OSNR, 514–521
preamplification and receiver Q-value, 510–511
saturation, 86–89
semiconductor amplifiers, 89–100
gain dynamics, 91–100
SRS for, 455–456
static and dynamic gain tilt, 311–314
optical attenuation. *See* attenuation in optical fibers
optical bandwidth, LEDs, 12–13
optical circulators, 356–361
optical coherent tomography (OCT), 232
optical communications. *See* fiber-optic transmission systems
optical couplers (2 x 2), transfer matrix for
balanced coherent detection, 203, 205–206
Mach-zender interferometers (MZIs), 161–162
Michelson interferometers, 170–171
optical double-sideband modulation, 123
optical feedback, 13–14
external, 28–29. *See also* external cavity laser diodes
intensity noise and, 262–263
optical fiber directional couplers, 330–334
optical fibers, 44–85, 366–367. *See also entries at* fiber
attenuation, 62–67, 382–394
cutback measurement technique, 382–384
optical time-domain interferometers (OTDRs). *See* OTDRs
birefringence. *See* birefringence in optical fibers
dispersion measurements, 394–409
chromatic dispersion, 400–409
intermodal dispersion, 395–399
fiber types, 367–374
specialty fibers, 370–374
standard fibers for transmission, 367–369
group velocity and dispersion, 67–76
chromatic dispersion, 70–73
group velocity dispersion, 69–70
modal dispersion, 73–74
polarization mode dispersion (PMD), 74–76
mode-field distribution, measuring, 374–382
far-field techniques, 378–380
near-field techniques, 380–382
nonlinear effects, 77–85, 446–476
Kerr effect nonlinearity, 79–85, 459–476
stimulated Brillouin scattering (SBS), 77–78, 447–453
stimulated Raman scattering (SRS), 78–79, 453–459
PMD sources and emulators, 441–446
polarization-dependent loss. *See* PDL
polarization-mode dispersion. *See* PMD
propagation modes, 49–62
electromagnetic field theory, 54–58
geometric optics analysis, 51–54
numerical aperture, 58–62
reflection and refraction, 44–49
Brewster angle, 49
Fresnel reflection coefficients, 45–47
optical field phase shift, 48–49
special cases, 47–48
systems of. *See* fiber-optic transmission systems
optical field transfer function, external modulators, 121–122
optical filters
misalignment of, R-OSNR due to, 615–616
MZIs as, 164–168
Michelson interferometers with, 174–175, 303
transfer functions, characterizing, 345–353
interferometer technique, 350–353
modulation phase-shift technique, 346–350
optical gain
adjusting in recirculating loops, 622–628
automatic gain control (AGC), 111–114
avalanche photodiodes, 42. *See also* avalanche photodiodes
gain flattening, 114–115
measurement of, 306–310
optical amplifiers, in general, 86–89. *See also* erbium-doped fiber

amplifiers (EDFAs); optical amplifiers; semiconductor optical amplifiers (SOAs)
optical input aperture for optical spectrum analyzers, 142–143
optical intensity modulators, 117–118
optical intensity noise. *See* relative intensity noise (RIN)
optical interface, 44
optical interferometric technique
  measuring fiber dispersion, 405–409
  measuring polarization-mode dispersion, 415–418
  measuring transfer functions with, 350–353
optical isolators, 353–356
optical low-coherent interferometry, 232–245
  Fourier-domain reflectometry (FDR), 240–245
  OLCR (optical low-coherent reflectometry), 232–240
optical network analyzers, 246–256
  S-parameters, 246–249
  scalar analyzers, 250–251
  vector analyzers, 251–256
optical phase. *See entries at* phase
optical phase noise, 25–26, 266–275
  coherent detection, 269–272
  optical isolators and, 354
  self-homodyne detection, 272–275
optical phonons, 78
optical polarimeters, 188–196
  lightwave polarization, description of, 188–190
  Stokes parameters for optical signals, 190–192
optical polarization. *See entries at* polarization
optical power, 111
  automatic power control (APC), 111–114
  fiber attenuation. *See* attenuation in optical fibers
  gain of optical amplifiers and, 306
  incident, with optical low-power reflectivity, 233
  LEDs (light-emitting diodes), 10–11
  maximum, for OSAs, 132
  NEP (noise-equivalent power), 40
  photon density and, 20
  receiver sensitivity, 485, 509–513
    BER vs. decision threshold voltage, 521–523
    OSNR margin and R-OSNR, 514–521
  reflected optical fields, 47
  relative intensity noise. *See* RIN
  small-signal modulation response, 23–24
  spectral density of optical field, 593
  Stokes parameters for optical signals, 190–192
  Stokes photons and, 77–78
  turn-on delay, 22–23
optical pulse measurement, 224–231
  optical receiver characterization, 304–306
optical pumping (EDFAs), 101, 110–111
optical receivers, wideband characterization of, 296–306
  frequency domain characterization, 299–301
  responsivity and linearity, 297–298
  short optical pulse technique, 304–306
  spontaneous-spontaneous emission noise beating technique, 301–304
optical recirculating loops, 616–628
  measurement procedure and time control, 618–622
  operating principle, 617–618
  optical gain adjustment in, 622–628
optical reflection. *See entries at* reflection
optical sampling, 216–219
  linear sampling, 220–221
  nonlinear sampling, 221–223
optical signal-to-noise ratio (OSNR), 514–521
  OSNR margin, 516–518
  required (R-OSNR), 516–521, 606–616
    diue to fiber nonlinearity, 610–614
    due to chromatic dispersion, 607–610
    due to filter misalignment, 615–616
optical signal waveform. *See* waveform measurement
optical single-side modulation. *See* single-sideband modulation
optical spectrum analyzers (OSAs), 345–353
  based on FPI, 153–157
    and gratings together, 159–160
  coherent, based on swept frequency lasers, 207–211
  gratings-based. *See* grating-based optical spectrum analyzers (OSAs)
  for optical gain measurement, 309–310
  with optical recirculating loops, 625
optical time-domain interferometers (OTDRs), 232
optical wavelength meters, 179–188
  based on Fizeau wedge interferometer, 186–188

optical wavelength meters (*Continued*)
based on Michelson interferometer, 180–183
signal coherence length, 185
spectral resolution, 184
wavelength calibration, 185–186
wavelength coverage, 183, 187
OSAs (optical spectrum analyzers)
based on FPI, 153–157
and gratings together, 159–160
coherent, based on swept frequency lasers, 207–211
gratings-based. *See* grating-based optical spectrum analyzers (OSAs)
for optical gain measurement, 309–310
with optical recirculating loops, 625
oscillation frequency, small-signal modulation response, 24
oscilloscopes
digital, 216–219
high-speed electric ADCs, 223–224
high-speed sampling of optical signals, 216–219
linear sampling, 220–221
nonlinear sampling, 220–221
jitter measurement, 532–534
operating principle of, 212–216
short optical pulse measurement, 224–231
optical receiver characterization, 304–306
time-base calibration in waveform measurement, 525–526
OSNR (optical signal-to-noise ratio), 514–521
required (R-OSNR), 516–521, 606–616
diue to fiber nonlinearity, 610–614
due to chromatic dispersion, 607–610
due to filter misalignment, 615–616
OSNR margin, 516–518
OSSB. *See* single-sideband modulation
OTDRs (optical time-domain interferometers), 232, 384–391
improvement considerations, 391–394
output focusing lens, optical spectrum analyzers, 143–144
output slit in optical spectrum analyzers, 144

P

P-I curves, 10–11, 18–19
P-J relationship, 19–20
p-type semiconductors, 5
APD structure, 41
pn-junction photodiodes, 32–34
Panda fibers, 371
parallel measurement technique, optical polarimeters, 193–194
parametric gain, 474
partial PDL, 558
passive optical components, 329–361
fiber-Bragg grating filters, 335–340
fiber-optic couplers, 330–334
optical circulators, 356–361
optical filter transfer functions characterization, 345–353
interferometer technique, 350–353
modulation phase-shift technique, 346–350
optical isolators, 353–356
WDM multiplexers and demultiplexers, 340–345
pattern generators in BER test sets, 501–504
in bit error detection, 504–508
jitter measurement, 531–535
pattern triggering, 215
PCFs (photonic crystal fibers), 372–374, 460
PDL (polarization-dependent loss), 359–360, 431–432, 438–441
in-situ monitoring, 551–556
partial, 558
percent of unavailability, 500
performance of fiber-optic systems. *See* fiber-optic transmission systems
phase
control, in Michelson interferometers, 176–177
modulation of. *See* chirp
nonlinear shift, 80–81
polarization, general description of, 188–190
reflected optical fields, 45–47
shift between incident and reflected beams, 48–49
phase condition, laser diodes, 14
phase detectors, measuring jitter with, 534–535
phase diversity, coherent homodyne systems, 204–207
phase modulation, XPM-induced, 572–575
phase noise, 25–26, 266–275
coherent detection, 269–272
optical isolators and, 354
self-homodyne detection, 272–275
phase shift key (PSK), 484
phase shift method for measuring chromatic dispersion

phase shift method for measuring chromatic dispersion (*Continued*)
  optical fiber chromatic dispersion, 400–402
  optical filter transfer functions, characterizing, 346–350
phase velocity, optical fibers, 67–69
photodetectors, 32–43
  autocorrelators, 225–226
  avalanche photodiodes (APDs), 41–43
    noise and SNR, 43
    Q function, 492
  bandwidth, 34–36
    short optical pulse technique, 304–306
    spontaneous-spontaneous emission noise beating technique, 301–304
  electrical characteristics, 36–37
  noise and SNR, 37–40
  with optical spectrum analyzers, 144
  pn-junction photodiodes, 32–34
  responsivity, 34–36
    avalanche photodiodes, 41–42
    characterization of, 297–298
    frequency domain characterization, 299–301
    short optical pulse technique, 304–306
    spontaneous-spontaneous emission noise beating technique, 301–304
  wideband characterization of, 296–306
    frequency domain characterization, 299–301
    responsivity and linearity, 297–298
    short optical pulse technique, 304–306
    spontaneous-spontaneous emission noise beating technique, 301–304
photodiodes, 32
  in coherent detection, 198
  in grating-based optical spectrum analyzers, 139, 144–145
  with sampling oscilloscopes, 219
photon density rate equation, 17, 21
photon lifetime, 17
photonic crystal fibers (PCFs), 372–374, 460
physical layer, optical transmission system, 483
physics constants, xvii
P-I curves, 10–11, 18–19
piezo-electric transducers (PZTs), 154
pin-hole output, optical spectrum analyzers, 144
PIN structure (photodetectors), 32–34
  preamplification and receiver Q-value, 510–511
  Q function, 492
planar lightwave circuits, 343
plano-mirror configuration, FPIs, 157
plastic optical fibers (POFs), 374
PM (phase modulation), XPM-induced, 572–575
PM fibers, 370–372, 445
PMD (polarization-mode dispersion), 74–76, 359, 409–438, 497
  birefringence representation, 409–413
  in-situ monitoring, 541–551
    fiber DGD vs. experienced DGD, 547–551
    using coherent detection, 543–547
  measurement of, 413–438
    fixed analyzer method, 420–424
    interferometric method, 415–418
    Jones Matrix method, 424–431
    Mueller Matrix method (MMM), 431–438
    Poincare arc method, 418–420
    pulse delay method, 413–415
  sources and emulators, 441–446
PMD parameter, 413
pn junctions, 5–6
  photodetectors, 32–34
POFs (plastic optical fibers), 374
Poincare arc method, 418–420
Poincare sphere, 191–192, 411
polarization
  general description of, 188–190
  optical circulators, 356–361
  optical isolators, 355–356
  spontaneous-spontaneous beat noise, calculating, 322
polarization average technique, 142
polarization dependence, OSAs, 133
polarization-dependent loss (PDL), 359–360, 431–432, 438–441
  partial, 558
polarization diversity configuration, OSA, 141
polarization diversity in coherent receivers, 202–204
polarization-maintaining (PM) fibers, 370–372, 445
polarization-mode dispersion (PMD), 74–76, 359, 409–438, 497
  birefringence representation, 409–413
  in-situ monitoring, 541–551
    fiber DGD vs. experienced DGD, 547–551
    using coherent detection, 543–547

polarization-mode dispersion (PMD) (*Continued*)
  measurement of, 413–438
    fixed analyzer method, 420–424
    interferometric method, 415–418
    Jones Matrix method, 424–431
    Mueller Matrix method (MMM), 431–438
    Poincare arc method, 418–420
    pulse delay method, 413–415
  sources and emulators, 441–446
polarization sensitivity compensation, OSAs, 140–142
polarization states (polarization rotation), 410
  OLCR (optical low-coherent reflectometry), 233, 236
  PDL (polarization-dependent loss) measurement, 438–441
positive slope triggering, 215
power, 111
  automatic power control (APC), 111–114
  fiber attenuation. *See* attenuation in optical fibers
  gain of optical amplifiers and, 306
  incident, with optical low-power reflectivity, 233
  LEDs (light-emitting diodes), 10–11
  maximum, for OSAs, 132
  NEP (noise-equivalent power), 40
  photon density and, 20
  receiver sensitivity, 485, 509–513
    BER vs. decision threshold voltage, 521–523
    OSNR margin and R-OSNR, 514–521
  reflected optical fields, 47
  relative intensity noise. *See* RIN
  small-signal modulation response, 23–24
  spectral density of optical field, 593
  Stokes parameters for optical signals, 190–192
  Stokes photons and, 77–78
  turn-on delay, 22–23
power loss
  fiber couplers, 330–331
  optical circulators, 359
  optical fibers. *See* attenuation in optical fibers
power margin, optical system, 513
power reflectivity, fiber Bragg grating filters, 337
power transfer functions
  external modulators, 121–122
  intermodal dispersion measurements, 396–399
  linewidth measurement using fiber dispersion, 287–288
  Mach-zender interferometers, 162–164
  Michelson interferometers, 172–174
    with FP optical filters, 174–175, 303
  of optical filters, characterizing, 345–353
    interferometer technique, 350–353
    modulation phase-shift technique, 346–350
  scanning FP interferometers, 147–148, 154
  thin-film filters, 341
PRBS test sequences, 501–504
  in bit error detection, 504–508
preamplification and receiver Q-value, 510–511
principal state of polarization (PSP), 411, 418–419
  PMD (polarization-mode dispersion) and, 587
propagating lightwave, polarization of. *See entries at* polarization
propagation constant, 52–53
  wavelength dependence of, 70–73
propagation equations. *See* rate equations
propagation modes of optical fibers, 49–62
  electromagnetic field theory, 54–58
  geometric optics analysis, 51–54
  modal dispersion, 73–74, 394, 395–399
  numerical aperture, 58–62
propagation speed. *See* group velocity, optical fibers
pseudo-random binary sequences (PRBSs), 501–504
  in bit error detection, 504–508
  jitter measurement, 531–535
pseudorandom waveforms, 281
PSK (phase shift key), 484
PSP (principal state of polarization), 411, 418–419
  PMD (polarization-mode dispersion) and, 587
pulse delay method, to measure DGD, 413–415
pulse distortion method for measuring fiber dispersion, 396–397
pulses (optical), measuring, 224–231
  optical receiver characterization, 304–306
pump absorption efficiency, 102
pump-probe technique, 466–468

pump source, EDFAs. *See* optical pumping (EDFAs)
pump waves, 447–448

Q

Q-factor (ring resonators), 167
Q-margin, 523
quality factor (Q-value), 483, 486–494
  decision threshold voltage vs., 521–523
  receiver sensitivity and power margin, 509–513
  for systems, 487, 490
    estimating with eye diagram parameterization, 494–498
quantum efficiency, LED, 11
quantum noise, optical amplifiers, 323

R

R-OSNR (required OSNR), 516–521, 606–616
  due to chromatic dispersion, 607–610
  due to fiber nonlinearity, 610–614
  due to filter misalignment, 615–616
radiation loss in optical fibers, 64
radiative recombination, 6
  direct and indirect semiconductors, 6–7
  erbium-doped fiber amplifiers (EDFAs), 101
  spontaneous vs. stimulated emission, 8–9
Raman crosstalk. *See* interchannel crosstalk
Raman scattering. *See* stimulated Raman scattering (SRS)
random jitter, 529
random waveform sampling, 218–219
range of wavelength meters, 183
rate equations
  for erbium-doped fiber amplifiers (EDFAs), 104–110
  for laser diodes, 16–18
Rayleigh scattering in optical fibers, 63, 65–67
  optical time-domain interferometers. *See* OTDRs
real-time sampling, 216
receivers, in optical transmission systems, 483, 508–523
  BER (bit error rate), 486–494
    decision threshold voltage vs., 521–523
  jitter tolerance, 530–531
  noise impact of modulation instability (MI), 594–597
  OSNR margin and R-OSNR, 514–521
  sensitivity of, 485, 509–513
receivers, wideband characterization of, 296–306
  frequency domain characterization, 299–301
  responsivity and linearity, 297–298
  short optical pulse technique, 304–306
  spontaneous-spontaneous emission noise beating technique, 301–304
recirculating loops, 616–628
  measurement procedure and time control, 618–622
  operating principle, 617–618
  optical gain adjustment in, 622–628
reflection in fiber couplers, 331
reflection in optical fibers
  Brewster angle, 49
  Fresnel reflection coefficients, 45–47
  minimizing (total transmission). *See* Brewster angle
  optical field phase shift, 48–49
  special cases, 47–48
reflectivity
  avoiding with optical isolators, 353–356
  fiber Bragg grating filters, 337
  as function of distance. *See* optical low-coherent interferometry
refraction in optical fibers
  Brewster angle, 49
  Fresnel reflection coefficients, 45–47
  index as function of electrical field. *See* electro-optic modulation response, measuring; electro-optic modulators
  nonlinear refractive index measurement, 459–476
    cross-phase modulation (XPM), 468–473
    four-wave mixing (FWM), 463–468
    modulation instability (MI), 473–476
    self-phase modulation (SPM), 460–463, 472
  optical field phase shift, 48–49
  special cases, 47–48
    critical angle, 48
    normal incidence, 47
refractive index (nonlinear), measuring, 460
relative calibration, FPI spectrum analyzers, 156–157
relative intensity noise (RIN), 25
  impact of modulation instability (MI). *See* modulation instability (MI)

relative intensity noise (RIN) (*Continued*)
measurement of, 261–266
optical amplifiers, 319, 322–323
optical isolators and, 354
relative wavelength accuracy, OSAs, 132
relaxation oscillation, 24
intensity noise and, 262
required OSNR (R-OSNR), 516–521, 606–616
diue to fiber nonlinearity, 610–614
due to chromatic dispersion, 607–610
due to filter misalignment, 615–616
resolution bandwidth of OSAs, 132, 139
responsivity, photodetectors, 34–36
avalanche photodiodes, 41–42
characterization of, 297–298
frequency domain characterization, 299–301
short optical pulse technique, 304–306
spontaneous-spontaneous emission noise beating technique, 301–304
responsivity, photodiodes, 34–36
retarders with optical polarimeters, 194–196
return loss, optical circulators, 359–360
return-to-zero (RZ) modulation format, 505–506
reverse-biased pn junctions, 32, 37
RF coherent detection, 200
RF direct detection, 200
RF network analyzers, 246–249
frequency domain measurement, 278–279
intermodal dispersion measurements, 398–399
self-homodyne measurement, 274
RIN (relative intensity noise), 25
impact of modulation instability (MI). *See* modulation instability (MI)
measurement of, 261–266
optical amplifiers, 319, 322–323
optical isolators and, 354
ring resonators, 166–167

S

S-parameters, 246–249
sampling oscilloscopes. *See* oscilloscopes
saturation, optical amplifiers, 88–89
SBS (stimulated Brillouin scattering), 77–78, 447–453
frequency dithering and, 580
scalar optical network analyzers, 250–251
scanning FP interferometers, 146–160
basic configuration and transfer function, 146–153
contrast, 151–152
finesse, 151
FSR (free spectral range), 150
HPBW (alf-power bandwidth), 151
optical configurations, 157–158
spectrum analyzers, 153–157
grating-based OSAs vs., 153
scattering
loss from, in fiber couplers, 330–331
loss from, in optical fibers, 63, 65–67
nonlinear processes of, 447. *See also* SBS; SRS
Rayleigh scattering in optical fibers, 63, 65–67
optical time-domain interferometers. *See* OTDRs
Scholow-Towns formula (modified), 268
Schrödinger equation, nonlinear, 79–85, 462–463, 473
modulation instability (MI), 590
second-order harmonic generation (SGH) crystals, 225, 227, 228
self-homodyne detection, 272–275
self-phase modulation (SPM), 81–82, 472
nonlinear index measurement using, 460–463, 472
R-OSNR due to fiber nonlinearity, 610–614
semiconductor lasers, characterization of, 260–275
phase noise and linewidth, 266–275
coherent detection, 269–272
self-homodyne detection, 272–275
RIN (relative intensity noise), 261–266
semiconductor optical amplifiers (SOAs), 89
EDFAs vs., 100–101
gain dynamics, 91–100
phase modulation, 99–100
wavelength conversion using cross-gain saturation, 95
wavelength conversion using FWM, 96–98
sensitivity of OSAs, 132
sensitivity of receivers, 485, 509–513
BER vs. decision threshold voltage, 521–523
OSNR margin and R-OSNR, 514–521
sensitivity requirement for MZI chirp measurement, 295
sequential measurement technique, optical polarimeters, 194
sequential waveform sampling, 217

SGH (second-order harmonic generation) crystals, 225, 227, 228
short fibers, PSP of, 418
short optical pulse measurement, 224–231
  optical receiver characterization, 304–306
shot noise, 38–39, 490
  avalanche photodiodes, 43
  optical amplifiers, 318–319
signal amplifiers. *See* optical amplifiers
signal BER. *See* BER (bit error rate)
signal coherence length, optical wavelength meters, 185
signal-dependent noise, observing, 492–493
signal error detection, 504–508. *See also* BER (bit error rate)
signal-independent noise, observing, 492–493
signal linewidth. *See* spectral linewidth
signal processing
  ADCs (analog-to-digital converters), 223–224
  with electro-optical modulators, 121–123
  to improve network performance, 486
  optical filter transfer function characterization, 353
  in optical time-domain interferometers, 385
signal-spontaneous emission beat noise, 318, 319–321
  optical signal-to-noise ratio (OSNR) and, 514–515
signal-to-noise ratio (SNR)
  avalanche photodiodes, 43
  in coherent detection, 201–202
    coherent OSAs, 210
  noise figure definition, 323–326
  OLCR (optical low-coherent reflectometry), 234
  optical (OSNR), 514–521
  photodetectors, 37–40
silicon-based CCDs, 145
single-frequency lasers, 26–31
  DFB laser diodes, 27–28, 278
  external cavity laser diodes, 28–31
single-mode optical fibers, 58, 367–368
  attenuation coefficient, 387
  attenuation measurement, 382–384
  chromatic dispersion, 72–73, 400
  fiber attenuation measurement, 391
  numerical aperture, 59–60
  polarization-mode dispersion (PMD), 409
single-sideband modulation, 121, 123–125
single sweep button (oscilloscopes), 216
skew rays, 51
SLED (superluminescence LED), 402
slit output, optical spectrum analyzers, 144
SLMs (spatial light modulators), 114–115
slow axis, polarization maintaining (PM) fibers, 371
small-signal gain tilt, 311
small-signal modulation response, 23–24
SMFs. *See* single-mode optical fibers
Snell's law, 45
SNR. *See* signal-to-noise ratio
SOAs. *See* semiconductor optical amplifiers
space charged region. *See* depletion region
spatial frequency, 377
spatial light modulators (SLMs), 114–115
spatial resolution, OTDR, 394
specialty optical fibers, 370–374
spectral density of noise power, 263–264
spectral density of optical field, 593
spectral linewidth, 26
  external cavity laser diodes, 30–31
  measurement of, 266–275
    coherent detection, 269–272
    self-homodyne detection, 272–275
spectral resolution, optical wavelength meters, 184
splitting ratio, fiber couplers, 332–333
SPM. *See* self-phase modulation
spontaneous emission, 8–9
  erbium-doped fiber amplifiers (EDFAs), 104, 107–109
  phase noise from. *See* phase noise
  rate of, laser diodes, 17
  relative intensity noise. *See* RIN
spontaneous recombination, 104
spontaneous-spontaneous emission beat noise, 301–304, 318, 321–323
  optical signal-to-noise ratio (OSNR) and, 514–515
  waveform measurement, 524
SRS (stimulated Raman scattering), 78–79, 453–459
  crosstalk caused by. *See also* interchannel crosstalk
  for optical amplification, 455–456
  Raman crosstalk with wide channel separation, 581–589
stability of power amplitude, OSAs, 133
standard optical fibers, for transmission, 367–369
standard single-mode fiber. *See* single-mode optical fibers

Stark effect, 126
state of polarization, 410
  OLCR (optical low-coherent reflectometry), 233, 236
  PDL (polarization-dependent loss) measurement, 438–441
static gain, optical amplifiers, 309
static gain tilt, optical amplifiers, 311–314
step-index fibers, 50, 56
stimulated Brillouin scattering (SBS), 77–78, 447–453
  frequency dithering and, 580
stimulated emission, 8–9
stimulated Raman scattering (SRS), 78–79, 453–459
  crosstalk caused by. *See also* interchannel crosstalk
  for optical amplification, 455–456
  Raman crosstalk with wide channel separation, 581–589
Stokes parameters for optical signals, 190–192
  measuring. *See* optical polarimeters
Stokes photons, 77
Stokes waves, 447–448. *See also* stimulated Brillouin scattering (SBS); stimulated Raman scattering (SRS)
stress birefringence, 75, 409
superluminescence LED (SLED), 402
surface-emitting diodes, 9. *See also* LEDs
sweep rate, OSAs, 133, 137
sweep speed control, oscilloscopes, 215–216
swept frequency lasers
  coherent OSAs based on, 207–211
  Fourier-domain reflectometry (FDR), 241–244
switchable retarders with optical polarimeters, 194–196
sync loss, 508
synchronization for bit error detection, 506–508
system quality factor, 483, 486–494
  decision threshold voltage vs., 521–523
  receiver sensitivity and power margin, 509–513
  for systems, 487, 490
    estimating with eye diagram parameterization, 494–498

T

T/R test sets, 247–248. *See also* optical network analyzers
TE mode, 57
Teralight fibers, 369
testing BER (bit error rate), 499–508
  error detection, 504–508
  pattern generators, 501–504
THD (total harmonic distortion), 280
thermal noise, 38–39
  avalanche photodiodes, 43
thin film–based interference filters, 340–342
threshold carrier density, 18–26
  MSR (mode suppression rate), 20–22
  P-J relationship, 19–20
  turn-on delay, 22–23
threshold condition, laser diodes, 14
threshold current density, 18–19
TIAs (transimpedance amplifiers), 297
time-base calibration in waveform measurement, 525–526
time delay, network analyzers, 249
time-domain characterization
  chromatic dispersion, 400–402
  of electro-optic modulation response, 281–282
  of erbium-doped fiber amplifiers (EDFAs), 327–329
  intermodal dispersion, 396–397
  of modulation-induced chirp, 292–296
time jitter. *See* jitter measurement
TM mode, 57
total harmonic distortion (THD), 280
total reflection, 48
  acceptance angle. *See* numerical aperture
  phase shift, 48–49
total transmission (optical fibers), 49
T/R test sets, 247–248. *See also* optical network analyzers
transfer functions
  external modulators, 121–122
  intermodal dispersion measurements, 396–399
  linewidth measurement using fiber dispersion, 287–288
  Mach-zender interferometers, 162–164
  Michelson interferometers, 172–174
    with FP optical filters, 174–175, 303
  of optical filters, characterizing, 345–353
    interferometer technique, 350–353
    modulation phase-shift technique, 346–350
  scanning FP interferometers, 147–148, 154
  thin-film filters, 341
transfer matrix
  2 × 2 optical couplers

transfer matrix (*Continued*)
balanced coherent detection, 203, 205–206
Mach-zender interferometers (MZIs), 161–162
Michelson interferometers, 170–171
FBGs, 337
modulation instability and, 590–600
transimpedance amplifiers (TIAs), 297
transmission systems. *See* fiber-optic transmission systems
transmission.reflection test sets, 247–248. *See also* optical network analyzers
transmitted power at reflective surfaces, 47
transmitters in optical transmission systems, 483
jitter generation, 531
transverse electric-field propagation mode, 57
transverse magnetic-field propagation mode, 57
tree configuration, Michelson interferometers, 177–179
trigger level, oscilloscope, 213, 214
trigger timeout (oscilloscopes), 216
Truewave fibers (TWFs), 369, 404–405
turn-on delay, 22–23
TWFs (Truewave fibers), 369, 404–405
2 x 2 optical couplers
balanced coherent detection, 203, 205–206
Mach-zender interferometers (MZIs), 161–162
Michelson interferometers, 170–171

U

ultrahigh-speed ADCs, 224
ultrashort optical pulses, measuring, 224–231
optical receiver characterization, 304–306
unblanking gate, oscilloscope, 213–214
unit interval (UI), 528
units conversion table, xvii

V

V-number, 57–58
numerical aperture and, 60
variable optical attenuator (VOA), 307, 465
vector optical network analyzers, 251–256
VOA (variable optical attenuator), 307, 465

W

water absorption peaks, 63
waterfall curve, 512
wave propagation. *See entries at* propagation
waveform jitter. *See* jitter measurement
waveform measurement, 211–231, 524
distortion sources, 492–493
high-speed electric ADC using optical techniques, 223–224
high-speed optical sampling, 216–219
linear sampling, 221–223
nonlinear sampling, 220–221
measuring distortion in, 524–527
observing distortion in eye diagrams, 492–493
oscilloscopes, operating principle of, 212–216
receiver sensitivity and power margin, 513
short optical pulse measurement, 224–231
optical receiver characterization, 304–306
waveguide dispersion, 71–73
wavelength accuracy of OSAs, 132
wavelength calibration
accuracy of, with OSAs, 133
FPI spectrum analyzers, 156–157, 185
optical wavelength meters, 185–186
wavelength components in fiber propagation. *See* dispersion, optical fibers
wavelength conversion in SOAs
using cross-gain saturation, 95
using FWM, 96–98
wavelength coverage, optical wavelength meters, 183, 187
wavelength meters, 179–188
based on Fizeau wedge interferometer, 186–188
based on Michelson interferometer, 180–183
signal coherence length, 185
spectral resolution, 184
wavelength calibration, 185–186
wavelength coverage, 183, 187
wavelength of laser operation. *See* operation wavelength
wavelength range of OSAs, 132
wavelength-selective switch, characterizing impact of, 615–616
wavelength-swept lasers
coherent OSAs based on, 207–211
Fourier-domain reflectometry (FDR), 241–244
wavelength vs. optical spectrum density. *See* grating-based optical spectrum analyzers (OSAs)

WDM combiner with EDFAs, 101
WDM optical system, 2–3
  EDFA gain flattening, 114–115
  impact of modulation instability, 590–605
    amplified multispan fiber systems, 600–601
    characterization of MI, 601–605
    transfer matrix formulation, 590–600
  multiplexers and demultiplexers, 340–345
  SRS and interchannel crosstalk, 79
wideband characterization of optical receivers, 296–306
  frequency domain characterization, 299–301
  responsivity and linearity, 297–298
  short optical pulse technique, 304–306
  spontaneous-spontaneous emission noise beating technique, 301–304
Wiener-Khintchine theorem, 593
WSS, characterizing impact of, 615–616

X

XPM. *See* cross-phase modulation

Z

zero-chirp modulators, 120
zero substitution patterns, 504

Lightning Source UK Ltd.
Milton Keynes UK
UKHW02n0051090218
317584UK00003B/154/P